INTRODUCTION TO BIOMEDICAL ENGINEERING

ACADEMIC PRESS SERIES IN BIOMEDICAL ENGINEERING

The focus of this series will be twofold. First, the series will produce a series of core text/references for biomedical engineering undergraduate and graduate courses. With biomedical engineers coming from a variety of engineering and biomedical backgrounds, it will be necessary to create a new cross-disciplinary teaching and self-study books. Secondly, the series will also develop handbooks for each of the major subject areas of biomedical engineering.

The series editor, Joseph Bronzino, is one of the most renowned biomedical engineers in the world. Joseph Bronzino is the Vernon Roosa Professor of Applied Science at Trinity College in Hartford, Connecticut.

This is a volume in
BIOMEDICAL ENGINEERING

JOSEPH BRONZINO, SERIES EDITOR
TRINITY COLLEGE, HARTFORD CONNECTICUT

INTRODUCTION TO BIOMEDICAL ENGINEERING

John D. Enderle
University of Connecticut
Storrs, Connecticut

Susan M. Blanchard
North Carolina State University
Raleigh, North Carolina

Joseph D. Bronzino
Trinity College
Hartford, Connecticut

ACADEMIC PRESS

A Harcourt Science and Technology Company

Amsterdam   Boston   Heidelberg   London   New York   Oxford
Paris   San Diego   San Francisco   Singapore   Sydney   Tokyo

Cover Image
A collage of the geometric model construction process. Starting from magnetic resonance images, a geometric model of the human thorax is constructed by segmenting the images, triangulating surfaces (such as the lung, shown in yellow), and tetrahedralizing the volume (shown in blue). A shaded view of the torso is shown in bronze, while a view of the ventricles of the heart are shown in red.

Photography courtesy of Chris Johnson, Rob MacLeod, and Mike Matheson
University of Utah

This book is printed on acid-free paper.

Academic Press
An imprint of Elsevier Science
525 B Street, Suite 1900, San Diego, California 92101-4495, USA
http://www.academicpress.com

Academic Press
84 Theobalds Road, London WC1X 8RR, UK
http://www.academicpress.com

Library of Congress Cataloging-in-Publication Data

Introduction to biomedical engineering/[editors] John D. Enderle, Susan M. Blanchard,
    Joseph D. Bronzino
        p.   cm.—(Academic Press series in biomedical engineering)
    International Standard Book Number: 0-12-238660-4
    1.   Biomedical engineering.   I. Enderle, John D.   II. Blanchard, Susan M.
    III.   Bronzino, Joseph D.
    IV.   Series
    R856.I47   1999
    610'.28—dc21   98-38976
                                                                CIP

PRINTED IN THE UNITED STATES OF AMERICA
        03   04   05   06   07   9   8   7   6   5   4   3

CONTENTS

DEDICATION

This book is dedicated to our families.

PREFACE

The purpose of this textbook is to serve as an introduction to and overview of the field of biomedical engineering. During the past 50 years, as the discipline of biomedical engineering has evolved, it has become clear that it is a diverse, seemingly all-encompassing field that includes such areas as biomechanics, biomaterials, bioinstrumentation, medical imaging, rehabilitation engineering, biosensors, biotechnology, and tissue engineering. Although it is not possible to cover all the biomedical engineering domains in this textbook, we have made an effort to focus on most of the major fields of activity in which biomedical engineers are engaged.

The text is written primarily for engineering students who have completed differential equations and basic courses in statics, dynamics, and linear circuits. Students in the biological sciences, including those in the fields of medicine and nursing, can also read and understand this material if they have the appropriate mathematical background.

Although we do attempt to be rigorous with our discussions and proofs, our ultimate aim is to help students grasp the nature of biomedical engineering. Therefore, we have compromised when necessary and have occasionally used less rigorous mathematics in order to be more understandable. A liberal use of illustrative examples amplifies concepts and develops problem-solving skills. Throughout the text, MATLAB (a matrix equation solver) and SIMULINK (an extension to MATLAB for simulating dynamic systems) are used as computer tools to assist with problem solving.

Chapters are written to provide some historical perspective of the major developments in a specific biomedical engineering domain as well as the fundamental principles that underlie biomedical engineering design, analysis, and modeling procedures in that domain. In addition, examples of some of the problems encountered, as well as the techniques used to solve them, are provided. Selected problems, ranging from simple to difficult, are presented at the end of each chapter in the same general order as covered in the text.

The material in this textbook has been designed for a one-semester, two-semester, or three-quarter sequence depending on the needs and interests of the instructor. Chapter 1 provides necessary background to understand the history and appreciate the field of biomedical engineering. Then, the text divides naturally into two parts, physiological systems and modeling (Chapters 2–10) and technology (Chapters 11–19). Serving as a capstone, Chapter 20 addresses the moral and ethical issues associated with the field of biomedical engineering.

John D. Enderle, Susan M. Blanchard,
Joseph D. Bronzino

ACKNOWLEDGMENTS

We would like to acknowledge Joel Claypool and Diane Grossman from Academic Press for their continued support and encouragement and to Kelly Ricci and Lana Traver from the PRD Group for their support.

We would also like to thank all those who have provided input to the development of the various chapters in this textbook.

CONTRIBUTORS

Numbers in parentheses indicate chapters of authors' contribution(s).

Susan M. Blanchard (2,5,6,13)
North Carolina State University
Raleigh, North Carolina

Joseph Bronzino (1,14,19,20)
Trinity College
Hartford, Connecticut

Stan Brown (11)
Food and Drug Administration
Gaithersburg, Maryland

Gerard Coté (17)
Texas A&M University
College Station, Texas

Roy B. Davis III (9)
Shriners Hospitals for Children
Greenville, South Carolina

John D. Enderle (3,7)
University of Connecticut
Storrs, Connecticut

R. J. Fisher (8)
University of Connecticut
Storrs, Connecticut

Carol Lucas (6)
University of North Carolina at Chapel Hill
Chapel Hill, North Carolina

Amanda Marley (5)
North Carolina State University
Raleigh, North Carolina

Yitzhak Mendelson (4)
Worcester Polytechnic Institute
Worcester, Massachusetts

Katherine Merritt (11)
Food and Drug Administration
Gaithersburg, Maryland

H. Troy Nagle (5)
North Carolina State University
Raleigh, North Carolina

Joseph Palladino (9)
Trinity College
Hartford, Connecticut

Bernhard Palsson (12)
University of California at San Diego
LaJolla, California

Sohi Rastegar (17)
Texas A&M University
College Station, Texas

Daniel Schneck (10)
Virginia Polytechnic Institute & State University
Blacksburg, Virginia

Kirk K. Shung (15)
Pennsylvania State University
University Park, Pennsylvania

Anne-Marie Stomp (13)
North Carolina State University
Raleigh, North Carolina

Andrew Szeto (18)
San Diego State University
San Diego, Carolina

LiHong Wang (17)
Texas A&M University
College Station, Texas

Steven Wright (16)
Texas A&M University
College Station, Texas

Melanie T. Young (6)
North Carolina State University
Raleigh, North Carolina

1 BIOMEDICAL ENGINEERING: A HISTORICAL PERSPECTIVE

Chapter Contents

At the conclusion of this chapter, readers will be able to:

- Identify the major role that advances in medical technology have played in the establishment of the modern health care system
- Define what is meant by the term biomedical engineering and the roles biomedical engineers play in the health care delivery system
- Explain why biomedical engineers are professionals

In the twentieth century, technological innovation has progressed at such an accelerated pace that it is has permeated almost every facet of our lives. This is especially true in the area of medicine and the delivery of health care services. Although the art of medicine has a long history, the evolution of a technologically based health care system capable of providing a wide range of effective diagnostic and therapeutic treatments is a relatively new phenomenon. Of particular importance in this evolutionary process has been the establishment of the modern hospital as the center of a technologically sophisticated health care system.

Since technology has had such a dramatic impact on medical care, engineering professionals have become intimately involved in many medical ventures. As a result, the discipline of biomedical engineering has emerged as an integrating medium for two dynamic professions, medicine and engineering, and has assisted in the struggle against illness and disease by providing tools (such as biosensors, biomaterials, image processing, and artificial intelligence) that can be utilized for research, diagnosis, and treatment by health care professionals.

Thus, biomedical engineers serve as relatively new members of the health care delivery team that seeks new solutions for the difficult problems confronting modern society. The purpose of this chapter is to provide a broad overview of technology's role in shaping our modern health care system, highlight the basic roles biomedical engineers play, and present a view of the professional status of the field as we begin the twenty-first century.

1.1 EVOLUTION OF THE MODERN HEALTH CARE SYSTEM

Primitive humans considered diseases to be "visitations," the whimsical acts of affronted gods or spirits. As a result, medical practice was the domain of the witch doctor and the medicine man and medicine woman. Yet even as magic became an integral part of the healing process, the cult and the art of these early practitioners were never entirely limited to the supernatural. These individuals, by using their natural instincts and learning from experience, developed a primitive science based on empirical laws. For example, through acquisition and coding of certain reliable practices, the arts of herb doctoring, bone setting, surgery, and midwifery were advanced. Just as primitive humans learned from observation that certain plants and grains were good to eat and could be cultivated, so the healers and shamans observed the nature of certain illnesses and then passed on their experiences to other generations.

Evidence indicates that the primitive healer took an active rather than simply intuitive interest in the curative arts, acting as a surgeon, a user of tools. For instance, skulls with holes made in them by trephiners have been collected in various parts of Europe, Asia, and South America. These holes were cut out of the bone with flint instruments to gain access to the brain. Although one can only speculate the purpose of these early surgical operations, magic and religious beliefs seem to be the most likely reasons. Perhaps this procedure liberated from the skull the malicious demons who were thought to be the cause of extreme pain (as in the case of migraine) or attacks of falling to the ground (as in epilepsy). That this procedure was carried out on living patients, some of whom actually survived, is evident from the rounded edges on the bone surrounding the hole which indicate that the bone had grown again after the operation. These survivors also achieved a special status of sanctity so that, after their death, pieces of their skull were used as amulets to ward off convulsive attacks. From these beginnings, the practice of medicine has become integral to all human societies and cultures.

It is interesting to note the fate of some of the most successful of these early practitioners. The Egyptians, for example, have held Imhotep, the architect of the first pyramid (3000 BC), in great esteem through the centuries, not as a pyramid builder but as a doctor. Imhotep's name signified "he who cometh in peace" because he visited the sick to give them "peaceful sleep." This early physician practiced his art so well that he was deified in the Egyptian culture as the god of healing.

Egyptian mythology, like primitive religion, emphasized the interrelationships between the supernatural and one's health. For example, consider the mystic sign Rx, which still adorns all prescriptions today. It has a mythical origin, the legend of the Eye of Horus. It appears that as a child Horus lost his vision after being viciously attacked by Seth, the demon of evil. Then Isis, the mother of Horus, called for assistance to Thoth, the most important god of health, who promptly restored the eye and its powers. Because of this intervention, the Eye of Horus became the Egyptian symbol of godly protection and recovery, and its descendant, Rx, serves as the most visible link between ancient and modern medicine.

The concepts and practices of Imhotep and the medical cult he fostered were duly recorded on papyri and stored in ancient tombs. One scroll (dated ca. 1500 BC), acquired by George Elbers in 1873, contains hundreds of remedies for numerous afflictions ranging from crocodile bite to constipation. A second famous papyrus (dated ca. 1700 BC), discovered by Edwin Smith in 1862, is considered to be the most important and complete treatise on surgery of all antiquity. These writings outline proper diagnoses, prognoses, and treatment in a series of surgical cases. These two papyri are certainly among the outstanding writings in medical history.

As the influence of ancient Egypt spread, Imhotep was identified by the Greeks with their own god of healing, Aesculapius. According to legend, Aesculapius was fathered by the god Apollo during one of his many earthly visits. Apparently, Apollo was a concerned parent, and, as is the case for many modern parents, he wanted his son to be a physician. He made Chiron, the centaur, tutor Aesculapius in the ways of healing. Chiron's student became so proficient as a healer that he soon surpassed his tutor and kept people so healthy that he began to decrease the population of Hades. Pluto, the god of the underworld, complained so violently about this course of events that Zeus killed Aesculapius with a thunderbolt and in the process promoted Aesculapius to Olympus as a god.

Inevitably, mythology has become entangled with historical facts, and it is not certain whether Aesculapius was in fact an earthly physician like Imhotep, the Egyptian. However, one thing is clear: By 1000 BC, medicine was already a highly respected profession. In Greece, the Aesculapia were temples of the healing cult and may be considered among the first hospitals. In modern terms, these temples were essentially sanatoriums that had strong religious overtones. In them, patients were received and psychologically prepared, through prayer and sacrifice, to appreciate the past achievements of Aesculapius and his physician priests. After the appropriate rituals, they were allowed to enjoy "temple sleep." During the night, "healers" visited their patients, administering medical advice to clients who were awake or interpreting dreams of those who had slept. In this way, patients became convinced that they would be cured by following the prescribed regimen of diet, drugs, or bloodletting. On the other hand, if they remained ill, it would be attributed to their lack of faith. With this approach, patients, not treatments, were at fault if they did not get well. This early use of the power of suggestion was effective then and is still important in medical treatment today. The notion of "healthy mind, healthy body" is still in vogue today.

One of the most celebrated of these "healing" temples was on the island of Cos, the birthplace of Hippocrates who as a youth became acquainted with the curative arts through his father, also a physician. Hippocrates was not so much an innovative physician as a collector of all the remedies and techniques that existed up to that time. Since he viewed the physician as a scientist instead of a priest, Hippocrates also injected an essential ingredient into medicine: its scientific spirit. For him, diagnostic observation and clinical treatment began to replace superstition. Instead of blaming disease on the gods, Hippocrates taught that disease was a natural process, one that developed in logical steps, and that symptoms were reactions of the body to disease. The body itself, he emphasized, possessed its own means of recovery, and the function of the physician was to aid these natural forces. Hippocrates treated each patient as an original case to be studied and documented. His shrewd descriptions of diseases are models for physicians even today. Hippocrates and the school of Cos trained a number of individuals who then migrated to the corners of the Mediterranean world to practice medicine and spread the philosophies of their preceptor. The work of Hippocrates and the school and tradition that stem from him constitute the first real break from magic and mysticism and the foundation of the rational art of medicine. However, as a practitioner, Hippocrates represented the spirit, not the science, of medicine, embodying the good physician: the friend of the patient and the humane expert.

As the Roman Empire reached its zenith and its influence expanded across half the world, it became heir to the great cultures it absorbed, including their medical advances. Although the Romans themselves did little to advance clinical medicine (the treatment of the individual patient), they did make outstanding contributions to public health. For example, they had a well-organized army medical service, which not only accompanied the legions on their various campaigns to provide "first aid" on the battlefield but also established "base hospitals" for convalescents at strategic points throughout the empire. The construction of sewer systems and aqueducts was truly a remarkable Roman accomplishment that provided their empire with the medical and social advantages of sanitary living. Insistence on clean drinking water and unadulterated foods affected the control and prevention of epidemics and, however primitive, made urban existence possible. Unfortunately, without adequate scientific knowledge about diseases, the preoccupation of the Romans with public health could not avert the periodic medical disasters, particularly the plague, that mercilessly befell its citizens.

Initially, the Roman masters looked upon Greek physicians and their art with disfavor. However, as the years passed, the favorable impression these disciples of Hippocrates made upon the people became widespread. As a reward for their service to the peoples of the Empire, Caesar (46 BC) granted Roman citizenship to all Greek practitioners of medicine in his empire. Their new status became so secure that when Rome suffered from famine that same year, these Greek practitioners were the only foreigners not expelled from the city. On the contrary, they were even offered bonuses to stay!

Ironically, Galen, who is considered the greatest physician in the history of Rome, was himself a Greek. Honored by the emperor for curing his "imperial fever," Galen became the medical celebrity of Rome. He was arrogant and a braggart and, unlike Hippocrates, reported only successful cases. Nevertheless, he was a remarkable physician. For Galen, diagnosis became a fine art; in addition to taking care of his own patients, he responded to requests for medical advice from the far reaches of the empire. He was so industrious that he wrote more than 300 books of anatomical observations which included selected case histories, the drugs he prescribed, and his boasts. His version of human anatomy, however, was misleading because he objected to human dissection and drew his human analogies solely from the studies of animals. However, because he so dominated the medical scene and was later endorsed by the Roman Catholic Church, Galen actually inhibited medical inquiry. His medical views and writings became both the "bible" and "the law" for the pontiffs and pundits of the ensuing Dark Ages.

With the collapse of the Roman Empire, the Church became the repository of knowledge, particularly of all scholarship that had drifted through the centuries into the Mediterranean. This body of information, including medical knowledge, was literally scattered through the monasteries and dispersed among the many orders of the Church.

The teachings of the early Roman Catholic Church and the belief in divine mercy made inquiry into the causes of death unnecessary and even undesirable. Members of the Church regarded curing patients by rational methods as sinful interference with the will of God. The employment of drugs signified a lack of faith by the doctor and patient, and scientific medicine fell into disrepute. Therefore, for almost 1000 years, medical research stagnated. It was not until the Renaissance in the 1500s that significant progress in the science of medicine occurred. Hippocrates had once taught that illness was not a punishment sent by the gods but a phenomenon of nature. Now, under the Church and a new God, the older views of the supernatural origins of disease were renewed and promulgated. Since disease implied demonic possession, the sick were treated by the monks and priests through prayer, laying on of hands, exorcism, penances, and exhibition of holy relics — practices officially sanctioned by the Church.

Although deficient in medical knowledge, the Dark Ages were not entirely lacking in charity toward the sick poor. Christian physicians often treated the rich and poor alike, and the Church assumed responsibility for the sick. Furthermore, the evolution of the modern hospital actually began with the advent of Christianity and is considered one of the major contributions of monastic medicine. With the rise in 335 AD of Constantine I, the first of the Roman emperors to embrace Christianity, all pagan temples of healing were closed, and hospitals were established in every cathedral city. (Note: The word hospital derives from the Latin *hospes,* meaning "host" or "guest". The same root has provided hotel and hostel.) These first hospitals were simply houses where weary travelers and the sick could find food, lodging, and nursing care. All these hospitals were run by the Church, and the art of healing was practiced by the attending monks and nuns.

As the Christian ethic of faith, humanitarianism, and charity spread throughout Europe and then to the Middle East during the Crusades, so did its "hospital system." However, trained "physicians" still practiced their trade primarily in the homes of their patients, and only the weary travelers, the destitute, and those considered hopeless cases found their way to hospitals. Conditions in these early hospitals varied widely. Although a few were well financed and well managed and treated their patients humanely, most were essentially custodial institutions to keep troublesome and infectious people away from the general public. In these establishments, crowding, filth, and high mortality among both patients and attendants were commonplace. Thus, the hospital was viewed as an institution to be feared and shunned.

The Renaissance and Reformation in the fifteenth and sixteenth centuries loosened the Church's stronghold on both the hospital and the conduct of medical practice. During the Renaissance, "true learning," the desire to pursue the true secrets of nature including medical knowledge, was again stimulated. The study of human anatomy was advanced and the seeds for further studies were planted by the artists Michelangelo, Raphael Durer, and, of course, the genius Leonardo da Vinci. They viewed the human body as it really was and not simply as a text passage from Galen. The Renaissance painters depicted people in sickness and pain, sketched in great detail, and in the process demonstrated amazing insight into the workings of the heart, lungs, brain, and muscle structure. They also attempted to portray the individual and to discover emotional as well as physical qualities. In this stimulating era, physicians began to approach their patients and the pursuit of medical knowledge in similar fashion. New medical schools, similar to the most famous of such institutions at Salerno, Bologna, Montpelier, Padua, and Oxford, emerged. These medical training centers once again embraced the Hippocratic doctrine that the patient was human, disease was a natural process, and commonsense therapies were appropriate in assisting the body to conquer its disease.

During the Renaissance, fundamentals received closer examination and the age of measurement began. In 1592, when Galileo visited Padua, Italy, he lectured on mathematics to a large audience of medical students. His famous theories and inventions (the thermoscope and the pendulum, in addition to the telescopic lens) were expounded upon and demonstrated. Using these devices, one of his students, Sanctorius, made comparative studies of the human temperature and pulse. A future graduate of Padua, William Harvey, later applied Galileo's laws of motion and mechanics to the problem of blood circulation. This ability to measure the amount of blood moving through the arteries helped to determine the function of the heart.

Galileo encouraged the use of experimentation and exact measurement as scientific tools that could provide physicians with an effective check against reckless speculation. Quantification meant theories would be verified before being accepted. Individuals involved in medical research incorporated these new methods into their activities. Body temperature and pulse rate became measures that could be related to other symptoms to assist the physician in diagnosing specific illnesses or disease. Concurrently, the development of the microscope amplified human vision, and an unknown world came into focus. Unfortunately, new scientific devices had little impact on the average physician, who continued to bloodlet and to disperse noxious ointments. Only in the universities did scientific groups band together to pool their instruments and their various talents.

In England, the medical profession found in Henry VIII a forceful and sympathetic patron. He assisted the doctors in their fight against malpractice and supported the establishment of the College of Physicians, the oldest purely medical institution in Europe. When he suppressed the monastery system in the early sixteenth century, church hospitals were taken over by the cities in which they were located. Consequently, a network of private, nonprofit, voluntary hospitals came into being. Doctors and medical students replaced the nursing sisters and monk physicians. Consequently, the professional nursing class became almost nonexistent in these public institutions. Only among the religious orders did "nursing" remain intact, further compounding the poor lot of patients confined within the walls of the public hospitals. These conditions were to continue until Florence Nightingale appeared on the scene years later.

Still another dramatic event was to occur. The demands made on England's hospitals, especially the urban hospitals, became overwhelming as the population of these urban centers continued to expand. It was impossible for the facilities to accommodate the needs of so many. Therefore, during the seventeenth century two of the major urban hospitals in London, St. Bartholomew's and St. Thomas, initiated a policy of admitting and attending to only those patients who could possibly be cured. The incurables were left to meet their destiny in other institutions such as asylums, prisons, or almshouses.

Humanitarian and democratic movements occupied center stage primarily in France and the American colonies during the eighteenth century. The notion of equal rights finally began, and as urbanization spread, American society concerned itself with the welfare of many of its members. Medical men broadened the scope of their services to include the "unfortunates" of society and helped to ease their suffering by advocating the power of reason and spearheading prison reform, child care, and the hospital movement. Ironically, as the hospital began to take up an active, curative role in medical care in the eighteenth century, the death rate among its patients did not decline but continued to be excessive. In 1788, for example, the death rate among the patients at the Hotel Dru in Paris, thought to be founded in the seventh century and the oldest hospital in existence today, was nearly 25%. These hospitals were lethal not only to patients but also to the attendants working in them, whose own death rate hovered between 6 and 12% per year.

Essentially, the hospital remained a place to avoid. In these circumstances, it is not surprising that the first American colonists postponed or delayed building hospitals. For example, the first hospital in America, the Pennsylvania Hospital, was not built until 1751, and the city of Boston took more than 200 years to erect its first hospital, the Massachusetts General, which opened its doors to the public in 1821.

Not until the nineteenth century could hospitals claim to benefit any significant number of patients. This era of progress was due primarily to the improved nursing practices fostered by Florence Nightingale on her return to England from the Crimean War. She demonstrated that hospital deaths were caused more frequently by hospital conditions than by disease. During the latter part of the nineteenth century, she was at the height of her influence, and few new hospitals were built anywhere in the world without her advice. During the first half of the nineteenth century, Nightingale forced medical attention to focus once more on the care of the patient. Enthusiastically and philosophically, she expressed her views on nursing: "Nursing is putting us in the best possible condition for nature to restore and preserve health. . . . The art is that of nursing the sick. Please mark, not nursing sickness."

Although these efforts were significant, hospitals remained, until this century, institutions for the sick poor. In the 1870s, for example, when the plans for the projected Johns Hopkins Hospital were reviewed, it was considered quite appropriate to allocate 324 charity and 24 pay beds. The hospital population before the turn of the century not only represented a narrow portion of the socioeconomic spectrum but also represented only a limited number of the type of diseases prevalent in the overall population. In 1873, for example, approximately half of America's hospitals did not admit contagious diseases, and many others would not admit incurables. Furthermore, in this period, surgery admissions in general hospitals constituted only 5% with trauma (injuries incurred by traumatic experience) making up a good portion of these cases.

American hospitals a century ago were simple in that their organization required no special provisions for research or technology and demanded only cooking and washing facilities. In addition, since the attending and consulting physicians were normally unsalaried, and the nursing costs were quite modest, the great bulk of the hospital's normal operation expenses was for food, drugs, and utilities. Not until the twentieth century did "modern medicine" come of age in the United States. As we shall see, technology played a significant role in its evolution.

1.2 THE MODERN HEALTH CARE SYSTEM

Modern medical practice is a relatively new phenomenon. Before 1900, medicine had little to offer the average citizen since its resources were mainly physicians, their education, and their little black bags. At this time physicians were in short supply, but for different reasons than exist today. Costs were minimal, demand small, and many of the services provided by the physician could also be obtained from experienced amateurs residing in the community. The individual's dwelling was the major site for treatment and recuperation, and relatives and neighbors constituted an able and willing nursing staff. Babies were delivered by midwives, and those illnesses not cured by home remedies were left to run their fatal course. Only in this century did the tremendous explosion in scientific knowledge and technology lead to the development of the American health care system with the hospital as its focal point and the specialist physician and nurse as its most visible operatives.

In the twentieth century, the advances made in the basic sciences (chemistry, physiology, pharmacology, etc.) began to occur much more rapidly. Indeed, this has been an era of intense interdisciplinary cross-fertilization. Discoveries in the physical sciences enabled medical researchers to take giant strides forward. For example, in 1903, William Einthoven devised the first electrocardiograph and measured the electrical changes that occurred during the beating of the heart. In the process, Einthoven initiated a new age for both cardiovascular medicine and electrical measurement techniques (Fig. 1.1).

Of all the new discoveries that now followed one another like intermediates in a chain reaction, the most significant for clinical medicine was the development of X rays. When W. K. Roentgen described his "new kinds of rays," the human body was opened to medical inspection. Initially these X rays were used in the diagnosis of bone fractures and dislocations. In the United States, X-ray machines brought this "modern technology" to most urban hospitals. In the process, separate departments of radiology were established, and the influence of their activities spread with almost every department of medicine (surgery, gynecology, etc.) advancing with the aid of this new tool. By the 1930s, X-ray visualization of practically all the organ systems of the body was possible by the use of barium salts and a wide variety of radiopaque materials.

The power this technological innovation gave physicians was enormous. The X ray permitted them to diagnose a wide variety of diseases and injuries accurately. In addition, being within the hospital, it helped trigger the transformation of the

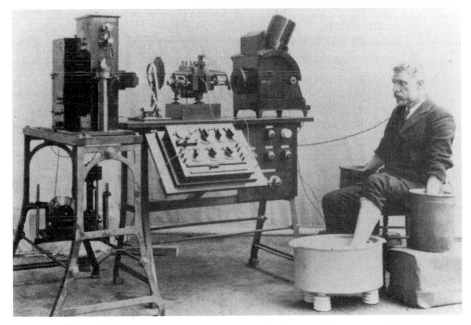

Fig. 1.1 Photograph depicting an early electrocardiogram machine.

hospital from a passive receptacle for the sick poor to an active curative institution for all the citizens of the American society.

When reviewing some of the most significant developments in health care practices, one is astounded to find that they have occurred fairly recently, that is, within the past 60 years. It was not until the introduction of sulfanilamide in the mid-1930s and penicillin in the early 1940s that the main danger of hospitalization, cross infection among patients, was significantly reduced. With these new drugs in their arsenals, surgeons were able to perform their operations without prohibitive morbidity and mortality due to infection. Also consider that, even though the different blood groups and their incompatibility were discovered in 1900 and sodium citrate was used in 1913 to prevent clotting, the full development of blood banks was not practical until the 1930s when technology provided adequate refrigeration. Until that time, "fresh" donors were bled, and the blood was transfused while it was still warm.

As technology in the United States blossomed so did the prestige of American medicine. From 1900 to 1929 Nobel prize winners in physiology or medicine came primarily from Europe, with no Americans among them. In the period 1930 to 1939, just before World War II, 7 Americans were honored by this award. During the postwar period, i.e., from 1945 to 1975, 39 American life scientists earned similar honors, and from 1975 to 1995, the number was 31. Most of these efforts were made possible by the advanced technology available to these clinical scientists.

The employment of the available technology assisted in advancing the development of complex surgical procedures. The Drinker respirator was introduced in 1927 and the first heart–lung bypass in 1939. In the 1940s, cardiac catheterization and angiography (the use of a cannula threaded through an arm vein and into the heart with the injection of radiopaque dye for the X-ray visualization of lung and heart vessels and valves) were developed. Accurate diagnoses of congenital and acquired heart disease (mainly valve disorders due to rheumatic fever) also became possible, and a new era of cardiac and vascular surgery began (Fig. 1.2).

Another child of this modern technology, the electron microscope, entered the medical scene in the 1950s and provided significant advances in visualizing relatively small cells. Body scanners to detect tumors arose from the same science that brought societies reluctantly into the atomic age. These "tumor detectives" used radioactive material and became commonplace in newly established departments of nuclear medicine in all hospitals.

The impact of these discoveries and many others was profound. The health care system that consisted primarily of the "horse and buggy" physician was gone forever, replaced by the doctor backed by and centered around the hospital, as medicine began to change to accommodate the new technology. Thus, it can be seen that the modern hospital in its contemporary, familiar form has been shaped during the twentieth century.

Following World War II, the evolution of comprehensive care greatly accelerated. The advanced technology that had been developed in the pursuit of military objectives now became available for peaceful applications with the medical profession benefiting greatly from this rapid surge of technological "finds." For instance, the

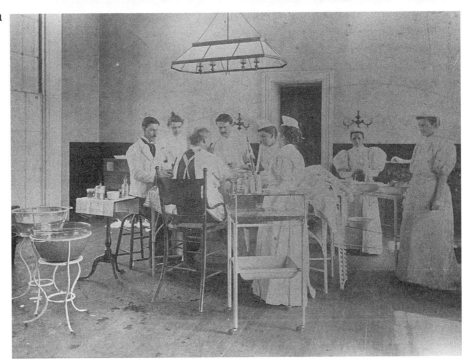

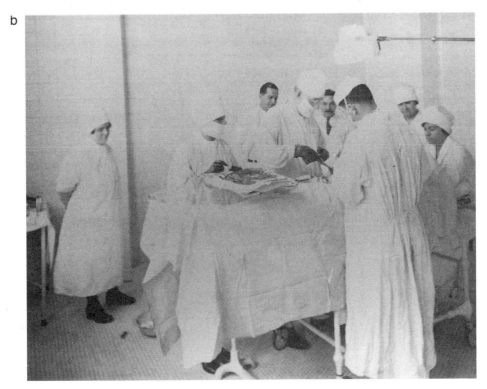

Fig. 1.2 Changes in the operating room: (a) the surgical scene at the turn of the century, (b) the surgical scene in the late 1920s and early 1930s, and (c) the surgical scene today (from Bronzino, 1977).

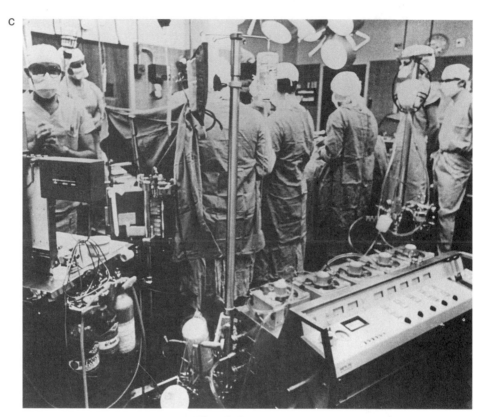

Fig. 1.2   *Continued*

realm of electronics came into prominence. The techniques for following enemy ships and planes, as well as providing aviators with information concerning altitude, air speed, and the like, were now used extensively in medicine to follow the subtle electrical behavior of the fundamental unit of the central nervous system, the neuron, or to monitor the beating heart of a patient.

Science and technology have leapfrogged past one another throughout recorded history. Anyone seeking a causal relation between the two was just as likely to find technology the cause and science the effect with the converse also holding true. As gunnery led to ballistics and the steam engine to thermodynamics, so powered flight led to aerodynamics. However, with the advent of electronics, this causal relation between technology and science changed to a systematic exploitation of scientific research and the pursuit of knowledge which were sometimes undertaken with technical uses in mind (Fig. 1.3).

The list becomes endless when one reflects on the devices produced by the same technology that permitted humans to stand on the moon. What was considered science fiction in the 1930s and the 1940s became reality. Devices continually changed

Fig. 1.3   Photograph of modern medical imaging facility.

to incorporate the latest innovations, which in many cases became outmoded in a very short period of time. Telemetry devices used to monitor the activity of a patient's heart freed both the physician and the patient from the wires that previously restricted them to the four walls of the hospital room. Computers, similar to those that controlled the flight plans of the *Apollo* capsules, now completely inundate our society. Since the 1970s, medical researchers have put these electronic brains to work performing complex calculations, keeping records (via artificial intelligence), and even controlling the very instrumentation that sustains life. The development of new medical imaging techniques such as computerized tomography and magnetic resonance imagining totally depended on a continually advancing computer technology. The citations and technological discoveries are so numerous that it is impossible to mention them all.

"Spare parts" surgery is now routine. With the first successful transplantation of a kidney in 1954, the concept of "artificial organs" gained acceptance and officially came into vogue in the medical arena (Fig. 1.4). Technology to provide prosthetic devices, such as artificial heart valves and artificial blood vessels, developed. Even an artificial heart program to develop a replacement for a defective or diseased human heart began. Although, to date, the results have not been satisfactory, this program has provided "ventricular assistance" for those who need it. These tech-

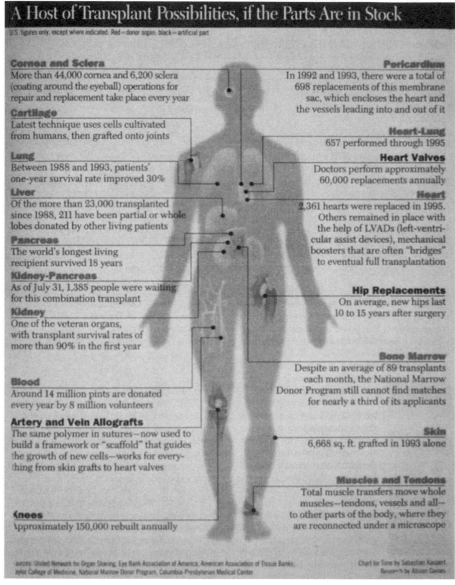

Fig. 1.4   Identification of various transplantation procedures (Chart by Sebastian Kaupert).

nological innovations radically altered surgical organization and utilization. The comparison of a hospital in which surgery was a relatively minor activity as it was a century ago to the contemporary hospital in which surgery plays a prominent role dramatically suggests the manner in which this technological effort has revolutionized the health profession and the institution of the hospital.

Through this evolutionary process, the hospital became the central institution that provided medical care. Because of the complex and expensive technology that could be based only in the hospital and the education of doctors oriented both as clinicians and investigators toward highly technological norms, both the patient and the physician were pushed even closer to this center of attraction. In addition, the effects of the increasing maldistribution and apparent shortage of physicians during the 1950s and 1960s also forced the patient and the physician to turn increasingly to the ambulatory clinic and the emergency ward of the urban hospital in time of need.

Emergency wards today handle not only an ever-increasing number of accidents (largely related to alcohol and the automobile) and somatic crises, such as heart attacks and strokes, but also problems resulting from the social environments that surround the local hospital. Respiratory complaints, cuts, bumps, and minor trauma constitute a significant number of the cases seen in a given day. In addition, there are individuals who live in the neighborhood of the hospital and simply cannot afford their own physician. Often such individuals enter the emergency ward for routine care of colds, hangovers, and even marital problems.

Because of these developments, the hospital has evolved as the focal point of the current system of health care delivery. The hospital, as currently organized, specializes in highly technical and complex medical procedures. This evolutionary process became inevitable as technology produced increasingly sophisticated equipment that private practitioners or even large group practices were economically unequipped to acquire and maintain. Only the hospital could provide this type of service. The steady expansion of scientific and technological innovations has not only necessitated specialization for all health professionals (physicians, nurses, and technicians) but also has required the housing of advanced technology within the walls of the modern hospital (Fig. 1.5).

In recent years, technology has struck medicine like a thunderbolt, providing far more advances in the past 50 years than in the previous 2000. In a culture steeped in science, it appears this trend will continue. However, the social and economic consequences of this vast outpouring of information and innovation must be fully understood if this technology is to be exploited effectively and efficiently.

For the future, technology offers great potential for affecting health care practices. It can provide health care for individuals in remote rural areas by means of closed-circuit television health clinics with complete communication links to a regional health center. Development of multiphasic screening systems can provide preventive medicine to the vast majority of our population and restrict admission to the hospital to those needing the diagnostic and treatment facilities housed there. Automation of patient and nursing records can inform physicians of the status of patients during their stay at the hospital and in their homes. With the creation of a cen-

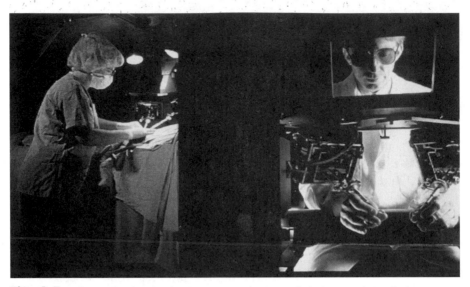

Fig. 1.5   The reach of technology — remote surgery techniques permit medical experts to assist others in the field (courtesy of Robert Holmgren/Zuma Syndication).

tral medical records system, anyone moving or becoming ill away from home can have records made available to the attending physician easily and rapidly. These are just a few of the possibilities that illustrate the potential of technology in creating the type of medical care system that will indeed be accessible, be of high quality, and be reasonably priced for all (Fig. 1.6).

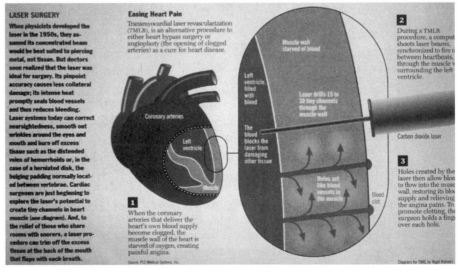

Fig. 1.6   Laser surgery — how it works (diagram by Nigel Holmes).

1.3 WHAT IS BIOMEDICAL ENGINEERING?

Many of the problems confronting health professionals today are of extreme importance to the engineer because they involve the fundamental aspects of device and systems analysis, design, and practical application — all of which lie at the heart of processes that are fundamental to engineering practice. These medically relevant design problems can range from very complex large-scale constructs, such as the design and implementation of automated clinical laboratories, multiphasic screening facilities (i.e., centers that permit many tests to be conducted), and hospital information systems, to the creation of relatively small and "simple" devices, such as recording electrodes and transducers that may be used to monitor the activity of specific physiological processes in either a research or a clinical setting. They encompass the many complexities of remote monitoring and telemetry and include the requirements of emergency vehicles, operating rooms, and intensive care units.

The American health care system encompasses many problems that represent challenges to certain members of the engineering profession called biomedical engineers. Since biomedical engineering involves applying the concepts, knowledge, and approaches of virtually all engineering disciplines (e.g., electrical, mechanical, and chemical engineering) to solve specific health care-related problems, the opportunities for interaction between engineers and health care professionals are many and varied.

Biomedical engineers may become involved, for example, in the design of a new medical imaging modality or development of new medical prosthetic devices to aid the disabled. Although what is included in the field of biomedical engineering is considered by many to be quite clear, many conflicting opinions concerning the field can be traced to disagreements about its definition. For example, consider the terms biomedical engineering, bioengineering, biological engineering, and clinical (or medical) engineer, which are defined in the Bioengineering Education Directory (Pacela, 1990). While Pacela defined "bioengineering" as the broad umbrella term used to describe this entire field, bioengineering is usually defined as a basic research-oriented activity closely related to biotechnology and genetic engineering, that is, the modification of animal or plant cells or parts of cells to improve plants or animals or to develop new microorganisms for beneficial ends. In the food industry, for example, this has meant the improvement of strains of yeast for fermentation. In agriculture, bioengineers may be concerned with the improvement of crop yields by treatment of plants with organisms to reduce frost damage. It is clear that in the future bioengineers will have tremendous impact on the quality of human life. The full potential of this specialty is difficult to image. Typical pursuits include the following:

- Development of improved species of plants and animals for food production
- Invention of new medical diagnostic tests for diseases
- Production of synthetic vaccines from clone cells
- Bioenvironmental engineering to protect human, animal, and plant life from toxicant and pollutants
- Study of protein–surface interactions

- Modeling of the growth kinetics of yeast and hybridoma cells
- Research in immobilized enzyme technology
- Development of therapeutic proteins and monoclonal antibodies

The term biomedical engineering appears to have the most comprehensive meaning. Biomedical engineers apply electrical, chemical, optical, mechanical, and other engineering principles to understand, modify, or control biological (i.e., human and animal) systems. When a biomedical engineer works within a hospital or clinic, he or she is more properly called a clinical engineer. However, this theoretical distinction is not always observed in practice since many professionals working within U.S. hospitals today continue to be called biomedical engineers.

The breadth of activity of biomedical engineers is significant. The field has moved significantly from being concerned primarily with the development of medical devices in the 1950s and 1960s to include a more wide-ranging set of activities. As illustrated in Fig. 1.7, the field of biomedical engineering now includes many new career areas. These areas include

- Application of engineering system analysis (physiologic modeling, simulation, and control to biological problems)
- Detection, measurement, and monitoring of physiologic signals (i.e., biosensors and biomedical instrumentation)
- Diagnostic interpretation via signal-processing techniques of bioelectric data
- Therapeutic and rehabilitation procedures and devices (rehabilitation engineering)
- Devices for replacement or augmentation of bodily functions (artificial organs)

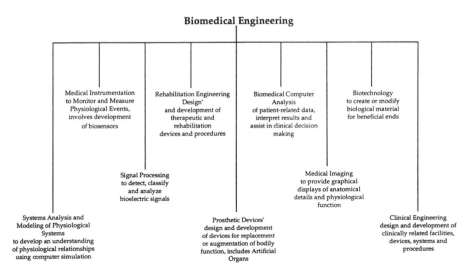

Fig. 1.7   Schematic diagram illustrating the various fields of activity within the discipline of biomedical engineering.

- Computer analysis of patient-related data and clinical decision making (i.e., medical informatics and artificial intelligence)
- Medical imaging, that is, the graphical display of anatomic detail or physiologic function
- The creation of new biologic products (i.e., biotechnology and tissue engineering)

Typical pursuits of biomedical engineers include

- Research in new materials for implanted artificial organs
- Development of new diagnostic instruments for blood analysis
- Computer modeling of the function of the human heart
- Writing software for analysis of medical research data
- Analysis of medical device hazards for safety and efficacy
- Development of new diagnostic imaging systems
- Design of telemetry systems for patient monitoring
- Design of biomedical sensors for measurement of human physiologic system variables
- Development of expert systems for diagnosis of diseases
- Design of closed-loop control systems for drug administration
- Modeling of the physiologic systems of the human body
- Design of instrumentation for sports medicine
- Development of new dental materials
- Design of communication aids for the disabled
- Study of pulmonary fluid dynamics
- Study of biomechanics of the human body
- Development of material to be used as replacement for human skin

The previous list is not intended to be all-inclusive. Many other applications use the talents and skills of the biomedical engineer. In fact, the list of activities of biomedical engineers depends on the medical environment in which they work. This is especially true for the clinical engineers, biomedical engineers employed in hospitals or clinical settings. Clinical engineers are essentially responsible for all the high-technology instruments and systems used in hospitals today, for the training of medical personnel in equipment safety, and for the design, selection, and use of technology to deliver safe and effective health care.

Biomedical engineering is thus an interdisciplinary branch of engineering heavily based both in engineering and in the life sciences. It ranges from theoretical, nonexperimental undertakings to state-of-the-art applications. It can encompass research, development, implementation, and operation. Accordingly, like medical practice, it is unlikely that any single person can acquire expertise that encompasses the entire field. As a result, there has been an explosion of biomedical engineering specialists to cover this broad spectrum of activity. However, because of the interdisciplinary nature of this activity, there is considerable interplay and overlapping of interest and effort between them. For example, biomedical engineers engaged in the development of biosensors may interact with those interested in prosthetic devices to develop a means to detect and use the same bioelectric signal to power a prosthetic device. Those engaged in automating the clinical chemistry laboratory may collaborate with those developing expert systems to assist clinicians in making clinical decisions based on specific laboratory data. The possibilities are endless.

Perhaps a greater potential benefit occurring from the utilization of biomedical engineers is the identification of problems and needs of our current health care delivery system that can be solved using existing engineering technology and systems methodology. Consequently, the field of biomedical engineering offers hope in the continuing battle to provide high-quality health care at a reasonable cost. If properly directed toward solving problems related to preventative medical approaches, ambulatory care services, and so on, biomedical engineers can provide the tools and techniques to make our health care system more effective and efficient.

1.4 ROLES PLAYED BY BIOMEDICAL ENGINEERS

In its broadest sense, biomedical engineering involves training essentially three types of individuals: (i) the clinical engineer in health care, (ii) the biomedical design engineer for industry, and (iii) the research scientist. Currently, one might also distinguish among three specific roles these biomedical engineers can play. Each is different enough to merit a separate description. The first type, the most common, might be called the "problem solver" (Fig. 1.8). This biomedical engineer (most likely the

Fig. 1.8   The role of the problem solver.

clinical engineer or biomedical design engineer) maintains the traditional service relationship with the life scientists who originate a problem that can be solved by applying the specific expertise of the engineer. For this problem-solving process to be efficient and successful, however, some knowledge of each other's language and a ready interchange of information must exist. Biomedical engineers must understand the biological situation to apply their judgment and contribute their knowledge toward the solution of the given problem as well as to defend their methods in terms that the life scientist can understand. If they are unable to do these things, they do not merit the "biomedical" prefix.

The second type, who is much rarer, might be called the "technological entrepreneur" (most likely a biomedical design engineer in industry). This individual assumes that the gap between the technological education of the life scientist or physician and current technological capability has become so great that the life scientist cannot pose a problem that will incorporate the application of existing technology. Therefore, technological entrepreneurs examine some portion of the biological or medical front and identify areas in which advanced technology might be advantageous. Thus, they pose their own problem and then proceed to provide the solution, at first conceptually and then in the form of hardware or software. Finally, these individuals must convince the medical community that they can provide a useful tool because, contrary to the situation in which problem solvers find themselves, the entrepreneur's activity is speculative at best and has no ready-made customer for the results. If the venture is successful, however, whether scientifically or commercially, then an advance has been made much earlier than it would have been through the conventional arrangement. Because of the nature of their work, technological entrepreneurs should have a great deal of engineering and medical knowledge as well as experience in numerous medical systems.

The third type of biomedical engineer, the "engineer scientist" (most likely found in academic institutions and industrial research labs), is primarily interested in applying engineering concepts and techniques to the investigation and exploration of biological processes (Fig. 1.9). The most powerful tool at their disposal is the construction of an appropriate physical or mathematical model of the specific biological system under study. Through simulation techniques and available computing machinery, they can use this model to understand features that are too complex for either analytical computation or intuitive recognition. In addition, this process of simulation facilitates the design of appropriate experiments that can be performed on the actual biological system. The results of these experiments can, in turn, be used to amend the model. Thus, increased understanding of a biological mechanism results from this iterative process.

This mathematical model can also predict the effect of these changes on a biological system in cases in which the actual experiments may be tedious, very difficult, or dangerous. The researchers are thus rewarded with a better understanding of the biological system, and the mathematical description forms a compact, precise language that is easily communicated to others. The activities of the engineer scientist inevitably involve instrument development because the exploitation of sophisticated measurement techniques is often necessary to perform the biological side of the experimental work. It is essential that engineer scientists work in a biological en-

Fig. 1.9   The biomedical engineer in the role of the engineer scientist.

vironment, particularly when their work may ultimately have a clinical application. It is not enough to emphasize the niceties of mathematical analysis while losing the clinical relevance in the process. This biomedical engineer is a true partner of the biological scientist and has become an integral part of the research teams being formed in many institutes to develop techniques and experiments that will unfold the mysteries of the human organism (Fig. 1.10).

Each of these roles envisioned for the biomedical engineer requires a different attitude and a specific degree of knowledge about the biological environment. However, each engineer must be a skilled professional with a significant expertise in engineering technology. Therefore, in preparing new professionals to enter this field at these various levels, biomedical engineering educational programs are continually

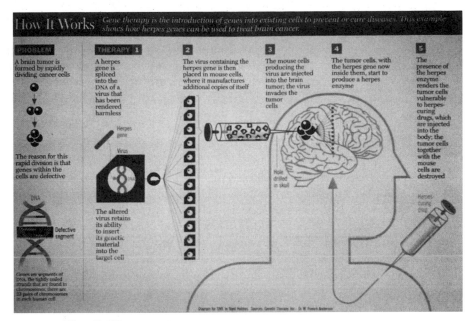

Fig. 1.10   Gene therapy — how it works (diagram by Nigel Holmes).

being challenged to develop curricula that will provide an adequate exposure to and knowledge about the environment without sacrificing essential engineering skills. As we continue to move into a period characterized by a rapidly growing aging population, rising social and economic expectations, and a need for the development of more adequate techniques for the prevention, diagnosis, and treatment of disease, development and employment of biomedical engineers have become a necessity. This is true not only because they may provide an opportunity to increase our knowledge of living systems but also because they constitute promising vehicles for expediting the conversion of knowledge to effective action.

The ultimate role of the biomedical engineer, like that of the nurse and physician, is to serve society. This is a profession, not just a skilled technical service. To use this new breed effectively, health care practitioners and administrators should be aware of the needs for these new professionals and the roles for which they are being trained. The great potential, challenge, and promise in this endeavor offer not only significant technological benefits but also humanitarian benefits.

1.5 PROFESSIONAL STATUS OF BIOMEDICAL ENGINEERING

Biomedical engineers are professionals. Professionals have been defined as an aggregate of people finding identity in sharing values and skills absorbed during a common course of intensive training. Parsons (1954) stated that one determines whether individuals are professionals by examining whether or not they have internalized certain given professional values. Friedson (1971) redefined Parson's definition by noting that a professional is someone who has internalized professional values and is to be recruited and licensed on the basis of his or her technical competence. Furthermore, he pointed out that professionals generally accept scientific standards in their work, restrict their work activities to areas in which they are technically competent, avoid emotional involvement, cultivate objectivity in their work, and put their clients' interests before their own.

The concept of a profession that is involved in the design, development, and management of medical technology encompasses three primary occupational models: science, business, and profession. Consider initially the contrast between science and profession. Science is seen as the pursuit of knowledge, its value hinging on providing evidence and communicating with colleagues. Profession, on the other hand, is viewed as providing a service to clients who have problems they cannot handle themselves. Science and profession have in common the exercise of some knowledge, skill, or expertise. However, while scientists practice their skills and report their results to knowledgeable colleagues, professionals, such as lawyers, physicians, and engineers, serve lay clients. To protect both the professional and the client from the consequences of the layperson's lack of knowledge, the practice of the profession is often regulated through such formal institutions as state licensing. Both professionals and scientists must persuade their clients to accept their findings. Professionals endorse and follow a specific code of ethics to serve society. On the other hand, scientists move their colleagues to accept their findings through persuasion.

Consider, for example, the medical profession. Its members are trained in caring for the sick, with the primary goal of healing them. These professionals not only have a responsibility for the creation, development, and implementation of that tradition but also are expected to provide a service to the public, within limits, without regard to self-interest. To ensure proper service, the profession itself closely monitors licensing and certification. Thus, medical professionals may be regarded as a mechanism of social control. However, this does not mean that other facets of society are not involved in exercising oversight and control of physicians in their practice of medicine.

A final attribute of professionals is that of integrity. Physicians tend to be both permissive and supportive in relationships with patients and yet are often confronted with moral dilemmas involving the desires of their patients and social interest. For example, how to honor the wishes of terminally ill patients while not facilitating the patients' deaths is a moral question that health professionals are forced to confront. A detailed discussion of the moral issues posed by medical technology is presented in Chapter 20.

One can determine the status of professionalization by noting the occurrence of six crucial events: (i) the first training school, (ii) the first university school, (iii) the first local professional association, (iv) the first national professional association, (v) the first state license law, and (vi) the first formal code of ethics.

The early appearances of the training school and the university affiliation underscore the importance of the cultivation of a knowledge base. The strategic innovative role of the universities and early teachers lies in linking knowledge to practice and creating a rationale for exclusive jurisdiction. Those practitioners pushing for prescribed training then form a professional association. The association defines the tasks of the profession: raising the quality of recruits; redefining their function to permit the use of less technically skilled people to perform the more routine, less involved tasks; and managing internal and external conflicts. In the process, internal conflict may arise between those committed to established procedures and newcomers committed to change and innovation. At this stage, some form of professional regulation, such as licensing or certification, surfaces because of a belief that it will ensure minimum standards for the profession, enhance status, and protect the layperson in the process.

The last area of professional development is the establishment of a formal code of ethics, which usually includes rules to exclude the unqualified and unscrupulous practitioners, rules to reduce internal competition, and rules to protect clients and emphasize the ideal service to society. A code of ethics usually is developed at the end of the professionalization process.

In biomedical engineering, all six critical steps mentioned previously have been clearly taken. The field of biomedical engineering, which originated as a professional group interested primarily in medical electronics in the late 1950s, has grown from a few scattered individuals to a very well-established organization. There are approximately 50 national societies throughout the world serving an increasingly expanding community of biomedical engineers. Today, the scope of biomedical engineering is enormously diverse. Over the years, many new disciplines, such as tissue engineering and artificial intelligence, which were once considered alien to the field are now an integral part of the profession.

Professional societies play a major role in bringing together members of this diverse community to share their knowledge and experience in pursuit of new technological applications that will improve the health and quality of life of human beings. Intersocietal cooperation and collaborations, both at national and international levels, are more actively fostered today through professional organizations, such as the International Federation of Medical and Biological Engineers (IFMBE), the American Institute of Medical and Biological Engineers (AIMBE), and Engineering in Medicine and Biology Society (EMBS) of the Institute of Electrical and Electronic Engineers (IEEE).

1.6 PROFESSIONAL SOCIETIES

1.6.1 American Institute for Medical and Biological Engineering

The United States has the largest biomedical engineering community in the world. Major professional organizations that address various cross sections of the field and serve more than 20,000 biomedical engineers include the American College of Clinical Engineering, the American Institute of Chemical Engineers, the American Medical Informatics Association, the American Society of Agricultural Engineers, the American Society for Artificial Internal Organs, the American Society of Mechanical Engineers, the Association for the Advancement of Medical Instrumentation, the Biomedical Engineering Society, the IEEE Engineering in Medicine and Biology Society, an interdisciplinary Association for the Advancement of Rehabilitation and Assistive Technologies, and the Society for Biomaterials. In an effort to unify all the disparate components of the biomedical engineering community in the United States as represented by these various societies, AIMBE was created in 1992. The primary goal of AIMBE is to serve as an umbrella organization in the United States for the purpose of unifying the bioengineering community, addressing public policy issues, and promoting the engineering approach in society's effort to enhance health and quality of life through the judicious use of technology. For information, contact the executive office, AIMBE, 1901 Pennsylvania Ave., N.W., Washington, DC 20005 (*http://bme.www.ecn.purdue.edu/BME/societies/AIMBE/aimbe.html*; Tel: 202-496-9661; Fax: 202-466-8489).

1.6.2 IEEE Engineering in Medicine and Biology Society

The IEEE is the largest international professional organization in the world and accommodates 37 different societies and councils under its umbrella structure. Of these 37, the EMBS represents the foremost international organization serving the needs of more than 8000 biomedical engineering members around the world. The field of interest of EMBS is application of the concepts and methods of the physical and engineering sciences in biology and medicine. Each year, the society sponsors a major international conference while cosponsoring a number of theme-oriented regional conferences throughout the world. Premier publications are a monthly journal (*Transactions on Biomedical Engineering*) and two quarterly journals (*Transactions on Rehabilitation Engineering* and *Transactions on Information Technology in Biomedicine*) and a bimonthly magazine, the *IEEE Engineering in Medicine and Biology Magazine*). For more information, contact the IEEE EMBS Secretariat, National Research Council of Canada, Room 220, Building M-19 Ottawa, Ontario K1A OR, 6 Canada (*http://www.bae.ncsu.edu/research/blanchard/www/embs/*; Tel: 613-954-4005; Fax: 613-954-2216; e-mail: *soc.emb@ieee.org*).

1.6.3 International Federation for Medical and Biological Engineering

Established in 1959, the IFMBE is an organization made up from an affiliation of national societies including members of transnational organizations. The current national affiliates are Argentina, Australia, Austria, Belgium, Brazil, Bulgaria, Canada, China, Cuba, Cyprus, Slovakia, Denmark, Finland, France, Germany, Greece, Hungary, Israel, Italy, Japan, Mexico, The Netherlands, Norway, Poland, South Africa, South Korea, Spain, Sweden, Thailand, United Kingdom, and the United States. The first transnational organization to become a member of the federation was the IEEE EMBS. Currently, the federation has an estimated 25,000 members from all of its constituent societies.

The primary goal of the IFMBE is to recognize the interests and initiatives of its affiliated member organizations and to provide an international forum for the exchange of ideas and dissemination of information. The major IFMBE activities include the publication of the federation's bimonthly journal, the *Journal of Medical and Biological Engineering and Computing,* and the MBEC *News;* establishment of close liaisons with developing countries to encourage and promote BME activities; and the organization of a major world conference every 3 years in collaboration with the International Organization for Medical Physics and the International Union for Physical and Engineering Sciences in Medicine. The IFMBE also serves as a consultant to the United Nations Industrial Development Organization and has nongovernmental status with the World Health Organization, the United Nations, and the Economic Commission for Europe. For more information, contact the Secretary General, International Federation for Medical and Biological Engineering, AMC, University of Amsterdam, Meibergdreef 15, 1105 AZ Amsterdam, The Netherlands (*http://vub.vub.ac.be/~ifmbe/ifmbe.html;* Tel: +31-205665200, ext. 5179; Fax: +31-20 6917233; e-mail: *ifmbe@amc.uva.nl*).

The activities of these biomedical engineering societies are critical to the continued advancement of the professional status of biomedical engineers. Therefore, all biomedical engineers, including students in the professon, are encouraged to become members of these societies and engage in the activities of true professionals.

EXERCISES

1. Select a specific medical technology from the historical periods indicated. Describe the fundamental principles of operation and discuss their impact on health care delivery: (a) 1900–1939, (b) 1945–1970, (c) 1970–1980, (d) 1980–1995.

2. Provide a definition for the term biomedical engineer and discuss the significance of the key words in the definition.

3. The term "genetic engineering" implies an engineering function. Is there one? Should this activity be included in the field of biomedical engineering?

4. Provide modern examples (i.e., names of individuals and their activities) of the three major roles played by biomedical engineers: (a) the problem solver, (b) the technological entrepreneur, and (c) the engineer scientist. (Hint: Review recent activities in one of the following journals: *Engineering in Medicine and Biology Magazine, Biomedical Instrumentation and Technology, Annals of Biomedical Engineering,* or *Journal of Clinical Engineering*).

5. Do the following groups fit the definition of a profession? Discuss how they do or do not: (a) registered nurses, (b) biomedical technicians, and (c) respiratory therapists.

6. List the areas of knowledge necessary to practice biomedical engineering. Identify where in the normal educational process one can acquire knowledge. How best can administrative skills be acquired?

7. Provide a copy of the home page for a biomedical engineering professional society and a list of their major activities for the coming year.

SUGGESTED READING

Atkinson, D. T. (1956). *Magic, Myth and Medicine.* Fawcett, New York.

Bronzino, J. D. (1971). The biomedical engineer — The roles he can play. *Science* **174,** 1001–1003.

Bronzino, J. D. (1977). *Technology for Patient Care.* Mosby, St. Louis.

Bronzino, J. D. (1986). *Biomedical Engineering and Instrumentation: Basic Concepts and Applications,* PWS Press, Boston.

Bronzino, J. D. (1991). *Biomedical Engineering Encyclopedia of Applied Physics,* Vol. 2, VCH Publishers, Inc. pp. 513–548.

Bronzino, J. D. (1995). *Biomedical Engineering Handbook.* CRC Press, Boca Raton, FL.

Calder, R. (1958). *Medicine and Man.* Allen & Unwin, London.

Crichton, M. (1970). *Five Patients — The Hospital Explained.* Knopf, New York.

Friedson, E. (1971). *Profession of Medicine.* Dodd, Mead, New York.

Johns, R. J. (1975). Current issues in biomedical engineering education. *IEEE Trans. Biomed. Eng.* **22,** 107–110.

Knowles, J. (1973). The hospital. *Sci. Am.* **229,** 128–137.

Marks, G., and Bealty, W. K. (1972). *Women in White.* Scribner's, New York.

Pacela, A. (1990). *Bioengineering Education Directory.* Quest, Brea, CA.

Parsons, T. (1954). *Essays in Sorrological Theories.* Free Press, Glencoe, IL.

Plonsey, R. (1973). New directions for biomedical engineering. *Eng. Education* **64,** 117–179.

Rushmer, R. (1972). *Medical Engineering Projections for Health Care Delivery.* Academic Press, New York.

Sigerest, H. E. (1951). *A History of Medicine.* Oxford Univ. Press, New York.

Susskind, C. (1973). *Understanding Technology.* Johns Hopkins Univ. Press, Baltimore, MD.

Wilensky, H. L. (1964). The professionalization of everyone. *Am. J. Soc.* **69,** 137–158.

Wolff, H. S. (1967). Bioengineering. *Biomed. Eng.* **2,** 547–549.

2 ANATOMY AND PHYSIOLOGY

Chapter Contents

At the conclusion of this chapter, the reader will be able to:

- Define anatomy and physiology and explain why they are important to biomedical engineering

- Define important anatomical terms
- Describe the cell theory
- List the major types of organic compounds and other elements found in cells
- Explain how the plasma membrane maintains the volume and internal concentrations of a cell
- Calculate the internal osmolarity and ionic concentrations of a model cell at equilibrium
- List and describe the functions of the major organelles found within mammalian cells
- Describe the similarities, differences, and purposes of replication, transcription, and translation
- List and describe the major components and functions of five organ systems — circulatory, respiratory, nervous, skeletal, and muscular
- Define homeostasis and describe how feedback mechanisms help maintain it

2.1   INTRODUCTION

Since biomedical engineering is an interdisciplinary field based in both engineering and the life sciences, it is important for biomedical engineers to have knowledge about and be able to communicate in both areas. Biomedical engineers must understand the basic components of the body and how they function well enough to exchange ideas and information with physicians and life scientists. Two of the most basic terms and areas of study in the life sciences are anatomy and physiology. Anatomy refers to the internal and external structures of the body and their physical relationships, whereas physiology refers to the study of the functions of those structures.

Figure 2.1a shows a male body in anatomical position. In this position, the body is erect and facing forward with the arms hanging at the sides and the palms facing outward. This particular view shows the anterior (ventral) side of the body, whereas Fig. 2.1c illustrates the posterior (dorsal) view of another male body that is also in anatomical position and Fig. 2.1b presents the lateral view of the female body. In clinical practice, directional terms are used to describe the relative positions of various parts of the body. Proximal parts are nearer to the trunk of the body or to the attached end of a limb than are distal parts (Fig. 2.1a). Parts of the body that are located closer to the head than other parts when the body is in anatomical position are said to be superior (Fig. 2.1b), whereas those located closer to the feet than other parts are termed inferior. Medial implies that a part is toward the midline of the body, whereas lateral means away from the midline (Fig. 2.1c). Parts of the body that lie in the direction of the head are said to be in the cranial direction, whereas those parts that lie in the direction of the feet are said to be in the caudal direction (Fig. 2.2).

Anatomical locations can also be described in terms of planes. The plane that divides the body into two symmetric halves along its midline is called the midsagittal plane (Fig. 2.2). Planes that are parallel to the midsagittal plane but do not divide the body into symmetric halves are called sagittal planes. The frontal plane is perpendicular to the midsagittal plane and divides the body into asymmetric anterior

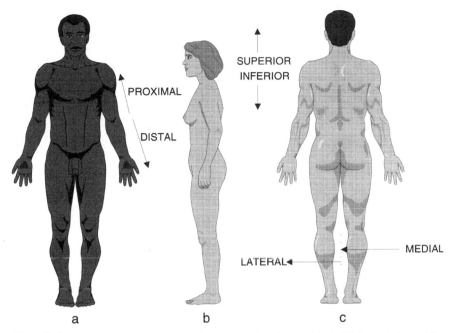

Fig. 2.1 (a) Anterior view of male body in anatomical position. (b) Lateral view of female body. (c) Posterior view of male body in anatomical position. Relative directions (proximal and distal, superior and inferior, and medial and lateral) are also shown.

and posterior portions. Planes that cut across the body and are perpendicular to the midsagittal and frontal planes are called transverse planes.

Human bodies are divided into two main regions, axial and appendicular. The axial part consists of the head, neck, thorax (chest), abdomen, and pelvis, whereas the appendicular part consists of the upper and lower extremities. The upper extremities, or limbs, include the shoulders, upper arms, forearms, wrists, and hands, whereas the lower extremities include the hips, thighs, lower legs, ankles, and feet. The abdominal region can be further divided into nine regions or four quadrants.

The cavities of the body hold the internal organs. The major cavities are the dorsal and ventral body cavities, whereas smaller ones include the nasal, oral, orbital (eye), tympanic (middle ear), and synovial (movable joint) cavities. The dorsal body cavity includes the cranial cavity that holds the brain and the spinal cavity that contains the spinal cord. The ventral body cavity contains the thoracic and abdominopelvic cavities that are separated by the diaphragm. The thoracic cavity contains the lungs and the mediastinum, which contains the heart and its attached blood vessels, the trachea, the esophagus, and all other organs in this region except for the lungs. The abdominopelvic cavity is divided by an imaginary line into the abdominal and pelvic cavities. The former is the largest cavity in the body and holds the stomach, small and large intestines, liver, spleen, pancreas, kidneys, and gall bladder. The latter contains the urinary bladder, the rectum, and the internal portions of the reproductive system.

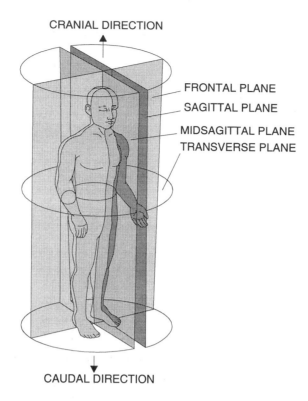

CRANIAL DIRECTION

FRONTAL PLANE
SAGITTAL PLANE
MIDSAGITTAL PLANE
TRANSVERSE PLANE

CAUDAL DIRECTION

Fig. 2.2   The body can be divided into sections by the frontal, sagittal, and transverse planes. The midsagittal plane goes through the midline of the body.

The anatomical terms described previously are used by physicians, life scientists, and biomedical engineers when discussing the whole human body or its major parts. Correct use of these terms is vital for biomedical engineers to communicate with health care professionals and to understand the medical problem of concern or interest. While it is important to be able to use the general terms that describe the human body, it is also important for biomedical engineers to have a basic understanding of some of the more detailed aspects of human anatomy and physiology.

2.2   CELLULAR ORGANIZATION

Although there are many smaller units such as enzymes and organelles that perform physiological tasks or have definable structures, the smallest anatomical and physiological unit in the human body that can, under appropriate conditions, live and reproduce on its own is the cell. Cells were first discovered more than 300 years ago shortly after Antony van Leeuwenhoek, a Dutch optician, invented the microscope. With his microscope, van Leeuwenhoek was able to observe "many very small animalcules, the motions of which were very pleasing to behold" in tartar scrapings from his teeth. Following the efforts of van Leeuwenhoek, Robert Hooke, a Curator of Instruments for the Royal Society of England, in the late 1600s further described cells when he used one of the earliest microscopes to look at the plant cell walls that remain in cork. These observations and others led to the cell theory developed by Theodor Schwann and Matthias Jakob Schleiden and formalized by Rudolf Virchow in the mid-1800s. The cell theory states that (i) all organisms are composed of one or more cells, (ii) the cell is the smallest unit of life, and (iii) all cells come from previously existing cells. Thus, cells are the basic building blocks of life.

Cells are composed mostly of organic compounds and water, with more than 60% of the weight in a human body coming from water. The organic compounds — carbohydrates, lipids, proteins, and nucleic acids — that cells synthesize are the molecules which are fundamental to sustaining life. These molecules function as energy packets, storehouses of energy and hereditary information, structural materials, and metabolic workers. The most common elements found in humans (in descending order based on percentage of body weight) are oxygen, carbon, hydrogen, nitrogen, calcium, phosphorus, potassium, sodium, chlorine, magnesium, sulfur, iron, and iodine. Carbon, hydrogen, oxygen, and nitrogen contribute more than 99% of all the atoms in the body. Most of these elements are incorporated into organic compounds, but some exist in other forms, such as phosphate groups and ions.

Carbohydrates are used by cells not only as structural materials but also to transport and store energy. There are three classes of carbohydrates: monosaccharides (e.g., glucose), oligosaccharides (e.g., lactose, sucrose, and maltose), and polysaccharides (e.g., glycogen). Lipids are greasy or oily compounds that will dissolve in each other but not in water. They form structural materials in cells and are the main reservoirs of stored energy. Proteins are the most diverse form of biological molecules. Specialized proteins, called enzymes, make metabolic reactions proceed at a faster rate than would occur if the enzymes were not available and enable cells to produce the organic compounds of life. Other proteins provide structural elements in the body, act as transport channels across plasma membranes, function as signals for changing activities, and provide chemical weapons against disease-carrying bacteria. These diverse proteins are built from a small number (20) of essential amino acids.

Nucleotides and nucleic acids make up the last category of important biological molecules. Nucleotides are small organic compounds that contain a five-carbon sugar (ribose or deoxyribose), a phosphate group, and a nitrogen-containing base that has a single or double carbon ring structure. Adenosine triphosphate (ATP) is the energy currency of the cell and plays a central role in metabolism. Other nucleotides are subunits of coenzymes which are enzyme helpers. The two nucleic acids are DNA (deoxyribonucleic acid) and RNA (ribonucleic acid). DNA (Fig. 2.3) is a unique, helical molecule that contains chains of paired nucleotides that run in opposite directions. Each nucleotide contains either a pyrimidine base — thymine (T) or cytosine (C) — with a single ring structure or a purine base — adenine (A) or guanine (G) — with a double ring. In the double helix of DNA, thymine always

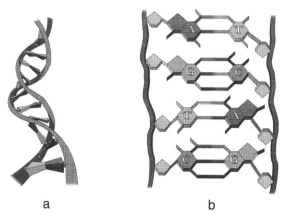

a   b

Fig. 2.3 (a) DNA consists of two chains of paired nucleotides that run in opposite directions and form a helical structure. (b) Thymine pairs with adenine (T–A) and cytosine pairs with guanine (C–G) due to hydrogen bonding between the bases.

pairs with adenine (T–A) and cytosine always pairs with guanine (C–G). RNA is similar to DNA except that it consists of a single helical strand, contains ribose instead of deoxyribose, and has uracil (U) instead of thymine.

All cells are surrounded by a plasma membrane that separates, but does not isolate, the cell's interior from its environment. Animal cells, such as those found in humans, are eukaryotic cells. A generalized animal cell is shown in Fig. 2.4. In addition to the plasma membrane, eukaryotic cells contain membrane-bound organelles and a membrane-bound nucleus. Prokaryotic cells, e.g., bacteria, lack membrane-bound structures other than the plasma membrane. In addition to a plasma membrane, all cells have a region that contains DNA (which carries the hereditary instructions for the cell) and cytoplasm (which is a semifluid substance that includes everything inside the plasma membrane except for the DNA).

2.2.1 Plasma Membrane

The plasma membrane performs several functions for the cell. It gives mechanical strength, provides structure, helps with movement, and controls the cell's volume and its activities by regulating the movement of chemicals in and out of the cell. The plasma membrane is composed of two layers of phospholipids interspersed with proteins and cholesterol (Fig. 2.5). The proteins in the plasma membranes of mammalian cells provide binding sites for hormones, recognition markers for identifying cells as one type or another, adhesive mechanisms for binding adjacent cells to each other, and channels for transporting materials across the plasma membrane. The phospholipids are arranged with their "water-loving" (hydrophilic) heads pointing outward and their "water-fearing" (hydrophobic) tails pointing inward. This double-layer arrangement of phospholipids interspersed with protein channels helps maintain the internal environment of a cell by controlling the substances that move

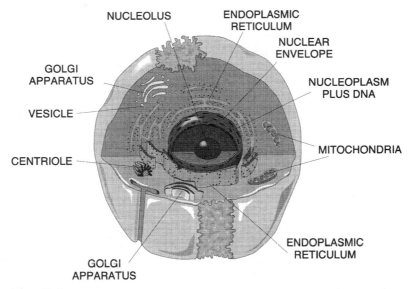

Fig. 2.4 Animal cells are surrounded by a plasma membrane. They contain a membrane-bound region, the nucleus, which contains DNA. The cytoplasm lies outside of the nucleus and contains several types of organelles that perform specialized functions.

across the membrane while the cholesterol molecules act as stabilizers to prevent extensive lateral movement of the lipid molecules.

Some molecules, e.g., oxygen, carbon dioxide, and water, can easily cross the plasma membrane, whereas other substances, e.g., large molecules and ions, must move through the protein channels. Osmosis is the process by which substances move across a selectively permeable membrane such as a cell's plasma membrane,

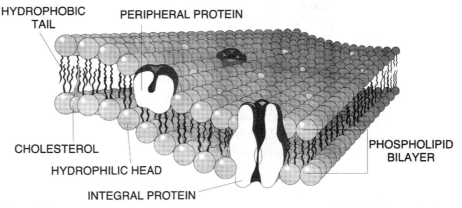

Fig. 2.5 The plasma membrane surrounds all cells. It consists of a double layer of phospholipids interspersed with proteins and cholesterol.

whereas diffusion refers to the movement of molecules from an area of relatively high concentration to an area of relatively low concentration. Substances that can easily cross the plasma membrane achieve diffusion equilibrium when there is no net movement of these substances across the membrane, i.e., the concentration of the substance inside the cell equals the concentration of the substance outside of the cell. Active transport, which requires an input of energy usually in the form of ATP, can be used to move ions and molecules across the plasma membrane and is often used to move them from areas of low concentration to areas of high concentration. This mechanism helps maintain concentrations of ions and molecules inside a cell that are different from the concentrations outside the cell. A typical mammalian cell has internal sodium ion (Na^+) concentrations of 12 mM (12 moles of Na^+ per 1000 liters of solution) and extracellular Na^+ concentrations of 120 mM, whereas intracellular and extracellular potassium ion (K^+) concentrations are on the order of 125 and 5 mM, respectively. In addition to positively charged ions (cations), cells also contain negatively charged ions (anions). A typical mammalian cell has intracellular and extracellular chloride ion (Cl^-) concentrations of 5 and 125 mM and internal anion (e.g., proteins, charged amino acids, sulfate ions, and phosphate ions) concentrations of 108 mM. These transmembrane ion gradients are used to make ATP, to drive various transport processes, and to generate electrical signals.

The plasma membrane plays an important role in regulating cell volume by controlling the internal osmolarity of the cell. Osmolarity is defined in terms of concentration of dissolved substances. A 1 osmolar (1 Osm) solution contains 1 mol of dissolved particles per liter of solution, whereas a 1 milliosmolar (1 mOsm) solution has 1 mol of dissolved particles per 1000 liters of solution. Thus, solutions with high osmolarity have low concentrations of water or other solvents. For biological purposes, solutions with 0.1 Osm glucose and 0.1 Osm urea have essentially the same concentrations of water. It is important to note that a 0.1 M solution of sodium chloride (NaCl) will form a 0.2 Osm solution since NaCl dissociates into Na^+ and Cl^- ions and thus has twice as many dissolved particles as a solution of a substance (e.g., glucose) that does not dissociate into smaller units. Two solutions are isotonic if they have the same osmolarity. One solution is hypotonic to another if it has a lower osmolarity and hypertonic to another if it has a higher osmolarity.

Consider a simple model cell that consists of a plasma membrane and cytoplasm. The cytoplasm in this model cell contains proteins that cannot cross the plasma membrane and water which can. At equilibrium, the total osmolarity inside the cell must equal the total osmolarity outside the cell. If the osmolarity inside and the osmolarity outside of the cell are out of balance, there will be a net movement of water from the side of the plasma membrane where it is more highly concentrated to the other until equilibrium is achieved. For example assume that a model cell (Fig. 2.6) contains 0.2 M protein and is placed in a hypotonic solution that contains 0.1 M sucrose. The plasma membrane of this model cell is impermeable to proteins and sucrose but freely permeable to water. The volume of the cell, 1 nl, is very small relative to the volume of the solution. In other words, changes in the cell's volume have no measurable effect on the volume of the external solution. What will happen to the volume of the cell as it achieves equilibrium?

At equilibrium, the osmolarity inside the cell must equal the osmolarity outside

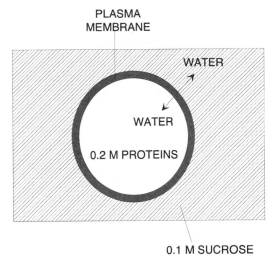

Fig. 2.6 A simple model cell which consists of cytoplasm, containing 0.2 *M* proteins, and a plasma membrane is placed in a solution of 0.1 *M* sucrose. The plasma membrane is insoluble to proteins and sucrose but allows water to pass freely in either direction. The full extent of the extracellular volume is not shown and is much larger than the cell's volume of 1 nl.

the cell. The initial osmolarity inside the cell is 0.2 Osm since the proteins do not dissociate into smaller units. The osmolarity outside the cell is 0.1 Osm due to the sucrose solution. A 0.2 Osm solution has 0.2 mol of dissolved particles per liter of solution, whereas a 0.1 Osm solution has half as many moles of dissolved particles per liter. The osmolarity inside the cell must decrease by a factor of 2 in order to achieve equilibrium. Since the plasma membrane will not allow any of the protein molecules to leave the cell, this can only be achieved by doubling the cell's volume. Thus, there will be a net movement of water across the plasma membrane until the cell's volume increases to 2 nl and the cell's internal osmolarity is reduced to 0.1 Osm — the same as the osmolarity of the external solution. The water moves down its concentration gradient by diffusing from where it is more highly concentrated in the 0.1 *M* sucrose solution to where it is less concentrated in the 0.2 *M* protein solution in the cell. What would happen to this same model cell if it was placed in pure water? (Hint: Think about how equilibrium could be achieved.)

Real cells are much more complex than the simple model described previously. In addition to achieving osmotic balance at equilibrium, real cells must also achieve electrical balance with regard to the ions that are present in the cytoplasm. The principle of electrical neutrality requires that the overall concentration of cations in a biological compartment (e.g., a cell) must equal the overall concentration of anions in that compartment. Consider another model cell (Fig. 2.7) with internal and external cation and anion concentrations similar to those of a typical mammalian cell. Is the cell at equilibrium if the plasma membrane is freely permeable to K^+ and

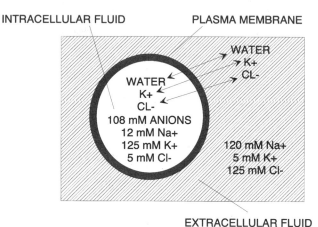

Fig. 2.7   A model cell with internal and external concentrations similar to those of a typical mammalian cell. The full extent of the extracellular volume is not shown and is much larger than the cell's volume.

Cl^- but impermeable to Na^+ and the internal anions? The total osmolarity inside the cell is 250 mOsm (12 mM Na^+, 125 mM K^+, 5 mM Cl^-, 108 mM anions) and the total osmolarity outside the cell is also 250 mOsm (120 mM Na^+, 5 mM K^+, 125 mM Cl^-), so the cell is in osmotic balance, i.e., there will be no net movement of water across the plasma membrane. If the average charge per molecule of the anions inside the cell is considered to be -1.2, then the cell is also approximately in electrical equilibrium (12 + 125 positive charges for Na^+ and K^+; 5 + 1.2 × 108 negative charges for Cl^- and the other anions). Real cells, however, cannot maintain this equilibrium without expending energy since real cells are permeable to Na^+. In order to maintain equilibrium and keep Na^+ from accumulating intracellularly, mammalian cells must actively pump Na^+ out of the cell against its diffusion and electrical gradients. Since Na^+ is pumped out through specialized protein channels at a rate equivalent to the rate at which it leaks in through other channels, it behaves osmotically as if it cannot cross the plasma membrane. Thus, mammalian cells exist in a steady state rather than at equilibrium since energy in the form of ATP must be used to prevent a net movement of ions across the plasma membrane.

One of the consequences of the distribution of charged particles in the intracellular and extracellular fluids is that an electrical potential exists across the plasma membrane. The value of this electrical potential depends on the intracellular and extracellular concentrations of ions that can cross the membrane and will be described more fully in Chapter 3.

In addition to controlling the cell's volume, the plasma membrane also provides a route for moving large molecules and other materials into and out of the cell. Substances can be moved into the cell by means of endocytosis (Fig. 2.8a) and out of the cell by means of exocytosis (Fig. 2.8b). In endocytosis, material (e.g., a bac-

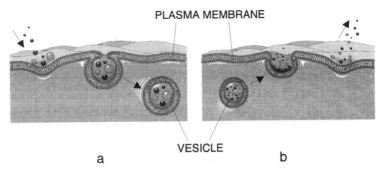

Fig. 2.8 Substances that are too large to pass through the integral proteins in the plasma membrane can be moved into the cell by means of endocytosis (a) and out of the cell by means of exocytosis (b).

terium) outside of the cell is engulfed by a portion of the plasma membrane which encircles it to form a vesicle. The vesicle then pinches off from the plasma membrane and moves its contents to the inside of the cell. In exocytosis, material within the cell is surrounded by a membrane to form a vesicle. The vesicle then moves to the edge of the cell where its membrane fuses with the plasma membrane and its contents are released to the exterior of the cell.

2.2.2 Cytoplasm and Organelles

The cytoplasm contains fluid (cytosol) and organelles. Ions (such as Na^+, K^+, and Cl^-) and molecules (such as glucose) are distributed through the cytosol via diffusion. Membrane-bound organelles include the nucleus, rough and smooth endoplasmic reticulum, the Golgi apparatus, lysosomes, and mitochondria. Nonmembranous organelles include nucleoli, ribosomes, centrioles, microvilli, cilia, flagella, and the microtubules, intermediate filaments, and microfilaments of the cytoskeleton.

The nucleus (Fig. 2.4) consists of the nuclear envelope (a double membrane) and the nucleoplasm (a fluid which contains ions, enzymes, nucleotides, proteins, DNA, and small amounts of RNA). Within its DNA, the nucleus contains the instructions for life's processes. Nuclear pores are protein channels that act as connections for ions and RNA, but not protein or DNA, to leave the nucleus and enter the cytoplasm and for some proteins to enter the nucleoplasm. Most nuclei contain one or more nucleoli. Each nucleolus contains DNA, RNA, and proteins and synthesizes the components of the ribosomes that cells use to make proteins.

The smooth and rough endoplasmic reticulum (ER), Golgi apparatus, and assorted vesicles (Figs. 2.4, 2.9a, and 2.9b) make up the cytomembrane system which delivers proteins and lipids for manufacturing membranes and accumulates and stores proteins and lipids for specific uses. The ER also acts as a storage site for calcium ions. The rough ER differs from the smooth ER in that it has ribosomes attached to its exterior surface. Ribosomes provide the platforms for synthesizing

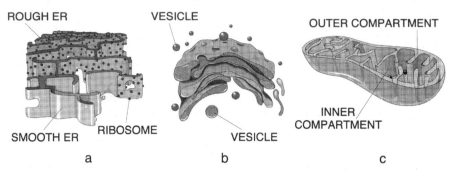

Fig. 2.9   Subcellular organelles. The endoplasmic reticulum (a), the Golgi apparatus (b), and vesicles (b) make up the cytomembrane system in the cell. The small circles on the endoplasmic reticulum (ER) represent ribosomes. The area containing ribosomes is called the rough ER, whereas the area that lacks ribosomes is called the smooth ER. The mitochondria (c) have a double membrane system which divides the interior into two compartments that contain different concentrations of enzymes, substrates, and hydrogen ions (H^+). Electrical and chemical gradients between the inner and outer compartments provide the energy needed to generate ATP.

proteins. Those that are synthesized on the rough ER are passed into its interior where nonproteinaceous side chains are attached to them. These modified proteins move to the smooth ER where they are packaged in vesicles. The smooth ER also manufactures and packages lipids into vesicles and is responsible for releasing stored calcium ions. The vesicles leave the smooth ER and become attached to the Golgi apparatus where their contents are released, modified, and repackaged into new vesicles. Some of these vesicles, called lysosomes, contain digestive enzymes which are used to break down materials that move into the cells via endocytosis. Other vesicles contain proteins, such as hormones and neurotransmitters, that are secreted from the cells by means of exocytosis.

The mitochondria (Figs. 2.9c and 2.10) contain two membranes — an outer membrane which surrounds the organelle and an inner membrane that divides the organelle's interior into two compartments. Approximately 95% of the ATP required by the cell is produced in the mitochondria in a series of oxygen-requiring reactions which produce carbon dioxide as a by-product. Mitochondria are different from most other organelles in that they contain their own DNA. The majority of the mitochondria in sexually reproducing organisms, such as humans, come from the mother's egg cell since the father's sperm contributes little more than the DNA in a haploid (half) set of chromosomes to the developing offspring.

Microtubules, intermediate filaments, and microfilaments provide structural support and assist with movement. Microtubules are long, hollow, cylindrical structures that radiate from microtubule organizing centers and, during cell division, from centrosomes, a specialized region of the cytoplasm that is located near the nucleus and contains two centrioles (Figs. 2.4 and 2.11a) oriented at right angles to each other. Microtubules consist of spiraling subunits of a protein called tubulin, whereas centrioles consist of nine triplet microtubules that radiate from their centers like the spokes of a wheel. Intermediate filaments are hollow and provide structure to the

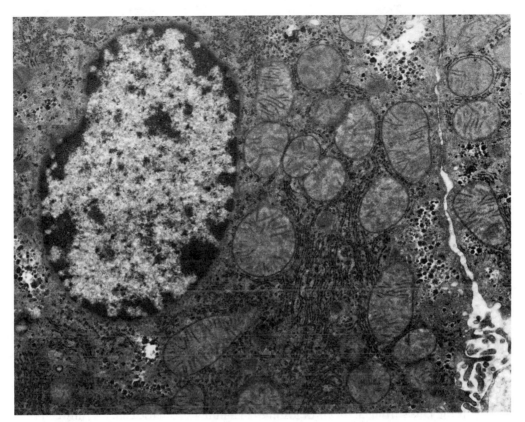

Fig. 2.10 Scanning electron micrograph of a normal mouse liver at 8000× magnification. The large round organelle on the left is the nucleus. The smaller round and oblong organelles are mitochondria that have been sliced at different angles. The narrow membranes in parallel rows are endoplasmic reticula. The small black dots on the ERs are ribosomes (photo courtesy of Valerie Knowlton, Center for Electron Microscopy, North Carolina State University.)

plasma membrane and nuclear envelope. They also aid in cell-to-cell junctions and in maintaining the spatial organization of organelles. Myofilaments are found in most cells and are composed of strings of protein molecules. Cell movement can occur when actin and myosin, protein subunits of myofilaments, interact. Microvilli (Fig. 2.11b) are extensions of the plasma membrane that contain microfilaments. They increase the surface area of a cell to facilitate absorption of extracellular materials.

Cilia (Fig. 2.11c) and flagella are parts of the cytoskeleton that have shafts composed of nine pairs of outer microtubules and two single microtubules in the center. Both types of shafts are anchored by a basal body which has the same structure as a centriole. Flagella function as whiplike tails that propel cells such as sperm. Cilia are generally shorter and more profuse than flagella and can be found on specialized cells such as those that line the respiratory tract. The beating of the cilia helps move mucus-trapped bacteria and particles out of the lungs.

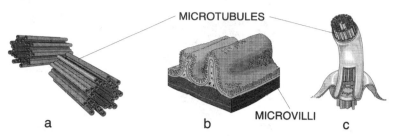

Fig. 2.11   Centrioles (a) contain microtubules and are located at right angles to each other in the cell's centrosome. These organelles play an important part in cell division by anchoring the microtubules that are used to divide the cell's genetic material. Microvilli (b), which are extensions of the plasma membrane, line the villi, tiny finger-like protrusions in the mucosa of the small intestine, and help increase the area available for the absorption of nutrients. Cilia (c) line the respiratory tract. The beating of these organelles helps move bacteria and particles trapped in mucus out of the lungs.

2.2.3   DNA and Gene Expression

DNA (Fig. 2.3) is found in the nucleus and mitochondria of eukaryotic cells. In organisms that reproduce sexually, the DNA in the nucleus contains information from both parents while that in the mitochondria comes from the organism's mother. In the nucleus, the DNA is wrapped around protein spools, called nucleosomes, and is organized into pairs of chromosomes. Humans have 22 pairs of autosomal chromosomes and two sex chromosomes, XX for females and XY for males (Fig. 2.12). If the DNA from all 46 chromosomes in a human somatic cell, i.e., any cell that does not become an egg or sperm cell, was stretched out end to end, it would be about 2 nm wide and 2 m long. Each chromosome contains thousands of individual genes that are the units of information about heritable traits. Each gene has a particular location in a specific chromosome and contains the code for producing one of the three forms of RNA (ribosomal RNA, messenger RNA, and transfer RNA). The Human Genome Project was begun in 1990 and has as its goal to first identify the

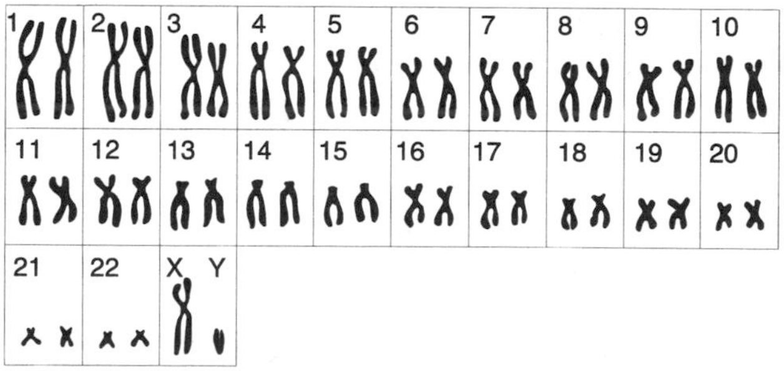

Fig. 2.12   This karyotype of a normal human male shows the 22 pairs of autosomal chromosomes in descending order based on size, as well as the X and Y sex chromosomes.

location of at least 3000 specific human genes and then to determine the sequence of nucleotides (about 3 billion!) in a complete set of haploid human chromosomes (one chromosome from each of the 23 pairs).

DNA replication occurs during cell division (Fig. 2.13). During this semiconservative process, enzymes unzip the double helix, deliver complementary bases to the nucleotides, and bind the delivered nucleotides into the developing complementary strands. Following replication, each strand of DNA is duplicated so that two double helices now exist, each consisting of one strand of the original DNA and one new strand. In this way, each daughter cell gets the same hereditary information that

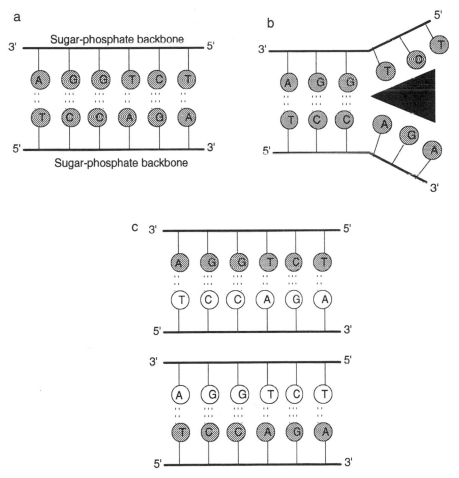

Fig. 2.13 During replication, DNA helicase (shown as a black wedge in b) unzips the double helix (a). Another enzyme, DNA polymerase, then copies each side of the unzipped chain in the 5′ to 3′ direction. One side of the chain (5′ to 3′) can be copied continuously, whereas the opposite side (3′ to 5′) is copied in small chunks in the 5′ to 3′ direction that are bound together by another enzyme, DNA ligase. Two identical double strands of DNA are produced as a result of replication.

was contained in the original dividing cell. During replication, some enzymes check for accuracy while others repair pairing mistakes so that the error rate is reduced to approximately one per billion.

Since DNA remains in the nucleus where it is protected from the action of the cell's enzymes and proteins are made on ribosomes outside of the nucleus, a method (transcription) exists for transferring information from the DNA to the cytoplasm. During transcription (Fig. 2.14), the sequence of nucleotides in a gene that codes for a protein is transferred to messenger RNA (mRNA) through complementary base pairing of the nucleotide sequence in the gene. For example, a DNA sequence of TACGCTCCGATA would become AUGCGAGGCUAU in the mRNA. The process is somewhat more complicated since the transcript produced directly from the DNA contains sequences of nucleotides, called introns, that are removed before the final mRNA is produced. The mRNA also has a tail, called a poly-A tail, of about 100–200 adenine nucleotides attached to one end. A cap with a nucleotide that has a methyl group and phosphate groups bonded to it is attached at the other end of the mRNA. Transcription differs from replication in that (i) only a certain stretch of

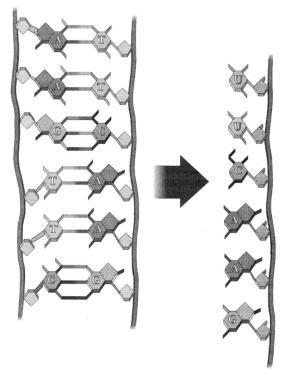

Fig. 2.14 During transcription, RNA is formed from genes in the cell's DNA by complementary base pairing to one of the strands. RNA contains uracil (U) rather than thymine (T) so the T in the first two pairs of the DNA become Us in the single-stranded RNA.

DNA acts as the template and not the whole strand, (ii) different enzymes are used, and (iii) only a single strand is produced.

After being transcribed, the mRNA moves out into the cytoplasm through the nuclear pores and binds to specific sites on the surface of the two subunits which make up a ribosome (Fig. 2.15). In addition to the ribosomes, the cytoplasm contains amino acids and another form of RNA, transfer RNA (tRNA). Each tRNA contains a triplet of bases, called an anticodon, and binds at an area away from the triplet to an amino acid that is specific for that particular anticodon. The mRNA that was produced from the gene in the nucleus also contains bases in sets of three. Each triplet in mRNA is called a codon. The four possibilities for nucleotides (A, U, C, and G) in each of the three places give rise to 64 (4^3) possible codons. These 64 codons make up the genetic code. Each codon codes for a specific amino acid, but some amino acids are specified by more than one codon. For example, AUG is the only mRNA codon for methionine (the amino acid that always signals the starting place for translation — the process by which the information from a gene is used to produce a protein), whereas UUA, UUG, CUU, CUC, CUA, and CUG are all codons for leucine. The anticodon on the tRNA that delivers the methionine to the ribo-

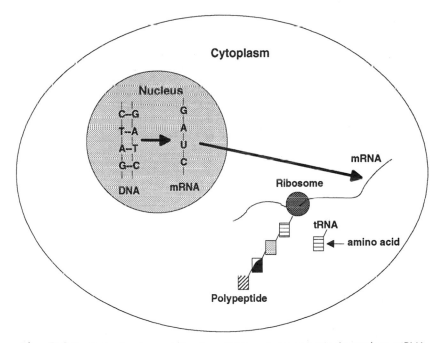

Fig. 2.15 Following transcription from DNA and processing in the nucleus, mRNA moves from the nucleus to the cytoplasm. In the cytoplasm, the mRNA joins with a ribosome to begin the process of translation. During translation, tRNA delivers amino acids to the growing polypeptide chain. Which amino acid is delivered depends on the three-base codon specified by the mRNA. Each codon is complementary to the anticodon of a specific tRNA. Each tRNA binds to a particular amino acid at a site that is opposite the location of the anticodon. For example, the codon CUG in mRNA is complementary to the anticodon GAC in the tRNA that carries leucine and will result in adding the amino acid leucine to the polypeptide chain.

some is UAC, whereas tRNAs with anticodons of AAU, AAC, GAA, GAG, GAU, and GAC deliver leucine.

During translation, the mRNA binds to a ribosome and tRNA delivers amino acids to the growing polypeptide chain in accordance with the codons specified by the mRNA. Peptide bonds are formed between each newly delivered amino acid and the previously delivered one. When the amino acid is bound to the growing chain, it is released from the tRNA, and the tRNA moves off into the cytoplasm where it joins with another amino acid that is specified by its anticodon. This process continues until a stop codon (UAA, UAG, or UGA) is reached on the mRNA. The protein is then released into the cytoplasm or into the rough ER for further modifications.

2.3   TISSUES

Groups of cells and surrounding substances that function together to perform one or more specialized activities are called tissues (Fig. 2.16). There are four primary types of tissue in the human body: epithelial, connective, muscle, and nervous. Epithelial tissues are either composed of cells arranged in sheets that are one or more layers thick or are organized into glands that are adapted for secretion. Typical functions of epithelial tissue include absorption (lining of the small intestine), secretion (glands), transport (kidney tubules), excretion (sweat glands), protection (skin; Fig. 2.16a), and sensory reception (taste buds). Connective tissues are the most abundant and widely distributed. Connective tissue proper can be loose (loosely woven fibers found around and between organs), irregularly dense (protective capsules around organs), and regularly dense (ligaments and tendons), whereas specialized connective tissue includes blood (Fig. 2.16b), bone, cartilage, and adipose tissue. Muscle tissue provides movement for the body through its specialized cells that can shorten in response to stimulation and then return to their uncontracted state. Figure 2.16c shows the three types of muscle tissue: skeletal (attached to bones), smooth (found in the walls of blood vessels), and cardiac (found only in the heart). Nervous tissue consists of neurons (Fig. 2.16d) that conduct electrical impulses and glial cells that protect, support, and nourish neurons.

2.4   MAJOR ORGAN SYSTEMS

Combinations of tissues that perform complex tasks are called organs, and organs that function together form organ systems. The human body has 11 major organ systems: integumentary, endocrine, lymphatic, digestive, urinary, reproductive, circulatory, respiratory, nervous, skeletal, and muscular. The integumentary system (skin, hair, nails, and various glands) provides protection for the body. The endocrine system (ductless glands such as the thyroid and adrenals) secretes hormones that regulate many chemical actions within cells. The lymphatic system (glands, lymph nodes, lymph, and lymphatic vessels) returns excess fluid and protein to the blood and helps defend the body against infection and tissue damage. The digestive

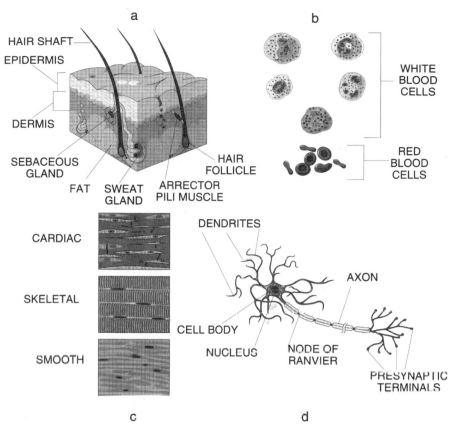

Fig. 2.16 Four tissue types. Skin [a] is a type of epithelial tissue that helps protect the body. Blood (b) is a specialized connective tissue. There are three types of muscle tissue (c): cardiac, skeletal, and smooth. Motor neurons (d) are a type of nervous tissue that conduct electrical impulses from the central nervous system to effector organs such as muscles.

system (stomach, intestines, and other structures) ingests food and water, breaks food down into small molecules which can be absorbed and used by cells, and removes solid wastes. The urinary system (kidneys, ureters, urinary bladder, and urethra) maintains the fluid volume of the body, eliminates metabolic wastes, and helps regulate blood pressure and acid-base and water–salt balances. The reproductive system (ovaries, testes, reproductive cells, and accessory glands and ducts) produces eggs or sperm and provides a mechanism for the production and nourishment of offspring. The circulatory system (heart, blood, and blood vessels) serves as a distribution system for the body. The respiratory system (airways and lungs) delivers oxygen to the blood from the air and carries away carbon dioxide. The nervous system (brain, spinal cord, peripheral nerves, and sensory organs) regulates most of the body's activities by detecting and responding to internal and external stimuli. The skeletal system (bones and cartilage) provides protection and support as well as sites for muscle attachments, the production of blood cells, and calcium and phosphorus storage. The muscular system (skeletal muscle) moves the body and its internal parts, maintains posture, and produces heat. Although biomedical engineers have made major contributions to understanding, maintaining, and/or replacing components in each of the 11 major organ systems, only the last 5 listed previously will be examined in greater detail.

2.4.1 Circulatory System

The circulatory system (Fig. 2.17) delivers nutrients and hormones throughout the body, removes waste products from tissues, and provides a mechanism for regulating temperature and removing the heat generated by the metabolic activities of the body's internal organs. Every living cell in the body is no more than 10–100 μm from a capillary (small blood vessels with walls only one cell thick that are 8 μm in diameter, approximately the same size as a red blood cell). This close proximity allows oxygen, carbon dioxide, and most other small solutes to diffuse from the cells into the capillary or from the capillary into the cells with the direction of diffusion determined by concentration and partial pressure gradients.

The heart (Fig. 2.18), the pumping station that moves blood through the blood vessels, consists of two pumps — the right side and the left side. Each side has one chamber (the atrium) that receives blood and another chamber (the ventricle) that

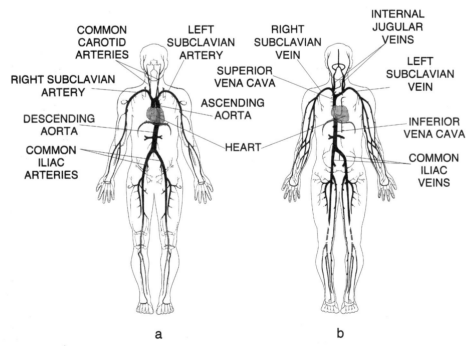

Fig. 2.17   (a) The distribution of the main arteries in the body which carry blood away from the heart. (b) The distribution of the main veins in the body which return the blood to the heart.

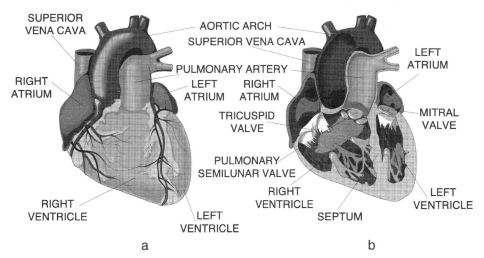

Fig. 2.18 (a) The outside of the heart as seen from its anterior side. (b) The same view after the exterior surface of the heart has been removed. The four interior chambers — right and left atria and right and left ventricles — are visible as are several valves.

pumps the blood away from the heart. The right side moves deoxygenated blood that is loaded with carbon dioxide from the body to the lungs, and the left side receives oxygenated blood that has had most of its carbon dioxide removed from the lungs and pumps it to the body. The vessels that lead to and from the lungs make up the pulmonary circulation, and those that lead to and from the rest of the tissues in the body make up the systemic circulation (Fig. 2.19). Blood vessels that carry blood away from the heart are called arteries, whereas those that carry blood toward the heart are called veins. The pulmonary artery is the only artery that carries deoxygenated blood, and the pulmonary vein is the only vein that carries oxygenated blood. The average adult has about 5 liters of blood with 80–90% in the systemic circulation at any one time; 75% of the blood is in the systemic circulation in the veins, 20% is in the arteries, and 5% is in the capillaries. Cardiac output is the product of the heart rate and the volume of blood pumped from the heart with each beat, i.e., the stroke volume. Each time the heart beats, about 80 ml of blood leave the heart. Thus, it takes about 60 beats for the average red blood cell to make one complete cycle of the body.

In the normal heart, the cardiac cycle, which refers to the repeating pattern of contraction (systole) and relaxation (diastole) of the chambers of the heart, begins with a self-generating electrical pulse in the pacemaker cells of the sinoatrial node (Fig. 2.20). This rapid electrical change in the cells is the result of the movement of ions across their plasma membranes. The permeability of the plasma membrane to Na^+ changes dramatically and allows these ions to rush into the cell. This change in the electrical potential across the plasma membrane from one in which the interior of the cell is more negative than the extracellular fluid (approximately -90 mV) to one in which the interior of the cell is more positive than the extracellular fluid (approximately 20 mV) is called depolarization. After a very short period of time

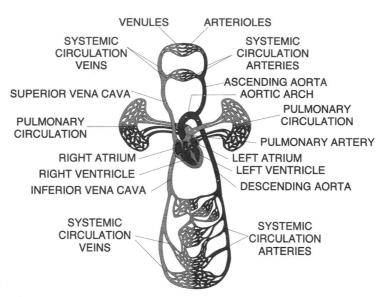

Fig. 2.19 Oxygenated blood leaves the heart through the aorta. Some of the blood is sent to the head and upper extremities and torso, whereas the remainder goes to the lower torso and extremities. The blood leaves the aorta and moves into other arteries, then into smaller arterioles, and finally into capillary beds where nutrients, hormones, gases, and waste products are exchanged between the nearby cells and the blood. The blood moves from the capillary beds into venules and then into veins. Blood from the upper part of the body returns to the right atrium of the heart through the superior vena cava, whereas blood from the lower part of the body returns through the inferior vena cava. The blood then moves from the right atrium to the right ventricle and into the pulmonary system through the pulmonary artery. After passing through capillaries in the lungs, the oxygenated blood returns to the left atrium of the heart through the pulmonary vein. It moves from the left atrium to the left ventricle and then out to the systemic circulation through the aorta to begin the same trip over again.

($<$0.3 s), changes in the membrane and activation of the sodium–potassium pumps result in repolarization, the restoration of the original ionic balance in the cells. The entire electrical event in which the polarity of the potential across the plasma membrane rapidly reverses and then becomes reestablished is called an action potential. The cells in the sinoatrial node depolarize on the average of every 0.83 s in a typical adult at rest. This gives a resting heart rate of 72 beats per minute with about $\frac{5}{8}$ of each beat spent in diastole and $\frac{3}{8}$ in systole.

Cardiac cells are linked and tightly coupled so that action potentials spread from one cell to the next. Activation wavefronts move across the atria at a rate of about 1 m/s. When cardiac cells depolarize, they also contract. This contraction process in the atria, atrial systole, moves blood from the right atrium to the right ventricle and from the left atrium to the left ventricle (Fig. 2.21). The activation wavefront then moves to the atrioventricular node where it slows to a rate of about 0.05 m/s to allow time for the ventricles to completely fill with the blood from the atria. After

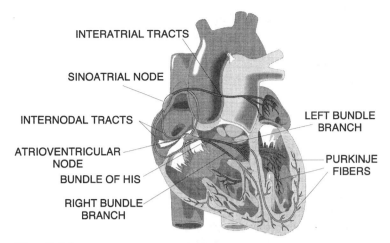

Fig. 2.20 Pacemaker cells in the sinoatrial node (SA node) depolarize first and send an activation wavefront through the atria. The propagating action potential slows down as it passes through the atrioventricular node (AV node) and then moves through the Bundle of His and Purkinje system very rapidly until it reaches the cells of the ventricles.

leaving the atrioventricular node, the activation wavefront moves to specialized conduction tissue, the Purkinje system, which spreads the wavefront very rapidly (at about 3 m/s) to many cells in both ventricles. The activation wavefront spreads through ventricular tissue at about 0.5 m/s. This results in the simultaneous con-

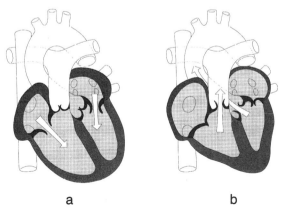

a   b

Fig. 2.21 (a) During the first part of the cardiac cycle, the atria contract (atrial systole) and move blood into the ventricles. (b) During the second part of the cardiac cycle, the atria relax (diastole), and the ventricles contract (ventricular systole) and move blood to the lungs (pulmonary circulation) and to the rest of the body (systemic circulation).

traction of both ventricles, ventricular systole, so that blood is forced from the heart into the pulmonary artery from the right ventricle and into the aorta from the left ventricle.

The electrocardiogram (ECG; Fig. 2.22) is an electrical measure of the sum of these ionic changes within the heart. The P wave represents the depolarization of the atria, and the QRS represents the depolarization of the ventricles. Ventricular repolarization shows up as the T wave, and atrial repolarization is masked by ventricular depolarization. Changes in the amplitude and duration of the different parts of the ECG provide diagnostic information for physicians. Many biomedical engineers have worked on methods for recording and analyzing ECGs.

During atrial and ventricular systole, special one-way valves (Fig. 2.23a) keep the blood moving in the correct direction. When the atria contract, the atrioventricular valves (tricuspid and mitral) open to allow blood to pass into the ventricles. During ventricular systole, the semilunar valves (aortic and pulmonary) open to allow blood to leave the heart while the atrioventricular valves close and prevent blood from flowing backwards from the ventricles to the atria. The aortic and pulmonary valves prevent blood from flowing back from the pulmonary artery and aorta into the right and left ventricles, respectively. If a valve becomes calcified or diseased or is not properly formed during embryonic development, it can be replaced by an artificial valve (Fig. 2.23b), a device that has been developed by cooperative work between biomedical engineers and physicians.

Blood pressure can be measured directly or indirectly (noninvasively). Direct blood pressure measurements are made by introducing a catheter or needle which is coupled to a pressure transducer into a vein or artery. Indirect methods include sphygmomanometry in which a cuff is used to apply sufficient pressure to an artery, usually in the arm, to prevent the flow of blood through the artery, and a stethoscope is used to listen to the change in sounds as the cuff is slowly deflated. The first

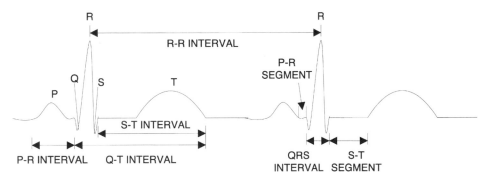

Fig. 2.22 Typical Lead II ECG. This electrocardiogram is typical of one that would be recorded from the body's surface by having a positive electrode on the left leg and a negative electrode on the right arm. The vertical direction represents voltage and the horizontal direction represents time. The P, R, and T waves are easily identified and are the result of the movement of ions in cells in different parts of the heart. Different intervals and segments have been identified which provide information about the health of the heart and its conduction system. The R–R interval can be used to determine heart rate.

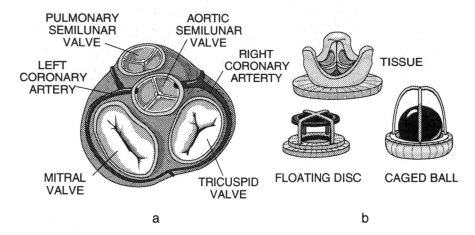

Fig. 2.23 (a) The atrioventricular (tricuspid and mitral) and semilunar (pulmonary and aortic) valves. (b) Three types of artificial valves — tissue, floating disc, and caged ball — that can be used to replace diseased or malformed human valves.

Korotkoff sounds occur when the systolic pressure, the highest pressure reached when the ventricles contract and eject blood, first exceeds the pressure in the cuff so that blood once again flows through the artery beneath the stethoscope. The Korotkoff sounds become muffled and disappear when the pressure in the cuff drops below the diastolic pressure, the minimum pressure that occurs at the end of ventricular relaxation. Another indirect measurement is the oscillometric method which uses a microprocessor to periodically inflate and slowly deflate a cuff. When blood breaks through the occlusion caused by the cuff, the walls of the artery begin to vibrate slightly due to the turbulent nature of the blood flow. The onset of these oscillations in pressure correlates with the systolic pressure. The oscillations decrease in amplitude over time with the diastolic pressure event corresponding to the point at which the rate of amplitude decrease suddenly changes slope. A third indirect measurement, the ultrasonic method, depends on the Doppler shift of sound waves that hit red blood cells that are flowing with the blood.

Blood in the systemic circulation leaves the heart through the aorta with an average internal pressure of about 100 mmHg (maximum systolic pressure of about 120 mmHg with a diastolic pressure of about 80 mmHg in a normal adult) and moves to medium-sized arteries (Fig. 2.17a) and arterioles. Arterioles lead to capillaries (average internal pressure of about 30 mmHg) which are followed by venules. Venules lead to medium-sized veins, then to large veins, and finally to the venae cavae (average internal pressure of about 10 mmHg) which return blood to the heart at the right atrium. Blood in the pulmonary circulation (Fig. 2.19) leaves the pulmonary artery and moves to arterioles and then to the capillary beds within the lungs. It returns to the heart through the left atrium. Blood flow is highest in the large arteries and veins (30–40 cm/s in the aorta; 5 cm/s in the vena cavae) and slowest in the capillary beds (1 mm/s) where the exchange of nutrients, metabolic wastes, gases, and hormones takes place. Pressures in the pulmonary circulation are lower

(25/10 mmHg) than in the systemic circulation due to the decreased pumping power of the smaller right ventricle compared to the left and to the lower resistance of blood vessels in the lungs.

2.4.2   Respiratory System

The respiratory system (Fig. 2.24a) moves air to and from the gas-exchange surfaces in the body where diffusion can occur between air and the circulating blood. It includes the conduction zone and the respiratory zone. In the conduction zone (mouth, nose, sinuses, pharynx, trachea, bronchi, and bronchioles), the air that enters the body is warmed, humidified, filtered, and cleaned. Mucus is secreted by cells in the conduction zone and traps small particles (>6 μm) before they can reach the respiratory zone. Epithelial cells that line the trachea and bronchi have cilia that beat in a coordinated fashion to move mucus toward the pharynx where it can be swallowed or expectorated. The respiratory zone, consisting of respiratory bronchioles with outpouchings of alveoli and terminal clusters of alveolar sacs, is where gas exchange between air and blood occurs (Fig. 2.24b). The respiratory zone comprises most of the mass of the lungs.

There are certain physical properties — compliance, elasticity, and surface tension — which are characteristic of lungs. Compliance refers to the ease with which lungs can expand under pressure. A normal lung is about 100 times more distensible than a toy balloon. Elasticity refers to the ease with which the lungs and other thoracic structures return to their initial sizes after being distended. This aids in pushing air out of the lungs during expiration. Surface tension is exerted by the thin

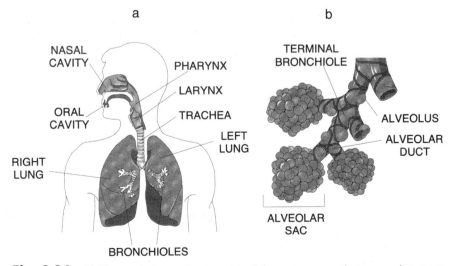

Fig. 2.24   (a) The respiratory system consists of the passageways that are used to move air into and out of the body and the lungs. (b) The terminal bronchioles and alveolar sacs within the lungs have alveoli where gas exchange occurs between the lungs and the blood in the surrounding capillaries.

film of fluid in the alveoli and acts to resist distention. It creates a force that is directed inward and creates pressure in the alveolus which is directly proportional to the surface tension and inversely proportional to the radius of the alveolus (Law of Laplace). Thus, the pressure inside an alveolus with a small radius would be higher than the pressure inside an adjacent alveolus with a larger radius and would result in air flowing from the smaller alveolus into the larger one. This could cause the smaller alveolus to collapse. This does not happen in normal lungs because the fluid inside the alveoli contains a phospholipid that acts as a surfactant. The surfactant lowers the surface tension in the alveoli and allows them to become smaller during expiration without collapsing. Premature babies often suffer from respiratory distress syndrome because their lungs lack sufficient surfactant to prevent their alveoli from collapsing. These babies can be kept alive with mechanical ventilators or surfactant sprays until their lungs mature enough to produce surfactant.

Breathing, or ventilation, is the mechanical process by which air is moved into (inspiration) and out of (expiration) the lungs. A normal adult takes about 15–20 breaths per minute. During inspiration, the inspiratory muscles contract and enlarge the thoracic cavity, the portion of the body where the lungs are located. This causes the alveoli to enlarge and the alveolar gas to expand. As the alveolar gas expands, the partial pressure within the respiratory system drops below atmospheric pressure by about 3 mmHg so that air easily flows in (Boyle's Law). During expiration, the inspiratory muscles relax and return the thoracic cavity to its original volume. Since the volume of the gas inside the respiratory system has decreased, its pressure increases to a value that is about 3 mmHg above atmospheric pressure. Air now moves out of the lungs and into the atmosphere.

Lung mechanics refers to the study of the mechanical properties of the lung and chest wall, whereas lung statics refers to the mechanical properties of a lung in which the volume is held constant over time. Understanding lung mechanics requires knowledge about the volumes within the lungs. Lung capacities contain two or more volumes. The tidal volume is the amount of air that moves in and out of the lungs during normal breathing (Fig. 2.25). The total lung capacity (TLC) is the amount of gas contained within the lungs at the end of a maximum inspiration. The vital capacity is the maximum amount of air that can be exhaled from the lungs after inspiration to TLC. The residual volume is the amount of gas remaining in the lungs after maximum exhalation. The amount of gas that can be inhaled after inhaling during tidal breathing is called the inspiratory reserve volume. The amount of gas that can be expelled by a maximal exhalation after exhaling during tidal breathing is called the expiratory reserve volume. The inspiratory capacity is the maximum amount of gas that can be inspired after a normal exhalation during tidal breathing, and the functional residual capacity is the amount of gas that remains in the lungs at this time.

All the volumes and capacities except those that include the residual volume can be measured with a spirometer. The classic spirometer is an air-filled container that is constructed from two drums of different sizes. One drum contains water and the other air-filled drum is inverted over an air-filled tube and floats in the water. The tube is connected to a mouthpiece used by the patient. When the patient inhales, the level of the floating drum drops. When the patient exhales, the level of the floating

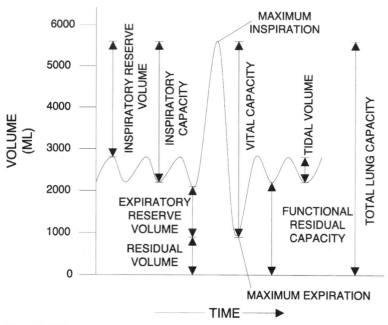

Fig. 2.25   Lung volumes and capacities, except for residual volume, functional residual capacity, and total lung capacity, can be measured using spirometry.

drum rises. These changes in floating drum position can be recorded and used to measure lung volumes.

Since spirograms record changes in volume over time, flow rates can be determined for different maneuvers. For example, if a patient exhales as forcefully as possible to residual volume following inspiration to TLC, then the forced expiratory volume ($FEV_{1.0}$) is the total volume exhaled at the end of 1 s. The $FEV_{1.0}$ is normally about 80% of the vital capacity. Restrictive diseases, in which inspiration is limited by reduced compliance of the lung or chest wall or by weakness of the inspiratory muscles, result in reduced values for $FEV_{1.0}$ and vital capacity but their ratio remains about the same. In obstructive diseases, such as asthma, the $FEV_{1.0}$ is reduced much more than the vital capacity. In these diseases, the TLC is abnormally large but expiration ends prematurely. Another useful measurement is the forced expiratory flow rate ($FEF_{25-75\%}$) which is the average flow rate measured over the middle half of the expiration, i.e., from 25 to 75% of the vital capacity. Flow-volume loops provide another method for analyzing lung function by relating the rate of inspiration and expiration to the volume of air that is moved during each process.

The TLC can be measured using the gas dilution technique. In this method, patients inspire to TLC from a gas mixture containing a known amount of an inert tracer gas, such as helium, and hold their breaths for 10 s. During this time, the inert gas becomes evenly distributed throughout the lungs and airways. Due to conservation of mass, the product of initial tracer gas concentration (which is known) times the amount inhaled (which is measured) equals the product of final tracer gas concentration (which is measured during expiration) times the TLC. Body plethysmography, which provides the most accurate method for measuring lung volumes, uses an airtight chamber in which the patient sits and breathes through a mouthpiece. This method makes use of Boyle's Law that states that the product of pressure and volume for gas in a chamber is constant under isothermal conditions. Changes in lung volume and pressure at the mouth when the patient pants against a closed shutter can be used to calculate the functional residual capacity. Since the expiratory reserve volume can be measured, the residual volume can be calculated by subtracting it from the functional residual capacity.

External respiration occurs in the lungs when gases are exchanged between the blood and the alveoli (Fig. 2.26). Each adult lung contains about 3.5×10^8 alveoli, which results in a large surface area (60–70 m^2) for gas exchange to occur. Each alveolus is only one cell layer thick so the air–blood barrier is only two cells thick (an alveolar cell and a capillary endothelial cell) which is about 2 μm. The partial pressure of oxygen in the alveoli is higher than the partial pressure of oxygen in the blood so oxygen moves from the alveoli into the blood. The partial pressure of carbon dioxide in the alveoli is lower than the partial pressure of carbon dioxide in the blood so carbon dioxide moves from the blood into the alveoli. During internal respiration, carbon dioxide and oxygen move between the blood and the extracellular fluid surrounding the body's cells. The direction and rate of movement of a gas depends on the partial pressures of the gas in the blood and the extracellular fluid, the surface area available for diffusion, the thickness of the membrane that the gas must

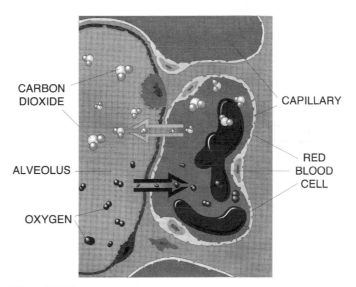

Fig. 2.26 During external respiration, oxygen moves from the alveoli to the blood and carbon dioxide moves from the blood to the air within the alveoli.

pass through, and a diffusion constant that is related to the solubility and molecular weight of the gas (Fick's Law).

Mechanical ventilators can be used to deliver air or oxygen to a patient. They can be electrically or pneumatically powered and can be controlled by microprocessors. Negative pressure ventilators, such as iron lungs, surround the thoracic cavity and force air into the lungs by creating a negative pressure around the chest. This type of ventilator greatly limits access to the patient. Positive pressure ventilators apply high-pressure gas at the entrance to the patient's lungs so that air or oxygen flows down a pressure gradient and into the patient. These ventilators can be operated in control mode to breathe for the patient at all times or in assist mode to help with ventilation when the patient initiates the breathing cycle. This type of ventilation changes the pressure within the thoracic cavity to positive during inspiration that affects venous return to the heart and cardiac output (the amount of blood the heart moves with each beat). High-frequency jet ventilators deliver very rapid (60–900 breaths per minute) low-volume bursts of air to the lungs. Oxygen and carbon dioxide are exchanged by molecular diffusion rather than by the mass movement of air. This method causes less interference with cardiac output than does positive pressure ventilation. Extracorporeal membrane oxygenation uses the technology that was developed for cardiopulmonary bypass machines. Blood is removed from the patient and passed through an artificial lung where oxygen and carbon dioxide are exchanged. It is warmed to body temperature before being returned to the patient. This technique allows the patient's lungs to rest and heal themselves and has been used successfully on some cold-water drowning victims and on infants with reversible pulmonary disease.

2.4.3 Nervous System

The nervous system, which is responsible for the integration and control of all the body's functions, has two major divisions: the central nervous system and the peripheral nervous system (Fig. 2.27). The former consists of all nervous tissue enclosed by bone (e.g., the brain and spinal cord), whereas the latter consists of all nervous tissue not enclosed by bone which enable the body to detect and respond to both internal and external stimuli. The peripheral nervous system consists of the 12 pairs of cranial and 31 pairs of spinal nerves with afferent (sensory) and efferent (motor) neurons.

The nervous system has also been divided into the somatic and autonomic nervous systems. Each of these systems consists of components from both the central and peripheral nervous systems. For example, the somatic peripheral nervous system consists of the sensory neurons, which convey information from receptors for pain, temperature, and mechanical stimuli in the skin, muscles, and joints to the central nervous system, and the motor neurons, which return impulses from the central nervous system to these same areas of the body. The autonomic nervous system is concerned with the involuntary regulation of smooth muscle, cardiac muscle, and glands and consists of the sympathetic and parasympathetic divisions. The sympathetic division causes blood vessels in the viscera and skin to constrict, vessels in the skeletal muscles to dilate, and heart rate to increase, whereas the parasympathetic

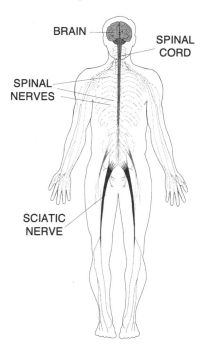

Fig. 2.27 The central nervous system (CNS) consists of all nervous tissue that is enclosed by bone, i.e., the brain and spinal cord, whereas the peripheral nervous system (PNS) consists of the nervous tissue that is not encased by bone.

division has the opposite effect on the vessels in the viscera and skin, provides no innervation to the skeletal muscles, and causes heart rate to decrease. Thus, the sympathetic division prepares the body for "fight or flight" and the parasympathetic division returns the body to normal operating conditions.

Specialized cells that conduct electrical impulses (neurons) or protect, support, and nourish neurons (glial cells) make up the different parts of the nervous system. The cell body of the neuron (Fig. 2.16d) gives rise to and nourishes a single axon and multiple, branching dendrites. The dendrites are the main receptor portion of the neuron although the cell body can also receive inputs from other neurons. Dendrites usually receive signals from thousands of contact points (synapses) with other neurons. The axon extends a few millimeters (in the brain) to a meter (from the spinal cord to the foot) and carries nerve signals to other nerve cells in the brain or spinal cord or to glands and muscles in the periphery of the body. Some axons are surrounded by sheaths of myelin that are formed by specialized, nonneural cells called Schwann cells. Each axon has many branches, called presynaptic terminals, at its end. These knoblike protrusions contain synaptic vesicles that hold neurotransmitters. When the neuron is stimulated by receiving a signal at its dendrites, the permeability of the cell's plasma membrane to sodium increases, just as occurred in cardiac cells, and an action potential moves from the dendrite to the cell body and then on to the axon. Gaps, called nodes of Ranvier, in the myelin sheaths of some axons allow the action potential to move more rapidly by essentially jumping from one node to the next. The vesicles in the presynaptic terminals release their neurotransmitter into the space between the axon and an adjacent neuron, muscle cell, or gland. The neurotransmitter diffuses across the synapse and causes a response (Fig. 2.28).

Neurons interconnect in several different types of circuits. In a divergent circuit, each branch in the axon of the presynaptic neuron connects with the dendrite of a different postsynaptic neuron. In a convergent circuit, axons from several presynaptic neurons meet at the dendrite(s) of a single postsynaptic neuron. In a simple feedback circuit, the axon of a neuron connects with the dendrite of an interneuron that connects back with the dendrites of the first neuron. A two-neuron circuit is one in which a sensory neuron synapses directly with a motor neuron, whereas a three-neuron circuit consists of a sensory neuron, an interneuron in the spinal cord, and a motor neuron. Both of these circuits can be found in reflex arcs (Fig. 2.29). The reflex arc is a special type of neural circuit that begins with a sensory neuron at a receptor (e.g., a pain receptor in the finger tip) and ends with a motor neuron at an effector (e.g., a skeletal muscle). Withdrawal reflexes are elicited primarily by stimuli for pain and heat great enough to be painful and are also known as protective or escape reflexes. They allow the body to respond quickly to dangerous situations without taking additional time to send signals to and from the brain and to process the information.

The brain is a large soft mass of nervous tissue and has three major parts: (i) cerebrum, (ii) diencephalon, and (iii) brain stem and cerebellum. The cerebrum (Fig.

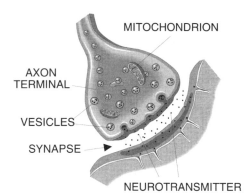

MITOCHONDRION

AXON TERMINAL

VESICLES

SYNAPSE

NEUROTRANSMITTER

Fig. 2.28 Following stimulation, vesicles in the axon terminal move to the synapse by means of exocytosis and release neurotransmitters into the space between the axon and the next cell which could be the dendrite of another neuron, a muscle fiber, or a gland. The neurotransmitters diffuse across the synapse and elicit a response from the adjacent cell.

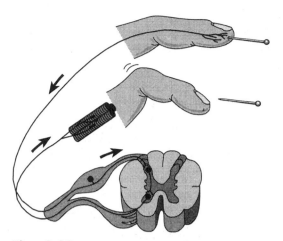

Fig. 2.29 This reflex arc begins with a sensory neuron in the finger that senses pain when the finger tip is pricked by the pin. An action potential travels from the sensory neuron to an interneuron and then to a motor neuron that synapses with muscle fibers in the finger. The muscle fibers respond to the stimulus by contracting and removing the fingertip from the pin.

2.30), which is divided into two hemispheres, is the largest and most obvious portion of the brain and consists of many convoluted ridges (gyri), narrow grooves (sulci), and deep fissures which result in a total surface area of about 2.25 m². The outer layer of the cerebrum, the cerebral cortex, is composed of gray matter (neurons with unmyelinated axons) that is 2–4 mm thick and contains over 50 billion neurons and 250 billion glial cells called neuroglia. The thicker inner layer is the

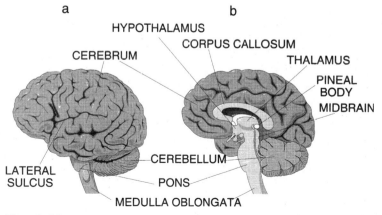

Fig. 2.30 (a) The exterior surface of the brain. (b) A midsagittal section through the brain.

white matter that consists of interconnecting groups of myelinated axons that project from the cortex to other cortical areas or from the thalamus (part of the diencephalon) to the cortex. The connection between the two cerebral hemispheres is called the corpus callosum (Fig. 2.30b). The left side of the cortex controls motor and sensory functions from the right side of the body, whereas the right side controls the left side of the body. Association areas that interpret incoming data or coordinate a motor response are connected to the sensory and motor regions of the cortex.

Fissures divide each cerebral hemisphere into a series of lobes that have different functions. The functions of the frontal lobes include initiating voluntary movement of the skeletal muscles, analyzing sensory experiences, providing responses relating to personality, and mediating responses related to memory, emotions, reasoning, judgment, planning, and speaking. The parietal lobes respond to stimuli from cutaneous (skin) and muscle receptors throughout the body. The temporal lobes interpret some sensory experiences, store memories of auditory and visual experiences, and contain auditory centers that receive sensory neurons from the cochlea of the ear. The occipital lobes integrate eye movements by directing and focusing the eye and are responsible for correlating visual images with previous visual experiences and other sensory stimuli. The insula is a deep portion of the cerebrum that lies under the parietal, frontal, and temporal lobes. Little is known about its function, but it seems to be associated with gastrointestinal and other visceral activities.

The diencephalon is the deep part of the brain that connects the midbrain of the brain stem with the cerebral hemispheres. Its main parts are the thalamus, hypothalamus, and epithalamus (Fig. 2.30b). The thalamus is involved with sensory and motor systems, general neural background activity, and the expression of emotion and uniquely human behaviors. Due to its two-way communication with areas of the cortex, it is linked with thought, creativity, interpretation and understanding of spoken and written words, and identification of objects sensed by touch. The hypothalamus is involved with integration within the autonomic nervous system, temperature regulation, water and electrolyte balance, sleep–wake patterns, food intake, behavioral responses associated with emotion, endocrine control, and sexual responses. The epithalamus contains the pineal body that is thought to have a neuroendocrine function.

The brain stem connects the brain with the spinal cord and automatically controls vital functions such as breathing. Its principal regions include the midbrain, pons, and medulla oblongata (Fig. 2.30b). The midbrain connects the pons and cerebellum with the cerebrum and is located at the upper end of the brain stem. It is involved with visual reflexes, the movement of eyes, focusing of the lenses, and the dilation of the pupils. The pons is a rounded bulge between the midbrain and medulla oblongata which functions with the medulla oblongata to control respiratory functions, acts as a relay station from the medulla oblongata to higher structures in the brain, and is the site of emergence of cranial nerve V. The medulla oblongata is the lowermost portion of the brain stem and connects the pons to the spinal cord. It contains vital centers that regulate heart rate, respiratory rate, constriction and dilation of blood vessels, blood pressure, swallowing, vomiting, sneezing, and coughing. The cerebellum is located behind the pons and is the second largest part of the brain. It processes sensory information that is used by the motor systems and is involved with coordinating skeletal muscle contractions and impulses for voluntary muscular movement that originate in the cerebral cortex. The cerebellum is a processing center that is involved with coordination of balance, body positions, and the precision and timing of movements.

2.4.4 Skeletal System

The average adult skeleton contains 206 bones, but the actual number varies from person to person and decreases with age as some bones become fused. Like the body, the skeletal system is divided into two parts: the axial skeleton and the appendicular skeleton (Fig. 2.31). The axial skeleton contains 80 bones (skull, hyoid bone, vertebral column, and thoracic cage), whereas the appendicular skeleton contains 126 (pectoral and pelvic girdles and upper and lower extremities). The skeletal system protects and supports the body, helps with movement, produces blood cells, and stores important minerals. It is made up of strong, rigid bones that are composed of

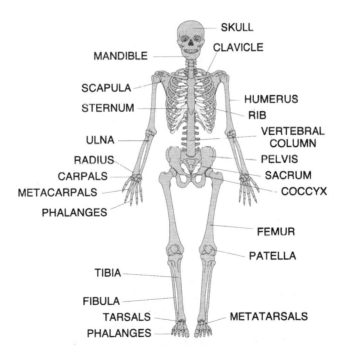

Fig. 2.31 The skull, hyoid bone (not shown), vertebral column, and thoracic cage (ribs, cartilage, and sternum) make up the axial skeleton, whereas the pectoral (scapula and clavicle) and pelvic girdles and upper and lower extremities make up the appendicular skeleton.

specialized connective tissue, bear weight, and form the major supporting elements of the body. Some support also comes from cartilage that is a smooth, firm, resilient, nonvascular type of connective tissue. Since the bones of the skeleton are hard, they protect the organs, such as the brain and abdominal organs, that they surround.

There are 8 cranial bones that support, surround, and protect the brain. Fourteen facial bones form the face and serve as attachments for the facial muscles that primarily move skin rather than bone. The facial bones, except for the lower jaw (mandible), are joined with each other and with the cranial bones. There are 6 auditory ossicles, 3 in each ear, that transmit sound waves from the external environment to the inner ear. The hyoid bone, which is near the skull but not part of it, is a small U-shaped bone that is located in the neck just below the lower jaw. It is attached to the skull and larynx (voice box) by muscles and ligaments and serves as the attachment for several important neck and tongue muscles.

The vertebral column starts out with approximately 34 bones, but only 26 independent ones are left in the average human adult. There are 7 cervical bones including the axis, which acts as a pivot around which the head rotates, and the atlas, which sits on the axis and supports the "globe" of the head. These are followed by 5 cervical, 12 thoracic, and 5 lumbar vertebrae and then the sacrum and the coccyx. The last two consist of 5 fused vertebrae. The vertebral column supports the weight of and allows movement of the head and trunk, protects the spinal cord, and provides places for the spinal nerves to exit from the spinal cord. There are four major curves (cervical, thoracic, lumbar, and sacral/coccygeal) in the adult vertebral column which allow it to flex and absorb shock. While movement between any two adjacent vertebrae is generally quite limited, the total amount of movement provided by the vertebral column can be extensive. The thoracic cage consists of 12 thoracic vertebrae (which are counted as part of the vertebral column), 12 pairs of ribs and their associated cartilage, and the sternum (breastbone). It protects vital organs and prevents the collapse of the thorax during ventilation.

Bones are classified as long, short, flat, or irregular according to their shape. Long bones, such as the femur and humerus, are longer than they are wide. Short bones, such as those found in the ankle and wrist, are as broad as they are long. Flat bones, such as the sternum and the bones of the skull, have a relatively thin and flattened shape. Irregular bones do not fit into the other categories and include the bones of the vertebral column and the pelvis.

Bones make up about 18% of the mass of the body and have a density of 1.9 g/cm^3. There are two types of bone: spongy and compact (cortical). Spongy bone forms the ends (epiphyses) of the long bones and the interior of other bones and is quite porous. Compact bone forms the shaft (diaphysis) and outer covering of bones and has a tensile strength of 120 N/mm^2, compressive strength of 170 N/mm^2, and Young's modulus of 1.8×10^4 N/mm^2. The medullary cavity, a hollow space inside the diaphysis, is filled with fatty, yellow marrow or red marrow that contains blood-forming cells.

Bone is a living organ that is constantly being remodeled. Old bone is removed by special cells, osteoclasts, and new bone is deposited by osteoblasts. Bone remodeling occurs during bone growth and in order to regulate calcium availability. The average skeleton is totally remodeled about three times during a person's lifetime. Osteoporosis is a disorder in which old bone is broken down faster than new bone is produced so that the resulting bones are weak and brittle.

The bones of the skeletal system are attached to each other at fibrous, cartilaginous, or synovial joints (Fig. 2.32). The articulating bones of fibrous joints are bound tightly together by fibrous connective tissue. These joints can be rigid and relatively immovable to slightly movable. This type of joint includes the suture joints in the skull. Cartilage holds together the bones in cartilaginous joints. These joints allow limited motion in response to twisting or compression and include the joints of the vertebral system and the joints that attach the ribs to the vertebral column and to the sternum. Synovial joints, such as the knee, are the most complex and varied and have fluid-filled joint cavities, cartilage that covers the articulating bones, and ligaments that help hold the joints together.

Synovial joints are classified into six types depending on their structure and the type of motion they permit. Gliding joints (Fig. 2.33) are the simplest type of synovial joint, allow back-and-forth or side-to-side movement, and include the intercarpal articulations in the wrist. Hinge joints, such as the elbow, permit bending in only one plane and are the most common type of synovial joint. The atlas and axis provide an example of a pivot joint that permits rotation. In condyloid articulations, an oval, convex surface of one bone fits into a concave depression on another bone. Condyloid joints, which include the metacarpophalangeal joints (knuckles) of the fingers, permit flexion–extension and rotation and are considered to be biaxial because rotation is limited to two axes of movement. The saddle joint, represented by the joint at the base of the thumb, is a modified condyloid joint that permits movement in several directions (multiaxial). Ball-and-socket joints allow motion in many directions around a fixed center. In these joints, the ball-shaped head of one bone fits into a cuplike concavity of another bone. This multiaxial joint is the most freely movable of all and includes the shoulder and hip joints. Biomedical engineers have helped develop artificial joints that are routinely used as replacements in diseased or injured hips, shoulders, and knees (Fig. 2.34).

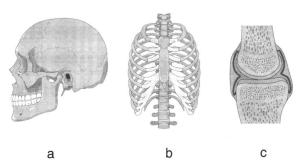

a   b   c

Fig. 2.32 Bones of the skeletal system are attached to each other at fibrous (a), cartilaginous (b), or synovial (c) joints.

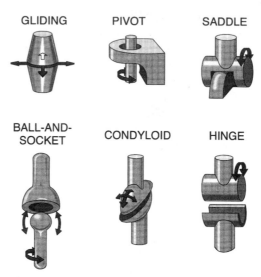

GLIDING   PIVOT   SADDLE

BALL-AND-
SOCKET   CONDYLOID   HINGE

Fig. 2.33   Synovial joints have fluid-filled cavities and are the most complex and varied types of joints. Each synovial joint is classified into one of six types depending on its structure and type of motion.

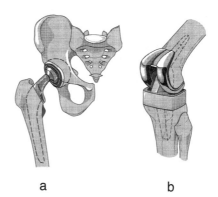

a   b

Fig. 2.34   Diseased or damaged hip (a) and knee (b) joints that are nonfunctional or extremely painful can be replaced by prostheses. Artificial joints can be held in place by a special cement [polymethylmethacrylate (PMMA)] and by bone ingrowth. Special problems occur at the interfaces due to the different elastic moduli of the materials (110 GPa for titanium, 2.2 GPa for PMMA, and 20 GPa for bone).

2.4.5 **Muscular System**

The muscular system (Fig. 2.35) is composed of 600–700 skeletal muscles, depending on whether certain muscles are counted as separate or as pairs, and makes up 40% of the body's mass. The axial musculature makes up about 60% of the skeletal muscles in the body and arises from the axial skeleton (Fig. 2.31). It positions the head and spinal column and moves the rib cage during breathing. The appendicular musculature moves or stabilizes components of the appendicular skeleton.

The skeletal muscles in the muscular system maintain posture, generate heat to maintain the body's temperature, and provide the driving force that is used to move the bones and joints of the body and the skin of the face. Muscles which play a major role in accomplishing a movement are called prime movers, or agonists. Muscles which act in opposition to a prime mover are called antagonists, whereas muscles which assist a prime mover in producing a movement are called synergists. The continual contraction of some skeletal muscles helps maintain the body's posture. If all these muscles relax, which happens when a person faints, the person collapses.

A system of levers, which consist of rigid lever arms that pivot around fixed points, is used to move skeletal muscle (Fig. 2.36). Two different forces act on

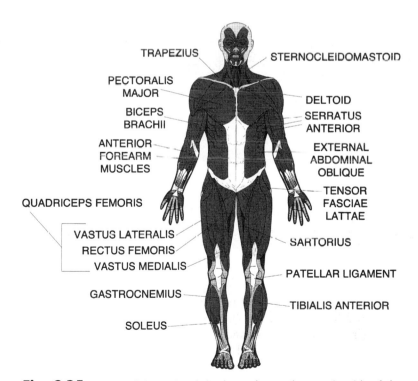

Fig. 2.35 Some of the major skeletal muscles on the anterior side of the body are shown.

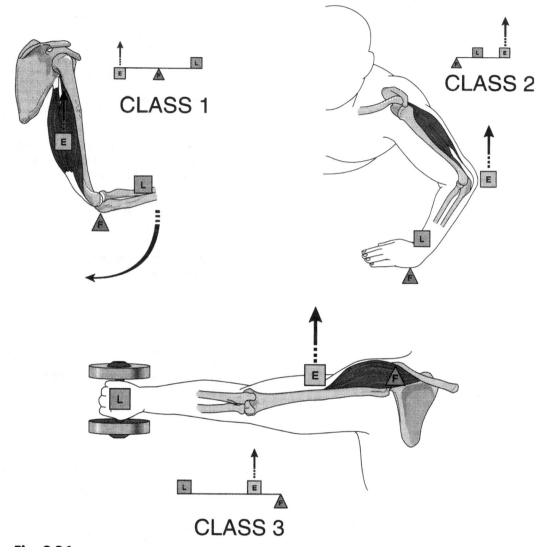

Fig. 2.36   Depending on the muscle in use, the location of the load, and the location of the fulcrum, the humerus can act as a class 1 Lever, a class 2 lever, or a class 3 Lever.

every lever: the weight to be moved, i.e., the resistance to be overcome, and the pull or effort applied, i.e., the applied force. Bones act as lever arms and joints provide a fulcrum. The resistance to be overcome is the weight of the body part which is moved and the applied force is generated by the contraction of a muscle or muscles at the insertion, the point of attachment of a muscle to the bone it moves. An example of a first-class lever, one in which the fulcrum is between the force and the weight, is the movement of the facial portion of the head when the face is tilted upwards. The fulcrum is formed by the joint between the atlas and the occipital bone of the skull and the vertebral muscles inserted at the back of the head generate the applied force which moves the weight, the facial portion of the head. A second-class lever is one in which the weight is between the force and the fulcrum. This can be found in the body when a person stands on "tip toe." The ball of the foot is the fulcrum and the applied force is generated by the calf muscles on the back of the leg. The weight that is moved is that of the whole body. A third-class lever is one in which the force is between the weight and the fulcrum. When a person has a bent elbow and holds a ball in front of the body, the applied force is generated by the contraction of the biceps brachii muscle. The weight to be moved includes the ball and the weight of the forearm and hand, and the elbow acts as the fulcrum.

The three types of muscle tissue — cardiac, skeletal, and smooth — share four important characteristics: (i) contractility, the ability to shorten; (ii) excitability, the capacity to receive and respond to a stimulus; (iii) extensibility, the ability to be stretched; and (iv) elasticity, the ability to return to the original shape after being stretched or contracted. Cardiac muscle tissue is found only in the heart, whereas smooth muscle tissue is found within almost every other organ where it forms sheets, bundles, or sheaths around other tissues. Skeletal muscles are composed of skeletal muscle tissue, connective tissue, blood vessels, and nervous tissue.

Each skeletal muscle is surrounded by a layer of connective tissue (collagen fibers) that separates the muscle from surrounding tissues and organs. These fibers come together at the end of the muscle to form tendons which connect the skeletal muscle to bone, skin (face), or the tendons of other muscles (hand). Other connective tissue fibers divide the skeletal muscles into compartments called fasicles that contain bundles of muscle fibers. Within each fascicle, additional connective tissue surrounds each skeletal muscle fiber and ties adjacent ones together. Each skeletal muscle fiber has hundreds of nuclei just beneath the cell membrane. Multiple nuclei provide multiple copies of the genes that direct the production of enzymes and structural proteins needed for normal contraction so that contraction can occur faster.

In muscle fibers, the plasma membrane is called the sarcolemma and the cytoplasm is called the sarcoplasm (Fig. 2.37). Transverse tubules (T tubules) begin at the sarcolemma and extend into the sarcoplasm at right angles to the surface of the sarcolemma. The T tubules, which play a role in coordinating contraction, are filled with extracellular fluid and form passageways through the muscle fiber. They make close contact with expanded chambers, cisternae, of the sarcoplasmic reticulum, a specialized form of the ER. The cisternae contain high concentrations of calcium ions which are needed for contraction to occur.

The sarcoplasm contains cylinders 1 or 2 μm in diameter that are as long as the entire muscle fiber and are called myofibrils. The myofibrils are attached to the sarcolemma at each end of the cell and are responsible for muscle fiber contraction. Myofilaments — protein filaments consisting of thin filaments (primarily actin) and thick filaments (mostly myosin) — are bundled together to make up myofibrils. Repeating functional units of myofilaments are called sarcomeres (Fig. 2.38). The sarcomere is the smallest functional unit of the muscle fiber and has a resting length of about 2.6 μm. The thin filaments are attached to dark bands, called Z lines, which form the ends of each sarcomere. Thick filaments containing double-headed myosin molecules lie between the thin ones. It is this overlap of thin and thick filaments that

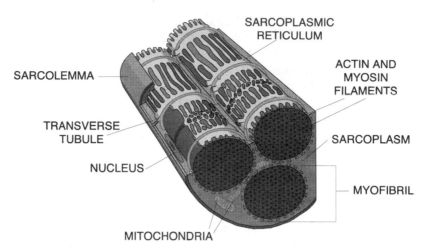

Fig. 2.37   Skeletal muscles are composed of muscle fascicles that are composed of muscle fibers such as the one shown here. Muscle fibers have hundreds of nuclei just below the plasma membrane, the sarcolemma. Transverse tubules extend into the sarcoplasm, the cytoplasm of the muscle fiber, and are important in the contraction process because they deliver action potentials that result in the release of stored calcium ions. Calcium ions are needed to create active sites on actin filaments so that cross-bridges can be formed between actin and myosin and the muscle can contract.

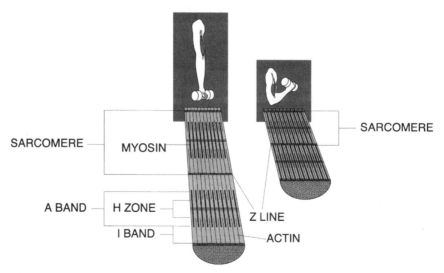

Fig. 2.38   The sarcomere is the basic functional unit of skeletal muscles and extends from one Z line to the next. Actin filaments are attached to the Z lines and extend into the A band where they overlap with the thicker myosin filaments. The H zone is the portion of the A band that contains no overlapping actin filaments. When the muscle changes from its extended, relaxed position (left) to its contracted state (right), the myosin filaments use cross-bridges to slide past the actin filaments and bring the Z lines closer together. This results in shorter sarcomeres and a contracted muscle.

gives skeletal muscle its banded, striated appearance. The I band is the area in a relaxed muscle fiber that just contains actin filaments, whereas the H zone is the area that just contains myosin filaments. The H zone and the area in which the actin and myosin overlap form the A band.

When a muscle contracts, myosin molecules in the thick filaments form cross-bridges at active sites in the actin of the thin filaments and pull the thin filaments toward the center of the sarcomere. The cross-bridges are then released and reformed at a different active site further along the thin filament. This results in a motion that is similar to the hand-over-hand motion that is used to pull in a rope. This action, the sliding filament mechanism, is driven by ATP energy and results in shortening of the muscle. Shortening of the muscle components (contraction) results in bringing the muscle's attachments (e.g., bones) closer together (Fig. 2.38).

Muscle fibers have connections with nerve. Sensory nerve endings are sensitive to length, tension, and pain in the muscle and send impulses to the brain via the spinal cord, whereas motor nerve endings receive impulses from the brain and spinal cord that lead to excitation and contraction of the muscle. Each motor axon branches and supplies several muscle fibers. Each of these axon branches loses its myelin sheath and splits up into a number of terminals that make contact with the surface of the muscle. When the nerve is stimulated, vesicles in the axon terminals release a neurotransmitter, acetylcholine, into the synapse between the neuron and the muscle. Acetylcholine diffuses across the synapse and binds to receptors in a special area, the motor end plate, of the sarcolemma. This causes the sodium channels in the sarcolemma to open up, and an action potential is produced in the muscle fiber. The resulting action potential spreads over the entire sarcolemmal surface and travels down all of the T tubules where it triggers a sudden massive release of calcium by the cisternae. Calcium triggers the production of active sites on the thin filaments so that cross-bridges with myosin can form and contraction occurs. Acetylcholinesterase breaks down the acetylcholine while the contraction process is under way so that the original relatively low permeability of the sarcolemma to sodium is restored.

A motor unit is a complex consisting of one motor neuron and the muscle fibers which it innervates. All the muscle fibers in a single motor unit contract at the same time, whereas muscle fibers in the same muscle but belonging to different motor units may contract at different times. When a contracted muscle relaxes, it returns to its original (resting) length if another contracting muscle moves it or if it is acted on by gravity. During relaxation, ATP is expended to move calcium back to the cisternae. The active sites that were needed for cross-bridge formation become covered so that actin and myosin can no longer interact. When the cross-bridges disappear, the muscle returns to its resting length, i.e., it relaxes.

The human body contains two different types of skeletal muscle fibers, fast and slow. Fast fibers can contract in 10 ms or less following stimulation and make up most of the skeletal muscle fibers in the body. They are large in diameter and contain densely packed myofibrils, large glycogen reserves (used to produce ATP), and relatively few mitochondria. These fibers produce powerful contractions that use up massive amounts of ATP and fatigue (can no longer contract despite continued neural stimulation) rapidly. Slow fibers take about three times as long to contract as fast fibers. They can continue to contract for extended periods of time because they contain (i) a more extensive network of capillaries so that they can receive more oxygen, (ii) a special oxygen-binding molecule called myoglobin, and (iii) more mitochondria which can produce more ATP than fast fibers. Muscles contain different amounts of slow and fast fibers. Those that are dominated by fast fibers (e.g., chicken breast muscles) appear white, whereas those that are dominated by slow fibers, (e.g., chicken legs) appear red. Most human muscles appear pink because they contain a mixture of both. Genes determine the percentage of fast and slow fibers in each muscle, but the ability of fast muscle fibers to resist fatigue can be increased through athletic training.

2.5   HOMEOSTASIS

Organ systems work together to maintain a constant internal environment within the body. Homeostasis is the process by which physical and chemical conditions within the internal environment of the body are maintained within tolerable ranges even when the external environment changes. Body temperature, blood pressure, and breathing and heart rates are some of the functions that are controlled by homeostatic mechanisms that involve several organ systems working together.

Extracellular fluid — the fluid that surrounds and bathes the body's cells — plays an important role in maintaining homeostasis. It circulates throughout the body and carries materials to and from the cells. It also provides a mechanism for maintaining optimal temperature and pressure levels, the proper balance between acids and bases, and concentrations of oxygen, carbon dioxide, water, nutrients, and many of the chemicals that are found in the blood.

Three components — sensory receptors, integrators, and effectors — interact to maintain homeostasis (Fig. 2.39). Sensory receptors, which may be cells or cell parts, detect stimuli, i.e., changes to their environment, and send information about the stimuli to integrators. Integrators are control points which pull together information from one or more sensory receptors. Integrators then elicit a response from effectors. The brain is an integrator that can send messages to muscles or glands or both. The messages result in some type of response from the effectors. The brain receives information about how parts of the body are operating and can compare this to information about how parts of the body should be operating.

Positive feedback mechanisms are ones in which the initial stimulus is reinforced by the response. There are very few examples of this in the human body since it disrupts homeostasis. Childbirth provides one example. Pressure from the baby's head in the birth canal stimulates receptors in the cervix which send signals to the hypothalamus. The hypothalamus responds to the stimulus by releasing oxytocin that enhances uterine contractions. Uterine contractions increase in intensity and force the baby further into the birth canal that causes additional stretching of the receptors in the cervix. The process continues until the baby is born, the pressure on the cervical stretch receptors ends, and the hypothalamus is no longer stimulated to release oxytocin.

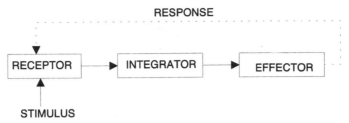

Fig. 2.39 Feedback mechanisms are used to help maintain homeostatis. A stimulus is received by a receptor which sends a signal (messenger) to an effector or to an integrator which sends a signal to an effector. The effector responds to the signal. The response feeds back to the receptor and modifies the effect of the stimulus. In negative feedback, the response subtracts from the effect of the stimulus on the receptor. In positive feedback, the response adds to the effect of the stimulus on the receptor.

Negative feedback mechanisms result in a response that is opposite in direction to the initiating stimulus. For example, receptors in the skin and elsewhere in the body detect the body's temperature. Temperature information is forwarded to the hypothalamus in the brain, which compares the body's current temperature to what the temperature should be (approximately 37°C). If the body's temperature is too low, messages are sent to contract the smooth muscles in blood vessels near the skin (reducing the diameter of the blood vessels and the heat transferred through the skin), to skeletal muscles to start contracting rapidly (shivering), and to the arrector pili muscles (Fig. 2.16a) to erect the hairs and form "goose bumps." The metabolic activity of the muscle contractions generates heat and warms the body. If the body's temperature is too high, messages are sent to relax the smooth muscles in the blood vessels near the skin (increasing the diameter of the blood vessels and the amount of heat transferred through the skin) and to sweat glands to release moisture and thus increase evaporative cooling of the skin. When the temperature of circulating blood changes enough in the appropriate direction that it reaches the set point of the system, the hypothalamus stops sending signals to the effector muscles and glands.

Another example of a negative feedback mechanism in the body involves the regulation of glucose in the bloodstream by clusters of cells, the pancreatic islets (Fig. 2.40). There are between 2×10^5 and 2×10^6 pancreatic islets scattered throughout the adult pancreas. When glucose levels are high, beta cells in the islets produce insulin which facilitates glucose transport across plasma membranes and into cells and enhances the conversion of glucose into glycogen which is stored in the liver. During periods of fasting or whenever the concentration of blood glucose drops below normal (70–110 mg/dl), alpha cells produce glucagon which stimulates the liver to convert glycogen into glucose and the formation of glucose from noncarbohydrate sources such as amino acids and lactic acid. When glucose levels return to normal, the effector cells in the pancreatic islets stop producing their respective hormone, i.e., insulin or glucagon. Some biomedical engineers are working on controlled drug delivery systems that can sense blood glucose levels and emulate the responses of the pancreatic islet cells, whereas other biomedical engineers are trying

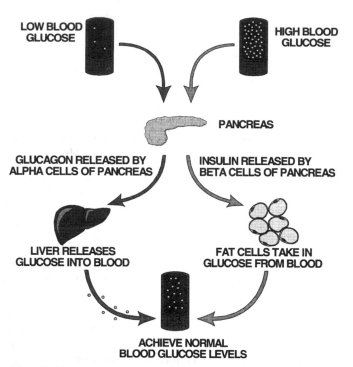

Fig. 2.40 Two negative feedback mechanisms help control the level of glucose in the blood. When blood glucose levels are higher than the body's set point (stimulus), beta cells in the pancreatic islets (receptors) produce insulin (messenger) which facilitates glucose transport across plasma membranes and enhances the conversion of glucose into glycogen for storage in the liver (effector). This causes the level of glucose in the blood to drop. When the level equals the body's set point, the beta cells stop producing insulin. When blood glucose levels are lower than the body's set point (stimulus), alpha cells in the pancreatic islets (receptors) produce glucagon (messenger) which stimulates the liver (effector) to convert glycogen into glucose. This causes the level of glucose in the blood to increase. When the level equals the body's set point, the alpha cells stop producing glucagon.

to develop an artificial pancreas which would effectively maintain appropriate blood glucose levels.

EXERCISES

1. Using as many appropriate anatomical terms as apply, write sentences which describe the positional relationship between your right knee and (i) your left ear, (ii) your nose, and (iii) the big toe on your right foot.

2. Using as many appropriate anatomical terms as apply, describe the position of the heart in the body and its position relative to the stomach.

3. Search the World Wide Web (WWW) to find a transverse section of the body that was imaged using computerized axial tomography or magnetic resonance imaging. Print the image and indicate its uniform resource locator (URL) or WWW address.

4. Name and give examples of the four classes of biologically important organic compounds. What are the major functions of each of the four groups?

5. What are the molarity and osmolarity of a 1-liter solution that contains half a mole of sodium bicarbonate?

6. Consider a simple model cell, such as the one in Fig. 2.6, that consists of cytoplasm and a plasma membrane. The cell's initial volume is 2 nl and it contains 0.2 M protein. The cell is placed in a large volume of 0.2 M NaCl. Neither Na^+ nor Cl^- can cross the plasma membrane and enter the cell. Is the 0.2 M NaCl solution hypotonic, isotonic, or hypertonic relative to the osmolarity inside the cell? Describe what happens to the cell as it achieves equilibrium in this new environment. What will be the final osmolarity of the cell? What will be its final volume?

7. Briefly describe how the principle of electrical neutrality affects the concentration of ions within a cell.

8. Consider the same model cell that was used in Exercise 6, but instead of being placed in 0.2 M NaCl, the cell is placed in 0.2 M urea. Unlike Na^+ and Cl^-, urea can cross the plasma membrane and enter the cell. Describe what happens to the cell as it achieves equilibrium in this environment. What will be the final osmolarity of the cell? What will be its final volume?

9. Briefly describe the path that a protein (e.g., a hormone) which is manufactured on the rough ER would take in order to leave the cell.

10. What major role do mitochondria have in the cell? Why might it be important to have this process contained within an organelle?

11. List and briefly describe three organelles that provide structural support and assist with cell movement.

12. Find a location on the WWW that describes the Human Genome Project. Print its home page and indicate its URL. Find and print an ideogram of a chromosome that shows a gene that causes a form of muscular dystrophy.

13. Briefly describe the major differences between replication and transcription.

14. Describe how the hereditary information contained in genes within the cell's DNA is expressed as proteins which direct the cell's activities.

15. Six different codons code for leucine, whereas only one codes for methionine. Why might this be important for regulating translation and producing proteins?

16. Copy the title page and abstract of five peer-reviewed journal articles that discuss engineering applications for five different organ systems in the body (one article per organ system). Review articles, conference proceeding

papers, copies of keynote addresses and other speeches, book chapters, and editorials are not acceptable. Good places to look are the *IEEE Transactions on Biomedical Engineering,* the *IEEE Transactions on Medical Imaging,* the *IEEE Transactions on Biomedical Information Technology,* the *IEEE Engineering in Medicine and Biology Magazine,* and the *Annals of Biomedical Engineering.* What information in the article indicates that it was reviewed?

17. Trace the path of a single red blood cell from a capillary bed in your big toe to the capillary beds of your right lung and back. What gases are exchanged and where are they exchanged during this process?

18. Draw and label a block diagram of pulmonary and systemic blood flow that includes the chambers of the heart, valves, major veins and arteries which enter and leave the heart, the lungs, and the capillary bed of the body. Use arrows to indicate the direction of flow through each component.

19. Describe how a 10-s ECG could be used to determine heart rate.

20. The total lung capacity of a patient is 5.9 liters. If the patient's inspiratory capacity was 3.3 liters, what would be the patient's functional residual capacity? What would you need to measure in order to determine the patient's residual volume?

21. Briefly describe the functions and major components of the central, peripheral, somatic, automatic, sympathetic, and parasympathetic nervous systems. Which ones are subsets of others?

22. Explain how sarcomeres shorten and how this results in muscle contraction.

23. How do the muscular and skeletal systems interact to produce movement?

24. Draw a block diagram to show the negative feedback mechanisms that help regulate glucose levels in the blood. Label the inputs, sensors, integrators, effectors, and outputs.

SUGGESTED READING

Alberts, B., Bray, D., Lewis, J., Raff, M., Roberts, K., and Watson, J. D. (1983). *Molecular Biology of the Cell.* Garland, New York.

Deutsch, S., and Deutsch, A. (1993). *Understanding the Nervous System — An Engineering Perspective.* IEEE Press, New York.

Guyton, A. C. (1991). *Basic Neuroscience. Anatomy & Physiology.* Saunders, Philadelphia.

Jenkins, D. B. (1991). *Hollinshead's Functional Anatomy of the Limbs and Back,* 6th ed. Saunders, Philadelphia.

Katz, A. M. (1986). *Physiology of the Heart.* Raven Press, New York.

Keynes, R. D., and Aidley, D. J. (1991). *Nerve & Muscle,* 2nd ed. Cambridge Univ. Press, Cambridge, UK.

Leff, A. R., and Schumacker, P. T. (1993). *Respiratory Physiology. Basics and Applications.* Saunders, Philadelphia.

Martini, F. H., and Bartholomew, E. F. (1997). *Essentials of Anatomy & Physiology.* Prentice Hall, Upper Saddle River, NJ.

Matthews, G. G. (1991). *Cellular Physiology of Nerve and Muscle.* Blackwell, Boston.

Starr, C. (1997). *BIOLOGY — Concepts and Applications.* Wadsworth, Belmont, CA.

Tate, P., Seeley, R. R., and Stephens, T. R. (1994). *Understanding the Human Body.* Mosby, St. Louis.

Van De Graaff, K. M., Fox, S. I., and LaFleur, K. M. (1997). *Synopsis of Human Anatomy & Physiology.* Brown, Dubuque, IA.

Van Wynsberghe, D., Noback, C. R., and Carola, R. (1995). *Human Anatomy and Physiology.* McGraw-Hill, New York.

West, J. B. (1990). *Respiratory Physiology — The Essentials,* 4th ed. Williams & Wilkins, Baltimore.

3   BIOELECTRIC PHENOMENA

Chapter Contents

At the conclusion of this chapter, the reader will be able to:

- Describe the history of bioelectric phenomenon

- Qualitatively explain how signaling occurs among neurons

- Calculate the membrane potential due to one or more ions

- Compute the change in membrane potential due to a current pulse through a cell membrane

- Describe the change in membrane potential with distance after stimulation

- Explain the voltage clamp experiment and an action potential

- Simulate an action potential using the Hodgkin–Huxley model

3.1   INTRODUCTION

Chapter 2 briefly described the nervous system and the concept of a neuron. Here the description of a neuron is extended by examining its properties at rest and during excitation. The concepts introduced here are basic and allow further investigation of more sophisticated models of the neuron or groups of neurons by using GENESIS (a general neural simulation program; Bower and Beeman, 1995) or extensions of the Hodgkin–Huxley model by using more accurate ion channel descriptions and neural networks. The models introduced here are an important first step in understanding the nervous system and how it functions.

Models of the neuron presented in this chapter have a rich history of development. This history continues today as new discoveries unfold that supplant existing theories and models. Much of the physiological interest in models of a neuron involves the neuron's use in transferring and storing information, while much engineering interest involves the neuron's use as a template in computer architecture and neural networks. To fully appreciate the operation of a neuron, it is important to understand the properties of a membrane at rest by using standard biophysics, biochemistry, and electric circuit tools. In this way, a more qualitative awareness of signaling via the generation of the action potential can be better understood.

The Hodgkin and Huxley theory that was published in 1952 described a series of experiments that allowed the development of a model of the action potential. This work was awarded a Nobel prize in 1963 (shared with John Eccles) and is discussed in Section 3.6. It is reasonable to question the usefulness of covering the Hodgkin–Huxley model in a textbook today given all the advances since 1952. One simple answer is that this model is one of the few timeless classics and should be covered. Another is that all current, and perhaps future, models have their roots in this model.

Section 3.2 describes a brief history of bioelectricity and can be easily omitted on first reading of the chapter. Section 3.3 describes the structure and provides a qualitative description of a neuron. Biophysics and biochemical tools useful in understanding the properties of a neuron at rest are presented in Section 3.4. An equivalent circuit model of a cell membrane at rest consisting of resistors, capacitors, and voltage sources is described in Section 3.5. Finally, Section 3.6 describes the Hodgkin-Huxley model of a neuron and includes a brief description of their experiments and the mathematical model describing an action potential.

3.2 HISTORY

3.2.1 The Evolution of a Discipline: The Galvani–Volta Controversy

In 1791, an article appeared in the *Proceedings of the Bologna Academy* reporting experimental results that, it was claimed, proved the existence of animal electricity. This now famous publication was the work of Luigi Galvani. At the time of its publication, this article caused a great deal of excitement in the scientific community and sparked a controversy that ultimately resulted in the creation of two separate and distinct disciplines — electrophysiology and electrical engineering. The controversy arose from the different interpretations of the data presented in this article. Galvani was convinced that the muscular contractions he observed in frog legs were due to some form of electrical energy emanating from the animal. On the other hand, Allesandro Volta, a professor of physics at the University of Padua, was convinced that the "electricity" described in Galvani's experiments originated not from the animal but from the presence of the dissimilar metals used in Galvani's experiments. Both of these interpretations were important. The purpose of this section, therefore, is to discuss them in some detail, highlighting the body of scientific knowledge available at the time these experiments were performed, the rationale behind the interpretations that were formed, and their ultimate effect.

3.2.2 Electricity in the Eighteenth Century

Before 1800, a considerable inventory of facts relating to electricity in general and bioelectricity in particular had accumulated. The Egyptians and Greeks had known that certain fish could deliver substantial shocks to an organism in their aqueous environment. Static electricity had been discovered by the Greeks, who produced it by rubbing resin (amber or, in Greek, *elektron*) with cat's fur or by rubbing glass with silk. For example, Thales of Miletus reported in 600 BC that a piece of amber when vigorously rubbed with a cloth responded with an "attractive power." Light particles, such as chaff, bits of papyrus, and thread, jumped from a distance and adhered to the amber. The production of static electricity at that time became associated with an aura.

More than 2000 years elapsed before the English physician, William Gilbert, picked up where Thales left off. Gilbert showed that not only amber but also glass, agate, diamond, sapphire, and many other materials when rubbed exhibited the same attractive power described by the Greeks. However, Gilbert did not report that particles could also be repelled. It was not until a century later that electrostatic repulsion was noted by Charles DuFay (1698–1739) in France.

The next step in the progress of electrification was an improvement of the friction process. Rotating rubbing machines were developed to give continuous and large-scale production of electrostatic charges. The first of these frictional electric machines was developed by Otto von Guericke (1602–1685) in Germany. In the eighteenth century, electrification became a popular science and experimenters discovered many new attributes of electrical behavior. In England, Stephen Gray (1666–1736) proved that electrification could flow hundreds of feet through ordinary twine when suspended by silk threads. Thus, he theorized that electrification was a "fluid." Substituting metal wires for the support threads, he found that the charges would quickly dissipate. Thus, the understanding that different materials can either conduct or insulate began to take shape. The "electrics," such as silk, glass, and resin, held charge. The "non-electrics," such as metals and water, conducted charges. Gray also found that electrification could be transferred by proximity of one charged body to another without direct contact. This was evidence of electrification by induction, a principle that was used later in machines that produced electrostatic charges.

In France, Charles F. DuFay, a member of the French Academy of Science, was intrigued by Gray's experiments. DuFay showed by extensive tests that practically all materials with the exception of metals and those too soft or fluid to be rubbed could be electrified. Later, however, he discovered that if metals were insulated they could hold the largest electric charge of all. DuFay found that rubbed glass would repel a piece of gold leaf, whereas rubbed amber, gum, or wax attracted it. He concluded that there were two different kinds of electric "fluids," which he labeled "vitreous" and "resinous." He found that while unlike charges attracted each other, like charges repelled. This indicated that there were two kinds of electricity.

In the American colonies, Benjamin Franklin (1706–1790) became interested in electricity and performed experiments that led to his hypothesis regarding the "one-fluid theory." Franklin stated that there was but one type of electricity and that the electrical effects produced by friction reflected the separation of electric fluid so that one body contained an excess and the other a deficit. He argued that "electrical fire" is a common element in all bodies and is normally in a balanced or neutral state. Excess or deficiency of charge, such as that produced by the friction between materials, created an imbalance. Electrification by friction was thus a process of separation rather than a creation of charge. By balancing a charge gain with an equal charge loss, Franklin had implied a law, namely, that the quantity of the electric charge is conserved. Franklin guessed that when glass was rubbed the excess charge appeared on the glass, and he called that positive electricity. He thus established the direction of conventional current from positive to negative. It is now known that the electrons producing a current move in the opposite direction.

Out of this experimental activity came an underlying philosophy or law. Up to the end of the eighteenth century, the knowledge of electrostatics was mainly qualitative. There were means for detecting, but not for measurement, and relationships between the charges had not been formulated. The next step was to quantify the phenomena of electrostatic charge forces.

For this determination, the scientific scene shifted back to France and the engineer-turned-physicist Charles A. Coulomb (1726–1806). Coulomb demonstrated that a force is exerted when two charged particles are placed in the vicinity of one another. However, he went a step beyond experimental observation by deriving a general relationship that completely expressed the magnitude of this force. His inverse-square law for the force of attraction or repulsion between charged bodies became one of the major building blocks in understanding the effect of a fundamental property of matter-charge. However, despite this wide array of discoveries, it is important to note that before the time of Galvani and Volta, there was no source that could deliver a continuous flow of electric fluid, a term that we now know implies both charge and current.

In addition to a career as statesman, diplomat, publisher, signer of the Declaration of Independence and the Constitution, Franklin was an avid experimenter and inventor. In 1743 at the age of 37, Franklin witnessed with excited interest a demonstration of static electricity in Boston and resolved to pursue the strange effects with investigations of his own. Purchasing and devising various apparati, Franklin became an avid electrical enthusiast. He launched into many years of experiments with electrostatic effects.

Franklin the scientist is most popularly known for his kite experiment during a thunderstorm in June 1752 at Philadelphia. Although various European investigators had surmised the identity of electricity and lightning, Franklin was the first to prove by an experimental procedure and demonstration that lightning was a giant electrical spark. Having previously noted the advantages of sharp metal points for drawing electrical fire, Franklin put them to use as "lightning rods." Mounted vertically on rooftops they would dissipate the thundercloud charge gradually and harmlessly to the ground. This was the first practical application in electrostatics.

Franklin's work was well received by the Royal Society in London. The origin of such noteworthy output from remote and colonial America made Franklin especially marked. In his many trips to Europe as statesman and experimenter, Franklin was lionized in social circles and eminently regarded by scientists.

3.2.3 Galvani's Experiments

Against such a background of knowledge of the electric fluid and the many powerful demonstrations of its ability to activate muscles and nerves, it is readily understandable that biologists began to suspect that the "nervous fluid" or the "animal spirit" postulated by Galen to course in the hollow cavities of the nerves and mediate muscular contraction, and indeed all the nervous functions, was of an electrical nature. Galvani, an obstetrician and anatomist, was by no means the first to hold such a view, but his experimental search for evidence of the identity of the electric and nervous fluids provided the critical breakthrough.

Speculations that the muscular contractions in the body might be explained by some form of animal electricity were common. By the eighteenth century, experimenters were familiar with the muscular spasms of humans and animals that were subjected to the discharge of electrostatic machines. As a result, electric shock was viewed as a muscular stimulant. In searching for an explanation of the resulting muscular contractions, various anatomical experiments were conducted to study the possible relationship of "metallic contact" to the functioning of animal tissue. In 1750, Johann Sulzer (1720–1779), a professor of physiology at Zurich, described a chance discovery that an unpleasant acid taste occurred when the tongue was put between two strips of different metals, such as zinc and copper, whose ends were in contact. With the metallic ends separated, there was no such sensation. Sulzer ascribed the taste phenomenon to a vibratory motion set up in the metals that stimulated the tongue and used other metals with the same results. However, Sulzer's reports went unheeded for a half-century until new developments called attention to his findings.

The next fortuitous and remarkable discovery was made by Luigi Galvani (1737–1798), descendant of a very large Bologna family, who at age 25 was made Professor of Anatomy at the University of Bologna. Galvani had developed an ardent interest in electricity and its possible relation to the activity of the muscles and nerves. Dissected frog legs were convenient specimens for investigation, and in his laboratory Galvani used them for studies of muscular and nerve activity. In these experiments, he and his associates were studying the responses of the animal tissue to various stimulations. In this setting, Galvani observed that, while a freshly prepared frog leg was being probed by a scalpel, the leg jerked convulsively whenever a nearby frictional electrical machine gave off sparks.

According to Green's translation (1953) Galvani, in writing of his experiments said:

> I had dissected and prepared a frog, and laid it on a table, on which there was an electrical machine. It so happened by chance that one of my assistants touched the point of his scalpel to the inner crural nerve of the frog; the muscles of the limb were suddenly and violently convulsed. Another of those who were helping to make the experiments in electricity thought that he noticed this happening only at the instant a spark came from the electrical machine. He was struck with the novelty of the action. I was occupied with other things at the time, but when he drew my attention to it I immediately repeated the experiment. I touched the other end of the crural nerve with the point of my scalpel, while my assistant drew sparks from the electrical machine. At each moment when sparks occurred, the muscle was seized with convulsions.

With an alert and trained mind, Galvani designed an extended series of experiments to resolve the cause of the mystifying muscle behavior. On repeating the experiments, he found that touching the muscle with a metallic object while the specimen lay on a metal plate provided the condition that resulted in the contractions.

Having heard of Franklin's experimental proof that a flash of lightning was of the same nature as the electricity generated by electric machines, Galvani set out to determine whether atmospheric electricity might produce the same results observed with his electrical machine. By attaching the nerves of frog legs to aerial wires and the feet to another electrical reference point known as electrical ground, he noted the same muscular response during a thunderstorm that he observed with the electrical machine. It was another chance observation during this experiment that lead to further inquiry, discovery, and controversy.

Galvani also noticed that the prepared frogs, which were suspended by brass hooks through the marrow and rested against an iron trellis, showed occasional convulsions regardless of the weather. In adjusting the specimens, he pressed the brass hook against the trellis and saw the familiar muscle jerk occurring each time he completed the metallic contact. To check whether this jerking might still be from some atmospheric effect, he repeated the experiment inside the laboratory. He found that the specimen, laid on an iron plate, convulsed each time the brass hook in the spinal marrow touched the iron plate. Recognizing that some new principle was involved, he varied his experiments to find the true cause. In the process, he found that by substituting glass for the iron plate, the muscle response was not observed but using a silver plate restored the muscle reaction. He then joined equal lengths of two different metals and bent them into an arc. When the tips of this bimetallic arc touched the frog specimens, the familiar muscular convulsions were obtained. As a result, he concluded not only that metal contact was a contributing factor but also that the intensity of the convulsion varied according to the kind of metals joined in the arc pair.

Galvani was now faced with trying to explain the phenomena he was observing. He had encountered two electrical effects for which his specimens served as indicator — one from the sparks of the electrical machine and the other from the contact of dissimilar metals. The electricity responsible for the action either resided in the anatomy of the specimens with the metals serving to release it or the effect was produced by the bimetallic contact with the specimen serving only as an indicator.

Galvani was primarily an anatomist and seized on the first explanation. He ascribed the results to "animal electricity" that resided in the muscles and nerves of the organism itself. Using a physiological model, he compared the body to a Leyden jar in which the various tissues developed opposite electrical charges. These charges flowed from the brain through nerves to the muscles. Release of electrical charge by metallic contact caused the convulsions of the muscles. He wrote:

> The idea grew that in the animal itself there was an indwelling electricity. We were strengthened in such a supposition by the assumption of a very fine nervous fluid that during the phenomena flowed into the muscle from the nerve, similar to the electric current of a Leyden Jar.

Galvani's hypothesis reflected the prevailing view of his day that ascribed the body activation to a flow of "spirits" residing in the various body parts.

In 1791, Galvani published his paper, *De Viribus Electricitatis In Motu Musculari,* in the proceedings of the Academy of Science in Bologna. This paper set forth his experiments and conclusions. Galvani's report created a sensation and implied to many a possible revelation of the mystery of the life force. Men of science and laymen alike, both in Italy and elsewhere in Europe, were fascinated and challenged by these findings. However, no one pursued Galvani's findings more assiduously and used them as a stepping stone to greater discovery than Allesandro Volta.

3.2.4   Volta's Interpretation

Galvani's investigations aroused a virtual furor of interest. Wherever frogs were found, scientists repeated his experiments with routine success. Initially, Galvani's explanation for the muscular contractions was accepted without question — even by the prominent physician Allesandro Volta, who had received a copy of Galvani's paper and verified the phenomenon.

Volta was a respected scientist in his own right. At age 24, Volta published his first scientific paper, *On the Attractive Force of the Electric Fire,* in which he speculated about the similarities between electric force and gravity. Engaged in studies of physics and mathematics and busy with experimentation, Volta's talents were so evident that before the age of 30 he was named the Professor of Physics at the Royal School of Como. Here he made his first important contribution to science with the invention of the electrophorus or "bearer of electricity." This was the first device to provide a replenishable supply of electric charge by induction rather than by friction.

In 1782, Volta was called to the professorship of physics at the University of Padua. There he made his next invention, the condensing electrophorus, a sensitive instrument for detecting electric charge. Earlier methods of charge detection employed the "electroscope," which consisted of an insulated metal rod that had pairs of silk threads, pith balls, or gold foil suspended at one end. These pairs diverged by repulsion when the rod was touched by a charge. The amount of divergence indicated the strength of the charge and thus provided quantitative evidence for Coulomb's Law.

By combining the electroscope with his electrophorus, Volta provided the scientific community with a detector for minute quantities of electricity. Volta continued to innovate and made his condensing electroscope a part of a mechanical balance that made it possible to measure the force of an electric charge against the force of gravity. This instrument was called an electrometer and was of great value in Volta's later investigations of the electricity created by contact of dissimilar metals.

Volta expressed immediate interest on learning of Galvani's 1791 report to the Bologna Academy on the "Forces of Electricity in Their Relation to Muscular Motion." Volta set out quickly to repeat Galvani's experiments and initially confirmed Galvani's conclusions on animal electricity as the cause of the muscular reactions. Along with Galvani, he ascribed the activity to an imbalance between electricity of the muscle and that of the nerve, which was restored to equilibrium when a metallic connection was made. On continuing his investigations, however, Volta began to have doubts about the correctness of that view. He found inconsistencies in the balance theory. In his experiments, muscles would convulse only when the nerve was in the electrical circuit made by metallic contact.

In an effort to find the true cause of the observed muscle activity, Volta went back to an experiment previously performed by Sulzer. When Volta placed a piece of tinfoil on the tip and a silver coin at the rear of his tongue and connected the two with a copper, he got a sour taste. When he substituted a silver spoon for the coin and omitted the copper wire, he got the same result as when he let the handle of the spoon touch the foil. When using dissimilar metals to make contact between the tongue and the forehead, he got a sensation of light. From these results, Volta came to the conclusion that the sensations he experienced could not originate from the metals as conductors but must come from the ability of the dissimilar metals themselves to generate electricity.

After 2 years of experimenting, Volta published his conclusions in 1792. While crediting Galvani with a surprising original discovery, he disagreed with him on what produced the effects. By 1794, Volta had made a complete break with Galvani. He became an outspoken opponent of the theory of animal electricity and proposed the theory of "metallic electricity." Galvani, by nature a modest individual, avoided any direct confrontation with Volta on the issue and simply retired to his experiments on animals.

Volta's conclusive demonstration that Galvani had not discovered animal electricity was a blow from which the latter never recovered. Nevertheless, he persisted in his belief in animal electricity and conducted his third experiment, which definitely proved the existence of bioelectricity. In this experiment, he held one foot of the frog nerve-muscle preparation and swung it so that the vertebral column and the sciatic nerve touched the muscles of the other leg. When this occurred or when the vertebral column was made to fall on the thigh, the muscles contracted vigorously. According to most historians, it was his nephew Giovanni Aldini (1762–1834) who championed Galvani's cause by describing this important experiment in which he probably collaborated. The experiment conclusively showed that muscular contractions could be evoked without metallic conductors. According to Fulton and Cushing (1936), Aldini wrote:

> Some philosophers, indeed, had conceived the idea of producing contractions in a frog without metals; and ingenious methods, proposed by my uncle Galvani, induced me to pay attention to the subject, in order that I might attain to greater simplicity. He made me sensible of the importance of the experiment and therefore I was long ago inspired with a desire of discovering that interesting process. It will be seen in the Opuscoli of Milan (No. 21), that I showed publicly, to the Institute of Bologna, contractions in a frog without the aid of metals so far back as the year 1794. The experiment, as described in a memoir addressed to M. Amorotti [sic] is as follows: I immersed a prepared frog in a strong solution of muriate of soda. I then took it from the solution, and, holding one extremity of it in my hand, I suffered the other to hang freely down. While in this position, I raised up the nerves with a small glass rod, in such a manner that they did not touch the muscles. I then suddenly removed the glass rod, and every time that the spinal marrow and nerves touched the muscular parts, contractions were excited. Any idea of a stimulus arising earlier from the action of the salt, or from the impulse produced by the fall of the nerves, may be easily removed. Nothing will be necessary but to apply the same nerves to the muscles of another prepared frog, not in a Galvanic circle; for, in this case, neither the salt, nor the impulse even if more violent, will produce muscular motion (Geddes and Hoff, p. 44, 1971).

The claims and counterclaims of Volta and Galvani developed rival camps of supporters and detractors. Scientists swayed from one side to the other in their opinions and loyalties. Although the subject was complex and not well understood, it was on the verge of an era of revelation. The next great contribution to the field was made by Carlo Matteucci, who both confirmed Galvani's third experiment and made a new discovery. Matteucci showed that the action potential precedes the contraction of skeletal muscle. In confirming Galvani's third experiment, which demonstrated the injury potential, Matteucci noted:

> I injure the muscles of any living animal whatever, and into the interior of the wound I insert the nerve of the leg, which I hold, insulated with glass tube. As I move this nervous filament in the interior of the wound, I see immediately strong contractions in the leg. To always obtain them, it is necessary that one point of the nervous filament touches the depths of the wound, and that another point of the same nerve touches the edge of the wound. (Geddes and Hoff, p. 45, 1971).

By using a galvanometer, Matteucci found that the difference in potential between an injured and uninjured area was diminished during a tetanic contraction. The study of this phenomenon occupied the attention of all succeeding electrophysiologists. More than this, however, Matteucci made another remarkable discovery — that a transient bioelectric event, now designated the action potential, accompanies the contraction of intact skeletal muscle. He demonstrated this by showing that a contracting muscle is able to stimulate a nerve that, in turn, causes contraction of the muscle it innervates. The existence of a bioelectric potential was established through the experiments of Galvani and Matteucci. Soon thereafter, the presence of an action potential was discovered in cardiac muscle and nerves.

Volta, on the other hand, advocated that the source of the electricity was due to the contact of the dissimilar metals only, with the animal tissue acting merely as the indicator. His results differed substantially depending on the pairs of metals used. For example, Volta found that the muscular reaction from dissimilar metals increased in vigor depending on the metals that were used.

In an effort to obtain better quantitative measurements, Volta dispensed with the use of muscles and nerves as indicators. He substituted instead his "condensing electroscope." He was fortunate in the availability of this superior instrument because the contact charge potential of the dissimilar metals was minute, far too small to be detected by the ordinary gold-leaf electroscope. Volta's condensing electroscope used a stationary disk and a removable disk separated by a thin insulating layer of shellac varnish. The thinness of this layer provided a large capacity for accumulation of charge. When the upper disk was raised after being charged, the condenser capacity was released to give a large deflection of the gold leaves.

Volta proceeded systematically to test the dissimilar metal contacts. He made disks of various metals and measured the quantity of the charge on each disk combination by the divergence of his gold foil condensing electroscope. He then determined whether the charge was positive or negative by bringing a rubbed rod of glass or resin near the electroscope. The effect of the rod on the divergence of the gold foil indicated the polarity of the charge.

Volta's experiments led him toward the idea of an electric force or electrical "potential." This, he assumed, resided in contact between the dissimilar metals. As Volta experimented with additional combinations, he found that an electrical potential also existed when there was contact between the metals and some fluids. As a result, Volta added liquids, such as brine and dilute acids, to his conducting system and classified the metal contacts as "electrifiers of the first class" and the liquids as electrifiers of the "second class."

Volta found that there was only momentary movement of electricity in a circuit composed entirely of dissimilar metals. However, when he put two dissimilar metals in contact with a separator soaked with a saline or acidified solution, there was a steady indication of potential. In essence, Volta was assembling the basic elements of an electric battery — two dissimilar metals and a liquid separator. Furthermore, he found that the overall electric effect could be enlarged by multiplying the elements. Thus, by stacking metal disks and the moistened separators vertically he constructed an "electric pile," the first electric battery. This was the most practical discovery of his career.

3.2.5 The Final Result

Considerable time passed before true explanations became available for what Galvani and Volta had done. Clearly, both demonstrated the existence of a difference in electric potential — but what produced it eluded them. The potential difference present in the experiments carried out by both investigators is now clearly understood. Although Galvani thought that he had initiated muscular contractions by discharging animal electricity resident in a physiological capacitor consisting of the nerve (inner conductor) and muscle surface (outer conductor), it is now known that the stimulus consists of an action potential which in turn causes muscular contractions.

It is interesting to note that the fundamental unit of the nervous system — the neuron — has an electric potential between the inside and outside of the cell even at rest. This membrane resting potential is continually affected by various inputs to the cell. When a certain potential is reached, an action potential is generated along its axon to all its distant connections. This process underlies the communication mechanisms of the nervous system. Volta's discovery of the electrical battery provided the scientific community with the first steady source of electrical potential, which when connected in an electric circuit consisting of conducting materials or liquids results in the flow of electrical charge, i.e., electrical current. This device launched the field of electrical engineering.

3.3 NEURONS

A reasonable estimate of the human brain is that it contains about 10^{12} neurons partitioned into fewer than 1000 different types in an organized structure of uniform appearance. While not important in this chapter, it is important to note that there are two classes of neuron, the nerve cell and the neuroglial cell. Even though there are 10–50 times as many neuroglial cells as nerve cells in the brain, attention is focused here on the nerve cell since the neuroglial cells are not involved in signaling and primarily provide a support function for the nerve cell. Therefore, the terms neuron and nerve cell are used interchangeably since the primary focus here is to better understand the signaling properties of a neuron. Overall, the complex abilities of the brain are best described by virtue of a neuron's interconnections with other neurons or the periphery and not a function of the individual differences among neurons.

A typical neuron, as shown in Fig. 3.1, is defined with four major regions: cell body, dendrites, axon, and presynaptic terminals. The cell body of a neuron contains the nucleus and other apparatus needed to nourish the cell and is similar to other cells. Unlike other cells, however, the neuron's cell body is connected to a number of branches called dendrites and a long tube called the axon that connects the cell body to the presynaptic terminals. Dendrites are the receptive surfaces of the neuron that receive signals from thousands of other neurons passively and without amplification. Located on the dendrite and cell body are receptor sites that receive input from presynaptic terminals from adjacent neurons. Neurons typically have 10^4 to 10^5 synapses. Communication between neurons, as described in Chapter 2, is through a neurotransmitter that changes membrane properties. Also connected to the cell body is a single axon that ranges in length from 1 m in the human spinal cord to a few millimeters in the brain. The diameter of the axon also varies from <1 to 500 μm. In general, the larger the diameter of the axon, the faster the signal travels. Signals traveling in the axon range from 0.5 to 120 m/s. The purpose of an axon is to serve as a transmission line to move information from one neuron to another at great speeds. Large axons are surrounded by a fatty insulating material called the myelin sheath and have regular gaps, called the nodes of Ranvier, that allow the action potential to jump from one node to the next. The action potential is most easily envisioned as a pulse that travels the length of the axon without decreasing in amplitude. Most of the remainder of this chapter is devoted to understanding this process. At the end of the axon is a network of up to 10,000 branches with endings called the presynaptic terminals. A diagram of the presynaptic terminal is shown in Fig. 2.28. All action potentials that move through the axon propa-

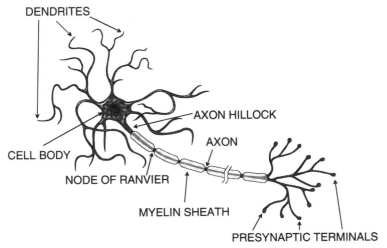

Fig. 3.1   Diagram of a typical neuron.

gate through each branch to the presynaptic terminal. The presynaptic terminals are the transmitting unit of the neuron, which when stimulated release a neurotransmitter that flows across a gap of approximately 20 nm to an adjacent cell where it interacts with the postsynaptic membrane and changes its potential.

3.3.1 Membrane Potentials

The neuron, like other cells in the body, has a separation of charge across its external membrane. The cell membrane is positively charged on the outside and negatively charged on the inside as illustrated in Fig. 3.2. This separation of charge, due to the selective permeability of the membrane to ions, is responsible for the membrane potential. In the neuron, the potential difference across the cell membrane is approximately 60–90 mV depending on the specific cell. By convention, the outside is defined as 0 mV (ground), and the resting potential is $V_m = v_i - v_o = -60$ mV. This charge differential is of particular interest since most signaling involves changes in this potential across the membrane. Signals such as action potentials are a result of electrical perturbations of the membrane. By definition, if the membrane is more negative than resting potential (i.e., -60 to -70 mV), it is called **hyperpolarization,** and an increase in membrane potential from resting potential (i.e., -60 to -50 mV) is called **depolarization.**

To create a membrane potential of -60 mV does not require the separation of many positive and negative charges across the membrane. The actual number, however, can be found from the relationship $Cdv = dq$, or $C\Delta v = \Delta q$ (Δq is the number

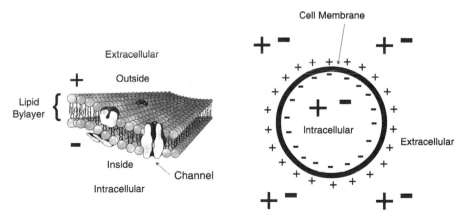

Fig. 3.2 Diagrams illustrating separation of charges across a cell membrane. (Left) A cell membrane with positive ions along the outer surface of the cell membrane and negative ions along the inner surface of the cell membrane. (Right) This further illustrates separation of charge by showing that only the ions along the inside and outside of the cell membrane are responsible for membrane potential (negative ions along the inside and positive ions along the outside of the cell membrane). Elsewhere negative and positive ions are approximately evenly distributed as indicated with the large plus/minus symbols for the illustration on the right. Overall, there is a net excess of negative ions inside the cell and a net excess of positive ions in the immediate vicinity outside the cell. For simplicity, the membrane shown on the right is drawn as the solid circle and ignores the axon and dendrites.

of charges times the electron charge of 1.6022×10^{-19} C). Therefore, with C = 1 μF/cm^2 and $\Delta v = 60 \times 10^{-3}$, the number of charges equals approximately 1×10^8 per cm^2. These charges are located within 1 μm of the membrane.

Graded Response and Action Potentials

A neuron can change the membrane potential of another neuron to which it is connected by releasing its neurotransmitter. The neurotransmitter crosses the synaptic cleft or gap, interacts with receptor molecules in the postsynaptic membrane of the dendrite or cell body of the adjacent neuron, and changes the membrane potential of the receptor neuron (Fig. 3.3).

The change in membrane potential at the postsynaptic membrane is due to a transformation from neurotransmitter chemical energy to electrical energy. The change in membrane potential depends on how much neurotransmitter is received and can be depolarizing or hyperpolarizing. This type of change in potential is typically called a **graded response** since it varies with the amount of neurotransmitter received. Another way of envisaging the activity at the synapse is that the neurotransmitter received is integrated or summed, which results in a graded response in the membrane potential. Note that while a signal from a neuron is either inhibitory or excitatory, specific synapses may be excitatory and others inhibitory, providing the nervous system with the ability to perform complex tasks.

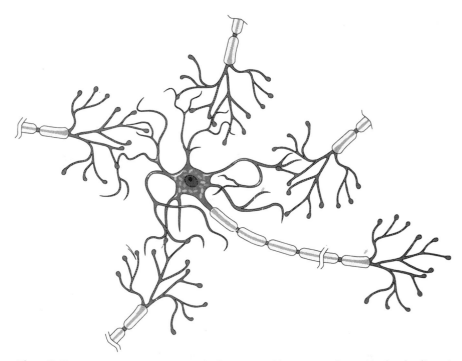

Fig. 3.3 Diagram illustrating a typical neuron with presynaptic terminals of adjacent neurons in the vicinity of its dendrites.

The net result of activation of the nerve cell is the action potential. The action potential is a large depolarizing signal of up to 100 mV that travels along the axon and lasts approximately 1–5 ms. Figure 3.4 illustrates a typical action potential. The action potential is an all or none signal that propagates actively along the axon without decreasing in amplitude. When the signal reaches the end of the axon at the presynaptic terminal, the change in potential causes the release of a packet of neurotransmitter. This is a very effective method of signaling over large distances. Additional details about the action potential are described throughout the remainder of this chapter after some tools for better understanding this phenomenon are introduced.

3.3.2 Resting Potential, Ionic Concentrations, and Channels

A resting membrane potential exists across the cell membrane because of the differential distribution of ions in and around the membrane of the nerve cell. The cell maintains these ion concentrations by using a selectively permeable membrane and, as described later, an active ion pump. A selectively permeable cell membrane with ion channels is illustrated in Fig. 3.2. The neuron cell membrane is approximately 10 nm thick and, because it consists of a lipid bilayer (i.e., two plates separated by an insulator), has capacitive properties. The extracellular fluid is composed of primarily Na^+ and Cl^-, and the intracellular fluid (cytoplasm) is composed primarily of K^+ and A^-. The large organic anions (A^-) are primarily amino acids and proteins and do not cross the membrane. Almost without exception, ions cannot pass through the cell membrane except through a channel.

Channels allow ions to pass through the membrane, are selective, and are either passive or active. Passive channels are always open and are ion specific. Figure 3.5

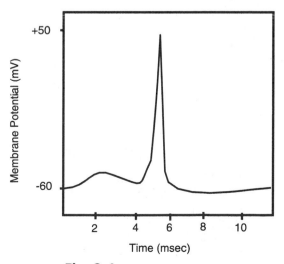

Fig. 3.4 An action potential.

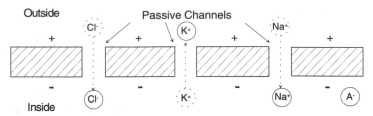

Fig. 3.5 Idealized cross section of a selectively permeable membrane with channels for ions to cross membrane. The thickness of membrane and size of the channels are not drawn to scale. When the diagram is drawn to scale, the cell membrane thickness is 20 times the size of the ions and 10 times the size of the channels, and the spacing between the channels is approximately 10 times the cell membrane thickness. Note that a potential difference exists between the inside and outside of the membrane as illustrated with the plus and minus signs. The membrane is selectively permeable to ions through ion-specific channels; that is, each channel shown here only allows one particular ion to pass through it.

illustrates a cross section of a cell membrane with passive channels only. As shown, a particular channel allows only one ion type to pass through the membrane and prevents all other ions from crossing the membrane through that channel. Passive channels exist for Cl^-, K^+, and Na^+. In addition, a passive channel exists for Ca^{2+}, which is important in the excitation of the membrane at the synapse. Active channels, or gates, are either opened or closed in response to an external electrical or chemical stimulation. The active channels are also selective and allow only specific ions to pass through the membrane. Typically, active gates open in response to neurotransmitters and an appropriate change in membrane potential. Figure 3.6 illustrates the concept of an active channel. Here, K^+ passes through an active channel and Cl^- passes through a passive channel. As will be shown, passive channels are

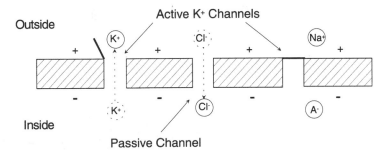

Fig. 3.6 Passive and active channels provide a means for ions to pass through the membrane. Each channel is ion specific. As shown, the active channel on the left allows K^+ to pass through the membrane, but the active channel on the right is not open, preventing any ion from passing through the membrane. Also shown is a passive Cl^- channel.

responsible for the resting membrane potential, and active channels are responsible for the graded response and action potentials.

3.4 BASIC BIOPHYSICS TOOLS AND RELATIONSHIPS

3.4.1 Basic Laws

Two basic biophysics tools and a relationship are used to characterize the resting potential across a cell membrane by quantitatively describing the impact of the ionic gradients and electric fields.

Fick's Law

The flow of particles due to diffusion is along the concentration gradient with particles moving from high-concentration areas to low ones. Specifically, for a cell membrane, the flow of ions across a membrane is given by

$$J(\text{diffusion}) = -D\frac{d[I]}{dx} \tag{3.1}$$

where J is the flow of ions due to diffusion, $[I]$ is the ion concentration, dx is the membrane thickness, and D is the diffusivity constant in m^2/s. The negative sign indicates that the flow of ions is from higher to lower concentration, and $\frac{d[I]}{dx}$ represents the concentration gradient.

Ohm's Law

Charged particles in a solution experience a force resulting from other charged particles and electric fields present. The flow of ions across a membrane is given by:

$$J(\text{drift}) = -\mu Z[I]\frac{dv}{dx} \tag{3.2}$$

where J is the flow of ions due to drift in an electric field $\vec{E}$, μ is the mobility in m^2/sV, Z is the ionic valence, $[I]$ is the ion concentration, v is the voltage across the membrane, and dv/dx is $(-\vec{E})$. Note Z is positive for positively charged ions (e.g., $Z = 1$ for Na^+ and $Z = 2$ for Ca^{2+}) and negative for negatively charged ions (e.g., $Z = -1$ for Cl^-). Positive ions drift down the electric field and negative ions drift up the electric field.

Figure 3.7 illustrates a cell membrane that is permeable to only K^+ and shows the forces acting on K^+. Assume that the concentration of K^+ is that of a neuron with a higher concentration inside than outside and that the membrane resting potential is negative from inside to outside. Clearly, only K^+ can pass through the membrane, and Na^+, Cl^-, and A^- cannot move since there are no channels for them to pass through. Depending on the actual concentration and membrane potential, K^+ will pass through the membrane until the forces due to drift and diffusion are balanced. The chemical force due to diffu-

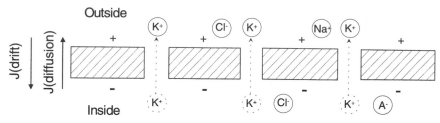

Fig. 3.7 Diagram illustrating the direction of the flow of K$^+$ due to drift and diffusion across a cell membrane that is only permeable to K$^+$.

sion from inside to outside decreases as K$^+$ moves through the membrane, and the electric force increases as K$^+$ accumulates outside the cell until the two forces are balanced.

Einstein Relationship

The relationship between the drift of particles in an electric field under osmotic pressure, that is, relationship between diffusivity and mobility, is given by

$$D = \frac{KT\mu}{q} \qquad (3.3)$$

where D is the diffusivity constant, μ is mobility, K is Boltzmann's constant, T is the absolute temperature in degrees Kelvin, and q is the magnitude of the electric charge (i.e., 1.60186×10^{-19} C).

3.4.2   Resting Potential of a Membrane Permeable to One Ion

The flow of ions in response to concentration gradients is limited by the selectively permeable nerve cell membrane and the resultant electric field. As described, ions pass through channels that are selective for that ion only. For clarity, the case of a membrane permeable to one ion only is considered first and then the case of a membrane permeable to more than one ion follows. It is interesting to note that neuroglial cells are permeable to only K$^+$ and that nerve cells are permeable to K$^+$, Na$^+$, and Cl$^-$. As will be shown, the normal ionic gradient is maintained if the membrane is permeable only to K$^+$ as in the neuroglial cell.

Consider the cell membrane shown in Fig. 3.7 that is permeable only to K$^+$ and assume that the concentration of K$^+$ is higher in the intracellular fluid than in the extracellular fluid. For this situation, the flow due to diffusion (concentration gradient) tends to push K$^+$ outside of the cell and is given by

$$J_K(\text{diffusion}) = -D\frac{d[\text{K}^+]}{dx} \qquad (3.4)$$

The flow due to drift (electric field) tends to push K^+ inside the cell and is given by

$$J_K(\text{drift}) = -\mu Z[K^+]\frac{dv}{dx} \tag{3.5}$$

which results in a total flow

$$J_K = J_K(\text{diffusion}) + J_K(\text{drift}) = -D\frac{d[K^+]}{dx} - \mu Z[K^+]\frac{dv}{dx} \tag{3.6}$$

Using the Einstein relationship $D = \dfrac{KT\mu}{q}$, the total flow is now given by

$$J_K = -\frac{KT}{q}\mu\frac{d[K^+]}{dx} - \mu Z[K^+]\frac{dv}{dx} \tag{3.7}$$

From Eq. (3.7), the flow of K^+ is found at any time for any given set of initial conditions. In the special case of steady state, that is, at equilibrium when the flow of K^+ into the cell is exactly balanced by the flow out of the cell or $J_K = 0$, Eq. (3.7) reduces to

$$0 = -\frac{KT}{q}\mu\frac{d[K^+]}{dx} - \mu Z[K^+]\frac{dv}{dx} \tag{3.8}$$

With $Z = 1$, Eq. (3.8) simplifies to

$$dv = -\frac{KT}{q[K^+]}d[K^+] \tag{3.9}$$

Integrating Eq. (3.9) from outside the cell to inside yields

$$\int_{v_o}^{v_i} dv = -\frac{KT}{q}\int_{[K^+]_o}^{[K^+]_i}\frac{d[K^+]}{[K^+]} \tag{3.10}$$

where v_o and v_i are the voltages outside and inside the membrane and $[K^+]_o$ and $[K^+]_i$ are the concentrations of potassium outside and inside the membrane. Thus,

$$v_i - v_o = -\frac{KT}{q}\ln\frac{[K^+]_i}{[K^+]_o} = \frac{KT}{q}\ln\frac{[K^+]_o}{[K^+]_i} \tag{3.11}$$

Equation (3.11) is known as the **Nernst equation,** named after a German physical chemist Walter Nernst, and $E_K = v_i - v_o$ is known as the Nernst potential for K^+. At room temperature, $\dfrac{KT}{q} = 26\text{mV}$, and thus the Nernst equation for K^+

$$E_K = v_i - v_o = 26\ln\frac{[K^+]_o}{[K^+]_i}\text{mV} \tag{3.12}$$

While Eq. (3.12) is specifically written for K^+, it can be easily derived for any permeable ion. At room temperature, the Nernst potential for Na^+ is

$$E_{Na} = v_i - v_o = 26\ln\frac{[Na^+]_o}{[Na^+]_i}\text{mV} \tag{3.13}$$

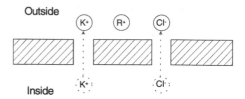

Fig. 3.8 Membrane is permeable to both K^+ and Cl^-, but not to a large cation R^+.

and the Nernst potential for Cl^- is

$$E_{Cl} = v_i - v_o = -26 \ln \frac{[Cl^-]_o}{[Cl^-]_i} \text{ mV} = 26 \ln \frac{[Cl^-]_i}{[Cl^-]_o} \text{ mV} \qquad (3.14)$$

The negative sign in Eq. (3.14) is due to $Z = -1$ for Cl^-.

3.4.3 Donnan Equilibrium

In a neuron at steady state (equilibrium) that is permeable to more than one ion, for example K^+, Na^+, and Cl^-, the Nernst potential for each ion is calculated using Eq. (3.12)–(3.14), respectively. The membrane potential, $V_m = v_i - v_o$, however, is due to the presence of all ions and is influenced by the concentration and permeability of each ion. In this section, the case in which two ions are permeable is presented. In the next section, the case in which any number of permeable ions are present is considered.

Suppose a membrane is permeable to both K^+ and Cl^-, but not to a large cation, R^+, as shown in Fig. 3.8. For equilibrium, the Nernst potentials for both K^+ and Cl^- must be equal, that is, $E_K = E_{Cl}$, or

$$E_K = \frac{KT}{q} \ln \frac{[K^+]_o}{[K^+]_i} = E_{Cl} = \frac{KT}{q} \ln \frac{[Cl^-]_i}{[Cl^-]_o} \qquad (3.15)$$

After simplifying,

$$\frac{[K^+]_o}{[K^+]_i} = \frac{[Cl^-]_i}{[Cl^-]_o} \qquad (3.16)$$

Equation (3.16) is known as the **Donnan equilibrium**. An accompanying principle is **space charge neutrality,** which states that the number of cations in a given volume is equal to the number of anions. Thus, in the equilibrium state, ions still diffuse across the membrane, but each K^+ that crosses the membrane must be accompanied by a Cl^- for space charge neutrality to be satisfied. If in Fig. 3.8 R^+ were not present, then at equilibrium the concentration of K^+ and Cl^- on both sides of the membrane would be equal. With R^+ in the extracellular fluid, the concentrations of [KCl] on both sides of the membrane are different as shown in Example Problem 3.1.

Example Problem 3.1

A membrane is permeable to K^+ and Cl^-, but not to a large cation R^+. Find the steady-state equilibrium concentration for the following initial conditions.

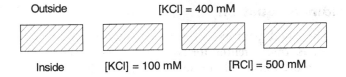

Outside

[KCl] = 400 mM

Inside   [KCl] = 100 mM   [RCl] = 500 mM

Solution

By conservation of mass,

$$[K^+]_i + [K^+]_o = 500$$

$$[Cl^-]_i + [Cl^-]_o = 1000$$

and by space charge neutrality,

$$[K^+]_i + 500 = [Cl^-]_i$$

$$[K^+]_o = [Cl^-]_o$$

From the Donnan equilibrium,

$$\frac{[K^+]_o}{[K^+]_i} = \frac{[Cl^-]_i}{[Cl^-]_o}$$

Substituting for $[K^+]_o$ and $[Cl^-]_o$ from the conservation of mass equations into the Donnan equilibrium equation gives

$$\frac{500 - [K^+]_i}{[K^+]_i} = \frac{[Cl^-]_i}{1000 - [Cl^-]_i}$$

and eliminating $[Cl^-]_i$ by using the space charge neutrality equations gives

$$\frac{500 - [K^+]_i}{[K^+]_i} = \frac{[K^+]_i + 500}{1000 - [K^+]_i - 500} = \frac{[K^+]_i + 500}{500 - [K^+]_i}$$

Solving the previous equation yields $[K^+]_i = 167$ mM at steady state . Using the conservation of mass equations and space charge neutrality equations gives $[K^+]_o = 333$ mM, $[Cl^-]_i = 667$ mM, and $[Cl^-]_o = 333$ mM at steady state. At steady state and at room temperature, the Nernst potential for either ion is 18 mV, as shown for $[K^+]$

$$E_K = v_i - v_o = 26 \ln\frac{333}{167} = 18 \text{ mV}$$

Summarizing, at steady state

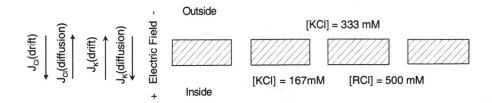

J_{Cl}(drift) J_{Cl}(diffusion) J_K(drift) J_K(diffusion)   Electric Field

− Outside

[KCl] = 333 mM

+ Inside   [KCl] = 167mM   [RCl] = 500 mM

3.4.4   Goldman Equation

The squid giant axon resting potential is -60 mV, which does not correspond to the Nernst potential for Na^+ or K^+. As a general rule, when V_m is affected by two or more ions, each ion influences V_m as determined by its concentration and membrane permeability. The Goldman equation quantitatively describes the relationship between V_m and permeable ions but applies only when the membrane potential or electric field is constant. This situation is a reasonable approximation for a resting membrane potential. Here the Goldman equation is first derived for K^+ and Cl^- and then extended to include K^+, Cl^-, and Na^+. The Goldman equation is used by physiologists to calculate the membrane potential for a variety of cells and, in fact, was used by Hodgkin, Huxley, and Katz in studying the squid giant axon.

Consider the cell membrane shown in Fig. 3.9. To determine V_m for both K^+ and Cl^-, flow equations for each ion are derived separately under the condition of a constant electric field and then combined using space charge neutrality to complete the derivation of the Goldman equation.

Potassium Ions

The flow equation for K^+ with mobility μ_K is

$$J_K = -\frac{KT}{q}\mu_K\frac{d[K^+]}{dx} - \mu_K Z_K[K^+]\frac{dv}{dx} \tag{3.17}$$

Under a constant electric field,

$$\frac{dv}{dx} = \frac{\Delta v}{\Delta x} = \frac{V}{\delta} \tag{3.18}$$

Substituting Eq. (3.18) into Eq. (3.17) with $Z_K = 1$ gives

$$J_K = -\frac{KT}{q}\mu_K\frac{d[K^+]}{dx} - \mu_K Z_K[K^+]\frac{V}{\delta} \tag{3.19}$$

Let the permeability for K^+, P_K, equal

$$P_K = \frac{\mu_K KT}{\delta q} = \frac{D_K}{\delta} \tag{3.20}$$

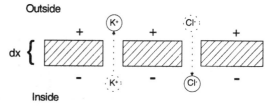

Fig. 3.9 Diagram illustrating a cell membrane permeable to both K^+ and Cl^-. The width of the membrane is $dx = \delta$.

Therefore, using Eq. (3.20) in Eq. (3.19) gives

$$J_K = \frac{-P_K q}{KT} V[K^+] - P_K \delta \frac{d[K^+]}{dx} \tag{3.21}$$

Rearranging the terms in Eq. (3.21) yields

$$dx = \frac{d[K^+]}{\dfrac{-J_K}{P_K \delta} - \dfrac{q V[K^+]}{KT\delta}} \tag{3.22}$$

Taking the integral of both sides, while assuming that J_K is independent of x, gives

$$\int_0^\delta dx = \int_{[K^+]_i}^{[K^+]_o} \frac{d[K^+]}{\dfrac{-J_K}{P_K \delta} - \dfrac{q V[K^+]}{KT\delta}} \tag{3.23}$$

resulting in

$$x \Big|_0^\delta = -\frac{KT\delta}{qV} \ln\left(\frac{J_K}{P_K\delta} + \frac{q V[K^+]}{KT\delta} \right) \Big|_{[K^+]_i}^{[K^+]_o} \tag{3.24}$$

and

$$\delta = \frac{KT\delta}{qV} \ln\left(\frac{\dfrac{J_K}{P_K\delta} + \dfrac{q V[K^+]_o}{KT\delta}}{\dfrac{J_K}{P_K\delta} + \dfrac{q V[K^+]_i}{KT\delta}} \right) \tag{3.25}$$

Removing δ from both sides of Eq. (3.25), bringing the term $-\dfrac{KT}{qV}$ to the other side of the equation, and then taking the exponential of both sides yields

$$e^{\frac{-qV}{KT}} = \frac{\dfrac{J_K}{P_K\delta} + \dfrac{q V[K^+]_o}{KT\delta}}{\dfrac{J_K}{P_K\delta} + \dfrac{q V[K^+]_i}{KT\delta}} \tag{3.26}$$

Solving for J_K in Eq. (3.26) gives

$$J_K = \frac{qV P_K}{KT} \left(\frac{[K^+]_o - [K^+]_i e^{\frac{-qV}{KT}}}{e^{\frac{-qV}{KT}} - 1} \right) \tag{3.27}$$

Chlorine Ions

The same derivation carried out for K^+ can be repeated for Cl^-, which yields

$$J_{Cl} = \frac{qV P_{Cl}}{KT} \left(\frac{[Cl^-]_o\, e^{\frac{-qV}{KT}} - [Cl^-]_i}{e^{\frac{-qV}{KT}} - 1} \right) \tag{3.28}$$

where P_{Cl} is the permeability for Cl^-.

Summarizing for Potassium and Chlorine Ions

From space charge neutrality, $J_K = J_{Cl}$ with Eqs. (3.27) and (3.28) gives

$$P_K\left([K^+]_o - [K^+]_i \, e^{\frac{-qV}{KT}}\right) = P_{Cl}\left([Cl^-]_o \, e^{\frac{-qV}{KT}} - [Cl^-]_i\right) \tag{3.29}$$

Solving for the exponential terms yields

$$e^{\frac{-qV}{KT}} = \frac{P_K[K^+]_o + P_{Cl}[Cl^-]_i}{P_K[K^+]_i + P_{Cl}[Cl^-]_o} \tag{3.30}$$

Solving for V gives

$$V = v_o - v_i = -\frac{KT}{q} \ln\left(\frac{P_K[K^+]_o + P_{Cl}[Cl^-]_i}{P_K[K^+]_i + P_{Cl}[Cl^-]_o}\right) \tag{3.31}$$

or in terms of V_m

$$V_m = \frac{KT}{q} \ln\left(\frac{P_K[K^+]_o + P_{Cl}[Cl^-]_i}{P_K[K^+]_i + P_{Cl}[Cl^-]_o}\right) \tag{3.32}$$

This equation is called the **Goldman equation**. Since sodium is also important in membrane potential, the Goldman equation for K^+, Cl^-, and Na^+ can be derived as

$$V_m = \frac{KT}{q} \ln\left(\frac{P_K[K^+]_o + P_{Na}[Na^+]_o + P_{Cl}[Cl^-]_i}{P_K[K^+]_i + P_{Na}[Na^+]_i + P_{Cl}[Cl^-]_o}\right) \tag{3.33}$$

where P_{Na} is the permeability for Na^+. To derive Eq. (3.33), first find J_{Na} and then use space charge neutrality $J_K + J_{Na} = J_{Cl}$. Equation (3.33) then follows. In general, when the permeability to one ion is exceptionally high compared with the other ions, then V_m predicted by the Goldman equation is very close to the Nernst equation for that ion.

Tables 3.1 and 3.2 contain the important ions across the cell membrane, the ratio of permeabilities, and Nernst potentials for the squid giant axon and frog skeletal muscle. The reason that the squid giant axon is extensively reported and used in experiments is due to its large size, lack of myelination, and ease of use. In general, the intracellular and extracellular concentration of ions in vertebrate neurons is approximately three or four times less than that in the squid giant axon.

Example Problem 3.2

Calculate V_m for the squid giant axon at 6.3° C.

Solution:

Using Equation 3.33 and the data in Table 3.1, gives

$$V_m = 25.3 \times \ln\left(\frac{1 \times 20 + 0.04 \times 440 + 0.45 \times 52}{1 \times 400 + 0.04 \times 50 + 0.45 \times 560}\right) \text{mV} = -60\text{mV}$$

■

TABLE 3.1 Approximate Intracellular and Extracellular Concentrations of the Important Ions across a Squid Giant Axon, Ratio of Permeabilities, and Nernst Potentials[a].

Ion	Cytoplasm (mM)	Extracellular fluid (mM)	Ratio of permeabilities	Nernst potential (mV)
K^+	400	20	1	−74
Na^+	50	440	0.04	55
$Cl-$	52	560	0.45	−60

[a]Note that the permeabilities are relative, that is, P_K: P_{Na}: P_{Cl}, and not absolute. Data were recorded at 6.3°C, resulting in KT/q approximately equal to 25.3 mV.

3.4.5 Ion Pumps

At rest, separation of charge and ionic concentrations across the cell membrane must be maintained, otherwise V_m changes. That is, the flow of charge into the cell must be balanced by the flow of charge out of the cell. For Na^+, the concentration and electric gradient creates a force that drives Na^+ into the cell at rest. At V_m, the K^+ force due to diffusion is greater than that due to drift and results in an efflux of K^+ out of the cell. Space charge neutrality requires the influx of Na^+ be equal to the flow of K^+ out of the cell. Although these flows cancel each other and space charge neutrality is maintained, this process cannot continue unopposed. Otherwise, $[K^+]_i$ goes to zero as $[Na^+]_i$ increases, with subsequent change in V_m as predicted by the Goldman equation.

Any change in the concentration gradient of K^+ and Na^+ is prevented by the Na–K pump. The pump transports a steady stream of Na^+ out of the cell and K^+ into the cell. Removal of Na^+ from the cell is against its concentration and electric gradient and is accomplished with an active pump that consumes metabolic energy. Figure 3.10 illustrates a Na–K pump along with an active and passive channel.

The Na–K pump has been found to be electrogenic; that is, there is a net transfer of charge across the membrane. Nonelectrogenic pumps operate without any net transfer of charge. For many neurons, the Na–K ion pump removes three Na^+ ions

TABLE 3.2 Approximate Intracellular and Extracellular Concentrations of the Important Ions across a Frog Skeletal Muscle, Ratio of Permeabilities, and Nernst Potentials[a].

Ion	Cytoplasm (mM)	Extracellular fluid (mM)	Ratio of permeabilities	Nernst potential (mV)
K^+	140	2.5	1.0	−105
Na^+	13	110	0.019	56
$Cl-$	3	90	0.381	−89

[a]Data were recorned at room temperature, resulting in KT/q approximately equal to 26mV.

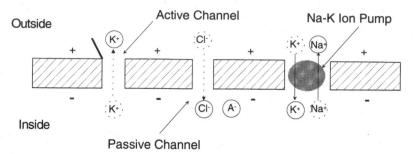

Fig. 3.10 An active pump is illustrated along with a passive and active channel.

for every two K^+ ions moved into the cell, which makes V_m slightly more negative than predicted with only passive channels.

In general, when the cell membrane is at rest, the active and passive ion flows are balanced and a permanent potential exists across a membrane only if

1. The membrane is impermeable to some ion(s).
2. An active pump is present.

The presence of the Na–K pump forces V_m to a given potential based on the K^+ and Na^+ concentrations that are determined by the active pump. Other ion concentrations are determined by V_m. For instance, since Cl^- moves across the membrane only through passive channels, the Cl^- concentration ratio at rest is determined from the Nernst equation with $E_{Cl} = V_m$, or

$$\frac{[Cl^-]_i}{[Cl^-]_o} = e^{\frac{-qV}{KT}}$$

(3.34)

Example Problem 3.3

Consider a membrane in which there is an active K^+ pump, passive channels for K^+ and Cl^-, and a nonequilibrium initial concentration of [KCl] on both sides of the membrane. Find an expression for the active K^+ pump.

Solution

From space charge neutrality, $J_{Cl} = J_K$, or

$$J_K = J_p - \frac{KT}{q}\mu_K\frac{d[K^+]}{dx} - \mu_K Z_K[K^+]\frac{dv}{dx}$$

$$J_{Cl} = -\frac{KT}{q}\mu_{Cl}\frac{d[Cl^-]}{dx} - \mu_{Cl}Z_{Cl}[Cl^-]\frac{dv}{dx}$$

where J_p is the flow due to the active K^+ pump.

Solving for $\frac{dv}{dx}$ using the J_{Cl} equation with $Z_{Cl} = -1$ gives

$$\frac{dv}{dx} = \frac{KT}{q[Cl^-]}\frac{d[Cl^-]}{dx}$$

By space charge neutrality, $[Cl^-] = [K^+]$, which allows rewriting the previous equation as

$$\frac{dv}{dx} = \frac{KT}{q[K^+]}\frac{d[K^+]}{dx}$$

At equilibrium, both flows are zero and with $Z_K = 1$, the J_K equation with $\frac{dv}{dx}$ substitution is given as

$$J_K = 0 = J_p - \mu_K[K^+]\frac{dv}{dx} - \frac{KT\mu_K}{q}\frac{d[K^+]}{dx}$$

$$= J_p - \mu_K[K^+]\frac{KT}{q[K^+]}\frac{d[K^+]}{dx} - \frac{KT\mu_K}{q}\frac{d[K^+]}{dx} = J_p - \frac{2KT\mu_K}{q}\frac{d[K^+]}{dx}$$

Moving J_p to the left side of the equation, multiplying both sides by dx, and then integrating yields

$$-\int_0^\delta J_p dx = -\frac{2KT\mu_K}{q}\int_{[K^+]_i}^{[K^+]_o} d[K^+] \qquad (3.10)$$

or

$$J_p = \frac{2KT\mu_K}{q\delta}([K^+]_o - [K^+]_i)$$

Note: In this example, if no pump was present, then at equilibrium the concentration on both sides of the membrane would be the same. ∎

3.5 EQUIVALENT CIRCUIT MODEL FOR THE CELL MEMBRANE

In this section, an equivalent circuit model is developed using the tools previously developed. Creating a circuit model is helpful when discussing the Hodgkin–Huxley model of an action potential in the next section, a model that introduces voltage- and time-dependent ion channels. As previously described in Sections 3.3 and 3.4, the nerve cell has three types of passive electrical characteristics: electromotive force, resistance, and capacitance. The nerve membrane is a lipid bilayer that is pierced by a variety of different types of ion channels, where each channel is characterized as being passive (always open) or active (gates that can be opened). Each ion channel is also characterized by its selectivity. In addition, there is the active Na–K pump that maintains V_m across the cell membrane.

3.5.1 Electromotive, Resistive, and Capacitive Properties

Electromotive Force Properties

The three major ions, K^+, Na^+, and Cl^-, are differentially distributed across the cell membrane at rest and across the membrane through passive ion channels as illustrated in Fig. 3.5. This separation of charge exists across the membrane and results in a voltage potential V_m as described by the Goldman equation (Eq. 3.33).

Across each ion-specific channel, a concentration gradient exists for each ion that creates an electromotive force, a force that drives that ion through the channel at a constant rate. The Nernst potential for that ion is the electrical potential difference across the channel and is easily modeled as a battery as is illustrated in Fig. 3.11 for K^+. The same model is applied for Na^+ and Cl^- with values equal to the Nernst potentials for each.

Resistive Properties

In addition to the electromotive force, each channel also has resistance; that is, it resists the movement of electrical charge through the channel. This is mainly due to collisions with the channel wall where energy is given up as heat. The term conductance, G, measured in Siemens (S), which is the ease with which the ions move through the membrane, is typically used to represent resistance. Since the conductances (channels) are in parallel, the total conductance is the total number of channels, N, times the conductance for each channel, G'

$$G = N \times G'$$

It is usually more convenient to write the conductance as resistance $R = \dfrac{1}{G}$, measured in ohms (Ω). An equivalent circuit for the channels for a single ion is now given as a resistor in series with a battery as shown in Fig. 3.12.

Conductance is related to membrane permeability, but they are not interchangeable in a physiological sense. Conductance depends on the state of the membrane, varies with ion concentration, and is proportional to the flow of ions through a membrane. Permeability describes the state of the membrane for a particular ion. Consider the case in which there are no ions on either side of the membrane. No matter how many channels are open, $G = 0$ because there are no ions available to flow across the cell membrane (due to a potential difference). At the same time, ion permeability is constant and determined by the state of the membrane.

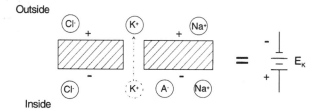

Fig. 3.11 A battery is used to model the electromotive force for a K^+ channel with a value equal to the K^+ Nernst potential. The polarity of the battery is given with the ground on the outside of the membrane in agreement with convention. From Table 3.1, note that the Nernst potential for K^+ is negative, which reverses the polarity of the battery, driving K^+ out of the cell.

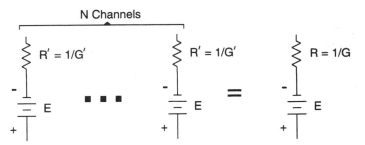

Fig. 3.12 The equivalent circuit for N ion channels is a single resistor and battery.

Equivalent Circuit for Three Ions

Each of the three ions, K^+, Na^+, and Cl^- are represented by the same equivalent circuit, as shown in Fig. 3.12, with Nernst potentials and appropriate resistances. Combining the three equivalent circuits into one circuit with the extracellular fluid and cytoplasm connected by short circuits completely describes a membrane at rest (Fig. 3.13).

Example Problem 3.4

Find V_m for the frog skeletal muscle (Table 3.2) if the Cl^- channels are ignored. Use $R_K = 1.7\,k\Omega$ and $R_{Na} = 15.67\,k\Omega$.

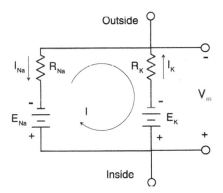

Solution

The diagram depicts the membrane circuit with mesh current I, current I_{Na} through the sodium channel, and current I_K through the potassium channel. Current I is found using mesh analysis:

$$E_{Na} + IR_{Na} + IR_K - E_K = 0$$

and solving for I,

$$I = \frac{E_K - E_{Na}}{R_{Na} + R_K} = \frac{(-105 - 56) \times 10^{-3}}{(15.67 + 1.7) \times 10^3} = -9.27\,\mu A$$

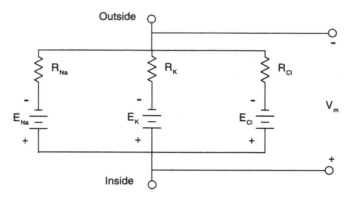

Fig. 3.13 Model of the passive channels for a small area of nerve at rest with each ion channel represented by a resistor in series with a battery.

yields

$$V_m = E_{Na} + I R_{Na} = -89\,\text{mV}$$

Notice that $I = -I_{Na}$ and $I = -I_K$, or $I_{Na} = I_K$ as expected. Physiologically, this implies that the inward Na^+ current is exactly balanced by the outward bound K^+ current. ■

Example Problem 3.5

Find V_m for the frog skeletal muscle if $R_{Cl} = 3.125\ \text{k}\Omega$.

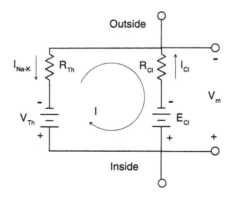

Solution

To solve, first find a Thévenin equivalent circuit for the circuit in Example Problem 3.4:

$$V_{Th} = V_m = -89\ \text{mV}$$

and

$$R_{Th} = \frac{R_{Na} \times R_K}{R_{Na} + R_K} = 1.534 \text{ k}\Omega$$

Since $E_{Cl} = V_{Th}$ according to Table 3.2, no current flows. This is the actual situation in most nerve cells. The membrane potential is determined by the relative conductances and Nernst potentials for K^+ and Na^+. The Nernst potentials are determined by the active pump that maintains the concentration gradient. Cl^- is usually passively distributed across the membrane. ∎

Na–K Pump

As shown in Example Problem 3.4 and Section 3.4, there is a steady flow of K^+ ions out of the cell and Na^+ ions into the cell even when the membrane is at the resting potential. Left unchecked, this would drive E_K and E_{Na} toward 0. To prevent this, current generators, the Na–K pump, are used that are equal and opposite to the passive currents and incorporated into the model as shown in Fig. 3.14.

3.5.2 Capacitive Properties

Capacitance occurs whenever electrical conductors are separated by an insulating material. In the neuron, the cytoplasm and extracellular fluid are the electrical conductors and the lipid bilayer of the membrane is the insulating material (Fig. 3.3). Capacitance for a neuron membrane is approximately 1 μF/cm². Membrane capacitance implies that ions do not move through the membrane except through ion channels.

The membrane can be modeled using the circuit in Fig. 3.15 by incorporating membrane capacitance with the electromotive and resistive properties. A consequence of membrane capacitance is that changes in membrane voltage are not im-

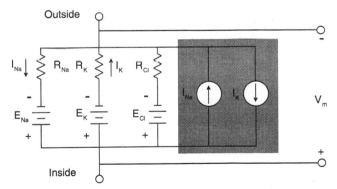

Fig. 3.14 Circuit model of the three passive channels for a small area of the nerve at rest with each ion channel represented by a resistor in series with a battery. The Na–K active pump is modeled as two current sources within the shaded box.

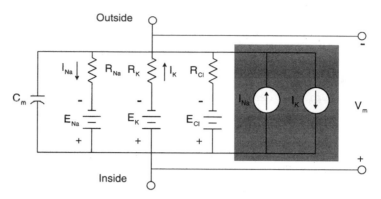

Fig. 3.15 Circuit model of a small area of the nerve at rest with all of its passive electrical properties. The Na–K active pump is within the shaded box as two current sources.

mediate but follow an exponential time course due to first-order time constant effects. To appreciate the effect of capacitance, the circuit in Fig. 3.15 is reduced to that shown in Fig. 3.16 by using a Thévenin equivalent for the batteries and the resistors with R_{Th} and V_{Th} given in Eqs. (3.35) and (3.36).

$$R_{Th} = \frac{1}{\dfrac{1}{R_K} + \dfrac{1}{R_{Na}} + \dfrac{1}{R_{Cl}}} \tag{3.35}$$

$$V_{Th} = \frac{R_{Na} R_{Cl} E_K + R_K R_{Cl} E_{Na} + R_K R_{Na} E_{Cl}}{R_{Na} R_{Cl} + R_K R_{Cl} + R_K R_{Na}} \tag{3.36}$$

The time constant for the membrane circuit model is $\tau = R_{Th} \times C_m$, and at 5τ the response is within 1% of steady state. The range for τ is from 1 to 20 ms in a

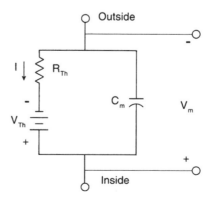

Fig. 3.16 Thévenin equivalent circuit of the model in Fig. 3.15.

typical neuron. In addition, at steady state, the capacitor acts as an open circuit and $V_{Th} = V_m$ as it should.

Example Problem 3.6

Compute the change in V_m due to a current pulse through the cell membrane.

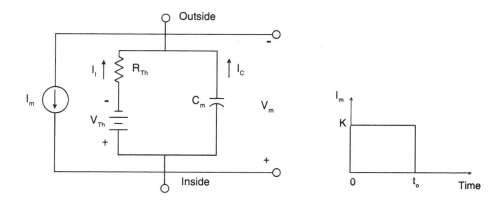

Solution

Experimentally, the stimulus current is a pulse passed through the membrane from an intracellular electrode to an extracellular electrode as depicted in the above circuit diagram.

The membrane potential, V_m, due to a current pulse, I_m, with amplitude K and duration t_0 applied at $t = 0$, is found by applying Kirchhoff's current law at the cytoplasm, yielding

$$-I_m + \frac{V_m - V_{Th}}{R_{Th}} + \frac{C_m d V_m}{dt} = 0$$

The Laplace transform of the node equation is

$$-I_m(s) + \frac{V_m(s)}{R_{Th}} - \frac{V_{Th}}{s R_{Th}} + s C_m V_m(s) - C_m V_m(0^+) = 0$$

Combining common terms gives

$$\left(s + \frac{1}{C_m R_{Th}}\right) V_m(s) = V_m(0^+) + \frac{I_m(s)}{C_m} + \frac{V_{Th}}{s C_m R_{Th}}$$

The Laplace transform of the current pulse is $I_m(s) = \frac{K}{s}(1 - e^{-t_0 s})$. Substituting $I_m(s)$ into the node equation and rearranging terms yields

$$V_m(s) = \frac{V_m(0^+)}{\left(s + \dfrac{1}{C_m R_{Th}}\right)} + \frac{K(1 - e^{-t_0 s})}{s C_m \left(s + \dfrac{1}{C_m R_{Th}}\right)} + \frac{V_{Th}}{s C_m R_{Th} \left(s + \dfrac{1}{C_m R_{Th}}\right)}$$

Performing a partial fraction expansion, and noting that $V_m(0^+) = V_{Th}$, gives

$$V_m(s) = KR_{Th}\left(\frac{1}{s} - \frac{1}{s + \dfrac{1}{C_m R_{Th}}}\right)(1 - e^{-t_o s}) + \frac{V_{Th}}{s}$$

Transforming back into the time domain yields the solution.

$$V_m(t) = V_{Th} + R_{Th}K\left(1 - e^{-\frac{t}{R_{Th}C_m}}\right)u(t) - R_{Th}K\left(1 - e^{-\frac{t-t_0}{R_{Th}C_m}}\right)u(t - t_0)$$

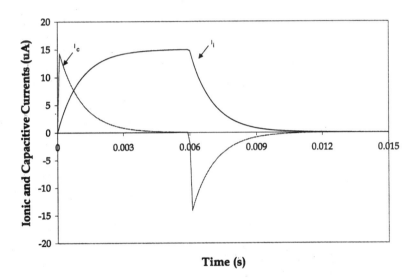

The ionic current (I_i) and capacitive current (I_c) are shown in the figure above, where

$$I_i = \frac{V_{Th} - V_m}{R_{Th}}$$

$$I_c = C_m\frac{dV_m}{dt}$$

Shown in the following figures are graphs of V_m in response to a 15 μA current pulse of 6 ms (upper) and 2 ms (lower) using parameters for the frog skeletal muscle. The time constant is approximately 1 ms. Note that in the figure on the left, V_m reaches steady state before the current pulse returns to zero, and in the figure on the lower, V_m falls short of the steady-state value reached on the upper.

The value of the time constant is important in the integration of currents (packets of neurotransmitter) at the synapse. Each packet of neurotransmitter acts as a current pulse. Note, the longer the time constant, the more time the membrane is excited. Most excitations are not synchronous, but because of τ, a significant portion of the stimulus is added together to cause signal transmission.

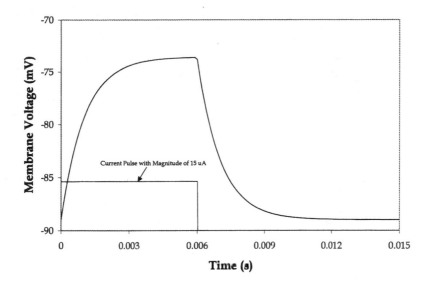

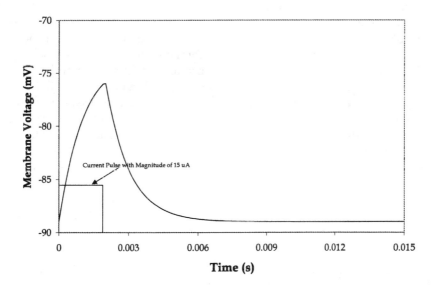

Shown in the following figure is the membrane voltage due to a series of 15μA current pulses of 6 ms duration with the onsets occurring at 0, 2, 4, 6, and 8 ms. Since the pulses occur within 5τ of the previous pulses, the effect of each on V_m is additive, allowing the membrane to depolarize to approximately -45 mV. If the pulses are spaced at intervals $>5\tau$, then V_m would be a series of pulse responses as previously illustrated.∎

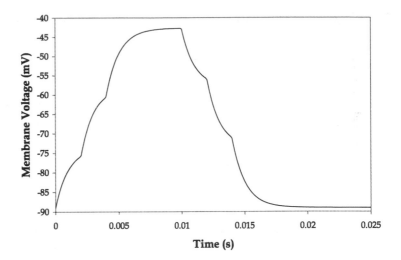

3.5.3   Change in Membrane Potential with Distance

The circuit model in Figs. 3.15 or 3.16 describes a small area or section of the membrane. In Example Problem 3.6, a current pulse was injected into the membrane and resulted in a change in V_m. The change in V_m in this section of the membrane causes current to flow into the adjacent membrane sections, which causes a change in V_m in each section and so on continuing throughout the surface of the membrane. Since the volume inside the dendrite is much smaller than the extracellular space, there is significant resistance to the flow of current in the cytoplasm from one membrane section to the next compared with the flow of current in the extracellular space. The larger the diameter of the dendrite, the smaller the resistance to the spread of current from one section to the next. To model this effect, a resistor, R_a, is placed in the cytoplasm connecting each section together as shown in the Fig. 3.17. This model is actually a three-dimensional surface and continues in the x, y, and z directions. The

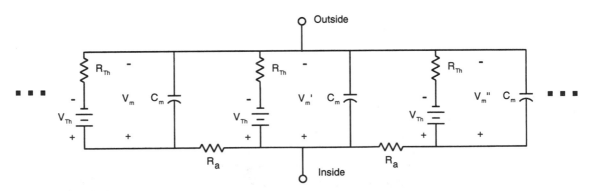

Fig. 3.17   Equivalent circuit of series of membrane sections connected with axial resistance, R_a.

outside resistance is negligible since it has a greater volume and is modeled as a short circuit.

Suppose a current is injected into a section of the dendrite similar to the situation in Example Problem 3.6 where t_0 is large. At steady state, the transient response due to C_m has expired and only current through the resistance is important. Most of the current flows out through the section into which the current was injected since it has the smallest resistance (R_{Th}) in relation to the other sections. The next largest current flowing out of the membrane occurs in the next section since it has the next smallest resistance, $R_{Th} + R_a$. The change in V_m, ΔV_m, from the injection site is independent of C_m and depends solely on the relative values of R_{Th} and R_a. The resistance seen n sections from the injection site is $R_{Th} + n \times R_a$. Since current decreases with distance from the injection site, then ΔV_m also decreases with distance from the injection site because it equals the current through that section times R_{Th}. The change in membrane potential, ΔV_m, decreases exponentially with distance and is given by

$$\Delta V_m = V_0 e^{-\frac{x}{\lambda}}$$

where $\lambda = \sqrt{\dfrac{R_{Th}}{R_a}}$ is the membrane length constant, x is the distance away from the injection site, and V_0 is the change in membrane potential at the injection site. The range of values for λ is 0.1 to 1 mm. The larger the value of λ, the greater the effect of the stimulation along the length of the membrane.

3.6 HODGKIN–HUXLEY MODEL OF THE ACTION POTENTIAL

Hodgkin and Huxley published five papers in 1952 that described a series of experiments and an empirical model of an action potential in a squid giant axon. Their first four papers described the experiments that characterized the changes in the cell membrane that occurred during the action potential. The last paper presented the empirical model. The empirical model they developed is not a physiological model based on the laws and theory developed in this chapter but a model based on curve fitting by using an exponential function. In this section, highlights of the Hodgkin–Huxley experiments are presented along with the empirical model. All the figures presented in this section were simulated using SIMULINK and the Hodgkin–Huxley empirical model parameterized with their squid giant axon data.

3.6.1 Action Potentials and the Voltage Clamp Experiment

The ability of nerve cells to conduct action potentials makes it possible for signals to be transmitted over long distances within the nervous system. An important feature of the action potential is that it does not decrease in amplitude as it is conducted away from its site of initiation. An action potential occurs when V_m reaches a value called the **threshold potential** at the axon hillock (see Fig. 3.1). Once V_m reaches threshold, time- and voltage-dependent conductance changes occur in the active

Na$^+$ and K$^+$ gates that drive V_m toward E_{Na}, then back to E_K, and finally to the resting potential. These changes in conductance were first described by Hodgkin and Huxley (1952a–d) [Katz was a coauthor on one paper (Hodgkin *et al.*, 1952) and a collaborator on several others]. Figure 3.18 illustrates a stylized action potential with the threshold potential at approximately -40 mV.

Stimulation of the postsynaptic membrane along the dendrite and cell body must occur for V_m to rise to the threshold potential at the axon hillock. As previously described, the greater the distance from the axon hillock, the smaller the contribution of postsynaptic membrane stimulation to the change in V_m at the axon hillock. Also, because of the membrane time constant, there is a time delay in stimulation at the postsynaptic membrane and the resultant change in V_m at the axon hillock. Thus, time and distance are important functions in describing the graded response of V_m at the axon hillock.

Once V_m reaches threshold, active Na$^+$ conductance gates are opened and an inward flow of Na$^+$ ions results, causing further depolarization. This depolarization increases Na$^+$ conductance, consequently inducing more Na$^+$ current. This iterative cycle continues driving V_m to E_{Na} and concludes with the closure of the Na$^+$ gates. A similar but slower change in K$^+$ conductance occurs that drives V_m back to the

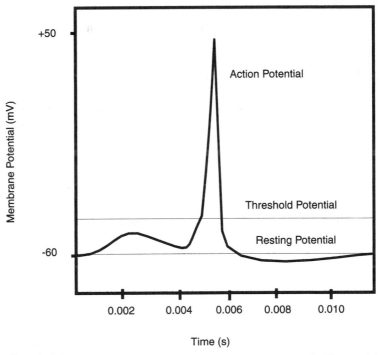

Fig. 3.18   Stylized diagram of an action potential once threshold potential is reached at approximately 5 ms. The action potential is due to voltage- and time-dependent changes in conductance. The action potential rise is due to Na$^+$ and the fall is due to K$^+$ conductance changes.

resting potential. Once an action potential is started, it continues until completion. This is called the "all or none" phenomenon. The active gates for Na^+ and K^+ are both functions of V_m and time.

The action potential moves through the axon at high speeds and appears to jump from one Node of Ranvier to the next in myelinated neurons. This occurs because the membrane capacitance of the myelin sheath is very small, making the membrane appear only resistive with almost instantaneous changes in V_m possible.

To investigate the action potential, Hodgkin and Huxley used an unmyelinated squid giant axon in their studies because of its large diameter (up to 1 mm) and long survival time of several hours in seawater at 6.3°C. Their investigations examined the then existing theory that described an action potential as due to enormous changes in membrane permeability that allowed all ions to freely flow across the membrane, driving V_m to zero. As they discovered, this was not the case. The success of the Hodgkin–Huxley studies was based on two new experimental techniques, the space clamp and voltage clamp, and collaboration with Cole and Curtis from Columbia University.

The space clamp allowed Hodgkin and Huxley to produce a constant V_m over a large region of the membrane by inserting a silver wire inside the axon and thus eliminating R_a. The voltage clamp allowed the control of V_m by eliminating the effect of further depolarization due to the influx of I_{Na} and efflux of I_K as membrane permeability changed. Selection of the squid giant axon was fortunate for two reasons: It was large and survived a very long time in sea water and it had only two types of voltage–time-dependent permeable channels. Other types of neurons have more than two voltage–time-dependent permeable channels that would have made the analysis extremely difficult or even impossible.

Voltage Clamp

To study the variable voltage–time resistance channels for K^+ and Na^+, Hodgkin and Huxley used a voltage clamp to separate these two dynamic mechanisms so that only the time dependent features of the channel were examined. Figure 3.19 illustrates the voltage clamp experiment by using the equivalent circuit model previously described. The channels for K^+ and Na^+ are represented using variable voltage–time resistances, and the passive gates for Na^+, K^+, and Cl^- are given by a leakage channel with resistance R_l (that is, the Thévenin equivalent circuit of the passive channels). The function of the voltage clamp is to suspend the interaction between Na^+ and K^+ channel resistance and the membrane potential. If the membrane voltage is not clamped, then changes in Na^+ and K^+ channel resistance modify membrane voltage, which then changes Na^+ and K^+ channel resistance and so on as previously described.

A voltage clamp is created by using two sets of electrodes. In an experiment, one pair injects current, I_m, to keep V_m constant and another pair is used to observe V_m. To estimate the conductance in the Na^+ and K^+ channels, I_m is also measured during the experiment. Meters for recording V_m and I_m are illustrated in Fig. 3.19. They are placed outside the seawater bath. Today, these would be connected to an analog-to-digital converter (ADC) with data stored in a hard disk of a computer. In 1952, these meters were strip chart recorders. The application of a clamp voltage,

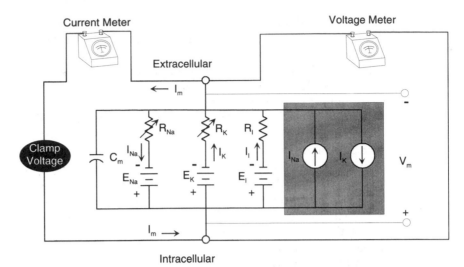

Fig. 3.19 Equivalent circuit model of an unmyelinated section of squid giant axon under voltage clamp conditions. The channels for K^+ and Na^+ are now represented using variable voltage–time resistances, and the passive gates for Na^+, K^+, and Cl^- are given by a leakage channel with resistance R_l. The Na–K pump is illustrated within the shaded area of the circuit. In the experiment, the membrane is immersed in the seawater bath.

V_c, causes a change in Na^+ conductance that results in an inward flow of Na^+ ions. This causes the membrane potential to be more positive than V_c. The clamp removes positive ions from inside the cell, which results in no net change in V_m. The current, I_m, is the dependent variable in the voltage clamp experiment and V_c is the independent variable.

To carry out the voltage clamp experiment, the investigator first selects a clamp voltage and then records the resultant membrane current, I_m, that is necessary to keep V_m at the clamp voltage. Figure 3.20 shows the resulting I_m due to a clamp voltage of -20 mV. Initially, the step change in V_m causes a large current to pass through the membrane that is primarily due to the capacitive current. The clamp voltage also creates a constant leakage current through the membrane that is equal to

$$I_i = \frac{V_c - E_l}{R_l} \tag{3.38}$$

Subtracting both the capacitive and leakage current from I_m leaves only the Na^+ and K^+ currents. To separate the Na^+ and K^+ currents, Hodgkin and Huxley substituted a large impermeable cation for Na^+ in the external solution. This eliminated the Na^+ current and left only the K^+ current. Returning the Na^+ to the external solution allowed the Na^+ current to be estimated by subtracting the capacitive, leakage, and K^+ currents from I_m. The Na^+ and K^+ currents due to a clamp voltage of -20 mV are illustrated in Fig. 3.21. Since the clamp voltage in Fig. 3.21 is above threshold, the Na^+ and K^+ channel resistances are engaged and

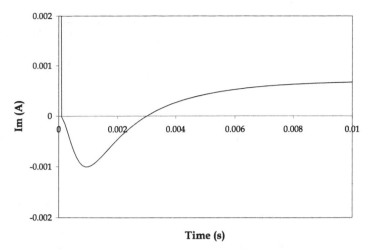

Fig. 3.20 Diagram illustrating membrane current I_m due to a -20 mV voltage clamp.

follow a typical profile. The Na^+ current rises to a peak first and then returns to zero as the clamp voltage is maintained. The K^+ current falls to a steady-state current well after the Na^+ current peaks and is maintained at this level until the clamp voltage is removed. This general pattern holds for both currents for all clamp voltages above threshold.

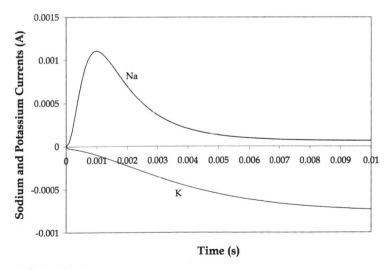

Fig. 3.21 Diagram illustrating sodium and potassium currents due to a -20 mV voltage clamp.

The Na^+ and K^+ channel resistance or conductance is easily determined applying Ohm's law to the circuit in Fig. 3.19 and the current waveforms in Fig. 3.21:

$$I_K = \frac{V_m - E_K}{R_K} = G_K \, (V_m - E_K) \tag{3.39}$$

$$I_{Na} = \frac{V_m - E_{Na}}{R_{Na}} = G_{Na} \, (V_m - E_{Na}) \tag{3.40}$$

These conductances are plotted as a function of clamp voltages ranging from -50 to 20 mV in Fig. 3.22.

For all clamp voltages above threshold, the rate of onset for opening Na^+ channels is more rapid than for K^+ channels, and the Na^+ channels close after a period of time while K^+ channels remain open while the voltage clamp is maintained. Once the Na^+ channels close, they cannot be opened until the membrane has been hyperpolarized to its resting potential. The time spent in the closed state is called the **refractory period.** If the voltage clamp is turned off before the time course for Na^+ is complete (returns to zero), G_{Na} almost immediately returns to zero, and G_K returns to zero slowly regardless of whether or not the time course for Na^+ is complete.

Example Problem 3.7

Compute I_c and I_l through a cell membrane for a subthreshold clamp voltage.

Solution

Assume that the Na^+ and K^+ voltage–time-dependent channels are not activated because the stimulus is below threshold. This eliminates these gates from the analysis although this is not actually true as shown in Example Problem 3.9. The cell membrane circuit is given by

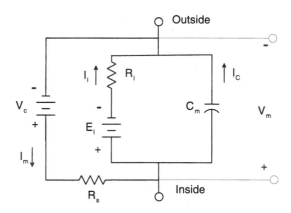

where R_s is the resistance of the wire. Applying Kirchhoff's current law at the cytoplasm gives

$$C_m \frac{dV_m}{dt} + \frac{V_m - E_l}{R_l} + \frac{V_m - V_c}{R_s} = 0$$

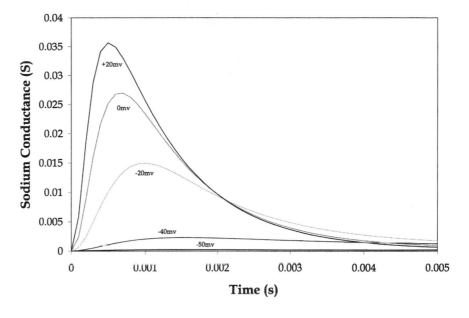

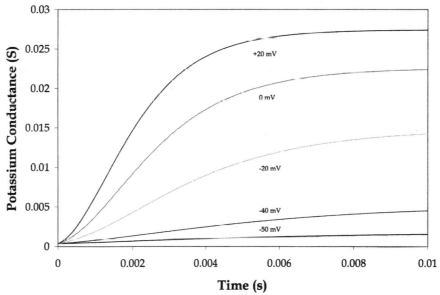

Fig. 3.22 Diagram illustrating the change in Na$^+$ and K$^+$ conductance with clamp voltage ranging from -50 mV (below threshold) to 20 mV. Note that the time scales are different in the two current plots.

Rearranging the terms in the previous equation yields

$$C_m \frac{dV_m}{dt} + \frac{R_l + R_s}{R_l R_s} V_m = \frac{R_l V_c + R_s E_l}{R_l R_s}$$

With the initial condition $V_m(0) = E_l$, the solution is given by

$$V_m = \frac{R_l V_c + R_s E_l}{R_l + R_s} + \frac{R_l(E_l - V_c)}{R_l + R_s} e^{-\frac{(R_l + R_s)t}{R_l R_s C_m}}$$

Now

$$I_c = C_m \frac{dV_m}{dt} = \frac{E_l - V_c}{R_s} e^{-\frac{(R_l + R_s)t}{R_l R_s C_m}}$$

and $I_t = \dfrac{V_m - E_l}{R_l}$. At steady state, $I_l = \dfrac{V_c - E_l}{R_l + R_s}$.   ■

Reconstruction of the Action Potential

By analyzing the estimated G_{Na} and G_K from voltage clamp pulses of various amplitudes and durations, Hodgkin and Huxley were able to obtain a complete set of nonlinear empirical equations that described the action potential. Simulations using these equations accurately describe an action potential in response to a wide variety of stimulations. Before presenting these equations, it is important to qualitatively understand the sequence of events that occur during an action potential by using previously described data and analyses. An action potential begins with a depolarization above threshold that causes an increase in G_{Na} and results in an inward Na^+ current. The Na^+ current causes a further depolarization of the membrane, which then increases the Na^+ current. This continues to drive V_m to the Nernst potential for Na^+. As shown in Fig. 3.22, G_{Na} is a function of both time and voltage and peaks and then falls to zero. During the time it takes for G_{Na} to return to zero, G_K continues to increase, which hyperpolarizes the cell membrane and drives V_m from E_{Na} to E_K. The increase in G_K results in an outward K^+ current. The K^+ current causes further hyperpolarization of the membrane, which then increases K^+ current. This continues to drive V_m to the Nernst potential for K^+, which is below resting potential. Figure 3.23 illustrates the changes in V_m, G_{Na}, and G_K during an action potential.

The circuit shown in Fig. 3.16 is a useful tool for modeling the cell membrane during small subthreshold depolarizations. This model assumes that the K^+ and Na^+ currents are small enough to neglect. As illustrated in Example Problem 3.6, a current pulse sent through the cell membrane briefly creates a capacitive current, which decays exponentially and creates an exponentially increasing I_l. Once the current pulse is turned off, capacitive current flows again and exponentially decreases to zero. The leakage current also exponentially decays to zero.

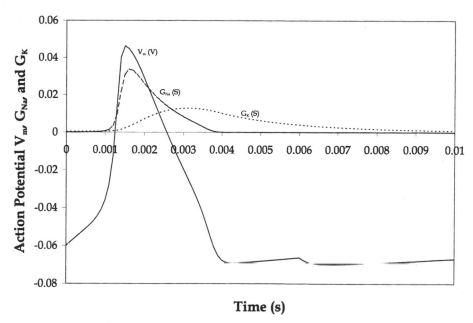

Fig. 3.23 V_m, G_{Na}, and G_K during an action potential.

As the current pulse magnitude is increased, depolarization of the membrane increases, causing activation of the Na$^+$ and K$^+$ voltage–time-dependent channels. For sufficiently large depolarizations, the inward Na$^+$ current exceeds the sum of the outward K$^+$ and leakage currents ($I_{Na} > I_K + I_l$). The value of V_m at this current is called **threshold**. Once the membrane reaches threshold, the Na$^+$ and K$^+$ voltage–time channels are engaged and run to completion as shown in Fig. 3.23.

If a slow-rising stimulus current is used to depolarize the cell membrane, then threshold will be higher. During the slow approach to threshold, inactivation of G_{Na} channels occurs and activation of G_K channels develops before threshold is reached. The value of V_m, where $I_{Na} > I_K + I_l$ is satisfied, is much larger than if the approach to threshold occurs quickly.

3.6.2 Equations Describing G_{Na} and G_K

The empirical equation used by Hodgkin and Huxley to model G_{Na} and G_K is of the form

$$G(t) = (A + Be^{-Ct})^D \qquad (3.41)$$

Values for the parameters A–D were estimated from the voltage clamp data that were collected on the squid giant axon. Not evident in Eq. 3.41 is the voltage dependence of the conductance channels. The voltage dependence is captured in the parameters as described in this section. In each of the conductance models, D is selected as 4 to give a best fit to the data. Figure 3.23 was actually calculated using SIMULINK, a simulation package that is part of MATLAB, and the parameter estimates found by Hodgkin and Huxley. Details concerning the simulation are discussed later in this section.

Potassium

The potassium conductance waveform shown in Fig. 3.22 is described by a rise to a peak while the stimulus is applied. This aspect is easily included in a model of G_K by using the general Hodgkin–Huxley expression as follows:

$$G_K = \overline{G}_K n^4 \tag{3.42}$$

where $\overline{G}_K$ is maximum K^+ conductance and n is thought of as a rate constant and given as the solution to the following differential equation:

$$\frac{d_n}{dt} = \alpha_n(1 - n) - \beta_n n \tag{3.43}$$

where

$$\alpha_n = 0.01 \frac{V + 10}{e^{\frac{V+10}{10}} - 1}$$

$$\beta_n = 0.125 e^{\frac{V}{80}}$$

$$V = V_{rp} - V_m$$

V_{rp} is the membrane potential at rest without any membrane stimulation, Note that V is the displacement from resting potential and should be negative. Clearly, G_K is a time-dependent variable since it depends on Eq. (3.43) and a voltage-dependent variable since n depends on voltage because of α_n and β_n.

Sodium

The sodium conductance waveform in Fig. 3.22 is described by a rise to a peak with a subsequent decline. These aspects are included in a model of G_{Na} as the product of two functions, one describing the rising phase and the other describing the falling phase, and modeled as

$$G_{Na} = \overline{G}_{Na} m^3 h \tag{3.44}$$

where $\overline{G}_{Na}$ is maximum Na^+ conductance and m and h are thought of as rate constants and given as the solutions to the following differential equations:

$$\frac{dm}{dt} = \alpha_m(1 - m) - \beta_m m \tag{3.45}$$

where

$$\alpha_m = 0.1 \frac{V + 25}{e^{\frac{V+25}{10}} - 1}$$

$$\beta_m = 4 e^{\frac{V}{18}}$$

and

$$\frac{dh}{dt} = \alpha_h(1 - h) - \beta_h h \tag{3.46}$$

where

$$\alpha_h = 0.07e^{\frac{V}{20}}$$

$$\beta_h = \frac{1}{e^{\frac{V + 30}{10}} + 1}$$

Note that m describes the rising phase and h describes the falling phase of G_{Na}. The units for the α_i's and β_i's in Eqs. (3.43), (3.45), and (3.46) are s^{-1}, whereas n, m, and h are dimensionless and range in value from 0 to 1. ∎

Example Problem 3.8

Calculate G_K and G_{Na} at resting potential for the squid giant axon using the Hodgkin–Huxley model. Parameter values are $\overline{G}_K - 36 \times 10^{-3}$ S and $\overline{G}_{Na} = 120 \times 10^{-3}$ S.

Solution

At resting potential, G_K and G_{Na} are constant with values dependent on n, m, and h. Since the membrane is at steady state $\frac{dn}{dt} = 0$, $\frac{dm}{dt} = 0$, and $\frac{dh}{dt} = 0$. Using Eqs. (3.43), (3.45), and (3.46), at resting potential and steady state

$$n = \frac{\alpha_n^0}{\alpha_n^0 + \beta_n^0}$$

$$m = \frac{\alpha_m^0}{\alpha_m^0 + \beta_m^0}$$

$$h = \frac{\alpha_h^0}{\alpha_h^0 + \beta_h^0}$$

where α_i^0 is α at $V = 0$ for $i = n$, m and h, and β_i^0 is β at $V = 0$ for $i = n$, m and h. Calculations yield $\alpha_n^0 = 0.0582$, $\beta_n^0 = 0.125$, $n = 0.31769$, $\alpha_m^0 = 0.2236$, $\beta_m^0 = 4$, $m = 0.05294$, $\alpha_h^0 = 0.07$, $\beta_h^0 = 0.04742$, and $h = 0.59615$. Therefore, at resting potential and steady state

$$G_K = \overline{G}_K n^4 = 36.0 \times 10^{-3}(0.31769)^4 = 0.3667 \times 10^{-3} \text{ S}$$

and

$$G_{Na} = \overline{G}_{Na} m^3 h = 120.0 \times 10^{-3}(0.05294)^3 \times 0.59615 = 0.010614 \times 10^{-3} \text{ S}$$

∎

3.6.3 Equation for the Time Dependence of the Membrane Potential

Figure 3.24 shows a model of the cell membrane that is stimulated via an external stimulus, I_m, which is appropriate for simulating action potentials. Applying Kirchhoff's current law at the cytoplasm yields

$$I_m = G_K(V_m - E_K) + G_{Na}(V_m - E_{Na}) + \frac{(V_m - E_l)}{R_l} + C_m\frac{dV_m}{dt} \quad (3.47)$$

where G_K and G_{Na} are the voltage–time-dependent conductances given by Eqs. (3.42) and (3.44).

Example Problem 3.9

For the squid giant axon, compute the size of the current pulse (magnitude and pulse width) necessary to raise the membrane potential from its resting value of -60 to -40 mV and then back to its resting potential. Neglect any changes in K^+ and Na^+ conductances from resting potential but include G_K and G_{Na} at resting potential in the analysis. Hodgkin–Huxley parameter values for the squid giant axon are $G_l = \frac{1}{R_l} = 0.3 \times 10^{-3}$ S, $\overline{G}_K = 36 \times 10^{-3}$ S, $\overline{G}_{Na} = 120 \times 10^{-3}$ S, $E_K = -72 \times 10^{-3}$ V, $E_l = -49.4 \times 10^{-3}$ V, $E_{Na} = 55 \times 10^{-3}$ V, and $C_m = 1 \times 10^{-6}$ F.

Solution

Let current I_m be given by

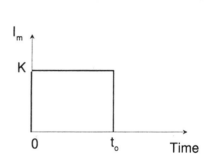

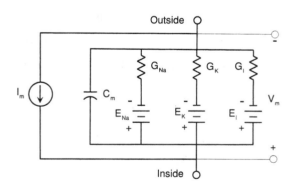

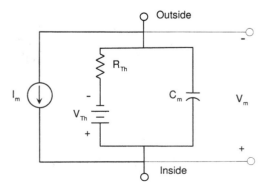

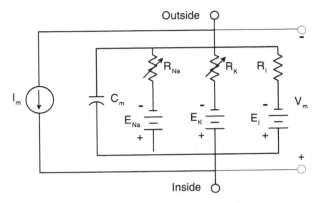

Fig. 3.24 Circuit model of an unmyelinated section of squid giant axon. The channels for K^+ and Na^+ are represented using the variable voltage–time conductances given in Eqs. (3.42) and (3.44). The passive gates for Na^+, K^+, and Cl^- are given by a leakage channel with resistance, R_l, and Nernst potential, E_l. The Na–K pump is not drawn for ease in analysis since it does not contribute any current to the rest of the circuit.

In Example Problem 3.8, the conductances at resting potential were calculated as $G_K = 0.3667 \times 10^{-3}$ S and $G_{Na} = 0.010614 \times 10^{-3}$ S. Since G_K and G_{Na} remain constant for a subthreshold current stimulus in this problem, the circuit in Fig. 3.24 reduces to the first circuit on the previous page. For ease in analysis, this circuit is replaced by the Thévenin equivalent circuit shown on the previous page with

$$R_{Th} = \frac{1}{G_{Na} + G_K + G_l} = 1.4764 \text{ k}\Omega$$

and $V_{Th} = -60$ mV.

Since the solution in Example Problem 3.6 is the same as the solution in this problem,

$$V_m(t) = V_{Th} + R_{Th}K\left(1 - e^{-\frac{t}{R_{Th}C_m}}\right)u(t) - R_{Th}K\left(1 - e^{-\frac{t-t_0}{R_{Th}C_m}}\right)u(t - t_0)$$

For convenience, assume the current pulse $t_0 > 5\tau$. Therefore, for $t \leq t_0$, $V_m = -40$ mV according to the problem statement at steady state, and from the previous equation, V_m reduces to

$$V_m = -0.040 = V_{Th} + K \times R_{Th} = -0.060 + K \times 1476.4$$

which yields $K = 13.6 \, \mu\text{A}$. Since $\tau = R_{Th}C_m = 1.47$ ms, any value for $t_0 > 5\tau = 7.35$ ms brings V_m to -40 mV with $K = 13.6 \, \mu\text{A}$. Naturally, a larger current pulse magnitude is needed for an action potential because as V_m exponentially approaches threshold (reaching it with a duration of infinity), the Na^+ conductance channels become active and shut down. ∎

To find V_m during an action potential, four differential equations (Eqs. (3.43, 3.45, 3.46 and 3.47) and six algebraic equations (α_i's and β_i's in Eqs. 3.43, 3.45, and 3.46) need to be solved. Since the system of equations is nonlinear due to the n^4 and m^3 conductance terms, an analytic solution is not possible. To solve for V_m, it is therefore necessary to simulate the solution. There are many computer tools that allow a simulation solution of nonlinear systems. SIMULINK, a general purpose toolbox in MATLAB that simulates solutions for linear and nonlinear and continuous and discrete dynamic systems, is used in this textbook. SIMULINK is a popular and widely used simulation program with a user-friendly interface that is fully integrated within MATLAB. SIMULINK is interactive and works on most computer platforms. Analogous to an analog computer, programs for SIMULINK are developed based on a block diagram of the system.

The SIMULINK program for an action potential is shown in Figs. 3.25–3.28. The block diagram is created by solving for the highest derivative term in Eq. (3.47),

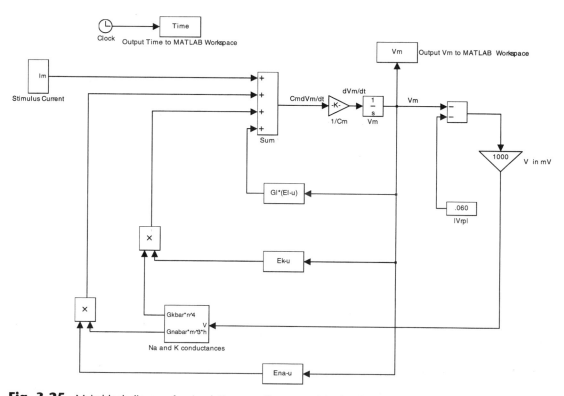

Fig. 3.25 Main block diagram for simulating an action potential using SIMULINK. The stimulus current is a pulse created by subtracting two step functions as described in Fig. 3.26. The Na$^+$ and K$^+$ conductance function blocks are described in Figs. 3.27 and 3.28.

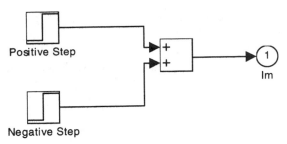

Fig. 3.26 The stimulus current.

which yields Eq. (3.48). The SIMULINK program is then created by using integrators, summers, etc.

$$\frac{dV_m}{dt} = \frac{1}{C_m}(I_m + G_K(E_K - V_m) + G_{Na}(E_{Na} - V_m) + G_l(E_l - V_m)) \quad (3.48)$$

Figure 3.25 shows the main block diagram. Figures 3.26–3.28 are subsystems that were created for ease in analysis. The Workspace output blocks were used to pass simulation results to MATLAB for plotting. Parameter values used in the simulation were based on the empirical results from Hodgkin and Huxley, with $G_l = \dfrac{1}{R_l} =$

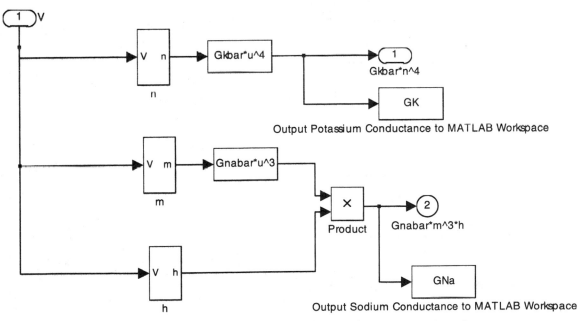

Fig. 3.27 SIMULINK program for the K^+ and Na^+ conductance channels.

0.3×10^{-3} S, $\overline{G}_K = 36 \times 10^{-3}$ S, $\overline{G}_{Na} = 120 \times 10^{-3}$ S, $E_K = -72 \times 10^{-3}$ V, $E_l = -49.4 \times 10^{-3}$ V, $E_{Na} = 55 \times 10^{-3}$ V, and $C_m = 1 \times 10^{-6}$ F. Figure 3.23 is a SIMULINK simulation of an action potential. The blocks $G_l*(El-u)$, Ek-u, and Ena-u are function blocks that were used to represent the terms $G_l(E_l - V_m)$, $(E_K - V_m)$ and $(E_{Na} - V_m)$ in Eq. (3.48), respectively.

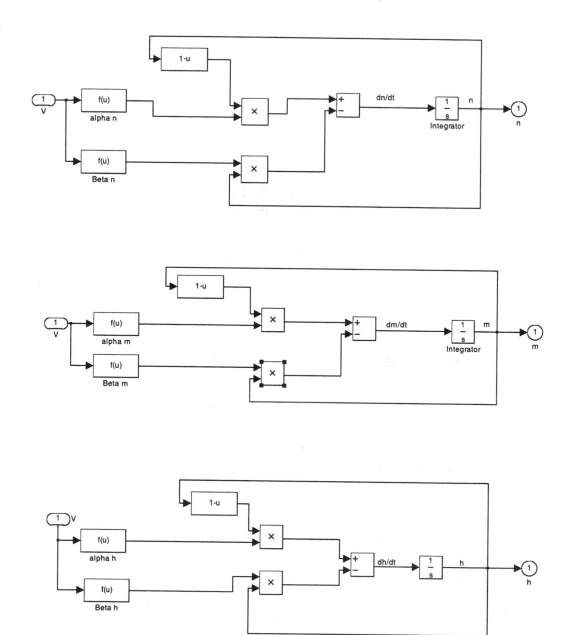

Fig. 3.28   SIMULINK program for the α and β terms in Eqs. (3.43), (3.45) and (3.46).

The stimulus pulse current was created by using the SIMULINK step function as shown in Fig. 3.26. The first step function starts at $t = 0$ with magnitude K, and the other one starts at $t = t_0$ with magnitude $-K$. The current pulse should be sufficient to quickly bring V_m above threshold.

Figure 3.27 illustrates the SIMULINK program for the conductance channels for Na^+ and K^+. Function blocks Gkbar*u^4, Gnabar*u^3, and Gnabar*m^3*h represent $\overline{G}_K n^4$, $\overline{G}_{Na} m^3$, and $\overline{G}_{Na} m^3 h$, respectively. The subsystems n, m and h are described in Fig. 3.28 and are based on six algebraic equations for α_i's and β_i's in Eqs. (3.43), (3.45), and (3.46).

Although this chapter has focused on the neuron, it is important to note that numerous other cells have action potentials that involve signaling or triggering. Many of the principles discussed in this chapter apply to these other cells as well, but the action potential-defining equations are different. For example, the cardiac action potential can be defined with a DiFranceso–Noble, Luo–Rudy, or other models rather than a Hodgkin–Huxley model of the neuron.

EXERCISES

1. Assume a membrane is permeable to only Ca^{2+}. (a) Derive the expression for the flow of Ca^{2+}; (b) Find the Nernst potential for Ca^{2+}.

2. Assume that a membrane is permeable to Ca^{2+} and Cl^- but not to a large cation R^+. The inside concentrations are $[RCl] = 100$ mM and $[CaCl_2] = 200$ mM, and the outside concentration is $[CaCl_2] = 300$ mM. (a) Derive the Donnan equilibrium. (b) Find the steady-state equilibrium concentration for Ca^{2+}.

3. Assume that a membrane is permeable to Ca^{2+} and Cl^-. The initial concentrations on the inside are different from those on the outside, and these are the only ions in the solution. (a) Write an equation for J_{Ca} and J_{Cl}. (b) Write an expression for the relationship between J_{Ca} and J_{Cl}. (c) Find the equilibrium voltage. (d) Find the relationship between the voltage across the membrane and $[CaCl_2]$ before equilibrium.

4. Assume that a membrane is permeable to only ion R^{3+}. The inside concentration is $[RCl_3] = 2$ mM and the outside concentration is $[RCl_3] = 1.4$ mM. (a) Write an expression for the flow of R^{3+}. (b) Derive the Nernst potential for R^{3+} at equilibrium.

5. Derive the Goldman equation for a membrane in which Na^+, K^+, and Cl^- are the only permeable ions.

6. Calculate V_m for the frog skeletal muscle at room temperature.

7. The following steady-state concentrations and permeabilities are given for a membrane.

Ion	Cytoplasm (mM)	Extracellular fluid (mM)	Ratio of permeabilities
K^+	140	2.5	1.0
Na^+	13	110	0.019
Cl^-	3	90	0.381

(a) Find the Nernst potential for K^+. (b) What is the resting potential predicted by the Goldman equation? (c) Explain whether space charge neutrality is satisfied. (d) Explain why the equilibrium membrane potential does not equal zero.

8. A membrane has the following concentrations and permeabilities.

Ion	Cytoplasm (mM)	Extracellular fluid (mM)	Ratio of permeabilities
K^+	?	4	?
Na^+	41	276	0.017
Cl^-	52	340	0.412

The resting potential of the membrane is -52 mV at room temperature. Find the K^+ cytoplasm concentration.

9. The following steady-state concentrations and permeabilities are given for a membrane. Note that A^+ is not permeable.

Ion	Cytoplasm (mM)	Extracellular fluid (mM)	Ratio of permeabilities
K^+	136	15	1.0
Na^+	19	155	0.019
Cl^-	78	112	0.381
A^+	64	12	—

(a) Find the Nernst potential for Cl^-. (b) What is the resting potential predicted by the Goldman equation? (c) Explain whether space charge neutrality is satisfied. (d) Explain why the equilibrium membrane potential does not equal zero. (e) Explain why the resting potential does not equal the Nernst potential of any of the ions.

10. A membrane is permeable to B^{3+} and Cl^-, but not to a large cation R^+. The following initial concentrations are given:

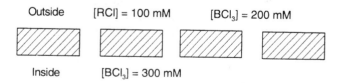

Outside   [RCl] = 100 mM   [BCl$_3$] = 200 mM

Inside   [BCl$_3$] = 300 mM

(a) Derive the Donnan equilibrium. (b) Find the steady-state equilibrium concentration for B^{3+}.

11. The following membrane is permeable to Ca^{2+} and Cl^-. (a) Write expressions for the flow of Ca^{2+} and Cl^- ions. (b) Write an expression for the relationship between J_{Ca} and J_{Cl}. (c) Find the equilibrium voltage. (d) Find the relationship between voltage across the membrane and the concentration of $CaCl_2$ before equilibrium is reached.

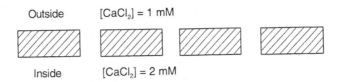

Outside $[CaCl_2]$ = 1 mM

Inside $[CaCl_2]$ = 2 mM

12. The following membrane has an active Ca^{2+} pump. Assume that the membrane is permeable to both Ca^{2+} and Cl^-, and the Ca^{2+} pump flow is J_p. The width of the membrane is δ. Find the pump flow as a function of $[Ca^{2+}]$.

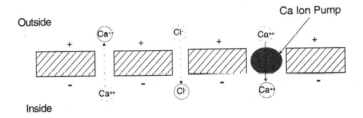

13. The membrane shown is permeable to K^+ and Cl^-. The active pump transports K^+ from the outside to the inside of the cell. The width of the membrane is δ.

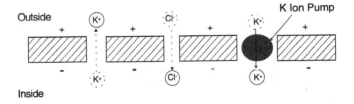

(a) Write an equation for the flow of each ion. (b) Find the flows at equilibrium. (c) Find the pump flow as a function of $([K^+]_i - [K^+]_o)$. (d) Qualitatively describe the ion concentration on each side of the membrane.

14. The following membrane is given with two active pumps. Assume that the membrane is permeable to Na^+, K^+, and Cl^- and $J_p(K) = J_p(Na) = J_p$. The width of the membrane is δ. Solve for the quantity $([Cl^-]_i - [Cl^-]_o)$ as a function of J_p.

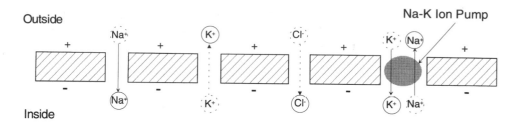

Outside
Na-K Ion Pump
Inside

15. The following steady-state concentrations and permeabilities are given for a membrane. Note that A^+ is not permeable. The ion channel resistances are $R_K = 1.7$ kΩ, $R_{Na} = 9.09$ kΩ, and $R_{Cl} = 3.125$ kΩ.

Ion	Cytoplasm (mM)	Extracellular fluid (mM)	Ratio of permeabilities
K^+	168	6	1.0
Na^+	50	337	0.019
Cl^-	41	340	0.381
A^+	64	12	—

(a) Find the Nernst potential for each ion. (b) Draw a circuit model for this membrane (Hint: See Fig. 3.13). (c) Find the membrane resting potential using the circuit in part (b). (d) Find the Thévenin equivalent circuit for the circuit in (b).

16. Suppose the membrane in Fig. 3.13 is given with $R_K = 0.1$ kΩ, $R_{Na} = 2$ kΩ, $R_{Cl} = 0.25$ kΩ, $E_K = -74$ mV, $E_{Na} = 55$ mV, and $E_{Cl} = -68$ mV. (a) Find V_m. (b) Find the Thévenin equivalent circuit.

17. Suppose a membrane has an active Na–K pump with $R_K = 0.1$ kΩ, $R_{Na} = 2$ kΩ, $R_{Cl} = 0.25$ kΩ, $E_K = -74$ mV, $E_{Na} = 55$ mV, and $E_{Cl} = -68$ mV, as shown in Fig. 3.14. Find I_{Na} and I_K for the active pump.

18. The following steady-state concentrations and permeabilities are given for a membrane. Note that A^+ is not permeable.

Ion	Cytoplasm (mM)	Extracellular fluid (mM)	Ratio of permeabilities
K^+	140	2.5	1.0
Na^+	13	110	0.019
Cl^-	3	90	0.381
A^+	64	12	—

(a) If $R_K = 1.7$ kΩ and $R_{Cl} = 3.125$ kΩ, then find R_{Na}. (b) Find the Thévenin equivalent circuit model.

19. Suppose that a membrane that has an active Na–K pump with $R_K = 0.1$ kΩ, $R_{Na} = 2$ kΩ, $R_{Cl} = 0.25$ kΩ, $E_K = -74$ mV, $E_{Na} = 55$ mV, $E_{Cl} = -68$ mV, and $C_m = 1$ μF as shown in Fig. 3.15 is stimulated by a current pulse of 10 μA for 6 ms. (a) Find V_m. (b) Find the capacitive current.

(c) Calculate the size of the current pulse applied at 6 ms for 1 ms necessary to raise V_m to -40 mV. (d) If the threshold voltage is -40 mV and the stimulus is applied as in (c), explain whether an action potential occurs.

20. Suppose that a membrane that has an active Na–K pump with $R_K = 2.727$ kΩ, $R_{Na} = 94.34$ kΩ, $R_{Cl} = 3.33$ kΩ, $E_K = -72$ mV, $E_{Na} = 55$ mV, $E_{Cl} = -49.5$ mV, and $C_m = 1$ μF as is shown in Fig. 3.15 is stimulated by a current pulse of 13 μA for 6 ms. Find (a) V_m, (b) I_K, and (c) the capacitive current.

21. Suppose a membrane has an active Na–K pump with $R_K = 1.75$ kΩ, $R_{Na} = 9.09$ kΩ, $R_{Cl} = 3.125$ kΩ, $E_K = -85.9$ mV, $E_{Na} = 54.6$ mV, $E_{Cl} = -9.4$ mV, and $C_m = 1$ μF as shown in Fig. 3.15. (a) Find the predicted resting membrane potential. (b) Find V_m if a small subthreshold current pulse is used to stimulate the membrane.

22. Suppose a membrane has an active Na–K pump with $R_K = 2.727$ kΩ, $R_{Na} = 94.34$ kΩ, $R_{Cl} = 3.33$ kΩ, $E_K = -72$ mV, $E_{Na} = 55$ mV, $E_{Cl} = -49.5$ mV, and $C_m = 1$ μF as shown in Fig. 3.15. Design a stimulus that will drive V_m to threshold at 3 ms. Assume that the threshold potential is -40 mV. (a) Find the current pulse magnitude and duration. (b) Find and sketch V_m.

23. Suppose a current pulse of 20 μA is passed through the membrane of a squid giant axon. The Hodgkin–Huxley parameter values for the squid giant axon are $G_l = \dfrac{1}{R_l} = 0.3 \times 10^{-3}$ S, $\bar{G}_K = 36 \times 10^{-3}$ S, $\bar{G}_{Na} = 120 \times 10^{-3}$ S, $E_K - -72 \times 10^{-3}$ V, $E_l = -49.4 \times 10^{-3}$ V, $E_{Na} = 55 \times 10^{-3}$ V, and $C_m = 1 \times 10^{-6}$ F. Simulate the action potential. Plot (a) V_m, G_{Na}, and G_K versus time; (b) V_m, n, m, and h versus time; and (c) V_m, I_{Na}, I_K, I_c, and I_l versus time.

24. Suppose a current pulse of 20 μA is passed through an axon membrane. The parameter values for the axon are $G_l = \dfrac{1}{R_l} = 0.3 \times 10^{-3}$ S, $\bar{G}_K = 36 \times 10^{-3}$ S, $\bar{G}_{Na} = 120 \times 10^{-3}$ S, $E_K = -12 \times 10^{-3}$ V, $E_l = 10.6 \times 10^{-3}$ V, $E_{Na} = 115 \times 10^{-3}$ V, and $C_m = 1 \times 10^{-6}$ F. Assume that Eqs. (3.41)–(3.48) describe the axon. Simulate the action potential. Plot (a) V_m, G_{Na}, and G_K versus time; (b) V_m, n, m, and h versus time; and (c) V_m, I_{Na}, I_K, I_c, and I_l versus time.

25. This exercise examines the effect of the threshold potential on an action potential for the squid giant axon. The Hodgkin–Huxley parameter values for the squid giant axon are $G_l = \dfrac{1}{R_l} = 0.3 \times 10^{-3}$ S, $\bar{G}_K = 36 \times 10^{-3}$ S, $\bar{G}_{Na} = 120 \times 10^{-3}$ S, $E_K = -72 \times 10^{-3}$ V, $E_l = -49.4 \times 10^{-3}$ V, $E_{Na} = 55 \times 10^{-3}$ V, and $C_m = 1 \times 10^{-6}$ F. (a) Suppose a current pulse of -10 μA, which hyperpolarizes the membrane, is passed through the membrane of a squid giant axon for a very long time. At time $t = 0$, the current pulse is removed. Simulate the resultant action potential. (b) The value of the threshold potential is defined as when $I_{Na} > I_K + I_l$. Changes in threshold potential can be easily implemented by changing the value of 25 in the equation for α_m to a lower value. Suppose the value of 25 is changed to 10 in the equation defining α_m and a current pulse of $-10\ \mu A$ (hyperpolarizes the membrane) is passed through the membrane of a squid giant axon for a very long time. At time $t = 0$, the current pulse is removed. Simulate the resultant action potential.

26. Simulate the plots shown in Figs. 3.20–3.22 in the voltage clamp mode.

27. Select an input current waveform necessary to investigate the refractory period that follows an action potential. (Hint: Use a two-pulse current input). What is the minimum refractory period? If the second pulse is applied before the minimum refractory period, how much larger is the stimulus magnitude that is needed to generate an action potential?

28. Explain whether the following circuit allows the investigator to conduct a voltage clamp experiment. The unmyelinated axon is sealed at both ends. A silver wire is inserted inside the axon that eliminates R_a. Clearly state any assumptions.

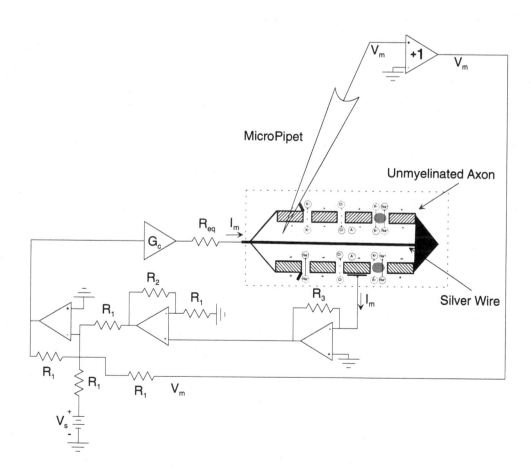

SUGGESTED READING

Bahill, A. T. (1981). *Bioengineering: Biomedical, Medical and Clinical Engineering.* Prentice-Hall, Englewood Cliffs, NJ.

Bronzino, J. D. (1995). *The Biomedical Engineering Handbook.* CRC Press, Boca Raton, FL.

Bower, J. M. and Beeman, D. (1995). *The Book of Genesis: Exploring Realistic Neural Models with the General Neural Simulation System.* Springer-Verlag, New York.

Deutsch, S. and Deutsch, A. (1993). *Understanding the Nervous System — An Engineering Perspective.* IEEE Press, New York.

Dibner, B. (1952). Galvani-Volta: A Controversy that Led to the Discovery of Useful Electricity. Burndy Library, New Haven, CT.

DiFrancesco, D. and Noble, D. (1985). A model of cardiac electrical activity incorporating ionic pumps and concentration changes. *Philos. Trans. R. Soc. London B*, **307**, 307–353.

Fulton, J. F. and Cushing, H. (1936). A bibliographical study of the Galvani and Aldini writings on animal electricity. *Ann. Sci.* **1**, 239–268.

Galvani, L. (1953). Commentary on the Effect of Electricity on Muscular Motion. (Translated by R. M. Green). Licht, New Haven, CT.

Galvani, L. (1953). De Viribus Electricitatis. (In English). Burnty Library, Norwalk, CT.

Guyton, A. C. and Hall, J. E. (1995). *Textbook on Medical Physiology,* 9th ed. Saunders, Philadelphia.

Hille, B. (1992). *Ionic Channels of Excitable Membranes,* 2nd ed. Sinauer, Sunderland, MA.

Hodgkin, A. and Huxley, A. (1952a). Currents carried by sodium and potassium ions through the membrane of the giant axon of *Loligo. J. Physiol. London* **116**, 449–472.

Hodgkin, A. and Huxley, A. (1952b). The components of membrane conductance in the giant axon of *Loligo. J. Physiol. London* **116**, 473–496.

Hodgkin, A. and Huxley, A. (1952c). The dual effect of membrane potential on sodium conductance in the giant axon of *Loligo. J. Physiol. London* **116**, 497–506.

Hodgkin, A. and Huxley, A. (1952d). A quantitative description of membrane current and its application to conduction and excitation in nerve. *J. Physiol. London* **117**, 500–544.

Hodgkin, A. Huxley, A. and Katz, B. (1952). Measurement of current-voltage relations in the membrane of the giant axon of *Loligo. J. Physiol. London* **116**, 424–448.

Kandel, E. R. Schwartz, J. H. and Jessell, T. M. (1992) *Principles of Neural Science,* 3rd ed. Appleton & Lange, Norwalk, CT.

Luo, C. and Rudy, Y. (1994). A dynamic model of the cardiac ventricular action potential: I. Simulations of ionic currents and concentration changes. *Circ. Res.* **74**, 1071.

Matthews, G. G. (1991). *Cellular Physiology of Nerve and Muscle.* Blackwell, Boston.

Nernst, W. (1889). Die elektromotorishe Wirksamkeit der Jonen. *Z. Physik. Chem.* **4**, 129–188.

Plonsey, R. and Barr, R. C. (1982). *Bioelectricity: A Quantitative Approach.* New York: Plenum Press.

Rinzel, J. (1990) Electrical excitability of cells, theory and experiment: Review of the Hodgkin–Huxley foundation and an update, in M. Mangel (ed.), Bull. Math. *Biology: Classics of Theoretical Biology.* **52**, 5–23.

4   BIOMEDICAL SENSORS

Chapter Contents

At the conclusion of this chapter, the reader will be able to:

- Describe the different classifications of biomedical sensors
- Describe the characteristics that are important for packaging materials associated with biomedical sensors
- Calculate the half-cell potentials generated by different electrodes immersed in an electrolyte solution
- Describe the electrodes that are used to record the ECG, EEG, and EMG and those that are used for intracellular recordings
- Describe how displacement transducers, airflow transducers, and thermistors are used to make physical measurements
- Describe how blood gases and blood pH are measured
- Describe how enzyme-based and microbial biosensors work and some of their uses
- Explain how optical biosensors work and describe some of their uses

4.1   INTRODUCTION

Biomedical sensors are used in clinical medicine and biological research for measuring a wide range of physiological variables. They are often called biomedical transducers and are the main building blocks of diagnostic medical instrumentation found in physician's offices, clinical laboratories, and hospitals. These sensors are routinely used *in vivo* to perform continuous invasive and noninvasive monitoring of critical physiological variables as well as *in vitro* to help clinicians in various diagnostic procedures. Some biomedical sensors are also used in nonmedical applications such as in environmental monitoring, agriculture, bioprocessing, food processing, and the petrochemical and pharmacological industries.

Increasing pressures to lower health care costs, optimize efficiency, and provide better care in less expensive settings without compromising patient care are shaping the future of clinical medicine. As part of this ongoing trend, clinical testing is rapidly being transformed by the introduction of new tests that will revolutionize the way physicians will diagnose and treat diseases in the future. Among these changes, patient self-testing and physician office screening are the two most rapidly expanding areas. This trend is driven by the desire of patients and physicians alike to have the ability to perform some types of instantaneous diagnosis and to move the testing apparatus from an outside central clinical laboratory closer to the point of care.

Biomedical sensors play an important role in a range of diagnostic medical applications. Depending on the specific needs, some sensors are used primarily in clinical laboratories to measure *in vitro* physiological quantities such as electrolytes, enzymes, and other biochemical metabolites in blood. Other biomedical sensors for measuring pressure, flow, and the concentrations of gases, such as oxygen and carbon dioxide, are used *in vivo* to follow continuously (monitor) the condition of a patient. For real-time continuous *in vivo* sensing to be worthwhile, the target analytes must vary rapidly and most often unpredictably.

The need for accurate medical diagnostic procedures places stringent requirement on the design and use of biomedical sensors. Usually, the first step in developing a biomedical sensor is to assess *in vitro* the accuracy, precision, range, response time, and drift of the sensor. Later, depending on the intended application, similar *in vivo* tests may be required to confirm the specifications of the device and to ensure that the measurement remains reliable and safe.

4.1.1 Sensor Classifications

Biomedical sensors are usually classified as physical, electrical, or chemical depending on their specific applications. Biosensors, which can be considered a special subclassification of biomedical sensors, refers to a group of sensors that have two distinct components: (i) a biological recognition element, such as a purified enzyme, antibody, or receptor, which functions as a mediator and provides the selectivity that is needed to detect the analyte of interest and (ii) a supporting structure, which also acts as a transducer and is positioned in intimate contact with the biological component. The purpose of the transducer is to convert the biochemical reaction into a quantifiable measurement, typically in the form of an optical, electrical, or physical signal.

4.1.2 Sensor Packaging

Packaging of certain biomedical sensors, primarily sensors for *in vivo* applications, is an important consideration during the design, fabrication, and use of the device. The sensor must be safe and remain functionally reliable. In the development of implantable biosensors, an additional key issue is the long operational lifetime and biocompatibility of the sensor. Whenever a sensor comes into contact with body fluids, the host itself may affect the function of the sensor or the sensor may affect the site in which it is implanted. For example, protein absorption and cellular deposits can alter the permeability of the sensor packaging that is designed to both protect the sensor and allow free chemical diffusion of certain analytes between the body fluids and the biosensor. Improper packaging of implantable biomedical sensors could lead to drift and a gradual loss of sensor sensitivity over time. Furthermore, inflammation of tissue, infection, or clotting in a vascular site may produce harmful adverse effects (see Chapter 11). Hence, the materials used in the construction of the sensor's outer body must be nonthrombogenic and nontoxic since they play a critical role in determining the overall performance and longevity of an implantable sensor. One convenient strategy is to utilize various polymeric covering materials and barrier layers to minimize leaching of potentially toxic sensor components into the body. It is also important to keep in mind that once the sensor is manufactured, common sterilization practices by steam, ethylene oxide, or gamma radiation must not alter the chemical diffusion properties of the sensor packaging material.

This chapter will examine the operating principles of different types of biomedical sensors including examples of invasive and noninvasive sensors for measuring biopotentials and other physical and biochemical variables encountered in different clinical and research applications.

4.2 BIOPOTENTIAL MEASUREMENTS

Biopotential measurements are made using different kinds of specialized electrodes. The function of these electrodes is to couple the ionic potentials generated inside the body to an electronic instrument. Biopotential electrodes are generally classified either as noninvasive (skin surface) or invasive (e.g., microelectrodes or wire electrodes).

4.2.1 The Electrolyte/Metal Electrode Interface

When a metal is placed in an electrolyte (i.e., an ionizable) solution, a charge distribution is created next to the metal/electrolyte interface as illustrated in Fig. 4.1. This localized charge distribution causes an electric potential, called a half-cell potential, to be developed across the interface between the metal and the electrolyte solution.

The half-cell potentials of several important metals are listed in Table 4.1. Note that the hydrogen electrode is considered to be the standard electrode against which the half-cell potentials of other metal electrodes are measured.

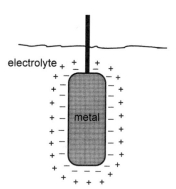

Fig. 4.1 Distribution of charges at a metal/electrolyte interface.

TABLE 4.1   Half-Cell Potentials

Primary metal and chemical reaction			Half-cell potential
Al	$\rightarrow$	$Al^{3+} + 3e^-$	−1.706
Cr	$\rightarrow$	$Cr^{3+} + 3e^-$	−0.744
Cd	$\rightarrow$	$Cd^{2+} + 2e^-$	−0.401
Zn	$\rightarrow$	$Zn^{2+} + 2e^-$	−0.763
Fe	$\rightarrow$	$Fe^{2+} + 2e^-$	−0.409
Ni	$\rightarrow$	$Ni^{2+} + 2e^-$	−0.230
Pb	$\rightarrow$	$Pb^{2+} + 2e^-$	−0.126
H_2	$\rightarrow$	$2H^+ + 2e^-$	−0.000 (standard by definition)
Ag	$\rightarrow$	$Ag^+ + e^-$	0.799
Au	$\rightarrow$	$Au^{3+} + 3e^-$	1.420
Cu	$\rightarrow$	$Cu^{2+} + 2e^-$	0.340
$Ag + Cl^-$	$\rightarrow$	$AgCl + 2e^-$	0.223

Example Problem 4.1

Silver and zinc electrodes are immersed in an electrolyte solution. Calculate the potential drop between these two electrodes.

Solution

From Table 4.1, the half-cell potentials for the silver and zinc electrodes are 0.799 and −0.763 V, respectively. Therefore, the potential drop between these two metal electrodes is equal to

$$0.799 - (-0.763) = 1.562 \text{ V}$$

Typically, biopotential measurements are made by utilizing two similar electrodes composed of the same metal. Therefore, the two half-cell potentials for these electrodes would be equal in magnitude. For example, two similar biopotential electrodes can be taped to the chest near the heart to measure the electrical potentials generated by the heart [electrocardiogram (ECG)]. Ideally, assuming that the skin-to-electrode interfaces are electrically identical, the differential amplifier attached to these two electrodes would amplify the biopotential (ECG) signal but the half-cell potentials would cancel out. In practice, however, disparity in electrode material or skin contact resistance could cause a significant DC offset voltage that would cause current to flow through the two electrodes. This current will produce a voltage drop across the body. The offset voltage will appear superimposed at the output of the amplifier and may cause instability or baseline drift in the recorded biopotential. ∎

Example Problem 4.2

Silver and aluminum electrodes are placed in an electrolyte solution. Calculate the current that will flow through the electrodes if the equivalent resistance of the solution is equal to 2 kΩ.

Solution

$$0.799 - (-1.706) = 2.505 \text{ V}$$

$$2.505 \text{ V/2 k}\Omega = 1.252 \text{ mA}$$ ■

4.2.2 ECG Electrodes

Examples of several types of noninvasive biopotential electrodes used primarily for ECG recording are shown in Fig. 4.2.

Rigid metal plate electrodes, which are usually made from an alloy of zinc, nickel, and copper, are usually used for short-term recording of biopotentials from the skin. These electrodes are attached to the skin either by an elastic strap (Fig. 4.2a) or a double-sided peel-off adhesive tape. A thin layer of electrolyte gel must be applied

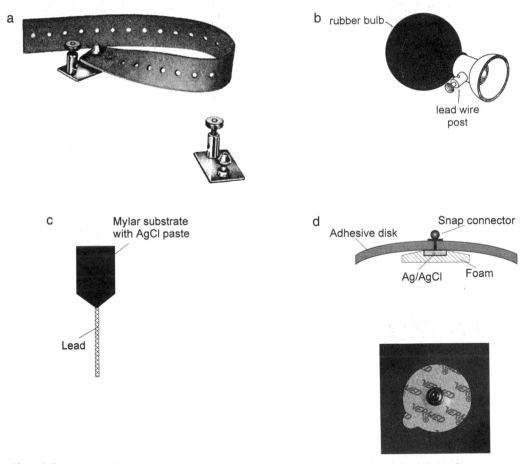

Fig. 4.2 Biopotential skin surface ECG electrodes: (a) Rigid metal plate electrode and attachment strap, (b) suction-type metal electrode, (c) flexible Mylar electrode, and (d) disposable snap-type Ag/AgCl electrode (courtesy of Vermont Medical, Inc., Bellows Falls, VT).

between the metal and the skin in order to establish good electrical contact with the skin during the recording of electrical potentials from the body. The main ingredients of these electrolytes are water and additional ionic salts, such as sodium chloride and potassium chloride.

Suction electrodes (Fig. 4.2b) are very convenient as ECG chest leads since they can be easily moved from one location to another. These electrodes are attached to the skin by using a rubber suction bulb to create a vacuum inside the hollow metallic cup.

A third type of biopotential sensor is a flexible electrode made of certain types of polymers or elastomers that are made electrically conductive by the addition of a fine carbon or metal powder. These electrodes are available with prepasted AgCl gel for quick and easy application to the skin (Fig. 4.2c).

The fourth and most common type of biopotential electrode is the "floating" silver/silver chloride electrode (Ag/AgCl) which is formed by electrochemically depositing a very thin layer of silver chloride onto a silver electrode (Fig. 4.2d). These electrodes are recessed and imbedded in foam that has been soaked with an electrolyte paste and provides good electrical contact with the skin. The electrolyte saturated foam is also known to reduce motion artifacts which could be produced, for example, during stress testing when the layer of the skin moves relative to the surface of the Ag/AgCl electrode. This motion artifact could cause large interference in the recorded biopotential and, in extreme cases, could severely degrade the measurement.

4.2.3 Electromyographic Electrodes

A number of different types of biopotential electrodes are used in recording electromyographic (EMG) signals from different muscles in the body. The shape and size of the recorded EMG signals depends on the electrical property of these electrodes and the recording location. For noninvasive recordings, proper skin preparation, which normally involves cleansing the skin with alcohol, or the application of a small amount of an electrolyte paste helps to minimize the impedance of the skin-electrode interface and improve the quality of the recorded signal considerably. The most common electrodes used for surface EMG recording and nerve conduction studies are circular discs, about 1 cm in diameter, that are made of silver or platinum.

For direct recording of electrical signals from nerves and muscle fibers, a variety of percutaneous needle electrodes are available as illustrated in Fig. 4.3.

The most common type of needle electrode is the concentric bipolar electrode shown in Fig. 4.3a. This electrode is made from thin metallic wires encased inside a larger canula or hypodermic needle. The two wires serve as the recording and reference electrodes. Another type of percutaneous EMG electrode is the unipolar needle electrode (Fig. 4.3b). This electrode is made of a thin wire that is mostly insulated by a thin layer of Teflon except for about 0.3 mm near the distal tip. Unlike a bipolar electrode, this electrode requires a second unipolar reference electrode to close the electrical circuit. The second recording electrode is normally placed either adjacent to the recording electrode or attached to the surface of the skin.

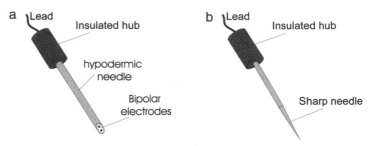

Fig. 4.3 Intramascular biopotentail electrodes: (a) bipolar and (b) unipolar configuration.

4.2.4 Electroencephalographic Electrodes

The most commonly used electrodes for recording signals from the brain [electroencephalograms (EEG)] are cup electrodes and subdermal electrodes. Cup electrodes are made of platinum or tin approximately 5–10 mm in diameter. These cups are filled with a conducting electrolyte gel and are attached to the scalp with an adhesive tape. Subdermal EEG electrodes are basically fine platinum or stainless-steel needle electrodes, about 10 mm long by 0.5 mm wide, which are inserted under the skin to provide a better electrical contact.

4.2.5 Microelectrodes

A microelectrode is a biopotential electrode with an ultra-fine tapered tip that can be inserted into individual biological cells. These electrodes serve an important role in recording action potentials from single cells and are commonly used in neurophysiological studies. The tip of these electrodes must be small with respect to the dimensions of the biological cell to avoid cell damage and at the same time sufficiently strong to penetrate the cell wall. Figure 4.4 illustrates the construction of three typical types of microelectrodes: glass micropipettes, metal microelectrodes, and solid-state microprobes.

In Fig. 4.4a, a hollow glass capillary tube, typically 1 mm in diameter, is heated and softened in the middle inside a small furnace and then quickly pulled apart from both ends. This process creates two similar microelectrodes with an open tip that has a diameter in the order of 0.1–10 μm. The larger end of the glass tube (the stem) is then filled with a 3 M KCl electrolyte solution. A small piece of Ag/AgCl wire is inserted through the stem to provide an electrical contact with the electrolyte solution. When the tip of the microelectrode is inserted into an electrolyte solution, such as the intracellular cytoplasm of a biological cell, ionic current can flow through the fluid junction at the tip of the microelectrode. This establishes a closed electrical circuit between the Ag/AgCl wire inside the microelectrode and the biological cell.

A different kind of microelectrode made from a small-diameter strong metal wire (e.g., tungsten or stainless steel) is illustrated in Fig. 4.4b. The tip of this microelectrode is usually sharpened to a diameter of a few micrometers by an electrochemical etching process. The wire is then insulated up to its tip.

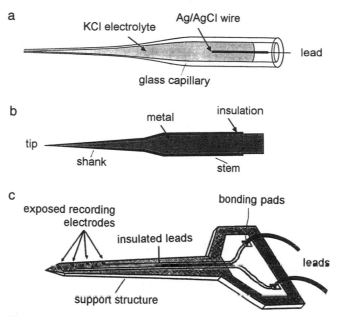

Fig. 4.4 Biopotential microelectrodes: (a) capillary glass microelectrode, (b) insulated metal microelectrode, and (c) solid-state multisite recording microelectrode.

Solid-state microfabrication techniques commonly used in the production of integrated circuits can be used to produce microprobes for multichannel recordings of biopotentials or for electrical stimulation of neurons in the brain or spinal cord. An example of such a microsensor is shown in Fig. 4.4c. This probe consists of a precisely micromachined silicon substrate with four exposed recording sites. One of the major advantages of this fabrication technique is the ability to mass produce very small and highly sophisticated microsensors with highly reproducible electrical and physical properties.

4.3 PHYSICAL MEASUREMENTS

4.3.1 Displacement Transducers

Inductive displacement transducers are based on the inductance L of a coil given by

$$L = n^2 G \mu \qquad (4.1)$$

where G is a geometric form constant, n is the number of coil turns, and μ is the permeability of the magnetically susceptible medium inside the coil. These types of transducers measure displacement by changing either the self-inductance of a single coil or the mutual inductance coupling between two or more stationary coils, typically by the displacement of a ferrite or iron core in the bore of the coil assembly. A widely used inductive displacement transducer is the linear variable differential transformer (LVDT) illustrated in Fig. 4.5.

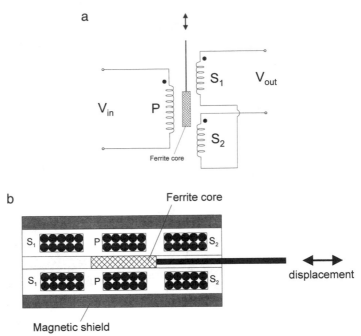

Fig. 4.5 LVDT transducer: (a) electric diagram and (b) cross-section view.

This device is essentially a three-coil mutual inductance transducer that is composed of a primary coil (P) and two secondary coils (S_1 and S_2) connected in series but in opposite polarity in order to achieve a wider linear output range. The mutual inductance coupled between the coils is changed by the motion of a high-permeability slug. The primary coil is usually excited by passing an AC current. When the slug is centered symmetrically with respect to the two secondary coils, the primary coil induces an alternating magnetic field on the secondary coils. This produces equal voltages (but of opposite polarities) across the two secondary coils. Therefore, the positive voltage excursions from one secondary coil will cancel out the negative voltage excursions from the other secondary coil, resulting in a zero net output voltage. When the core moves toward one coil, the voltage induced in that coil is increased in proportion to the displacement of the core while the voltage induced in the other coil is decreased proportionally, leading to a typical voltage-displacement diagram as illustrated in Fig. 4.6. Since the voltages induced in the two secondary coils are out of phase, special phase-sensitive electronic circuits must be used to detect both the position and the direction of the core's displacement.

Blood flow through an exposed vessel can be measured by means of an electromagnetic flow transducer. It is used extensively in research studies to measure blood flow in major blood vessels near the heart, including the aorta at the point where it exits from the heart.

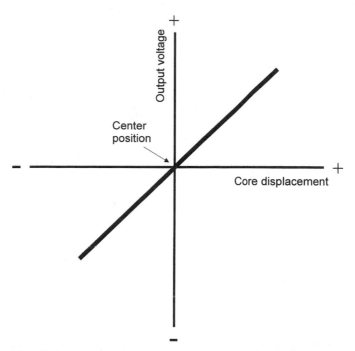

Fig. 4.6 Output voltage versus core displacement of a typical LVDT transducer.

Consider a blood vessel of diameter l filled with blood flowing with a uniform velocity u. If the blood vessel is placed in a uniform magnetic field $\vec{B}$ that is perpendicular to the direction of blood flow, the negatively charged anion and positively charged cation particles in the blood will experience a force $\vec{F}$, which is normal to both the magnetic field and blood flow directions and is given by

$$\vec{F} = q\,(\vec{u} \times \vec{B}) \tag{4.2}$$

where q is the elementary charge $(1.6 \times 10^{-19}\ \text{C})$. As a result, these charged particles will be deflected in opposite directions and will move along the diameter of the blood vessels according to the direction of the force vector $\vec{F}$. This movement will produce an opposing force $\vec{F}_0$ that is equal to

$$\vec{F}_0 = q\,\vec{E} = q\!\left(\frac{V}{l}\right) \tag{4.3}$$

where $\vec{E}$ is the net electrical field produced by the displacement of the charged particles and V is the potential produced across the blood vessel. At equilibrium, these two forces will be equal. Therefore, the potential difference, V, is given by

$$V = Blu \tag{4.4}$$

and is proportional to the velocity of blood through the vessel.

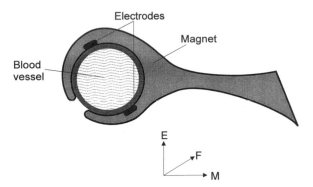

Fig. 4.7   Electromagnetic blood-flow probe.

Example Problem 4.3

Calculate the voltage in a magnetic flow probe if the probe is applied across a blood vessel with a diameter of 0.5 cm and the flow rate of blood is 5 cm/s. Assume that the magnitude of the magnetic field is 1.5×10^{-5} tesla (T).

Solution

From Eq. (4.4),

$$V = Blu = (1.5 \times 10^{-5}\text{T})(0.5 \text{ cm})(5 \text{ cm/s}) = 37.5 \text{ } \mu\text{V} \qquad \blacksquare$$

Practically, this device consists of a clip-on probe that fits snugly around the blood vessel as illustrated in Fig. 4.7. The probe contains electrical coils to produce an electromagnetic field transverse to the direction of blood flow. The coil is usually excited by an AC current. A pair of very small biopotential electrodes are attached to the housing and rest against the wall of the blood vessel to pick up the induced potential. The flow-induced voltage is an AC voltage at the same frequency as the

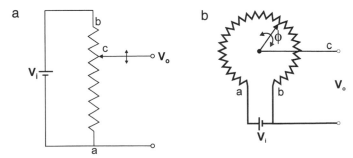

Fig. 4.8   Linear translational (a) and angular (b) displacement transducers.

excitation voltage. Using an AC method instead of DC excitation helps to remove any offset potential error due to the contact between the vessel wall and the biopotential electrodes.

A potentiometer is a resistive-type transducer that converts either linear or angular displacement into an output voltage by moving a sliding contact along the surface of a resistive element. Figure 4.8 illustrates linear and angular-type potentiometric transducers. A voltage, V_i is applied across the resistor R. The output voltage, V_o between the sliding contact and one terminal of the resistor is linearly proportional to the displacement. Typically, a constant current source is passed through the variable resistor and the small change in output voltage is measured by a sensitive voltmeter using Ohm's law (i.e., $I = V/R$).

Example Problem 4.4

Calculate the change in output voltage of a linear potentiometer transducer that undergoes a 20% change in displacement.

Solution

Assume that the current flowing through the transducer is constant. From Ohm's law,

$$\Delta V = I \times \Delta R$$

Hence, since the resistance between the sliding contact and one terminal of the resistor is linearly proportional to the displacement, a 20% change in displacement will produce a 20% change in the output voltage of the transducer. ■

In certain clinical situations, it is desirable to measure changes in the peripheral volume of a leg when the venous outflow of blood from the leg is temporarily occluded by a blood pressure cuff. This volume-measuring method is called plethysmography and can indicate the presence of venous clots in the legs. The measurement can be performed by wrapping an elastic resistive transducer around the leg and measuring the rate of change in resistance of the transducer as a function of time. This change corresponds to relative changes in the blood volume of the leg. If a clot is present, it will take more time for the blood stored in the leg to flow out through the veins after the temporary occlusion is removed. A similar transducer can be used to follow a patient's breathing patterns by wrapping the elastic band around the chest.

An elastic resistive transducer consists of a thin elastic tube filled with an electrically conductive material as illustrated in Fig. 4.9. The resistance of the conductor inside the flexible tubing is given by

$$R = \rho\left(\frac{l}{A}\right) \tag{4.5}$$

where ρ is the resistivity of the electrically conductive material, l is the length, and A the cross-sectional area of the conductor.

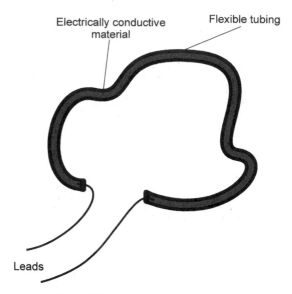

Electrically conductive material

Flexible tubing

Leads

Fig. 4.9 Elastic resistive transducer.

Example Problem 4.5

A 10-cm long elastic resistive transducer with a resting resistance of 0.5 kΩ is wrapped around the chest. Assume that the chest diameter during exhalation is 33 cm. Calculate the resistance of the transducer after it has been applied to the chest.

Solution

After the transducer is stretched around the chest, its length will increase from 10 to 103.7 cm. Assuming that the cross-sectional area of the transducer remains unchanged after it is stretched, the resistance will increase to

$$R_{\text{stretched}} = 0.5 \times \left(\frac{103.7}{10} \right) = 5.18 \text{ k}\Omega \qquad \blacksquare$$

Example Problem 4.6

Calculate the change in voltage that is induced across the elastic transducer in Example Problem 4.5. Assume that normal breathing produces a 10% change in chest circumference and a constant current of 5 mA is flowing through the transducer.

Solution

From Ohm's law ($V = IR$),

$$V = 5 \text{ mA} \times 5.18 \text{ k}\Omega = 25.9 \ V$$

If R changes by 10%, then

$$V = 5 \text{ mA} \times 1.1 \times 5.18 \text{ k}\Omega = 28.5 \text{ V}$$

$$\Delta V = 2.59 \text{ V} \qquad \blacksquare$$

Strain gauges are displacement-type transducers that measure changes in the length of an object as a result of an applied force. These transducers produce a resistance change that is proportional to the fractional change in the length of the object, also called strain, S, which is defined as

$$S = \frac{\Delta l}{l} \tag{4.6}$$

where Δl is the fractional change in length and l is the initial length of the object. Examples include resistive wire elements and certain semiconductor materials.

To understand how a strain gauge works, consider a fine wire conductor of length l, cross-sectional area A, and resistivity ρ. The resistance of the unstretched wire is given by Eq. (4.5). Now suppose that the wire is stretched within its elastic limit by a small amount, Δl, such that its new length becomes $(l + \Delta l)$. Because the volume of the stretched wire must remain constant, the increase in the wire length results in a smaller cross-sectional area, $A_{\text{stretched}}$. Thus,

$$lA = (l + \Delta l)A_{\text{stretched}} \tag{4.7}$$

The resistance of the stretched wire is given by

$$R_{\text{stretched}} = \rho \frac{l + \Delta l}{A_{\text{stretched}}} \tag{4.8}$$

The increase in the resistance of the stretched wire ΔR is

$$\Delta R = R_{\text{stretched}} - \rho \frac{l}{A} \tag{4.9}$$

Substituting Eq. (4.8) into Eq. (4.9) gives

$$\Delta R = \rho \frac{(l + \Delta l)^2}{lA} - \rho \frac{l}{A} = \frac{\rho(l^2 + 2l\Delta l + \Delta l^2 - l^2)}{Al} \tag{4.10}$$

Assume that for small changes in length, $\Delta l << l$, this relationship simplifies to

$$\Delta R = \rho \frac{2\Delta l}{A} = \frac{2\Delta l}{l} R \tag{4.11}$$

The fractional change in resistance $(\Delta R/R)$ divided by the fractional change in length $(\Delta l/l)$ is called the gauge factor, G. For a metal wire strain gauge made of constantan, G is approximately equal to 2. Semiconductor strain gauges made of silicon have a gauge factor about 70–100 times higher and are therefore much more sensitive than metallic wire strain gauges.

Example Problem 4.7

Calculate the strain in a metal wire gauge for a fractional change in resistance of 10%.

Solution

Combine Eq. (4.6) and (4.11) to obtain

$$\frac{\Delta R}{R} = \frac{2\Delta l}{l} = 2S$$

$$\frac{0.1}{2R} = S$$

$$\frac{0.05}{R} = S$$

■

Strain gauges typically are classified into two categories, bonded or unbonded. A bonded strain gauge has a folded thin wire cemented to a semiflexible backing material, as illustrated in Fig. 4.10.

An unbonded strain gauge consists of multiple resistive wires (typically four) stretched between fixed and movable rigid frames. In this configuration, when a deforming force is applied to the structure, two of the wires are stretched and the other two are shortened proportionally. This configuration is used in blood pressure transducers and is illustrated in Fig. 4.11. In this arrangement, a diaphragm is coupled directly by an armature to a movable frame that is inside the transducer. Blood in a peripheral vessel is coupled through a thin fluid-filled (saline) catheter to a disposable dome that is sealed by the flexible diaphragm. Changes in blood pressure during the pumping action of the heart apply a force on the diaphragm that causes the movable frame to move from its resting position. This movement causes the strain gauge wires to stretch or compress and results in a cyclical change in resistance that is proportional to the pulsatile blood pressure measured by the transducer.

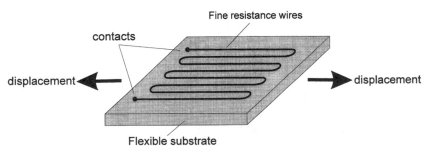

Fig. 4.10 Bonded-type strain gauge transducer.

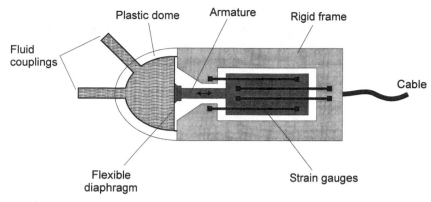

Fig. 4.11 Resistive strain gauge (unbonded type) blood pressure transducer.

In general, the change in resistance of a strain gauge is typically quite small. In addition, changes in temperature can also cause thermal expansion of the wire and subsequently lead to large changes in the resistance of a strain gauge. Therefore, very sensitive electronic amplifiers with special temperature compensation circuits are used in most applications.

The capacitance C between two equal-size parallel plates of cross-sectional area A separated by a distance d is given by

$$C = \varepsilon_0 \varepsilon_r \left(\frac{A}{d} \right) \qquad (4.12)$$

where ε_0 is the dielectric constant of free space (8.85×10^{-14} F/cm) and ε_r is the relative dielectric constant of the insulating material placed between the two plates. The method that is most commonly used to measure displacement in capacitance transducers involves changing the separation distance d between a fixed and a movable plate as illustrated in Fig. 4.12.

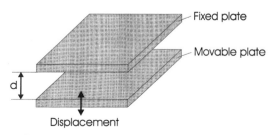

Fig. 4.12 Capacitive displacement transducer.

Example Problem 4.8

Two metal plates with an area of 3 cm^2 and separation distance of 0.1 mm are used to form a capacitance transducer. If the material between the two plates has a dielectric constant of 2×10^{-2}, calculate the capacitance of the transducer.

Solution

$$C = \varepsilon_0 \varepsilon_r \frac{A}{d} = (8.85 \times 10^{-14} \text{ F/cm} \times 2 \times 10^{-2} \times 3 \text{ cm}^2) / 0.01 \text{ cm} = 0.53 \text{ pF} \quad \blacksquare$$

Capacitive displacement transducers can used to measure respiration or movement of a patient by attaching multiple transducers to a mat that is placed on a bed. A capacitive displacement transducer can also be used as a pressure transducer by attaching the movable plate to a thin diaphragm that is in contact with a fluid or air. By applying a voltage across the capacitor and amplifying the small AC signal generated by the vibrations of the diaphragm, it is possible to obtain a signal that is proportional to the applied external pressure source.

Piezoelectric transducers are used extensively in cardiology to listen to heart sounds (phonocardiography), in automated blood pressure measurements, and for measurement of physiological forces and accelerations. They are also commonly employed in generating ultrasonic waves (high-frequency sound waves typically above 20 kHz) which are used for measuring blood flow or imaging internal soft structures in the body.

A piezoelectric transducer consists of a small crystal (usually quartz) that contracts if an electric field (usually in the form of a short voltage impulse) is applied across its plates as illustrated in Fig. 4.13. Conversely, if the crystal is mechanically strained, it will generate a small electric potential. Besides quartz, several other ceramic materials, such as barium titanate and lead zirconate titanate, are also known to produce a piezoelectric effect.

The piezoelectric principle is based on the phenomenon that when an asymmetrical crystal lattice is distorted by an applied force F, the internal negative and positive charges are reoriented. This causes an induced surface charge Q on the opposite sides of the crystal. This charge is directly proportional to the applied force and is given by

$$Q = kF \tag{4.13}$$

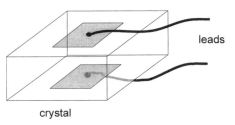

crystal

Fig. 4.13   Ultrasonic transducer.

where k is a proportionality constant for the specific piezoelectric material. By assuming that the piezoelectric crystal acts like a parallel plate capacitor, the voltage across the crystal, V, is given by

$$V = \frac{Q}{C} \tag{4.14}$$

where C is the equivalent capacitance of the crystal.

Example Problem 4.9

Derive a relationship for calculating the output voltage across a piezoelectric transducer that has a thickness, d, and an area, A, in terms of an applied force, F.

Solution

The capacitance of a piezoelectric transducer can be approximated by Eq. (4.12). Equation (4.14) is combined with the relationships given by Eq. (4.13) and (4.14) to give

$$V = \frac{Q}{C} = \frac{kF}{C} = \frac{kFd}{\varepsilon_0 \varepsilon_r A} \qquad\blacksquare$$

Since the crystal has an internal leakage resistance, any steady charge produced across its surfaces will eventually be dissipated. Consequently, these piezoelectric transducers are not suitable for measuring a steady or low-frequency DC force. Instead, they are used either as variable force transducers or as mechanically resonating devices to generate high frequencies (typically from 1 to 10 MHz) either in crystal-controlled oscillators or as ultrasonic pulse transducers.

Piezoelectric transducers are commonly used in biomedical applications to measure the thickness of an object or in noninvasive blood pressure monitors. For instance, if two similar crystals are placed across an object (e.g., a blood vessel), one crystal can be excited to produce a short burst of ultrasound. The time it takes for this sound to reach the other transducer can be measured. Assuming that the velocity of sound propagation in soft tissue, c_t, is known (typically 1540 m/s), the time, t, it takes the ultrasonic pulse to propagate across the object can be measured and used to calculate the separation distance, d, of the two transducers from the following relationship

$$d = c_t\, t \tag{4.15}$$

4.3.2 Airflow Transducers

One of the most common airflow transducers is the Fleish pneumotachometer illustrated in Fig. 4.14. The device consists of a straight short tube section with a fixed screen obstruction in the middle that produces a slight pressure drop as the air is passed through it. The pressure drop created across the screen is measured by a differential pressure transducer. The signal produced by the pressure transducer is proportional to the velocity of the air. The tube is normally cone shaped to generate a laminar flow pattern. A small heater heats the screen so that water vapor does not

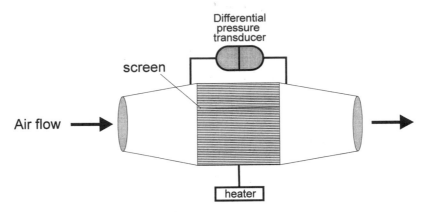

Fig. 4.14   Fleish airflow transducer.

condense on it over time and produce an artificially high pressure drop. Fleish pneumotachometers are used to monitor volume, flow, and breathing rates of patients on mechanical ventilators.

4.3.3   Temperature Measurement

Body temperature is one of the most tightly controlled physiological variables and one of the four basic vital signs used in the daily assessment of patients. The interior (core) temperature in the body is remarkably constant, about 37°C for a healthy person, and is normally maintained within ±0.5°C. Therefore, elevated body temperature is a sign of disease or infection, whereas a significant drop in skin temperature may be a good clinical indication of shock.

There are two distinct areas in the body where temperature is measured routinely — the surface of the skin under the armpit and inside a body cavity such as the mouth or the rectum. The two most commonly used devices to measure body temperature are thermistors, which require direct contact with the skin or mucosal tissues, and noncontact thermometers, which measure body core temperature inside the auditory canal.

Thermistors are temperature-sensitive transducers made of compressed sintered metal oxides (such as nickel, manganese, or cobalt) that change their resistance with temperature. Commercially available thermistors range in shape from small beads to large disks as illustrated in Fig. 4.15.

Fig. 4.15   Common forms of thermistors.

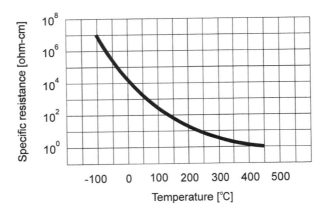

Fig. 4.16 Resistivity versus temperature characteristics of a typical thermistor.

Mathematically, the resistance-temperature characteristic of a thermistor can be approximated by

$$R_T = R_0 \exp\left[\beta\left(\frac{1}{T} - \frac{1}{T_0}\right)\right] \tag{4.16}$$

where R_0 is the resistance at a reference temperature T_0, (in degrees K), R_T is the resistance at temperature T (in degrees K), and β is a material constant, typically between 2500 and 5500 K. A typical resistance-temperature characteristic of a thermistor is shown in Fig. 4.16. Note that unlike metals and conventional resistors which have a positive temperature coefficient (as the temperature increases, the resistance increases), thermistors have a nonlinear relationship between temperature and resistance and a negative temperature coefficient. Increasing the temperature decreases the resistance of the thermistor.

Example Problem 4.10

A thermistor with a material constant β of 4500 K is used as a thermometer. Calculate the resistance of this thermistor at 25°C. Assume that the resistance of this thermistor at body temperature (37°C) is equal to 85Ω.

Solution

Using the resistance-temperature characteristic of a thermistor (Eq. 4.16) gives

$$R_T = 85 \exp\left[4500\left(\frac{1}{298} - \frac{1}{310}\right)\right] = 152.5 \ \Omega$$

∎

The size and mass of a thermistor probe in a medical thermometer must be small in order to give a rapid response time to temperature variations. The probe is normally covered with a very thin sterile plastic that is also disposable to prevent cross-contamination between patients.

A thermistor sensor is utilized in a thermodilution technique for measuring cardiac output (the volume of blood ejected by the heart each minute) as illustrated in Fig. 4.17. The technique involves the bolus injection of a cold indicator solution, usually a saline solution kept at 0–5°C, via a pulmonary artery catheter. The catheter is inserted into either the femoral or jugular veins. The tip of the flexible catheter is passed through the right side of the heart into the pulmonary artery with the aid of a small inflatable balloon. The cold liquid mixes with the venous blood in

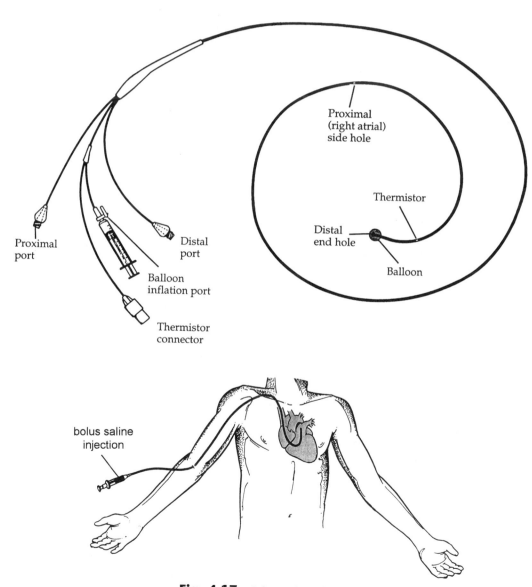

Fig. 4.17 A Swan–Ganz thermodilution catheter.

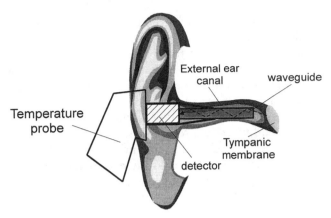

Fig. 4.18   Non-contact-type infrared ear thermometer.

the right atrium of heart and causes the blood to cool slightly. The cooled blood is ejected by the right ventricle into the pulmonary artery, where it contacts a thermistor located in the wall near the tip of a Swan–Ganz catheter. The thermistor measures the change in blood temperature as the blood passes on to the lungs. An instrument measures the extent of blood cooling which is inversely proportional to cardiac output.

Noncontact thermometers measure the temperature of the ear canal wall near the tympanic membrane, which is known to track the core temperature by about 0.5–1.0°C. Basically, as illustrated in Fig. 4.18, infrared radiation from the tympanic membrane is channeled to a heat-sensitive detector through a metal waveguide that has a gold-plated inner surface for better reflectivity. The detector, which is either a thermopile or a pyroelectric sensor that converts heat flow into an electric current, is normally maintained in a constant temperature environment to minimize inaccuracies due to fluctuation in ambient temperature. A disposable speculum is used on the probe to protect patients from cross-contamination.

4.4   BLOOD GASES AND pH SENSORS

Measurements of arterial blood gases (pO_2 and pCO_2) and pH are frequently performed on critically ill patients in both the operating room and the intensive care unit and are used by physicians to determine the need for adjusting mechanical ventilation or administering pharmacological agents. These measurements provide information about the respiratory and metabolic imbalances in the body and reflect the adequacy of blood oxygenation and CO_2 elimination.

Traditionally, blood gas analysis has been performed by withdrawing blood from a peripheral artery. The blood sample is then transported to a clinical laboratory for analysis. The need for rapid test results in the management of unstable, critically ill patients has led to the development of newer methods for continuous noninvasive blood gas monitoring. This allows the physician to follow trends in the patient's condition as well as receive immediate feedback on the adequacy of certain therapeutic interventions.

Noninvasive sensors for measuring O_2 and CO_2 in arterial blood are based on the discovery that gases, such as O_2 and CO_2, can easily diffuse through the skin. Diffusion occurs due to a partial pressure difference between the blood in the superficial layers of the skin and the outermost surface of the skin. This concept has been used to develop two types of noninvasive electrochemical sensors for monitoring pO_2 and pCO_2 transcutaneously. Furthermore, the discovery that blood changes its color depending on the amount of oxygen chemically bound to the hemoglobin in the erythrocytes has led to the development of several optical methods to measure the oxygen saturation in blood.

4.4.1 Oxygen Measurement

A quantitative method for measuring blood oxygenation is of great importance in assessing the circulatory and respiratory condition of a patient. Oxygen is transported by the blood from the lungs to the tissues in two distinct states. Under normal physiological conditions, approximately 2% of the total amount of oxygen carried by the blood is dissolved in the plasma. This amount is linearly proportional to the blood pO_2. The remaining 98% is carried inside the erythrocytes in a loose reversible chemical combination with hemoglobin (Hb) as oxyhemoglobin (HbO_2). Thus, there are two options for measuring blood oxygenation — either using a pO_2 sensor or measuring oxygen saturation (the relative amount of HbO_2 in the blood) by means of an oximeter.

A polarographic pO_2 sensor, also widely known as a Clark electrode, is used to measure the partial pressure of O_2 gas in a sample of air or blood. The measurement is based on the principle of polarography as illustrated in Fig. 4.19.

The electrode utilizes the ability of O_2 molecules to react chemically with H_2O in the presence of electrons to produce hydroxyl (OH^-) ions. This electrochemical reaction, called an oxidation/reduction or redox reaction, requires an externally applied constant polarizing (bias) voltage source of about 0.6 V.

Oxygen is reduced (consumed) at the surface of a noble metal (e.g., platinum or gold) cathode (the electrode connected to the negative side of the voltage source) according to the following chemical reaction:

$$O_2 + 2H_2O + 4e^- \leftrightarrow 4OH^-$$

In this reduction reaction, an O_2 molecule takes four electrons and reacts with two water molecules, generating four hydroxyl ions. The resulting OH^- ions migrate and react with a reference Ag/AgCl anode (the electrode connected to the positive side of the voltage source), causing a two-step oxidation reaction to occur as follows:

$$Ag \leftrightarrow Ag^+ + e^-$$

$$Ag^+ + Cl^- \leftrightarrow AgCl_\downarrow$$

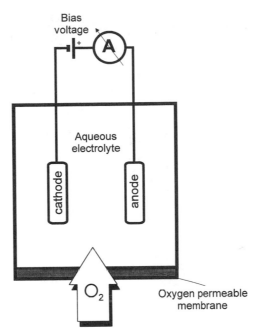

Fig. 4.19 Principle of a polarographic Clark-type pO_2 sensor.

In this oxidation reaction, silver from the electrode is first oxidized to silver ions and electrons are liberated to the anode. These silver ions are immediately combined with chloride ions to form silver chloride that precipitates on the surface of the anode. The current flowing between the anode and the cathode in the external circuit produced by this reaction is directly (i.e., linearly) proportional to the number of O_2 molecules constantly reduced at the surface of the cathode. The electrodes in the polarographic cell are immersed in an electrolyte solution of potassium chloride and surrounded by an O_2-permeable Teflon or polypropylene membrane that permits gases to diffuse slowly into the electrode. Thus, by measuring the change in current between the cathode and the anode, the amount of oxygen that is dissolved in the solution can be determined.

With a rather minor change in the configuration of a polarographic pO_2 sensor, it is also possible to measure the pO_2 transcutaneously. Figure 4.20 illustrates a cross section of a Clark-type transcutaneous pO_2 sensor. This sensor is essentially a standard polarographic pO_2 electrode that is attached to the surface of the skin by double-sided adhesive tape. It measures the partial pressure of oxygen that diffuses from the blood through the skin into the Clark electrode similar to the way it measures the pO_2 in a sample of blood. However, since the diffusion of O_2 through the skin is normally very low, a miniature heating coil is incorporated into the housing of this electrode to cause gentle vasodilatation (increased local blood flow) of the capillaries in the skin. By raising the local skin temperature to about 43°C, the pO_2

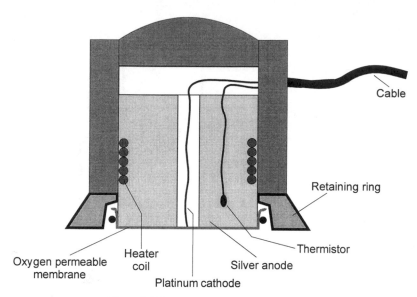

Fig. 4.20   Transcutaneous pO_2 sensor.

measured by the transcutaneous sensor approximates that of the underlying arterial blood. This electrode has been used extensively in monitoring newborn babies in the intensive care unit. However, as the skin becomes thicker and matures in adult patients, the gas diffusion properties of the skin change significantly and cause large errors and inconsistent readings.

Various methods for measuring the oxygen saturation, SO_2 (the relative amount of oxygen carried by the hemoglobin in the erythrocytes), of blood *in vitro* or *in vivo* in arterial blood (S_aO_2) or mixed venous blood (S_vO_2), have been developed. These methods, referred to as oximetry, are based on the light absorption properties of blood and, in particular, the relative concentration of Hb and HbO_2 since the characteristic color of deoxygenated blood is blue, whereas fully oxygenated blood has a distinct bright red color.

The measurement is performed at two specific wavelengths: a red wavelength, λ_1, where there is a large difference in light absorbance between Hb and HbO_2 (e.g., approximately 660 nm), and a second wavelength, λ_2, in the near-infrared region of the spectrum, typically chosen between 805 and 960 nm. The second wavelength can be either isobestic (a region of the spectrum around 805 nm where the absorbance of light is independent of blood oxygenation) or around 940–960 nm, where the absorbance of Hb is slightly smaller than that of HbO_2. Figure 4.21 shows the optical absorption spectra of blood in the visible and near-infrared region.

The measurement is based on Beer–Lamber's law that relates the transmitted light power, P_t to the incident light power, P_0, according to the following relationship:

$$P_t = P_0 \times 10^{-abc} \qquad (4.17)$$

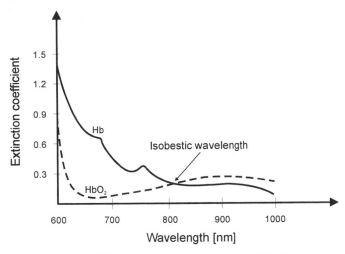

Fig. 4.21 Optical properties of Hb and HbO$_2$.

where a is a wavelength-dependent constant called the extinction coefficient (or molar absorptivity) of the sample, b is the light path length through the sample, and c is the concentration of the sample.

Assuming for simplicity that (i) $\lambda_1 = 660$ nm and $\lambda_2 = 805$ nm (i.e., isobestic), (ii) the hemolyzed blood sample (blood in which the erythrocytes have been ruptured, i.e., the hemoglobin has been released and uniformly mixed with the plasma) consists of a two-component mixture of Hb and HbO$_2$ and (iii) the total light absorbance by the mixture of these two components is additive, a simple mathematical relationship can be derived for computing the oxygen saturation of blood:

$$SO_2 = A - B\left[\frac{OD(\lambda_1)}{OD(\lambda_2)}\right] \tag{4.18}$$

where A and B are two coefficients that are functions of the specific absorptivity of Hb and HbO$_2$, OD is defined as the optical density, i.e., $\log_{10}(1/T)$ (where T is the light transmission given by P_t/P_0) and SO_2 is defined as $C_{HBO_2} / (C_{HB} + C_{HBO_2})$.

The measurement of SO_2 in blood can be performed either *in vitro* or *in vivo*. *In vitro* measurement using a bench-top oximeter requires a sample of hemolyzed blood, usually drawn from a peripheral artery. The sample is then injected into an optical cuvette (a parallel-wall glass container) which holds the sample while it is being illuminated sequentially by light from an intense white source after proper wavelength selection using narrow-band optical filters.

SO_2 can also be measured *in vivo* using a pulse oximeter. Noninvasive optical sensors for measuring S_aO_2 by a pulse oximeter consist of a pair of small and inexpensive light emitting diodes (LEDs) — typically a red LED around 660 nm and an infrared LED around 960 nm — and a single, highly sensitive silicon photodetector. These components are mounted inside a reusable spring-loaded clip or a disposable adhesive wrap. The sensor is usually attached either to the fingertip or earlobe such

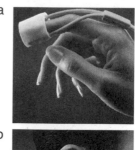

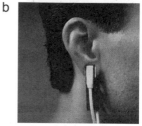

Fig. 4.22 Fingertip (a) and earlobe (b) pulse oximeter probes (courtesy of Criticare Systems, Inc., Waukesha, WI).

that the tissue is sandwiched between the light source and the photodetector as shown in Fig. 4.22. Electronic circuits inside the pulse oximeter generate signals to turn on the two LEDs in a sequential manner and synchronously measure the photodetector output when the corresponding LEDs are activated.

Pulse oximetry relies on the detection of the photoplethysmographic signal, as illustrated in Fig. 4.23. This signal is caused by changes in arterial blood volume

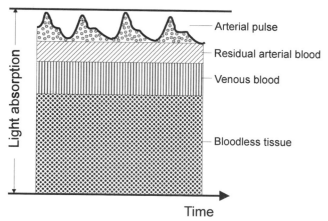

Fig. 4.23 Time dependence of light absorption by a peripheral vascular tissue bed illustrating the effect of arterial pulsation.

associated with periodic contractions of the heart during systole. The magnitude of this signal depends on the amount of blood ejected from the heart into the peripheral vascular bed with each cardiac cycle, the optical absorption of the blood, skin, and tissue, and the wavelength used to illuminate the blood. S_aO_2 is derived by analyzing the magnitude of the red and infrared photoplethysmograms measured by the photodetector. Electronic circuits separate the photopletysmogram into its pulsatile (AC) and nonpulsatile (DC) signal components. An algorithm inside the pulse oximeter performs a mathematical normalization by which the AC signal at each wavelength is divided by the corresponding DC component that results mainly from the light absorbed by the bloodless tissue, residual arterial blood when the heart is in diastole, venous blood, and skin pigmentation. Since it is assumed that the AC portion in the photoplethysmogram results only from the arterial blood component, this scaling process provides a normalized red/infrared ratio, R, which is highly dependent on the color of the arterial blood (i.e., S_aO_2) but is largely independent of the volume of arterial blood entering the tissue during systole, skin pigmentation, skin thickness, and vascular structure. Hence, the instrument does not need to be recalibrated for measurements on different patients. The mathematical relationship between S_aO_2 and R is programmed by the manufacturer into the pulse oximeter.

4.4.2 pH Electrodes

pH describes the balance between acid and base in a solution. Acidic solutions have an excess of hydrogen ions (H^+), whereas basic solutions have an excess of hydroxyl ions (OH^-). In a dilute solution, the product of these ion concentrations is a constant (1.0×10^{-14}). Therefore, the concentration of either ion can be used to express the acidity or alkalinity of a solution. All neutral solutions have a pH of 7.0.

The measurement of blood pH is fundamental to many diagnostic procedures. In normal blood, pH is maintained under tight control and is typically approximately 7.40 (slightly basic). By measuring the pH of the blood, it is possible to determine whether the lungs are removing sufficient CO_2 gas from the body or how well the kidneys regulate the acid-base balance.

A pH electrode essentially consists of two separate electrodes: a reference electrode and an active (indicator) electrode as illustrated in Fig. 4.24. The two electrodes are typically made of an Ag/AgCl wire dipped in a KCl solution and encased in a glass container. A salt bridge, which is essentially a glass tube containing an electrolyte enclosed in a membrane that is permeable to all ions, maintains the potential of the reference electrode at a constant value regardless of the solution under test. Unlike the reference electrode, the active electrode is sealed with hydrogen-impermeable glass except at the tip. The reference electrode may also be combined with the indicator electrode in a single glass housing.

The boundary separating two solutions has a potential proportional to the hydrogen ion concentration of one solution and, at a constant temperature of 25°C, is given by

$$V = -59 \text{ mV} \times \log_{10} [H^+] + C \qquad (4.19)$$

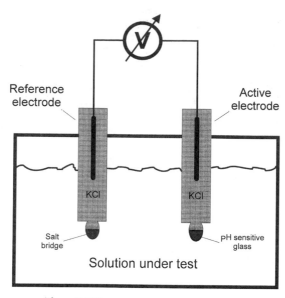

Fig. 4.24   Principle of a pH electrode.

where C is a constant. Since pH is defined as

$$pH = -\log_{10}[H^+] \qquad (4.20)$$

the potential of the active pH electrode V is proportional to the pH of the solution under test and is equal to

$$V = 59 \times pH + C \qquad (4.21)$$

The value of C is usually compensated for electronically when the pH electrode is calibrated by placing the electrode inside buffer solutions with known pH values.

4.4.3   Carbon Dioxide Sensors

Electrodes for measurement of partial pressure of CO_2 in blood or other liquids are based on measuring the pH as illustrated in Fig. 4.25. The measurement is based on the observation that, when CO_2 is dissolved in water, it forms a weakly dissociated carbonic acid (H_2CO_3) that subsequently forms free hydrogen and bicarbonate ions according to the following chemical reaction:

$$CO_2 + H_2O \leftrightarrow H_2CO_3 \leftrightarrow H^+ + HCO_3^-$$

As a result of this chemical reaction, the pH of the solution is changed. This change generates a potential between the glass pH and a reference (e.g., Ag/AgCl) electrode that is proportional to the negative logarithm of the pCO_2.

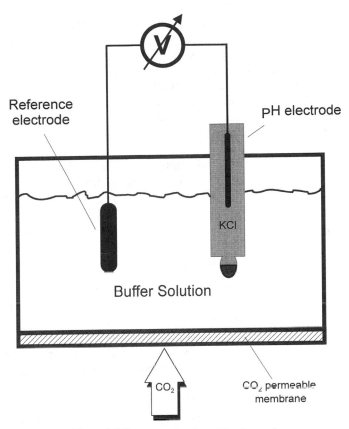

Fig. 4.25 Principle of a pCO$_2$ electrode.

4.5 BIOANALYTICAL SENSORS

The number of analytes that can be measured with electrochemical sensors can be increased significantly by adding biologically specific mediators (reagents that either undergo reactions or act as catalysts) to the semipermeable membrane structure. Several biosensors that have been constructed and used mainly for research applications have different enzymes and microorganisms as the primary sensing elements. Although these biosensors have been used successfully *in vitro* to demonstrate unique medical and industrial applications, further technical improvements are necessary to make these sensors robust and reliable enough to fulfill the demanding requirements of routine analytical and clinical applications. Examples of some interesting sensor designs are given in the following sections.

4.5.1 Enzyme-Based Biosensors

Enzymes constitute a group of more than 2000 proteins having so-called biocatalytic properties. These properties give the enzymes their unique and powerful ability to accelerate chemical reactions inside biological cells. Most enzymes react only with specific substrates even though they may be contained in a complicated mixture with other substances. It is important to keep in mind, however, that soluble enzymes are very sensitive to both temperature and pH variations and they can be inactivated by many chemical inhibitors. For practical biosensor applications, these enzymes are normally immobilized by insolubilizing the free enzymes via entrapment into an inert and stable matrix such as starch gel, silicon rubber, or polyacrylamide. This process is important to ensure that the enzyme retains its catalytic properties and can be reusable.

The action of specific enzymes can be utilized to construct a range of different biosensors. A typical example of an enzyme-based sensor is a glucose sensor that uses the enzyme glucose oxidase. An immobilized enzyme which acts as a catalyst, such as glucose oxidase (g.o.), is commonly used to detect glucose by measuring electrochemically either the amount of hydrogen peroxide (H_2O_2) or gluconic acid produced or the amount of oxygen consumed according to the following reaction:

$$\text{Glucose} + O_2 \xrightleftharpoons{\text{g.o.}} \text{gluconic acid} + H_2O_2$$

Biocatalytic enzyme-based sensors generally consist of an electrochemical gas-sensitive transducer or an ion-selective electrode with an enzyme immobilized in or on a membrane that serves as the biological mediator. The analyte diffuses from the bulk sample solution into the biocatalytic layer where an enzymatic reaction takes place. The electroactive product that is formed (or consumed) is usually detected by an ion-selective electrode. A membrane separates the basic sensor from the enzyme if a gas is consumed (such as O_2) or is produced (such as CO_2 or NH_3). Although the concentration of the bulk substrate drops continuously, the rate of consumption is usually negligible. The decrease is detected only when the test volume is very small or when the area of the enzyme membrane is large enough. Thus, this electrochemical analysis is nondestructive and the sample can be reused. Measurements are usually performed at a constant pH and temperature either in a stirred medium solution or in a flowthrough solution.

4.5.2   Microbial Biosensors

A number of microbial sensors have been developed mainly for on-line control of biochemical processes in various environmental, agricultural, food, and pharmaceutical applications. Microbial biosensors typically involve the assimilation of organic compounds by the microorganisms, followed by a change in respiration activity (metabolism), or the production of specific electrochemically active metabolites, such as H_2, CO_2, or NH_3, that are secreted by the microorganism.

A microbial biosensor is composed of immobilized microorganisms that serve as specific recognition elements and an electrochemical or optical sensing device that is used to convert the biochemical signal into an electronic signal that can be processed. The operation of a microbial biosensor is a five-step process: (i) The substrate is transported to the surface of the sensor, (ii) the substrate diffuses through the membrane to the immobilized microorganism, (iii) a reaction occurs at the immobilized organism, (iv) the products formed in the reaction are transported through the membrane to the surface of the detector, and (v) the products are measured by the detector.

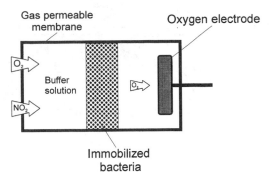

Fig. 4.26 Principle of a NO_2 microbial-type biosensor.

Examples of microbial biosensors include ammonia (NH_3) and nitrogen dioxide (NO_2) sensors that utilize nitrifying bacteria as the biological sensing component. Ammonia biosensors are based on nitrifying bacteria, such as *Nitrosomonas* sp., that use ammonia as a source of energy and oxidize ammonia as follows:

$$NH_3 + 1.5O_2 \xrightarrow[\textit{Nitrosomonas sp.}]{} NO_2 + H_2O + H^+$$

The oxidation process proceeds at a high rate, and the amount of oxygen consumed by the immobilized bacteria can be measured directly by a polarographic oxygen electrode placed behind the bacteria.

Nitric oxide (NO) and NO_2 are the two principal pollution gases of nitrogen in the atmosphere. The principle of a NO_2 biosensor is shown in Fig. 4.26. When a sample of NO_2 gas diffuses through the gas-permeable membrane, it is oxidized by the *Nitrobacter* sp. bacteria as follows:

$$2NO_2 + O_2 \xrightarrow[\textit{Nitrosomonas sp.}]{} 2NO_3$$

Similar to an ammonia biosensor, the consumption of O_2 around the membrane is determined by an electrochemical oxygen electrode.

The use of microbial cells in electrochemical sensors offers several advantages over enzyme-based electrodes, the principal one being the increased electrode lifetime of several weeks. On the other hand, microbial sensors may be less favorable compared with enzyme electrodes with respect to specificity and response time.

4.6 OPTICAL BIOSENSORS

4.6.1 Optical Fibers

Optical fibers are used to transmit light from one location to another. They are made from two concentric and transparent glass or plastic materials as illustrated in Fig. 4.27. One is known as the core and the second layer, which serves as a coating material, is called the cladding.

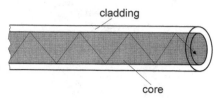

Fig. 4.27   Principle of optical fibers.

The core and cladding of an optical fiber have a different index of refraction, n. The index of refraction is a number that expresses the ratio of the light velocity in free space to its velocity in a specific material. For instance, the refractive index for air is equal to 1.0, whereas the refractive index for water is equal to 1.33. Assuming that the refractive index of the core material is n_1 and the refractive index of the cladding is n_2 ($n_1 > n_2$), Snell's law gives

$$n_1\sin\phi_1 = n_2\sin\phi_2 \tag{4.22}$$

where ϕ is the angle of incidence as illustrated in Fig. 4.28.

Accordingly, any light passing from a lower refractive index to a higher refractive index is bent toward the line that is perpendicular to the interface of the two materials. For small incident angles, ϕ_1, the light ray enters the fiber core and bends inwards at the first core/cladding interface. For larger incident angles, ϕ_2, the ray exceed a minimum angle required to bend it back into the core when it reaches the corecladding boundary. Consequently, the light escapes into the cladding. By setting $\sin\phi_2 = 1.0$, the critical angle, ϕ_{cr}, is given by

$$\sin\phi_{cr} = \frac{n_2}{n_1} \tag{4.23}$$

Any light rays that enter the optical fiber with incidence angles greater than ϕ_{cr} are internally reflected inside the core of the fiber by the surrounding cladding. Conversely, any entering light rays with incidence angles smaller than ϕ_{cr} escape through the cladding and are therefore not transmitted by the core.

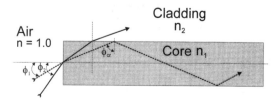

Fig. 4.28   Optical fiber illustrating the incident and refracted light rays. The solid line shows the light ray escaping from the core into the cladding. The dashed line shows the ray undergoing total internal reflection inside the core.

Example Problem 4.11

Assume that a beam of light passes from a layer of glass with a refractive index of $n_1 = 1.47$ into a second layer of glass with a refractive index of $n_2 = 1.44$. Using Snell's law, calculate the critical angle for the boundary between these two glass layers.

Solution

$$\phi_{cr} = \arcsin\left(\frac{n_2}{n_1}\right) = \arcsin(0.9796)$$

$$\phi_{cr} = 78.4°$$

Therefore, light that strikes the boundary between these two glasses at an angle greater than 78.4° will be reflected back into the first layer. ∎

4.6.2 Sensing Mechanisms

Optical fibers can be used to develop a whole range of sensors for biomedical applications. These sensors are small, flexible, and free from electrical interference. They can produce an instantaneous response to microenvironments that surround their surface.

Commercial fiber optic sensors for blood gas monitoring became available during the past decade. While many different approaches have been taken, they all have some features in common as illustrated in Fig. 4.29. First, all sensors are interfaced with an optical module. The module supplies the excitation light, which may be

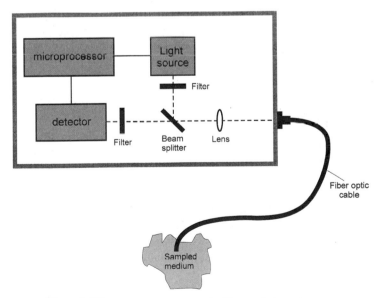

Fig. 4.29 General principle of a fiber optic-based sensor.

from a monochromatic source such as a diode laser or from a broadband source (e.g., quartz-halogen) that is filtered to provide a narrow bandwidth of excitation. Typically, two wavelengths of light are used: One wavelength is sensitive to changes in the species to be measured, whereas the other wavelength is unaffected by changes in the analyte concentration. This wavelength serves as a reference and is used to compensate for fluctuations in source output and detector stability. The light output from the optic module is coupled into a fiber optic cable through appropriate lenses and an optical connector.

Several sensing mechanisms can be utilized to construct optical fiber sensors. In fluorescence-based sensors, the incident light excites fluorescence emission, which changes in intensity as a function of the concentration of the analyte to be measured. The emitted light travels back down the fiber to the monitor where the light intensity is measured by a photodetector. In other types of fiber optic sensors, the light-absorbing properties of the sensor chemistry change as a function of analyte chemistry. In the absorption-based design, a reflective surface near the tip or some scattering material within the sensing chemistry itself is usually used to return the light back through the same optical fiber.

4.6.3 Indicator-Mediated Fiber Optic Sensors

Since only a limited number of biochemical substances have an intrinsic optical absorption or fluorescence property that can be measured directly with sufficient selectivity by standard spectroscopic methods, indicator-mediated sensors have been developed to use specific reagents that are immobilized either on the surface or near the tip of an optical fiber. In these sensors, light travels from a light source to the end of the optical fiber where it interacts with a specific chemical or biological recognition element. These transducers may include indicators and ion-binding compounds (ionophores) as well as a wide variety of selective polymeric materials. After the light interacts with the biological sample, it returns through either the same optical fiber (in a single-fiber configuration) or a separate optical fiber (in a dual-fiber configuration) to a detector, which correlates the degree of light attenuation with the concentration of the analyte.

Typical indicator-mediated sensor configurations are shown schematically in Fig. 4.30. The transducing element is a thin layer of chemical material that is placed near the sensor tip and is separated from the blood medium by a selective membrane. The chemical-sensing material transforms the incident light into a return light signal with a magnitude that is proportional to the concentration of the species to be measured. The stability of the sensor is determined by the stability of the photosensitive material that is used and also by how effective the sensing material is protected from leaching out of the probe. In Fig. 4.30a the indicator is immobilized directly on a membrane positioned at the end of the fiber. An indicator in the form of a powder can also be physically retained in position at the end of the fiber by a special permeable membrane as illustrated in Fig. 4.30b or a hollow capillary tube as illustrated in Fig. 4.30c.

Optical fibers have been used to construct immunoassay biosensors. The principle for this relies on the propagation of light along the optical fiber. When light trav-

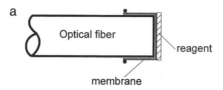

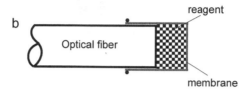

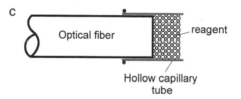

Fig. 4.30 Different indicator-mediated fiber optic sensor configurations.

els along the fiber, it is not just confined to the core region. A small fraction (referred to as an evanescent wave) penetrates a characteristic distance (typically on the order of one wavelength or less) beyond the core boundary into the surrounding cladding. The evanescent wave is attenuated exponentially, typically following Beer–Lamber's law. This concept has been exploited to construct several biosensors in which a portion of the cladding near the distal tip of an optical fiber has been removed and replaced by an optically absorbing compound. The light, which propagates inside the optical fiber core, undergoes several multiple internal reflections along the sides of the unclad portion of the fiber tip. Usually, an optically reflecting material is coated on the distal tip of the fiber to divert the beam back through the same fiber where it is detected by a sensitive photodetector. In some designs, a stable fluorophore can be used instead of an absorbing material. The excitation light, which is typically generated either by a stable laser source or by a combination of a broadband light source and a narrow-width optical filter, is absorbed by the fluorophore and emits a detectable fluorescent light at a slightly higher wavelength than the excitation light. This concept provides improved sensitivity because narrow-

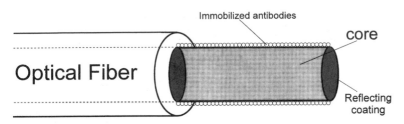

Fig. 4.31 Principle of a fiber optic immunoassay biosensor.

width optical filters can be used to separate the incident light from the fluorescent light components that reach the photodetector.

Evanescent-type biosensors can be used in immunological diagnostics to detect antibodyantigen binding. Figure 4.31 shows a conceptual diagram of an immunoassay biosensor. The immobilized antibody on the surface of the unclad portion of the fiber captures the antigen from the sample solution, which is normally introduced into a small flowthrough chamber where the fiber tip is located. The sample solution is then removed and labeled antibody is added into the flow chamber. A fluorescent signal is excited and measured when the labeled antibody binds to the antigen that is already immobilized by the antibody.

EXERCISES

1. Two identical silver electrodes are placed in an electrolyte solution. Calculate the potential drop between the two electrodes.
2. By how much would the inductance of an inductive displacement transducer coil change if the number of coil turns increases by a factor of three?
3. Determine the ratio between the cross-sectional areas of two blood vessels. Assume that the voltage ratio induced in identical magnetic flow probes is equal to $1:5$ and the ratio of blood flow through these vessels is $2:3$.
4. A 10 kΩ linear rotary transducer is used to measure the angular displacement of the knee joint. Calculate the change in output voltage for a $135°$ change in the angle of the knee if a constant current of 1 mA is supplied to the transducer.
5. Provide a step-by-step derivation of Eq. (4.11).
6. An elastic resistive transducer with an initial resistance, R_0, and length, l_0, is stretched to a new length. Assuming that the cross-sectional area of the transducer changes during stretching, derive a mathematical relationship for the change in resistance, ΔR, as a function of the initial length, l_0; the change in length, Δl; the volume of the transducer, V; and the resistivity, ρ.
7. Plot the capacitance (y axis) versus displacement (x axis) characteristics of a capacitance transducer.
8. Calculate the sensitivity of a capacitive transducer (i.e., $\Delta C/\Delta d$) for small changes in displacements.
9. A capacitive transducer is used in a mattress to measure changes in breath-

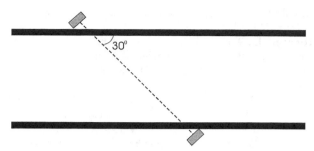

Fig. 4.32 Two identical ultrasonic transducers are positioned across a blood vessel.

ing patterns of an infant. During inspiration and expiration, the rate of change (i.e., dV/dt) in voltage across the capacitor is equal to ± 1 V/s, and this change can be modeled by a triangular waveform. Plot the corresponding changes in current flow through this transducer.

10. Two identical ultrasonic transducers are positioned across a blood vessel as shown in Fig. 4.32. Calculate the diameter of the blood vessel if it takes 400 ns for the ultrasonic sound to propagate from one transducer to the other.

11. Calculate the resistance of a thermistor at 98°F assuming that the resistance of this thermistor at 15°C is equal to 1.5 kΩ and $\beta = 2750$.

12. The resistance of a thermistor with a $\beta = 4000$ measured at 20°C is equal to 500 Ω. Find the temperature of the thermistor when the resistance is doubled.

13. Calculate the β of a thermistor if it has a resistance of 3 kΩ at 72°F (room temperature) and a resistance of 2.46 kΩ when the room temperature increases by 15%.

14. Sketch the current (y axis) versus pO_2 (x axis) characteristics of a polarographic Clark electrode.

15. A pH electrode is attached to sensitive voltmeter that reads 0.348 V when the electrode is immersed in a buffer solution with a pH of 4.50. After the pH electrode is moved to an unknown buffer solution, the reading of the voltmeter is decreased by 20%. Calculate the pH of the unknown buffer.

16. Plot the optical density, OD, of an absorbing solution (y axis) versus the concentration (x axis) of this solution. What is the slope of this curve?

17. If an incident light ray passing from air into water has a 55° angle with respect to the normal, calculate the angle of the refracted light ray.

SUGGESTED READING

Allocca, J. A., and Stuart A. (1984). *Transducers: Theory and Applications.* Reston, VA.

Aston, R. (1990), *Principles of Biomedical Instrumentation and Measurement.* Macmillan, New York.

Cobbold R. S. C. (1974), *Transducers for Biomedical Measurement: Principles and Applications*. Wiley, New York.

Cromwell, L., Weibell, F. J., and Pfeiffer, E. J. (1980). *Biomedical Instrumentation and Measurements*. Prentice Hall, Englewood Cliffs, NJ.

Geddes, L. A., and Baker, L. E. (1968). *Principles of Applied Biomedical Instrumentation*. Wiley, New York.

Hall, E. A. H. (1991). *Biosensors*. Prentice Hall, Englewood Cliffs.

Neuman M. R. (1995). V. Biomedical sensors. In *The Biomedical Engineering Handbook* (J. D. Bronzino, Ed.). CRC Press, Boca Raton, FL.

Webster, J. G. (1988). *Encyclopedia of Medical Devices and Instrumentation*. Wiley, New York.

Webster, J. G. (1998). *Medical Instrumentation, Application and Design*. John Wiley & Sons.

Wise, D. L. (1991). *Bioinstrumentation and Biosensors*. Dekker, New York.

5 BIOINSTRUMENTATION

Chapter Contents

After completing this chapter, the reader will be able to:

- Discuss some of the major milestones that occurred during the evolution of medical instrumentation
- Analyze the Wheatstone bridge circuit
- Describe the components of a basic instrumentation system
- Describe a variety of circuits that use operational amplifiers to condition analog signals and calculate the effect that those circuits have on the signal that is being monitored
- Determine the transfer function for cascaded simple circuits
- Explain the effects of different filters (low-pass, high-pass, band-pass, and band-stop) on analog input signals
- Describe some of the ways in which noise is handled when designing instruments
- Design a differential biopotential amplifier system
- Explain how analog signals are converted into digital signals
- Describe the components of a computer
- Describe the role of computers in medical instrumentation

5.1 INTRODUCTION

Since the beginning of recorded history, humans have been interested in fashioning medical tools and finding new ways to heal the sick. When people lived in caves, medical technology consisted of primitive tools, such as stones, roots, herbs, and branches, but there is evidence that some surgical procedures were performed. Fossilized skulls have been uncovered that indicate ancient healers drilled holes in the heads of the sick. This procedure, called trephining, was probably performed as a cure for seizures, epilepsy, or severe head pain. The growth of new bone around the holes in the skull is evidence that the individuals who underwent this treatment survived for some period of time.

Although medical technology continued to evolve, it remained quite primitive for thousands of years. By the year 1000 BC, the field of medicine and its technology represented a highly respected profession with temples of healing as the precursors of the first hospitals. Hippocrates (~400 BC), the most notable of these early persons, is said to be the founder of Western medicine because he introduced the scientific spirit. As a result of his contributions, diagnostic observation and clinical treatment began to replace superstition.

The scientific method had its most dramatic impact during the twentieth century. Before 1900, medicine had little to offer the typical citizen because its resources were mainly the education and little black bag of the physician. The origins of the changes that occurred within medical science are found in several developments that took place in the applied sciences. During the early nineteenth century, diagnosis was based on physical examination, and treatment was designed to heal the structural abnormality. By the late nineteenth century, diagnosis was based on laboratory tests, and treatment was designed to remove the cause of the disorder. The trend toward the use of technology accelerated throughout the twentieth century. During this period, hospitals became institutions of research and technology. Professionals in the areas of chemistry, physics, mechanical engineering, and electrical engineering began to work in conjunction with the medical field, and biomedical engineering became a recognized profession. As a result, medical technology advanced more in the twentieth century than it had in the rest of history combined (Fig. 5.1).

During this period, the area of electronics had a significant impact on the development of new medical technology. Men such as Richard Caton and Augustus Desire proved that the human brain and heart depended on bioelectric events. In 1903, William Einthoven expanded on these ideas after he created the first string galvanometer. Einthoven placed two skin sensors on a man and attached them to the ends of a silvered wire that was suspended through holes drilled in both ends of a large permanent magnet. The suspended silvered wire moved rhythmically as the subject's heart beat. By projecting a tiny light beam across the silvered wire, Einthoven was able to record the movement of the wire as waves on a scroll of moving photographic paper. Thus, the invention of the string galvanometer led to the creation of the electrocardiogram (ECG), which is routinely used today to measure and record the electrical activity of abnormal hearts and to compare those signals to normal ones.

In 1929, Hans Berger created the first electroencephalogram (EEG), which is used to measure and record electrical activity of the brain. In 1935, electrical amplifiers were used to prove that the electrical activity of the cortex had a specific rhythm, and in 1960 electrical amplifiers were used in devices such as the first implantable pacemaker that was created by William Chardack and Wilson Greatbatch. These are just a small sample of the many examples in which the field of electronics has been used to significantly advance medical technology.

Many other advancements that were made in medical technology originated from research in basic and applied physics. In 1895, the X-ray machine, one of the most important technological inventions in the medical field, was created when W. K. Roentgen found that X rays could be used to give pictures of the internal structures of the body. Thus, the X-ray machine was the first imaging device to be created. (Radiation imaging is discussed in detail in Chapter 14.)

Another important addition to medical technology was provided by the invention of the computer, which allowed much faster and more complicated analyses and functions to be performed. One of the first computer-based instruments in the field of medicine, the sequential multiple analyzer plus computer (SMAC; Fig. 5.2), was used to store a vast amount of data pertaining to clinical laboratory information. The invention of the computer made it possible for laboratory tests to be performed and analyzed faster and more accurately.

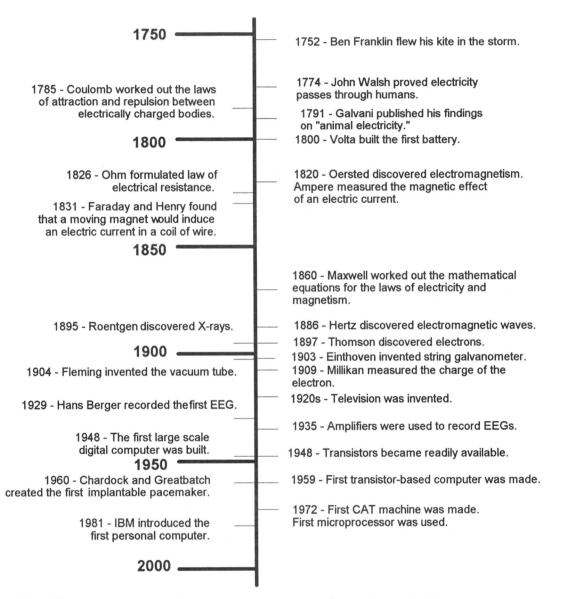

1750

1752 - Ben Franklin flew his kite in the storm.

1785 - Coulomb worked out the laws of attraction and repulsion between electrically charged bodies.

1774 - John Walsh proved electricity passes through humans.

1791 - Galvani published his findings on "animal electricity."

1800

1800 - Volta built the first battery.

1826 - Ohm formulated law of electrical resistance.

1820 - Oersted discovered electromagnetism. Ampere measured the magnetic effect of an electric current.

1831 - Faraday and Henry found that a moving magnet would induce an electric current in a coil of wire.

1850

1860 - Maxwell worked out the mathematical equations for the laws of electricity and magnetism.

1895 - Roentgen discovered X-rays.

1886 - Hertz discovered electromagnetic waves.

1900

1897 - Thomson discovered electrons.

1903 - Einthoven invented string galvanometer.

1904 - Fleming invented the vacuum tube.

1909 - Millikan measured the charge of the electron.

1929 - Hans Berger recorded the first EEG.

1920s - Television was invented.

1935 - Amplifiers were used to record EEGs.

1948 - The first large scale digital computer was built.

1948 - Transistors became readily available.

1950

1960 - Chardock and Greatbatch created the first implantable pacemaker.

1959 - First transistor-based computer was made.

1972 - First CAT machine was made. First microprocessor was used.

1981 - IBM introduced the first personal computer.

2000

Fig. 5.1 Time line for major inventions and discoveries that led to modern medical instrumentation.

The first large-scale computer-based medical instrument was created in 1972 when the computerized axial tomography (CAT) machine was invented. The CAT machine created an image that showed all the internal structures that lie in a single plane of the body. This new type of image made it possible to have more accurate

Fig. 5.2 SMAC.

and easier diagnosis of tumors, hemorrhages, and other internal damage from information that was obtained noninvasively (for details, see Chapter 14).

Telemedicine, which uses computer technology to transmit information from one medical site to another, is being explored to permit access to health care for patients in remote locations. A specialist in a major hospital can use telemedicine to receive information on a patient in a rural area and send back a plan of treatment specific for that patient.

Today, there is a wide variety of medical devices and instrumentation systems. Some are used to monitor patient conditions or acquire information for diagnostic purposes, e.g., ECG and EEG machines, whereas others are used to control physiological functions, e.g., pacemakers and ventilators. Some devices, such as pacemakers, are implantable, whereas many others are used noninvasively. This chapter will focus on those features that are common to devices that are used to acquire and process physiological data. Section 5.2 will describe the components of a basic instrumentation system. Section 5.3 provides information about analog circuits and their elements, and Section 5.4 discusses the ways in which analog signals can be processed. Section 5.5 discusses instrumentation design with emphasis on amplifying the signal, reducing noise, preventing signal aliasing, and converting the signal from analog to digital format. These concepts are introduced in the context of designing a biopotential amplifier system. Section 5.6 provides some basic information about computer-based instrumentation systems that handle digital signals.

5.2 BASIC INSTRUMENTATION SYSTEM

The quantity, property, or condition that is measured by an instrumentation system is called the measurand (Fig. 5.3). This can be a bioelectric signal, such as those generated by muscles or the brain, or a chemical or mechanical signal that is converted to an electrical signal. As explained in Chapter 4, sensors are used to convert physical measurands into electric outputs. The outputs from these biosensors are analog signals, i.e., continuous signals, that are sent to the analog processing and digital conversion block. There, the signals are amplified, filtered, conditioned, and converted to digital form. Methods for modifying analog signals, e.g., amplification and filtering, are discussed in Sections 5.4 and 5.5. Biosignal processing (signal conditioning) is used to help compensate for undesirable sensor characteristics. Once the analog signals have been digitized and converted to a form that can be stored and processed by digital computers, many more methods of signal conditioning can be applied (for details see Chapter 6).

Basic instrumentation systems also include output display devices that enable human operators to view the signal in a format that is easy to understand. These displays may be numerical or graphical, discrete or continuous, and permanent or temporary. Most output display devices are intended to be observed visually, but some also provide audible output, e.g., a beeping sound with each heartbeat.

In addition to displaying data, many instrumentation systems have the capability of storing data. In some devices, the signal is stored briefly so that further processing can take place or so that an operator can examine the data. In other cases, the signals are stored permanently so that different signal processing schemes can be applied at a later time. Holter monitors, for example, acquire 24 h of ECG data that are later processed to determine arrhythmic activity and other important diagnostic characteristics.

With the invention of the telephone and now with the internet, signals can be acquired with a device in one location, perhaps in a patient's home, and transmitted to another device for processing and/or storage. This has made it possible, for example, to provide quick diagnostic feedback if a patient has an unusual heart rhythm while at home. It has also allowed medical facilities in rural areas to transmit diagnostic images to tertiary care hospitals so that specialized physicians can help general practitioners make more accurate diagnoses.

Two other components play important roles in instrumentation systems. The first is the calibration signal. A signal with known amplitude and frequency content is applied to the instrumentation system at the sensor's input. The calibration signal allows the components of the system to be adjusted so that the output and input have a known, measured relationship. Without this information, it is impossible to convert the output of an instrument system into a meaningful representation of the measurand.

Another important component, a feedback element, is not a part of all instrumentation systems. These devices include pacemakers and ventilators that stimulate the heart or the lungs. Some feedback devices collect physiological data and stimulate a response (e.g., a heartbeat or breath) when needed or are part of biofeedback systems in which the patient is made aware of a physiological measurement (e.g., blood pressure) and uses conscious control to change the physiological response.

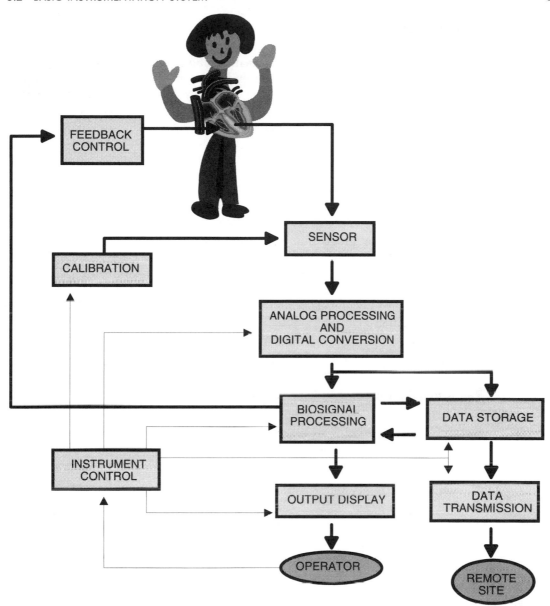

Fig. 5.3   Basic instrumentation systems contain sensors that convert the measurand, e.g., cardiac electrical activity, into an electrical signal, signal processing components that condition (i.e., filter, amplify, and in some cases digitize) the analog signal, and components that are used to store and display the analog and/or digital data. In some cases, the data may be transmitted to another site for interpretation or may be used as part of a control circuit that feeds back to the patient. Thick lines indicate signal paths, whereas thin lines represent instrument control.

5.3 ANALOG CIRCUITS

This section provides some basic information about electric circuit theory. It begins with a brief description of simple circuits containing several elements. One of these circuits, the Wheatstone bridge, is widely used in medical instrumentation. Phasors, the Laplace domain, and superposition are introduced in preparation for solving more complex circuits that are typical of those found in medical instrumentation.

5.3.1 Multiple Analog Elements

Simple Circuits

If the same current flows from one circuit element to another, the two (or more) elements are said to be in series (Figs. 5.4a and 5.4c). In other words, the same source current flows in all of the elements. Two or more elements are in parallel if the same voltage appears across each of the elements (Figs. 5.4b and 5.4d). An equivalent resistor (R_{EQ}) is determined for N resistors in series (Fig. 5.4a) from the following equation:

$$R_{EQ} = \sum_{i=1}^{N} R_i \qquad (5.1)$$

R_{EQ} is found for N resistors in parallel (Fig. 5.4B) from Eq. (5.2):

$$\frac{1}{R_{EQ}} = \sum_{i=1}^{N} \frac{1}{R_i}$$

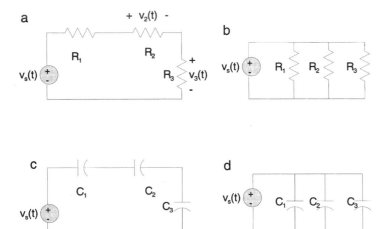

Fig. 5.4 Simple circuits. Voltage source with resistors in series (a), resistors in parallel (b), capacitors in series (c), and capacitors in parallel (d).

or

$$G_{EQ} = \sum_{i=1}^{N} G_i \qquad (5.2)$$

For two resistors, R_1 and R_2, in parallel, Eq. (5.2) reduces to

$$R_{EQ} = \frac{R_1 R_2}{R_1 + R_2}$$

Inductors follow the same rules as those for resistors. On the other hand, capacitances in parallel (Fig. 5.4.d) add like resistors in series, whereas capacitances in series (Fig. 5.4.c) combine in the same manner as resistances in parallel.

Voltage Divider

Due to Kirchoff's voltage law (KVL), the sum of the voltages across the three resistors in Fig. 5.4A equals the voltage supplied by the voltage source. Thus, the resistors divide the total voltage into several parts that are proportional to the resistance in the different resistors. The voltage across the second resistor (R_2) is given by

$$v_2(t) = \frac{R_2}{R_1 + R_2 + R_3} v(t) \qquad (5.3)$$

whereas the voltage across R_3 is given by

$$v_3(t) = \frac{R_3}{R_1 + R_2 + R_3} v(t) \qquad (5.4)$$

The **Wheatstone bridge** is a resistive circuit that uses two voltage dividers. Its general form is shown in Fig. 5.5. In this circuit, $v_s(t)$, R_1, R_2 and R_3 are known parameters, whereas R_x is the variable sensor element. The voltage difference between a and b can be measured and then used to solve for the unknown resistance, R_x, which may be a strain gauge, a thermistor, or some other resistive element.

$$v_{ab}(t) = v_{R_2}(t) - v_{R_x}(t) = v_s(t)\left(\frac{R_2}{R_1 + R_2} - \frac{R_x}{R_3 + R_x}\right) \qquad (5.5)$$

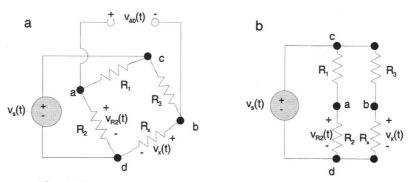

Fig. 5.5 Two different ways to draw the Wheatstone bridge circuit.

Example Problem 5.1

What is the voltage that would be measured at $v_{ab}(t)$ if the sensor resistance, R_x, is 5 kΩ, and $R_1 = 20$ kΩ, $R_2 = 10$ kΩ, $R_3 = 10$ kΩ, and $v_s(t) = 20$ V?

Solution

From Eq. (5.5),

$$v_{ab}(t) = 20\text{V}\left(\frac{10 \text{ k}\Omega}{10 \text{ k}\Omega + 20 \text{ k}\Omega} - \frac{5 \text{ k}\Omega}{10 \text{ k}\Omega + 5 \text{ k}\Omega}\right)$$

Thus,

$$v_{ab}(t) = 0 \text{ V}$$

This is known as the null condition and represents a balanced Wheatstone bridge.

■

5.3.2 Time-Varying Signals

Most of the electric power that is used for industrial and household applications worldwide is time varying and is either 50- or 60-Hz sinusoidal voltage and current. Sinusoidal voltage sources can be represented mathematically as

$$v_1(t) = V_1 \cos(\omega t + \theta) \tag{5.6}$$

where the voltage is defined by its frequency (ω in rad/s), its phase angle (θ in rad), and its peak magnitude (V_1). Unlike resistors, capacitors and inductors are dynamic elements that have values that exhibit derivative or integral relationships when time-varying voltage sources are used. Frequency can also be expressed in cycles per second (hertz or Hz) where

$$f = \frac{\omega}{2\pi} \tag{5.7}$$

Hence, a power-line frequency of 60 Hz may also be expressed as $\omega = 2\pi 60 = 377$ rad/s.

Phasors

Phasors, consisting of a real part [Re] and an imaginary part [Im], are used to express purely sinusoidal voltages and currents in the form of Eq. (5.6) and are manipulated mathematically by complex-number algebra. Equation (5.6) in phasor notation is

$$\hat{V}_1 = V_1 e^{j\theta} = V_1 \angle \theta \tag{5.8}$$

where $e^{j\theta} = \cos\theta + j\sin\theta$ and j is the square root of -1. Ohm's law for phasors is

$$\hat{V}_1 = Z\hat{I} \tag{5.9}$$

where $\hat{V}$ is the phasor voltage, $\hat{I}$ is the phasor current, and Z is the impedance. These are all complex numbers that contain real and imaginary parts.

Example Problem 5.2

Given an impedance of $Z = 25 + j10 \, \Omega$ and a phasor current of

$$\hat{I} = 5 - j3$$

Find the phasor voltage across the impedance.

Solution

From Eq. (5.9),

$$\hat{V} = Z\hat{I} = (25 + j10)(5 - j3) = 125 + j50 - j75 - j^2 30 = 155 - j25$$

In magnitude/angle notation

$$|\hat{V}| = \sqrt{(155)^2 + (25)^2} = 157.0$$

$$\angle\theta = \arctan\left(-\frac{25}{155}\right) = -10.18°$$

$$\hat{V} = 157.0\angle - 10.18°$$ ■

In the phasor domain, the impedances for circuit elements R, C, and L are defined as follows for sinusoidal steady state:

$$Z_R = R \tag{5.10}$$
$$Z_L = j\omega L \tag{5.11}$$
$$Z_C = \frac{1}{j\omega C} \tag{5.12}$$

In general, Z is a complex number, $Z = R + jX$, where R is resistance and X is called reactance. The quantities in Eqs. (5.10)–(5.12) are therefore called **resistance, inductive reactance,** and **capacitive reactance,** whereas R, L, and C have the same meaning that was given previously. Impedances behave like resistors when they are in series or parallel (see Eqs. 5.1 and 5.2). For example, the impedance of a resistor, capacitor, and inductor that are connected in series is

$$Z = R + j\omega L + \frac{1}{j\omega C} = R + j\left(\omega L - \frac{1}{\omega C}\right) \tag{5.13}$$

Laplace Domain

The Laplace domain provides another method for analyzing complex circuits. The Laplace transform is defined as

$$\mathcal{L}\{f(t)\} = F(s) = \int_{0^-}^{\infty} f(t)e^{-st}dt \tag{5.14}$$

where

$$s = \sigma + j\omega \tag{5.15}$$

Example Problem 5.3

Find the Laplace transform of a step function, $f(t) = u(t)$.

$$f(t) = 0, t < 0$$

$$f(t) = 1, t \geq 0$$

Solution

$$F(s) = \int_0^\infty 1 e^{st} dt = -\frac{1}{s} e^{-st} \Big|_0^\infty = \frac{1}{s}$$

∎

The Laplace transform can be used to convert functions of time, t, to functions of the Laplace variable, s. Once the transformation has been accomplished, it is possible to solve for unknown variables in the Laplace domain by using algebraic equations. To complete the solution process, the inverse Laplace transform is taken to give the variables in the time domain.

In the Laplace domain (Table 5.1), Ohm's law is

$$V(s) = I(s)\, R \tag{5.16}$$

so

$$Z_R(s) = \frac{V(s)}{I(s)} = R \tag{5.17}$$

Equation (5.8) becomes

$$I(s) = C\, s\, V(s) \tag{5.18}$$

so

$$Z_C(s) = \frac{V(s)}{I(s)} = \frac{1}{sC} \tag{5.19}$$

Equation (5.10) becomes

$$V(s) = L\, s\, I(s) \tag{5.20}$$

TABLE 5.1   Comparison of the Time Domain and the Laplace Domain

	Time domain	Laplace domain
	$v(t)$	$V(s)$
	$i(t)$	$I(s)$
	R	Z_R
	L	Z_L
	C	Z_C
	Differential equations	Algebraic equations

so

$$Z_L(s) = \frac{V(s)}{I(s)} = sL \tag{5.21}$$

KVL and Kirchoff's circuit law (KCL) apply in both the phasor domain and the Laplace domain. These principles are used when analyzing complex circuits. One of the advantages to working in the phasor or Laplace domains is that circuits can be analyzed by means of algebraic manipulations rather than differential equations (Fig. 5.6).

Figure 5.7a shows two elements in series, whereas Fig. 5.7b shows two elements in parallel. The equivalent impedance, $Z_{EQ}(s)$, in the Laplace domain for elements in series is given by

$$Z_{EQ}(s) = \sum_{i=1}^{N} Z_i(s) \tag{5.22}$$

Thus, the equivalent impedance for the elements shown in Fig. 5.7b would be

$$Z_{EQ}(s) = Ls + \frac{1}{sC}$$

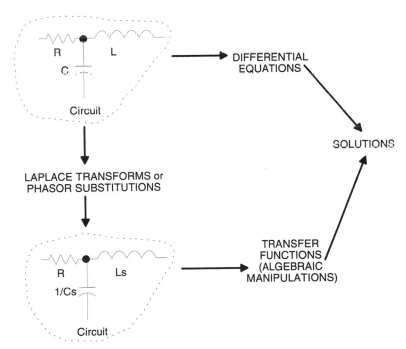

Fig. 5.6 Laplace transforms and phasor substitutions help simplify circuit analysis.

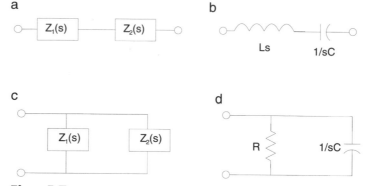

Fig. 5.7 Diagrammatic representation of Laplace domain impedances in series (a and b) and in parallel (c and d). The boxes labeled $Z_x(s)$
represent any basic circuit element.

The equivalent impedance for elements in parallel is given by

$$\frac{1}{Z_{EQ}(s)} = \sum_{i=1}^{N} \frac{1}{Z_i(s)} \tag{5.23}$$

If there are only two elements in parallel as in Fig. 5.7c, this reduces to

$$Z_{EQ}(s) = \frac{Z_1(s)Z_2(s)}{Z_1(s) + Z_2(s)} \tag{5.24}$$

The equivalent impedance for the circuit shown in Fig. 5.7d would be

$$Z_{EQ}(s) = \frac{\dfrac{R}{sC}}{R + \dfrac{1}{sC}}$$

and can be written as

$$Z_{EQ}(s) = \frac{1}{C} \frac{1}{s + \dfrac{1}{RC}}$$

5.3.3 Superposition

In linear circuits with multiple sources, the combination of the sources is the sum of
each of them taken one at a time. This analysis is carried out by removing all the
sources (i.e., setting them equal to zero) except one. After the circuit is analyzed
with the first source, it is set to zero and another source is used. When all the sources
have been analyzed one at a time, the final result is obtained by summing the individual solutions.

Example Problem 5.4

Use superposition to find the current, i_x, in Fig. 5.8a.

Solution

The circuit contains two current sources, 3 A and 5 A. The first loop (Fig. 5.8b) is analyzed by replacing the 5-A source with an open circuit so that no current will flow through that branch. The current, $(i_x)_1$, through the 2 Ω resistor is 3 A. When the 3-A source is replaced with an open circuit so that current will flow through the other branch (Fig. 5.8c), the current, $(i_x)_2$, through the 2 Ω resistor is -5 A. Summing $(i_x)_1$ and $(i_x)_2$ gives $i_x = -2$ A since $i_x = (i_x)_1 + (i_x)_2$. ∎

5.4 SIGNAL CONDITIONING

Analog circuits provide the building blocks for instruments used in biomedical applications. This section provides an overview of analog circuits that are used to condition physiological signals.

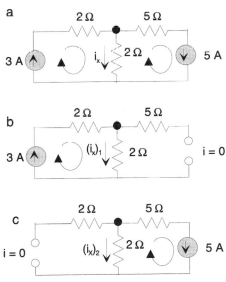

Fig. 5.8 Superposition is used to solve circuits that contain more than one source. (a) Circuit with two current sources; (b) the 5-A source is replaced by an open circuit; (c) the 3-A source is replaced by an open circuit.

5.4.1 Operational Amplifiers

Operational amplifiers, or op-amps, are integrated circuits that contain many components in a relatively small package. An ideal op-amp is shown in Fig. 5.9. A_V represents the **open-loop voltage gain** of the amplifier and is typically around 10^6. The output voltage, v_{out}, of an op-amp is

$$v_{out} = A_V v_{pn} = Av(v_p - v_n)$$ (5.25)

a

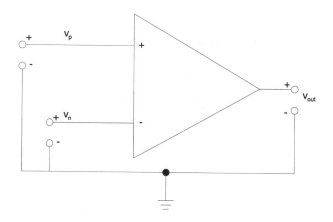

b

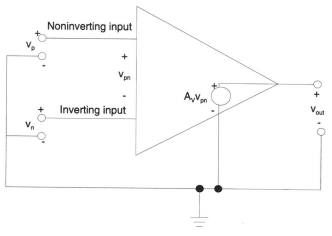

Fig. 5.9 (a) Circuit model of an ideal operational amplifier (op-amp) with two inputs and one output. (b) In the source model, the noninverting input is attached to the positive input of the op-amp, whereas the inverting input is attached to the negative input. A_V is called the open-loop gain of the op-amp. $v_{pn} = v_p - v_n$.

Op-amps are often used in circuits that include feedback loops, i.e., a signal path from the output to one of its input terminals. An **inverting amplifier**, which has a negative feedback loop, is shown in Fig. 5.10. In op-amp circuits, it is often convenient to mark certain nodes (connection points) in the circuit with voltage labels, e.g., v_n. These node voltage labels refer to the voltage value at the point in the circuit with respect to ground — an important point to remember.

Certain rules apply in the analysis of all ideal op-amps: (i) No current flows into the v_p or v_n terminals and (ii) $v_p = v_n$ since the open-loop gain is infinite. In the circuit shown in Fig. 5.10,

$$i_n = 0 \text{ so } i_2 = i_1 \qquad \text{Rule 1}$$

and

$$v_n = v_p = 0 \qquad \text{Rule 2}$$

since the noninverting input is connected to ground. Thus,

$$\frac{v_{in} - v_n}{R_1} = \frac{v_n - v_{out}}{R_2} \Rightarrow \frac{v_{in}}{R_1} = -\frac{v_{out}}{R_2}$$

and

$$v_{out} = v_{in}\left(-\frac{R_2}{R_1}\right)$$

For an inverting amplifier, the gain, G, of the amplifier is given by

$$G = \frac{v_{out}}{v_{in}} = -\frac{R_2}{R_1} \qquad (5.26)$$

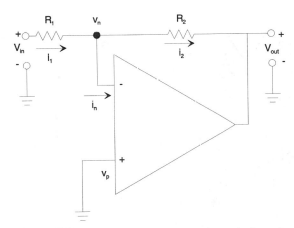

Fig. 5.10 Inverting amplifier. Note that v_n is the voltage from the negative terminal to ground (see Fig. 5.9).

Example Problem 5.5

What would be the output voltage for the inverting amplifier in Fig. 5.10 if $R_2 = 100$ kΩ, $v_{in} = 0.025 \cos(25t)$, and R_1 is replaced by two 20 kΩ resistors in parallel?

Solution

First, reduce the parallel resistors to a single equivalent element:

$$R_{EQ} = \frac{(20 \text{ k}\Omega)(20 \text{ k}\Omega)}{20 \text{ k}\Omega + 20 \text{ k}\Omega} = 10 \text{k}\Omega$$

From Eq. (5.26),

$$G = -\frac{100 \text{ k}\Omega}{10 \text{ k}\Omega} = -10$$

Thus, $v_{out} = -0.25 \cos(25t)$ as is shown in Fig. 5.11. Note that the minus sign results in an output voltage that is 180° (π rad) out of phase with the input voltage since $\cos(-\theta) = -\cos(\theta)$. ■

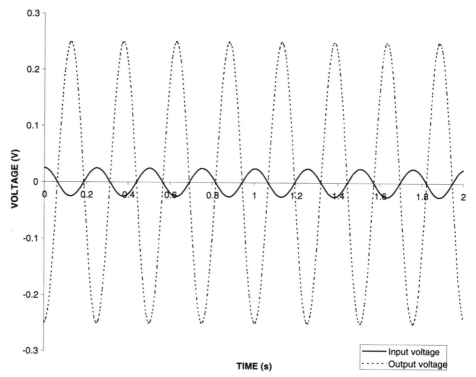

Fig. 5.11　The input and output voltages for Example Problem 5.5.

Another important type of op-amp circuit is called the **unity buffer** or **voltage follower** (Fig. 5.12). In this type of circuit, $G = 1$. Unity buffers drive a current into a load without drawing any current from the input source since $i_{in} = i_p$ and $i_p = 0$ (rule 1). This is a particularly important feature for physiological measurements. In the unity buffer, $v_p = v_{in}$ and $v_n = v_{out}$. Since rule 2 states that $v_p = v_n$, $v_{in} = v_{out}$. Thus, $G = 1$. Note that any current supplied to circuits connected to the buffer comes from within the op-amp.

A unity buffer can be used in conjunction with a voltage divider circuit (Fig. 5.13). Since the op-amp takes no current at its inputs,

$$v_p = v_{in} \frac{R_2}{R_1 + R_2}$$

but $v_p = v_n = v_{out}$ since this is a unity buffer. Thus,

$$v_{out} = v_{in} \frac{R_2}{R_1 + R_2}$$

which gives a gain of

$$G = \frac{R_2}{R_1 + R_2} \tag{5.27}$$

Several additional important op-amp configurations and the relationship between their input and output voltages are shown in Fig. 5.14. Figure 5.14a shows a configuration with a single input voltage, whereas Figs. 5.14b–5.14d show configurations with two input voltages. **Instrumentation amplifiers** (Fig. 5.14d) are often used in biomedical applications because they have an extremely high input impedance, which means that they will draw very little current from the system that is being measured. With instrumentation amplifiers, it is also possible to obtain a high gain with low resistor values.

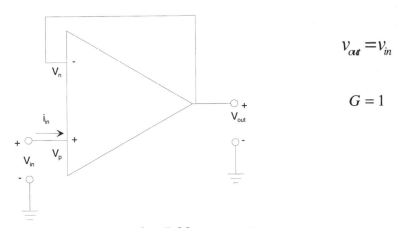

Fig. 5.12 Voltage follower.

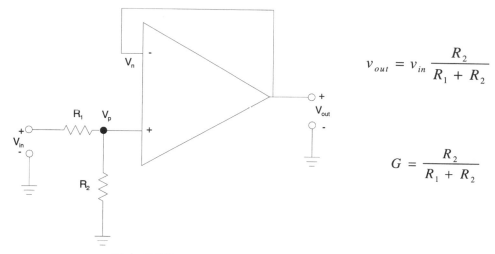

$$v_{out} = v_{in} \frac{R_2}{R_1 + R_2}$$

$$G = \frac{R_2}{R_1 + R_2}$$

Fig. 5.13   Voltage follower in a voltage divider circuit.

Instrumentation amplifiers also reject common-mode signals, i.e., those signals that are most likely due to environmental noise rather than to some aspect of the signal that is being measured. The response of a differential amplifier, such as the one shown in Fig. 5.14d, can be described as having differential-mode and common-mode components, e.g.,

$$v_o(t) = A_{dm}\big(v_p(t) - v_n(t)\big) + A_{cm}\big(v_n(t) + v_p(t)\big) \tag{5.28}$$

Differential amplifiers are designed so that A_{dm} is very large and A_{cm} is very small. A parameter that describes the relationship between A_{dm} and A_{cm} is called the common-mode rejection ratio (CMRR) and may be defined as

$$CMRR = \frac{|A_{dm}|}{A_{cm}} \tag{5.29}$$

or

$$(CMRR)_{dB} = 20\log_{10}\left|\frac{A_{dm}}{A_{cm}}\right| \tag{5.30}$$

5.4.2   Transfer Functions and Complex Impedance

Transfer functions describe the relationship between the input and output of an instrument or system (Fig. 5.15a). Unlike the gain, which indicates the relationship between the amplitude of the input and output, the transfer function also provides information on how the instrument or system affects the frequency content and phase

a   Non-Inverting Amplifier (Gain > 1)

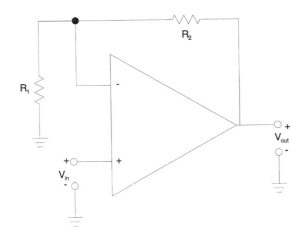

$$v_{out} = \left(\frac{R_1 + R_2}{R_1} \right) v_{in}$$

$$G = \frac{R_1 + R_2}{R_1}$$

b   Inverting Summer

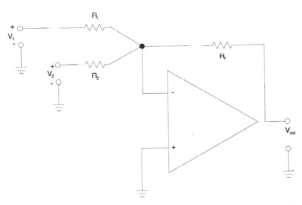

$$v_{out} = \left(\frac{-R_f}{R_1} \right) v_1 + \left(\frac{-R_f}{R_2} \right) v_2$$

Fig. 5.14 Some additional important op-amp configurations. (a) Noninverting amplifier (gain > 1); (b) inverting summer; (c) subtractor; (d) instrumentation amplifier.

C   Subtractor

$$v_{out} = \left(\frac{-R_2}{R_1} \right) v_1 + \left[\left(\frac{R_1 + R_2}{R_1} \right) \left(\frac{R_4}{R_3 + R_4} \right) \right] v_2$$

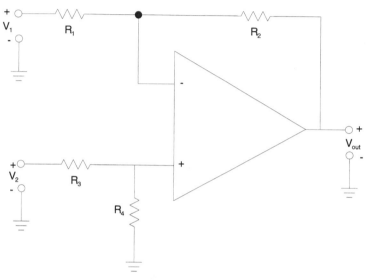

Fig. 5.14 *Continued*

of the output relative to the input. The general equation for the transfer function in the Laplace domain is

$$T(s) = \frac{V_{out}(s)}{V_{in}(s)} \tag{5.31}$$

Figure 5.15b shows an op-amp circuit that has elements with complex impedances, i.e., $Z_i(s)$ and $Z_f(s)$, rather than linear resistance. $I_n(s)$, $V_p(s)$, and $V_n(s)$ all equal zero due to the rules that apply to ideal op-amps. Thus, $I_i(s)$ must equal $I_f(s)$. It follows that

$$\frac{V_{in}(s) - V_p(s)}{Z_i(s)} = \frac{V_p(s) - V_{out}(s)}{Z_f(s)} \tag{5.32}$$

and

$$\frac{V_{in}(s)}{Z_i(s)} = \frac{-V_{out}(s)}{Z_f(s)}$$

d   Instrumentation Amplifier

$$v_{out} - v_{ref} = G(v_2 - v_1)$$

$$G = \frac{R_4}{R_3}\left(1 + 2\frac{R_2}{R_1}\right)$$

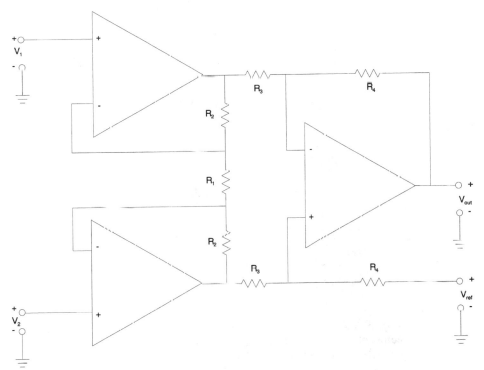

Fig. 5.14 *Continued*

or

$$T(s) = \frac{V_{out}(s)}{V_{in}(s)} = \frac{-Z_f(s)}{Z_i(s)} \tag{5.33}$$

Example Problem 5.6

What is the transfer function for the circuit shown in Fig. 5.16?

a

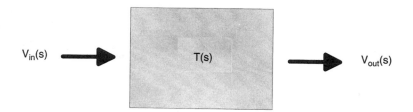

b

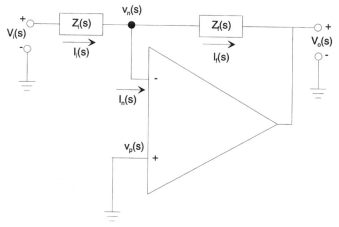

Fig. 5.15   (a) The transfer function expresses the relationship between the input and the output of a system. (b) An op-amp circuit that has elements with complex impedances, i.e., $Z_i(s)$ and $Z_f(s)$, rather than linear resistance.

Solution

$Z_i(s)$ has two series elements (Eq. 5.22) so

$$Z_i(s) = R_1 + \frac{1}{sC_1} = \frac{R_1C_1 + 1}{sC_1}$$

$Z_f(s)$ has two parallel elements (Eq. 5.24) so

$$Z_f(s) = \frac{R_2\left(\dfrac{1}{sC_2}\right)}{R_2 + \dfrac{1}{sC_2}} = \frac{R_2}{R_2C_2s + 1}$$

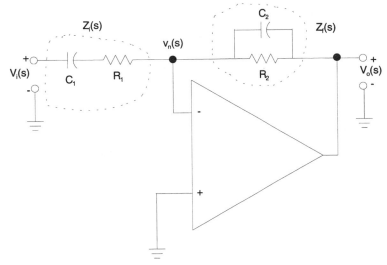

Fig. 5.16 Example Problem 5.6.

From Eq. (5.32),

$$T(s) = -\frac{\left(\dfrac{R_2}{R_2 C_2 s + 1}\right)}{\left(\dfrac{R_1 C_1 s + 1}{s C_1}\right)}$$

which reduces to

$$T(s) = \frac{1}{R_1 C_2 (s)} \frac{s}{\left(s + \dfrac{1}{R_1 C_1}\right)\left(s + \dfrac{1}{R_2 C_2}\right)}$$ ∎

Transfer functions that operate one after another on a signal are said to be cascaded (Fig. 5.17). The overall transfer function for the instrument or system is the product of the intermediate transfer functions:

$$T(s) = \frac{V_{\text{out}}(s)}{V_{\text{in}}(s)} = T_1(s) T_1(s) \cdots T_n(s) \tag{5.34}$$

A requirement for Eq. (5.34) to be valid is that either (i) the input of a cascaded circuit block must drive only an op-amp (i.e., the input current is zero) or (ii) the output of the previous cascaded circuit block has its output driven by an op-amp, which means that it can supply any needed current without changing its output voltage.

Example Problem 5.7

Find the transfer function for the circuit shown in Fig. 5.18.

Fig. 5.17   Cascaded transfer functions.

Solution

At the interface between the two cascaded blocks, the output of the first block is driven by an op-amp (rule 2), so

$$T(s) = T_1(s) \, T_2(s)$$

The transfer function for the first block can be found from that of the noninverting amplifier shown in Fig. 5.14, whereas the transfer function for the second block is given by Eq. (5.20):

$$T_1(s) = \frac{Z_2(s) + Z_1(s)}{Z_1(s)} = \frac{R_1 + \dfrac{1}{sC_1}}{\dfrac{1}{sC_1}} = R_1 C_1 s + 1$$

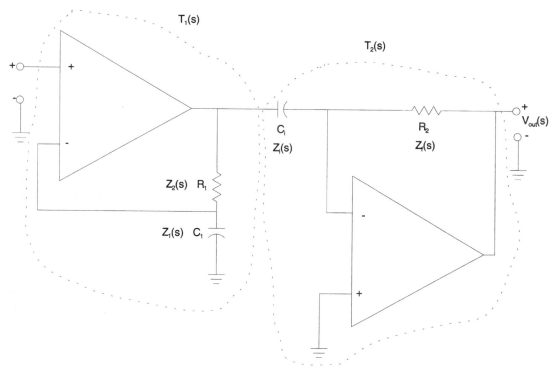

Fig. 5.18   Example of cascaded transfer functions.

$$T_2(s) = \frac{Z_f(s)}{Z_i(s)} = -\frac{R_2}{\dfrac{1}{sC_2}} = -R_2C_2s$$

$$T(s) = T_1(s)T_2(s) = (R_1C_1s + 1)(-R_2C_2s)$$ ■

5.4.3 Filters and Frequency Response

Filters are used to modify the frequency content of input signals. Low-pass filters attenuate the frequency content of a signal that is above a certain cutoff frequency, f_c, whereas high-pass filters attenuate frequencies below the specified cutoff frequency. Band-pass filters are formed by cascading a high-pass and a low-pass filter and attenuate the frequencies above and below the cutoff frequencies of the cascaded filters. Band-stop filters are formed by cascading a low-pass and a high-pass filter and attenuate the frequencies between the cutoff frequencies of the cascaded filters. First-order low-pass and high-pass filters are shown in Fig. 5.19 and frequency responses of the four filter types are shown in Fig. 5.20.

The frequency response of a filter is found by exciting the circuit with a sinusoid of known magnitude and frequency ($V_{in}\cos \omega t$) and computing or measuring the circuit's output response, $V_{out}\cos(\omega t + \theta)$, as shown in Fig. 5.21a. This results in a magnitude (v_{out}/v_{in}) versus frequency ω plot and a phase θ versus frequency ω plot (Fig. 5.21b). One easy way to calculate the frequency response is to find the transfer function, $T(s)$, and manipulate its form as shown below.

Transfer functions can be factored into poles and zeroes, which are either real or occur in complex conjugate pairs:

$$T_2(s) = K\,\frac{(s + z_1)(s + z_2)\cdots}{(s + p_1)(s + p_2)\cdots} \tag{5.35}$$

where the complex conjugate pairs take the form

$$(s + \alpha + j\beta)(s + \alpha - j\beta) = (s + \alpha)^2 + \beta^2 \tag{5.36}$$

The form of Eq. (5.34) can be changed as follows

$$T(s) = \frac{Kz_1z_2\cdots}{p_1p_2\cdots}\,\frac{\left(1 + \dfrac{s}{z_1}\right)\left(1 + \dfrac{s}{z_2}\right)\cdots}{\left(1 + \dfrac{s}{p_1}\right)\left(1 + \dfrac{s}{p_2}\right)\cdots} \tag{5.37}$$

where K_D is defined as

$$K_d = \frac{Kz_1z_2\cdots}{p_1p_2\cdots} \tag{5.38}$$

Since the complex Laplace variable has the form $s = \sigma + j\omega$, the transfer function becomes the frequency response of the circuit if σ is set equal to zero. Hence,

$$s = j\omega = j2\pi f \tag{5.39}$$

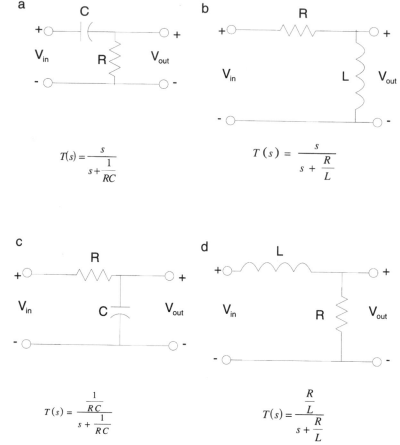

Fig. 5.19 Voltage dividers used to produce filters: (a and b) first-order high-pass filters; (c and d) first-order low-pass filters. These circuits assume that the input is driven by an op-amp and that the output, V_{out}, drives another op-amp ($i_{out} = 0$). The transfer functions are given for each circuit.

where ω is in radians per second and f in cycles per second or Hertz. Thus,

$$T(j\omega) = K_d \frac{\left(1 + \dfrac{j\omega}{z_1}\right)\left(1 + \dfrac{j\omega}{z_2}\right)\cdots}{\left(1 + \dfrac{j\omega}{p_1}\right)\left(1 + \dfrac{j\omega}{p_2}\right)\cdots} \qquad (5.40)$$

$T(j\omega)$ is a complex function,

$$T(j\omega) = |T(j\omega)|e^{j\theta_T(\omega)} \qquad (5.41)$$

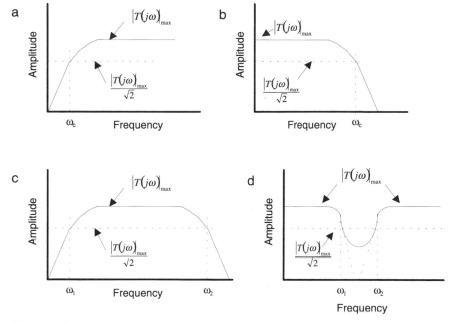

Fig. 5.20 Frequency response plots for (a) high-pass, (b) low-pass, (c) band-pass, and (d) band-stop filters. The band-pass and band-stop filters have two cutoff frequencies, ω_1 and ω_2.

where θ_T is the angle or phase and the magnitude is $|T(j\omega)|$. These variables are related to the frequency response shown in Fig. 5.21:

$$|T(j\omega)| = \frac{V_{out}}{V_{in}} = \text{magnitude response}$$

$$\theta_{T(j\omega)} = \theta = \text{phase response}$$

At $\omega = 0$, the circuit is in the DC state since there are no oscillations and

$$T(j0) = K_d \tag{5.42}$$

Thus, K_d represents the DC gain. The cutoff frequency, f_c, is defined as the frequency at which

$$|T(j\omega_c)| = \frac{|T(j\omega)|_{max}}{\sqrt{2}} = 0.707|T(j\omega)|_{max} \tag{5.43}$$

and

$$\omega_c = 2\pi f_c$$

from Eq. (5.7). Note that the frequency function, $T(j\omega)$, is related to the continuous Fourier transform as discussed in Chapter 6.

a

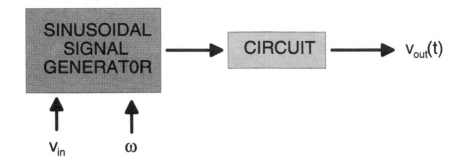

b

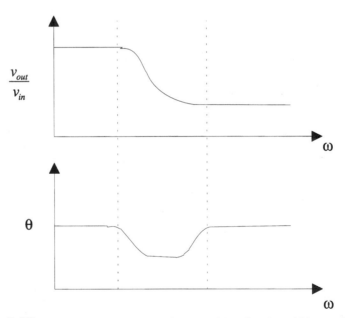

Fig. 5.21   The frequency response of a circuit (a) with a sinusoid input of known magnitude and frequency ($V_{in}\cos \omega t$) can be described in (b) a magnitude (v_{out}/v_{in} versus frequency ω plot and a phase θ versus frequency ω plot.

Example Problem 5.8

Assume that the transfer function for a circuit is given by

$$T(s) = K\frac{s + \alpha}{s + \beta}$$

where $\alpha = 10$, $\beta = 1$, and $K = 0.1$. What is the cutoff frequency for this filter? What type of filter is this? What is its DC gain?

Solution

The answers to this problem can be determined by examining the magnitude of the frequency response, $|T(j\omega)|$. Factoring α from the numerator and β from the denominator gives

$$T(s) = K\frac{\alpha}{\beta}\frac{\left(1 + \dfrac{s}{\alpha}\right)}{\left(1 + \dfrac{s}{\beta}\right)} \tag{5.44}$$

Substituting the given values for K, α, and β and $s = j\omega$ gives

$$T(j\omega) = \frac{\left(1 + \dfrac{j\omega}{10}\right)}{\left(1 + \dfrac{j\omega}{1}\right)}$$

Table 5.2 was produced by solving for ω at different values:

$$|T(j0)| = 1$$

$$|T(j1)| = \frac{(1 + 0.1j)}{(1 + j)} = \frac{1\angle 5.7°}{\sqrt{2}\angle 45°} = 0.707$$

$$|T(j10)| = \frac{(1 + 1j)}{(1 + 10j)} = \frac{\sqrt{2}\angle 45°}{10\angle 84.3°} = 0.141$$

$$|T(j\infty)| = \frac{\left(\dfrac{j\omega}{10}\right)}{\left(\dfrac{j\omega}{1}\right)} = \frac{1}{10} = 0.1$$

TABLE 5.2 Values for $|T(j\omega)|$ at Different Values of ω for the Transfer Function in Example Problem 5.8

ω	$|T(j\omega)|$
0	1 DC gain, $K_d = 1$
1	0.707
10	0.141
∞	0.1

From the table, $|T(j\omega)|_{max} = 1$. The cutoff frequency occurs at $0.707 |T(j\omega)|_{max} = 0.707$. Therefore, $\omega_c = 1$ and $f_c = 1/2\pi$. This is a low-pass filter since the magnitude is highest when ω has low-frequency values. It matches the waveform in Fig. 5.20b. ∎

In the previous example, the cutoff frequency of a first-order low-pass filter was found by using Eq. (5.44). For this equation with $\alpha \geq 10 \beta$,

$$|T(j\omega)|_{max} = |T(j0)| = K_{dc} = K\frac{\alpha}{\beta}$$

and

$$\frac{1}{\sqrt{2}}\left(K\frac{\alpha}{\beta}\right) = \left(K\frac{\alpha}{\beta}\right)\left|\frac{1 + j\frac{\omega_c}{\alpha}}{1 + j\frac{\omega_c}{\beta}}\right|$$

$$\frac{1}{\sqrt{2}}\left|1 + j\frac{\omega_c}{\beta}\right| = \left|1 + j\frac{\omega_c}{\alpha}\right|$$

$$\frac{1}{\sqrt{2}}\sqrt{1 + \frac{\omega_c^2}{\beta^2}} = \sqrt{1 + \frac{\omega_c^2}{\alpha^2}}$$

With $\alpha \geq 10 \beta$, this equation has the approximate solution, $\omega_c = \beta$, so the cutoff frequency can be found by inspection of Eq. (5.44):

$$\frac{1}{\sqrt{2}}\sqrt{1 + \frac{\beta^2}{\beta^2}} = \sqrt{1 + \frac{\beta^2}{\alpha^2}} = 1$$

5.5 INSTRUMENTATION DESIGN

Figure 5.3 illustrated the various functional blocks that are needed in a medical instrumentation system. The purpose of this type of instrument is to monitor the output of a sensor or sensors and to extract information from the signals that are produced by the sensors. Biomedical sensors were described in Chapter 4 and some of the signals that are recorded from biomedical sensors are described in Chapter 6.

This information acquired from a biomedical instrumentation system is used to control or diagnose abnormal conditions in a patient or to obtain data for basic research. General-purpose computer systems are readily available with analog-to-digital (A/D) conversion included as part of the unit. The A/D converter typically requires an input signal in the range of ±5 V. However, the sensor output signals for biological applications can range from a few microvolts to several hundred millivolts (Table 5.3). Thus, a signal conditioning block is needed to amplify the signal. In addition to amplification, the signal conditioning block must also reduce noise and prevent frequency aliasing during the sampling process. Each of these three functions of signal conditioning is examined in more detail in the following sections.

TABLE 5.3 Voltage and Frequency Ranges for Some Important Parameters That Are Measured in the Human Body

Parameter sensor location	Voltage range	Frequency range (Hz)
Electrocardiography (ECG) Skin electrodes	0.5–4 m V	0.01–250
Electroencephalography (EEG) Scalp electrodes	5–200 μV	DC–150
Electrogastrography (EGG)		
Skin–surface electrodes	10–1000 μV	DC–1
Stomach–surface electrodes	0.5–80 mV	DC–1
Electromyography (EMG) Needle electrodes	0.1–5 mV	DC–10,000
Electrooculography (EOG) Contact electrodes	50–3500 μV	DC–50
Electroretinography (ERG) Contact electrodes	0–900 μV	DC–50
Nerve potentials Surface or needle electrodes	0.01–3 mV	DC–10,000

5.5.1 Signal Amplification

Most biological amplifiers contain op-amps. Since a signal gain of up to 10^7 may be required, the final amplifier design often consists of a cascade of amplifier modules, each of which has a gain of 10–10,000. When cascading these amplifier modules, great care must be used to ensure that small DC offset voltages generated by the op-amp circuit stages are removed from the signal path. A DC offset of only 5 μV in a sensor's output would translate into a 50 V signal at the A/D converter if the amplifier gain was set at 10^7! High-pass filters with cutoff frequencies around 1.0 Hz are commonly used at various points in the cascade of amplifier stages to remove any small DC offsets that may be present.

5.5.2 Noise Reduction

There are two ways in which signals can be corrupted in a measurement system. First, interference noise can occur when unwanted signals are introduced into the system by outside sources, e.g., power lines and transmitted radio and television electromagnetic waves. This kind of noise can be reduced by careful attention to the circuit's wiring configuration to minimize coupling effects.

Interference noise is introduced by power lines (50 or 60 Hz), fluorescent lights, AM/FM radio broadcasts, computer clock oscillators, laboratory equipment, cellular phones, etc. Electromagnetic energy radiating from noise sources is injected into the amplifier circuit or into the patient by capacitive and/or inductive coupling. Even the action potentials from nerve conduction in the patient can generate noise at the sensor/amplifier interface. Filters are used to reduce the noise and to maximize the signal-to-noise ratio at the input of the A/D converter.

Low-frequency noise (amplifier DC offsets, sensor drift, temperature fluctuations, etc.) can be eliminated by a high-pass filter with the cutoff frequency set above the noise frequencies and below the biological signal frequencies (Fig. 5.20a). High-frequency noise (nerve conduction, radio broadcasts, computers, cellular phones, etc.) can be reduced by a low-pass filter with the cutoff set below the noise frequencies and above the frequencies of the biological signal that is being monitored (Fig. 5.20b).

Power-line noise can be a very difficult problem in biological monitoring since the 50- or 60-Hz frequency is usually within the frequency range of the biological signal that is being measured. Band-stop filters are commonly used to reduce power-line noise (Fig. 5.20d). The notch frequency in these band-stop filters is set to the power-line frequency of 50 or 60 Hz with the cutoff frequencies located a few Hertz to either side. Figure 5.22 illustrates how an amplifier system can be used to reduce noise.

The second type of corrupting signal is called inherent noise. Inherent noise arises from random processes that are fundamental to the operation of the circuit's elements and, hence, can be reduced by good circuit design practice. While inherent noise can be reduced, it can never be eliminated. Both interference and inherent noise can be modeled in the frequency domain. Low-pass filters can be used to reduce high-frequency components. However, noise signals within the frequency range of the biosignal being amplified cannot be eliminated by this filtering approach. The performance of the amplifier circuit within the frequency range of interest can be characterized by the signal-to-noise ratio (SNR). For an amplifier output $v_o(t)$

$$v_o(t) = v_s(t) + v_n(t) \tag{5.45}$$

where $v_s(t)$ is the desired signal and $v_n(t)$ is the corrupting noise. Then,

$$\text{SNR} = 20\log_{10}\left[\frac{V_{s(\text{rms})}}{V_{n(\text{rms})}}\right] \tag{5.46}$$

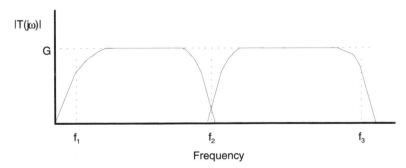

Fig. 5.22 Frequency response of an amplifier system. f_1 is typically approximately 1.0 Hz, f_2 is the power-line frequency, and f_3 is the high-frequency noise cutoff frequency. G is the passband amplifier gain.

where

$$V_{i(\text{rms})} = \left[\frac{1}{T}\int v_i^2(t)dt\right]^{\frac{1}{2}} \tag{5.47}$$

5.5.3 Frequency Aliasing

According to the Nyquist theorem, which is described more fully in Chapter 6, signals must be sampled at more than twice the highest frequency of interest (f_{max}) in order to adequately represent the original signal with interpolation between sample points. For example, the frequency range of interest in electrocardiograms is from very low frequencies to around 100 Hz, with much of the power of the signal occurring around 20 Hz. To adequately sample the higher frequency components of the ECG, the sampling rate, f_s, for the A/D converter must be at least 200 Hz, i.e., 200 samples/s. The sampling rate for intracellular recordings is much higher due to the fast upstroke of the action potential. Systems that are designed to record from intracellular cardiac electrodes often have sampling rates higher than 20 kHz. In practice, sampling rates are generally selected that are 5–10 times the highest frequency content of the signal.

Shannon's sampling theorem, which provides the basis for the Nyquist theorem, also indicates that any frequency in the signal that is being sampled which is greater than 0.5 f_s, i.e., $f_{\text{max}} > 0.5\ f_s$, will be reflected down into the biosignal frequency range and will corrupt the recorded signal's information content. This problem can be eliminated by using a low-pass filter with $f_c < f_{\text{max}}$ to limit the frequency content of the signal that is being sampled. This process is illustrated by a specific example in Fig. 5.23, which expands the concepts presented in Fig. 5.22. For this example, the range for gain can be 10 to 10^7. With the antialiasing filter cutoff set to 1500 Hz (twice the highest frequency of interest in this particular biosignal), frequencies above 2 kHz can be neglected. Thus, a sampling frequency of 4 kHz will

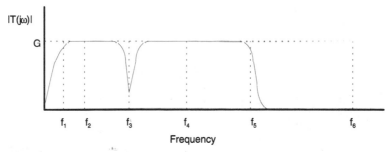

Fig. 5.23 Amplifier with antialiasing filter for a cardiac application. f_1, low-frequency cutoff; f_2, lower limit of biosignal frequency range; f_3, power-line frequency; f_4, upper limit of biosignal frequency range; f_5, antialiasing filter cutoff and high-frequency noise reduction; f_6, A/D converter sampling frequency; G, passband amplifier gain.

produce very clean signals in the digital computer. The antialiasing filter serves the additional function of reducing high-frequency noise.

5.5.4 Design of a Differential Biopotential Amplifier System

The design for a high-gain analog signal conditioning circuit used to take differential biopotential measurements is presented in this section. The system is constructed by cascading several op-amp circuits to produce the desired transfer function. The input signals for this system range from 0.5 μV to 500 mV and must be amplified to a range of ±5 V to be in the appropriate range for the A/D converter which follows the signal conditioning block.

The positive input to the instrumentation amplifier (V_2 in Fig. 5.14d) is generally an electrode that is placed on the tissue that is being monitored, whereas the negative input electrode (V_1 in Fig. 5.14d) can be placed on adjacent tissue that is also actively changing its electrical characteristics or on inactive tissue. When ECGs are recorded from the body's surface, both electrodes respond to the changing electrical currents generated by the heart. When unipolar electrograms are recorded from the ventricular surface of the heart, the positive electrode is placed on ventricular muscle tissue while the negative, or reference, electrode is placed on nonactivating tissue, e.g., the aorta.

The first step in designing the signal conditioning circuit involves determining the minimum and maximum gains that are needed:

$$\text{Largest gain: } \frac{5\text{ V}}{0.5\text{ μV}} = 10^7$$

$$\text{Smallest gain: } \frac{5\text{ V}}{5 \times 10^{-3}\text{V}} = 10$$

An instrumentation amplifier is used to provide gains of 1, 10, 100, 1000, and 10,000. An output stage is added to extend this by a factor of 1000 to give 10^7.

From Fig. 5.14d, the gain equation for an instrumentation amplifier is given by

$$G = \frac{v_{\text{out}} - v_{\text{ref}}}{v_2 - v_1} = \frac{R_4}{R_3}\left(1 + 2\frac{R_2}{R_1}\right) \tag{5.48}$$

The Burr–Brown INA118 model instrumentation amplifier is a good choice for this design and has the following specifications:

$$R_1 = R_G \text{ (the gain adjusting resistor)}$$

$$R_2 = 25\text{ k}\Omega$$

$$R_3 = R_4 = 60\text{ k}\Omega$$

Thus,

$$G = 1 + \frac{50{,}000}{R_G}$$

Figure 5.24a illustrates the use of the INA118 with a second fixed amplifier stage to achieve the design specifications. With this arrangement, Table 5.4 gives the gain settings achieved with different values of R_G.

Figure 5.24b gives a more detailed view of the amplifier blocks. Unity gain buffers are used to ensure that the amplifier draws no current from the sensor electrodes. Next, a high-pass filter is used to remove any DC offset before the signal is amplified by the instrumentation amplifier stage. The output of the instrumentation amplifier is fed back into v_{ref} by an integrator. This serves as an additional high-pass filter stage to remove DC offset in the instrumentation amplifier output. The final two blocks are a low-pass filter for high-frequency noise reduction and antialiasing and a high-pass filter for low-frequency noise and DC offset removal. Each of these blocks is discussed in more detail in the following sections.

Unity Buffer and High-Pass Filter

The details of the unity buffer and high-pass filter are shown in Fig. 5.25. Note that the voltages v_A and v_B are measured with respect to ground. KCL is used to determine the gain of the cascaded circuit. From the rules for ideal op-amps, $v_B = v_{out}$.

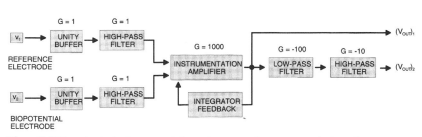

Fig. 5.24 (a) Block diagram of the gain stages for a system designed to measure the difference between a biopotential electrode and a reference electrode, (b).

TABLE 5.4 Gain Settings for the Burr–Brown INA118 Model Instrumentation Amplifier with Different Values of R_G

	Gain	
R_G	$(V_{out})_1$	$(V_{out})_2$
∞	1	10^3
5 kΩ	10	10^4
500 Ω	100	10^5
50 Ω	10^3	10^6
5 Ω	10^4	10^7

KCL at v_A :

$$i_1 + i_3 = i_2$$

$$\frac{V_{in}(s) - V_A(s)}{\dfrac{1}{sC_1}} + \frac{V_{out}(s) - V_A(s)}{R_2} = \frac{V_A(s) - V_{out}(s)}{\dfrac{1}{sC_2}} \qquad (5.49)$$

$$(R_2 C_1 s + R_2 C_2 s + 1) V_A(s) = (R_2 C_1 s) V_{in}(s) + (R_2 C_2 s + 1) V_{out}(s)$$

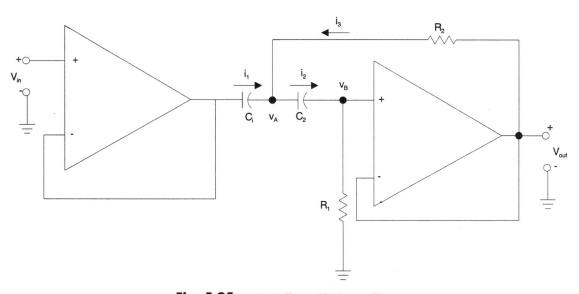

Fig. 5.25 Unity buffer and high-pass filter.

KCL at v_B:

$$i_2 = i_4$$

$$\frac{V_A(s) - V_{out}(s)}{\dfrac{1}{sC_2}} = \frac{V_{out}(s)}{R_1} \tag{5.50}$$

$$R_1C_2sV_A(s) = (R_1C_2s + 1)V_{out}(s)$$

Substituting Eq. (5.50) into Eq. (5.49) gives

$$(R_2C_1s + R_2C_2s + 1)\frac{(R_1C_2s + 1)}{R_1C_2s}V_{out}(s) = (R_2C_2s)V_{in}(s) + (R_2C_2s + 1)V_{out}(s)$$

which can be solved to give

$$\frac{V_{out}(s)}{V_{in}(s)} = \frac{s^2}{\dfrac{1}{R_1C_1R_2C_2}(R_1C_1R_2C_2s^2 + (R_2C_1 + R_2C_2)s + 1)} \tag{5.51}$$

$$\frac{V_{out}(s)}{V_{in}(s)} = \frac{s^2}{s^2 + 2\zeta\omega_n s + \omega_n^2}$$

which is a second-order high-pass filter with

$$\omega_n = \frac{1}{\sqrt{R_1C_1R_2C_2}}$$

and

$$\xi = \frac{R_2C_1 + R_2C_2}{2\sqrt{R_1C_1R_2C_2}}$$

Example Problem 5.9

What is the cutoff frequency, f_c, for the filter described by Eq. (5.51) if $C_1 = C_2 = 0.1\ \mu F$ and $R_1 = R_2 = 1\ M\Omega$?

Solution

Substituting the given values for the resistors and capacitors into Eq. (5.51) gives

$$\frac{V_{out}(s)}{V_{in}(s)} = \frac{s^2}{s^2 + \left(\dfrac{0.1 + 0.1}{0.1 \cdot 0.1}\right)s + \dfrac{1}{0.1 \cdot 0.1}}$$

$$\frac{V_{out}(s)}{V_{in}(s)} = \frac{s^2}{(s + 10^2)}$$

$$\frac{V_{out}(s)}{V_{in}(s)} = \frac{s^2}{s^2 + 20s + 100}$$

Thus, $\omega_c = 10$. From Eq. (5.7), $f_c = 1.59$ Hz. ∎

Instrumentation Amplifier with Integrator Feedback

The equation for the gain of an instrumentation amplifier is given in Eq. (5.48). For the circuit shown in Fig. 5.26,

$$\frac{V_{\text{ref}}(s)}{V_{\text{out}}(s)} = -\frac{\left(\dfrac{1}{sC}\right)}{R} = -\frac{1}{sRC}$$

Substituting for $v_{\text{ref}}(s)$ in Eq. (5.48) gives

$$\frac{V_{\text{out}}(s) - \left(-\dfrac{V_{\text{out}}(s)}{sRC}\right)}{V_2(s) - V_1(s)} = G$$

$$V_{\text{out}}(s)\left(1 + \frac{1}{sRC}\right) = G\big(V_2(s) - V_1(s)\big)$$

$$\frac{V_{\text{out}}(s)}{V_2(s) - V_1(s)} = \frac{G}{1 + \dfrac{1}{sRC}} = \frac{GRCs}{RCS + 1}$$

$$\frac{V_{\text{out}}(s)}{V_2(s) - V_1(s)} = \frac{Gs}{s + \dfrac{1}{RC}} \tag{5.52}$$

This is a first-order high-pass filter with cutoff frequency of $1/RC$.

Example Problem 5.10

What is the cutoff frequency, f_c, for the filter described by Eq. (5.52) if $C = 0.1\ \mu\text{F}$ and $R = 1\ \text{M}\Omega$?

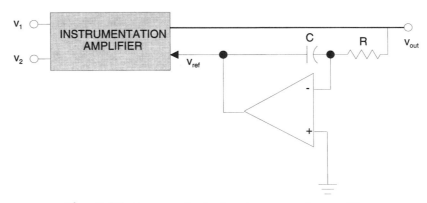

Fig. 5.26   Integrator feedback on instrumentation amplifier.

Solution

Substituting in Eq. (5.52) gives

$$\frac{V_{\text{out}}(s)}{V_2(s) - V_1(s)} = \frac{Gs}{s + 10}$$

Thus, $\omega_c = 10$. From Eq. (5.7), $f_c = 1.59$ Hz. ■

Low-Pass Filter

The low-pass filter in Fig. 5.27 is similar to the op-amp circuit that was shown in Fig. 5.15B. In Fig. 5.27, the feedback impedance consists of a resistor in parallel with a capacitor. $Z_f(s)$ can be determined from Eq. (5.23), whereas $Z_i(s)$ is equal to R_1:

$$Z_f(s) = \frac{R_2\left(\dfrac{1}{sC}\right)}{R_2 + \dfrac{1}{sC}} = \frac{R_2}{R_2 Cs + 1} = \frac{\dfrac{1}{C}}{s + \dfrac{1}{R_2 C}}$$

The transfer function can be found from Eq. (5.33):

$$\frac{V_{\text{out}}(s)}{V_{\text{in}}(s)} = \frac{Z_f(s)}{Z_i(s)} = -\frac{1}{R_1 C}\left(\frac{1}{s + \dfrac{1}{R_2 C}}\right) \tag{5.53}$$

Example Problem 5.11

What would be the cutoff frequency for the circuit shown in Fig. 5.27 if $R_1 = 1$ kΩ, $R_2 = 100$ kΩ, and $C = 1$ nF?

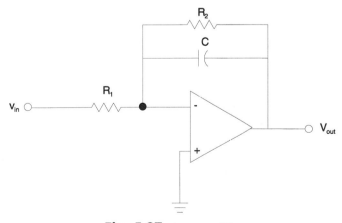

Fig. 5.27 Low-pass filter.

Solution

From the given values, $R_1C = 10^{-6}$ and $R_2C = 10^{-4}$. Using Eq. (5.53),

$$\frac{V_{out}(s)}{V_{in}(s)} = -\frac{10^6}{s + 10^4}$$

$$K_D = \text{DC gain} = -\frac{10^6}{10^4} = -100$$

$$fc = \frac{10^4}{2\pi} = 1592 \text{ Hz}$$

■

Final High-Pass Filter Stage

The final high-pass circuit is shown in Fig. 5.28. $Z_f(s)$ is equal to R_2. The input consists of elements in series so Eq. (5.22) applies:

$$Z_i(s) = R_1 + \frac{1}{sC} = \frac{R_1Cs + 1}{sC}$$

The transfer function can be found from Eq. (5.33)

$$\frac{V_{out}(s)}{V_{in}(s)} = -\frac{R_2}{\left(\dfrac{R_1Cs + 1}{sC}\right)} = -\frac{R_2}{R_1}\left(\frac{s}{s + \dfrac{1}{R_1C}}\right) \qquad (5.54)$$

Example Problem 5.12

What would be the high-frequency gain and cutoff frequency for the circuit shown in Fig. 5.28 if $R_1 = 100 \text{ k}\Omega$, $R_2 = 1 \text{ M}\Omega$, and $C = 1 \text{ }\mu\text{F}$?

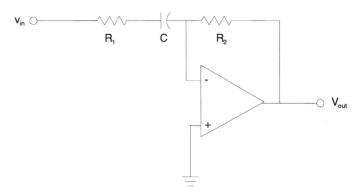

Fig. 5.28 High-pass filter.

Solution

From the given values, $R_1C = 0.1$. Using Eq. (5.54), the transfer function for the circuit in Fig. 5.28 is

$$\frac{V_{out}(s)}{V_{in}(s)} = -\frac{10^6}{10^5}\frac{s}{s + \frac{1}{0.1}} = -\frac{10s}{s + 10} = -\frac{10}{1 + \frac{10}{s}}$$

Evaluating the final term at $s = \infty$ gives the gain at high frequencies. The cutoff frequency for this circuit is

$$G\Big|_{s = \infty} = -\frac{10}{1 + \frac{10}{\infty}} = -10$$

$$\omega_c = \frac{1}{R_1C_1} = 10 \text{ rad} \qquad f_c = \frac{10}{2\pi} = 1.59 \text{ Hz} \qquad \blacksquare$$

Completing the Design

Figure 5.29 shows the final design of the differential amplifier system that incorporates the unity buffer, high-pass filter, instrumentation amplifier with integrator feedback, low-pass filter, and final high-pass filter. The transfer function for each stage can be determined from Example Problems 5.9–5.12. In the overall design, $R = 1$ MΩ, $C = 0.1$ μF, and R_G is selected from Table 5.3 as 50 Ω.

The transfer function for the unity buffer is 1, and the transfer function for the first high-pass filter was determined by using Eq. (5.51). Thus,

$$\frac{V_3(s)}{V_1(s)} = \frac{V_4(s)}{V_2(s)} = \frac{s^2}{s + 10}$$

$$V_3(s) = \frac{s^2}{(s + 10)^2} V_1(s)$$

$$V_4(s) = \frac{s^2}{(s + 10)^2} V_2(s)$$

$$V_4(s) - V_3(s) = \left(\frac{s^2}{(s + 10)^2}\right)(V_2(s) - V_1(s)) \qquad (5.55)$$

With $R_G = 50$ Ω, the gain for the INA118 instrumentation amplifier is 1000. From Example Problem 5.10, the transfer function for the instrumentation amplifier with integrator feedback is

$$\frac{V_5(s)}{V_4(s) - V_3(s)} = \frac{1000s}{s + 10}$$

From Example Problem 5.11, the transfer function for the low-pass filter is

$$\frac{V_6(s)}{V_5(s)} = -\frac{10^6}{s + 10^4}$$

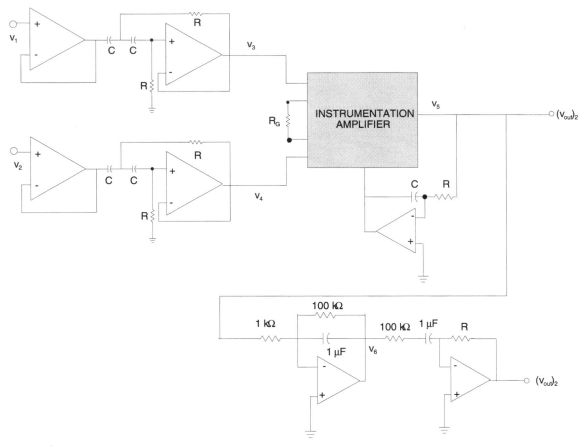

Fig. 5.29 v_1 is the reference electrode, and v_2 is the biopotential electrode. $R = 1\ \text{M}\Omega$, $C = 0.1\ \mu\text{F}$, and $R_G = 50\ \Omega$.

The transfer function for the final high-pass filter was found in Example Problem 5.12:

$$\frac{(V_{\text{out}}(s))_2}{V_6(s)} = -\frac{10s}{s + 10}$$

Using Eq. (5.34), the transfer function for the parts of the circuit between $(v_{\text{out}})_2$, v_4, and v_3 is given by

$$\frac{(V_{\text{out}}(s))_2}{V_4(s) - V_3(s)} = \left(-\frac{10s}{s + 10}\right)\left(-\frac{10^6}{s + 10^4}\right)\left(\frac{1000s}{s + 10}\right) = \frac{10^{10}s^2}{(s + 10)^2(s + 10^4)} \qquad (5.56)$$

Substituting Eq. (5.55) into Eq. (5.56) gives

$$\frac{(V_{\text{out}}(s))_2}{V_3(s) - V_1(s)} = \left(\frac{10^{10}s^2}{(s + 10)^2\ (s + 10^4)}\right)\left(\frac{s^2}{(s + 10)^2}\right) = \frac{10^{10}s^4}{(s + 10)^4(s + 10^4)} \qquad (5.57)$$

This is the transfer function for the total circuit.

This overall circuit operates as a band-pass filter with fourth-order low-frequency cutoff at 1.59 Hz ($\omega = 10$) and first-order high-frequency cutoff at 1592 Hz ($\omega = 10^4$). Equation (5.57) can be used to determine the overall gain for the circuit in the pass band (1.59 Hz $< f <$ 1592 Hz):

$$T(s) = \frac{10^{10}s^4}{(s + 10)^4(s + 10^4)} = \frac{10^2 s^4}{\left(1 + \dfrac{s}{10}\right)^4\left(1 + \dfrac{s}{10^4}\right)} \tag{5.58}$$

Substituting $j\omega$ for s (Eq. 5.39) in Eq. (5.58) gives

$$T(j\omega) = \frac{10^2\omega^4}{\left(1 + \dfrac{j\omega}{10}\right)^4\left(1 + \dfrac{j\omega}{10^4}\right)} \tag{5.59}$$

Selecting a frequency of 159 Hz in the pass band gives $\omega = 1000$ (Eq. 5.7). Substituting this value for ω in Eq. (5.59) gives

$$T(j\omega) = \frac{10^2(10^3)^4}{\left(j\dfrac{10^3}{10}\right)^4} = \frac{10^{14}}{(10^2)^4} = 10^6$$

Thus, the gain in the pass band for this example design is 10^6.

5.5.5 A/D Conversion

A/D converters convert the analog signal received from sensors to a digital form. This is analogous to taking snapshots of the voltage value of the signal at specified time intervals. The time period between snapshots is determined by the sampling rate of the system, whereas the voltage values of the signal depend on the number of bits in the A/D converter and its full-scale voltage range.

Example Problem 5.13

A data acquisition system has a sampling rate of 2 kHz. How often will the acquired voltage value be updated?

Solution

A sampling rate of 2 kHz is equivalent to taking 2000 samples every second. Since the samples are taken at regularly spaced intervals, one sample is taken every 1/2000 s, which is the equivalent of every 0.5 ms. ∎

A/D converters are characterized by the number of bits they produce in the digitized output, typically 8–12. A **bit** is a single piece of information that has the value of 0 or 1. Models with more bits have better resolution, i.e., each bit represents a smaller division of the total voltage range of the A/D converter. The A/D converter's resolution determines the smallest unit of voltage into which an analog signal can be divided. To make effective use of an A/D converter, low-amplitude signals must be amplified to several orders of magnitude larger than the resolution of the A/D converter.

Example Problem 5.14

How would the decimal number 179 be represented in an eight-bit A/D converter?

Solution

A/D converter values are based on powers of 2. The first bit (the least significant bit) in eight bits represents 2^0 (1) and the last bit represents 2^7 (128). Thus, the eight-bit binary value of 179 is 10110011:

Example Problem 5.15

What would be the resolution of a 10-bit A/D converter with an input range of 0–10 V?

Solution

The resolution is the full-scale voltage divided by the total number of possible values minus 1 since the number of intervals into which the voltage range can be divided is one less than the total number of values represented by the number of bits:

$$\frac{10 - 0 \text{ V}}{(2^{10} - 1) \text{ bits}} = 9.78 \frac{\text{mV}}{\text{bit}} \quad \blacksquare$$

Figure 5.30 shows an analog signal that has been sampled by an eight-bit A/D converter with a full-scale range of ± 2.5V. The eight dashed horizontal lines represent the eight bits and the dashed vertical lines represent the times at which samples were taken. The dark lines represent the output from the A/D converter. Note that many of the important characteristics of the waveform (e.g., peaks and notches) are not represented in the output from the A/D converter. The degree to which the output resembles the input waveform can be improved by increasing the sampling rate, which decreases the time interval between samples, and by using an A/D converter with more bits and a smaller full-scale voltage range, which improves the resolution of the A/D converter.

Example Problem 5.16

How much data will be generated if a system with 12-bit A/D converters records for 10 min from 16 electrodes at 200 Hz? Can this amount of information be stored on a 1.44 Mb floppy disk?

Solution

Each electrode will acquire the following amount of data in each second:

$$\left(\frac{12 \text{ bits}}{\text{sample}} \right) \left(\frac{200 \text{ samples}}{\text{s}} \right) = 2400 \frac{\text{bits}}{\text{s}}$$

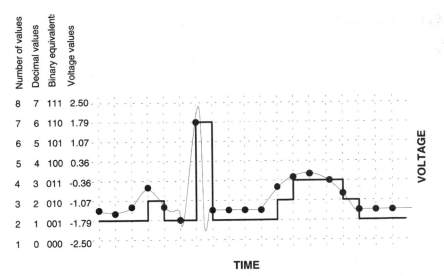

Fig. 5.30 A three-bit ±2.5 V A/D converter has eight (2³) binary values and divides the full-scale voltage into seven equal intervals. The resolution of this A/D converter would be 0.714 V/interval. The dashed horizontal lines represent the transition from one binary value to the next in the A/D converter. The dashed vertical lines represent the times at which samples were taken. The solid circles represent the actual voltage value of the waveform at each sample time. The solid dark line represents the output from the A/D converter.

During the sampling time of 10 min, all 16 electrodes will acquire the following amount of data:

$$\left(16 \text{ electrodes}\right)\left(10 \text{ min}\right)\left(\frac{60 \text{ s}}{\text{min}}\right)\left(\frac{2400 \text{ bits}}{\text{s} \times \text{electrode}}\right) = 23{,}040{,}000 \text{ bits}$$

This is the equivalent of 2.88 kb of data and could easily be stored on a 1.44 Mb floppy disk. However, many research studies in cardiac electrophysiology acquire data from more than 2000 electrodes at many different times during a multihour experiment. In addition, sampling rates of 1000–2000 samples/s are used to acquire data from the heart's surface. The combination of more electrodes, faster sampling rates, and frequent sampling during long studies can result in large increases in data storage requirements. ∎

5.6 COMPUTER-BASED INSTRUMENTATION SYSTEMS

Computers are found in many different instrument systems. They are used for real-time data acquisition systems, intelligent patient monitoring, automated specimen analysis, knowledge-based and expert diagnosis systems, digital signal processing systems, intelligent prosthetic devices, and intelligent implantable devices.

5.6.1   Computers and Programming Languages

Computers consist of three basic units: the central processing unit (CPU), the arithmetic and logic unit (ALU), and memory (Fig. 5.31). The CPU directs the functioning of all other units and controls the flow of information among the units during processing procedures. It is controlled by program instructions. The ALU performs all arithmetic calculations (add, subtract, multiply, and divide) as well as logical operations (AND, OR, and NOT) that compare one set of information to another.

Computer memory consists of read-only memory (ROM) and random-access memory (RAM). ROM is permanently programmed into the integrated circuit that forms the basis of the CPU and cannot be changed by the user. RAM stores information temporarily and can be changed by the user. RAM is where user-generated programs, input data, and processed data are stored.

Computers are binary devices that use the presence of an electrical signal to represent 1 and the absence of an electrical pulse to represent 0. The signals are combined in groups of 8 bits, a **byte**, to code information. A **word** is made up of two bytes. Most desktop computers that are available today are 32-bit systems, which means that they can address 4.29×10^9 locations in memory. The first microcomputers were 8-bit devices that could interact with only 256 memory locations.

Programming languages relate instructions and data to a fixed array of binary bits so that the specific arrangement has only one meaning. Letters of the alphabet and other symbols (e.g., punctuation marks) are represented by special codes. The American Standard Code for Information Exchange (ASCII) provides a common

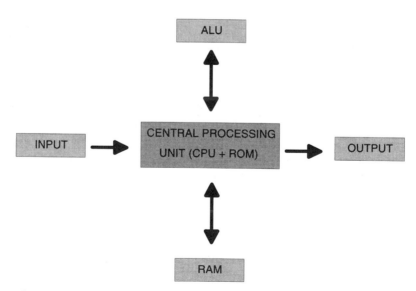

Fig. 5.31   The basic components of a computer are the central processing unit (CPU), the arithmetic and logic unit (ALU), and memory. Memory consists of read-only memory (ROM) and random access memory (RAM).

standard that allows different types of computers to exchange information. When word processing files are saved as text files, they are saved in ASCII format. Ordinarily, word processing files are saved in special program-specific binary formats, but almost all data analysis programs can import and export data in ASCII files.

The lowest level of computer language is machine language which consists of the 0s and 1s that the computer interprets. Machine language represents the natural language of a particular computer. At the next level, assembly languages use English-like abbreviations for binary equivalents. Programs written in assembly language can manipulate memory locations directly. These programs run very quickly and are often used in data acquisition systems that must rapidly acquire a large number of samples, perhaps from an array of sensors, at a very high sampling rate.

Higher level languages (e.g., FORTRAN, PERL, and C^{++}) contain statements that accomplish tasks that require many machine or assembly language statements. Instructions in these languages often resemble English and contain commonly used mathematical notations. Higher level languages are easier to learn than machine and assembly languages. Program instructions are designed to tell computers when and how to use various hardware components to solve specific problems. These instructions must be delivered to the CPU of a computer in the correct sequence in order to give the desired result.

When computers are used to acquire physiological data, programming instructions tell the computer when data acquisition should begin, how often samples should be taken from how many sensors, how long data acquisition should continue, and where the digitized data should be stored. The rate at which a system can acquire samples is dependent on the speed of the computer's clock (e.g., 233 MHz) and the number of computer instructions that must be completed in order to take a sample. Some computers can also control the gain on the input amplifiers so that signals can be adjusted during data acquisition. In other systems, the gain of the input amplifiers must be manually adjusted.

5.7 SUMMARY

In this chapter, a brief discussion of the history of biomedical instruments was followed by a short review of multielement analog circuits. Operational amplifiers were then introduced due to their important contributions to almost every type of bioinstrument. When signals are acquired from biological sources, they must be conditioned. Signal conditioning includes amplification, filtering, noise reduction, and sampling at the correct rate in order to avoid aliasing. A differential biopotential amplifier system for acquiring cardiac signals was designed in order to illustrate the various components of signal conditioning. After an analog signal has been appropriately conditioned, it can be digitized and stored in a computer. Most bioinstrumentation systems that are available today are computer-based systems. Digitized signals can undergo additional processing. Biosignal processing with computers is discussed in Chapter 6.

EXERCISES

1. The voltage, v_{ab} (t), in a Wheatstone bridge circuit (Fig. 5.5) is measured as 0.25 V. Find the sensor resistance, R_x, if $R_1 = R_2 = R_3 = 25$ kΩ and v_s (t) = 10 V.

2. Find the phasor current through the impedance if $Z = 10 + j15$ and $V = 150 + j0$. Give the answer in magnitude/angle notation.

3. Find the impedance at 60 Hz for a capacitor of 2 μF, an inductor of 5 mH, and a resistor of 10 kΩ that are connected in series.

4. Find the impedence at 60 Hz for a capacitor of 2 μF, an inductor of 5 mH, and a resistor of 10 kΩ that are connected in parallel.

5. Find the impedance in the Laplace domain for a capacitor of 10 nF, an inductor of 25 mH, and a resistor of 5 MΩ that are connected in series.

6. Find the impedance in the Laplace domain for a capacitor of 10 nF, an inductor of 25 mH, and a resistor of 5 MΩ that are connected in parallel.

7. Use superposition to find the current i_x, in the 2 Ω resistor in the following circuit:

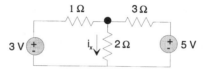

8. Design a circuit that contains inverting amplifiers and will have a gain of 20 with no phase shift between the input and output voltages. Use an input voltage of 0.025 cos(25t). Graph 2 s of input and output data.

9. Derive the gain for a noninverting amplifier by using the rules that apply to ideal op-amps.

10. Find the output voltage in the following circuit if $v_1 = 50$ mV, $v_2 = 175$ mV, and $v_3 = 300$ mV:

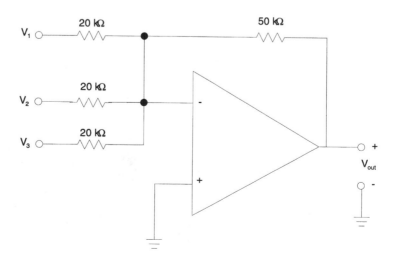

11. Find v_{out} in Fig. 5.14C if $R_1 = R_4 = 50$ kΩ, $R_2 = 100$ kΩ, and $R_3 = 25$ kΩ with $v_1 = 10 \sin(5t)$ and $v_2 = 5 \cos(10t)$. Graph v_1, v_2, and v_{out}. Show at least one complete cycle of each.

12. For the instrumentation amplifier in Fig. 5.14D, select a resistor combination that will satisy the following equation:

$$v_{out} = 500 \ (v_2 - v_1)$$

13. Find the transfer function for the following circuit. Is this filter a high-, band-, or low-pass filter?

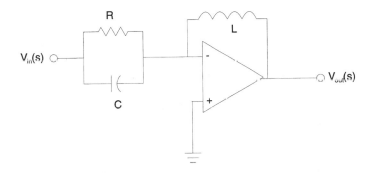

14. Find the transfer function for the following circuit. Is this filter a high-, band-, or low-pass filter?

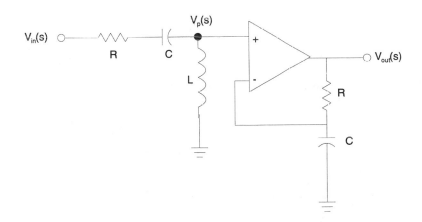

15. Find the transfer function for the following circuit. Is this filter a high-, band-, or low-pass filter?

16. Design a low-pass filter with a gain of 10 and a cutoff frequency of 100 Hz that does not change the phase.

17. Design a differential biopotential amplifier system using an instrumentation amplifier with a passband gain of 500 over the frequency range of 0.05–100 Hz.

18. What would be the change in resolution if an 8-bit A/D converter with a full-scale voltage range of ± 5 V was replaced by a 12-bit A/D converter with a full-scale voltage range of 0–5 V?

19. Consider a data acquisition system with 12-bit accuracy that acquires electrophysiological signals from 1024 electrodes simultaneously. **(a)** If a sampling rate of 2.5 kHz is used to acquire 15 min of data, how much disk storage space (in bytes) will be required? **(b)** Aliasing will occur if frequencies in the original analog signal exceed what value? Show how you determined your answer. **(c)** How could you design the system to prevent aliasing? Be specific.

20. A multichannel data acquisition system with a 10-bit A/D converter stores data in groups of 5 bytes for every four channels. The first byte contains the 8 least significant bits for the first channel with the second through the fourth bytes containing the 8 least significant bits for the second through fourth channels. The fifth byte contains the two most significant beats for all four channels with bits 8 and 7 containing bits 10 and 9 for the first channel, bits 6 and 5 containing bits 10 and 9 for the second channel, bits 4 and 3 containing bits 10 and 9 for the third channel, and bits 2 and 1 containing bits 10 and 9 for the fourth channel. **(a)** What would be the decimal values recorded from each of four channels if the 5 bytes in the A/D converter were 11110001, 10010010, 11111011, 00101011, 10001100? **(b)** Assume that the A/D converter has a full-scale range of 0–10 V. What would be the voltage value that was recorded for each of the four channels? **(c)** If the system had a gain of 10^4, what would be the voltage value at the signal source for each of the four channels?

SUGGESTED READING

Aston, R. (1990). *Principles of Biomedical Instrumentation and Measurement.* Macmillan, New York.

Bronzino, J. D. (1986). *Biomedical Engineering and Instrumentation: Basic Concepts and Applications.* PWS Engineering, Boston.

Bronzino, J. D., Smith, V. H., and Wade, M. L. (1990). *Medical Technology and Society: An Interdisciplinary Perspective.* MIT Press, Cambridge, MA.

Carr, J. J., and Brown, J. M. (1998). *Introduction to Biomedical Equipment Technology,* 3rd ed. Prentice Hall, Upper Saddle River, NJ.

Davis, A. B. (1981). *Medicine and Its Technology.* Greenwood, Westport, CT.

Dempster, J. (1993). *Computer Analysis of Electrophysiological Signals.* Academic Press, (1993), London.

Dubin, D. (1989). *Rapid Interpretation of EKG's — A Programmed Course,* 4th ed. Cover, Tampa, FL.

Goode, P. V., Jr. Cardiac wavefront control using low-energy stimulation. Doctoral dissertation, North Carolina State University, Raleigh.

Horenstein, M. N. (1990). *Microelectronic Circuits and Devices,* Prentice Hall, Upper Saddle River, NJ.

Johns, D. A., and Martin, K. (1997). *Analog Integrated Circuit Design.* Wiley, New York.

Nagel, J. (1995). Biopotential amplifiers. In *The Biomedical Engineering Handbook* (J. D. Bronzino Ed.). CRC Press, Boca Raton, FL.

Poole, G., and Poole, L. (1964). *Electronics in Medicine.* McGraw–Hill, New York.

Rizzoni G. (1993). *Principles and Applications of Electrical Engineering.* Irwin, Homewood, IL.

Rosen, A., and Rosen, H. D. (Eds.), (1995). *New Frontiers in Medical Device Technology.* Wiley New York.

Spencer, D. D. (1997). *The Timetable of Computers.* Camelot, Ormond Beach, FL.

Webster, J. G. (Ed.), (1992). *Medical Instrumentation — Application and Design,* 2nd ed. Houghton Mifflin, Boston.

Welkowitz, W., Deutsch, S., and Akay M. (1992). *Biomedical Instruments — Theory and Design,* 2nd ed. Academic Press, San Diego.

Wise, D. E. (Ed.), (1990). *Bioinstrumentation: Research, Developments, and Applications.* Butterworth, Stoneham, MA.

Wise, D. E. (Ed.), (1991). *Bioinstrumentation and Biosensors.* Dekker, New York.

6 BIOSIGNAL PROCESSING

Chapter Contents

At the end of this chapter, the reader will be able to:

- Describe the different origins and types of biosignals
- Distinguish between deterministic, periodic, transient, and random signals
- Explain the process of analog-to-digital conversion
- Define the sampling theorem
- Describe the main purposes and uses of the wavelet and Fourier transforms
- Define the *Z* transform
- Explain the basic concepts and advantages of fuzzy logic
- Describe the basic concepts of artificial neural networks

6.1 INTRODUCTION

Biosignals are space, time, or space–time records of a biological event such as a beating heart or a contracting muscle. The electrical, chemical, and mechanical activity that occurs during this biological event often produces signals that can be measured and analyzed. Biosignals, therefore, contain information that can be used to explain the underlying physiological mechanisms of a specific biological event or system.

Biosignals can be acquired in a variety of ways, e.g., by a physician who uses a stethoscope to listen to a patient's heart sounds or by the use of highly complex and technologically advanced biomedical instruments. In most cases, it is not sufficient to simply acquire a biosignal. They must be analyzed to retrieve the most relevant information from them. The basic methods of signal analysis, e.g., amplification, filtering, digitization, processing, and storage, can be applied to many biosignals. These techniques are generally accomplished by using digital computers. In addition to these common procedures, different digital methods have been developed for analyzing biosignals. These include signal averaging, wavelet analysis, and artificial intelligence techniques.

6.2 PHYSIOLOGICAL ORIGINS OF BIOSIGNALS

6.2.1 Bioelectric Signals

Nerve and muscle cells generate bioelectric signals that are the result of electrochemical changes within and between cells (see Chapter 3). If a nerve or muscle cell is stimulated by a stimulus that is strong enough to reach a necessary threshold, the cell will generate an action potential. The complete action potential, which represents the flow of ions across the cell membrane, can be measured by using intracellular electrodes. The action potential generated by an excited cell can be transmitted from one cell to adjacent cells. When many cells become excited, an electric field is generated and propagates through the biological medium. Changes in extracellular potential can be measured on the surface of the organ or organism by using surface electrodes. The electrocardiogram (ECG), the electroencephalogram (EEG), and the electromyogram (EMG) are all examples of this phenomenon (Fig. 6.1).

6.2.2 Biomagnetic Signals

Different organs, including the heart, brain, and lungs, generate magnetic fields that are weak compared to other events as electrical changes occur in them. Biomagnetism is the measurement of the magnetic signals that are associated with specific physiological activity. Biomagnetic signals therefore can provide valuable additional information that is not usually contained in bioelectric signals. Furthermore, they can be used to obtain additional information about intracellular activity.

6.2.3 Biochemical Signals

Biochemical signals contain information about the levels and changes in various chemicals in the body. For example, the concentration of various ions, such as calcium and potassium, in cells can be measured and recorded as can the changes in the partial pressures of oxygen (pO_2) and carbon dioxide (pCO_2) in the respiratory system or blood. All these constitute biochemical signals. These biochemical signals can be used for a variety of purposes, such as determining levels of glucose, lactate, and metabolites and providing information about the function of various physiological systems.

6.2.4 Biomechanical Signals

Mechanical functions of biological systems, which include motion, displacement, tension, force, pressure, and flow, also produce biosignals. Blood pressure, for example, is a measurement of the force that blood exerts against the walls of blood vessels. Changes in blood pressure can be recorded as a waveform (Fig. 6.2). The upstrokes in the waveform represent the contraction of the ventricles of the heart as blood is ejected from the heart into the body and blood pressure increases to the systolic pressure, the maximum blood pressure. The downward portion of the waveform depicts ventricular relaxation as the blood pressure drops to the minimum value that is called the diastolic pressure.

6.2.5 Bioacoustic Signals

Bioacoustic signals are a special subset of biomechanical signals that involve vibration (motion). Many biological events produce acoustic noise. For instance, the flow of blood through the valves in the heart has a distinctive sound. Measurements of the bioacoustic signal of a heart valve can be used to help determine whether or not

a

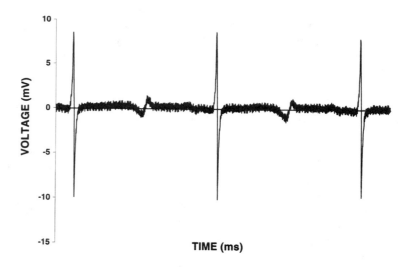

b

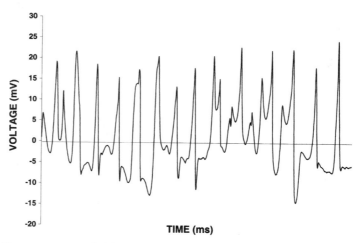

Fig. 6.1   (a) Electrogram recorded from the surface of a pig's heart during normal sinus rhythm. (b) Electrogram recorded from the surface of the same pig's heart during ventricular fibrillation.

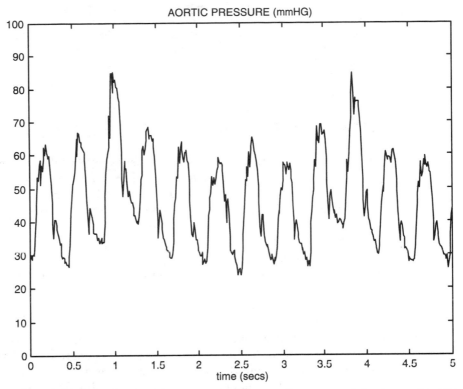

Fig. 6.2 Blood pressure waveform recorded from the aortic arch of a 4-year-old child (sampled at 200 samples/s).

it is operating properly. The respiratory system, joints, and muscles also generate bioacoustic signals that propagate through the biological medium and can often be measured at the skin surface by using acoustic transducers such as microphones and accelerometers.

6.2.6 Biooptical Signals

Biooptical signals are generated by the optical attributes of biological systems. Biooptical signals may occur naturally or, in some cases, the signals may be induced using a biomedical technique. For example, information about the health of a fetus may be obtained by measuring the fluorescence characteristics of the amniotic fluid. Estimations of cardiac output can be made by using the dye dilution method that involves monitoring the concentration of a dye as it recirculates through the bloodstream.

Example Problem 6.1

What type of biosignals would the muscles in your lower legs produce if you were to sprint across a paved street?

Solution

Biomechanical signals would be produced by the motion of the muscles and the forces imposed as your feet hit the pavement. Bioelectric signals would also be produced as the muscles were stimulated by nerves and the muscle cells contracted.   ■

6.3 CHARACTERISTICS OF BIOSIGNALS

There are two general ways to classify biosignals by signal characteristic — continuous and discrete. Continuous signals are defined over a continuum of time or space and are described by continuous variable functions. Discrete signals are defined only at discrete points in time or space and are represented as sequences of numbers. Signals that are produced by biological systems are almost always continuous signals.

Biosignals can also be classified as being either deterministic or random signals. Deterministic signals can be described by mathematical functions or rules. Periodic signals and transient signals are deterministic. Periodic signals are usually composed of the sum of different sine waves or sinusoid components and can be expressed as

$$x(n) = x(n + aT) \tag{6.1}$$

where $x(n)$ is the signal, a is an integer, and T is the period. Periodic signals have a basic waveshape with a duration of T. Transient signals are nonzero or vary over only a finite duration and then decay to a constant value. The sine wave, shown in Fig. 6.3a, is a simple example of a periodic signal. The product of a decaying exponential and a sine wave, as shown in Fig. 6.3b, is a transient signal.

Real biosignals almost always have some unpredictable noise or change in parameters and, therefore, are not deterministic. The ECG of a normal heart rate at rest is an example of a signal that appears to be almost periodic. The basic waveshape consists of the P wave, QRS complex, and T wave and repeats (see Fig. 2.22). However, the precise shapes of the P waves, QRS complexes, and T waves vary over time. The length of time between QRS complexes, which is known as the R–R interval, also changes over time as a result of heart rate variability (HRV). HRV is used as a diagnostic tool to predict the health of a heart that has experienced a heart attack. The extended outlook for patients with low HRV is generally worse than it is for patients with high HRV.

Random signals, also called stochastic signals, are highly correlated only at phase zero. Mathematical functions cannot be used to precisely describe random signals. Random signals often show distribution probabilities and can be expressed in terms of statistical properties. Consequently, statistical techniques are often used for the analysis of random signals. The EMG, an electrical recording of electrical activity in skeletal muscle that is used for the diagnosis of neuromuscular disorders, is a random signal. Stationary random signals are signals for which the statistics or frequency spectra remain the same over time. Conversely, nonstationary random signals have statistical properties or frequency spectra that vary with time. The identification of stationary segments of random signals is important for signal processing and pattern analysis.

a

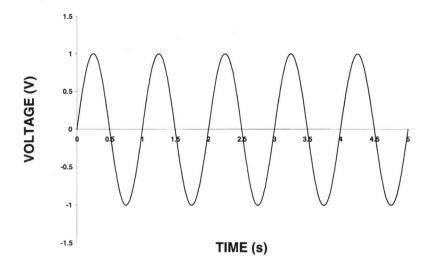

b

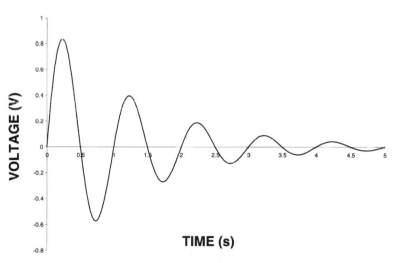

Fig. 6.3 (a) Periodic sine wave signal $x(t) = \sin(\omega t)$ with period of 1 Hz. (b) Transient signal $y(t) = e^{-0.75t}\sin(\omega t)$ for the same 1-Hz sine wave.

Example Problem 6.2

Ventricular fibrillation (VF) is a cardiac arrhythmia in which there are no regular QRS complexes, T waves, or rhythmic contractions of the heart muscle (Fig. 6.1b). VF often leads to sudden cardiac death, which is one of the leading causes of death in the United States. What type of biosignal would most probably be recorded by an ECG when a heart goes into VF?

Solution

An ECG recording of a heart in ventricular fibrillation will be a random, bioelectric signal. ∎

6.4   SIGNAL ACQUISITION

6.4.1   Overview of Biosignal Data Acquisition

Biosignals are often very small, contain unwanted noise, and can even be masked by other biosignals from different biological phenomena. In order to extract the information from a biosignal that may be crucial to understanding a particular biological system or event, sophisticated data acquisition techniques and equipment are commonly used. A diagram of the basic components in a bioinstrumentation system is shown in Fig. 6.4.

Throughout the data acquisition procedure, it is critical that the information in the original biosignal be preserved. Since these signals are often used to aid the diagnosis of pathological disorders, the procedures of amplification, analog filtering and analog-to-digital (A/D) conversion should not cause misleading or impercepti-

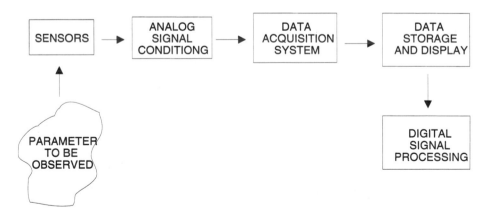

Fig. 6.4 Sensors adapt the signal that is being observed into an electrical analog signal that can be measured with a data acquisition system. The data acquisition system converts the analog signal into a calibrated digital signal that can be stored. Digital signal processing techniques are applied to the stored signal to reduce noise and extract additional information that can improve understanding of the physiological meaning of the original parameter.

ble distortions in the biosignal. Distortions in the biosignal could lead to an improper diagnosis.

6.4.2 Sensors, Amplifiers, and Analog Filters

Signals are first detected in the biological medium, such as a cell or on the skin's surface, by using a sensor. A sensor converts a physical measurand into an electric output and provides an interface between biological systems and electrical recording instruments. The type of biosignal determines what type of sensor will be used. ECGs, for example, are detected by electrodes that have a silver/silver chloride interface that converts the original signal, which is created by the movement of ions, into an electrical signal. Arterial blood pressure is measured by a sensor that detects changes in pressure. It is very important that the sensor used to detect the biosignal does not affect the properties and characteristics of the signal it is measuring.

After the biosignal has been detected by using a sensor, it is usually amplified and filtered. Operational amplifiers are electronic circuits that are used primarily to increase the amplitude of biosignals. An analog filter may then be used to remove noise or to compensate for distortions caused by the sensor. Amplification and filtering of the biosignal may also be necessary in order to meet the hardware specifications of the data acquisition system. Continuous signals may need to be limited to a certain band of frequencies before the signal can be digitized by using A/D conversion.

6.4.3 A/D Conversion

An A/D converter is used to change the biosignal from a continuous analog waveform to a digital signal. An A/D converter is a computer-controlled voltmeter, which measures an input analog signal and gives a numeric representation of the signal as its output. Figure 6.5a shows an analog signal and Fig. 6.5b shows a digital version of the same signal. The analog waveform, originally detected by the sensor and subsequently amplified and filtered, is a continuous signal. The A/D converter transforms the continuous analog signal into a discrete digital signal. The discrete signal consists of a sequence of numbers that can easily be stored and processed on a digital computer. A/D conversion is particularly important because, due to advances in computer technology, the storage and analysis of biosignals is becoming increasingly computer based.

The digital conversion of an analog biosignal does not produce an exact copy of the original signal. The discrete digital signal is a digital approximation of the original analog signal that is generated by repeatedly sampling the amplitude level of the original signal at fixed time intervals. As a result, the original analog signal is represented as a sequence of numbers — the digital signal.

A/D converters are characterized by the number of bits they use to generate the numbers in the digital signal, and their resolution increases as the number of bits increases. A 16-bit A/D converter will have better resolution than an 8-bit A/D converter. The resolution of an A/D converter is determined by the voltage range of the input analog signal divided by the numeric range of the A/D converter. The numeric range is calculated as $2^{(\text{number of bits})}$ minus 1.

a

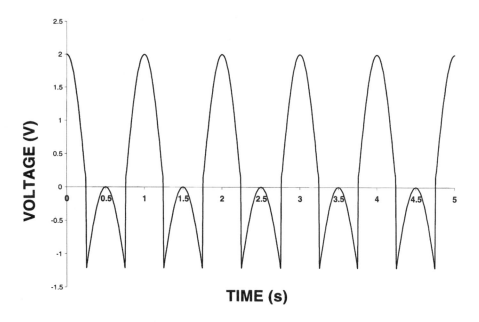

b

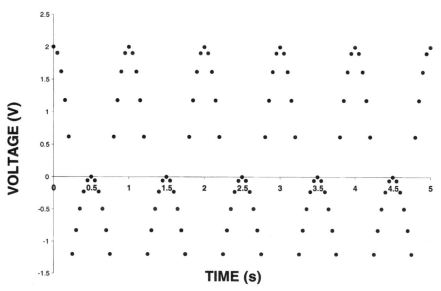

Fig. 6.5 (a) Analog version of a periodic signal; (b) digital version of the analog signal.

Example Problem 6.3

Find the resolution of an eight-bit A/D converter when an input signal with a 10 V range is digitized.

Solution

$$\frac{\text{input voltage range}}{2^{(\text{number of bits})} - 1} = \frac{10 \text{ V}}{255} = 0.0392 \text{ V/bit} = 39.2 \text{ mV/bit} \quad \blacksquare$$

The two main processes involved in A/D conversion are sampling and quantization. If $x(t)$ is an analog signal, then sampling involves recording the amplitude value of $x(t)$ every T seconds. The amplitude value is denoted as $x(kT)$, where k denotes the position or number of the sample in the sample set or data sequence. Finite data sequences are generally used in digital signal processing. Therefore, the range of a data sequence is $k = 0, 1, \cdots N - 1$, where N is the total number of samples in the data sequence. T is the sampling interval or the time between the samples. The sampling frequency, or the sampling rate, is equal to $1/T$ Hz.

Sampling an analog signal must be accomplished so that the digitized signal provides an accurate representation of the original analog signal. Thus, the sampling rate is critical for the generation of an accurate digital signal. If the sampling rate is too low, distortions will occur in the digital signal. Nyquist's theorem states that the minimum sampling rate should be twice the highest frequency in the original signal. The Nyquist rate can be calculated as

$$f_{\text{nyquist}} = 2 \times f_{\text{max}} \quad (6.2)$$

where f_{max} is the highest frequency present in the analog signal.

Example Problem 6.4

The frequency content of an analog EEG signal is 0.5–100 Hz. What is the lowest rate at which the signal can be sampled to produce an accurate digital signal?

Solution

Highest frequency in analog signal = 100 Hz

$$f_{\text{nyquist}} = 2 \times f_{\text{max}} = 2 \times 100 \text{ Hz} = 200 \text{ samples/s} \quad \blacksquare$$

Another problem often encountered is determining what happens if a signal is not sampled at a rate high enough to produce an accurate signal. One form of the sampling theorem is the following statement: All frequencies of the form $[f - kf_s]$, where $-\infty \leq k \leq \infty$ and $f_s = 1/T$, look the same once they are sampled.

Example Problem 6.5

A 360-Hz signal is sampled at 200 samples/s. What frequency will the "aliased" digital signal look like?

Solution

According to the formula, $f_s = 200$, and the pertinent set of frequencies that look alike is in the form of $[360 - k200] = [\cdots 360 \; 160 \; -40 \; -240 \cdots]$. The only signal in this group that will be accurately sampled is the signal at 40 Hz since the sampling rate is more than twice this value. -40 Hz and 40 Hz look alike for real signals, i.e., $\cos(-\omega t) = \cos(\omega t)$ and $\sin(-\omega t) = -\sin(\omega t)$. Thus, the sampled signal will look the same as if it had been a 40-Hz signal. The process is illustrated in Fig. 6.6.   ■

Since digitized samples are usually stored and analyzed on computers, which operate on binary numbers, every sample generated by the sampling process must be quantized. Quantization is the process by which the series of samples is transformed into binary numbers that are limited in size by the number of bits available in the A/D converter. The samples must be approximated in order to fit into the 8, 12, 16, or more bits of the A/D converter. If the number of bits is not large enough, the approximation of the sample to fit in the required number of bits may introduce quantization errors. Quantization errors occur when the approximated binary number representation of the sample is significantly different from the original value of the sample.

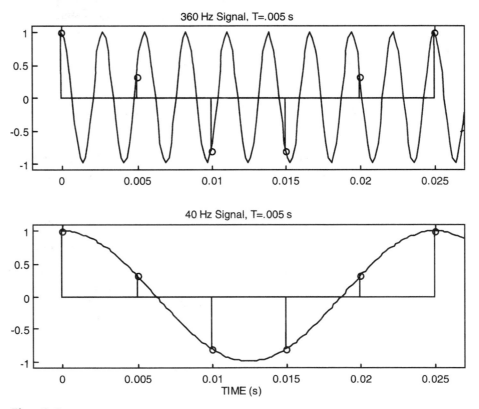

Fig. 6.6   A 360-Hz sine wave is sampled every 5 ms, i.e., at 200 samples/s. This sampling rate will adequately sample a 40-Hz sine wave but not a 360-Hz sine wave.

6.5 FREQUENCY DOMAIN REPRESENTATION OF BIOSIGNALS

Biosignals can be represented in both the time domain and the frequency domain. The graphs of the signals shown thus far in this chapter have been in the time domain. The Fourier transform is a basic operation that is used to transform signals from the time domain into the frequency domain. Frequency components of signals, which may be difficult to discern in the time domain, can often be separated and analyzed more easily in the frequency domain. For many different biosignal processing methods and applications, including filtering and spectral analysis, it is useful to have the frequency domain representations of the biosignals. Filtering operations can, in some cases, be applied more efficiently in the frequency domain, and spectral information about biosignals can be obtained from frequency representations of biosignals.

The Fourier transform, which changes continuous signals from the time domain into the frequency domain, is given by

$$X(\omega) = \int_{-\infty}^{\infty} x(t)e^{-j\omega t}dt \tag{6.3}$$

Tables of Fourier transforms for many common signals can be found in most digital signal processing textbooks.

Example Problem 6.6

Find the Fourier transform (FT) of the rectangular pulse signal

$$x(t) = 1, |t| < a$$

$$0, |t| > a$$

Solution

Equation (6.3) is used:

$$X(\omega) = \int_{-a}^{a} e^{-j\omega t}dt = \frac{2 \sin \omega a}{\omega}$$ ∎

The inverse Fourier transform (IFT) is expressed as Eq. (6.4) and is used to transform continuous signals from the frequency domain into the time domain:

$$x(t) = \frac{1}{2\pi} \int_{-\infty}^{\infty} X(\omega)e^{j\omega t}d\omega \tag{6.4}$$

Signals that are periodic with the length of the period, T, can be represented by the Fourier series, which has many different forms when the signal is a real signal:

$$x(t) = \sum_{m=-\infty}^{+\infty} c_m e^{jk\omega_o t}, \text{ where } \omega_o = \frac{2\pi}{T}$$

or

$$x(t) = \frac{A_0}{2} + \sum_{m=1}^{\infty} \left(a_m \cos m\omega_o t + b_m \sin m\omega_o t\right)$$

or

$$x(t) = \frac{A_0}{2} + \sum_{m=1}^{\infty} \left(A_m \cos(m\omega_o t + \phi_m) \right)$$

where

$$c_m = \frac{1}{T} \int_{-\frac{T}{2}}^{\frac{T}{2}} x(t) e^{-jm\omega_o t} dt; \quad a_m = \frac{2}{T} \int_{-\frac{T}{2}}^{\frac{T}{2}} x(t) \cos m\omega_o t dt; \quad b_m = \frac{2}{T} \int_{-\frac{T}{2}}^{\frac{T}{2}} x(t) \sin m\omega_o t dt$$

and the relationship between the coefficients is

$$c_m = \frac{a_m - jb_m}{2} = \frac{A_m}{2} e^{j\phi_m}; \quad A_m = \sqrt{a^2{}_m + b^2{}_m}; \quad \phi_m = \tan^{-1}\left(\frac{-b_m}{a_m}\right)$$

Each value for *m* represents a "harmonic" for the periodic signal. Many periodic biological signals can be represented with very few harmonics. Figures 6.7 and 6.8 illustrate a harmonic reconstruction of an aortic pressure waveform. Figure 6.7 plots

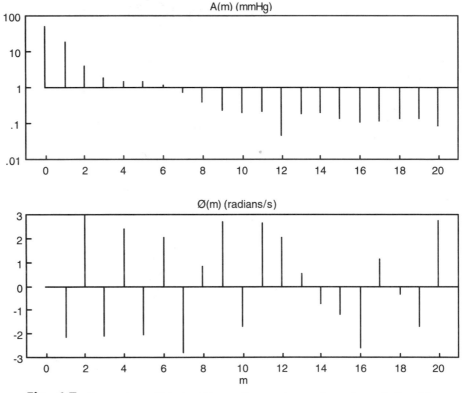

Fig. 6.7 Harmonic coefficients of the aortic pressure waveform shown in Fig. 6.2.

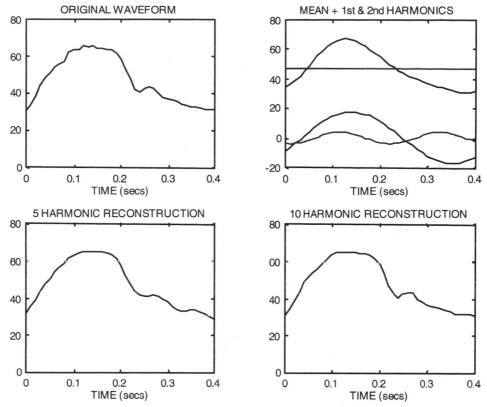

Fig. 6.8 Harmonic reconstruction of the aortic pressure waveform shown in Fig. 6.2.

the coefficients for a cosine series representation. The A_m's are plotted on a $\log_{10}$ scale so the smaller values are visible. Figure 6.8 shows several levels of harmonic reconstruction. The mean plus the first and second harmonics provide the basis for the general systolic and diastolic shape. Additional harmonics add details.

For discrete signals, the discrete Fourier transform (DFT) and inverse discrete Fourier transform are used to move between the time and frequency domains. The DFT is given by

$$X(m) = \sum_{k=0}^{N-1} x(k)e^{-j\frac{2\pi mk}{N}} \; ; m = 0, 1, \cdots, N/2 \qquad (6.5)$$

where N is the number of samples and is assumed to be an even number. The inverse DFT is given by

$$x(k) = \frac{1}{N}\sum_{m=0}^{N-1} X(m)e^{j\frac{2\pi mk}{N}} \; ; k = 0, 1, \cdots, N-1 \qquad (6.6)$$

If $x(k)$ is a real signal, the very nature of the DFT makes $x(k)$ a periodic signal with a period of N if the inverse transform is calculated for k outside the $[0: N - 1]$ range. [One can easily check that $x(k) = x(k + rN)$ in the above signal.] This leads to a discrete Fourier series (DFS) notation that is similar to the FS notation described previously where

$$c_m = \frac{1}{N}X(m); \; x(k) = \sum_{m=0}^{N-1} c_m e^{jkm\Omega_o}; \; where \; \Omega_o = \frac{2\pi}{N}$$

or

$$x(k) = \frac{A_0}{2} + \sum_{m=0}^{\frac{N}{2}-1} \left(a_m \cos m\Omega_o k + b_m \sin m\Omega_o k\right) = \frac{A_{N/2}}{2} \cos \frac{N}{2}\Omega_o k$$

or

$$x(k) = \frac{A_0}{2} + \sum_{m=1}^{\frac{N}{2}-1} A_m \cos\left(m\Omega_o k + \phi_m\right) + \frac{A_{N/2}}{2} \cos \frac{N}{2}\Omega_o k$$

The relationships between the coefficients are the same for the DFS as for the FS.

An efficient algorithm for calculating the DFT is the fast Fourier transform (FFT), which is not another transform but simply an algorithm. The outputs of the FFT and DFT algorithms are the same. However, the FFT has a much faster execution time than the DFT. The ratio of computing time for the FFT and DFT is

$$\frac{\text{FFT computing time}}{\text{DFT computing time}} = \frac{\log_2 N}{2N} \tag{6.7}$$

In order for the FFT to be efficient, the number of data samples, N, must be a power of 2. If N is not a power of 2, another DFT algorithm is usually used.

Figure 6.9 shows a signal and the corresponding DFT, which was calculated using the FFT algorithm. The signal shown in Fig. 6.9a is a sine wave with a frequency of 100 Hz. Figure 6.9b shows the FFT of the 100-Hz sine wave. Notice that the peak of the FFT occurs at 100 Hz. Figure 6.10a shows a 100-Hz sine wave that was corrupted with random noise. The frequency of the signal is not clear in the time domain. However, after transformation into the frequency domain, the signal, shown in Fig. 6.10b, reveals a definite frequency component at 100 Hz, which is marked by the large peak in the FFT.

The FT performs well for signals that have a constant frequency over time, but it does not adequately display the changes in frequency for signals with varying frequencies. An alternative method such as the short-time Fourier transform (STFT) or wavelet transform may provide better results. The STFT and wavelet transform are discussed in Section 6.9.

The FT has several properties that help make calculations involving frequency domain transformations easier. The DFT has very similar properties, but the equations vary slightly since the FT is used for continuous signals and the DFT is used for discrete signals. A few of the properties of the FT are discussed below. Let $x_1(t)$

a

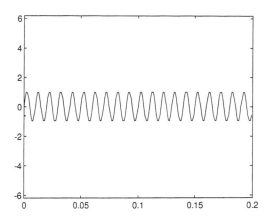

b

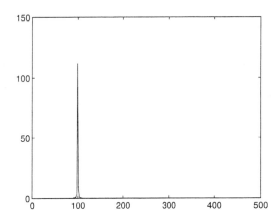

Fig. 6.9 (a) 100-Hz sine wave; (b) fast Fourier transform of 100-Hz sine wave.

and $x_2(t)$ be two signals in the time domain. The FTs of $x_1(t)$ and $x_2(t)$ are represented as $X_1(\omega) = F\{x_1(t)\}$ and $X_2(\omega) = F\{x_2(t)\}$.

Linearity: The FT is a linear operator. Therefore, for any constants a_1 and a_2,

$$F\{a_1x_1(t) + a_2x_2(t)\} = a_1 X_1(\omega) + a_2 X_2(\omega) \qquad (6.8)$$

Time shifting: If $x_1(t - t_0)$ is a signal in the time domain that is shifted in time, the FT can be represented as

$$F\{x_1(t - t_0)\} = e^{-j\omega t_0} X_1(\omega) \qquad (6.9)$$

a

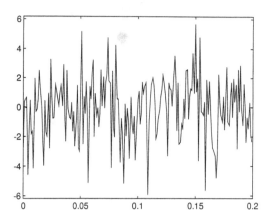

b

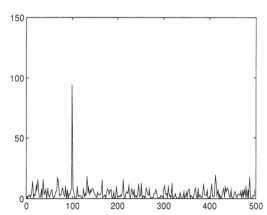

Fig. 6.10 (a) 100-Hz sine wave corrupted with noise; (b) fast Fourier transform of noisy 100-Hz sine wave.

Frequency shifting: If $X_1(\omega - \omega_0)$ is the FT of a signal, shifted in frequency, the IFT is

$$F^{-1}\{X_1(\omega - \omega_0)\} = e^{j\omega_0 t}x(t) \tag{6.10}$$

Convolution theorem: The convolution integral of two signals in the time domain is defined as

$$x(t) = \int_{-\infty}^{\infty} x_1(\tau)x_2(t - \tau)d\tau = x_1(t) * x_2(t) \tag{6.11}$$

The FT of $x(t)$, which is the convolution of two signals in the time domain, is

$$X(\omega) = F\{x(t)\} = F\{x_1(t) * x_2(t)\} = X_1(\omega)X_2(\omega) \tag{6.12}$$

Thus, the convolution in the time domain, which is a relatively complex operation, becomes simple multiplication in the frequency domain.

Next, consider the convolution of two signals, $X_1(\omega)$ and $X_2(\omega)$, in the frequency domain. The convolution integral in the frequency domain is expressed as

$$X(\omega) = \int_{-\infty}^{\infty} X_1(v)x_2(\omega - v)dv = X_1(\omega) * X_2(\omega) \tag{6.13}$$

The IFT of $X(\omega)$ is

$$x(t) = F^{-1}\{X(\omega)\} = F^{-1}\{X_1(\omega) * X_2(\omega)\} = 2\pi\, x_1(t)\, x_2(t) \tag{6.14}$$

Consequently, the convolution of two signals in the frequency domain is 2π times the product of the two signals in the time domain. Convolution is an important property for the filtering of biosignals.

Example Problem 6.7

What is the FT of $3\sin(25t) + 4\cos(50t)$? Express your answer only in a symbolic equation. Do not evaluate the result.

Solution

$$F\{3\sin(25t) + 4\cos(50t)\} = 3F\{\sin(25t)\} + 4F\{\cos(50t)\} \qquad \blacksquare$$

6.6 THE Z TRANSFORM

An alternative method for representing digital signals is the z transform. The z transform is used to transform discrete signals from the time domain into the z domain. For continuous signals, the Laplace transform, which transforms continuous signals into the s domain, has the same role as the z transform. The forward two-sided z transform $X(z)$ of a digital signal, which is used to transform a digital signal into the z domain, is defined as

$$X(z) = \sum_{k=-\infty}^{\infty} x_k z^{-k} \tag{6.15}$$

The one-sided z transform, where x_k is defined only for $k > 0$, is given by

$$X(z) = \sum_{k=0}^{\infty} x_k z^{-k} = x_0 + x_1 z^{-1} + x_2 z^{-2} + \cdots + x_k z^{-k} \tag{6.16}$$

In most practical applications of digital biosignal processing, the signal is a data sequence with N samples. Therefore, the one-sided z transform is used for $k = 0 \cdots N - 1$. Tables of common z transforms and their inverse transforms can be found in most digital signal processing textbooks.

The z transform of most sampled signals can be found quite easily. The data sequence of a sampled signal can be represented as

$$[y(0), y(T), y(2T), \cdots, y(kT)] \tag{6.17}$$

The z transform for the sampled signal given by Eq. (10.16) for the one-sided z transform

$$Y(z) = y(0) + y(T)z^{-1} + y(2T)z^{-2} + \cdots + y(kT)z^{-k} \qquad (6.18)$$

where T is the sampling period or interval.

A sampled signal is a data sequence with each number separated from the next by the sampling period. In the z transform, the value of the multiplier, $y(kT)$, is the value of the data sample. The value of the negative exponent of the z term indicates the number of sampling periods after the start of sampling at which the data sample occurs. The variable z^{-1} represents a time separation of one period, T, between each term in the data sequence of the digital signal. In Eq. (6.18), $y(0)$ is the value of the sampled data at $t = 0$, and $y(T)$ is the value of the sampled data that was sampled after the first sampling period. The z transform is an important method for describing the sampling process of a signal and is widely applied in digital filtering.

The z transform of the unit impulse function will be examined as an initial example. The unit impulse function may be represented as the sequence $[1, 0, 0, 0, \cdots, 0]$. The z transform of this sequence is

$$Y(z) = 1 + 0\,z^{-1} + 0\,z^{-2} + \cdots + 0\,z^{-k} = 1 + 0 + 0 + \cdots + 0 = 1 \qquad (6.19)$$

The unit impulse function and its z transform are often used in the study of filter performance.

Example Problem 6.8

An A/D converter is used to convert a recorded signal of the electrical activity inside a nerve into a digital signal. The first five samples are -60.0, -49.0, -36.0, -23.0, and -14.0 mV. What is the z transform of this data sequence of the biosignal? How many sample periods after the start of the sampling process was the data sample -23.0 recorded?

Solution

$$Y(z) = -60.0 + -49.0\,z^{-1} + -36.0\,z^{-2} + -23.0\,z^{-3} + -14.0\,z^{-4}$$

The value of the negative exponent of the -23.0 mV term is 3. Therefore, the data sample with the value of -23.0 was recorded three sampling periods after the start of sampling. ∎

The following are some of the properties of the z transform which can help simplify obtaining and analyzing z transforms of digital signals. Let $x_1(k)$ and $x_2(k)$ be two digital signals with corresponding z transforms, $X_1(z)$ and $X_2(z)$.

Linearity: The z transform is a linear operator. For any constants a_1 and a_2,

$$Z\{a_1x_1(k) + a_2x_2(k)\} = \sum_{k=0}^{\infty} [a_1x_1(k) + a_2x_2(k)]z^{-k} = a_1X_1(z) + a_2X_2(z) \qquad (6.20)$$

Delay: Let $x_1(k - n)$ be the original signal that is delayed by n. The z transform of the delayed signal is

$$Z\{x_1(k - n)\} = \sum_{k=0}^{\infty} x_1(k - n)z^{-k} = \sum_{k=0}^{\infty} x_1(k)z^{-(k+n)} = z^{-n}X_1(z) \qquad (6.21)$$

Convolution: As with the FT one of the important properties of the z transform is that convolution of two signals in the time domain becomes multiplication in the z domain. Let $x(k)$ be the convolution of $x_1(k)$ and $x_2(k)$; then

$$x(k) = x_1(k) * x_2(k)$$

$X(z)$, the z transform of $x(k)$, can be calculated as

$$X(z) = Z\{x(k)\} = X_1(z)X_2(z) \tag{6.22}$$

6.7 DIGITAL FILTERS

Digital systems are described by digital filters, or difference equations, just like analog systems are described by differential equations. The general form of a real-time digital filter is

$$y(k) = \sum_{m=0}^{M} b_m x(k - m) - \sum_{m=1}^{N} a_m y(k - m) \tag{6.23}$$

For example, if $M = 2$ and $N = 2$, then

$$y(k) = b_0 x(k) + b_1 x(k - 1) + b_2 x(k - 2) - a_1 y(k - 1) - a_2 y(k - 2)$$

where $x(k)$ represents the input sequence and $y(k)$ represents the output sequence. The following are digital sequences of particular importance:

The unit sample of impulse sequence:

$$\delta(k) = 1 \text{ if } k = 0$$
$$0 \text{ if } k \neq 0$$

The unit-step sequence:

$$u(k) = 1 \text{ if } k \geq 0$$
$$0 \text{ if } k < 0$$

The exponential sequence:

$$a^k u(k) = a^k \text{ if } k \geq 0$$
$$0 \text{ if } k < 0$$

Digital systems, like analog systems, are defined by their impulse responses. If the response has a finite number of nonzero points, the filter is called an FIR filter or a finite impulse response filter. If the response has an infinite number of points, the filter is called an IIR or infinite impulse response filter. A positive quality of digital filters is the ease with which the output for any input can be calculated.

Example Problem 6.9

Find the impulse response for the digital filter

$$y(k) = \frac{1}{2}x(k) + \frac{1}{2}y(k - 1)$$

Solution

Assume the system is at rest before input begins, i.e., $y(n) = 0$ for $n \leq 0$:

$$y(-2) = \frac{1}{2}\delta(-2) + \frac{1}{2}y(-3) = 0 + 0 = 0$$

$$y(-1) = \frac{1}{2}\delta(-1) + \frac{1}{2}y(-2) = 0 + 0 = 0$$

$$y(0) = \frac{1}{2}\delta(0) + \frac{1}{2}y(-1) = \frac{1}{2} + 0 = \frac{1}{2}$$

$$y(1) = \frac{1}{2}\delta(1) + \frac{1}{2}y(0) = 0 + \frac{1}{2}\left(\frac{1}{2}\right) = \left(\frac{1}{2}\right)^2$$

$$y(2) = \frac{1}{2}\delta(2) + \frac{1}{2}y(1) = 0 + \frac{1}{2}\left(\frac{1}{2}\right)^2 = \left(\frac{1}{2}\right)^3$$

$$y(3) = \frac{1}{2}\delta(3) + \frac{1}{2}y(2) = 0 + \frac{1}{2}\left(\frac{1}{2}\right)^3 = \left(\frac{1}{2}\right)^4$$

$$\cdots$$

$$y(k) = \left(\frac{1}{2}\right)^{k+1}u(k)$$

The impulse response for the filter is an exponential sequence. This is an IIR filter.

■

Example Problem 6.10

Find the impulse response for the digital filter

$$y(k) = \frac{1}{3}x(k) + \frac{1}{3}x(k-1) + \frac{1}{3}x(k-2)$$

Solution

Assume the system is at rest before input begins, i.e., $y(n) = 0$ for $n \leq 0$:

$$y(-2) = \frac{1}{3}\delta(-2) + \frac{1}{3}\delta(-3) + \frac{1}{3}\delta(-4) = 0 + 0 + 0 = 0$$

$$y(-1) = \frac{1}{3}\delta(-1) + \frac{1}{3}\delta(-2) + \frac{1}{3}\delta(-3) = 0 + 0 + 0 = 0$$

$$y(0) = \frac{1}{3}\delta(0) + \frac{1}{3}\delta(-1) + \frac{1}{3}\delta(-2) = \frac{1}{3} + 0 + 0 = \frac{1}{3}$$

$$y(1) = \frac{1}{3}\delta(1) + \frac{1}{3}\delta(0) + \frac{1}{3}\delta(-1) = 0 + \frac{1}{3} + 0 = \frac{1}{3}$$

$$y(2) = \frac{1}{3}\delta(2) + \frac{1}{3}\delta(1) + \frac{1}{3}\delta(0) = 0 + 0 + \frac{1}{3} = \frac{1}{3}$$

$$y(3) = \frac{1}{3}\delta(3) + \frac{1}{3}\delta(2) + \frac{1}{3}\delta(1) = 0 + 0 + 0 = 0$$

$$y(4) = \frac{1}{3}\delta(4) + \frac{1}{3}\delta(3) + \frac{1}{3}\delta(2) = 0 + 0 + 0 = 0$$

$$\cdots$$

$$y(k) = 0; \, k \geq 3$$

This is an FIR filter with only three nonzero points. ■

IIR filters are particularly useful for simulating analog systems. The main advantage of an IIR filter is that the desired job can usually be accomplished with fewer filter coefficients than it can with an FIR filter, i.e., IIR filters tend to be more efficient. The main disadvantage of an IIR filter is that signals may be distorted in an undesirable way. FIR filters can be designed with symmetry to prevent undesired signal distortion.

Digital filters, as the name implies, are most often designed to perform specific "filtering" operations: low-pass filters, high-pass filters, band-pass filters, band-stop filters, notch filters, etc. However, digital filters can be used to simulate most analog systems, e.g., to differentiate and to integrate. Many textbooks have been written on digital filter design. The key components of the process are illustrated below.

From digital filter to transfer function: The transfer function for the digital system, $H(z)$, can be obtained by rearranging the difference equation (Eq. 6.23) and applying Eq. (6.21). $H(z)$ is the quotient of the z transform of the output, $Y(z)$, divided by the z transform of the input, $X(z)$:

$$y(k) + a_1 y(k-1) + a_2 y(k-2) \cdots + a_N y(k-N)$$
$$= b_0 x(k) + b_1 x(k-1) + \cdots + b_M x(k-M)$$

$$Y(z) + a_1 z^{-1} Y(z) + a_2 z^{-2} Y(z) \cdots + a_N z^{-N} Y(z)$$
$$= b_0 X(z) + b_1 z^{-1} X(z) + b_2 z^{-2} X(z) \cdots + b_M z^{-M} X(z)$$

$$Y(z)(1 + a_1 z^{-1} + a_2 z^{-2} \cdots + a_N z^{-N}) = X(z)(b_0 + b_1 z^{-1} + b_2 z^{-2} \cdots + b_M z^{-M})$$

$$H(z) = \frac{Y(z)}{X(z)} = \frac{b_0 + b_1 z^{-1} + b_2 z^{-1} \cdots + b_M z^{-M}}{1 + a_1 z^{-1} + a_2 z^{-1} \cdots + a_N z^{-N}} \tag{6.24}$$

From transfer function to frequency response: The frequency response ($H'(\Omega)$) of a digital system can be calculated directly from $H(z)$, where Ω is in radians. If the data are samples of an analog signal as previously described, the relationship between ω and Ω is $\Omega = \omega T$:

$$H'(\Omega) = H(z)\Big|_{z = e^{j\Omega}} \tag{6.25}$$

For a linear system, an input sequence of the form

$$x(k) = A\sin(\Omega_0 k + \Phi)$$

will generate an output whose steady-state sequence will fit into the following form

$$y(k) = B\sin(\Omega_0 k + \Phi).$$

Values for B and Φ can be calculated directly

$$B = A|H'(\Omega_0)|$$

$$\Phi = \Phi + \text{angle}(H'(\Omega_0))$$

Example Problem 6.11

The input sequence for a digital filter is

$$x(k) = 100\sin\left(\frac{\pi}{2}k\right).$$

What is the steady state form of the output?

Solution

$$y(k) - \frac{1}{2}y(k-1) = \frac{1}{2}x(k)$$

$$H(z) = \frac{1/2}{1 - \frac{1}{2}z^{-1}}$$

$$H'\left(\frac{\pi}{2}\right) = H(e^{j\pi/2}) = \frac{1/2}{1 - \frac{1}{2}e^{-j\pi/2}} = \frac{1/2}{1 + \frac{1}{2}j} = 0.4 - j0.2 = 0.45e^{-j0.15\pi}$$

Therefore,

$$y(k) = 45\sin(\frac{\pi}{2}k - 0.15\pi)$$

Filter design problems begin with identifying the frequencies that are to be kept versus the frequencies that are to be removed from the signal. For ideal filters, $|H'(\Omega_{\text{keep}})| = 1$ and $|H'(\Omega_{\text{remove}})| = 0$. The filters in Example Problems 6.9 and 6.10 could both be considered low-pass filters. However, their frequency responses (Fig. 6.11) show that neither is a particularly good low-pass filter. An ideal low-pass filter that has a cutoff frequency of $\pi/4$ with $|H'(\Omega)| = 1$ for $|\Omega| < \pi/4$ and $|H'(\Omega)| = 0$ for $\pi/4$ and $|\Omega| < \pi$ is superimposed for comparison.

6.8 SIGNAL AVERAGING

Many biological signals are periodic in nature. Signals associated with the beating heart — blood pressure, blood velocity, and ECG — fall into this category. However, due to intrinsic variability (which is healthy), noise, and/or the influence of

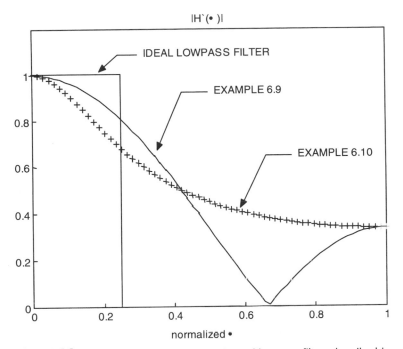

Fig. 6.11 A frequency domain comparison of low-pass filters described in Example Problems 6.9 and 6.10. An ideal low-pass filter with a cutoff frequency at $\pi/4$ radians or 0.25 when normalized by π radians is superimposed for comparison. The cutoff frequency of a low-pass filter is usually defined as the frequency at which the amplitude is equal to 1/sqrt(2) or approximately 0.71, which matches Example Problem 6.10. Both digital filters have the same amplitude at f_{max}, i.e., where they normalized $\Omega = 1$.

other functions (e.g., respiration), beat-to-beat differences are to be expected. Figure 6.2 is an example of a blood pressure signal that has all of the previously mentioned variability.

Blood pressure signals have many features that clinicians and researchers use to determine a patient's health. Examples of variables that are often measured include the peak pressure while the heart is ejecting blood (systolic phase), the lowest pressure achieved while the valve is closed (diastolic phase), the peak derivative (dP/dt) during the early part of the systolic phase (considered an indication of the strength of the heart), and the time constant of the exponential decay during diastole (a function of the resistance and compliance of the blood vessels).

One way to determine variables of interest is to calculate the variable for each beat in a series of beats and then report the means. Another approach is to average the signal such that a representative beat is obtained. The variables are then estimated based on the representative beat. Many acquisition systems are designed to calculate a signal-averaged beat as data are collected. The summation process is triggered by a signal or signal-related feature. The ECG signal, which has many sharp features, is often used for heartbeat-related data. Figure 6.12 shows a signal-averaged pressure waveform for the data shown in Fig. 6.2.

The previous blood pressure example illustrates signal averaging in the time domain. For signals that are random in nature, signal averaging in the frequency domain is preferable. Figure 6.13 illustrates an EEG signal sampled over the occipital lobe of a patient. The sampling rate was 16 Hz. EEG analysis is usually done in the frequency domain since the presence of different frequencies is indicative of different brain states, e.g., sleeping, resting, and alertness. The power at each frequency estimate, which can be approximated by the square of the FT, is the measurement of choice.

If a DFT is performed on the data to estimate the power of the frequencies in the signal, the expected variance in the measurement is equal to the measurement itself. To reduce the variance, a statistical approach must be taken. One popular approach is known as the Welch or periodogram averaging method. The signal is broken into L sections (disjoint if possible) of N points each. A DFT is performed on each of the

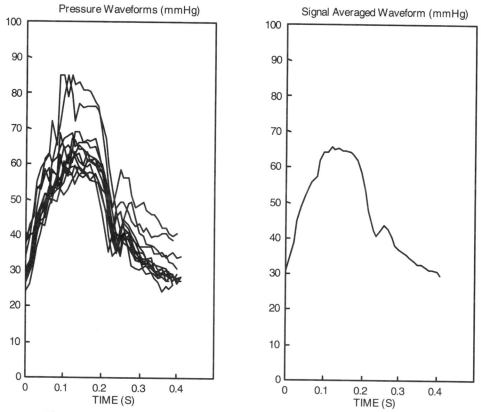

Fig. 6.12   A signal-averaged pressure waveform for the data shown in Fig. 6.2.

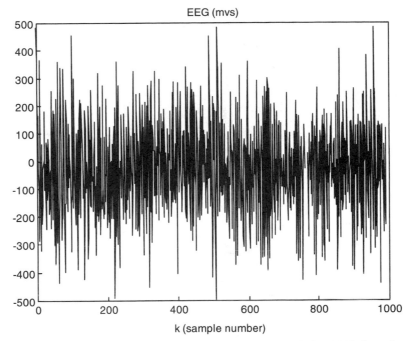

Fig. 6.13 An EEG signal containing 1000 samples sampled at 16 Hz from the occipital area.

L sections. The final result for the N frequencies is then the average at each frequency for the L sections. Notation for the process is given here.

The N data points in section i are denoted as

$$x_i(k) = x(k + (i-1)N) \qquad 0 \le k \le N - 1, 1 \le i \le L$$

if the segments are consecutive and disjoint. The power estimate based on the DFT of an individual segment i is

$$\hat{P}_i(m) = \frac{1}{N} \left| \sum_{k=0}^{N-1} x_i(k)e^{-j\frac{2\pi mk}{N}} \right|^2 \quad \text{for } 0 \le m \le N - 1 \qquad (6.26)$$

where m is associated with the power at $\Omega = 2\pi m/N$ radians. The averaged signal is calculated by taking the mean at each frequency

$$\hat{P}(m) = \frac{1}{L} \sum_{i=1}^{L} \hat{P}_i(m) \qquad (6.27)$$

The selection of N is very important since N determines the resolution in the frequency domain. For example, if data are sampled at 500 samples/s and the resolution is desired at the 1-Hz level, at least 1 s or 500 samples ($N = 500$) should be in-

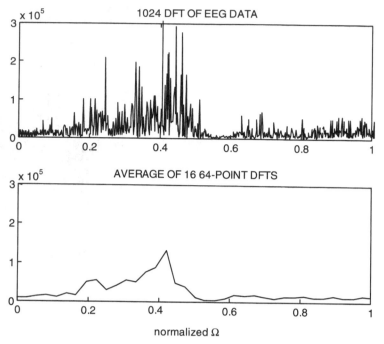

Fig. 6.14   DFT averaging of an EEG.

cluded in each of the L sections. If resolution at the 10-Hz level is sufficient, only 0.1 s or 50 data points need to be included in each section. This process decreases the variance by a factor of $1/L$. This averaging process is demonstrated for the EEG data in Fig. 6.14. Modifications to the procedure may include using overlapping segments if a larger value for L is needed and the number of available data points is not sufficient and/or multiplying each section by a window that forces continuity at the end points of the segments.

6.9   WAVELET TRANSFORM AND SHORT-TIME FOURIER TRANSFORM

The FT (Eq. 6.3) is a well-known signal processing tool for breaking a signal into constituent sinusoidal waveforms of different frequencies (Fig. 6.15). For many applications, particularly those that change little over time, knowledge of the overall frequency content may be all that is desired. The FT, however, does not delineate the nature or time of transitory changes.

The STFT and wavelet transform (WT) have been designed to help preserve the time-domain information. The STFT approach is to perform a FT on only a small section (window) of data at a time, thus mapping the signal into a two-

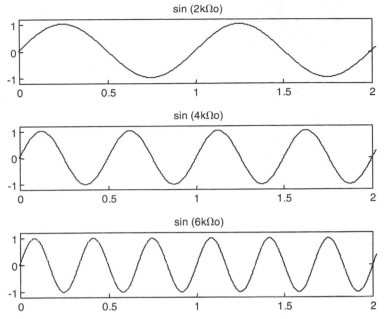

Fig. 6.15 The Fourier transform has been used to break a sinusoidal waveform into its constituent sinusoidal waveforms. Time has been normalized for the *x* axis.

dimensional (2-D) function of time and frequency. The transform is described mathematically as

$$X(\omega, a) = \int_{-\infty}^{\infty} x(t)g(t - a)e^{-j\omega t}dt \tag{6.28}$$

where $g(t)$ may define a simple box or pulse function. The inverse of the STFT is given as

$$X(\omega, a) = K_g \iint X(\omega, a)g(t - a)e^{j\omega t}dtda \tag{6.29}$$

where K_g is a function of the window used.

To avoid the "boxcar" or "rippling" effects associated with a sharp window, the box may be modified to have more gradually tapered sides. Both designs are shown in Fig. 6.16. The windows are superimposed on a totally periodic aortic pressure signal. For clarity, the windows have been multiplied by a factor of 100.

The STFT amplitudes for three box window sizes — one-half period, one period, and two periods — are illustrated in Fig. 6.17. The vertical lines in the top figure are indicative of longer periodicities than the window. The solid colored horizontal lines in the bottom two figures indicate that the frequency content is totally independent

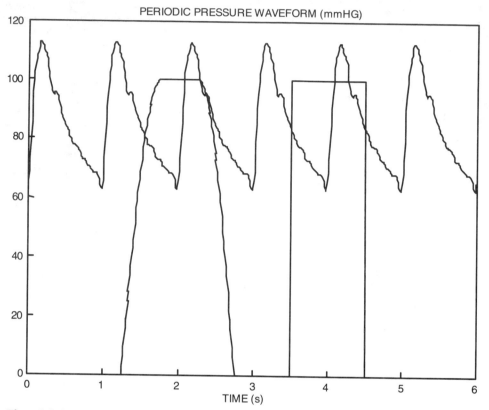

Fig. 6.16 An example of two windows that might be used to perform a STFT on a perfectly periodic aortic pressure waveform. Each window approximates the width of one pulse. The tapered window on the left can help avoid the "boxcar" or "rippling" effects associated with the sharp window on the right. For clarity, the windows have been multiplied by a factor of 100.

of time at that window size. This is expected since the window includes either one or two perfect periods. The dark (little or no frequency content) horizontal lines interspersed with the light lines in the bottom figure indicate that multiple periods exist within the window.

In contrast, Fig. 6.18 shows an amplitude STFT spectrum for the aperiodic pressure waveform shown in Fig. 6.2 with the window size matched as closely as possible to the heart rate. The mean has been removed from the signal so the variation in the lowest frequences (frequency level 0) reflects change with respiration. The level of the heart rate (level 1) is most consistent across time, and the variability increases with frequency.

The main disadvantage of the STFT is that the width of the window remains fixed throughout the analysis. Wavelet analysis represents a change from both the FT and STFT in that the constituent signals are no longer required to be sinusoidal

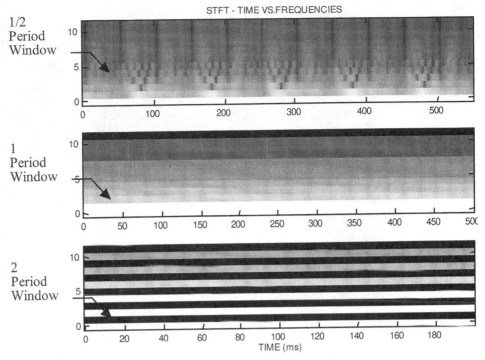

Fig. 6.17 A two-dimensional rendering of the STFT amplitude coefficients for three box window sizes — one-half period, one period, and two periods — applied to the perfectly periodic data shown in Fig. 6.16. The lighter the color, the higher the amplitude. For example, the 0th row corresponds to the mean term of the transform, which is the largest in all cases. Higher rows corresponds to harmonics of the data, which in general decrease with frequency. The vertical lines in the top figure are indicative of longer periodities than the window. The solid colored horizontal lines in the bottom two figures indicate that the frequency content is totally independent of time at that window size. The dark (little or no frequency content) horizontal lines interspersed with the light lines in the bottom figure indicate that multiple periods exist within the window.

and the windows are no longer of fixed length. In wavelet analysis, the signals are broken up into shifted and scaled versions of the original or "mother" wavelet, $\psi(t)$. Figure 6.19 shows examples of two wavelets, the Haar on the left and one from the Daubechies (db2) series on the right. The sharp corners enable the transform to match up with local details that are not possible to observe using a FT. The notation for the 2-D WT is

$$C(a,s) = \int_{-\infty}^{\infty} x(t)\varphi(a,s,t)dt \qquad (6.30)$$

where a = scale factor and s = the position factor. C can be interpreted as the correlation coefficient between the scaled, shifted wavelet and the data. Figure 6.20

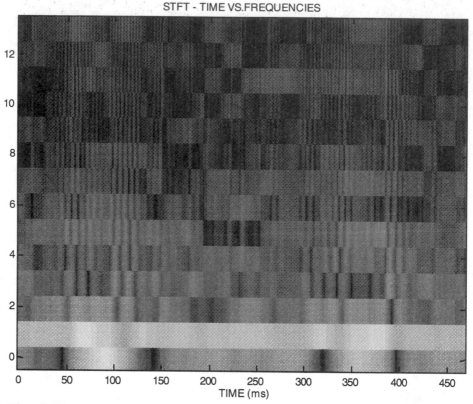

STFT - TIME VS.FREQUENCIES

Fig. 6.18 A two-dimensional rendering of the STFT of the aperiodic aortic pressure tracing shown in Fig. 6.2. The window size was matched as closely as possible to the heart rate. The mean was removed from the signal so the variation in the lowest frequencies, i.e., frequency level 0, reflects changes with respiration.

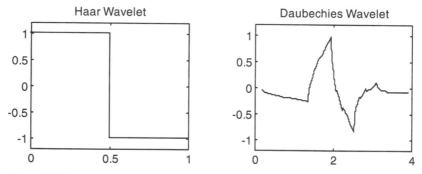

Fig. 6.19 The general shape of two wavelets commonly used in wavelet analysis. The sharp corners enable the transform to match up with local details not possible to observe when using a Fourier transform that matches only sinusoidal shapes.

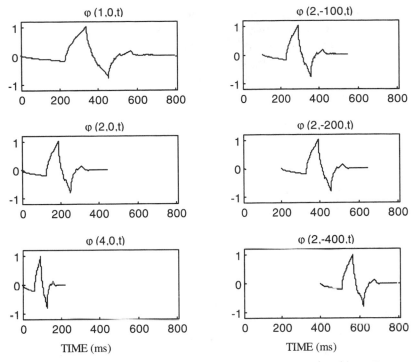

Fig. 6.20 Illustrations of the db2 wavelet at several scales and positions. The upper left-hand corner illustrates the basic waveform φ(t). The notation for the illustrations is given in the form φ (scale, delay, t). Thus, φ(t) = φ(1, 0, t), ψ(2t − 100) = φ(2, −100, t), etc.

illustrates the db2 ($\varphi(t)$) wavelet at different scales and positions, e.g., $\varphi(2, -100, t) = \varphi(2t - 100)$. The inverse transform is

$$x(t) = K_\varphi \iint C(a,s)\varphi(a,s,t)dtds \tag{6.31}$$

where K_φ is a function of the wavelet used.

In practice, wavelet analysis is performed on digitized signals using a subset of scales and positions (see MATLAB's Wavelet Toolbox). One computational process is to recursively break the signal into low-frequency ("high-scale" or "approximation") and high-frequency ("low scale" or "detail") components using digital low-pass and high-pass filters that are functions of the mother wavelet. The output of each filter will have the same number of points as the input. In order to keep the total number of data points the same at each level, every other data point of the output sequences is discarded. This is a process known as "downsampling." Using "upsampling" and a second set of digital filters, called reconstruction filters, the process can be reversed, and the original data set is reconstructed. Remarkably, the inverse discrete wavelet transform does exist!

While this process will rapidly yield wavelet transform coefficients, the power of discrete wavelet analysis lies in its ability to examine waveform shapes at different levels of decomposition and to selectively reconstruct waveforms using only the level of approximation and detail that are desired. Applications include detecting discontinuities and breakdown points, detecting long-term evolution, detecting self-similarity (e.g., fractal trees), identifying pure frequencies (similar to FT), and suppressing, denoising, and/or compressing signals.

For comparison purposes, DFTs and Discrete WTs are illustrated for the pressure waveforms shown in Fig. 6.2. Figure 6.21 shows details of the DFT on the entire record of data. The beat-to-beat differences are reflected by the widened and irregular values around the harmonics of the heart rate. The respiration influence is apparent at the very low frequencies.

Finally, an example from the MATLAB Wavelet Toolbox is shown that uses the same pressure waveform. Figure 6.22 is a 2-D rendering of the WT coefficients. The x axis shows the positions and the y axis shows the scales with the low scales on the bottom and the high scales on the top. The top scale clearly shows the two respiratory cycles in the signal. More informative than the transform coefficients, however, is a selective sample of the signal details and approximations. As the scale is changed from a1 to a7 (Fig. 6.23), the approximation goes from emphasizing the heart rate components to representing the respiration components. The details show that the noise at the heart rate levels is fairly random at the lower scales but becomes quite regular as the heart rate data become the noise.

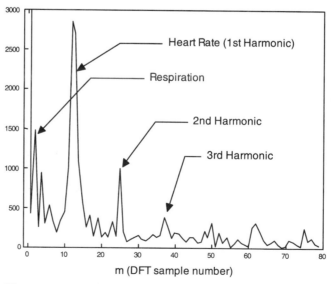

Fig. 6.21   DFT of pressure data from Fig. 6.2. The first, second, and third harmonics of the heart rate are clearly visible.

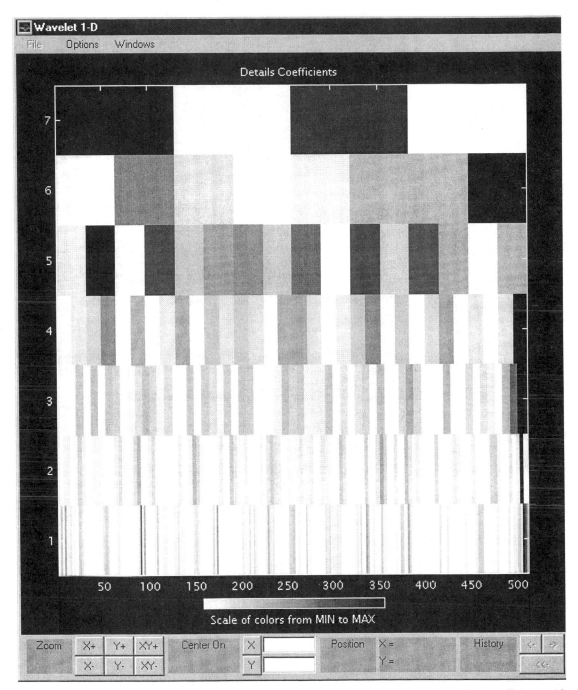

Fig. 6.22 MATLAB was used to produce a two-dimensional rendering of the wavelet transform coefficients with the Daubechies wavelet applied to the aortic pressure tracing in Fig. 6.2. The x axis shows the positions and the y axis shows the scales, with the low scales on the bottom and the high scales on the top. The associated waveforms at selected levels of these scales are shown in Fig. 6.23.

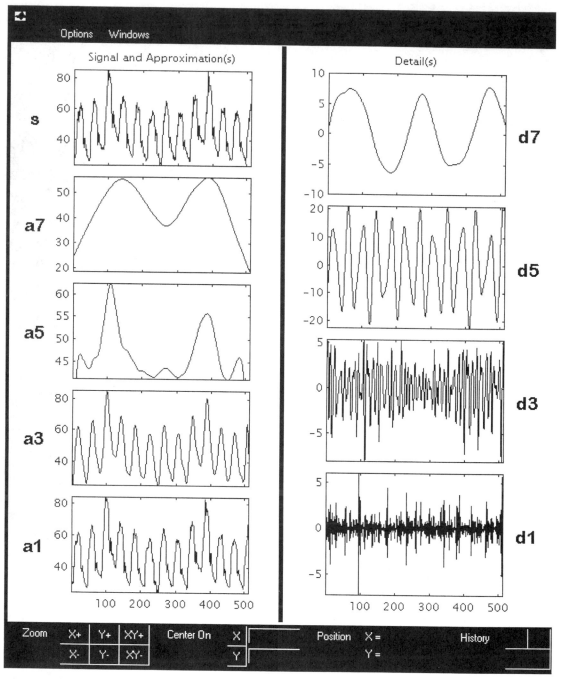

Fig. 6.23 A selective sample of the signal details and approximations generated by MATLAB as part of the wavelet transform process.

6.10 ARTIFICIAL INTELLIGENCE TECHNIQUES

Artificial intelligence (AI) is a broad topic that has many different fields, including fuzzy logic, neural networks, and expert systems. The principal aim of AI is to create intelligent machines. The term intelligent, in regard to machines, indicates computer-based systems that can interact with their environment and adapt to changes in the environment. The adaptation is accomplished through self-awareness and perceived models of the environment that are based on qualitative and quantitative information. In other words, the basic goal of AI techniques is to produce machines that are more capable of human-like reasoning, decision making, and adaptation.

The machine intelligence quotient (MIQ) is a measure of the intelligence level of machines. The higher the MIQ of a machine, the higher the capacity of the machine for automatic reasoning and decision making. The MIQ of a wide variety of machines has risen significantly during the past few years. Many computer-based consumer products, industrial machinery, and biomedical instruments and systems are using more sophisticated AI techniques. Advancements in the development of fuzzy logic, neural networks, and other soft computing techniques have contributed significantly to the improvement of the MIQ of many machines.

Soft computing is an alliance of complementary computing methodologies. These methodologies include fuzzy logic, neural networks, probabilistic reasoning, and genetic algorithms. Various types of soft computing often can be used synergistically to produce superior intelligent systems. The primary aim of soft computing is to allow for imprecision since many of the parameters that machines must evaluate do not have precise numeric values. Parameters of biological systems can be especially difficult to measure and evaluate precisely.

6.10.1 Fuzzy Logic

Fuzzy logic is based on the concept of using words, rather than numbers, for computing since words tend to be much less precise than numbers. Computing has traditionally involved calculations that use precise numerical values, whereas human reasoning generally uses words. Fuzzy logic attempts to approximate human reasoning by using linguistic variables. Linguistic variables are words that are used to describe a parameter. For body temperature, linguistic variables that might be used are high fever, above normal, normal, below normal, and frozen. The linguistic variables are more ambiguous than the number of degrees Fahrenheit, such as 105.0, 98.9, 98.6, 97.0, and 27.5.

In classical mathematics, numeric sets called crisp sets are defined, and the basic elements of fuzzy systems are fuzzy sets. An example of a crisp set is A = [0, 20]. Crisp sets have precisely defined, numeric boundaries. Fuzzy sets do not have sharply defined bounds. Consider the categorization of people by age. Using crisp sets, the age groups could be divided as A = [0, 20], B = [30, 50], and C = [60, 80]. Figure 6.24a shows the characteristic function for the sets A, B, and C. The value of the function is either 0 or 1 depending on whether or not the age of a person is within the bounds of set A, B, or C. The scheme using crisp sets lacks flexibility. If a person is 25 years old or 37 years old, he or she is not categorized.

a

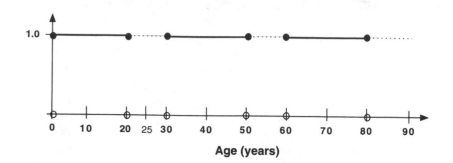

b

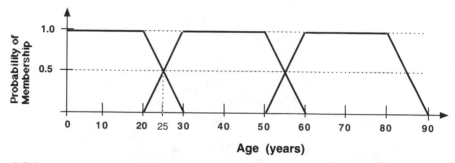

Fig. 6.24 (a) Crisp sets for the classification of people by age; (b) fuzzy sets for the classification of people by age.

If the age groups were instead divided into fuzzy sets, the precise divisions between the age groups would no longer exist. Linguistic variables, such as young, middle-aged, and old could be used to classify the individuals. Figure 6.24b shows the fuzzy sets for age categorization. Note the overlap between the categories. The words are basic descriptors, not precise measurements. A 30-year-old woman may seem old to a 6-year-old boy but quite young to an 80-year-old man. For the fuzzy sets, a value of 1 represents a 100% degree of membership to a set. A value of 0 indicates that there is no membership in the set. All numbers between 0 and 1 show the degree of membership to a group. A 25-year-old person, for instance, belongs 50% to the young set and 50% to the middle-aged set.

As with crisp sets from classical mathematics, operations are also defined for fuzzy sets. The fuzzy set operation of intersection is shown in Fig. 6.25a. Figure 6.25b shows the fuzzy union operator, and Fig. 6.25c shows the negation operator for fuzzy sets. The solid line indicates the result of the operator in Figs. 6.25a–6.25c.

Although it is easy to form fuzzy sets for a simple example such as age classification, fuzzy sets for more sophisticated applications are derived by using sophisticated calibration techniques. The linguistic variables are formulated mathematically

a

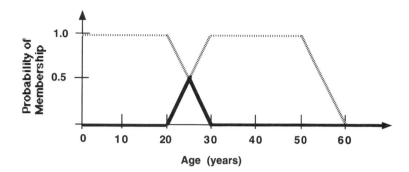

b

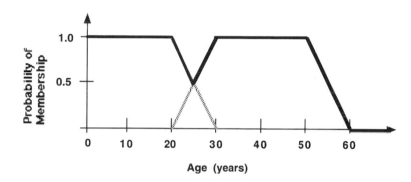

c

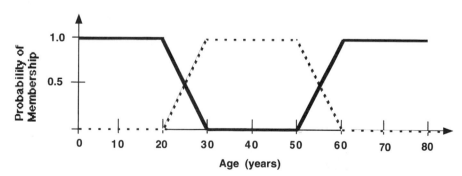

Fig. 6.25 (a) Intersection of fuzzy sets: YOUNG AND MIDDLE-AGED. (b) Union of fuzzy sets: YOUNG OR MIDDLE-AGED. (c). Negation of fuzzy sets: NOT MIDDLE-AGED.

and then can be processed by computers. Once the fuzzy sets have been established, rules are constructed. Fuzzy logic is a rule-based logic. Fuzzy systems are constructed by using a large number of rules. Most rules used in fuzzy logic computing are if/then statements that use linguistic variables. Two simple rules that use the fuzzy sets for age classification might be

> If patient is YOUNG, then use TREATMENT A.
> If patient is MIDDLE-AGED or OLD, then use TREATMENT B.

The degree of membership in a group helps determine which rule will be used and, consequently, the type of action that will be taken or, in the previous example, the type of treatment that will be used. Defuzzification methods are used to determine which rules will be used to produce the final output of the fuzzy system.

For many applications, fuzzy logic has significant advantages over traditional numeric computing methods. Fuzzy logic is particularly useful when information is too limited or too complex to allow for numeric precision since it tolerates imprecision. If an accurate mathematical model cannot be constructed, fuzzy logic may prove valuable. However, if a process can be described or modeled mathematically, then fuzzy logic will not generally perform better than traditional methods.

Biomedical engineering applications, which involve the analysis and evaluation of biosignals, often have attributes that confound traditional computing methods but are well suited to fuzzy logic. Biological phenomena are often not precisely understood and can be extremely complex. Biological systems also vary significantly between individuals. In addition, many key quantities in biological systems cannot be measured precisely due to limitations in existing sensors and other biomedical measuring devices. Sensors may have the capability to measure biological quantities intermittently or in combination with other parameters but not independently. Blood glucose sensors, for example, are sensitive not only to blood glucose but also to urea and other elements in the blood. Fuzzy logic can be used to help compensate for the limitations of sensors.

Fuzzy logic is being used in a variety of biomedical engineering applications. Closed-loop drug delivery systems, which are used to automatically administer drugs to patients, have been developed by using fuzzy logic. In particular, fuzzy logic may prove valuable in the development of drug delivery systems for anesthetic administration since it is difficult to precisely measure the amount of anesthetic that should be delivered to an individual patient by using conventional computing methods. Fuzzy logic is also being used to develop improved neuroprosthetics for paraplegics. Neuroprosthetics for locomotion use sensors controlled by fuzzy logic systems to electrically stimulate necessary leg muscles and will ideally enable the paraplegic patient to walk.

Example Problem 6.12

A fuzzy system is used to categorize people by heart rates. The system is used to help determine which patients have normal resting heart rates, bradychardia, or tachycardia. Bradychardia is a cardiac arrhythmia in which the resting heart rate is <60 beats per minute, whereas tachycardia is defined as a cardiac arrhythmia in which the resting heart rate is >100 beats per minute. A normal heart rate is considered to be in the range of 70–80 beats per minute. What are three linguistic variables that might be used to describe the resting heart rates of the individuals?

Solution

A variety of linguistic variables may be used. The names are important only because they offer a good description of the categories and problem. Slow, normal, and fast might be used. Another possibility is simply bradychardia, normal, and tachychardia. ∎

6.10.2 Artificial Neural Networks

Artificial neural networks (ANN) are theoretically based on biological neural networks. Biological neural networks, which are composed of biological neurons, are highly complex and may consist of billions of neurons, each connected to thousands of other neurons. The human brain is one of the most sophisticated biological neural networks. Highly developed biological neural networks are able to learn from experience, recognize patterns, and react to changes in the environment.

ANNs are simpler than biological neural networks. A sophisticated ANN contains only a few thousand neurons with several hundred connections. Although simpler than biological neural networks, the aim of ANNs is to build computer systems that have learning, generalized processing, and adaptive capabilities resembling those seen in biological neural networks. ANNs can learn to recognize certain inputs and to produce a particular output for a given input. Therefore, ANNs are commonly used for pattern detection and classification of biosignals.

ANNs consist of multiple, interconnected neurons. Different types of neurons can be represented in an ANN. Neurons are arranged in a layer, and the different layers of neurons, which are connected to other neurons, form the neural network. The manner in which the neurons are interconnected determines the architecture of the ANN. There are many different ANN architectures. Figure 6.26 shows a schematic of a simple ANN with three layers of neurons and a total of six neurons. The first layer is called the input layer and has two neurons, which accept the input to the network. The middle layer contains three neurons and much of the processing occurs here. The output layer has one neuron that provides the result determined by the ANN.

Mathematical equations are used to describe the connections between the neurons. The diagram in Fig. 6.26 represents a single neuron and a mathematical method for determining the output of the neuron. The equation for calculating the total input to the neuron is

$$x = (\text{input}_1 \times \text{weight}_1) + (\text{input}_2 \times \text{weight}_2) + \text{bias weight} \qquad (6.32)$$

The output for the neuron is determined by using a mathematical function, $g(x)$. Threshold functions and nonlinear sigmoid functions such as those shown in Fig. 6.27 are commonly used. The output y of a neuron using the sigmoid function is calculated from the following simple equation:

$$y = 1/(1 + e^{-x}) \qquad (6.33)$$

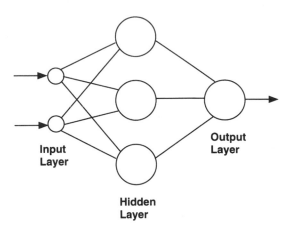

Fig. 6.26 Schematic of a simple artificial neural network with six neurons and three layers.

In biosignal processing applications, the inputs to the first layer or input layer of the ANN can be raw data, a preprocessed signal, or extracted features from a biosignal. Raw data are generally samples from a digitized signal. Preprocessed signals are biosignals that have been transformed, filtered, or processed using some other method before being input to the neural network. Features can also be extracted from biosignals and used as inputs for the neural network. Extracted features might include thresholds, a particular, reoccurring waveshape, or the period between waveforms.

The ANN must learn to recognize the features or patterns in an input signal, but this is not the case initially. In order for the ANN to learn, a training process must occur in which the user of the ANN presents the neural network with many differ-

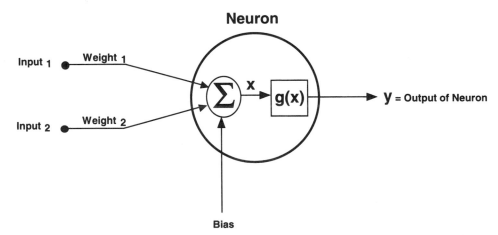

Fig. 6.27 Diagram of a single neuron showing mathematical input and output relationships.

ent examples of input. Each example is given to the ANN many times. Over time, after the ANN has been presented with all the input examples several times, the ANN learns to produce particular outputs for specific inputs.

There are a variety of types of learning paradigms for ANNs. Learning can be broadly divided into two categories — unsupervised learning and supervised learning. In unsupervised learning, the outputs for the given input examples are not known. The ANN must perform a sort of self-organization. During unsupervised learning, the ANN learns to recognize common features in the input examples and produces a specific output for each different type of input. Types of ANNs with unsupervised learning that have been used in biosignal processing include the Hopfield network and self-organizing feature maps networks.

In supervised learning, the desired output is known for the input examples. The output that the ANN produces for a particular input or inputs is compared against the desired output or output function. The desired output is known as the target. The difference between the target and the output of the ANN is calculated mathematically for each given input example. A common training method for supervised learning is backpropagation. The multilayered perceptron trained with backpropagation is a type of a network with supervised learning that has been used for biosignal processing.

Backpropagation is an algorithm that attempts to minimize the error of the ANN. The error of the ANN can be regarded as simply the difference between the output of the ANN for an input example and the target for that same input example. Backpropagation uses a gradient-descent method to minimize the network error. In other words, the network error is gradually decreased down an error slope that is in some respects similar to how a ball rolls down a hill. The name backpropagation refers to the way by which the ANN is changed to minimize the error. Each neuron in the network is "credited" with a portion of the network error. The relative error for each neuron is then determined, and the connection strengths between the neurons are changed to minimize the errors. The weights, such as those that were shown in Fig. 6.27, represent the connection strengths between neurons. The calculations of the neuron errors and weight changes propagate backwards through the ANN from the output neurons to the input neurons. Backpropagation is the method of finding the optimum weight values that produce the smallest network error.

ANNs are well suited for a variety of biosignal processing applications and may be used as a tool for nonlinear statistical analysis. They are often used for pattern recognition and classification. In addition, ANNs have been shown to perform faster and more accurately than conventional methods for signals that are highly complex or contain high levels of noise. ANNs also have the ability to solve problems that have no algorithmic solution; that is, problems for which a conventional computer program cannot be written. Since ANNs learn, algorithms are not required to solve problems.

As advances are made in artificial intelligence techniques, ANNs will be used more extensively in biosignal processing and biomedical instrumentation. The viability of ANNs for applications ranging from the analysis of ECG and EEG signals to the interpretation of medical images and the diagnosis of a variety of diseases has been investigated. In neurology, research has been conducted by using ANNs to characterize brain defects that occur in disorders such as epilepsy, Parkinson's disease, and Alzheimer's disease. ANNs have also been used to characterize and classify ECG signals of cardiac arrhythmias. One study used an ANN in the emergency room to diagnose heart attacks. The results of the study showed that, overall, the ANN was able to diagnose heart attacks better than were the emergency room physicians. ANNs have the advantage of not being affected by fatigue, distractions, or emotional stress. As artificial intelligence technologies advance, ANNs may provide a superior tool for many biosignal processing tasks.

Example Problem 6.13

A neuron in a neural network has three inputs and uses a sigmoid function to calculate the output of the neuron. The three values of the inputs are 0.1, 0.9, and 0.1. The weights associated with these three inputs are 0.39, 0.72, and 0.26, and the bias weight is 0.48 after training. What is the output of the neuron?

Solution

Using Eq. (6.32) to calculate the relative sum of the inputs gives

$$x = (\text{input}_1 \times \text{weight}_1) + (\text{input}_2 \times \text{weight}_2) + (\text{input}_3 \times \text{weight}_3) + \text{bias weight}$$

$$= (0.1)\,0.39 + (0.9)\,0.72 + (0.1)\,0.24 + 0.48$$

$$= 1.19$$

The output of the neuron is calculated using Eq. (6.33):

$$y = 1/(1 + e^{-X}) = 1/(1 + e^{-1.19}) = 0.77 \qquad \blacksquare$$

EXERCISES

1. What types of biosignals would the nerves in your legs produce during a sprint across the street?
2. What types of biosignals can be recorded with an EEG? Describe these in terms of both origins and characteristics of the signal.
3. Describe the biosignal that the electrical activity of a normal heart would generate during a bicycle race.
4. A 16-bit A/D converter is used to convert an analog biosignal with a minimum voltage of -30 mV and a maximum voltage of 90 mV. What is the resolution?
5. An EMG recording of skeletal muscle activity has been sampled at 200–250 Hz and correctly digitized. What is the highest frequency of interest in the original EMG signal?
6. Given $x(t) = e^{-at}u(t)$ and $h(t) = e^{-bt}u(t)$, where a and b are constants >0, explain why it would be easier to evaluate the convolution $x(t)^* h(t)$ in the frequency domain.

7. An ECG recording of the electrical activity of the heart during ventricular fibrillation is digitized and the signal begins with the following data sequence [$-90.0, 10.0, -12.0, -63.0, 7.0, -22.0$]. The units of the data sequence are given in mV. What is the z transform of this data sequence of the biosignal?

8. Examine the characteristics of the digital filter

$$y(k) = \frac{1}{4}x(k) + \frac{1}{4}x(k-1) + \frac{1}{2}y(k-1)$$

Find the impulse response, $H(z)$ and $H'(\Omega)$. Use MATLAB to calculate and plot $|H'(\Omega)|$ for $0 < \Omega < \pi$. Observe the difference between this filter and the filter in Example Problem 6.10. Why is this a better low-pass filter? What is the output if the input sequence is $x(k) = 100 \sin(\frac{\pi}{2}k + \frac{\pi}{8})$?

What is the output if the input sequence is $x(k) = 100\,u(k)$?

9. Accurate measurements of blood glucose levels are needed for the proper treatment of diabetes. Glucose is a primary carbohydrate, which circulates throughout the body and serves as an energy source for cells. In normal individuals, the hormone insulin regulates the levels of glucose in the blood by promoting glucose transport out of the blood to skeletal muscle and fat tissues. Diabetics suffer from improper management of glucose levels, and the levels of glucose in the blood can become too high. Describe how fuzzy logic might be used in the control of a system for measuring blood glucose levels. What advantages would the fuzzy logic system have over a more conventional system?

10. Describe three biosignal processing applications for which artificial neural networks might be used. Give at least two advantages of artificial neural networks over traditional biosignal processing methods for the applications you listed.

11. The fuzzy sets in Example Problem 6.12 have been calibrated so that a person with a resting heart rate of 95 beats per minute has a 75% degree of membership in the normal category and a 25% degree of membership in the tachycardia category. A resting heart rate of 65 beats per minutes indicates a 95% degree of membership in the normal category. Draw a graph of the fuzzy sets.

SUGGESTED READING

Akay, M. (1994). *Biomedical Signal Processing*. Academic Press, San Diego.

Akay, M. (Ed.), (1998). *Time Frequency and Wavelets in Biomedical Signal Processing*. IEEE Press, New York.

Bauer, P., Nouak, S., and Winkler, R. (1996). *A brief course in fuzzy logic and fuzzy control* (*http://www.fill.uni-linz.ac.at/fuzzy*). Fuzzy Logic Laboratorium Linz-Hagenberg, Linz, Austria.

Bishop, C. M. (1995). *Neural Networks for Pattern Recognition*. Oxford Univ. Press, New York.

Ciaccio, E. J., Dunn, S. M., and Akay, M. (1993a). Biosignal pattern recognition and interpretation systems: Part 1 of 4: Fundamental concepts. *IEEE Eng. Med. Biol. Mag.* **12**(3), 89–97.

Ciaccio, E. J., Dunn, S. M., and Akay, M. (1993b). Biosignal pattern recognition and interpretation systems: Part 2 of 4: Methods for feature extraction and selection. *IEEE Eng. Med. Biol. Mag.* **12**(4), 106–113.

Ciaccio, E. J., Dunn, S. M., and Akay, M. (1994a). Biosignal pattern recognition and interpretation systems: Part 3 of 4: Methods of classification. *IEEE Eng. Med. Biol. Mag.* **13**(1), 129–135.

Ciaccio, E. J., Dunn, S. M., and Akay, M. (1994b). Biosignal pattern recognition and interpretation systems: Part 4 of 4: Review of applications. *IEEE Eng. Med. Biol. Mag.* **13**(2), 269–273, 283.

Cohen, A. (1986a). *Biomedical Signal Processing: Volume I Time and Frequency Domain Analysis*. CRC Press, Boca Raton, FL.

Cohen, A. (1986b). *Biomedical Signal Processing: Volume II Compression and Automatic Recognition*. CRC Press, Boca Raton, FL.

Dempster, J., (1993). *Computer Analysis of Electrophysiological Signals,* Academic Press, San Diego.

Haykin, S. (1994). *Neural Networks — A Comprehensive Foundation*. Macmillan, New York.

Onaral, B. (Ed.) (1995). VI. Biomedical signal analysis. In *The Biomedical Engineering Handbook* (J. D. Bronzino, Ed.). CRC Press, Boca Raton, FL.

Oppenheim, A. V., and Schafer, R. W. (1975). *Digital Signal Processing*. Prentice Hall, Englewood Cliffs, NJ.

Oppenheim, A. V., Willsky, A. S., and Young, I. T. (1983). *Signals and Systems*. Prentice Hall, Englewood Cliffs, NJ.

Roberts, R. A., and Mullis, C. T. (1987). *Digital Signal Processing*. Addison-Wesley, Reading, MA.

Smith, M. (1996). *Neural Networks for Statistical Modeling*. International Thomson Computer Press, Boston.

Stearns, S. D., and David, R. A. (1993). *Signal Processing Algorithms in Fortran and C*. Prentice Hall, Englewood Cliffs, NJ.

Thompkins, W. J. (1993). *Biomedical Digital Signal Processing*. Prentice Hall, Englewood Cliffs, NJ.

Williams, C. S. (1993). *Designing Digital Filters*. Prentice Hall, Englewood Cliffs, NJ.

Zadeh, L. A. (1987). *Fuzzy Sets and Applications*. Wiley, New York.

Ziemer, R. E., Tranter, W. H., and Fannin, D. R. (1993). *Signals and Systems: Continuous and Discrete,* 3rd ed. Macmillan, New York.

7 PHYSIOLOGICAL MODELING

Chapter Contents

At the conclusion of this chapter, the reader will be able to:

- Describe the process used to build a mathematical physiological model
- Qualitatively describe a saccadic eye movement
- Describe the saccadic eye movement system with a second-order model
- Explain the importance of the pulse-step saccadic control signal
- Explain how a muscle operates using a nonlinear and linear muscle model
- Simulate a saccade with a fourth-order saccadic eye movement model
- Describe how the brain controls the saccadic eye movement system and the major neural sites
- Estimate the parameters of a model using system identification

7.1   INTRODUCTION

A **quantitative** physiological model is a mathematical representation that approximates the behavior of an actual physiological system. **Qualitative** physiological models, most often used by biologists, describe the actual physiological system without the use of mathematics. Quantitative physiological models, however, are much more useful and are the subject of this chapter. Physiological systems are almost always dynamic and mathematically characterized with differential equations. The modeling techniques developed in this chapter are intimately tied to many other interdisciplinary areas, such as physiology, biophysics, and biochemistry, and involve electrical and mechanical analogs. A model is usually constructed using basic and natural laws. This chapter extends this experience by presenting models that are more complex and involve larger systems.

Creating a model is always accompanied by carrying out an experiment and obtaining data. The best experiment is one that provides data that are related to variables used in the model. Consequently, the design and execution of an experiment is one of the most important and time-consuming tasks in modeling. A model constructed from basic and natural laws then becomes a tool for explaining the underlying processes that cause the experimental data and predicting the behavior of the system to other types of stimuli. Models serve as vehicles for thinking, organizing complex data, and testing hypotheses. Ultimately, modeling's most important goals are the generation of new knowledge, prediction of observations before they occur, and assistance in designing new experiments.

Figure 7.1 illustrates the typical steps in developing a model. The first step in-

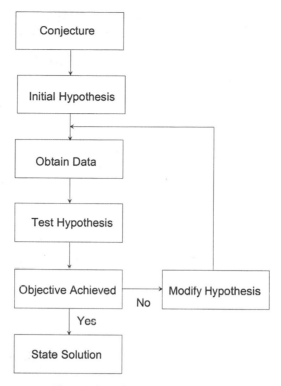

Fig. 7.1 Flow chart for modeling.

volves observations from an experiment or a phenomenon that leads to a conjecture or a verbal description of the physiological system. An initial hypothesis is formed via a mathematical model. The strength of the model is tested by obtaining data and testing the model against the data. If the model performs adequately, the model is satisfactory, and a solution is stated. If the model does not meet performance specifications, then the model is updated and additional experiments are carried out. Usually some of the variables in the model are observable and some are not. New experiments provide additional data that increase the understanding of the physiological system by providing information about previously unobservable variables, which improves the model. The process of testing the model against the data continues until a satisfactory solution is attained. Usually a statistical test is performed to test the goodness of fit between the model and the data. One of the characteristics of a good model is how well it predicts the future performance of the physiological system.

The introduction of the digital computer, programming languages, and simulation software have caused a rapid change in the use of physiological models. Before digital computers, mathematical models of biomedical systems were either oversimplified or involved a great deal of hand calculation as described in the Hodgkin–Huxley investigations published in 1952. Today, digital computers have become so common that the terms modeling and simulation have almost become synonymous. This has allowed the development of much more realistic or homeomorphic models that include as much knowledge as possible about the structure and interrelationships of the physiological system without any overriding concern about the number of calculations. While models can continue to be made more complex, it is important to evaluate the value added with each stage of complexity — the model should be made as simple as possible to explain the data but not so simple that it becomes meaningless. On the other hand, a model that is made too complex is also of little use.

7.1.1　Deterministic and Stochastic Models

A deterministic model is one that has an exact solution that relates the independent variables of the model to each other and to the dependent variable. For a given set of initial conditions, a deterministic model yields the same solution each and every time. A stochastic model involves random variables that are functions of time and include probabilistic considerations. For a given set of initial conditions, a stochastic model yields a different solution each and every time. Suffice it to say that solutions involving stochastic models are much more involved than the solution for a deterministic model.

It is interesting to note that all deterministic models include some measurement error. The measurement error introduces a probabilistic element into the deterministic model so that it might be considered stochastic. However, in this chapter, models are deterministic if their principle features lead to definitive predictions. On the other hand, models are stochastic if their principle features depend on probabilistic elements. This chapter is primarily concerned with deterministic models.

7.1.2　Solutions

There are two types of solutions available to the modeler. A closed-form solution exists for models that can be solved by analytic techniques such as solving a differential equation using the classical technique or by using Laplace transforms. For example, given the following differential equation

$$\ddot{x} + 4\dot{x} + 3x = 9$$

with initial conditions $x(0) = 0$ and $\dot{x}(0) = 1$, the solution is found as

$$x(t) = -4e^{-t} + e^{-3t} + 3$$

A numerical or simulation solution exists for models that have no closed-form solution. Consider the following function:

$$x = \int_{-20}^{20} \frac{1}{33\sqrt{2\pi}} e^{-\frac{1}{2}\left(\frac{t-7}{33}\right)^2} dt$$

This function (the area under a Gaussian curve) has no closed-form solution and must be solved using an approximation technique such as the trapezoidal rule for integration. Most nonlinear differential equations do not have an exact solution and must be solved via an iterative method or simulation package such as SIMULINK. This was the situation in Chapter 3, Section 3.6 when the Hodgkin–Huxley model was solved.

Inverse Solution

Engineers often design and build systems to a predetermined specification. They often use a model to predict how the system will behave because a model is efficient and economical. The model that is built is called a plant and consists of parameters that completely describe the system, the characteristic equation. The engineer selects the parameters of the plant to achieve a certain set of specifications, such as rise time, settling time, or peak overshoot time.

In contrast, biomedical engineers involved with physiological modeling do not build the physiological system but only observe the behavior of the system — the input and output of the system — and then characterize it with a model. Characterizing the model as illustrated in Fig. 7.1 involves identifying the form or structure of the model, collecting data, and then using the data to estimate the parameters of the model. The goal of physiological modeling is not to design a system but to identify the components (or parameters) of the system. Often, data needed for building the model are not the data that can be collected using existing biosensors and bioinstrumentation as discussed in Chapters 4 and 5. Typically, the recorded data are transformed from measurement data into estimates of the variables used in the model. Collecting appropriate data is usually the most difficult aspect of the discovery process.

Model building typically involves estimating the parameters of the model that optimize, in a mean square error sense, the output of the model or model prediction, $\hat{x}_i$, and the data, x_i. For example, one metric for estimating the parameters of a model, S, is given by minimizing the sum of squared errors between the model prediction and the data:

$$S = \sum_{i=1}^{n} \varepsilon_i^2 = \sum_{i=1}^{n} (x_i - \hat{x}_i)^2$$

where ε_i is the error between the data x_i and the model prediction $\hat{x}_i$. This technique provides an unbiased estimate with close correspondence between the model prediction and the data.

In order to provide a feeling for the modeling process described in Fig. 7.1, this chapter focuses on one particular system — the fast eye movement system, the modeling of which began with early muscle modeling experiences in the 1920s and continues today with neural network models for the control of the fast eye movement system. This physiological system is probably the best understood of all systems in the body. Some of the reasons for this success are the relative ease in obtaining data, the simplicity of the system in relation to others, and the lack of feedback during dynamic changes in the system. In this chapter, a qualitative description of the fast eye movement system is presented followed by the first model of the system by Westheimer, who used a second-order model, published in 1954. The 1964 model of the system by Robinson is presented next because of its fundamental advances in describing the input to the system. With the physical understanding of the system in place, a detailed presentation of muscle models is given with the early work of Levin and Wyman (1927) and Fenn and Marsh (1935). These muscle models are important in developing a realistic model that accurately depicts the system. Using the more accurate muscle models, the fast eye movement model is revisited by examining the model presented by Bahill and coworkers (1980) and then Enderle and coworkers (1984). Next, the control mechanism for this system is described from the basis of physiology, systems control theory, and neural networks based on anatomical pathways. System identification or parameter estimation closes the chapter. The literature on the fast eye movement system is vast, and the material covered in this chapter is not exhaustive but rather a representative sample from the field. Not covered in this chapter is how visual information is collected and processed by the body and how the body reacts to the information.

7.2 AN OVERVIEW OF THE FAST EYE MOVEMENT SYSTEM

A fast eye movement is usually referred to as a **saccade** and involves quickly moving the eye from one image to another image. Saccade is a French term that means to pull and originated from the jerk of the reins on a horse. This type of eye movement is very common and is observed most easily while reading. When the end of a line is reached, the eyes are moved quickly to the beginning of the next line. The saccade system is part of the oculomotor system that controls all movements of the eyes due to any stimuli. The eyes are moved within orbit by the oculomotor system to allow the individual to locate, see, and track objects in visual space. Each eye can be moved within the orbit in three directions: vertically, horizontally, and torsionally. These movements are due to three pairs of agonist–antagonist muscles. These muscles are called antagonistic pairs because their activities oppose each other and follow the principle of reciprocal innervation. Shown in Fig. 7.2 is a diagram illustrating the muscles of the eye. The overall strategy of the system is to keep the central portion of the retina, called the fovea, on the target of interest.

The oculomotor system responds to visual, auditory, and vestibular stimuli, which results in one of five types of eye movements: Fast eye movements are used to locate or acquire targets, smooth pursuit eye movements are used to track or follow a target, vestibular ocular movements are used to maintain the eyes on the target during head movements, vergence eye movements are used to track near and far targets, and optokinetic eye movements are reflex movements that occur when moving through a target-filled environment. Except for the vergence eye movement, each of the movements are conjugate; that is, the eyes move together in the same direction and distance. The vergence eye movement system uses nonconjugate eye movements to keep the eyes on the target. If the target moves closer, the eyes converge — farther away, they diverge. Each of these movements is controlled by a different neuronal system and uses the same final common pathway to the muscles of the eye. In addition to the five types of eye movements, these stimuli also cause head and body movements. Thus, the visual system is part of a multiple input–multiple output system.

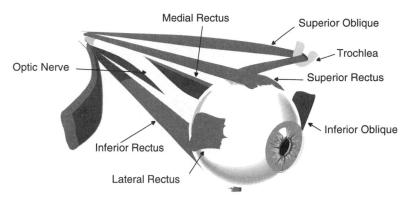

Fig. 7.2 Diagram illustrating the muscles and optic nerve of the right eye. The left eye is similar except the lateral and medial rectus muscles are reversed. The lateral and medial rectus muscles are used to move the eyes in a horizontal motion. The superior rectus, inferior rectus, superior oblique, and inferior oblique are used to move the eyes vertically and torsionally. The contribution from each muscle depends on the position of the eye. When the eyes are looking straight ahead, called primary position, the muscles are stimulated and under tension.

Regardless of the input, the oculomotor system is responsible for movement of the eyes so that targets are focused on the central $\frac{1}{2}°$ region of the retina, known as the fovea (Fig. 7.3). Lining the retina are photoreceptive cells that translate images into neural impulses. These impulses are then transmitted along the optic nerve to the central nervous system via parallel pathways to the superior colliculus and the cerebral cortex. The fovea is more densely packed with photoreceptive cells than the retinal periphery; thus, a higher resolution image (or higher visual acuity) is generated in the fovea than in the retinal periphery. The purpose of the fovea is to allow us to *clearly* see an object and the purpose of the retinal periphery is to allow us to *detect* a new object of interest. Once a new object of interest is detected in the periphery, the saccade system redirects the eyes, as fast as possible, to the new object. This type of saccade is typically called a goal-directed saccade.

During a saccade, the oculomotor system operates in an open-loop mode. After the saccade, the system operates in a closed-loop mode to ensure that the eyes reached the correct destination. The saccade system operates without feedback during a fast eye movement because information from the retina and muscle proprioceptors is not transmitted quickly enough during the eye movement for use in altering the control signal. The oculomotor plant and saccade generator are the basic elements of the saccadic system. The oculomotor plant consists of three muscle pairs and the eyeball. These three muscle pairs contract and lengthen to move the eye in horizontal, vertical, and torsional directions. Each pair of muscles acts in an antagonistic fashion due to reciprocal innervation by the saccade generator. For simplicity, the models described here involve only horizontal eye movements and one pair of muscles, the lateral and medial rectus muscle.

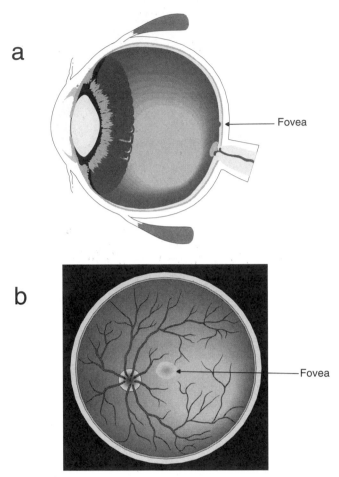

Fig. 7.3   (a) Diagram illustrating a side view of the eye. The rear surface of the eye is called the retina. The retina is part of the central nervous system and consists of two photoreceptors, rods and cones. (b) Front view looking at the rear inside surface (retina) of the eye. The fovea is located centrally and is approximately 1 mm in diameter. The oculomotor system maintains targets centered on the fovea.

7.2.1   Saccade Characteristics

Saccadic eye movements, among the fastest voluntary muscle movements the human is capable of producing, are characterized by a rapid shift of gaze from one point of fixation to another. Shown in Fig. 7.4 is a 10° saccade. The usual experiment for recording saccades is for a subject to sit before a horizontal target display of small light emitting diodes (LEDs). Subjects are instructed to maintain their eyes on the lit LED by moving their eyes as fast as possible to avoid errors. A saccade is made by the subject when the active LED is switched off and

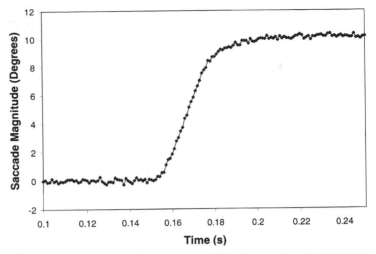

Fig.7.4 Sample saccadic eye movement of approximately 10°. Data were collected with a sampling rate of 1000 samples/s.

another LED is switched on. Saccadic eye movements are conjugate and ballistic with a typical duration of 30–100 ms and a latency of 100–300 ms. The subject was looking straight ahead when the target switched from the 0° position to the 10° position as illustrated in Fig. 7.4. The subject then executed a saccade 150 ms later and completed the saccade at 200 ms. The latent period in Fig. 7.4 is approximately 150 ms and thought to be the time interval during which the central nervous system (CNS) determines whether to make a saccade and, if so, calculates the distance the eyeball is to be moved, transforming retinal error into transient muscle activity.

Generally, saccades are extremely variable, with wide variations in the latent period, time to peak velocity, peak velocity, and saccade duration. Furthermore, variability is well coordinated for saccades of the same size. Saccades with lower peak velocity are matched with longer saccade durations, and saccades with higher peak velocity are matched with shorter saccade durations. Thus, saccades driven to the same destination usually have different trajectories.

To appreciate differences in saccade dynamics, it is often helpful to describe them with saccade main sequence diagrams. The main sequence diagrams plot saccade peak velocity–saccade magnitude, saccade duration–saccade magnitude, and saccade latent period–saccade magnitude. Shown in Fig. 7.5 are the main sequence characteristics for a subject executing 26 saccades. The subject actually executed 52 saccades in both the positive and negative directions; only the results of the saccades in the positive direction are displayed in Fig. 7.5 for simplicity. Notice that the peak velocity–saccade magnitude starts off as a linear function and then levels off to a constant for larger saccades. Many researchers have fit this relationship to an exponential function. The solid lines in Fig. 7.5A include an exponential fit

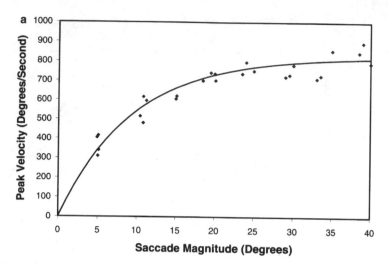

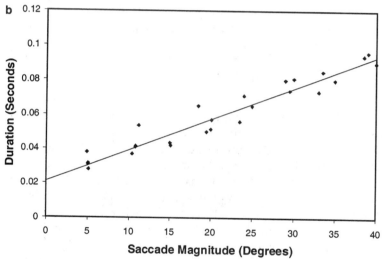

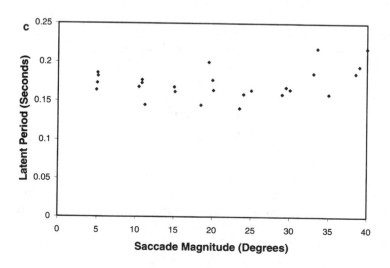

to the data for positive eye movements. The lines in the first graph are fitted to the equation

$$V = \alpha \left(1 - e^{-\frac{x}{\beta}}\right) \tag{7.1}$$

where V is the maximum velocity, x is the saccade size, and the constants α and β were evaluated to minimize the summed error squared between the model and the data. Note that α represents the steady state of the peak velocity–saccade magnitude curve and β represents the "time constant" for the peak velocity–saccade magnitude curve. For this data set, $\alpha = 825$, and $\beta = 9.3$.

A similar pattern is observed with eye movements moving in the negative direction, but the parameters α and β are typically different from the values computed for the positive direction. The exponential shape of the peak velocity–saccade amplitude relationship might suggest that the system is nonlinear if a step input to the system is assumed. A step input provides a linear peak velocity–saccade amplitude relationship. In fact, the saccade system is not driven by a step input but rather by a more complex pulse-step waveform. Thus, the saccade system cannot be assumed to be nonlinear solely based on the peak velocity–saccade amplitude relationship. The input to the saccade system is discussed more fully in Section 7.3.

Shown in Fig. 7.5B are data depicting a linear relationship between saccade duration–saccade magnitude. If a step input is assumed, then the dependence between saccade duration and saccade magnitude also might suggest that the system is nonlinear. A linear system with a step input always has a constant duration. Since the input is not characterized by a step waveform, the saccade system cannot be assumed to be nonlinear solely based on the saccade duration–saccade magnitude relationship. Shown in Fig. 7.5C is the latent period–saccade magnitude data. It is quite clear that the latent period does not show any linear relationship with saccade size, i.e., the latent period's value appears independent of saccade size. However, some investigators have proposed a linear relationship between the latent period and saccade magnitude. This feature is unimportant for the presentation in this chapter since in the development of the oculomotor plant models the latent period is implicitly assumed within the model.

Because of the complexity of the eye movement system, attention is restricted to horizontal fast eye movements. In reality, the eyeball is capable of moving horizontally, vertically, and torsionally. An appropriate model for this system would include a model for each muscle and a separate controller for each muscle pair. The horizontal eye movement models in this chapter are historical and are presented in increasing complexity with models of muscle introduced out of sequence so that their importance is fully realized. Not every oculomotor model is discussed. A few are presented for illustrative purposes.

Fig. 7.5 Main sequence diagrams for positive saccades. Similar shapes are observed for negative saccades. (a) Peak velocity–saccade magnitude, (b) saccade duration–saccade magnitude, and (c) latent period–saccade magnitude for 26 saccadic movements by a single subject (adapted from: Enderle 1988).

7.3 WESTHEIMER'S SACCADIC EYE MOVEMENT MODEL

The first quantitative saccadic horizontal eye movement model, illustrated in Fig. 7.6, was published by Westheimer in 1954. Based on visual inspection of a recorded 20° saccade and the assumption of a step controller, Westheimer proposed a second order model (Eq. 7.2) that follows directly from Fig. 7.6:

$$J\ddot{\theta} + B\dot{\theta} + K\theta = \tau(t) \tag{7.2}$$

Laplace analysis is used to analyze the characteristics of this model and compare it to data. Taking the Laplace transform of Eq. (7.2) with zero initial conditions yields

$$s^2 J\theta + sB\theta + K\theta = \tau(s)$$
$$(s^2 J + sB + K)\theta = \tau(s) \tag{7.3}$$

The transfer function of Eq. (7.3), written in standard form, is given by

$$H(s) = \frac{\theta(s)}{\tau(s)} = \frac{\dfrac{\omega_n^2}{K}}{s^2 + 2\zeta\omega_n s + \omega_n^2} \tag{7.4}$$

where $\omega_n = \sqrt{\dfrac{K}{J}}$, and $\zeta = \dfrac{B}{2\sqrt{KJ}}$. Based on the saccade trajectory for a 20° saccade, Westheimer estimated $\omega_n = 120$ rads per second, and $\zeta = 0.7$. With the input $\tau(s) = \gamma/s$, $\theta(t)$ is determined as

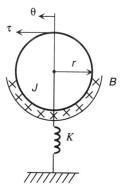

Fig. 7.6 Westheimer's second-order model of the saccade system. The parameters J, B, and K are rotational elements for moment of inertia, friction, and stiffness, respectively, and represent the eyeball and its associated viscoelasticity. The torque applied to the eyeball by the lateral and medial rectus muscles is given by $\tau(t)$, and θ is the angular eye position. The radius of the eyeball is r.

$$\theta(t) = \frac{\gamma}{K} \left[1 + \frac{e^{-\zeta\omega_n t}}{\sqrt{1 - \zeta^2}} \cos(\omega_d t + \phi) \right] \qquad (7.5)$$

where $\omega_d = \omega_n \sqrt{1 - \zeta^2}$ and $\phi = \pi + \tan^{-1} \frac{-\zeta}{\sqrt{1 - \zeta^2}}$.

Example Problem 7.1

Show the intermediate steps in going from Eq. (7.4) to Eq. (7.5).

Solution

Substituting the input, $\tau(s) = \frac{\gamma}{s}$, into Eq. (7.4) yields

$$\theta(s) = \frac{\gamma\omega_n^2}{Ks(s^2 + 2\zeta\omega_n s + \omega_n^2)}$$

Assuming a set of complex roots based on the estimates from Westheimer, a partial fraction expansion gives

$$\theta(s) = \frac{\gamma}{Ks} + \frac{\dfrac{\gamma}{2K\left((\zeta^2 - 1) - j\zeta\sqrt{1 - \zeta^2}\right)}}{s + \zeta\omega_n - j\omega_n\sqrt{1 - \zeta^2}} + \frac{\dfrac{\gamma}{2K\left((\zeta^2 - 1) + j\zeta\sqrt{1 - \zeta^2}\right)}}{s + \zeta\omega_n + j\omega_n\sqrt{1 - \zeta^2}}$$

$$= \frac{\gamma}{Ks} + \frac{|M|e^{j\phi}}{s + \zeta\omega_n - j\omega_n\sqrt{1 - \zeta^2}} + \frac{|M|e^{-j\phi}}{s + \zeta\omega_n + j\omega_n\sqrt{1 - \zeta^2}}$$

$|M|$ is the magnitude of the partial fraction coefficient (numerator of either of the complex terms); that is,

$$|M| = \frac{\gamma}{2K\sqrt{(\zeta^2 - 1)^2 + \zeta^2(1 - \zeta^2)}} = \frac{\gamma}{2K\sqrt{1 - \zeta^2}}$$

ϕ, the phase angle, is found by first removing the imaginary term from the denominator of the first partial fraction coefficient by multiplying by the complex conjugate and then rearranging terms; that is,

$$\frac{\gamma}{2K\left((\zeta^2 - 1) - j\zeta\sqrt{1 - \zeta^2}\right)}$$

$$= \frac{\gamma\left((\zeta^2 - 1) + j\zeta\sqrt{1 - \zeta^2}\right)}{2K\left((\zeta^2 - 1) - j\zeta\sqrt{1 - \zeta^2}\right)\left((\zeta^2 - 1) + j\zeta\sqrt{1 - \zeta^2}\right)}$$

$$= \frac{\gamma\left(-(\sqrt{1-\zeta^2})^2 + j\zeta\sqrt{1-\zeta^2}\right)}{2K\left((\zeta^2-1) - j\zeta\sqrt{1-\zeta^2}\right)\left((\zeta^2-1) + j\zeta\sqrt{1-\zeta^2}\right)}$$

$$= \gamma\sqrt{1-\zeta^2} \ \frac{(-\sqrt{1-\zeta^2} + j\zeta)}{2K\left((\zeta^2-1) - j\zeta\sqrt{1-\zeta^2}\right)\left((\zeta^2-1) + j\zeta\sqrt{1-\zeta^2}\right)}$$

As shown in the figure in this solution, the phase angle of the partial fraction coefficient from the previous equation is then given by

$$\phi = \pi + \psi = \pi + \tan^{-1}\frac{\zeta}{-\sqrt{1-\zeta^2}}$$

Note that the hypotenuse of the triangle in the figure is 1.

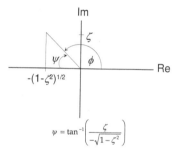

Returning to the time domain yields Eq. (7.5) by noting that the form of the solution for the complex terms is

$$2|M|e^{-\zeta\omega_n t}\cos\left(\omega_n\sqrt{1-\zeta^2}\ t + \phi\right)$$

It is always helpful to check analysis results and one easy point to check is usually at time zero:

$$\theta(0) = \frac{\gamma}{K}\left[1 + \frac{\cos(\phi)}{\sqrt{1-\zeta^2}}\right] = \frac{\gamma}{K}\left[1 + \frac{-\sqrt{1-\zeta^2}}{\sqrt{1-\zeta^2}}\right] = 0$$

since the saccade starts at primary position or $\theta(0) = 0$.   ∎

To fully explore the quality of a model, it is necessary to compare its performance against the data. For a saccade, convenient metrics are time to peak overshoot, which gives an indication of saccade duration, and peak velocity. These metrics were discussed in Section 7.2.1 when the main sequence diagram was described.

The time to peak overshoot of saccade model, T_p, is found by first calculating

$$\frac{\partial \theta}{\partial t} = \frac{\gamma \, e^{-\zeta \omega_n t}}{K\sqrt{1 - \zeta^2}} \left[-\zeta \omega_n \cos(\omega_d t + \phi) - \omega_d \sin(\omega_d t + \phi) \right] \quad (7.6)$$

using the chain rule on Eq. (7.5) and then determining T_p from $\left. \frac{\partial \theta}{\partial t} \right|_{t=T_p} = 0$, yielding

$$T_p = \frac{\pi}{\omega_n \sqrt{1 - \zeta^2}} \quad (7.7)$$

With Westheimer's parameter values, $T_p = 37$ ms for saccades of all sizes, which is independent of saccade magnitude and not in agreement with the experimental data that have a duration which increases as a function of saccade magnitude as presented in Fig. 7.5.

Example Problem 7.2

Show that Eq. (7.7) follows from Eq. (7.6) set equal to zero. Find the value of $\theta(T_p)$.

Solution:

With

$$\frac{\gamma \, e^{-\zeta \omega_n t}}{K\sqrt{1 - \zeta^2}} \left[-\zeta \omega_n \cos(\omega_d t + \phi) - \omega_d \sin(\omega_d t + \phi) \right] = 0$$

the terms multiplying the sinusoids are removed since they do not equal zero. Therefore,

$$-\zeta \omega_n \cos(\omega_d t + \phi) = \omega_d \sin(\omega_d t + \phi) = \omega_n \sqrt{1 - \zeta^2} \sin(\omega_d t + \phi)$$

which reduces to

$$\tan(\omega_d t + \phi) = \frac{-\zeta}{\sqrt{1 - \zeta^2}} = \tan(\phi)$$

The last term in the previous equation follows from Example Problem 7.1. Now

$$\tan(\omega_d t + \phi) = \tan(\phi)$$

only when $\omega_d t = n\pi$, or $t = \frac{n\pi}{\omega_d}$. The time to peak overshoot is the smallest value of n that satisfies $t = \frac{n\pi}{\omega_d}$, which is $n = 1$. Thus, with $t = T_p$

$$T_p = \frac{\pi}{\omega_d} = \frac{\pi}{\omega_n \sqrt{1 - \zeta^2}}$$

With T_p substituted into Eq. (7.5), the size of the saccade at time of peak overshoot is

$$\theta(T_p) = \frac{\gamma}{K}\left[1 + \frac{e^{-\zeta\omega_n \frac{\pi}{\omega_n\sqrt{1-\zeta^2}}}}{\sqrt{1-\zeta^2}}\cos\left(\omega_d \frac{\pi}{\omega_n\sqrt{1-\zeta^2}} + \phi\right)\right]$$

$$= \frac{\gamma}{K}\left[1 + \frac{e^{-\zeta\omega_n \frac{\pi}{\omega_n\sqrt{1-\zeta^2}}}}{\sqrt{1-\zeta^2}}\cos(\pi + \phi)\right]$$

$$= \frac{\gamma}{K}\left[1 + \frac{e^{-\zeta\omega_n \frac{\pi}{\omega_n\sqrt{1-\zeta^2}}}}{\sqrt{1-\zeta^2}}\sqrt{1-\zeta^2}\right]$$

$$= \frac{\gamma}{K}\left(1 + e^{-\zeta \frac{\pi}{\sqrt{1-\zeta^2}}}\right)$$

The predicted saccade peak velocity, $\dot{\theta}(t_{pv})$, is found by first calculating

$$\frac{\partial^2\theta}{\partial t^2} = \frac{-\gamma\, e^{-\zeta\omega_n t}}{K\sqrt{1-\zeta^2}}\left(-\zeta\omega_n(\zeta\omega_n\cos(\omega_d t + \phi) + \omega_d\sin(\omega_d t + \phi))\right.$$

$$\left. + (-\zeta\omega_n\omega_d\sin(\omega_d t + \phi) + \omega_d^2\cos(\omega_d t + \phi))\right) \tag{7.8}$$

and then determining time at peak velocity, t_{pv} from $\left.\dfrac{\partial^2\theta}{\partial t^2}\right|_{t=t_{pv}} = 0$, yielding

$$t_{pv} = \frac{1}{\omega_d}\tan^{-1}\left(\frac{\sqrt{1-\zeta^2}}{\zeta}\right) \tag{7.9}$$

Substituting t_{pv} into Eq. (7.6) gives the peak velocity $\dot{\theta}(t_{pv})$. Using Westheimer's parameter values with any arbitrary saccade magnitude given by $\Delta\theta = \dfrac{\gamma}{K}$ and Eq. (7.6) gives

$$\dot{\theta}(t_{pv}) = 55.02\Delta\theta \tag{7.10}$$

Equation (7.10) indicates that peak velocity is directly proportional to saccade magnitude. As illustrated in the main sequence diagram shown in Fig. 7.5, experimental peak velocity data have an exponential form and do not represent a linear function as predicted by the Westheimer model. Both Eqs. (7.7) and (7.10) are consistent with linear systems theory. That is, for a step input to a linear system, the duration (and time to peak overshoot) stays constant regardless of the size of the input, and the peak velocity increases with the size of the input.

Example Problem 7.3

Show that Eq. (7.9) follows from Eq. (7.8).

Solution:

With

$$\frac{-\gamma\, e^{-\zeta\omega_n t}}{K\sqrt{1-\zeta^2}}\left(-\zeta\omega_n(\zeta\omega_n \cos(\omega_d t + \phi) + \omega_d \sin(\omega_d t + \phi))\right.$$
$$\left. + (-\zeta\omega_n \omega_d \sin(\omega_d t + \phi) + \omega_d^2 \cos(\omega_d t + \phi))\right) = 0$$

the terms multiplying the sinusoids are removed since they do not equal zero. Therefore,

$$\left((\omega_d^2 - \zeta^2\omega_n^2)\cos(\omega_d t + \phi)\right) - (2\zeta\omega_n \omega_d \sin(\omega_d t + \phi)) = 0$$

which reduces to

$$\tan(\omega_d t + \phi) = \frac{\omega_d^2 - \zeta^2\omega_n^2}{2\zeta\omega_n \omega_d} = \frac{\omega_n^2(1 - \zeta^2) - \zeta^2\omega_n^2}{2\zeta\omega_n \omega_n \sqrt{1-\zeta^2}}$$

$$= \frac{1 - 2\zeta^2}{2\zeta\sqrt{1-\zeta^2}} = \left(1 - \frac{1}{2\zeta^2}\right)\tan(\psi)$$

The last term in the previous expression makes use of $\tan(\psi) = \dfrac{-\zeta}{\sqrt{1-\zeta^2}}$ as given

in Example Problem 7.1. Also note that $\tan(\psi) = \tan(\phi)$ as evident from the figure in Example Problem 7.1. Now the trigonometric identity

$$\tan(\omega_d t + \phi) = \frac{\tan(\omega_d t) + \tan(\phi)}{1 - \tan(\omega_d t)\tan(\phi)}$$

is used to simplify the previous expression involving $\tan(\omega_d t + \phi)$, which after substituting yields

$$\tan(\omega_d t) + \tan(\phi) = (1 - \tan(\omega_d t)\tan(\phi)) \times \left(1 - \frac{1}{2\zeta^2}\right)\tan(\psi)$$

$$= \tan(\psi) - \frac{1}{2\zeta^2}\tan(\psi) - \left(1 - \frac{1}{2\zeta^2}\right)\tan(\psi)\tan(\omega_d t)\tan(\phi)$$

Using $\tan(\psi) = \tan(\phi)$ and rearranging the previous equation gives

$$\tan(\omega_d t)\left[1 + (\tan(\phi))^2\left(1 - \frac{1}{2\zeta^2}\right)\right] = \frac{-\tan(\phi)}{2\zeta^2}$$

or

$$\tan(\omega_d t) = \frac{-\tan(\phi)}{2\zeta^2\left(1 + (\tan(\phi))^2\left(\frac{2\zeta^2 - 1}{2\zeta^2}\right)\right)} = \frac{-\tan(\phi)}{2\zeta^2 + (\tan(\phi))^2(2\zeta^2 - 1)}$$

With $\tan(\phi) = \dfrac{-\zeta}{\sqrt{1 - \zeta^2}}$, the previous expression reduces to

$$\tan(\omega_d t) = \frac{\dfrac{\zeta}{\sqrt{1 - \zeta^2}}}{2\zeta^2 + \dfrac{\zeta^2(2\zeta^2 - 1)}{1 - \zeta^2}} = \frac{\sqrt{1 - \zeta^2}}{\zeta}$$

or

$$t_{pv} = \frac{1}{\omega_d}\tan^{-1}\left(\frac{\sqrt{1 - \zeta^2}}{\zeta}\right)$$

Westheimer (1954) noted the differences between saccade duration–saccade magnitude and peak velocity–saccade magnitude in the model and the experimental data and inferred that the saccade system was not linear because the peak velocity–saccade magnitude plot was nonlinear. He also noted that the input was not an abrupt step function. Overall, this model provided a satisfactory fit to the eye position data for a saccade of 20° but not for saccades of other magnitudes. Interestingly, Westheimer's second-order model proves to be an adequate model for saccades of all sizes if a different input function, as described in the next section, is assumed. Due to its simplicity, the Westheimer model of the oculomotor plant is still popular today.

7.4   THE SACCADE CONTROLLER

One of the challenges in modeling physiological systems is the lack of data or information about the input to the system. For instance, in the fast eye movement system the input is the neurological signal from the CNS to the muscles connected to the eyeball. Information about the input is not available in this system since it involves thousands of neurons firing at a very high rate. Recording the signal would involve invasive surgery and instrumentation that is not yet available. The difficulty in modeling this system, as well as most physiological systems, is the lack of information about the input. Often, however, it is possible to obtain information about the input via indirect means as described in this section for the fast eye movement system.

In 1964, Robinson performed an experiment in an attempt to measure the input to the eyeballs during a saccade. To record the measurement, one eye was held fixed using a suction contact lens while the other eye performed a saccade from target to target. Since the same innervation signal is sent to both eyes during a saccade, Robinson inferred that the input, recorded through the transducer attached to the fixed eyeball, was the same input driving the other eyeball. He proposed that muscle tension driving the eyeballs during a saccade is a pulse plus a step, or simply, a pulse-step input (Fig. 7.7).

Microelectrode studies have been carried out to record the electrical activity in oculomotor neurons. Figure 7.8 illustrates a micropipet being used to record the activity in the oculomotor nucleus, an important neuron population responsible for driving a saccade. Additional experiments on oculomotor muscle have been carried out to learn more about the saccade controller since Robinson's initial study. In 1975, for instance, Collins and coworkers reported using a miniature "C"-gauge force transducer to measure muscle tension *in vivo* at the muscle tendon during unrestrained human eye movements. This type of study has allowed a better understanding of the tensions exerted by each muscle rather than the combined effect of both muscles as shown in Fig. 7.7.

It is important to distinguish between the tension or force generated by a muscle, called muscle tension, and the force generator within the muscle, called the **active-state tension generator.** The active-state tension generator creates a force within the muscle that is transformed through the internal elements of the muscle into the muscle tension. Muscle tension is external and measurable, and the active-state tension is internal and unmeasurable. Active-state tension follows most closely the neural input to the muscle. From Fig. 7.8, a pattern of neural activity is observed as follows:

1. The muscle that is being contracted (agonist) is stimulated by a pulse, followed by a step to maintain the eyeball at its destination.
2. The muscle that is being stretched (antagonist) is unstimulated during the saccade (stimulated by a pause or a negative pulse to zero), followed by a step to maintain the eyeball at its destination.

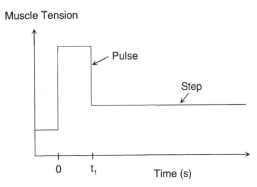

Fig. 7.7. Diagram illustrating the muscle tension recorded during a saccade.

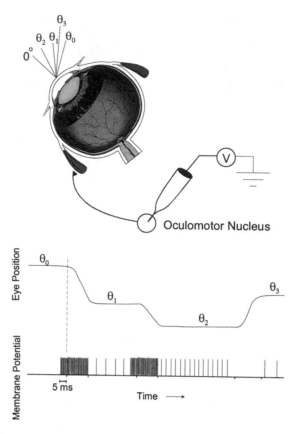

Fig. 7.8 Diagram of recording of a series of saccades using a micropipet and the resultant electrical activity in a single neuron. Spikes in the membrane potential indicate an action potential occurred with dynamics as described in Chapter 3. Saccade neural activity initiates with a burst of neural firing approximately 5 ms before the eye begins to move and continues until the eye has almost reached its destination, a process that is described in Section 7.9. Relative position of the eye is shown at the top with angles θ_0–θ_3. Initially the eye starts in position θ_0, a position in the extremity in which the muscle is completely stretched with zero input. To move the eye from θ_0 to θ_1, neural burst firing occurs. To maintain the eye at θ_1, a steady firing occurs in the neuron. The firing rate for fixation is in proportion to the shortness of the muscle. Next, the eye moves from θ_1 to θ_2. This saccade moves much more slowly than the first saccade with approximately the same duration as the first. The firing level is also approximately at the same level as the first. The difference in input corresponds to fewer neurons firing to drive the eye to its destination, which means a smaller input than the first saccade. Because the muscle is shorter after completing this saccade, the fixation firing rate is higher than that of the previous position at θ_1. Next, the eye moves in the opposite direction to θ_3. Since the muscle is lengthening, the input to the muscle is zero; that is, no action potentials are used to stimulate the muscle. The fixation firing level θ_3 is less than that for θ_1 because the muscle is longer.

Figure 7.9 quantifies these relationships for the agonist neural input, N_{ag}, and the antagonist neural input, N_{ant}. The pulse input is required to get the eye to the target as soon as possible, and the step is required to keep the eye at that location.

It has been reported that the active-state tensions are not identical to the neural controllers but can be described by low-pass filtered pulse-step waveforms. The active-state tensions are shown in Fig. 7.9 as dashed lines with time-varying time constants τ_{ac} and τ_{de}. It is thought that the low-pass filtering involves the movement of Ca^{2+} across the cell membrane. Some investigators have reported a different set of time constants for the agonist and antagonist activity, and others have noted a firing frequency-dependent agonist activation time constant. Others suggest that the agonist activation time constant is a function of saccade magnitude. For simplicity in this textbook, activation and deactivation time constants are assumed to be identical for both agonist and antagonist activity.

In 1964, Robinson described a model for fast eye movements (constructed from empirical considerations) which simulated saccades over a range of 5 to 40° by

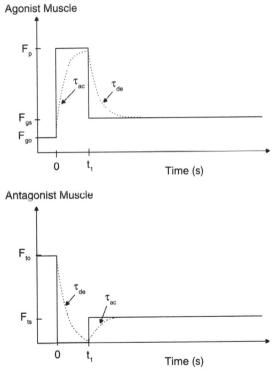

Fig. 7.9 Agonist, N_{ag}, and antagonist, N_{ant}, control signals (solid lines) and the agonist, F_{ag}, and antagonist, F_{ant}, active-state tensions (dashed lines). Note that the time constant for activation, τ_{ac}, is different from the time constant for deactivation, τ_{de}. The time interval, t_1, is the duration of the pulse.

changing the amplitude of the pulse-step input. These simulation results were adequate for the position–time relationship of a saccade, but the velocity–time relationship was inconsistent with physiological evidence. To correct this deficiency of the model, physiological and modeling studies of the oculomotor plant were carried out from the 1960s through the 1990s that allowed the development of a more homeomorphic oculomotor plant. Essential to this work was the construction of oculomotor muscle models.

7.5 DEVELOPMENT OF AN OCULOMOTOR MUSCLE MODEL

It is clear that an accurate model of muscle is essential in the development of a model of the horizontal fast eye movement system that is driven by a pair of muscles (lateral and medial rectus muscles). In fact, the Westheimer model does not include any muscle model and relies solely on the inertia of the eyeball, friction between the eyeball and socket, and elasticity due to the optic nerve and other attachments as the elements of the model. In this section, the historical development of a muscle model is discussed as it relates to the oculomotor system. Muscle model research involves a broad spectrum of topics, ranging from micro models that deal with the sarcomeres to macro models in which collections of cells are grouped into a lumped parameter system and described with ordinary mechanical elements. Here the focus is on a macro model of the oculomotor muscle based on physiological evidence from experimental testing. The model elements, as presented, consist of an active-state tension generator (input), elastic elements, and viscous elements. Each element is introduced separately and the muscle model is incremented in each subsection. It should be noted that the linear muscle model presented at the end of this section completely revises the subsections preceeding it. The subsections before the last are presented because of their historical significance and to gain an appreciation of the current muscle model.

7.5.1 Passive Elasticity

Consider the experiment of stretching an unexcited muscle and recording tension to investigate the passive elastic properties of muscle. The data curve shown in Fig. 7.10 is a typical recording of the tension observed in an eye rectus muscle. The tension required to stretch a muscle is a *nonlinear* function of distance. Thus, in order to precisely model this element, a nonlinear spring element should be used. Note that the change in length at 0 refers to the length of the muscle at primary position (looking straight ahead). Thus, the eye muscles are stretched, approximately 3 mm, when the eye movement system is at rest in primary position. At rest, the muscle length is approximately 37 mm.

To be useful in a linear model of muscle, Fig. 7.10 must be linearized in the vicinity of an operating point. The operating point should be approximately centered in the region in which the spring operates. In Fig. 7.10, a line tangent to the curve at primary position provides a linear approximation to the spring's behavior in this region. For ease in analysis, the following relationships hold for a sphere representing the eyeball radius of 11 mm:

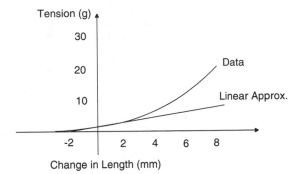

Fig. 7.10 Diagram illustrating the tension-displacement curve for unexcited muscle. The slope of the linear approximation to the data is muscle passive elasticity, K_{pe}.

$$1 \text{ g} = 9.806 \times 10^{-3} \, N$$

$$1° = 0.192 \text{ mm} = 1.92 \times 10^{-4} \text{ m}$$

The slope of the line, K_{pe}, is approximately

$$K_{pe} = 0.2 \frac{g}{°} = 0.2 \frac{g}{°} \times \frac{9.806 \times 10^{-3} \, N}{1 \text{ g}} \times \frac{1°}{1.92 \times 10^{-4} \, m} = 10.2 \frac{N}{m}$$

and represents the elasticity of the passive elastic element. The choice of the operating region is of vital importance concerning the slope of the curve. At this time, a point in the historical operating region of rectus muscle is used. In most of the oculomotor literature, the term K_{pe} is typically subtracted out of the analysis and is not used. Later, the operating point will be revisited and this element will be completely removed from the model.

7.5.2 Active-State Tension Generator

In general, a muscle produces a force in proportion to the amount of stimulation. The element responsible for the creation of force is the active-state tension generator. Note that this terminology is used so that there is no confusion with the force created within the muscle when the tension created by the muscle is discussed. The active-tension generator is included along with the passive elastic element in the muscle model as shown in Fig. 7.11. The relationship between tension, T, active-state tension, F, and elasticity is given by

$$T = F - K_{pe}x \tag{7.11}$$

Isometric (constant length) experiments have been performed on humans over the years to estimate the active-tension generator at different levels of stimulation. These experiments were usually performed in conjunction with strabismus surgery when muscles were detached and reattached to correct crossed eyes. Consider the tension created by a muscle when stimulated as a function of length as shown in Fig.

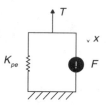

Fig. 7.11 Diagram illustrating a muscle model consisting of an active-state tension generator, F, and a passive elastic element, K_{pe}. Upon stimulation of the active-state tension generator a tension, T, is exerted by the muscle.

7.12. The data were collected from the lateral rectus muscle that was detached from one eyeball while the other unoperated eyeball fixated at different locations in the nasal (N) and temporal (T) directions from −45 to 45°. This experiment was carried out under the assumption that the same neural input is sent to each eyeball (Hering's law of equal innervation); thus the active-state tension in the freely moving eyeball should be the same as that in the detached lateral rectus muscle. At each

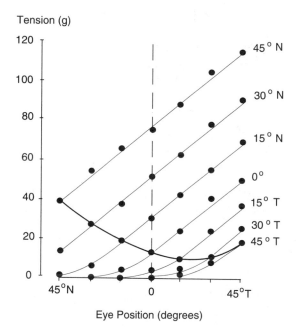

Fig. 7.12. Length–tension curves for lateral rectus muscle at different levels of activation. Dots represent tension data recorded from the detached lateral rectus muscle during strabismus surgery while the unoperated eyeball fixated at targets from –45 to 45° (adapted from Collins *et al.*, 1975).

fixation point, the detached lateral rectus muscle was stretched and tension data were recorded at each of the points indicated on the graph. The thick line represents the muscle tension at that particular eye position under normal conditions. The curve for 45° T is the zero stimulation case and represents the passive mechanical properties of muscle. The results are similar to those reported in Fig. 7.10. Note that the tension generated is a *nonlinear* function of the length of the muscle.

To compare the model in Fig. 7.11 against the data in Fig. 7.12, it is convenient to subtract the passive elasticity in the data (represented by the 45° T curve) from each of the data curves 30° T through 45° N, leaving only the hypothetical active-state tension. Shown in the graph on the left in Fig. 7.13 is one such calculation for 15° N, with the active-state tension given by the dashed line. The other curves in Fig. 7.12 give similar results and have been omitted because of the clutter. The dashed line should represent the active-state tension, which appears to be a function of length. If this was a *pure* active-state tension element, the subtracted curve should be a horizontal line indicative of the size of the input. One such input is shown for the active-state tension in the graph on the right in Fig. 7.13. The result in Fig. 7.13 implies that either the active-state tension's effect is a nonlinear element, i.e., there may be other nonlinear or linear elements missing in the model, or perhaps some of the assumptions made in the development of the model are wrong. For the moment, consider that the analysis is correct and assume that some elements are missing. This topic will be revisited later.

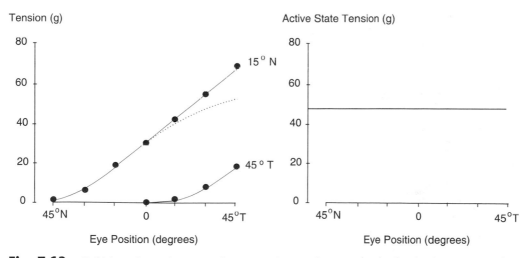

Fig. 7.13. (Left) Length–tension curves for extraocular muscle at two levels of activation corresponding to the 15° N and 45° T positions. Dots represent tension data recorded from the detached lateral rectus muscle during strabismus surgery while the unoperated eyeball fixated at targets. Dashed line is the 15° curve with the 45° curve subtracted from it. The resultant dashed curve represents the active state tension as a function of eye position. (Right) The theoretical graph for active-state tension vs eye position as given by Eq. (7.11) (adapted from Collins *et al.*, 1975).

7.5.3 Elasticity

With the effects of the passive elasticity removed from the tension data as shown in Fig. 7.13, a relationship still exists between length and tension as previously described. To account for the relationship between length and tension, a new elastic element is added into the model as shown in Fig. 7.14 and described by Eq. (7.12):

$$T = F - K_{pe}x - Kx \qquad (7.12)$$

The new elastic element, K, accounts for the slope of the subtracted curve shown with dashes in the graph on the left in Fig. 7.13. The slope of the line, K, at primary position is approximately 0.8 g/° = 40.86 N/m (a value typically reported in the literature). The slope for each of the curves in Fig. 7.12 can be calculated in the same manner at primary position, with the resultant slopes all approximately equal to the same value as the one for 15° N. At this time, the introduction of additional experiments will provide further insight to the development of the muscle model.

Series Elastic Element

Experiments carried out by Levin and Wyman (1927), and Collins (1975) indicated the need for a series elasticity element in addition to the other elements previously presented in the muscle model. The experimental setup and typical data from the experiment are shown in Fig. 7.15. The protocol for this experiment, called the quick release experiment, is as follows: (i) A weight is hung onto a muscle, (ii) the muscle is fully stimulated at time t_1, and (iii) the weight is released at time t_2. At time t_2, the muscle changes length almost instantaneously when the weight is released. The only element that can instantaneously change its length is a spring. Thus to account for this behavior, a spring, called the series elastic element, K_{se}, is connected in series to the active-state tension element. The updated muscle model is shown in Fig. 7.16.

Based on the experiment carried out by Collins (1975) on rectus eye muscle, an estimate for K_{se} was given as 125 N/m (2.5 g/°). Since the value of K_{se} does not equal the value of K, another elastic element is needed to account for this behavior.

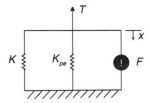

Fig. 7.14 Diagram illustrating a muscle model consisting of an active-state tension generator F, passive elastic element, K_{pe}, and elastic element, K. Upon stimulation of the active-state tension generator, a tension T is exerted by the muscle.

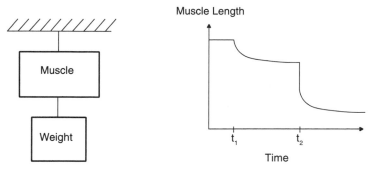

Fig. 7.15 Diagram illustrating the quick release experiment. (Left) The physical setup of the experiment. (Right) Typical data from the experiment. At time t_1 the muscle is fully stimulated and at time t_2 the weight is released.

Length Tension Elastic Element

Given the inequality between K_{se} and K, another elastic element, called the length–tension elastic element, K_{lt}, is placed in parallel with the active-state tension element as shown in the illustration on the left in Fig. 7.17. For ease of analysis, K_{pe} is subtracted out (removed) using the graphical technique shown in Fig. 7.13. To estimate a value for K_{lt}, the muscle model shown on the right in Fig. 7.17 is analyzed and reduced to an expression involving K_{lt}. Analysis begins by summing the forces acting on nodes 1 and 2.

$$T = K_{se}(x_2 - x_1) \tag{7.13}$$

$$F = K_{lt}x_2 + K_{se}(x_2 - x_1) \quad \rightarrow \quad x_2 = \frac{F + K_{se}x_1}{K_{se} + K_{lt}} \tag{7.14}$$

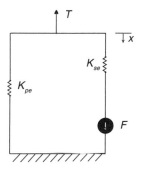

Fig. 7.16 Diagram illustrating a muscle model consisting of an active-state tension generator F, passive elastic element K_{pe}, and series elastic element K_{se}. Upon stimulation of the active-state tension generator F, a tension T is exerted by the muscle.

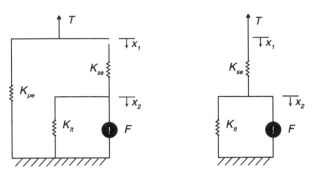

Fig. 7.17 (Left) A muscle model consisting of an active-state tension generator F in parallel to a length–tension elastic element K_{lt}, connected to a series elastic element K_{se}, all in parallel with the passive elastic element K_{pe}. Upon stimulation of the active-state tension generator F, a tension T is exerted by the muscle. (Right) The same muscle model except that K_{pe} has been removed.

Substituting x_2 from Eq. (7.14) into Eq. (7.13) gives

$$T = \frac{K_{se}}{K_{se} + K_{lt}}(F + K_{se}x_1) - K_{se}x_1 = \frac{K_{se}}{K_{se} + K_{lt}}F - \frac{K_{se}K_{lt}}{K_{se} + K_{lt}}x_1 \qquad (7.15)$$

Equation (7.15) is an equation for a straight line with y-intercept $\dfrac{K_{se}}{K_{se} + K_{lt}}F$ and slope $\dfrac{K_{se}K_{lt}}{K_{se} + K_{lt}}$. The slope of the length–tension curve in Fig. 7.11 is given by $K = 0.8\ g/° = 40.86$ N/m. Therefore,

$$K = \frac{K_{se}K_{lt}}{K_{se} + K_{lt}} = 40.86\ N/m \qquad (7.16)$$

Solving Eq. (7.16) for K_{lt} yields

$$K_{lt} = \frac{K_{se}K}{K_{se} - K} = 60.7\ N/m \qquad (7.17)$$

7.5.4　Force–Velocity Relationship

Early experiments indicated that muscle had elastic as well as viscous properties. Muscle was tested under isotonic (constant force) experimental conditions as shown in Fig. 7.18 to investigate muscle viscosity. The muscle and load were attached to a lever with a high lever ratio. The lever reduced the gravity force (mass × gravity) of the load at the muscle by one over the lever ratio and the inertial force (mass × acceleration) of the load by one over the lever ratio squared. With this arrangement,

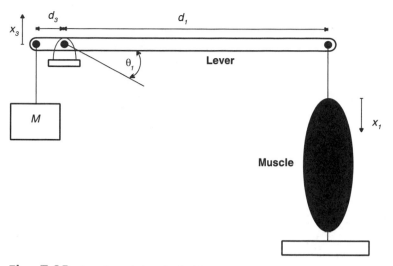

Fig. 7.18 Drawing of the classical isotonic experiment with inertial load and muscle attached to the lever. The muscle is stretched to its optimal length according to experimental conditions and attached to ground.

it was assumed that the inertial force exerted by the load during isotonic shortening could be ignored. The second assumption was that if mass was not reduced enough by the lever ratio (enough to be ignored), then taking measurements at maximum velocity provided a measurement at a time when acceleration is zero, and therefore inertial force equals zero. If these two assumptions are valid, then the experiment would provide data free of the effect of inertial force as the gravity force is varied.

According to the experimental conditions, the muscle is stretched to its optimal length at the start of the isotonic experiment. The isotonic experiment begins by attaching a load M, stimulating the muscle, and recording position. The two curves in Fig. 7.19 depict the time course for the isotonic experiment for a small and

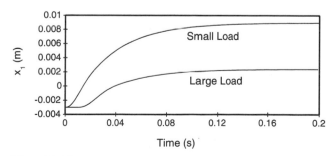

Fig. 7.19 Diagram illustrating typical response of a muscle stimulated with a large and small load.

large load. Notice that the durations of both responses are approximately equal regardless of the load, despite the apparent much longer time delay associated with the large load. Also notice that the heavier the load, the less the total shortening. Maximum velocity is calculated numerically from the position data. To estimate muscle viscosity, this experiment is repeated with many loads at the same stimulation level and maximum velocity is calculated. Figure 7.20 illustrates the typical relationship between load ratio (P/P_0) and maximum velocity, where $P = Mg$ and P_0 is the isometric tension (the largest weight that the muscle can move) for maximally stimulated muscle. This curve is usually referred to as the **force–velocity curve.**

Clearly, the force–velocity curve is nonlinear and follows a hyperbolic shape. If a smaller stimulus than maximum is used to stimulate the muscle, then a family of force–velocity curves results as shown in Fig. 7.21. Each curve is generated with a different active-state tension as indicated. The force–velocity characteristics in Fig. 7.21 are similar to those shown in Fig. 7.20. In particular, the slope of the force–velocity curve for a small value of active-state tension is quite different than that for a large value of the active-state tension in the operating region of the eye muscle (i.e., approximately 800°/s).

To include the effects of viscosity from the isotonic experiment in the muscle model, a viscous element is placed in parallel with the active-state tension generator and the length–tension elastic element as shown in Fig. 7.22. The impact of this element is examined by analyzing the behavior of the model in Example Problem 7.4 by simulating the conditions of the isotonic experiment. At this stage, it is assumed that the viscous element is linear in this example. For simplicity, the lever is removed along with the virtual acceleration term $M\ddot{x}_1$. A more thorough analysis including the lever is considered in Section 7.6.2. For simplicity, the passive elastic element K_{pe} is removed from the diagrams and analysis.

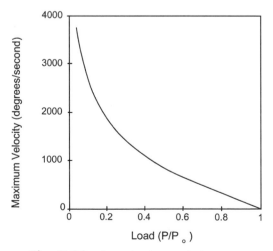

Fig. 7.20   Illustrative force–velocity curve.

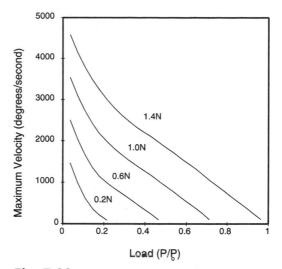

Fig. 7.21 Illustrative family of force–velocity curves for active-state tensions ranging from 1.4 to 0.2 N.

Example Problem 7.4

Consider the system shown in the figure that represents a model of the isotonic experiment. Assume that the virtual acceleration term $M\ddot{x}_1$ can be ignored. Calculate and plot maximum velocity as a function of load.

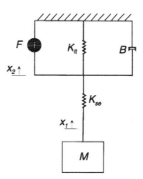

Solution

Assume that $\dot{x}_2 > \dot{x}_1$ and that the mass is supported so that $\dot{x}_1 > 0$. Let the term $K_{st} = K_{se} + K_{lt}$. Summing the forces acting on nodes 1 and 2 gives

$$Mg = K_{se}(x_2 - x_1) \quad \rightarrow \quad x_1 = x_2 - \frac{Mg}{K_{se}}$$

$$F = B\dot{x}_2 + K_{lt}x_2 + K_{se}(x_2 - x_1)$$

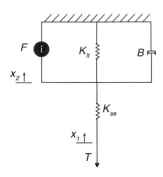

Fig. 7.22 Diagram illustrates a muscle model consisting of an active-state tension generator F in parallel with a length–tension elastic element K_{lt} and viscous element B, connected to a series elastic element K_{se}. The passive elastic element K_{pe} has been removed from the model for simplicity. Upon stimulation of the active-state tension generator F, a tension T is exerted by the muscle.

Substituting x_1 into the second equation yields

$$F = B\dot{x}_2 + K_{lt}x_2 + Mg$$

Solving previous equation for x_2 and $\dot{x}_2$ gives

$$x_2(t) = \frac{F - Mg}{K_{lt}}\left(1 - e^{\frac{-K_{lt}t}{B}}\right)$$

$$\dot{x}_2(t) = \frac{F - Mg}{B}\,e^{\frac{-K_{lt}t}{B}}$$

Maximum velocity, V_{max}, for all loads is given by $V_{max} = \dfrac{F - Mg}{B}$ and $\dot{x}_1 = \dot{x}_2$ since $\dot{x}_1 = \dfrac{d}{dt}\left(x_2 - \dfrac{Mg}{K_{se}}\right)$. The graph depicts a linear relationship between maximum velocity and load.

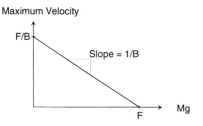

The assumption of a linear viscosity element appears to be in error since the analysis in Example Problem 7.4 predicts a linear relationship between load and maximum velocity (according to the assumptions of the solution) and the data from the isotonic experiment shown in Fig. 7.20 are clearly nonlinear. Thus, a reasonable assumption is that the viscosity element is nonlinear.

Traditionally, muscle viscosity is characterized by the nonlinear Hill hyperbola, given by

$$V_{max}(P + a) = b(P_0 - P) \tag{7.18}$$

where V_{max} is the maximum velocity, P is the external force, P_0 the isometric tension, and a and b are the empirical constants representing the asymptotes of the hyperbola. As described previously, P_0 represents the isometric tension, which is the largest weight that the muscle can move, and P is the weight Mg. Hill's data suggests that

$$a = \frac{P_0}{4} \quad \text{and} \quad b = \frac{V_{max}}{4}$$

Therefore, with these values for a and b, the Hill equation is rewritten from Eq. (7.18) as

$$P = P_0 - \frac{V_{max}(P_0 + a)}{b + V_{max}} = P_0 - BV_{max} \tag{7.19}$$

where

$$B = \frac{P_0 + a}{b + V_{max}} \tag{7.20}$$

The term B represents the viscosity of the element. Clearly, the force due to viscosity is nonlinear due to the velocity term, V_{max}, in the denominator of Eq. (7.20).

In oculomotor models, usually V_{max} is replaced by $\dot{x}_2$, P is replaced by muscle tension, T, and P_0 is replaced by the active-state tension, F, as defined from Fig. 7.22. Therefore, Eqs. (7.19) and (7.20) are rewritten as

$$T = F - B\dot{x}_2 \tag{7.21}$$

where

$$B = \frac{F + a}{b + \dot{x}_2} \tag{7.22}$$

Example Problem 7.5

Write an equation for the tension created by the muscle model shown in Fig. 7.22.

Solution

Assuming that $\dot{x}_2 > \dot{x}_1$, the forces acting at nodes 1 and 2 give

$$T = K_{se}(x_2 - x_1) \quad \rightarrow \quad x_2 = \frac{T}{K_{se}} + x_1$$

$$F = B\dot{x}_2 + K_{lt}x_2 + K_{se}(x_2 - x_1)$$

Eliminating x_2 in the second equation by substituting from the first equation yields

$$F = B\dot{x}_2 + (K_{se} + K_{lt}) \left(\frac{T}{K_{se}} + x_1 \right) - K_{se}x_1$$

Solving for T gives

$$T = \frac{K_{se}}{K_{se} + K_{lt}} F - \frac{K_{se}K_{lt}}{K_{se} + K_{lt}} x_1 - \frac{K_{se}B}{K_{se} + K_{lt}} \dot{x}_2$$

Equation (7.22) can be substituted for parameter B in the previous equation to give a nonlinear model of oculomotor muscle. ∎

Some oculomotor investigators have reported values for a and b in the Hill equation that depend on whether the muscle is being stretched or contracted. There is some evidence to suggest that stretch dynamics are different from contraction dynamics. However, the form of the viscosity expression for muscle shortening or lengthening is given by Eq. (7.22), with values for a and b parameterized appropriately. For instance, Hsu and coworkers (1976) described the viscosity for shortening and lengthening for oculomotor muscles as

$$B_{ag} = \frac{F_{ag} + AG_a}{\dot{x}_2 + AG_b} \tag{7.23}$$

$$B_{ant} = \frac{F_{ant} - ANT_a}{\dot{x}_2 + ANT_b} \tag{7.24}$$

where AG_a, AG_b, ANT_a, and ANT_b are parameters based on the asymptotes for contracting (agonist) or stretching (antagonist), respectively.

Eqs. (7.23) and (7.24) define B_{ag} and B_{ant} as nonlinear functions of velocity. To develop a linear model of muscle, both B_{ag} and B_{ant} can be linearized by making a straight-line approximation to the hyperbolic functions shown in Fig. 7.21. Antagonist activity is typically at the 5% level ($F = 0.2$ N) and agonist activity is at the 100% level ($F = 1.4$ N).

7.6 A LINEAR MUSCLE MODEL

This section reexamines the static and dynamic properties of muscle in the development of a linear model of oculomotor muscle. The updated linear model for oculomotor muscle is shown in Fig. 7.23. Each of the elements in the model is linear and supported with physiological evidence. The muscle is modeled as a parallel combination of viscosity B_2 and series elasticity K_{se}, connected to the parallel combination of active-state tension generator F, viscosity element B_1, and length–tension elastic element K_{lt}. Variables x_1 and x_2 describe the displacement from the equilibrium for the stiffness elements in the muscle model. The only structural difference between this model and the previous oculomotor muscle model is the addition of viscous element B_2 and the removal of passive elasticity K_{pe}. As will be described,

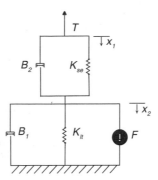

Fig. 7.23 Diagram illustrates an updated linear muscle model consisting of an active-state tension generator F in parallel with a length–tension elastic element K_{lt} and viscous element B_1, connected to a series elastic element K_{se} in parallel with a viscous element B_2. Upon stimulation of the active-state tension generator F, a tension T is exerted by the muscle.

the viscous element B_2 is vitally important to describe the nonlinear force–velocity characteristics of the muscle, and the elastic element K_{pe} is unnecessary.

The need for two elastic elements in the linear oculomotor muscle model is supported through physiological evidence. As described previously, the use and value of the series elasticity K_{se} was determined from the isotonic–isometric quick release experiment by Collins (1975). Length–tension elasticity K_{lt} was estimated in a slightly different fashion than before from the slope of the length–tension curve. Support for the two linear viscous elements is based on the isotonic experiment and estimated from simulation results presented in this section.

7.6.1 Length–Tension Curve

The basis for assuming nonlinear elasticity is the nonlinear length–tension relation for excited and unexcited muscle for tensions below 10 g as shown in Fig. 7.12. Using a miniature C-gauge force transducer, Collins (1975) measured muscle tension *in vivo* at the muscle tendon during unrestrained human eye movements. Data shown in Fig. 7.12 were recorded from the rectus muscle of the left eye by measuring the isometric tensions at different muscle lengths, ranging from eye positions of −45 to 45°, and different levels of innervation, established by directing the subject to look at the corresponding targets with the unhampered right eye from −45° T to 45° N. The change in eye position during this experiment corresponds to a change in muscle length of approximately 18 mm. Collins described the length–tension curves as "straight, parallel lines above about 10 g. Below the 10 g level, the oculorotary muscles begin to go slack." He also reported that the normal range of tensions for the rectus muscle during all eye movements never falls below 10 g into the slack region when the *in vivo* force transducer is used.

In developing a muscle model for use in the oculomotor system, it is imperative that the model accurately exhibits the static characteristics of rectus eye muscle within the normal range of operation. Thus, any oculomotor muscle model must have length–tension characteristics consisting of straight, parallel lines above 10-g tension. Since oculomotor muscles do not operate below 10 g, it is not important that the linear behavior of the model does not match this nonlinear portion in the length–tension curves observed in the data as was done in the development of the muscle model in Section 7.3. As demonstrated in this section, by concentrating on the operational region of the oculomotor muscles, accurate length–tension curves are obtained from the muscle model using series elastic and length–tension elastic elements, even when active-state tension is zero. Thus, there is no need to include a passive elastic element in the muscle model as previously required.

Since the rectus eye muscle is not in equilibrium at primary position (looking straight ahead, 0°) within the oculomotor system, it is necessary to define and account for the equilibrium position of the muscle. Equilibrium denotes the unstretched length of the muscle when the tension is zero, with zero input. It is assumed that the active-state tension is zero on the 45° T length–tension curve. Typically, the equilibrium position for rectus eye muscle is found from within the slack region, where the 45° T length–tension curve intersects the horizontal axis. Note that this intersection point was not shown in the data collected by Collins (1975) (Fig. 7.12) but is reported to be approximately 15° (3 mm short of primary position), a value that is typical of those reported in the literature.

Since the muscle does not operate in the slack region during normal eye movements, using an equilibrium point calculated from the operational region of the muscle provides a much more realistic estimate for the muscle. Here, the equilibrium point is defined according to the straight-line approximation to the 45° T length–tension curve above the slack region. The value at the intersection of the straight-line approximation with the horizontal axis gives an equilibrium point of $-19.3°$. By use of the equilibrium point at $-19.3°$, there is no need to include an additional elastic element K_{pe} to account for the passive elasticity associated with unstimulated muscle as others have done.

The tension exerted by the linear muscle model shown in Fig. 7.23 is given by

$$T = \frac{K_{se}}{K_{se} + K_{lt}} F - \frac{K_{se}K_{lt}}{K_{se} + K_{lt}} x_1 \tag{7.25}$$

With the slope of the length–tension curve equal to 0.8 g/° = 40.86 N/m in the operating region of the muscle (non-slack region), K_{se} = 2.5 g/° = 125 N/m, and Eq. (7.25) has a slope of

$$\frac{K_{se}K_{lt}}{K_{se} + K_{lt}} \tag{7.26}$$

K_{lt} is evaluated as 1.2 g/° = 60.7 N/m. To estimate the active-state tension for fixation at all locations, Eq. (7.25) is used to solve for F for each innervation level straight-line approximation, yielding

$$F = 0.4 + 0.0175\theta \ N \quad \text{for} \quad \theta > 0° \ (\text{N direction}) \tag{7.27}$$

and

$$F = 0.4 + 0.0120\theta N \quad \text{for} \quad \theta \leq 0° \text{ (T direction)} \qquad (7.28)$$

where θ is the angle that the eyeball is deviated from the primary position measured in degrees, and $\theta = 5208.7 \times (x_1 - 3.705)$. Note that $5208.7 = \dfrac{180}{\pi r}$, where r equals the radius of the eyeball.

Figure 7.24 displays a family of static length–tension curves obtained using Eqs. (7.25)–(7.28), which depicts the length–tension experiment. No attempt is made to describe the activity within the slack region since the rectus eye muscle does not normally operate in that region. The length–tension curves shown in Fig. 7.24 are in excellent agreement with the data shown in Fig. 7.12 within the operating region of the muscle.

7.6.2   Force-Velocity Relationship

The original basis for assuming nonlinear muscle viscosity is that the expected linear relation between external load and maximum velocity (see Example Problem 7.4) was not found in early experiments by Fenn and Marsh (1935). As Fenn and

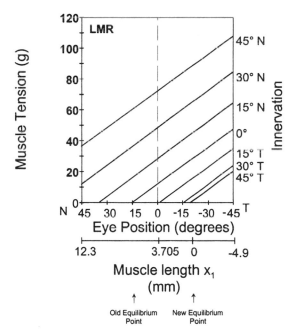

Fig. 7.24 Length–tension curve generated using Eqs. (7.25)–(7.28) derived from the linear muscle model and inputs: $F = 130$ g for 45° N, $F = 94.3$ g for 30° N, $F = 64.9$ g for 15° N, $F = 40.8$ g for 0°, $F = 21.7$ g for 15°T, $F = 5.1$ g for 30°T, and $F = 0$ g for 45°T. These lines were parameterized to match Fig. 7.12.

Marsh reported, "If the muscle is represented accurately by a viscous elastic system this force–velocity curve should have been linear, the loss of force being always proportional to the velocity. The slope of the curve would then represent the coefficient of viscosity." Essentially the same experiment was repeated for rectus eye muscle by Close and Luff (1974) with similar results.

The classical force–velocity experiment was performed to test the viscoelastic model for muscle as described in Section 7.5.4. Under these conditions, it was first assumed that the inertial force exerted by the load during isotonic shortening could be ignored. The second assumption was that if mass was not reduced enough by the lever ratio (enough to be ignored), then taking measurements at maximum velocity provided a measurement at a time when acceleration is zero, and therefore inertial force equals zero. If these two assumptions are valid, then the experiment would provide data free of the effect of inertial force as the gravity force is varied. Both assumptions are incorrect. The first assumption is wrong since the inertial force is never minimal (minimal would be zero) and therefore has to be taken into account. The second assumption is wrong since, given an inertial mass not equal to zero, maximum velocity depends on the forces that act prior to the time of maximum velocity. The force–velocity relationship is carefully reexamined with the inertial force included in the analysis in this section.

The dynamic characteristics for the linear muscle model are described with a force–velocity curve calculated via the lever system presented in Fig. 7.18 and according to the isotonic experiment. For the rigid lever, the displacements x_1 and x_3 are directly proportional to the angle θ_1 and to each other, such that

$$\theta_1 = \frac{x_1}{d_1} = \frac{x_3}{d_3} \tag{7.29}$$

The equation describing the torques acting on the lever is given by

$$Mgd_3 + Md_3^2\ddot{\theta}_1 = d_1 K_{se}(x_2 - x_1) + d_1 B_2(\dot{x}_2 - \dot{x}_1) \tag{7.30}$$

The equation describing the forces at node 2, inside the muscle, is given by

$$F = K_{lt}x_2 + B_1\dot{x}_2 + B_2(\dot{x}_2 - \dot{x}_1) + K_{se}(x_2 - x_1) \tag{7.31}$$

Equation (7.30) is rewritten by removing θ_1 using Eq. (7.29); hence

$$Mg\frac{d_3}{d_1} + M\left(\frac{d_3}{d_1}\right)^2 \ddot{x}_1 = K_{se}(x_2 - x_1) + B_2(\dot{x}_2 - \dot{x}_1) \tag{7.32}$$

Ideally, to calculate the force–velocity curve for the lever system, $x_1(t)$ is found first. Then $\dot{x}_1(t)$ and $\ddot{x}_1(t)$ are found from $x_1(t)$. Finally, the velocity is found from $V_{max} = \dot{x}_1(T)$, where time T is the time it takes for the muscle to shorten to the stop, according to the experimental conditions of Close and Luff (1974). While this velocity may not be maximum velocity for all data points, the symbol V_{max} is used to denote the velocities in the force–velocity curve for ease in presentation. Note that this definition of velocity differs from the Fenn and Marsh (1935) definition of velocity. Fenn and Marsh denoted maximum velocity as $V_{max} = \dot{x}_1(T)$, where time T is found when $\ddot{x}_1(T) = 0$.

It should be noted that this is a third-order system and the solution for $x_1(t)$ is not trivial and involves an exponential approximation (for an example of an exponential approximation solution for V_{max} from a fourth-order model, see Enderle and Wolfe, 1988). It is more expedient, however, to simply simulate a solution for $x_1(t)$ and then find V_{max} as a function of load.

Using a simulation to reproduce the isotonic experiment, elasticities estimated from the length–tension curves as previously described, and data from rectus eye muscle, parameter values for the viscous elements in the muscle model are found as $B_1 = 2.0$ Ns/m and $B_2 = 0.5$ Ns/m as demonstrated by Enderle and coworkers (1991). The viscous element B_1 is estimated from the time constant from the isotonic time course. The viscous element B_2 is calculated by trial and error so that the simulated force–velocity curve matches the experimental force–velocity curve.

Shown in Fig. 7.25 are the force–velocity curves using the model described in Eq. (7.32) (triangles), plotted along with an empirical fit to the data (solid line). It is clear that the force–velocity curve for the linear muscle model is hardly a straight line, and that this curve fits the data well.

The muscle lever model described by Eqs. (7.30) and (7.31) is a third-order linear system and characterized by three poles. Dependent on the values of the parameters, the eigenvalues (or poles) consist of all real poles or a real and a pair of complex conjugate poles. A real pole is the dominant eigenvalue of the system. Through a sensitivity analysis, viscous element B_1 is the parameter that has the greatest effect on the dominant eigenvalue or time constant for this system, whereas viscous element B_2 has very little effect on the dominant eigenvalue. Thus, viscous element B_1 is estimated so that the dominant time constant of the lever system

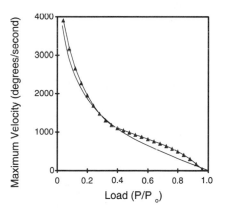

Fig. 7.25 Force–velocity curve derived from simulation studies with the linear muscle model with an end stop. Shown with triangles indicating simulation calculation points and an empirical fit to the force–velocity data (solid line) as described by Close and Luff (1974) (adapted from Enderle *et al.*, 1991).

model (approximately B_1/K_{lt} when $B_1 > B_2$) matches the time constant from the isotonic experimental data. For rectus eye muscle data, the duration of the isotonic experiment is approximately 100 ms. A value for $B_1 = 2.0$ Ns/m yields a simulated isotonic response with approximately the same duration. For skeletal muscle experimental data, the duration of the isotonic experiment is approximately 400 ms, and a value for $B_1 = 6.0$ Ns/m yields a simulated isotonic response with approximately the same duration. It is known that fast and slow muscles have differently shaped force–velocity curves and that the fast muscle force–velocity curve data has less curvature. Interestingly, the changes in the parameter values for B_1 as suggested here give differently shaped force–velocity curves consistent with fast (rectus eye muscle) and slow (skeletal muscle) muscle.

The parameter value for viscous element B_2 is selected by trial and error so that the shape of the simulation force–velocity curve matches the data. As the value for B_2 is decreased from 0.5 Ns/m, the shape of the force–velocity curves changes to a more linear-shaped function. Moreover, if the value of B_2 falls below approximately 0.3 Ns/m, strong oscillations appear in the simulations of the isotonic experiment, which are not present in the data. Thus, the viscous element B_2 is an essential component in the muscle model. Without it, the shape of the force–velocity curve is more linear and the time course of the isotonic experiment does not match the characteristics of the data.

Varying the parameter values of the lever muscle model changes the eigenvalues of the system. For instance, with $M = 0.5$ kg, the system's nominal eigenvalues (as defined with the parameter values previously specified) are a real pole at -30.71 and a pair of complex conjugate poles at $-283.9 \pm j221.2$. If the value of B_2 is increased, three real eigenvalues describe the system. If the value of B_2 is decreased, a real pole and a pair of complex conjugate poles continue to describe the system. Changing the value of B_1 does not change the eigenvalue composition but does significantly change the value of the dominant eigenvalue from -292 with $B_1 = 0.1$ to -10 with $B_1 = 6$.

7.7   A LINEAR HOMEOMORPHIC SACCADIC EYE MOVEMENT MODEL

Based on physiological evidence, Bahill and coworkers (1980) presented a linear fourth order model of the horizontal oculomotor plant that provides an excellent match between model predictions and horizontal eye movement data. This model eliminates the differences seen between velocity and acceleration predictions of the Westheimer and Robinson models and the data. For ease in this presentation, the 1984 modification of this model by Enderle and coworkers is used.

Figure 7.26 illustrates the mechanical components of the oculomotor plant for horizontal eye movements, the lateral and medial rectus muscle, and the eyeball. The agonist muscle is modeled as a parallel combination of an active-state tension generator F_{ag}, viscosity element B_{ag}, and elastic element K_{lt} connected to a series elastic element K_{se}. The antagonist muscle is similarly modeled as a parallel combination of an active-state tension generator F_{ant}, viscosity element B_{ant}, and elastic element K_{lt} connected to a series elastic element K_{se}. The eyeball is modeled as a

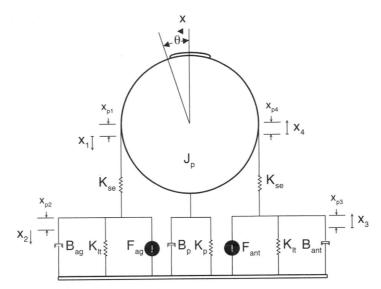

Fig. 7.26 This diagram illustrates the mechanical components of the oculomotor plant. The muscles are shown extended from equilibrium, a position of rest, at the primary position (looking straight ahead), consistent with physiological evidence. The average length of the rectus muscle at the primary position is approximately 40 mm, and at the equilibrium position it is approximately 37 mm. θ is the angle the eyeball is deviated from the primary position, and variable x is the length of arc traversed. When the eye is at the primary position, both θ and x are equal to zero. Variables x_1–x_4 are the displacements from equilibrium for the stiffness elements in each muscle. Values x_{p1} x_{p4} are the displacements from equilibrium for each of the variables x_1–x_4 at the primary position. The total extension of the muscle from equilibrium at the primary position is $x_{p1} + x_{p2}$ or $x_{p3} + x_{p4}$, which equals approximately 3 mm. It is assumed that the lateral and medial rectus muscles are identical, such that $x_{p1} = x_{p4}$ and $x_{p3} = x_{p2}$. The radius of the eyeball is r.

sphere with moment of inertia J_p, connected to viscosity element B_p and elastic element K_p. The passive elasticity of each muscle is included in spring K_p for ease in analysis. Each of the elements defined in the oculomotor plant is ideal and linear.

Physiological support for the muscle model is based on the model presented in Section 7.5 using linear approximations to the force–velocity curve and elasticity curves. Eyeball moment of inertia, elasticity, and viscosity in Fig. 7.26 are supported from the studies by Robinson and coworkers during strabismus surgery. Passive elasticity of the eyeball is usually from a combination of the effects due to the four other oculomotor muscles and optic nerve. Viscous effects are attributed to the friction of the eyeball within the eye socket. Moment of inertia is due to the eyeball.

By summing the forces at junctions 2 and 3 (the equilibrium positions for x_2 and x_3) and the torques acting on the eyeball, using Laplace variable analysis about the operating point, the linear homeomorphic model (as shown in Fig. 7.26), is derived as

$$\delta\left(K_{se}\left(K_{st}(F_{ag} - F_{ant}) + B_{ant}\dot{F}_{ag} - B_{ag}\dot{F}_{ant}\right)\right)$$
$$= \dddot{\theta} + C_3\,\dddot{\theta} + C_2\,\ddot{\theta} + C_1\,\dot{\theta} + C_0\,\theta \tag{7.33}$$

where

$$K_{st} = K_{se} + K_{lt},\; J = \frac{57.296 J_p}{r^2},\; B = \frac{57.296 B_p}{r^2},\; K = \frac{57.296 K_p}{r^2},\; \delta = \frac{57.296}{rJB_{ANT}B_{AG}}$$

$$C_3 = \frac{JK_{st}(B_{ag} + B_{ant}) + BB_{ant}B_{ag}}{JB_{ant}B_{ag}}$$

$$C_2 = \frac{JK_{st}^2 + BK_{st}(B_{ag} + B_{ant}) + B_{ant}B_{ag}(K + 2K_{se})}{JB_{ant}B_{ag}}$$

$$C_1 = \frac{BK_{st}^2 + (B_{ag} + B_{ant})(KK_{st} + 2K_{se}K_{st} - K_{se}^2)}{JB_{ant}B_{ag}}$$

$$C_0 = \frac{KK_{st}^2 + 2K_{se}K_{st}K_{lt}}{JB_{ant}B_{ag}}$$

The agonist and antagonist active-state tensions follow from Fig. 7.9, which assumes no latent period, and are given by the following low-pass filtered waveforms:

$$\dot{F}_{ag} = \frac{N_{ag} - F_{ag}}{\tau_{ag}} \quad \text{and} \quad \dot{F}_{ant} = \frac{N_{ant} - F_{ant}}{\tau_{ant}} \tag{7.34}$$

where N_{ag} and N_{ant} are the neural control inputs (pulse-step waveforms), and

$$\tau_{ag} = \tau_{ac}\,(u(t) - u(t - t_1)) + \tau_{de}\,u(t - t_1) \tag{7.35}$$

$$\tau_{ant} = \tau_{de}(u(t) - u(t - t_1)) + \tau_{ac}\,u(t - t_1) \tag{7.36}$$

are the time-varying time constants.

Based on an analysis of experimental evidence, a set of parameter estimates for the oculomotor plant are $K_{se} = 125$ N m^{-1}, $K_{lt} = 32$ N m^{-1}, $K = 66.4$ N m^{-1}, $B = 3.1$ Ns m^{-1}, $J = 2.2 = 10^{-3}$ Ns^2 m^{-1}, $B_{ag} = 3.4$ Ns m^{-1}, $B_{ant} = 1.2$ Ns m^{-1}, $\tau_{ac} = 0.009$ s, $\tau_{de} = 0.0054$ s, and $\delta = 5.80288 \times 10^5$, and the steady-state active-state tensions are:

$$F_{ag} = \begin{cases} 0.14 + 0.0185\theta & N \quad for \quad \theta < 14.23° \\ 0.0283\theta & N \quad for \quad \theta \geq 14.23° \end{cases} \tag{7.37}$$

$$F_{ant} = \begin{cases} 0.14 - 0.00980\theta & N \quad for \quad \theta < 14.23° \\ 0 & N \quad for \quad \theta \geq 14.23° \end{cases} \tag{7.38}$$

Since saccades are highly variable, estimates of the dynamic active-state tensions are usually carried out on a saccade by saccade basis. One method to estimate the active-state tensions is to use the system identification technique, a conjugate gradient search program carried out in the frequency domain discussed in Section 7.10. Estimates for agonist pulse magnitude are highly variable from saccade to saccade, even for saccades of the same size. Agonist pulse duration is closely coupled with pulse amplitude. As the pulse amplitude increases, the pulse duration decreases for saccades of the same magnitude. Reasonable values for the pulse amplitude for this model range from about 0.6 to 1.4 N. The larger the magnitude of the pulse amplitude, the larger the peak velocity of the saccade.

Example Problem 7.6

Show that Eq. (7.33) follows from Fig. 7.26.

Solution

From Fig. 7.26, the following relationships exist among variables x_1, x_4, θ, and x:

$$x_1 = x - x_{p1}$$

$$x_4 = x + x_{p4}$$

$$\theta = 57.296 \, x/r$$

Assuming that $\dot{x}_2 > \dot{x}_1 \geq \dot{x}_4 > \dot{x}_3$, the forces acting at nodes 2 and 3 and torques acting on the eyeball give

$$F_{ag} = B_{ag}\dot{x}_2 + K_{se}(x_2 - x_1) + K_{lt}x_2$$

$$rK_{se}(x_2 - x_1) \quad rK_{se}(x_4 - x_3) - J_p\ddot{\theta} + B_p\dot{\theta} + K_p\theta$$

$$K_{se}(x_4 - x_3) = F_{ant} + K_{lt}x_3 + B_{ant}\dot{x}_3$$

The previous three equations could be simplified and written as one equation by taking the Laplace transform of each equation and eliminating all variables except x. However, initial conditions must be considered in the transforms. An alternative method is to reduce the equations to include only changes from the initial eye position. Introduce

$$\hat{x} = x - x(0)$$

$$\hat{\theta} = \theta - \theta(0)$$

$$\hat{x}_1 = x_1 - x_1(0)$$

$$\hat{x}_2 = x_2 - x_2(0)$$

$$\hat{x}_3 = x_3 - x_3(0)$$

$$\hat{x}_4 = x_4 - x_4(0)$$

$$\hat{F}_{ag} = F_{ag} - F_{ag}(0)$$

$$\hat{F}_{ant} = F_{ant} - F_{ant}(0)$$

Node equations 2 and 3 and the equation for the torques acting on the eyeball are rewritten in terms of the new variables as

$$\hat{F}_{ag} = K_{st}\hat{x}_2 + B_{ag}\dot{\hat{x}}_2 - K_{se}\hat{x}$$

$$\hat{F}_{ant} = K_{se}\hat{x} - K_{st}\hat{x}_3 - B_{ant}\dot{\hat{x}}_3$$

$$K_{se}(\hat{x}_2 + \hat{x}_3 - 2\hat{x}) = J\ddot{\hat{x}} + B\dot{\hat{x}} + K\hat{x}$$

where

$$K_{st} = K_{se} + K_{lt}$$

$$J = \frac{57.296 J_p}{r^2}$$

$$B = \frac{57.296 B_p}{r^2}$$

$$K = \frac{57.296 K_p}{r^2}$$

Note that variables $\hat{x}_1$ and $\hat{x}_4$ are eliminated from the previous equations since they are both equal to $\hat{x}$ and $\theta = \frac{57.296\hat{x}}{r}$. The Laplace transform of node equations 2 and 3 and the equation for the torques acting on the eyeball are given by

$$L\{\hat{F}_{ag}\} = (K_{st} + sB_{ag})\hat{x}_2 - K_{se}\hat{x}$$

$$L\{\hat{F}_{ant}\} = K_{se}\hat{x} - (K_{st} + sB_{ag})\hat{x}_3$$

$$K_{se}(\hat{x}_2 + \hat{x}_3 - 2\hat{x}) = (Js^2 + Bs + K)\hat{x}$$

These three equations are written as one equation by first eliminating $\hat{x}_2$ and $\hat{x}_3$ from the node equations; that is,

$$\hat{x}_2 = \frac{L\{\hat{F}_{ag}\} + K_{se}\hat{x}}{K_{st} + sB_{ag}} \quad \text{and} \quad \hat{x}_3 = \frac{K_{se}\hat{x} - L\{\hat{F}_{ant}\}}{K_{st} + sB_{ant}}$$

and upon substituting into the torque equation this yields

$$K_{se}\left(\frac{L\{\hat{F}_{ag}\} + K_{se}\hat{x}}{K_{st} + sB_{ag}} + \frac{K_{se}\hat{x} - L\{\hat{F}_{ant}\}}{K_{st} + sB_{ant}} - 2\hat{x}\right) = (s^2 J + sB + K)\hat{x}$$

Multiplying the previous equation by $(sB_{ant} + K_{st}) \times (sB_{ag} + K_{st})$ reduces it to

$$K_{se}(sB_{ant} + K_{st})L\{\hat{F}_{ag}\} - K_{se}(sB_{ag} + K_{st})L\{\hat{F}_{ant}\} = (P_4 s^4 + P_3 s^3 + P_2 s^2 + P_1 s + P_0)\hat{x}$$

where

$$P_4 = JB_{ant}B_{ag}$$

$$P_3 = JK_{st}(B_{ag} + B_{ant}) + BB_{ant}B_{ag}$$

$$P_2 = JK_{st}^2 + BK_{st}(B_{ag} + B_{ant}) + B_{ant}B_{ag}(K + 2K_{se})$$

$$P_1 = BK_{st}^2 + (B_{ag} + B_{ant})(KK_{st} + 2K_{se}K_{st} - K_{se}^2)$$

$$P_0 = KK_{st}^2 + 2K_{se}K_{st}K_{lt}$$

Transforming back into the time domain yields

$$K_{se}\left(K_{st}(\hat{F}_{ag} - \hat{F}_{ant}) + B_{ant}\dot{\hat{F}}_{ag} - B_{ag}\dot{\hat{F}}_{ant}\right) = P_4\,\dddot{\hat{x}} + P_3\,\dddot{\hat{x}} + P_2\,\ddot{\hat{x}} + P_1\,\dot{\hat{x}} + P_0\,\hat{x}$$

Note that with $\dot{F}_{ag} = \dot{\hat{F}}_{ag}$, $\dot{F}_{ant} = \dot{\hat{F}}_{ant}$, and derivatives of $\hat{x}$ equal to x, the previous equation is rewritten as

$$K_{se}\left(K_{st}(\hat{F}_{ag} - \hat{F}_{ant}) + B_{ant}\dot{F}_{ag} - B_{ag}\dot{F}_{ant}\right) = P_4\,\dddot{x} + P_3\,\dddot{x} + P_2\,\ddot{x} + P_1\,\dot{x} + P_0\,\hat{x}$$

In general, it is not necessary for $F_{ag}(0) = F_{ant}(0)$. With $\hat{x} = x - x(0)$ and $\hat{F}_{ag} - \hat{F}_{ant} = F_{ag} - F_{ag}(0) - F_{ant} + F_{ant}(0)$, the steady-state difference of the two node equations is

$$F_{ag}(0) - F_{ant}(0) = K_{st}(x_2(0) + x_3(0)) - K_{se}(x_1(0) + x_4(0))$$

With $x_1(0) = x(0) - x_{p1}$ and $x_4(0) + x(0) - x_{p1}$,

$$F_{ag}(0) - F_{ant}(0) = K_{st}(x_2(0) + x_3(0)) - 2K_{se}x(0)$$

From the node 1 equation at time 0^-,

$$K_{se}(x_2(0) + x_3(0) - 2x(0)) = Kx(0)$$

or

$$x_2(0) + x_3(0) = \left(\frac{K}{K_{se}} + 2\right)x(0)$$

This gives

$$F_{ag}(0) - F_{ant}(0) = \left(K_{st}\left(\frac{K}{K_{se}} + 2\right) - 2K_{se}\right)x(0)$$

Therefore,

$$K_{se}\left(K_{st}(F_{ag} - F_{ant}) - K_{st}\,(F_{ag}(0) - F_{ant}(0)) + B_{ant}\dot{F}_{ag} - B_{ag}\dot{F}_{ant}\right)$$
$$= P_4\,\dddot{x} + P_3\,\dddot{x} + P_2\,\ddot{x} + P_1\,\dot{x} + P_0\,(x - x(0))$$

From the previous expression for $F_{ag}(0) - F_{ant}(0)$, it follows from multiplying both sides by $K_{se}K_{st}$ that

$$K_{se}K_{st}(F_{ag}(0) - F_{ant}(0)) = K_{se}K_{st}\left(K_{st}\left(\frac{K}{K_{se}} + 2\right) - 2K_{se}\right)x(0)$$

The right-hand side of the previous equation reduces to

$$K_{se}K_{st}(F_{ag}(0) - F_{ant}(0)) = (K_{st}^2K + 2K_{se}K_{st}^2 - 2K_{se}^2K_{st})x(0)$$
$$= (K_{st}^2K + 2K_{se}K_{st}K_{lt})x(0) = P_0\,x(0)$$

Substituting the previous equation into the differential equation describing the system gives

$$K_{se}\big(K_{st}(F_{ag} - F_{ant}) + B_{ant}\dot{F}_{ag} - B_{ag}\dot{F}_{ant}\big) = P_4\,\ddddot{x} + P_3\,\dddot{x} + P_2\,\ddot{x} + P_1\,\dot{x} + P_0 x$$

With $x = \theta r/57.296$ substituted into the previous result, Eq. (7.33) follows.   ■

7.8 A TRUER LINEAR HOMEOMORPHIC SACCADIC EYE MOVEMENT MODEL

The linear model of the oculomotor plant presented in Section 7.7 is based on a nonlinear oculomotor plant model by Hsu and coworkers (1976) using a linearization of the force–velocity relationship and elasticity curves. Using the linear model of muscle described in Section 7.6, it is possible to avoid the linearization and to derive a truer linear homeomorphic saccadic eye movement model.

The linear muscle model in Section 7.6 has the static and dynamic properties of rectus eye muscle, a model without any nonlinear elements. As presented, the model has a nonlinear force–velocity relationship that matches eye muscle data using linear viscous elements, and the length–tension characteristics are also in good agreement with eye muscle data within the operating range of the muscle. Some additional advantages of the linear muscle model are that a passive elasticity is not necessary if the equilibrium point $x_e = -19.3°$ rather than $15°$, and muscle viscosity is a constant that does not depend on the innervation stimulus level.

Figure 7.27 illustrates the mechanical components of the updated oculomotor plant for horizontal eye movements, the lateral and medial rectus muscle, and the eyeball. The agonist muscle is modeled as a parallel combination of viscosity B_2 and series elasticity K_{se}, connected to the parallel combination of active-state tension generator F_{ag}, viscosity element B_1, and length–tension elastic element K_{lt}. For simplicity, agonist viscosity is set equal to antagonist viscosity. The antagonist muscle is similarly modeled with a suitable change in active-state tension to F_{ant}. Each of the elements defined in the oculomotor plant is ideal and linear.

The eyeball is modeled as a sphere with moment of inertia J_p, connected to a pair of viscoelastic elements connected in series. The update of the eyeball model is based on observations by Robinson (1981), and the following discussion. In the model of the oculomotor plant described in Section 7.7, passive elasticity K_{pe} was combined with the passive elastic orbital tissues. In the new linear model muscle presented in Section 7.6, the elastic element K_{pe} was no longer included in the muscle model. Thus, the passive orbital tissue elasticity needs to be updated due to the elimination of K_{pe} and the new observations by Robinson. As reported by Robinson (1981), "When the human eye, with horizontal recti detached, is displaced and suddenly released, it returns rapidly about 61% of the way with a time constant of about 0.02 sec, and then creeps the rest of the way with a time constant of about 1 sec." As suggested from this observation, there are at least two viscoelastic elements. Here it is proposed that these two viscoelastic elements replace the single parallel viscoelastic element of the previous oculomotor plant. Connected to the sphere, $B_{p1} \parallel K_{p1}$ are connected in series to $B_{p2} \parallel K_{p2}$. As reported by Robinson, total orbital elasticity is equal to 7.8×10^{-7} g/°. Thus, with the time constants previously described, the orbital viscoelastic elements are evaluated as $K_{p1} = 1.28 \times 10^{-6}$ g/°, $K_{p2} = 1.98 \times$

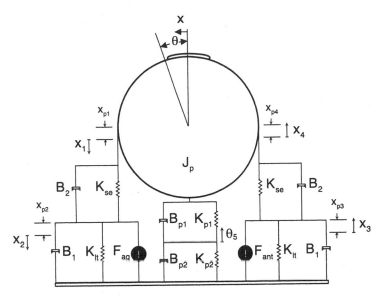

Fig. 7.27 This diagram illustrates the mechanical components of the updated oculomotor plant. The muscles are shown to be extended from equilibrium, a position of rest, at the primary position (looking straight ahead), consistent with physiological evidence. The average length of the rectus muscle at the primary position is approximately 40 mm, and at the equilibrium position it is approximately 37 mm. θ is the angle the eyeball is deviated from the primary position, and variable x is the length of arc traversed. When the eye is at the primary position, both θ and x are equal to zero. Variables x_1-x_4 are the displacements from equilibrium for the stiffness elements in each muscle, and θ_5 is the rotational displacement for passive orbital tissues. Values $x_{p1}-x_{p4}$ are the displacements from equilibrium for each of the variables x_1-x_4 at the primary position. The total extension of the muscle from equilibrium at the primary position is $x_{p1} + x_{p2}$ or $x_{p3} + x_{p4}$, which equals approximately 3 mm. It is assumed that the lateral and medial rectus muscles are identical, such that $x_{p1} = x_{p4}$ and $x_{p3} = x_{p2}$. The radius of the eyeball is r.

10^{-6}g/°, $B_{p1} = 2.56 \times 10^{-8}$gs/°, and $B_{p2} = 1.98 \times 10^{-6}$gs/°. For modeling purposes, θ_5 is the variable associated with the change from equilibrium for these two pairs of viscoelastic elements. Both θ and θ_5 are removed from the analysis for simplicity using the substitution $\theta = 57.296x/r$ and $\theta_5 = 57.296x_5/r$.

By summing the forces acting at junctions 2 and 3 and the torques acting on the eyeball and junction 5, a set of four equations is written to describe the oculomotor plant:

$$F_{ag} = K_{lt}x_2 + B_1\dot{x}_2 + K_{se}(x_2 - x_1) + B_2(\dot{x}_2 - \dot{x}_1)$$

$$B_2(\dot{x}_4 - \dot{x}_3) + K_{se}(x_4 - x_3) = F_{ant} + K_{lt}x_3 + B_1\dot{x}_3$$

$$B_2(\dot{x}_2 + \dot{x}_3 - \dot{x}_1 - \dot{x}_4) + K_{se}(x_2 + x_3 - x_1 - x_4) = J\ddot{x} + B_3(\dot{x} - \dot{x}_5) + K_1(x - x_5)$$

$$K_1(x - x_5) + B_3(\dot{x} - \dot{x}_5) = B_4\dot{x}_5 + K_2x_5$$

(7.39)

where

$$J = \frac{57.296}{r^2}J_p, \quad B_3 = \frac{57.296}{r^2}B_{p1}, \quad B_4 = \frac{57.296}{r^2}B_{p2},$$

$$K_1 = \frac{57.296}{r^2}K_{p1}, \quad K_2 = \frac{57.296}{r^2}K_{p2}$$

Using Laplace variable analysis about an operating point similar to the analysis used in Example Problem 7.6 yields

$$K_{se}K_{12}(F_{ag} - F_{ant}) + (K_{se}B_{34} + B_2K_{12})(\dot{F}_{ag} - \dot{F}_{ant}) + B_2B_{34}(\ddot{F}_{ag} - \ddot{F}_{ant})$$

$$= C_4\ddddot{x} + C_3\dddot{x} + C_2\ddot{x} + C_1\dot{x} + C_0x \tag{7.40}$$

where

$$B_{12} = B_1 + B_2, \quad B_{34} = B_3 + B_4, \quad K_{12} = K_1 + K_2$$

$$C_4 = JB_{12}B_{34}$$

$$C_3 = B_3B_4B_{12} + 2B_1B_2B_{34} + JB_{34}K_{st} + JB_{12}K_{12}$$

$$C_2 = 2B_1B_{34}K_{se} + JK_{st}K_{12} + B_3B_{34}K_{st} + B_3B_{12}K_{12} + K_1B_{12}B_{34} - B_3^2K_{st}$$
$$\qquad - 2K_1B_3B_{12} + 2B_2K_{lt}B_{34} + 2B_1K_{12}B_2$$

$$C_1 = 2K_{lt}B_{34}K_{se} + 2B_1K_{12}K_{se} + B_3K_{st}K_2 + K_1B_{34}K_{st} + K_1B_{12}K_{12}$$
$$\qquad - K_{st}K_1B_3 - K_1^2B_{12} + 2B_2K_{lt}K_{12}$$

$$C_0 = 2K_{lt}K_{se}K_{12} + K_1K_{st}K_2$$

Converting from x to θ gives

$$\delta\big(K_{se}K_{12}(F_{ag} - F_{ant}) + (K_{se}B_{34} + B_2K_{12})(\dot{F}_{ag} - \dot{F}_{ant}) + B_2B_{34}(\ddot{F}_{ag} - \ddot{F}_{ant})\big)$$

$$= \ddddot{\theta} + P_3\dddot{\theta} + P_2\ddot{\theta} + P_1\dot{\theta} + P_0\theta \tag{7.41}$$

where

$$\delta = \frac{57.296}{rJB_{12}B_4}, \quad P_3 = \frac{C_3}{C_4}, \quad P_2 = \frac{C_2}{C_4}, \quad P_1 = \frac{C_1}{C_4}, \quad P_0 = \frac{C_0}{C_4}$$

Based on the updated model of muscle and length–tension data presented in Section 7.6, steady-state active-state tensions are determined as

$$F = \begin{cases} 0.4 + 0.0175\theta & \text{N} \quad \text{for } \theta \geq 0° \\ 0.4 - 0.0125\theta & \text{N} \quad \text{for } \theta < 0° \end{cases} \tag{7.42}$$

The agonist and antagonist active-state tensions follow from Fig. 7.9, assume no latent period, and are given by the following low-pass filtered waveforms:

$$\dot{F}_{ag} = \frac{N_{ag} - F_{ag}}{\tau_{ag}} \quad \text{and} \quad \dot{F}_{ant} = \frac{N_{ant} - F_{ant}}{\tau_{ant}} \tag{7.43}$$

where N_{ag} and N_{ant} are the neural control inputs (pulse-step waveforms), and

$$\tau_{ag} = \tau_{ac}\big(u(t) - u(t - t_1)\big) + \tau_{de}\,u(t - t_1) \tag{7.44}$$

$$\tau_{ant} = \tau_{de}\big(u(t) - u(t - t_1)\big) + \tau_{ac}\,u(t - t_1) \tag{7.45}$$

are the time-varying time constants. Based on an analysis of experimental data, parameter estimates for the oculomotor plant are $K_{se} = 125 \text{ Nm}^{-1}$, $K_{lt} = 60.7 \text{ Nm}^{-1}$, $B_1 = 2.0 \text{ Nsm}^{-1}$, $B_2 = 0.5 \text{ Nsm}^{-1}$, $J = 2.2 \times 10^{-3} \text{ Ns}^2\text{m}^{-1}$, $B_3 = 0.538 \text{ Nsm}^{-1}$, $B_4 = 41.54 \text{ Nsm}^{-1}$, $K_1 = 26.9 \text{ Nm}^{-1}$, $K_2 = 41.54 \text{ Nm}^{-1}$, and $r = 0.0118 \text{ m}$.

Saccadic eye movements simulated with this model have characteristics that are in good agreement with the data, including position, velocity and acceleration, and the main sequence diagrams. As before, the relationship between agonist pulse magnitude and pulse duration is tightly coupled. Material presented in Section 7.9 further explores this relationship by examining the neural network responsible for saccades.

Example Problem 7.7

Using the oculomotor plant model described with Eq. (7.41), parameters given after this model, and the steady-state input from Eq. (7.42), simulate a 10° saccade. Plot agonist and antagonist active-state tension, position, velocity, and acceleration vs time. Compare the simulation with the main sequence diagram in Fig. 7.5.

Solution

The solution to this example problem involves selecting a set of parameters (F_p, t_1, τ_{ac}, and τ_{de}) that match the characteristics observed in the main sequence diagram shown in Fig. 7.5. There is a great deal of flexibility in simulating a 10° saccade. The only constraints for the 10° saccade simulation results are that the duration is approximately 40–50 ms and peak velocity is in the 500–600°/s range. For realism, a latent period of 150 ms has been added to the simulation results. A SIMULINK block diagram of Eq. (7.40) is shown in the following figures.

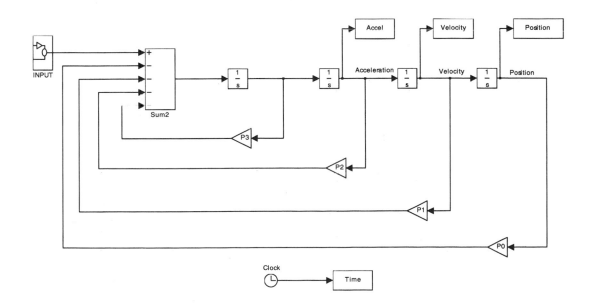

where the INPUT block is given by

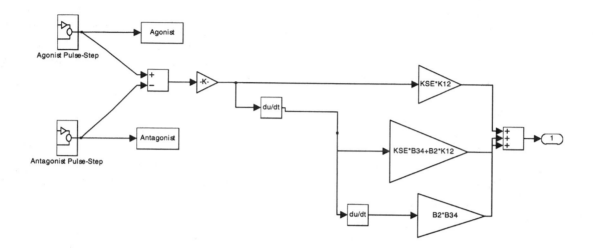

the Agonist Pulse-Step block is given by

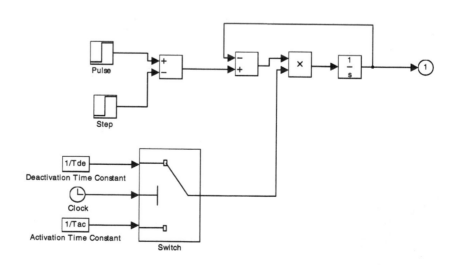

and the Antagonist Pulse-Step is given by

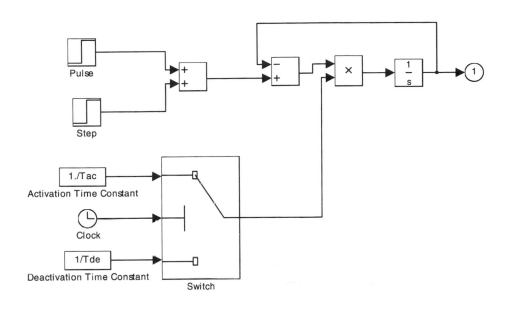

With $F_p = 1.3$ N, $t_1 = 0.10$ s, $\tau_{ac} = 0.018$ s and $\tau_{de} = 0.018$ s, the following simulation results

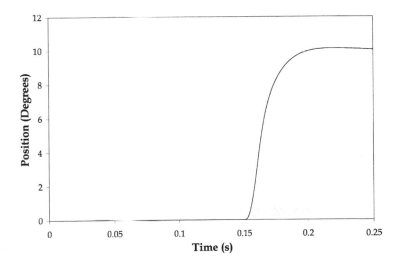

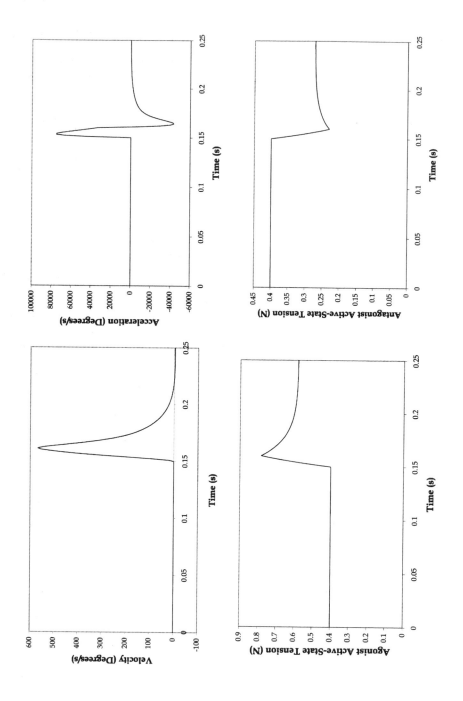

These 10° simulation results have the main sequence characteristics with a peak velocity of 568 °/s and a duration of 45 ms.

Many other parameter sets can also simulate a 10° saccade. For instance, consider reducing τ_{de} to .009 s. Because the antagonist active-state tension activity goes toward zero more quickly than in the previous case, a greater total active-state tension $(F_{ag} - F_{ant})$ results. Therefore, to arrive at 10° with the appropriate main sequence characteristics, F_p needs to be reduced to 1.0 N if τ_{ac} remains at 0.018 s and $t_1 = 0.0115$ s. This 10° simulation is shown in the following figures:

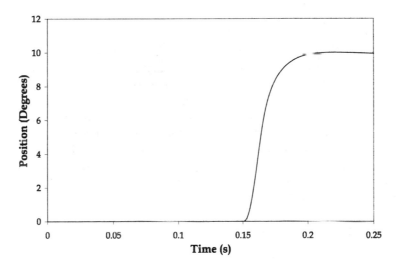

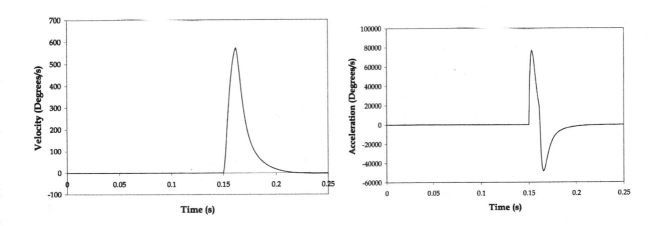

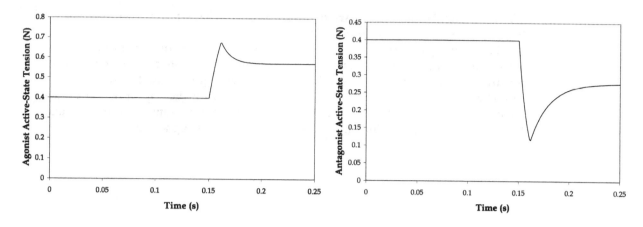

To simulate larger saccades with main sequence characteristics, the time constants for the agonist and antagonist active-state tensions can be kept at the same values as those for the 10° saccades or made functions of saccade amplitude [see Bahill (1981) for several examples of amplitude-dependent time constants]. Main sequence simulations for 15 and 20° saccades are obtained with $F_p = 1.3$ N and the time constants are both fixed at 0.018 s (the first case) by changing t_1 to 0.0155 and 0.0223 s, respectively. For example, the 20° simulation results are shown in the following figures with a peak velocity of 682 °/s and a duration of 60 ms:

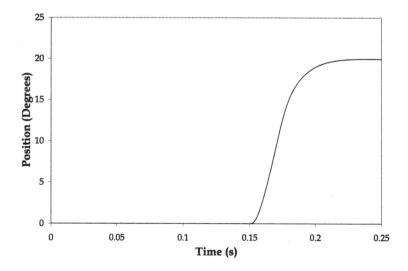

In general, as F_p increases, t_1 decreases to maintain the same saccade amplitude. Additionally, peak velocity increases as F_p increases. For the saccade amplitude to remain a constants as either or both time constants increase, F_p should also increase. ∎

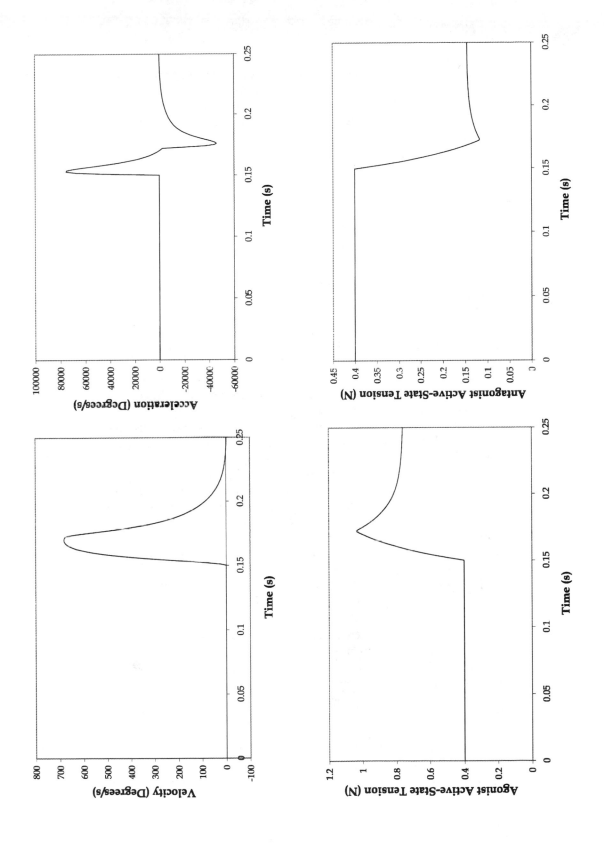

Note: **The following section is for advanced readers only. It presents a model of brain function that is a complex process, serving as an input to the physiological system just described. It is placed here to complete the model of the oculomotor system. For those readers not interested in the topic, advance to Section 7.10 without any loss in continuity.**

7.9　SACCADE PATHWAYS

Clinical evidence and lesion and stimulation studies point toward the participation of vitally important neural sites in the control of saccades, including the cerebellum, superior colliculus, thalamus, cortex, and other nuclei in the brain stem, and that saccades are driven by two parallel neural networks. From each eye, the axons of retinal ganglion cells exit and join other neurons to form the optic nerve. The optic nerves from each eye then join at the optic chiasm, where fibers from the nasal half of each retina cross to the opposite side. Axons in the optic tract synapse in the lateral geniculate nucleus (a thalamic relay) and continue to the visual cortex. This portion of the saccade neural network is concerned with the recognition of visual stimuli. Axons in the optic tract also synapse in the superior colliculus. This second portion of the saccade neural network is concerned with the location of visual targets and is primarily responsible for goal-directed saccades.

Saccadic neural activity of the superior colliculus and cerebellum, in particular, has been identified as the saccade initiator and terminator, respectively, for a goal-directed saccade. The frontal eye field and the thalamus, while very important, have less important roles in the generation of goal-directed saccades to visual stimuli. The frontal eye fields are primarily concerned with voluntary saccades, and the thalamus appears to be involved with corrective saccades. Shown in Fig. 7.28 is a diagram illustrating important sites and abbreviations for the generation of a conjugate goal-directed horizontal saccade in both eyes. Each of the sites and connections detailed are fully supported by physiological evidence. Some of these neural sites will be briefly described here.

The superior colliculus contains two major functional divisions: a superficial division and an intermediate or deep division. Inputs to the superficial division are almost exclusively visual and originate from the retina and the visual cortex. The deep layers provide a convergence site for sensory signals from several modalities and a source of efferent commands for initiating saccades. The superior colliculus is the initiator of the saccade and thought to translate visual information into motor commands.

The deep layers of the superior colliculus initiate a saccade based on the distance between the current position of the eye and the desired target. The neural activity in the superior colliculus is organized into movement fields that are associated with the direction and saccade amplitude and does not involve the initial position of the eyeball whatsoever. The movement field is shown in Fig. 7.29 for a 20° saccade. Neurons active during a particular saccade are shown as the dark circle, representing a desired 20° eye movement. Active neurons in the deep layers of the superior col-

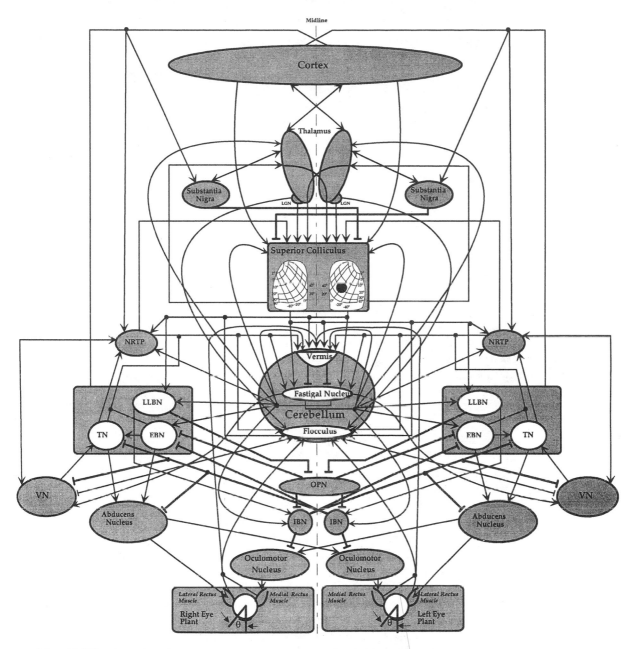

Fig. 7.28 Shown is a diagram illustrating important sites for the generation of a conjugate horizontal saccade in both eyes. It consists of the familiar premotor excitatory burst neurons (EBN), inhibitory burst neurons (IBN), long-lead burst neurons (LLBN), omnipause neurons (OPN), tonic neurons (TN), the vestibular nucleus (VN), Abducens nucleus, oculomotor nucleus, cerebellum, substantia nigra, nucleus reticularis tegmenti points (NRTP), the thalamus, the deep layers of the superior colliculus, and the oculomotor plant for each eye. Excitatory inputs are shown with an arrow, inhibitory inputs are shown with a ⊥. Consistent with current knowledge, the left and right structures of the neural circuit model are maintained. Since interest here is in goal-directed visual saccades, the cortex has not been partitioned into the frontal eye field and posterior eye field (striate, prestriate, and inferior parietal cortices). LGN, lateral geniculate nucleus.

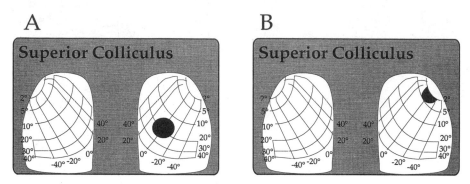

Fig. 7.29 Movement fields of the superior colliculus: (a) 20° horizontal saccade; (b) 2° horizontal saccade.

liculus generate a high-frequency burst of activity beginning 18–20 ms before a saccade and end sometime toward the end of the saccade. The exact timing for the end of the burst firing is quite random and can occur slightly before or slightly after the saccade ends. Each active bursting neuron discharges maximally, regardless of the initial position of the eye. Neurons discharging for small saccades have smaller movement fields, and those for larger saccades have larger movement fields. All the movement fields are connected to the same set of long-lead burst neurons (LLBNs).

The cerebellum is responsible for the coordination of movement and is composed of a cortex of gray matter, internal white matter, and three pairs of deep nuclei: the fastigial nucleus, the interposed and globose nucleus, and the dentate nucleus. The deep cerebellar nuclei and the vestibular nuclei transmit the entire output of the cerebellum. Output of the cerebellar cortex is carried through Purkinje cells. Purkinje cells send their axons to the deep cerebellar nuclei and have an inhibitory effect on these nuclei. The cerebellum is involved with both eye and head movements, and both tonic and phasic activity are reported in the cerebellum. The cerebellum is not directly responsible for the initiation or execution of a saccade but contributes to saccade precision. Sites within the cerebellum that are important for the control of eye movements include the oculomotor vermis, fastigial nucleus, and the flocculus. Consistent with the operation of the cerebellum for other movement activities, the cerebellum is postulated here to act as the coordinator for a saccade and act as a precise gating mechanism.

The cerebellum is included in the saccade generator as a time-optimal gating element, using three active sites during a saccade: the vermis, the fastigial nucleus, and the flocculus. The vermis is concerned with the absolute starting position of a saccade in the movement field and corrects control signals for initial eye position. Using proprioceptors in the oculomotor muscles and an internal eye position reference, the vermis is aware of the current position of the eye. The vermis is also aware of the signals (dynamic motor error) used to generate the saccade via the connection with the nucleus reticularis tegmenti pontis (NRTP) and the superior colliculus.

Concerning the oculomotor system, the cerebellum has inputs from superior colliculus (SC), lateral gebniculate nucleus, oculomotor muscle proprioceptors, and striate cortex via NRTP. The cerebellum sends inputs to the NRTP, LLBN, excitatory burst neurons (EBNs), vestibular nucleus, thalamus, and superior colliculus. The oculomotor vermis and fastigial nuclei are important in the control of saccade amplitude, and the flocculus, perihypoglossal nuclei of the rostral medulla, and possibly the pontine and mesencephalic reticular formation are thought to form the integrator within the cerebellum. One important function of the flocculus may be to increase the time constant of the neural integrator for saccades starting at locations different from primary position.

The fastigial nucleus receives input from the superior colliculus as well as from other sites. The output of the fastigial nucleus is excitatory and projects ipsilaterally and contralaterally as shown in Fig. 7.28. During fixation, the fastigial nucleus fires tonically at low rates. Twenty ms before a saccade, the contralateral fastigial nucleus bursts, and the ipsilateral fastigial nucleus pauses and then discharges with a burst. The pause in ipsilateral firing is due to Purkinje cell input to the fastigial nucleus. The sequential organization of Purkinje cells along beams of parallel fibers suggests that the cerebellar cortex might function as a delay, producing a set of timed pulses that could be used to program the duration of the saccade. For the case of nonprimary position saccades, different temporal and spatial schemes, via cerebellar control, are necessary to produce the same size saccade. It is postulated here that the cerebellum acts as a gating device that precisely stops a saccade based on the initial position of the eye in the orbit.

The paramedian pontine reticular formation (PPRF) has neurons that burst at frequencies up to 1000 Hz during saccades and are silent during periods of fixation and neurons that fire tonically during periods of fixation. Neurons that fire at steady rates during fixation are called tonic neurons (TNs) and are responsible for holding the eye steady. The TN firing rate depends on the position of the eye (presumably through a local integrator type network). The TNs are thought to provide the step component to the motoneuron. There are two types of burst neurons in the PPRF called the LLBN and a medium-lead burst neuron (MLBN). During periods of fixation, these neurons are silent. The LLBN bursts at least 12 ms before a saccade and the MLBN bursts <12 ms (typically 6–8 ms) before the saccade. The MLBNs are connected monosynaptically with the abducens nucleus.

There are two types of neurons within the MLBN: the EBNs and the inhibitory burst neurons (IBNs). The EBN and IBN labels describe the synaptic activity upon the motoneurons. The EBN excites and is responsible for the burst firing, and the IBNs inhibit and are responsible for the pause. A mirror image of these neurons exists on both sides of the midline. The IBNs inhibit the EBNs on the contralateral side.

Also within the brain stem is another type of saccade neuron called the omnipause neuron (OPN). The OPN fires tonically at approximately 200 Hz during periods of fixation and is silent during saccades. The OPN stops firing approximately 10–12 ms before a saccade and resumes tonic firing approximately 10 ms before the end of the saccade. OPNs are known to inhibit the MLBN and are inhibited by the LLBN. The OPN activity is responsible for the precise timing between groups of neurons that causes a saccade.

Qualitatively, a saccade occurs according to the following sequence of events within the PPRF. First, the ipsilateral LLBNs are stimulated by the SC, initiating the saccade. The LLBNs then inhibit the tonic firing of the OPNs. When the OPNs cease firing, the MLBN is released from inhibition and begins firing (these neurons fire spontaneously and are also stimulated by the fastigial nucleus). The ipsilateral IBNs are stimulated by the ipsilateral LLBNs and the contralateral fastigial nucleus of the cerebellum. The ipsilateral EBNs are stimulated by the contralateral fastigial nucleus of the cerebellum and fire spontaneously when released from inhibition. Except for the fastigial nucleus, there are no other accepted excitatory inputs to the EBNs. The burst firing in the ipsilateral IBNs inhibits the contralateral EBNs and abducens nucleus and the ipsilateral oculomotor nucleus. The burst firing in the ipsilateral EBNs causes the burst in the ipsilateral abducens nucleus, which stimulates the ipsilateral lateral rectus muscle and the contralateral oculomotor nucleus. With the stimulation of the ipsilateral lateral rectus muscle by the ipsilateral abducens nucleus and the inhibition of the ipsilateral medial rectus muscle via the oculomotor nucleus, a saccade occurs in the right eye.

Simultaneously, with the contralateral medial rectus muscle stimulated by the contralateral oculomotor nucleus and with the inhibition of the contralateral lateral rectus muscle via the abducens nucleus, a saccade occurs in the left eye. Thus, the eyes move conjugately under the control of a single drive center. The saccade is terminated with the resumption of tonic firing in the OPN via the fastigial nucleus.

7.9.1   Saccade Control Mechanism

Although the purpose for a saccadic eye movement is clear, that is, to quickly redirect the eyeball to the destination, that for the neural control mechanism is not. Until quite recently, saccade generator models involved a ballistic or preprogrammed control to the desired eye position based on retinal error alone. Today, an increasing number of investigators are putting forth the idea that visual goal-directed saccades are controlled by a local feedback loop that continuously drives the eye to the desired eye position. This hypothesis first gained acceptance in 1975 when Robinson suggested that saccades originate from neural commands that specify the desired position of the eye rather than the preprogrammed distance the eye must be moved. The value of the actual eye position is subtracted from the desired position to create an error signal that completes the local feedback loop that drives a high-gain burst element to generate the neural pulse. This neural pulse continuously drives the eyes until the error signal is zero. Subsequently, a number of other investigators have modified the local feedback mechanism proposed by Robinson to better describe the neural connections and firing patterns of brain stem neurons in the control of horizontal saccadic eye. In addition to the Robinson model (as modified by van Gisbergen *et al.*, 1981), two other models by Scudder (1988) and Enderle (1994) describe a saccade generator. All the models involve three types of premotor neurons (burst, tonic, and pause cells, as previously described) and involve a pulse-step change in firing rate at the motoneuron during a saccadic eye movement. While the general pattern of motoneuron activity is qualitatively accepted during a saccadic eye movement, there is little agreement on a quantitative discharge description. The saccade generator models by Robinson and Scudder are structured to provide a control signal that is proportionally weighted (or dependent) to the desired saccade size, as opposed to the saccade generator model, which is structured to provide a control signal that is independent of saccade amplitude.

The Enderle saccade generator model uses a first-order time optimal or a saccade amplitude-independent controller within the SC and cerebellum. The concepts underlying this hypothesis are that each muscle's active-state tension is described by a low-pass filtered pulse-step waveform in which the magnitude of the agonist pulse is a maximum regardless of the amplitude of the saccade and that only the duration of the agonist pulse affects the size of the saccade. The antagonist muscle is inhibited during the period of maximum agonist stimulation. The saccade generator illustrated in Fig. 7.28 operates under these principles and is discussed further in this section.

The saccade amplitude-independent controller and saccade amplitude-dependent controller yield trajectory simulations that are markedly different. Neural inputs from a saccade amplitude dependent controller yield saccades which almost immediately separate into separate position trajectories for saccades of increasing sizes. Neural inputs from a saccade amplitude-independent controller yield saccades that follow the same position trajectory during the pulse phase (the early stages of a saccade) and then separate for saccades of increasing sizes. Velocity characteristics for both types of controllers behave similarly.

Shown in Fig. 7.30 are five saccades elicited from target movements of 4, 8, 12, 16, and 20°. Notice that the 12, 16, and 20° eye movements follow the same trajectory before separating toward the end of the eye movement, consistent with a saccade amplitude-independent controller. The 4 and 8° eye movements initially start with the other eye movements but separate earlier than predicted by a time-optimal controller. This behavior is caused by a smaller population of neurons firing maximally (and thus time optimally) in the SC for small saccades. It is envisioned that the reduced number of neurons is based on the "time constant of the burster"; that is, once firing is initiated it takes approximately 10–20 ms to turn off. Thus, small saccades could never occur via all neurons firing maximally, as observed in the microelectrode recordings, if the system did not compensate by reducing the number of active neurons.

Significant differences exist among interpretations of microelectrode recordings, especially for small saccades. For large saccades (those >10°), all investigators support an identical high level of steady discharge in the agonist motoneurons. Most investigators support characteristics for the transition band between 5 and 10° saccades similar to those for 10° and greater saccades. For saccades <5°, investigators

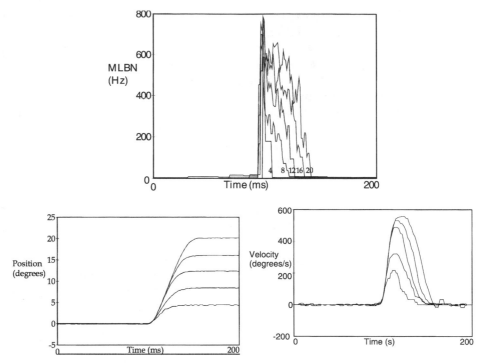

Fig. 7.30 Extracelluar single-unit recordings from within the abducens nucleus (MLBN), eye position, and velocity obtained from rhesus monkeys during saccadic eye movements. Eye position data were recorded using magnetic coils, and neural activity was recorded using tungsten microelectrodes. No filtering of the data was carried out, but the firing frequency was placed in the usual format of frequency of firing over 1-ms intervals rather than the electrical activity itself. Details of the experiment and training are reported elsewhere (Sparks *et al.,* data provided by Dr. David Sparks).

report a significantly lowered burst rate. Operationally, this may be predicted by a first-order time-optimal controller as implemented with the neural burst generator circuit for microsaccades. In general, whenever a retinal error exists, the SC fires maximally, driving the dynamic motor error to zero, initiating burst activity through contralateral long-lead burst, excitatory burst, and inhibitory burst neurons. Note that there is no instantaneous rise in the burst rate in the LLBNs and subsequent bursting neurons, but rather the rate changes exponentially according to physiological evidence. During very small saccades, the duration of the firing burst from the SC is extremely short and fewer neurons fire, resulting in activity within the bursting neurons returning to zero firing rates before reaching a peak firing rate. This activity during small saccades has the appearance of a controller that depends on the size of the saccade, but it is actually due to a time-optimal, saccade amplitude-independent controller.

A functional block diagram of the neural sites (nucleus) of the horizontal saccade generator model is shown in Figs. 7.31 and 7.32. Table 7.1 summarizes additional

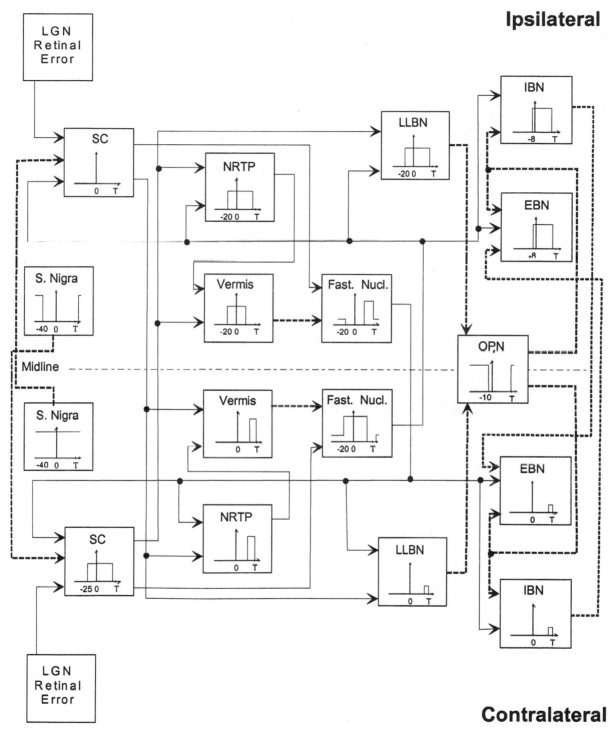

Fig. 7.31   A functional block diagram of the saccade generator model. Solid lines are excitatory and dashed lines are inhibitory. This figure illustrates the first half of the network.

Ipsilateral

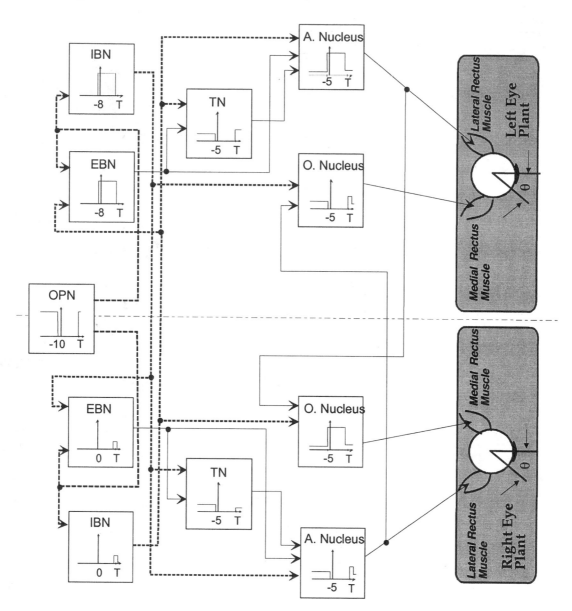

Contralateral

Fig. 7.32 A functional block diagram of the saccade generator model. Solid lines are excitatory and dashed lines are inhibitory. This figure illustrates the second half of the network.

TABLE 7.1 Activity in Neural Sites during a Saccade

Neural Site	Onset before Saccade (ms)	Peak Firing Rate (Hz)	End Time
Abducens nucleus	5	400–800	Ends approx 5 ms before saccade ends
Contralateral fastigial nucleus	20	200	Pulse ends with pause approx 10 ms before saccade ends, resumes tonic firing approx 10 ms after saccade ends
Contralateral superior colliculus	20–25	800–1000	Ends approx when saccade ends
Ipsilateral cerebellar vermis	20–25	600–800	Ends approx 25 ms before saccade ends
Ipsilateral EBN	6–8	600–800	Ends approx 10 ms before saccade ends
Ipsilateral fastigial nucleus	20	Pause during saccade and a burst of 200 Hz toward the end of the saccade	Pause ends with burst approx 10 ms before saccade ends, resumes tonic firing approx 10 ms after saccade ends
Ipsilateral FEF	>30	600–800	Ends approx when saccade ends
Ipsilateral IBN	6–8	600–800	Ends approx 10 ms before saccade ends
Ipsilateral LLBN	20	800–1000	Ends approx when saccade ends
Ipsilateral NRTP	20–25	800–1000	Ends approx when saccade ends
Ipsilateral substantia nigra	40	40–100	Resumes firing approx 40–150 ms after saccade ends
OPN	6–8	150–200 (before and after)	Ends approx when saccade ends

firing characteristics for the neural sites. The output of each block represents the firing pattern at each neural site observed during the saccade; time zero indicates the start of the saccade and T represents the end of the saccade. Naturally, the firing pattern observed for each block represents the firing pattern for a single neuron, as recorded in the literature, but the block represents the cumulative effect of all the neurons within that site. Consistent with a time-optimal control theory, neural activity is represented within each of the blocks as pulses and/or steps to reflect their operation as timing gates. The SC fires maximally as long as the dynamic motor error is greater than zero, in agreement with the first-order time-optimal controller and physiological evidence. Notice that the LLBNs are driven by the SC as long as there is a feedback error maintained by the cerebellar vermis. In all likelihood, the maximal firing rate by the SC is stochastic, depending on a variety of physiological factors such as the interest in tracking the target, anxiety, frustration, and stress.

7.10 SYSTEM IDENTIFICATION

In traditional applications of electrical, mechanical, and chemical engineering, the main application of modeling is as a design tool to allow the efficient study of the effects of parametric variation on system performance as a means of cost containment.

In modeling physiological systems, the goal is not to design a system but to identify the parameters and structure of the system. Ideally, the input and the output of the physiological system are known and some information about the internal dynamics of the system is available (Fig. 7.33). In many cases, either the input or the output is not measurable or observable but is estimated from a remote signal, and no information about the system is known. System identification is the process of creating a model of a system and estimating the parameters of the model. This section introduces the concept of system identification in both the frequency and the time domain.

A variety of signals are available to the biomedical engineer as described in Chapters 3–5. Those produced by the body include action potentials, EEGs, EKGs, EMGs, EOG, and pressure transducer output. Additional signals are available through ultrasound, X-ray tomography, MRI, and radiation. From these signals a model is built and parameters are estimated according to the modeling plan in Fig. 7.1. Before work on system identification begins, understanding the characteristics of the input and output signals is important; that is, knowing the voltage range, frequency range, whether the signal is deterministic or stochastic, and if coding (i.e., neural mapping) is involved. Most biologically generated signals are low-frequency and involve some coding. For example, EEGs have an upper frequency of 30 Hz and eye movements have an upper frequency under 100 Hz. The saccadic system uses neural coding that transforms burst duration into saccade amplitude. After obtaining the input and output signals, these signal must be processed. A fundamental block is the amplifier, which is characterized by its gain and frequency as described in Chapter 5. Note that the typical amplifier is designed as a low-pass filter since noise amplification is not desired. Interestingly, most amplifiers have storage elements (i.e., capacitors and inductors), so the experimenter must wait until the transient response of the amplifier has been completed before any useful information can be extracted. An important point to remember is that the faster the cutoff of the filter, the longer the transient response of the amplifier.

In undergraduate classes, a system (the transfer function or system description) and input are usually provided and a response or output is requested. While this seems difficult, it is actually much easier than trying to determine the parameters of a physiological system when the input (and perhaps not the direct input as described in the saccadic eye movement system) and noisy output characteristics of the model are known. In the ideal case, the desired result here is the transfer function, which can be determined from

$$H(s) = \frac{V_o(s)}{V_i(s)} \tag{7.46}$$

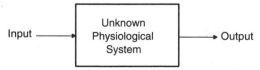

Fig. 7.33   Block diagram of typical physiological system without feedback.

7.10.1 Classical System Identification

The simplest and most direct method of system identification is sinusoidal analysis. A source of sinusoidal excitation is needed that usually consists of a sine wave generator, a measurement transducer, and a recorder to gather frequency response data. Many measurement transducers are readily available for changing physical variables into voltages as described in Chapter 4. Devices that produce the sinusoidal excitation are much more difficult to obtain and are usually designed by the experimenter. Recording the frequency response data can easily be obtained from an oscilloscope. Figure 7.34 illustrates the essential elements of sinusoidal analysis.

The experiment to identify model parameters using sinusoidal analysis is simple to carry out. The input is varied at discrete frequencies over the entire spectrum of interest and the output magnitude and phase are recorded for each input. To illustrate this technique, consider the following analysis. From Fig. 7.34, it is clear that the transfer function is given by

$$H(j\omega) = \frac{V_o(j\omega)}{V_i(j\omega)} \tag{7.47}$$

The Fourier transform of the input is

$$V_i(j\omega) = \Gamma\{A\cos(\omega_x t + \theta)\} = A\int e^{-j\omega\lambda}\cos(\omega_x\lambda t + \theta)d\lambda$$

$$= A\int e^{-j\omega\left(-\frac{\theta}{\omega_\lambda} + \tau\right)}\cos(\omega_x\tau)d\tau \tag{7.48}$$

by substituting $\lambda = \tau - \dfrac{\theta}{\omega_x}$. Factoring out the terms not involving τ gives

$$V_i(j\omega) = Ae^{\frac{j\omega\theta}{\omega_x}}\int e^{j\omega\tau}\cos\omega_x\tau\, d\tau = Ae^{\frac{j\omega\theta}{\omega_x}}\left[\pi\delta(\omega - \omega_x) + \pi\delta(\omega + \omega_x)\right] \tag{7.49}$$

Similarly,

$$V_o(j\omega) = Be^{\frac{j\omega\phi}{\omega_x}}\left[\pi\delta(\omega - \omega_x) + \pi\delta(\omega + \omega_x)\right] \tag{7.50}$$

According to Eq. (7.47)

$$H(j\omega) = \frac{V_o(j\omega)}{V_i(j\omega)} = \frac{B}{A}\frac{e^{\frac{j\omega\theta}{\omega x}}}{e^{\frac{j\omega\theta}{\omega x}}} = \frac{B}{A}e^{\frac{j\omega\theta(\phi-\theta)}{\omega_x}} \tag{7.51}$$

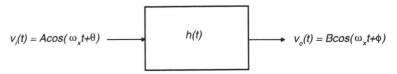

Fig. 7.34 Impulse response block diagram.

At steady state with $\omega = \omega_x$, Eq. (7.51) reduces to

$$H(j\omega) = \frac{B}{A} e^{j(\phi - \theta)} \tag{7.52}$$

Each of these quantities in Eq. (7.52) is known (i.e., B, A, ϕ, and θ), so the magnitude and phase angle of the transfer function is also known. Thus, ω_x can be varied over the frequency range of interest to determine the transfer function.

In general, a transfer function, $G(s)\big|_{s=j\omega} = G(j\omega)$, is composed of the following terms:

1. Constant term K
2. M poles or zeros at the origin of the form $(j\omega)^M$
3. P poles of the form $\prod_{p=1}^{P} (1 + j\omega\tau_p)$ or Z zeros of the form $\prod_{z=1}^{Z} (1 + j\omega\tau_z)$.

   Naturally, the poles or zeros are located at $-\dfrac{1}{\tau}$.

4. R complex poles of the form $\prod_{r=1}^{R}\left(1 + \left(\dfrac{2\zeta_r}{\omega_{nr}}\right)j\omega + \left(\dfrac{j\omega}{\omega_{nr}}\right)^2\right)$ or S zeros of

   the form $\prod_{s=1}^{S}\left(1 + \left(\dfrac{2\zeta_s}{\omega_{ns}}\right)j\omega + \left(\dfrac{j\omega}{\omega_{ns}}\right)^2\right)$.

5. Pure time delay $e^{-j\omega T_d}$

where M, P, Z, R, S, and T_d are all positive integers. Incorporating these terms, the transfer function is written as

$$G(j\omega) = \frac{K \times e^{-j\omega T_d} \times \left(\prod_{z=1}^{Z} (1 + j\omega\tau_z)\right) \times \left(\prod_{s=1}^{S}\left(1 + \left(\dfrac{2\zeta_s}{\omega_{ns}}\right)j\omega + \left(\dfrac{j\omega}{\omega_{ns}}\right)^2\right)\right)}{(j\omega)^M \times \left(\prod_{p=1}^{P} (1 + j\omega\tau_p)\right) \times \left(\prod_{s=1}^{R}\left(1 + \left(\dfrac{2\zeta_s}{\omega_{ns}}\right)j\omega + \left(\dfrac{j\omega}{\omega_{nr}}\right)^2\right)\right)} \tag{7.53}$$

This equation is used as a template when describing the data with a model. To determine the value of the unknown parameters in the model, the logarithm and asymptotic approximations to the transfer function are used. In general, the logarithmic gain (in dB), of the transfer function template is

$$20 \log\left|G(j\omega)\right| = 20 \log K + 20 \sum_{z=1}^{Z} \log\left|1 + j\omega\tau_z\right| + 20 \sum_{s=1}^{S} \log\left|1 + \left(\frac{2\zeta_s}{\omega_{ns}}\right)j\omega + \left(\frac{j\omega}{\omega_{ns}}\right)^2\right|$$

$$- 20 \log\left|G(j\omega)^M\right| - 20 \sum_{p=1}^{P} \log\left|1 + j\omega\tau_p\right| - 20 \sum_{r=1}^{R} \log\left|1 + \left(\frac{2\zeta_r}{\omega_{nr}}\right)j\omega + \left(\frac{j\omega}{\omega_{nr}}\right)^2\right| \tag{7.54}$$

and the phase, in degrees, is

$$\phi(\omega) = -\omega T_d + \sum_{z=1}^{Z} \tan^{-1}(\omega\tau_z) + \sum_{s=1}^{S} \tan^{-1}\left(\frac{2\zeta_s\omega_{ns}\omega}{\omega_{ns}^2 - \omega^2}\right) - M \times (90°)$$

$$- \sum_{p=1}^{P} \tan^{-1}(\omega\tau_p) - \sum_{r=1}^{R} \tan^{-1}\left(\frac{2\zeta_r\omega_{nr}\omega}{\omega_{nr}^2 - \omega^2}\right)$$

(7.55)

where the phase angle of the constant is 0° and the magnitude of the time delay is 1. Evident from these expressions is that each term can be considered separately and added together to obtain the complete Bode diagram. The asymptotic approximations to the logarithmic gain for the poles and zeros are given by

- **Poles at the origin**
  Gain: $-20 \log|(j\omega)| = -20 \log \omega$. The logarithmic gain at $\omega = 1$ is 0 (i.e., the line passes through 0 dB at $\omega = 1$ rad/s).
  Phase: $\phi = -90°$. If there is more than one pole, the slope of the gain changes by $M \times (-20)$ and the phase by $M \times (-90°)$.
- **Pole on the real axis**

$$\text{Gain: } -20 \log|1 + j\omega\tau_p| = \begin{cases} 0 & \text{for } \omega < \dfrac{1}{\tau_p} \\[2ex] -20 \log(\omega\tau_p) & \text{for } \omega > \dfrac{1}{\tau_p} \end{cases}$$

  Phase: An asymptotic approximation to $-\tan^{-1}(\omega\tau_p)$ is drawn with a straight line from 0° at 1 decade below $\omega = \dfrac{1}{\tau_p}$ to $-90°$ at 1 decade above $\omega = \dfrac{1}{\tau_p}$. The pole is located at $-\dfrac{1}{\tau_p}$.
- **Zero on the real axis**

$$\text{Gain: } 20 \log|1 + j\omega\tau_z| = \begin{cases} 0 & \text{for } \omega < \dfrac{1}{\tau_z} \\[2ex] 20 \log(\omega\tau_z) & \text{for } \omega \geq \dfrac{1}{\tau_z} \end{cases}$$

  Phase: An asymptotic approximation to $\tan^{-1}(\omega\tau_z)$ is drawn with a straight line from 0° at 1 decade below $\omega = \dfrac{1}{\tau_z}$ to 90° at 1 decade above $\omega = \dfrac{1}{\tau_z}$. The zero is located at $-\dfrac{1}{\tau_z}$.
- **Complex poles**

$$\text{Gain: } -20 \log\left|1 + \left(\frac{2\zeta_r}{\omega_{nr}}\right)j\omega + \left(\frac{j\omega}{\omega_{nr}}\right)^2\right| = \begin{cases} 0 & \text{for } \omega < \omega_{nr} \\[2ex] -40 \log\left(\dfrac{\omega}{\omega_{nr}}\right) & \text{for } \omega \geq \omega_{nr} \end{cases}$$

A graph of the actual magnitude–frequency is shown in Fig. 7.35 with $\omega_n = 1.0$ and ζ ranging from 0.05 to 1.0. Notice that as ζ decreases from 1.0, the magnitude peaks at correspondingly larger values. As ζ approaches zero, the magnitude approaches infinity at $\omega = \omega_{n_r}$. For values of $\zeta > 0.707$ there is no resonance.

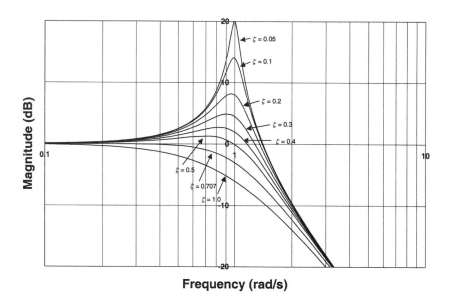

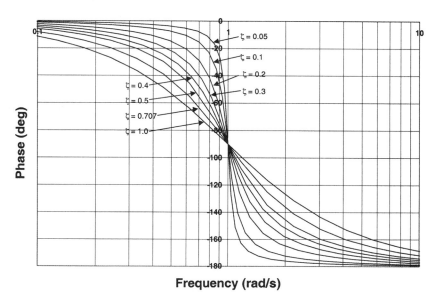

Fig. 7.35   Bode plot of complex poles with $\omega_n = 1.0$ rad/s

Phase: Depending on the value of ζ_r, the shape of the curve is quite variable but in general is $0°$ at 1 decade below $\omega = \omega_{n_r}$ and $-180°$ at 1 decade above $\omega = \omega_{n_r}$.

A graph of the actual phase–frequency is shown in Fig. 7.35 with $\omega_n = 1.0$ and ζ ranging from 0.05 to 1.0. Notice that as ζ decreases from 1.0, the phase changes more quickly from $0°$ to $180°$ over a smaller frequency interval.

The poles are located at $-\zeta_r\omega_{n_r} \pm j\omega_{n_r}\sqrt{1 - \zeta_r^2}$.

- **Complex zeros**

$$\text{Gain: } 20\log\left|1 + \left(\frac{2\zeta_s}{\omega_{n_s}}\right)j\omega + \left(\frac{j\omega}{\omega_{n_s}}\right)^2\right| = \begin{cases} 0 & \text{for } \omega < \omega_{n_s} \\ 40\log\left(\dfrac{\omega}{\omega_{n_s}}\right) & \text{for } \omega \geq \omega_{n_s} \end{cases}$$

Phase: Depending on the value of ζ_s, the shape of the curve is quite variable but in general is $0°$ at 1 decade below $\omega = \omega_{n_s}$ and $180°$ at 1 decade above $\omega = \omega_{n_s}$.

Both the magnitude and the phase follow the two graphs shown in Fig. 7.35 and the previous discussion with regard to the complex poles with the exception that the slope is 40 dB/decade rather than -40 dB/decade.

The zeros are located at $-\zeta_s\omega_{n_s} \pm j\omega_{n_s}\sqrt{1 - \zeta_s^2}$.

- **Time delay**
    Gain: 1 for all ω
    Phase: $-\omega T_d$
- **Constant K**
    Gain: $20\log K$
    Phase: 0

The frequency at which the slope changes in a Bode magnitude–frequency plot is called a break or corner frequency. The first step in estimating the parameters of a model involves identifying the break frequencies in the magnitude–frequency and/or phase–frequency responses. This simply involves identifying points at which the magnitude changes slope in the Bode plot. Poles or zeros at the origin have a constant slope of -20 or $+20$ dB/decade, respectively, from $-\infty$ to ∞. Real poles or zeros have a change in slope at the break frequency of -20 or 20 dB/decade, respectively. The value of the pole or zero is the break frequency. Estimating complex poles or zeros is much more difficult. The first step is to locate the break frequency ω_n, the point at which the slope changes by 40 dB/decade. To estimate ζ, use the actual magnitude–frequency (size of the peak) and phase–frequency (quickness of changing $180°$) curves in Fig. 7.35 to closely match the data.

The error between the actual logarithmic gain and straight-line asymptotes at the break frequency is 3 dB for a pole on the real axis (the exact curve equals the asymptote $(-3$ dB). The error drops to 0.3 dB 1 decade below and above the break frequency. The error between the real zero and the asymptote is similar except the exact curve equals the asymptote $+3$ dB. At the break frequency for the complex poles or zeros, the error between the actual logarithmic gain and straight-line asymptotes depends on ζ and can be quite large as observed from Fig. 7.35.

Example Problem 7.8

Sinusoids of varying frequencies were applied to an open–loop system and the following results were measured. Construct a Bode diagram and estimate the transfer function.

Frequency (rad/s)	20 log|G| (dB)	Phase (degrees)
0.01	58	−90
0.02	51	−91
0.05	44	−91
0.11	37	−93
0.24	30	−97
0.53	23	−105
1.17	15	−120
2.6	6	−142
5.7	−7	−161
12.7	−20	−171
28.1	−34	−176
62	−48	−178
137	−61	−179
304	−75	−180
453	−82	−180
672	−89	−180
1000	−96	−180

Solution

Bode plots of gain and phase vs frequency for the data given are shown in the following graphs. From the phase–frequency graph, it is clear that this system has two more poles than zeros because the phase angle approaches −180° as $\omega \rightarrow \infty$. Also, note that there is no peaking observed in the gain–frequency graph or sharp changes in the phase–frequency graph so there does not appear to be any lightly damped ($\zeta < 0.5$) complex poles. However, this does not imply that there are no heavily damped complex poles at this time.

For frequencies in the range 0.01–1 rad/s, the slope of the gain–frequency graph is −20 dB per decade. Since it is logical to assume that the slope remains at −20 dB for frequencies <0.01 rad/s, a transfer function with a pole at the origin provides such a response. If possible, it is important to verify that the magnitude–frequency response stays at −20 dB per decade for frequencies <0.01 rad/s to the limits of the recording instrumentation.

For frequencies in the range 2–1000 rad/s, the slope of the gain–frequency graph is −40 dB per decade. Since a pole has already been identified in the previous frequency interval, it is reasonable to conclude that there is another pole in this interval. Other possibilities exist, such as an additional pole and zero that are closely spaced or a complex pole and a zero. However, in the interest of simplicity and because these possibilities are not evident in the graphs, the existence of a pole is all that is required to describe the data.

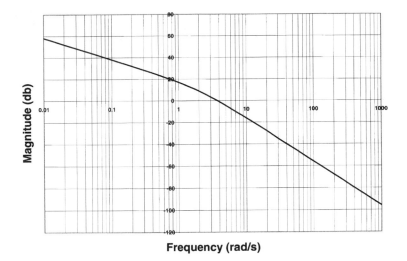

Frequency (rad/s)

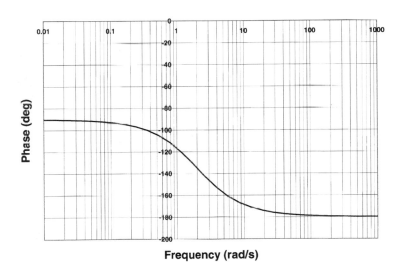

Frequency (rad/s)

The model contains a pole at the origin and another pole somewhere in the region above 1 rad/s. It is also safe to conclude that the model does not contain heavily damped complex poles since the slope of the gain–frequency graph is accounted for completely.

To estimate the unknown pole, straight lines are drawn tangent to the magnitude–frequency curve as shown in the following figure. The intersection of the two lines gives the break frequency for the pole at approximately 2 rad/s. Notice that the actual curve is approximately 3 dB below the asymptotes as discussed previously and observed in the following figure. The model developed thus far is

$$G(j\omega) = \frac{1}{j\omega\left(\dfrac{j\omega}{2} + 1\right)}$$

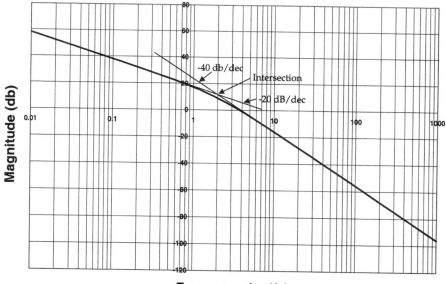

To determine the constant K, the magnitude at $\omega = 1$ rad/s is investigated. The pole at zero contributes a value of 0 toward the logarithmic gain. The pole at 2 contributes a value of

$$-1 \text{ dB } \left(-20 \log \left|\frac{j\omega}{2} + 1\right| = -20 \log\left(\sqrt{\left(\frac{1}{2}\right)^2 + 1^2}\right) = -1 \text{ dB}\right)$$

toward the logarithmic gain. The reason that the contribution of the pole at 2 was computed was that the point 1 rad/s was within the range of plus/minus a decade of the break frequency. At $\omega = 1$ rad/s, the nonzero terms are

$$20 \log|G(\omega)| = 20 \log K - 20 \log \left|\frac{j\omega}{2} + 1\right|$$

From the gain–frequency graph, $20 \log|G(\omega)| = 17$ dB. Therefore,

$$17 = 20 \log K - 1$$

or $K = 8$. The model now consists of

$$G(j\omega) = \frac{8}{j\omega\left(\dfrac{j\omega}{2} + 1\right)}$$

The last term to investigate is whether there is a time delay in the system. At the break frequency $\omega = 2.0$ rad/s, the phase angle from the current model should be

$$\phi(\omega) = -90 - \tan^{-1}\left(\frac{\omega}{2}\right) = -90 - 45 = -135$$

This value is approximately equal to the data, and thus there does not appear to be a time delay in the system. ∎

Example Problem 7.8 illustrated a process of thinking in determining the structure and parameters of a model. Carrying out an analysis in this fashion on complex systems is extremely difficult if not impossible. Software packages that automatically carry out estimation of poles, zeros, a time delay, and gain of a transfer function from data are available, such as the System Identification toolbox in MATLAB. There are also other programs that provide more flexibility than MATLAB in analyzing complex systems, such as the FORTRAN program written by Seidel (1975).

There are a variety of other inputs that one can use to stimulate the system to elicit a response. These include such transient signals as a pulse, step, and ramp and noise signals such as white noise and pseudo-random binary sequences. The reason these techniques might be of interest is that not all systems are excited via sinusoidal input. One such system is the fast eye movement system. Here we typically use a step input to analysis the system.

In analyzing the output data obtained from step input excitation to determine the transfer function, we use a frequency response method using the Fourier transform and the fast Fourier transform (FFT). The frequency response of the input is known (s^{-1}). The frequency response of the output is calculated via a numerical algorithm called the FFT. To calculate the FFT of the output, the data must first be digitized using an analog-to-digital converter and stored in disk memory. Care must be taken to antialias (low-pass) filter any frequency content above the highest frequency of the signal and sample at a rate of at least 2.5 times the highest frequency. The transfer function is then calculated according to Eq. (7.46).

7.10.2 Identification of a Linear First-Order System

Another type of identification technique used specifically for a first-order system is presented here using a time domain approach. Assume that the system of interest is a first-order system that is excited with a step input. The response to the input is

$$v_o(t) = v_{ss} + Ke^{-\frac{t}{\tau}}u(t) \tag{7.55}$$

where $K = -(v_{ss} - v(0))$, and the response is shown in Fig. 7.36. Note that at $t = \tau$, the response is 63% of the way from the initial to the steady-state value. Similarly, at $t = 4\tau$, the response is 98% of the way from the initial to the steady-state value.

Suppose step input data are collected from an unknown first-order system. To describe the system, the parameters of Eq. 7.55 need to be estimated. One way to estimate the system time constant is from the initial slope of the response and using a smoothed steady-state value (via averaging). That is, the time constant the time constant τ is found from

$$\dot{v}(t) = \frac{1}{\tau}(v_{ss} - v(0))e^{-\frac{t}{\tau}} \rightarrow \tau = \frac{v_{ss} - v(0)}{\dot{v}(t)}e^{-\frac{t}{\tau}} \tag{7.57}$$

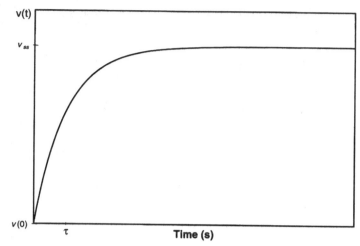

Fig. 7.36   First-order system response to a step input.

At $t = 0$, $\tau = \dfrac{v_{ss} - v(0)}{\dot{v}(0)}$, where $\dot{v}(0)$ is the initial slope of the response. The equation for estimating τ is nothing more than the equation of a straight line. This technique is illustrated in Fig. 7.37.

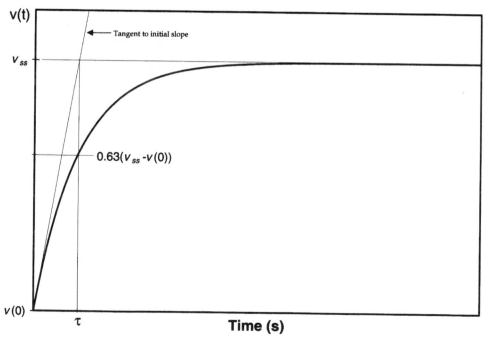

Fig. 7.37   Estimating the time constant from the initial slope of the response.

Example Problem 7.9

The following data were collected for the step response from an unknown first-order system. Find the parameters that describe the model.

t	v(t)
0	0.0
0.05	0.56
0.1	0.98
0.15	1.30
0.2	1.54
0.25	1.73
0.3	1.86
0.4	2.04
0.5	2.15
0.7	2.24
1.0	2.27
1.5	2.28
2.0	2.28

Solution

The model under consideration is described by Eq. (7.56), $v_o(t) = v_{ss} + Ke^{-\frac{t}{\tau}} u(t)$, with unknown parameters v_{ss}, K, and τ. Clearly, $v_{ss} = 2.28$ and $K = -2.28$. The data are plotted in the following figure along with the tangent to $\dot{v}(o)$. From the graph, $\tau = 1.7$ s.

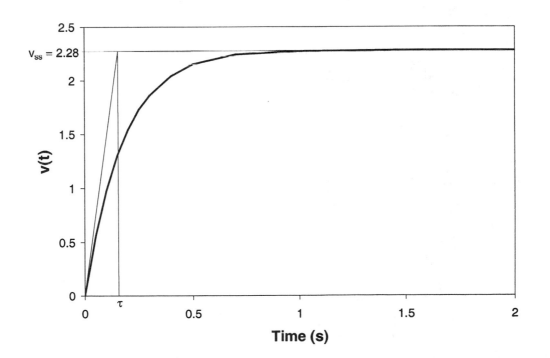

The model is given by

$$v(t) = 2.28\left(1 - e^{-\frac{t}{1.7}}\right)u(t)$$

or by

$$V(s) = \frac{v_{ss}}{s\tau + 1} = \frac{2.28}{0.17s + 1} = \frac{13.4}{s + 5.9}$$

∎

7.10.3 Identification of a Linear Second-Order System

Consider estimating the parameters of a second-order system using a time domain approach. For ease in analysis, suppose the system in Fig. 7.38 is given.

The differential equation describing the system is given by $f(t) = M\ddot{x} + B\dot{x} + Kx$. It is often convenient to rewrite the original differential equation in the standard form for ease in analysis:

$$\frac{f(t)}{M} = \ddot{x} + 2\zeta\omega_n\dot{x} + \omega_n^2 x \qquad (7.58)$$

where ω_n is the undamped natural frequency and ζ is the damping ratio, which for the system in Fig. 7.38 is given by

$$\zeta = \frac{B}{2\sqrt{KM}} \quad \text{and} \quad \omega_n = \sqrt{\frac{K}{M}}$$

The roots of the characteristic equation are

$$s_{1,2} = -\zeta\omega_n \pm \omega_n\sqrt{\zeta^2 - 1} = -\zeta\omega_n \pm j\omega_n\sqrt{1 - \zeta^2} = -\zeta\omega_n \pm j\omega_d \qquad (7.59)$$

By holding ω_n constant and varying ζ, the roots move about the complex plane as illustrated in Fig. 7.39. A system with $0 < \zeta < 1$ is called underdamped, with $\zeta = 1$ critically damped and $\zeta > 1$ overdamped. The natural or homogenous solution of this system can be solved using the classical approach via the roots of characteristic equation using the following equations:

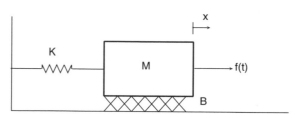

Fig. 7.38 A simple mechanical system.

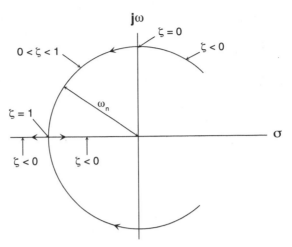

Fig. 7.39 Root trace as ζ is varied while ω_n is held constant. The circle radius is ω_n.

Damping	Natural Response Equation
Overdamped	$x(t) = x_{ss} + A_1 e^{s_1 t} + A_2 e^{s_2 t}$
Underdamped	$x(t) = x_{ss} + B_1 e^{-\zeta \omega_n t} \cos(\omega_d t + \phi)$
Critically damped	$x(t) = x_{ss} + (C_1 + C_2 t) e^{-\zeta \omega_n t}$

where x_{ss} is the steady-state value of $x(t)$ and A_1, A_2, B_1, ϕ, C_1, and C_2 are the constants which describe the system evaluated from the initial conditions of the system.

To estimate ζ and ω_n, a step input of magnitude γ is applied to the system, and the data are collected. After visually inspecting a plot of the data, one of the three types of responses is selected that describes the system. The parameters are then estimated from the plot. For instance, consider the following step response for an unknown system.

It appears that a suitable model for the system might be a second-order underdamped model (i.e., $0 < \zeta < 1$), with a solution similar to the one carried out in Example Problem 7.1:

$$x(t) = C \left[1 + \frac{e^{-\zeta \omega_n t}}{\sqrt{1 - \zeta^2}} \cos(\omega_d t + \phi) \right] \qquad (7.60)$$

where C is the steady-state response x_{ss}.

The following terms, illustrated in Fig. 7.40, are typically used to describe, quantitatively, the response to a step input.

- Rise time, T_r: The time for the response to rise from 10 to 90% of steady state.
- Settling time, T_s: The time for the response to settle within $\pm 5\%$ of the steady-state value.

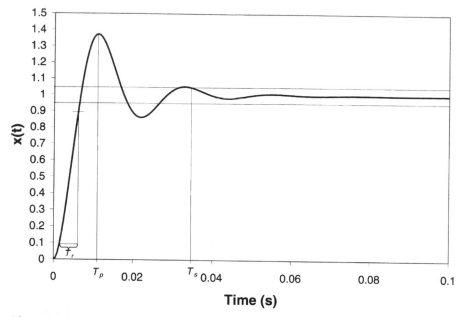

Fig. 7.40 Sample second-order response with graphical estimates of rise time, time to peak overshoot, and settling time.

- Peak overshoot time, T_p: The time for the response to reach the first peak overshoot.

Using values graphically determined from the data for the previous quantities as shown in Fig. 7.40 provides estimates for the parameters of the model. In Example Problem 7.2, the peak overshoot time was calculated for a second-order underdamped system as

$$T_p = \frac{\pi}{\omega_n \sqrt{1 - \zeta^2}} \tag{7.61}$$

and the response at T_p as

$$x(T_p) = C\left(1 + e^{-\zeta \frac{\pi}{\sqrt{1+\zeta^2}}}\right) \tag{7.62}$$

With the performance estimates calculated from the data and Eqs. (7.61) and (7.62), it is possible to estimate ζ and ω_n. First find ζ by using the Eq. (7.62). Then, using the solution for ζ, substitute this value into Eq. (7.62) to find ω_n. The phase angle ϕ in Equation in Equation 7.60 is determined using Equation 7.5 and the estimate for ζ, that is

$$\phi = \pi + \tan^{-1} \frac{-\zeta}{\sqrt{1 - \zeta^2}} \tag{7.63}$$

Example Problem 7.10

Find ζ and ω_n for the data in Fig. 7.40.

Solution

From the data in Fig. 7.40, $C = 1.0$, $T_p = 0.011$, and $x(T_p) = 1.37$. Therefore,

$$x(T_p) = C\left(1 + e^{-\zeta\frac{\pi}{\sqrt{1+\zeta^2}}}\right) \rightarrow \zeta = \sqrt{\frac{\frac{(\ln(x(T_p) - 1))^2}{\pi^2}}{1 + \frac{(\ln(x(T_p) - 1))^2}{\pi^2}}} = 0.3$$

$$T_p = \frac{\pi}{\omega_n\sqrt{1 - \zeta^2}} \rightarrow \omega_n = \frac{\pi}{T_p\sqrt{1 - \zeta^2}} = 300 \text{ rad/s} \quad \blacksquare$$

EXERCISES

1. What is the main sequence diagram? How do results from the Westheimer model in Section 7.3 compare with the main sequence diagram?
2. Simulate a 20° saccade with the Westheimer model in Section 7.3 with $\zeta = 0.7$ and $\omega_n = 120$ radians/s using SIMULINK. Assume that $K = 1$ N/m. Repeat the simulation for a 5, 10, and 15° saccade. Compare these results with the main sequence diagram in Figure 7.5.
3. Suppose the input to the Westheimer model in Section 7.3 is a pulse-step waveform as described in Section 7.4, and $\zeta = 0.7$, $\omega_n = 120$ radians/s and $K = 1$ N/m. (1) Estimate the size of the step necessary to keep the eyeball at 20°. (2) Using SIMULINK, find the pulse magnitude that matches the main sequence diagram in Figure 7.5 necessary to drive the eyeball to 20°. (3) Repeat part (2) for saccades of 5, 10 and 15°. (4) Compare these results with those of the Westheimer model.
4. A model of the saccadic eye movement system is characterized by the following equation.

$$\tau = 1.74 \times 10^{-3}\ddot{\theta} + 0.295\dot{\theta} + 25\theta$$

where τ is the applied torque. Suppose $\tau = 250u(t)$ and the initial conditions are zero. Use Laplace Transforms to solve for $\theta(t)$. Sketch $\theta(t)$.
5. Suppose a patient ingested a small quantity of radioactive iodine (I^{131}). A simple model describing the removal of I^{131} from the bloodstream into the urine and thyroid is given by a first-order differential equation. The rate of transfer of I^{131} from the bloodstream into the thyroid is given by K_1 and the urine by K_2. The time constant for the system is $\tau = \frac{1}{K_1 + K_2}$.
(a) Sketch the response of the system. (b) Suppose the thyroid is not functioning and does not take up any I^{131}, sketch the response of the system and compare to the result from (a).

6. Consider the system below defined with $M_1 = 2$ kg, $M_2 = 1$ kg, $B_2 = 1$ Ns/m, $B_2 = 2$ Ns/m, $K_1 = 1$ N/m, and $K_2 = 1$ N/m. Let $f(t)$ be the applied force, and x_1 and x_2 be the displacements from rest. (a) Find the transfer function. (b) Use MATLAB to draw the Bode diagram.

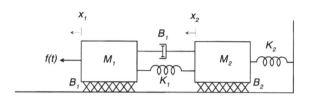

7. Consider the system illustrated in the following diagram defined with $M_1 = 1$ kg, $M_2 = 2$ kg, $B_2 = 2$ Ns/m, $K_1 = 1$ N/m, and $K_2 = 1$ N/m. Let $f(t)$ be the applied force, and x_1 and x_2 be the displacements from rest. The pulley is assumed to have no inertia or friction. (a) Write the differential equations that describe this system. (b) If the input is a step with magnitude 10 N, simulate the solution with SIMULINK.

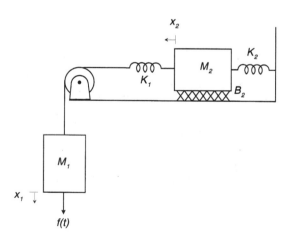

8. Consider the system in the following diagram, where there are two viscous elements B_2, K_1 is a translational element and K_2 is a rotational element. Let $\tau(t)$ be the applied torque, x_1 the displacement of M from rest and θ the angular displacement of the element J from rest (i.e., when the springs are neither stretched or compressed). The pulley has no inertia or friction, and the cable does not stretch. (a) Write the differential equations that describe the system. (b) Write the state variable equations that describe the system.

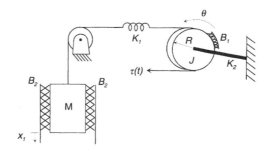

9. Consider the system in the following diagram, where there are two viscous elements B_2, K_1 is a translational element and K_2 is a rotational element. Let $f(t)$ be the applied force, x_1 the displacement of M_1 and x_2 the displacement of M_2 from rest, and θ the angular displacement of the element J from rest (i.e., when the springs are neither stretched nor compressed). The pulley has no inertia or friction, and the cable does not stretch. **(a)** Write the differential equations that describe the system. **(b)** Write the state variable equations that describe the system.

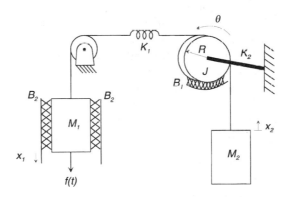

10. Consider the system in the following diagram defined with $M_3 = 0.5$ kg, $K_3 = 4$ N/m (translational), $J_1 = 0.5$ kg/m^2, $B_1 = 1$ N/s/m, $K_1 = 2$ N-m, $J_2 = 2$ kg/m^2, $K_2 = 1$ N/m, $R_1 = 0.2$ m, and $R_2 = 1.0$ m. Let $\tau(t)$ be the applied torque, x_3 be the displacement of M_3 from rest, and θ_1 and θ_2 be the angular displacement of the elements J_1 and J_2 from rest (i.e., when the springs are neither stretched nor compressed. **(a)** Write the differential equations that describe this system. **(b)** If the input is a step with magnitude 10 N, simulate the solution with SIMULINK.

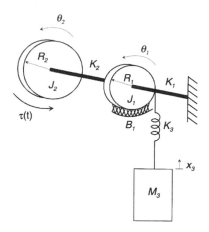

11. Given the Westheimer model described in Section 7.3 with $\zeta = 1/\sqrt{2}$, $\omega_n = 100$ rad/s and $K = 1$, solve for the general response with a pulse-step input as described in Fig. 7.7. Examine the change in the response as the pulse magnitude is increased and the duration of the pulse, t_1, is decreased while the steady state size of the saccade remains constant.

12. With the Westheimer model described in Section 7.3, separately estimate ζ and ω_n for a 5, 10, 15, and 20° saccades using information in the main sequence diagram in Fig. 7.5. Assume that peak overshoot, $x(T_p)$, is 1° greater than the saccade size. Simulate the four saccades. Develop a relationship between ζ and ω_n as a function of saccade size that matches the main sequence diagram. With these relationships, plot T_p and peak velocity as a function of saccade size. Compare these results to the original Westheimer main sequence results and those in Fig. 7.5.

13. Consider an unexcited muscle model as shown in Fig. 7.22 with $K_{lt} = 32$ N/m, $K_{se} = 125$ N/m and $B = 3.4$ Ns/m (F = 0 for the case of an unexcited muscle). **(a)** Find the transfer function $H(j\omega) = \dfrac{X_1}{T}$. **(b)** Use MATLAB to draw the Bode diagram.

14. Consider an unexcited muscle model in Fig. 7.23 with $K_{lt} = 60.7$ N/m, $K_{se} = 125$ N/m, $B_1 = 2$ Ns/m and $B_2 = 0.5$ Ns/m (F = 0 for the case of an unexcited muscle) **(a)** Find the transfer function $H(j\omega) = \dfrac{X_1}{T}$. **(b)** Use MATLAB to draw the Bode diagram.

15. Simulate a 20° saccade using the linear homeomorphic saccadic eye movement model from Section 7.7 using SIMULINK.

16. Simulate a 20° saccade using the linear homeomorphic saccadic eye movement model from Section 7.8 using SIMULINK.

17. Consider the linear homeomorphic saccadic eye movement model given in Eq. (7.41). **(a)** Find the transfer function. **(b)** Use MATLAB to draw the Bode diagram.

18. Verify the length–tension curves in Fig. 7.24.

19. Verify the force–velocity curve for the muscle model in Fig. 7.25. (Hint use SIMULINK to calculate peak velocity for each value of M.)

20. Consider the following model of the passive orbital tissues driven by torque τ with $Kp = 0.5$ g/°, $B_p = 0.06$ g s/°, and $J_p = 4.3 \times 10^{-5}$ g s²/°. All elements are rotational. **(a)** Find the transfer function $\theta(s)/\tau(s)$. **(b)** Use MATLAB to draw the Bode diagram.

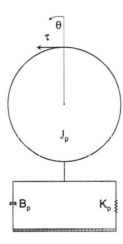

21. Consider the following model of the passive orbital tissues driven by torque τ with $J_p = 4.308 \times 10^{-5}$ gs²/°, $K_{p1} = 0.5267$ g/°, $K_{p2} = 0.8133$ g/°, $B_{p1} = 0.010534$ gs/°, and $B_{p2} = 0.8133$ gs/°. All elements are rotational. **(a)** Find the transfer function $\theta_1(s)/\tau(s)$. **(b)** Use MATLAB to draw the Bode diagram.

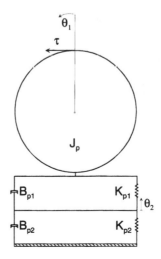

22. Consider the following model of the eye movement system. The elements are all rotational and $f_K(\theta) = K_1\theta^2$ (a nonlinear rotational spring).
(a) Write the nonlinear differential equation that describes this system.
(b) Write a linearized differential equation using a Taylor series 1st order approximation about an operating point.

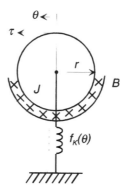

23. Suppose the passive elasticity of unexcited muscle is given by the following nonlinear translational force-displacement relationship

$$f(x) = |x|x$$

where x is the displacement from equilibrium position. Determine a linear approximation for this nonlinear element in the vicinity of the equilibrium point.

24. Sinusoids of varying frequencies were applied to an open-loop system and the following results were measured. Construct a Bode diagram and estimate the transfer function.

Frequency (rad/s)	Magnitude ratio, $|G|$	Phase (degrees)
0.6	2.01	−3.3
1.6	2.03	−8.9
2.6	2.09	−15.0
3.6	2.17	−21.8
5.5	2.37	−38.1
6.1	2.43	−44.8
7.3	2.49	−60.0
9.8	2.16	−93.6
12.7	1.39	−123.2
32.9	0.18	−164.2
62.1	0.05	−170.0
100	0.018	−175.0

25. Sinusoids of varying frequencies were applied to an open-loop system and the following results were measured. Construct a Bode diagram and estimate the transfer function.

Frequency (rad/s)	20 log|G| (dB)	Phase (degrees)
0.001	6.02	−0.086
0.356	6.02	−3.06
1.17	5.96	−10.0
2.59	5.74	−22.0
8.53	3.65	−64.9
12.7	1.85	−88.1
18.9	−0.571	−116.0
41.8	−6.64	−196.0
62.1	−9.95	−259.0
137	−16.8	−479.0
304	−23.6	959.0
1000	−34.0	−2950.0

26. Sinusoids of varying frequencies were applied to an open-loop system and the following results were measured. Construct a Bode diagram and estimate the transfer function.

Frequency (rads/s)	Magnitude ratio, |G|	Phase (degrees)
.011	2.0	−0.62
.024	2.0	−1.37
.053	2.0	−3.03
1.17	2.0	−6.7
2.6	1.93	−14.5
5.7	1.74	−29.8
12.7	1.24	−51.8
28.1	0.67	−70.4
62	0.32	−80.9
137	0.15	−85.8
304	0.07	−88.1
453	0.044	−88.7
672	0.03	−89.1
1000	0.02	−89.4

27. Sinusoids of varying frequencies were applied to an open-loop system and the following results were measured (data from Seidel, 1975). Construct a Bode diagram for the data.

Frequency (rads/s)	Magnitude ratio, $|G|$	Phase (radians)
1	1.	−.035
3	.95	−.227
7	.77	−.419
10	.7	−.541
15	.67	−.611
20	.63	−.768
25	.6	−.995
30	.53	−1.08
35	.48	−1.24
40	.44	−1.31
50	.35	−1.52
60	.31	−1.92
70	.33	−1.61
80	.35	−1.83
90	.32	−2.08
100	.32	−2.23
110	.3	−2.53
120	.29	−2.72
130	.27	−2.9
140	.26	−3.0

Estimate the transfer function if it consists of (a) two poles, (b) a pole and a complex pole pair, (c) two poles, a zero, and a complex pole pair, and (d) three poles, a zero, and a complex pole pair. (Hint: It may be useful to solve this program using the MATLAB System Identification toolbox or Seidel's program).

28. The following data were collected for the step response for an unknown first-order system. Find the parameters that describe the model.

t	$v(t)$
0.0	0.00
0.005	3.41
0.01	5.65
0.015	7.13
0.02	8.11
0.025	8.75
0.03	9.18
0.035	9.46
0.04	9.64
0.045	9.76
0.05	9.84
0.055	9.90
0.06	9.93
0.1	10.0

29. Suppose a second-order underdamped system response to a step is given by Equation 7.60 and has $C = 10$, $T_p = 0.050$ and $x(T_p) = 10.1$. Find ζ and ω_n.

30. A stylized 10° saccade is shown in the following figure. Estimate ζ and ω_n for the Westheimer model described in Section 7.3. Calculate the time to peak velocity and peak velocity.

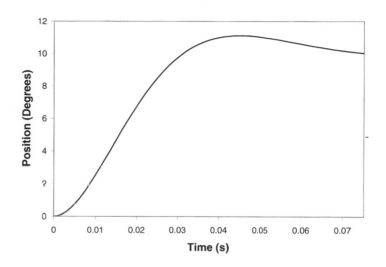

SUGGESTED READING

Bahill, A. T. (1981). *Bioengineering: Biomedical, Medical and Clinical Engineering.* Prentice Hall, Englewood Cliffs, NJ.

Bahill, A. T., Latimer, J. R., and Troost, B. T. (1980). Linear homeomorphic model for human movement. *IEEE Trans. Biomed. Eng.* **27**, 631–639.

Carpenter, R. H. H. (1988). *Movements of the Eyes.* Pion, London.

Close, R. I., and Luff, A. R. (1974). Dynamic properties of inferior rectus muscle of the rat. *J. Physiol.* **236**, 259–270.

Collins, C. C. (1975). The human oculomotor control system. In *Basic Mechanisms of Ocular Motility and Their Clinical Implications* (G. Lennerstrand and P. Bach-y-Rita, Eds.), pp. 145–180. Pergamon, Oxford.

Collins, C. C., O'Meara, D. M., and Scott, A. B. (1975). Muscle tension during unrestrained human eye movements. *J. Physiol.* **245**, 351.

Enderle, J. D. (1988). Observations on pilot neurosensory control performance during saccadic eye movements. *Aviation Space Environ. Med.* **59**, 309.

Enderle, J. D. (1994). A physiological neural network for saccadic eye movement control, Armstrong Laboratory/AO-TR-1994-0023. Air Force Material Command, Brooks Air Force Base, TX.

Enderle, J. D. (1995). The fast eye movement control system. In *The Biomedical Engineering Handbook* ed. (J. Bronzino, Ed.), CRC Press, Boca Raton, FL. pp. 2463–2483.

Enderle, J. D., and Wolfe, J. W. (1987). Time-optimal control of saccadic eye movements. *IEEE Trans. Biomed. Eng.* **BME-34**(1), 43–55.

Enderle, J. D., and Wolfe, J. W. (1988). Frequency response analysis of human saccadic eye movements: Estimation of stochastic muscle forces. *Computers in Biol. Med.* **18**(3), 195–219.

Enderle, J. D., Wolfe, J. W., and Yates, J. T. (1984). The linear homeomorphic saccadic eye movement model — A modification. *IEEE Trans. Biomed. Eng.* **31**(11), 717–720.

Enderle, J. D., Engelken, E. J., and Stiles, R. N. (1991). A comparison of static and dynamic characteristics between rectus eye muscle and linear muscle model predictions. *IEEE Trans. Biomed. Eng.* **BME-38**, 12, 1235–1245.

Fenn, W. O., and Marsh, B. S. (1935). Muscular force at different speeds of shortening. *J. Physiol. London* **35**, 277–297.

Hill, A. V. (1938). The heat of shortening and the dynamic constants of muscle. *Proc. R. Soc. London Ser. B* **126**, 136–195,

Hsu, F. K., Bahill, A. T., and Stark, L. (1976). Parametric sensitivity of a homeomorphic model for saccadic and vergence eye movements. *Computer Prog. Biomed.* **6**, 108–116.

Kuo, B. C. (1991). *Automatic Control Systems.* Prentice Hall, Englewood Cliffs, NJ.

Leigh, R. J., and Zee, D. S. (1983). *The Neurology of Eye Movements.* Davis, Philadelphia.

Levin, A. and Wyman, J. (1927). The viscous elastic properties of muscle. *Proc. R. Soc. London Ser. B***101**, 218–243.

Robinson, D. A. (1964). The mechanics of human saccadic eye movement. *J. Physiol. London* **174**, 245–264.

Robinson, D. A. (1975). Oculomotor control signals. In *Basic Mechanisms of Ocular Motility and Their Clinical Implication* (G. Lennerstrand and P. Bach-y-Rita, Eds.), pp. 337–374. Pergamon, Oxford.

Robinson, D. A. (1981). Models of mechanics of eye movements. In *Models of Oculomotor Behavior and Control* (B. L. Zuber), (Ed.), pp. 21–41. CRC Press, Boca Raton, FL.

Scudder, C. A. (1988). A new local feedback model of the saccadic burst generator. *J. Neurophysiol.* **59**(4), 1454–1475.

Seidel, R. C. (1975, September). Transfer-function-parameter estimation from frequency response data — A FORTRAN program, *NASA Technical Memorandum NASA TM X-3286.* Lewis Research Center, NASA, Cleveland OH 44135

Sparks, D. L. (1986). Translation of sensory signals into commands for control of saccadic eye movements: Role of the primate superior colliculus. *Physiol. Rev.* **66**, 118–171.

Sparks, D. L., Holland, R., and Guthrie, B. L. (1976). Size and distribution of movement fields in the monkey superior colliculus. *Brain Res.* **113**, 21–34.

van Gisbergen, J. A. M., Robinson, D. A., and Gielen, S. (1981). A quantitative analysis of generation of saccadic eye movements by burst neurons. *J. Neurophysiol.* **45**, 417–442.

Westheimer, G. (1954). Mechanism of saccadic eye movements. *AMA Arch. Ophthalmol.* **52**, 710–724.

Wilkie, D. R. (1968). *Muscle: Studies in Biology,* Vol. 11. Arnold, London.

8 COMPARTMENTAL ANALYSIS

Chapter Contents

At the conclusion of this chapter, the reader will be able to:

- Identify various types of models to meet specific goals
- Define various terms useful in compartmental analysis, such as kinetic homogeneity, and explain the concept of a compartment
- Select physiologic systems that are candidates for compartmental modeling using the five model postulates presented: existence, homogeneity, conservation, stationarity, and linearity

- Formulate simple models to simulate system responses based on all model postulates
- Obtain solutions for the model equations and evaluate their appropriateness to emulate system behavior
- Modify the simple models by relaxing one or more of the postulates and obtain solutions for simulation and design
- Incorporate more complex concepts other than zero- or first-order behavior for transport of solids, liquids, or gases between compartments
- Evaluate the need to seek alternatives from the classical linear compartment models

8.1   INTRODUCTION

The specific goal of compartmental analysis is to represent complicated physiologic systems with relatively simple mathematical models. Once the model is developed, system simulation can be readily accomplished to provide insights into system structure and performance.

In compartmental analysis, systems that are continuous and essentially nonhomogeneous are replaced with a series of discrete spatial regions, termed compartments, considered to be homogeneous. Thus, each subsequent compartment is modeled as a lumped parameter system. For example, a physiologic system requiring partial differential equations to describe transient spatial variations in the concentrations of desired components can be simulated using a series of ordinary differential equations using compartmental analysis.

Before proceeding, some definitions and concepts are required. A system can be defined by a class of dynamic models widely used in quantitative studies of the kinetics of materials in physiologic systems. Materials are considered to be either exogenous (such as a drug or tracer) or endogenous (such as a substrate like glucose or an enzyme or hormone like insulin). Kinetics refers to time-variant processes, such as production, distribution, transport, utilization, and substrate–hormone control interactions.

A compartment is an amount of material or spatial region that acts as though it is well mixed and kinetically homogeneous. The concept of well mixed is related to uniformity of information. This means that any samples taken from the compartment at the same time will have identical properties and are equally representative of the system. Kinetic homogeneity means that each particle within a chamber has the same probability of taking any exit pathway. A compartmental model then is defined as a finite number of compartments with specific interconnections among them, each representing a flux of material which physiologically represents transport from one location to another and/or a chemical transformation. When a compartment is a physical space, parts that are accessible for measurement must be distinguished from those that are inaccessible. The definition of a compartment is actually a theoretical construct which could combine material from several physical spaces within a system. Consequently, the ability to equate a compartment to a physical space depends on the system being studied and associated model assumptions.

There are several possible candidates for compartments in specific biological systems. Blood plasma can be considered a compartment as well as a substance such as glucose within the plasma. Zinc in bone and thyroxin in the thyroid can also be compartments. Since experiments can be conducted that follow different substances in plasma, such as glucose, lactate, and alanine, there can be more than one plasma compartment — one for each substance being studied. Extending this concept to other physiologic systems, glucose and glucose-6-phosphate can represent two different compartments within a liver cell. Thus, it should be apparent that a physical space may actually represent more than one compartment.

Compartmental analysis is the combining of material with similar characteristics into entities that are homogeneous that permits a complex physiologic system to be represented by a finite number of compartments and pathways. The actual number of compartments required depends on both the complexity of these large systems and the robustness of the experimental protocol. The associated model incorporates known and postulated physiology and biochemistry and thus is unique for each system that is studied. It provides the investigator with invaluable insights into system structure and performance but is only as good as the assumptions that were incorporated into its development.

The focus of this chapter is on the utilization of compartmental analysis to better understand the working of the human body. Other scientists/engineers also use compartmental analysis in studying pharmacokinetics, chemical reaction engineering, fluid transport analysis, and even semiconductor design and fabrication.

This chapter develops the ability to select physiologic systems as candidates for compartmental modeling, to formulate appropriate models using the five model postulates as presented, and to obtain solutions for the model equations to conduct system simulation and design studies.

8.2 MODEL POSTULATES

Classical compartmental analysis is defined by five postulates (existence, homogeneity, conservation, stationarity, and linearity) that must be valid. However, to better describe real systems it may be necessary to relax one or more of these postulates. In doing this, it is possible to lose the simplicity of the mathematical analysis. These modified forms generally require computer simulation and nonlinear parameter estimation. The consequence of these extended efforts is that these models can cover a much wider range of phenomena. Section 8.3 will address these issues in more detail. At this point, however, brief descriptions of the five postulates required for the classical form of compartmental analysis are presented.

Postulate 1: Existence — the ability to represent the system as one or more compartments. This of course need not be related in a straightforward fashion to anatomic structures since they may be purely logical constructs. It is important to note that not all biological systems lend themselves to compartmentalization; therefore, different analysis techniques must be employed.

Postulate 2: Homogeneity — requires that all substances are distributed uniformly and all entering substances are instantaneously mixed. However, since no real system can satisfy either condition, a less stringent approach must be taken. It is sufficient to require that mixing occurs within a time scale that is significantly shorter than the characteristic times for other phenomenological events to occur as well as the sampling times.

Postulate 3: Conservation — refers to the concept that mass is conserved; consequently, system model equations are obtained from mass balance considerations.

Postulate 4: Stationarity — means that all intrinsic properties of the system remain constant, including temperature, compartment volume, and rate constants. However, concentrations and other extrinsic properties are time dependent. Clearly, this is an oversimplification since variations occur due to biological rhythms, externally induced stresses, etc. Time averaging of cyclic property variations is a means of taking into account this nonstationarity.

Postulate 5: Linearity — assumes that all fluxes leaving a compartment are linearly proportional to the amount (concentration) in that compartment. Consequently, all input fluxes are either proportional to the concentration in the previous compartment or have a specified functionality if they are at the entry points of the system. Furthermore, there are no time delays between exit from one compartment and entrance at the next. These requirements permit the flux from compartment i to compartment j to be expressed by the linear relationship

$$F_{ij} = k_{ij}C_i \tag{8.1}$$

The parameter k_{ij} is termed a rate constant. Thus, the dynamics of these systems can be described by linear, first-order differential equations.

8.3 COMPARTMENTAL STRUCTURE

Identifying the number of compartments and the connections among them that describes the physiologic system under investigation may be the most difficult step in compartmental model building. The structure must reflect the following: (i) There may be some a priori knowledge about the system which can be incorporated in the structure, (ii) specific assumptions can be made about the system which are reflected in the structure, and (iii) testing via simulation must be conducted on alternate structures to determine what is needed to fit the available data. The result at this stage is a model which has a set of unknown parameters which must be determined using well-established parameter estimation techniques.

One major advantage associated with compartmentalization lies in the ability to reduce the model complexity through the use of lumped versus distributed systems. This permits the use of ordinary differential equations to describe system dynamics instead of more complicated partial differential equations. The following discussion will help clarify these points. Consider the need to remove a toxic compound from a body fluid, such as a xenobiotic drug from blood, by an external device such as an artificial liver. The detoxification may be accomplished by adsorption onto specific receptors bound to solid beads where further reaction can take place. In either event, it is assumed that a first-order process occurs uniformly across a given cross section of flow and that it varies with depth into the packed bed of beads. This system can be modeled as a flow reactor with a time-varying input, for example, as the feed composition of uremic toxins in blood being fed to a dialyzer varies because of the multiple-pass requirements. Mass balance considerations for the system, shown schematically in Fig. 8.1, generate the following partial differential equation:

$$\frac{\partial C}{\partial t} + v\frac{\partial C}{\partial x} = \frac{D\partial^2 C}{\partial x^2} - kC \tag{8.2}$$

where C is the concentration of the toxin, v is the linear velocity of blood in this tubular "reactor" (a volumetric flow, F, can be obtained by multiplying v by the cross-sectional area for flow), x is the flow direction, D is the axial diffusion coefficent, and k is the first-order rate constant. This k can represent either a mass transfer coefficient (transport to the bead surface) or a biochemical reaction parameter (on the bead surface), dependent on which mechanism is "controlling." These details will be discussed later; here the focus will be on how a compartmental system consistent with our model can be structured. Typically in these convective flow situations the axial dispersion term is negligible. Thus, the system becomes

$$\frac{\partial C}{\partial t} + v\frac{\partial C}{\partial x} + kC = 0 \tag{8.3}$$

Given an initial and a boundary condition for the situation of interest, an analytical solution for $C(x,t)$ is obtainable. An approach using deviation (perturbation) variables and Laplace transforms is a suggested exercise that is given at the end of the chapter. At this time, the distributed parameter system is represented as a finite number of well-mixed chambers connected in series (Fig. 8.2). The inlet to the first chamber is denoted as C_0 and the exit as C_1, which is also the inlet to the second and so forth along the pathway. For algebraic simplicity, the volume of each compartment is taken to be equal and given as $V_j = V/n$, where V is the total volume of the system and n is the number of compartments, which is not known a priori. The actual number required to emulate the reactor depends on the accuracy desired from the physical system being simulated:

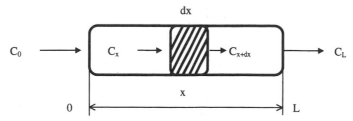

Fig. 8.1 Schematic of a packed bed (tubular) flow reactor, representing an approach for an artificial liver system.

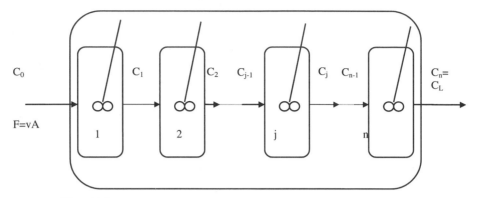

Fig. 8.2   Schematic of a finite number of well-mixed chambers in series.

For chamber 1: $\qquad V_1 \dfrac{dC_1}{dt} = FC_0 - FC_1 - k_1 C_1 V_1$

For chamber 2: $\qquad V_2 \dfrac{dC_2}{dt} = FC_1 - FC_2 - k_2 C_2 V_2 \qquad\qquad$ (8.4)

For chamber j: $\qquad V_j \dfrac{dC_j}{dt} = FC_{j-1} - FC_j - k_j C_j V_j$

The simplest way to observe the equivalence of these two approaches is to study the system at steady state. The dynamic response comparison yields the same conclusion and is studied as an exercise at the end of the chapter. For ease of illustration, the rate constant k will be considered to be independent of concentration, and the state variables, such as pressure and temperature, that affect its value are held constant. This allows the subscript on k to be dropped. Since all the volumes are equal, they can be referred to as Vn. The corresponding residence time for each chamber is $\theta_n = Vn/F$, whereas that for the tubular reactor is $\tau = L/v$, which is also equal to $n\theta_n$. The steady-state solution to Eq. (8.3) is

$$C(x) = C_0 \exp(-kx/v) \qquad\qquad (8.5)$$

where C_0 is the steady input to the system. This could be viewed as a "single-pass" analysis for the reactor. The compartment model equations simplify to an algebraic form. Represented as transfer functions, they become the following for the jth element:

$$\frac{C_j}{C_{j-1}} = \frac{1}{k\theta_n + 1} \qquad\qquad (8.6)$$

Writing this for the entire series of chambers, i.e., from C_0 to C_n gives

$$\frac{C_n}{C_0} = \left(\frac{C_n}{C_{n-1}}\right)\left(\frac{C_{n-1}}{C_{n-2}}\right) \cdots \left(\frac{C_2}{C_1}\right)\left(\frac{C_1}{C_0}\right) = \prod_{j=0}^{n} \frac{C_{j+1}}{C_j} \qquad\qquad (8.7)$$

Thus,

$$\frac{C_n}{C_0} = \left[\frac{1}{k\theta_n + 1} \right]^n \tag{8.8}$$

or

$$\theta_n = \left(\frac{1}{k} \right) \left[\left(\frac{C_0}{C_n} \right)^{\frac{1}{n}} - 1 \right] \tag{8.9}$$

The total residence time for the full length of the system must be considered for comparison to the distributed system. Given that $\tau = \dfrac{L}{v} = n\theta_n$ and noting that $c(L) = c_n$ gives

$$\frac{C_n}{C_0} = \exp(-k\tau) \quad \text{or} \quad \tau = \frac{1}{k} \ln \left(\frac{C_0}{C_n} \right) \tag{8.10}$$

To demonstrate the equivalence between the approaches, Eq. (8.9) is multiplied by n. Then the limit is taken as n gets large, i.e., $n \rightarrow \infty$.

$$\lim_{n \rightarrow \infty} (n\theta_n) = \lim_{n \rightarrow \infty} \left(\frac{n}{k} \right) \left[\left(\frac{C_0}{C_n} \right)^{\frac{1}{n}} - 1 \right] \tag{8.11}$$

Application of l'Hopitals' rule allows clarification of the limit process, i.e.,

$$\lim_{n \rightarrow \infty} \frac{\dfrac{d}{dn} \left[\left(\dfrac{C_0}{C_n} \right)^{\frac{1}{n}} - 1 \right]}{\dfrac{d}{dn} \left(\dfrac{k}{n} \right)} = \lim_{n \rightarrow \infty} \left[\frac{\left(\dfrac{C_0}{C_n} \right)^{\frac{1}{n}} \ln \left(\dfrac{C_0}{C_n} \right) \dfrac{d \dfrac{1}{n}}{dn}}{k \dfrac{d \dfrac{1}{n}}{dn}} \right] = \frac{1}{k} \ln \left(\frac{C_0}{C_n} \right) \tag{8.12}$$

This, of course, is identical to the result from the distributed system. Systems originally modeled by a complicated partial differential equation can be represented by a "large" number of simpler, lumped-parameter, ordinary differential equations. How large is large needs to be evaluated for each particular system studied while the model objectives are kept in focus.

Example Problem 8.1

Transform the following partial differential equation for $y(x,t)$ and the associated boundary and initial conditions into an ordinary differential equation in the Laplace domain. Use deviation variables to further simplify the equation and initial condition.

$$\frac{\partial y}{\partial t} + a \frac{\partial y}{\partial x} + b(y - y_i) = 0, \quad \text{with} \quad y(0,t) = y_0 \quad \text{and} \quad y(x,0) = y_i$$

Solution

Define $Y(x,t) = y(x,t) - y_i$; then the system equation becomes

$$\frac{\partial Y}{\partial t} + a\frac{\partial Y}{\partial x} + bY = 0, \quad \text{with} \quad Y(0,t) = y_0 - y_i \quad \text{and} \quad Y(x,0) = 0$$

Using Laplace transforms, the dependent variable is now $\hat{Y}(x,s)$ and the partial differential equation becomes an ordinary differential equation since the Laplace variable, s, is considered a constant. Using the initial condition, the result is

$$a\frac{d\hat{Y}}{dx} + (s + b)\hat{Y} = 0 \quad \text{and} \quad \hat{Y}(0,s) = (y_0 - y_i)/s \qquad \blacksquare$$

Example Problem 8.2

Hollow fiber cartridge systems, typically used in dialysis, are often used as bioreactors in tissue engineering studies for artificial organs. An application is the detoxification of blood by exposure to immobilized liver cells in the extracapillary space. The cells are sustained by nutrients fed through the fiber lumens. Mass transfer characterization tests have shown that the extracapillary space can be represented as multiple well-mixed compartments in series. Under flow conditions similar to those used for the reacting system, three compartments were adequate. For an extracapillary space of 360 ml and a blood flow of 200 ml/min, a fractional conversion (detoxification) of 0.4 was obtained for a given toxin that followed first-order kinetics. How many cartridges must be connected in series to obtain a 0.9 fractional conversion? If only one cartridge is available, what change in flow conditions would be necessary to obtain this 0.9 fractional conversion?

Solution

The rate constant for the reaction can be obtained from the given conversion and flow conditions, i.e., using Eq. (8.8) with $n = 3$,

$$\theta_n = \tau/n = \left(\frac{360 \text{ ml}}{200 \text{ ml/min}}\right)/3 = 0.6 \text{ min}$$

$$\text{and} \quad C_n/C_0 = 1 - 0.4 = 0.6 \quad \text{gives} \quad k = 0.309 \text{ min}^{-1}$$

To obtain a fractional conversion of 0.9, i.e., $C_n/C_0 = 0.1$, requires

$$n = \frac{\ln[C_n/C_0]}{\ln\left[\dfrac{1}{k\theta_n + 1}\right]} = \frac{\ln(0.1)}{\ln\left[\dfrac{1}{((0.309)(0.6) + 1)}\right]} = 13.5$$

compartments, equivalent to 4.5 cartridges, i.e., 5 since only whole cartridges can be used.

For a conversion of 0.9 in one cartridge, the residence time for each compartment, θ_n, must be increased as (from Eq. 8.8), $0.1 = \left[\dfrac{1}{(0.309\theta_n + 1)}\right]^3$, which gives $\theta_n =$

3.735 min or $\tau = 3\theta_n = 11.2$ min. Consequently, the flow must be reduced, as per the definition of τ,

$$Q = \frac{V}{\tau} = \left(\frac{360 \text{ ml}}{11.2 \text{ min}}\right) = 32.1 \text{ ml/min} \qquad \blacksquare$$

To further illustrate the usefulness of this compartmentalization approach, a complex physical situation can be taken to show how lumping due to significant differences in system response times (i.e., characteristic times, also referred to as system time constants) simplifies both the overall view of the system and its analysis. An example from pharmacokinetics has been selected that is concerned with the study and characterization of the time course of drug absorption, distribution, metabolism, and excretion. The purpose is to determine the relationship of these processes to the intensity and time course of therapeutic and toxicological effects of the substance in question. A schematic of the various steps in the transfer of a drug from its absorption site [e.g., the gastrointestinal (GI) tract] to the blood and its subsequent distribution and elimination in the body is given in Fig. 8.3.

Here k_{ij} is the rate constant of the species (the drug in this case) from compartment i to compartment j. The reversible step is obviously characterized by k_{ji}. Once the drug is absorbed into the blood, it quickly distributes itself between the plasma and erythrocytes. Within the plasma, it distributes between the water phase and the plasma proteins, particularly albumin (sometimes to α_1 acid glycoproteins but rarely to globulin). Most drugs are relatively small molecules that are readily transported through capillary walls and reach the extracellular fluids of essentially every organ in the body. They are also sufficiently lipid soluble to be distributed into the intracellular fluids of various tissues. In every location, the drug is partitioned between body water and proteins (or other macromolecules) dispersed in the body fluids or as components of the cells. The body can therefore be thought of as a collection of individual compartments, each containing a portion of the administered drug. The communication between the compartments is, as previously discussed, through a transport rate constant (k_{ij}). The transfer of the drug from the blood to the other body fluids and tissues is termed distribution. Typically, this process is extremely rapid and reversible so that a state of distribution equilibrium is assumed to exist between the plasma, erythrocytes, other body fluids, and tissue components. This concept permits the interpretation of the variations in drug concentration in the plasma as an indication of changes in drug levels at the other sites, including those of pharmacological interest.

The elimination process consists of three major components (or routes): (i) from the blood to urine via the kidneys; (ii) into other excretory fluids, such as bile and saliva; and (iii) enzymatic or biochemical transformation (i.e., metabolism) in the tissues/organs (e.g., the liver) or in the plasma itself. These processes are typically characterized as irreversible and are responsible for the physical and biochemical removal of the drug from the body.

The distribution and elimination processes occur concurrently, although at different rates. The instant a drug reaches the blood its distribution generally occurs more rapidly than elimination, primarily due to the difference in the rate constants.

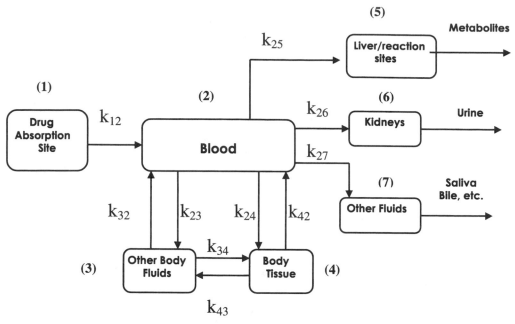

Fig. 8.3 Schematic representation of drug absorption, distribution, and elimination.

If the difference is significantly large, distribution equilibrium can be assumed, i.e., the drug can be distributed before any appreciable amount is eliminated. Under these conditions, the body can be characterized by a single compartment (Fig. 8.4). The compartments for blood, other body fluids, and tissues are "lumped" into one chamber with one inlet and three outlets, all considered irreversible. Since all three elimination rate processes are considered to be linear, they can be combined into one single transport rate constant by simply summing them. This is now an extremely simple representation for a complex system, i.e., a single compartment with one time variant input and one time variant output. It is apparent that the amount of drug at the absorption site decreases with time as the drug is distributed and eliminated. These dynamic processes are the object of investigation. After administration, there is a continual increase in the amount of drug converted to metabolites and/or physically eliminated. As a result, the amount of drug in the body at any

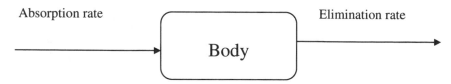

Fig. 8.4 Simplified version of Fig. 8.3 for drug dynamics.

time is the net transient response to these input and output processes. Compartmental analysis allows the system to be represented by three differential equations as follows:

$$\text{For absorption (input):} \qquad \frac{dA}{dt} = -k_0\, A \qquad (8.13)$$

$$\text{For body transients:} \qquad \frac{dB}{dt} = k_0\, A - k_1\, B \qquad (8.14)$$

$$\text{For elimination:} \qquad \frac{dE}{dt} = k_1\, B \qquad (8.15)$$

where A is the drug concentration from the absorption site, B is the drug concentration in the body, and E is the appearance of eliminated drug. The appropriate initial conditions are $E(0) = B(0) = 0$ and $A(0) = A_0$, the initial dose of drug at the absorption site as a consequence of either oral or intramuscular administration. The general form for the solution is represented in Fig. 8.5.

The analytic solutions to Eqs. (8.13)–(8.15) are obtained straightforwardly by use of Laplace transforms or by conventional techniques:

$$A(t) = A_0 \exp(-k_0\, t) \qquad (8.16)$$

$$B(t) = \left(\frac{k_0\, A_0}{k_1 - k_0}\right)[\exp(-k_0\, t) - \exp(-k_1\, t)] \qquad (8.17)$$

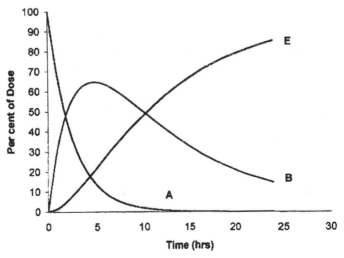

Fig. 8.5 Curve A is the time course of drug disappearance from the absorption site. Curve E is the appearance of eliminated drug in all forms. Curve B is the net result indicating drug transients in the body.

$$E(t) = A_0 - A(t) - B(t) = A_0\left\{1 - \left(\frac{1}{k_1 - k_0}\right)[k_1 \exp(-k_0 t) - k_0 \exp(-k_1 t)]\right\} \quad (8.18)$$

The time when the maximum in B occurs is

$$t_{max} = \frac{\ln(k_1/k_0)}{(k_1 - k_0)} \quad (8.19)$$

and the corresponding value for B at t_{max} is

$$B(t_{max}) = B_{max} = A_0\left(\frac{k_1}{k_0}\right)^{\frac{k_1}{(k_0 - k_1)}} \quad (8.20)$$

This model and variations of it provide excellent illustrations of the usefulness of single-compartment models. Applications other than those for pharmacokinetics are discussed in subsequent sections. However, this current model or modifications to it are useful in providing a quantitative index of the persistence of a drug in the body. It will be used to determine the duration of clinical effect and a dose administration schedule. From Eqs. (8.19) and (8.20) it is apparent that as the absorption rate increases compared to elimination processes, i.e., the ratio of $(k_1/k_0) \ll 1$, the maximum in drug plasma concentration increases and the time to reach this maximum is shortened. This maximum needs to be determined since any drug can be toxic at high concentrations. The absorption rate can be controlled by administration procedures. The current model reflects either an oral or an intramuscular injection with a known release/absorption rate.

Example Problem 8.3

Find the maximum "body" alcohol level and the time when it occurs for an 80-kg human subject who consumes 90 g of alcohol (orally) in less than 6 s. Prior tests indicate that $k_0/k_1 = 5.0$ and that the elimination rate constant $k_1 = 0.008$ min^{-1}.

Solution

The initial amount of alcohol at the distribution site is $A_0 = 90$ g and the absorption rate constant $k_0 = 0.008 (5.0) = 0.04$ min^{-1}. Thus, from Eq. (8.19)

$$t_{max} = \frac{\ln\left(\frac{k_0}{k_1}\right)}{(k_0 - k_1)} = \frac{\ln(5)}{(0.04 - 0.008)} = 50.3 \text{ min}$$

and from Eq. (8.20)

$$B_{max} = A_0\left(\frac{k_1}{k_0}\right)^{\frac{k_1}{(k_0 - k_1)}} = 90\left(\frac{1}{5}\right)^{0.25} = 60.2 \text{ g}$$

A delta function can be used for the initial condition for the drug concentration in the body to simulate a rapid intravenous injection (referred to as bolus). The model is simplified since only the change in the amount of drug in the body (or plasma) with time must be described. This is represented by

$$\frac{dB}{dt} = -k_1 B \quad \text{with} \quad B_0 = A_0 \tag{8.21}$$

and

$$\frac{dE}{dt} = -\frac{dB}{dt} \tag{8.22}$$

Therefore, only the time course for B must be monitored. This is given by

$$B(t) = B_0 \exp(-k_1 t) \tag{8.23}$$

Use of this model to analyze body response to a bolus injection of an antibiotic and to illustrate that its half-life is a quantitative index for drug evaluation/performance studies is shown in the following example.

Example Problem 8.4

A 2-g dose of a semisythetic penicillin with an apparent distribution volume of 10 liters is injected into healthy subjects. The average plasma concentration response curves are as shown in Fig. 8.6. These results are consistent with Eq. (8.23). From these data determine (a) the initial concentration, (b) the elimination rate, (c) the half-life of the antibiotic in the body, and (d) how often the drug must be given to ensure concentrations of 12.5% of the initial value.

Solution

Extrapolation to $t = 0$ gives $B_0 = 0.2$ g/liter, which is consistent with that calculated from the apparent volume of distribution and the amount of drug injected. The elimination rate constant is readily obtained from the slope: $k_1 = 0.693 \text{ h}^{-1}$. Using the definition of half-life, i.e., the time required to reduce the concentration by 50%, Eq. (8.23) gives $t_{1/2} = \dfrac{\ln 2}{k_1}$, thus $t_{1/2} = 1$ h. If the objective is to ensure that drug plasma concentrations will not be lower than 12.5% of the initial value then injections must be repeated every 3 h. It should be noted that an alternative performance parameter could be the rate constant, k_1. From the solution $B = B_0 \exp(-k_1 t)$, observe that k_1 is equal to the reciprocal of the system time constant. It is well known that a system perturbation is reduced to 5.0% of its initial value after three time constants. However, since dealing with powers of 0.5 is less cumbersome than with the exponential function, the drug half-life was selected. ■

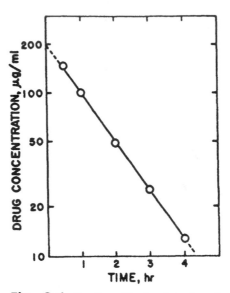

Fig. 8.6 Semilogarithmic plot of penicillin concentration in plasma after a 2-g intravenous dose.

Due to the possible harmful effects related to either excessive dosages or lack of level maintenance from mistimed drug administration, many drugs are encapsulated for a long-term controlled-release schedule. This can lead to a constant infusion rate (i.e., zero-order process) and thus Eq. (8.13) for the absorption rate is replaced by the following:

$$\frac{dA}{dt} = -k_0 \tag{8.24}$$

The rate of change of the amount of drug in the body (B) during infusion is given by

$$\frac{dB}{dt} = k_0 - k_1 B \tag{8.25}$$

and the solution is obtained straightforwardly:

$$B(t) = \left(\frac{k_0}{k_1}\right)[1 - \exp(-k_1 t)] \tag{8.26}$$

The concentration in the body (and thus plasma concentration) increases during infusion and is at its maximum when infusion is terminated at time T. After this, it will decline according to

$$B(t) = B_{max} \exp[-k_1(t - T)] \tag{8.27}$$

where B_{max} is obtained from Eq. (8.26) with $t = T$. Two important points must be noted. First, the maximum drug plasma concentration after intravenous infusion is always lower than that after bolus injection of the same dose (i.e., $k_0 T$). Second, since B_{max} is linearly proportional to k_0, doubling the dose (i.e., doubling the infusion rate over the same time period) doubles the maximum concentration.

It is apparent that various administration strategies can be simulated using Eqs. (8.13) and (8.14) and subsequent modifications. Some of these were described previously. This simulation capability is imperative since many drugs will not produce a pharmacological effect or the desired response unless a minimum concentration is transported to the site of action. This requires that a threshold level be exceeded in the plasma due to the distribution equilibrium that is established. A therapeutic plasma concentration range can be predicted and experimentally validated. By prescribing a drug in an appropriate dosing regime, the physician expects to elicit the desired clinical response promptly. Books on pharmacokinetics can be consulted for more detailed discussions and examples.

8.3.1 Single-Compartment Models

An introduction to the single-compartment concept was given in the previous section. It is the simplest model and may be considered as a special case of a two-compartment model because the substance in question is transported by one mechanism or another from one compartment and can enter another. A dilute suspension of red blood cells is one of the simplest examples of this concept, in which the erythrocytes themselves form the compartment. A tracer such as radioactive potassium can be put into the medium, and changes in radioactivity within the cells can be monitored. It is also possible to start with prelabeled erythrocytes and measure their decay rate. Since the potassium transport rate is quite slow, measurements can be performed at moderate rates.

It is pertinent to discuss how this system meets the five modeling postulates presented earlier. The existence and conservation postulates are met trivially. Homogeneity implies that all red cells are assumed to behave identically. This is a fair assumption even though differences between cells can be demonstrated. An "average" or representative cell behavior can be specified. Stationarity presents a greater challenge. The potassium transport experiments may be 48 h or longer in duration, hence justifying the constant cell properties assumption becomes problematic albeit tolerable.

The most troublesome problem in applying compartmental analysis to this system is the issue of linearity. This requires, among other things, that all unidirectional fluxes from a compartment are linearly proportional to the concentration therein. Of course, zero-order behavior is an acceptable alternative. To meet this requirement, a Taylor series expansion could be performed about a "steady-state value" (e.g., a stable isotope). Discarding all higher order terms in this perturbation variable will provide an appropriate linearized form with an apparent first-order rate constant. To illustrate this point, the efflux is considered to be proportional to the square of the internal potassium concentration (c). This flux can be symbolically represented as

$$F_1 = k_1 \, c^2 \tag{8.28}$$

If a small amount of labeled tracer (δc^*) is added and the $(\delta c^*)^2$ terms and higher are discarded,

$$F_1 = F_{1s} + F_1^* \cong k_1 \, c_s^2 + 2k_1 c_s \, \delta c^* \tag{8.29}$$

where s refers to a stable isotope. From Eq. (8.29) the apparent rate constant k_{1a} is defined as

$$k_{1a} = 2k_1 \, c_s \tag{8.30}$$

instead of k_1 itself. It should be noted that there are several mechanisms for transport of potassium ions between the red blood cells and the surrounding medium, and that some are actually zero- or first-order naturally.

It is appropriate at this point to establish a basis for zero-order and first-order kinetic behavior in biological systems. This is easily demonstrated for both chemical kinetics and diffusion processes. This will be illustrated for chemical kinetics since a similar analysis is applied for the other processes and nothing new is revealed. Not only do biochemical reactions govern metabolism but also they are coupled to many diffusion and adsorption phenomena. Most of these reactions are catalyzed by specific enzymes and thus are often represented by the Michaelis–Menton reaction network scheme:

$$E + S \underset{k_{-1}}{\overset{k_1}{\rightleftharpoons}} ES^* \xrightarrow{k_2} E + P \tag{8.31}$$

where the enzyme E is viewed as reacting with a specific substrate/reactant S in a reversible step to form an activated enzyme–substrate complex ES^*, which subsequently is converted to the product P and the original form of the enzyme. If the kinetics are assumed to be controlled by the rate of complex decomposition to product, then the reversible steps can be considered to be in equilibrium, or at least at a quasi-steady state. The functional form obtained is similar in either situation so the equilibrium assumption can be selected due to the simpler algebra and ease of physical interpretation of the model parameters. Thus,

$$\frac{[E][S]}{[ES^*]} = K_{eq} = \text{an equilibrium constant} \tag{8.32}$$

The total amount of enzyme E is invariant so

$$[E_0] = [E] + [ES^*] = \text{initial amount of enzyme} \tag{8.33}$$

It is easily shown using Eqs. (8.32) and (8.33) that

$$[ES^*] = \frac{[E_0][S]}{([S] + K_{eq})} \tag{8.34}$$

This is difficult to measure directly, so through use of Eq. (8.34) the rate of product formation can be written as

$$r = \frac{k_2[E_0][S]}{([S] + K_{eq})} \tag{8.35}$$

From this form, it becomes clear that both zero-order and first-order kinetics can be observed in various substrate concentration regimes. For example, when $[S] \gg K_{eq}$ Eq. (8.35) becomes $r = k_2 [E_0]$, i.e., a constant, and thus zero-order kinetics. In this concentration regime, the rate of reaction is limited by the quantity of enzyme present. Consequently, the rate is at its maximum value. When $K_{eq} \gg [S]$ Eq. (8.35) becomes $r = k_2[E_0][S]/K_{eq}$, which yields first-order kinetics with an apparent rate constant, $k_a = k_2[E_0]/K_{eq}$.

This section is concluded with two examples to illustrate compartment selection, model formulation, parameter estimation from experimental results, and predictive and/or design capabilities.

Modeling of the Patient–Artificial Kidney System

The physiology of the kidneys was described in Chapter 2. A brief discussion of a few concepts pertinent to renal insufficiency is appropriate to set the basis for evaluating artificial kidney performance. Renal insufficiency elicits the clinical picture of uremia, a condition in which urea concentration in blood is chronically elevated. Urea is a metabolic end product in the catabolism of proteins that is relatively nontoxic even at high concentrations. However, it mirrors the impaired renal elimination and the resulting accumulation in body fluids of other toxic substances. Some of these have been identified (e.g., phenols, guanidine, and diverse polypeptides), whereas others remain unknown and are therefore referred to as uremic toxins (or middle molecules, approximately 1–30 kDa). Uremia is often expressed as blood urea nitrogen (BUN), which expresses the nitrogen content of the urea and is actually 0.47 times (approximately half) the urea concentration and serves as an indicator of the severity of renal disease. Each gram of protein consumed produces about 250 mg of urea. For example, a 70-kg patient who consumes a typical 1.0 g of protein per kilogram of body weight per day would produce 18 g of urea distributed over a fluid volume of 42 liters (a typical distribution volume for this size patient). A 3.5-h treatment with a blood dialysis unit (hemodialyzer) would lead to a 64% reduction in urea concentration.

Dialysis is a membrane separation process in which one or more dissolved molecular species diffuse across a selective barrier in response to a difference in concentration. The "stripping" solution, referred to as the dialysate, is a buffered electrolyte solution that usually contains glucose at or above physiologic concentration that circulates through the water compartment of the hemodialyzer to control diffusional transport of small molecules across the membrane and achieve the desired blood concentration. Figure 8.7 is a simplified model representation of this system. Note that the patient is modeled as a single compartment and the artificial kidney as a two-compartment model. This is not of consequence at this point since the current focus is on the patient response. A more complicated multicompartmental approach is given later. The performance of the artificial kidney will be taken as a known and for a single pass through the device is represented as

$$C_{Bo} = C_{Bi} \exp\left(\frac{-KA}{Q_B}\right) \tag{8.36}$$

where K is the appropriate transport parameter for urea in the device, A is the transport area, Q_B is the blood flow rate, and C_{Bi} and C_{Bo} are the urea concentrations in the blood at the device inlet and outlet points, respectively. This equation is valid only for a certain instant in time since C_{Bi} will vary with time during the treatment period. To determine how the urea content of the patient varies over a specified time interval, Eq. (8.36) must be combined with a mass balance on urea for the patient. A 50-liter well-stirred tank of fluid of uniform concentration is taken as the compartment representing the patient, V_b. Neglecting urea production by the body during the treatment cycle (relaxed in later sections), this mass balance yields the following equation:

$$V_b \frac{dC_{Bi}}{dt} = Q_B\left(C_{Bo} - C_{Bi}\right) \tag{8.37}$$

Substituting Eq. (8.36) for C_{Bo} into Eq. (8.37) and integrating, with the initial condition that $C_{Bi}(0) = C_{Bi}^0$, gives

$$C_{Bi} = C_{Bi}^0 \exp\left(\frac{Q_B(\beta - 1)t}{V_b}\right) \tag{8.38}$$

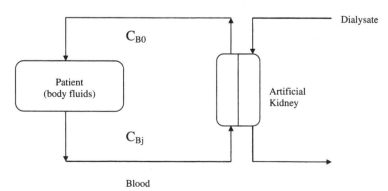

Fig. 8.7 Model of a patient–artificial kidney system.

where $\beta = \exp(-KA/Q_B)$. This result can be used to predict the required treatment time. However, treatment time is not the only performance specification. The blood must not be cleared of urea at a rate faster than that which the other body fluids (e.g., cerebrospinal fluid) can transfer their urea to the blood. This would induce an osmotic shift of water into the brain and other regions, causing potentially dangerous swelling. This simple compartment model cannot address these issues. This requires modified multicompartmental analysis.

Example Problem 8.5

Evaluate a flat plate dialyzer with the following specifications: $K = 16.7 \times 10^{-3}$ cm/min, $A = 1.0$ m^2, blood is processed at a flow rate $Q_B = 200$ cm^3/min, and there is a BUN content of 150 mg% when initially connected to the device. If treatment is considered complete when the patient's BUN is 50 mg%, how long must the patient be connected to the device?

Solution

$$\left|\frac{KA}{Q_B}\right| = 16.7 \times 10^{-3}\frac{\text{cm}}{\text{min}} \cdot \frac{10,000 \text{ cm}^2}{200 \dfrac{\text{cm}^3}{\text{min}}} = 0.833$$

$$\beta = \exp\left[\frac{-KA}{Q_B}\right] = \exp[-0.833] = 0.435$$

Use Eq. (8.38) to find time of treatment:

$$t = \left(\frac{V_b}{Q_B(\beta - 1)}\right)\ln\left(\frac{C_{Bi}}{C_{Bi}^0}\right) = \frac{50,000 \text{ cm}^3}{200 \dfrac{\text{cm}^3}{\text{min}}(0.435 - 1)}\ln\left(\frac{50}{150}\right) = 486 \text{ min, or } 8.1 \text{ h}$$

This is consistent with actual clinical treatment time using flat plate dialyzers with membrane transfer areas of 1.0 m^2. ∎

Data Analysis of a Sustained-Release Medication

A one-compartment model is used to study the pharmacokinetics of an experimental drug (XDI) and to interpret the data via parameter estimation techniques. Two different formulations of this drug are used, with one in solution and the other as a tablet. Both are administered orally. Average plasma concentrations from six test animals are plotted versus time in Fig. 8.8. The initial dose (A_0) was 79 mg/kg for the tablet (◆) formulation and 28.8 mg/kg for the solution (■). For the sustained-release tablet, a fraction f_i of this initial dose is released immediately and a fraction f_r is released by a first-order process with a rate constant k_r. The consequence of this split-release mechanism is that the body compartment free stream (from the GI tract) is slightly more complex than the absorption rate expression used earlier. The elimination process is once again the sum of the contributions from metabolism and urination and is represented by a single rate parameter k_e. Since the absorption rate into the body is proportional to the amount in the GI tract (G) with

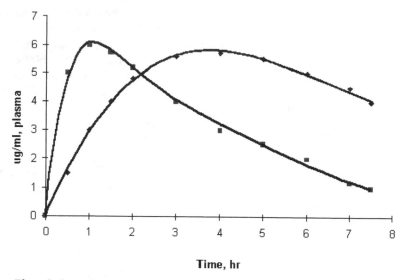

Fig. 8.8 Average plasma concentration from six test animals after oral administration as tablets (◆, 79 mg/kg) and solution (■, 28.8 mg/kg). The solid lines are the model predictions.

rate constant k_a, a mathematical description of the time course of G is needed. This requires knowledge of the amount immediately released into the GI tract since it will be the initial condition on G. The subsequent input rate from the tablet is thus proportional to the amount remaining, A. The amount remaining in the tablet is thus found from Eq. (8.13):

$$\frac{dA}{dt} = -k_r A$$

Noting that $A_0 f_i$ is released immediately, the initial condition for A is thus $A_0 f_r$, i.e., the amount remaining in tablet form. This integrates to Eq. (8.16):

$$A(t) = A_0 \, f_r \, \exp(-k_r t)$$

Since the input to the GI tract is $k_r A$, a balance for G gives

$$\frac{dG}{dt} = A_0 f_r k_r \exp(-k_r t) - k_a G \tag{8.39}$$

with the initial condition that $G(0) = A_0 f_i$ (amount released immediately). This integrates to

$$G(t) = \left(\frac{A_0 f_r k_r}{(k_a - k_r)}\right)[\exp(-k_r t) - \exp(-k_a t)] + A_0 f_i \exp(-k_a t) \tag{8.40}$$

Now the amount in the body is represented by

$$\frac{dB}{dt} = k_a G - k_e B \tag{8.41}$$

With the initial condition, $B(0) = 0$, the solution is

$$B(t) = \frac{A_0 f_r k_a k_r}{(k_a - k_r)(k_e - k_r)}[\exp(-k_r t) - \exp(-k_e t)]$$

$$+ \left\{ \frac{A_0 f_i k_a - \left(\frac{A_0 f_r k_a k_r}{(k_a - k_r)}\right)}{(k_e - k_a)} \right\}[\exp(-k_a t) - \exp(-k_e t)] \quad (8.42)$$

Since B is the amount of drug in the body, the concentration in the plasma can be obtained by dividing by the distribution volume V_d, which is yet undetermined, i.e., a model parameter. To obtain values for k_r, f_r, and f_i, tablets are dissolved in simulated gastric juice, a standard solution of HCl, NaCl, and pepsin. These experiments yielded values of $k_r = 0.25$ h^{-1}, $f_r = 0.88$, and $f_i = 0.12$. The data analysis procedure is easier to describe using the solution data since the algebra is simplified by the fact that $f_r = 0$ and hence $f_i = 1.0$. Equation (8.42) divided by V_d becomes

$$C_B = \frac{A_0 k_a}{V_d (k_e - k_a)}[\exp(-k_a t) - \exp(-k_e t)] \quad (8.43)$$

For the case in which $k_a > k_e$ (the most common case), and for large times (i.e., when the amount of $A \rightarrow 0$),

$$C_B \simeq \alpha \exp(-k_e t) \quad (8.44)$$

where $\alpha = A_0 k_a / V_d (k_a - k_e)$. A plot of $\ln C_B$ versus time will be linear in this large time regime with a slope of $-k_e$ and an extrapolated intercept of $\ln \alpha$. Figure 8.9 is a plot of the "solution" data and clearly produces the desired result: $k_e = 0.28$ h^{-1} and $\ln \alpha = 2.22$, i.e., $\alpha = 9.21$. From the definition of α it can be seen that there are two parameters, k_a and V_d, yet to be determined. An additional "equa-

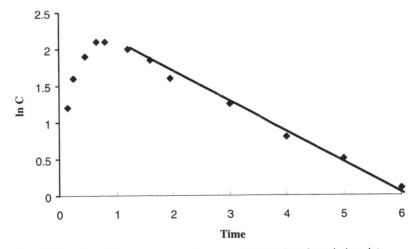

Fig. 8.9 Natural log of concentration versus time using the solution data for the "long" time regime analysis.

tion" is needed to remove one of these "degrees of freedom." Rearrangement of Eq. (8.43) with the preexponential term of $-\alpha$ gives

$$\exp(-k_a t) = \exp(-k_e t) - \frac{C_B}{\alpha} \qquad (8.45)$$

The right-hand side of Eq. (8.45) contains the known parameters k_e and α. A plot of $\ln\left(\exp(-k_e t) - \dfrac{C_B}{\alpha}\right)$ versus time can be constructed (as seen in Fig. 8.10). It produces a straight line with slope $= -k_a$. The value obtained is $k_a = 2.0$ h^{-1}. From these results (i.e., $k_e = 0.28$, $k_a = 2.0$, $\alpha = 9.21$, and $A_0 = 28.8$), $V_d = 3.5$ liters/kg. For model validation, Eq. (4.42) can be used to predict the plasma concentrate profile for the tablet case. From Fig. 8.8 it can be seen that the model predictions, i.e., the solid curves, are quite good for both cases.

8.3.2 Multicompartment Models

The concepts developed in the preceding sections can be applied to models with any number of compartments, with each characterized by differential equations derived from mass balance considerations. Such models are somewhat more complex and thus can be used to represent a greater variety of systems. The model of thyroxin metabolism in the isolated perfused liver will be developed as an example.

Thyroxin is 3, 5, 3′, 5′-tetraiodothyronine and is also known as T4. It plays a critical role in regulating metabolism and is formed in normal humans in the thyroid, an endocrine gland. The flow pathways consist of transport into plasma where it is partially bound reversibly to carrier proteins, partial degradation in the liver and subsequent movement on to bile, and as free (inorganic) iodine. During metabolism, iodines are removed successively, and the compounds T3, T2, and T1 (3, 2, and 1 iodine atoms per "thyroxin" molecule) are formed. These compounds are physiologically active in a similar mode to T4 but are more rapidly degraded.

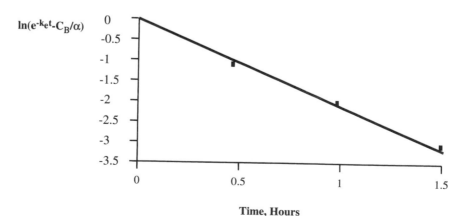

Fig. 8.10   Natural log $(\exp(-k_e t) - C_B/\alpha)$ versus time to recover k_a from Eq. (8.45).

Therefore, T4 is easier to monitor. Various models have been proposed to represent iodine and thyroxin metabolism in the intact organism. Some are very complex yet oversimplified if compared to the real animal. Simpler systems for isolated components, e.g., the liver, were considered as shown in Fig. 8.1.

Rat blood with radiolabeled-iodine T4 can be perfused into isolated rat liver, and radioactivity of the liver and blood are monitored with time. Cumulative radioactivity of bile, collected from the canulated bile duct, is also measured to complete the mass balance and determine the fate of the radiolabeled iodine. Some inorganic iodine is returned to the blood after metabolism by the liver. Therefore, when conducting these experiments, the transient concentration of inorganic iodine as well as changes in the perfusate as aliquots for analysis are removed periodically must be taken into account. Experiments of this nature have successfully proved that the system represented in Fig. 8.11 is too simple. Model predictions were completely unacceptable, and a more complex compartmental system was needed. The first modification was subdivision of the liver compartment. This increased the connective pathways to the "outside" and treated these paths as connections to open boxes. Later examples will further develop these points. Continuation with this thyroxin model's complex modifications will become too cumbersome for this introduction. There is, however, intrinsic merit in returning to the simpler model, even though it was proven ineffective in its predictive capabilities, to demonstrate the analysis techniques and establishment of baseline considerations.

Writing the mass balance equations for the two major compartments (blood and liver) in terms of the mass of T4 (m_1 and m_2, respectively) gives

$$\frac{dm_1}{dt} = -k_{12}m_1 + k_{21}m_2 \qquad (8.46a)$$

$$\frac{dm_2}{dt} = k_{12}m_1 - (k_{21} + k_{23})m_2 \qquad (8.46b)$$

Note from Fig. 8.11 that the radioactive label, ^{131}I, is excreted by the liver in the bile and as the iodine of the blood. These are lumped together as compartment 3. It is important to note that this excretion is not a compartment in the restricted sense defined earlier but rather is often termed an open compartment to emphasize that its volume has no meaning. This also renders C_3 meaningless, and thus it is not included in the Eq. (8.46a) and (8.46b). However, m_3 is measured experimentally and can be predicted from this model once m_2 and k_{23} are known.

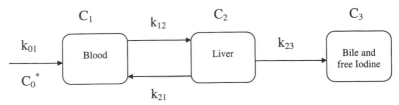

Fig. 8.11 Multicompartment system representing metabolism of thyroxin by the liver.

These more complex models can be solved, as in the single-compartment model, by guessing a sum of exponentials as the appropriate form and substituting into the differential equations. A solution is sought from

$$m_1 = \alpha_{11} \exp(-\lambda_1 t) + \alpha_{12} \exp(-\lambda_2 t) \tag{8.47a}$$

$$m_2 = \alpha_{21} \exp(-\lambda_1 t) + \alpha_{22} \exp(-\lambda_2 t) \tag{8.47b}$$

These six solution constants (αs and λs) will be determined in terms of the rate constants (k_{ij}), i.e., the model parameters for this system. The details are as follows. Substitute Eqs. (8.47a) and (8.47b) into Eqs. (8.46a) and (8.46b) to obtain

$$[(k_{12} - \lambda_1)\alpha_{11} - k_{21}\alpha_{21}]\exp(-\lambda_1 t) \\ + [(k_{12} - \lambda_2) - k_{21}\alpha_{22}]\exp(-\lambda_2 t) = 0 \tag{8.48a}$$

$$[(k_{21} + k_{23} - \lambda_1)\alpha_{21} - k_{21}\alpha_{11}]\exp(-\lambda_1 t) \\ + [(k_{21} + k_{23} - \lambda_2)\alpha_{22} - k_{12}\alpha_{12}]\exp(-\lambda_2 t) = 0 \tag{8.48b}$$

Since solutions of the form given in Eqs. (8.47a) and (8.47b) are being sought — that is, the functions $\exp(-\lambda_1 t)$ and $\exp(-\lambda_2 t)$ must be independent of each other — all coefficients in Eqs. (8.48a) and (8.48b) must be zero. Two sets of homogeneous equations, rewritten in matrix form to illustrate further requirements, are obtained:

$$\begin{bmatrix} (k_{12} - \lambda_1) & -k_{21} \\ -k_{12} & (k_{12} + k_{23} - \lambda_1) \end{bmatrix} \begin{pmatrix} \alpha_{11} \\ \alpha_{21} \end{pmatrix} = \begin{pmatrix} 0 \\ 0 \end{pmatrix} \tag{8.49a}$$

$$\begin{bmatrix} (k_{12} - \lambda_2) & -k_{21} \\ -k_{12} & (k_{12} + k_{23} - \lambda_2) \end{bmatrix} \begin{pmatrix} \alpha_{12} \\ \alpha_{22} \end{pmatrix} = \begin{pmatrix} 0 \\ 0 \end{pmatrix} \tag{8.49b}$$

For there to be a nontrivial solution to these sets, the determinant of the coefficient matrices must be zero. This "eigenvalue" problem only yields nonunique solutions, i.e., components of the α vectors are not determined uniquely and are not independent of each other. Noting, of course, that the coefficient matrices are identical except for the subscript on the λs, the following characteristic equation can be solved for the two values of λ, which are the roots of this equation:

$$(k_{12} - \lambda_j)(k_{21} + k_{23} - \lambda_j) - k_{12}k_{21} = 0 = \lambda_j^2 - (k_{12} + k_{21} + k_{23})\lambda_j + k_{12}k_{21} \tag{8.50}$$

The two roots are

$$\lambda_1, \lambda_2 = \frac{(k_{12} + k_{21} + k_{23})}{2} \pm \sqrt{\left[\left(\frac{(k_{12} + k_{21} + k_{23})}{2} \right)^2 - k_{12}k_{21} \right]} \tag{8.51}$$

Substituting into Eqs. (8.49a) and (8.49b), i.e. selecting either equation from each set, gives the necessary consistency relationships:

$$\alpha_{21} = \left[\frac{k_{12} - \lambda_1}{k_{21}} \right] \alpha_{11}$$

$$\alpha_{12} = \left[\frac{k_{21}}{k_{12} - \lambda_2} \right] \alpha_{22} \tag{8.52}$$

Substituting into Eqs. (8.47a) and (8.47b) gives

$$m_1 = \alpha_{11}\exp(-\lambda_1 t) + \left[\frac{k_{21}}{k_{12} - \lambda_2}\right]\alpha_{22}\exp(-\lambda_2 t)$$

$$m_2 = \alpha_{11}\left[\frac{k_{12} - \lambda_1}{k_{21}}\right]\exp(-\lambda_1 t) + \alpha_{22}\exp(-\lambda_2 t)$$

(8.53)

Specification of the initial conditions on m_1 and m_2 will give α_{11} and α_{22} uniquely. For T4, $m_1(0)$ and $m_2(0)$ are 1 and 0, respectively; thus,

$$1 = \alpha_{11} + \alpha_{22}\left[\frac{k_{21}}{k_{12} - \lambda_2}\right]$$

$$0 = \alpha_{11}\left[\frac{k_{12} - \lambda_1}{k_{21}}\right] + \alpha_{22}$$

or

(8.54)

$$\alpha_{11} = \left[\frac{k_{12} - \lambda_2}{\lambda_1 - \lambda_2}\right]$$

$$\alpha_{22} = \left[\frac{(k_{12} - \lambda_1)(k_{12} - \lambda_2)}{k_{21}(\lambda_2 - \lambda_1)}\right]$$

All the parameters in the solution set for m_1 and m_2, as given in Eqs. (8.47a) and (8.47b) or (8.53), are specified in terms of the model parameters, k_{ij}, through use of Eqs. (8.51), (8.52), and (8.54). As mentioned earlier, this model failed to fit the experimental data so a more complex model was needed. The first modification should be to subdivide the liver into multiple compartments. This does not eliminate the problems of poor fit. This example will not be considered further due to the level of algebraic complexities that it involves. Instead, a more straightforward system will be selected to demonstrate the advantages of this compartmental subdivision technique.

The drug distribution system represented by Eqs. (8.13)–(8.15) is modified. The body compartment is divided into two compartments, blood and tissue. This is consistent with actual behavior since most drugs tend to distribute differently between blood and tissue. In fact, they enter various tissues, such as fat and muscle, to differing extents. Since the therapeutic effect generally depends on tissue concentration, blood–tissue distributions must be taken into account by constructing the model system as given in Fig. 8.12. Note that this is simply the next level of complexity since there is no drug elimination by metabolism directly from the tissue space. It is lumped into the elimination pathway as was done previously. The following are the governing differential equations:

$$\frac{dA}{dt} = -k_0 A$$

(8.13)

$$\frac{dP}{dt} = k_0 A + k_{21}T - (k_{12} + k_1)P$$

(8.55)

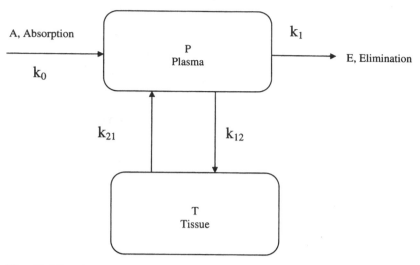

Fig. 8.12 Two-compartment model divided into plasma and tissue for drug absorption, distribution, and elimination in the body.

$$\frac{dT}{dt} = -k_{21}T + k_{12}P \tag{8.56}$$

$$\frac{dE}{dt} = k_1 P \tag{8.57}$$

The solution for P, with $A(0) = A_0$ and $P(0) = T(0) = 0$, is

$$P(t) = k_0 A_0 \left[\frac{(k_{21}-\alpha)\exp(-\alpha t)}{(k_0-\alpha)(\beta-\alpha)} + \frac{(k_{21}-\beta)\exp(-\beta t)}{(k_0-\beta)(\alpha-\beta)} + \frac{(k_{21}-k_0)\exp(-k_0 t)}{(\alpha-k_0)(\beta-k_0)} \right] \tag{8.58}$$

where

$$\alpha = 0.5[a + \sqrt{(a^2 - b)},] \qquad \beta = 0.5[a - \sqrt{(a^2 - b)}]$$

$$a = k_{12} + k_1 + k_{21}, \qquad\qquad b = 4k_{21}k_1$$

The solution for T is similar in form to Eq. (8.58):

$$T(t) = k_0 k_{12} A_0 \left[\frac{\exp(-\alpha t)}{(k_0 - \alpha)(\beta - \alpha)} + \frac{\exp(-\beta t)}{(k_0 - \beta)(\alpha - \beta)} + \frac{\exp(-k_0 t)}{(\alpha - k_0)(\beta - k_0)} \right] \tag{8.59}$$

Note that the solutions have a triexponential form that is consistent with the fact that there are three independent equations for the four species present. The value of the fourth is set once the other three are determined. This is readily seen from the addition of the four equations to yield $dA + dP + dT + dE = 0$.

Two-Compartment Model of a Patient–Artificial Kidney System

This model considers the patient's body as two compartments and explicitly accounts for urea generation by metabolism. A schematic is given in Fig. 8.13.

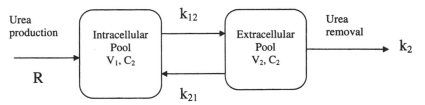

Fig. 8.13 Schematic of a two-compartment model for a patient undergoing dialysis.

Mass balances on both compartments yield

$$V_1 \frac{dC_1}{dt} = R - k_{12}C_1 + k_{21}C_2 \qquad (8.60a)$$

$$V_2 \frac{dC_2}{dt} = k_{12}C_1 - k_{21}C_2 - k_2C_2 \qquad (8.60b)$$

where k_{12} and k_{21} are actually interpool mass transport parameters that are identical to a mass transfer coefficient times an interfacial area and typically are equal to each other, k_2 is the clearance rate constant for the dialysis system. Since the extracellular pool includes the blood as well as interstitial fluid, the blood urea concentration is the same as that in the extracellular pool as a whole. A value for R is determined for each patient from BUN measurements on the patient's blood during the periods between dialysis. A value of 0.18 g/h is a reasonable choice for this example. The clearance constant is a function of the mass transfer characteristics of the device and the operating conditions. A value of 4.6 liters/h is typical. The total fluid volume in a patient's body can be estimated from physiological correlations and the ratio of V_1 to V_2 is also obtainable in this manner. The data in Table 8.1 were obtained from a patient whose V_T was estimated as 33.2 liters and $V_1/V_2 = 2.33$.

The system equations are coupled so they can be solved with a numerical procedure. (Solving this sytem using Laplace transforms is given as an exercise at the end of the chapter). All model parameters except $k_{12} = k_{21} = k$ were set as described

TABLE 8.1 BUN Data as a Function of Dialysis Time for Three Runs with a 21-h "Recovery" Period between Runs

	BUN (mg%)		
Time (h)	Run No. 1	Run No. 2	Run No. 3
0	63	54	47
1	53	44	38
2	47	39	32
3	43	35	29

previously. To find k, an initial guess was made, the equation was numerically integrated, and the output was compared to the data sets given in Table 8.1. At each point a deviation between actual and simulated values was obtained. The absolute values were totaled and used as the objective function in an optimization search, i.e., new values of k were selected by the optimization search algorithm, and a new model output and a new "error" point were obtained. This continued until a minimum was reached. The corresponding value for k obtained in this manner was $k = 33.1$ liters/h. This model predicted patient urea dynamics superbly. Actual performance data, both while on the device and during recovery periods, can be obtained from more extensive reports given in Suggested Reading.

8.4   MODIFIED COMPARTMENTAL ANALYSIS

In the previous sections, it was shown that compartmental analysis generates models that are characterized by simplicity and relative ease of manipulation. These can provide excellent first approximations and may prove to be sufficient if the selected goals are achieved. When they are not adequate, one or more of the postulates given previously that constitute the basis for using these simpler models must be relaxed. The consequence is that the mathematical analysis became more complicated but, fortunately, not prohibitive due to the advances in computational techniques available with modern computers. It is possible to modify and extend compartmental analysis when one or more postulates are relaxed to cover a wider group of phenomena. Most require simulation and nonlinear parameter estimation, which emphasize the role of computer technology. In this section the requirement that the rate of loss of a substance must be proportional to the amount in the compartments will be relaxed. The input or the output or both may be nonlinear relationships in the conservation of mass equation. The following example describing blood glucose regulation illustrates these points.

8.4.1   Blood Glucose Regulation

Glucose concentration varies from a typical value (immediately after eating) of between 90 and 200 mg/100 ml to 40 mg/100 ml (3 h after eating). Two main mechanisms for the control of glucose concentration, automatic feedback involving hormonal secretion (insulin) by the pancreas and storage in the liver, will be considered.

The pancreas, in addition to its digestive functions, secretes insulin directly into the blood. Insulin facilitates glucose diffusion (transport) across the cell membrane. In its absence, glucose transport falls to 25% of normal (referred to as diabetes). With excess amounts of insulin, the blood glucose level can be driven as low as 20 mg/100 ml. When too low (hypoglycemia), symptoms of distress, dizziness, and unconsciousness may occur. Levels that are too high (hyperglycemic) may lead to ketoacidosis, excessive glucose in the urine, and eventually coma if the condition is not treated. Therefore, the rate of insulin secretion is regulated so that glucose can be maintained at a constant level.

The liver acts as a storage compartment for glucose. When excess amounts of glucose are present, two-thirds of this excess is stored in the liver almost immediately. Conversely, when the level in the plasma is too low, this stored glucose is released by the liver to replenish the supply in the blood. Insulin has a moderating effect on liver function. Figure 8.14 illustrates normal and abnormal (diabetes) responses to a bolus injection of glucose. A compartmental model is shown schematically in Fig. 8.15.

The balance equations are

$$\frac{dG}{dt} = R - F_E(G, I) \tag{8.61a}$$

$$\frac{dI}{dt} = F_P(G, I) \tag{8.61b}$$

where G is glucose concentration (mg/ml) in the plasma, I is the insulin concentration (μU/ml), R is the injected glucose rate (mg/ml s), $F_E(G, I)$ is the nonlinear elimination rate (loss) of glucose to the liver and tissues (mg/ml s), and $F_p(G, I)$ is the nonlinear rate of insulin into the plasma (μU/ml s).

It is convenient to expand the functions F_E and F_p with regard to their "fasting" values, G_F and I_F. Because the system is in a quasi-steady state after the subject fasts overnight, R is zero and so are the derivative terms. This gives $F_E(G_F, I_F) = F_p(G_F, I_F) = 0$. Furthermore, the variables are changed by using differences from the fasting values, i.e., g and p defined as $g = G - G_F$ and $p = I - I_F$. The functions are linearized about G_F and I_F:

$$F_E(G, I) = F_E(G_F, I_F) + g\frac{\partial F_E}{\partial G}\bigg|_{(G_F, I_F)} + p\frac{\partial F_E}{\partial I}\bigg|_{(G_F, I_F)} \tag{8.62a}$$

$$F_P(G, I) = F_P(G_F, I_F) + g\frac{\partial F_P}{\partial G}\bigg|_{(G_F, I_F)} + p\frac{\partial F_P}{\partial I}\bigg|_{(G_F, I_F)} \tag{8.62b}$$

Furthermore, we define the following transport rate constants: $K_1 = \frac{\partial}{\partial G}F_E\big|_{(G_F, I_F)}$, independent elimination transfer rate constant in which the system returns glucose to its fasting value; $K_2 = \frac{\partial}{\partial I}F_E\big|_{(G_F, I_F)}$, elimination transfer rate constant in which the system returns insulin to its fasting value; $K_3 = \frac{\partial}{\partial G}F_P\big|_{(G_F, I_F)}$, pancreas transfer rate constant in which the system returns glucose to its fasting value; and $K_4 = \frac{\partial}{\partial I}F_P\big|_{(G_F, I_F)}$, pancreas metabolic rate constant in which the system returns insulin to its fasting value.

Substituting into Eqs. (8.61a) and (8.61b) gives

$$\frac{dg}{dt} = -K_1 g - K_2 p + R \tag{8.63a}$$

$$\frac{dp}{dt} = K_3 g - K_4 p \tag{8.63b}$$

Normal

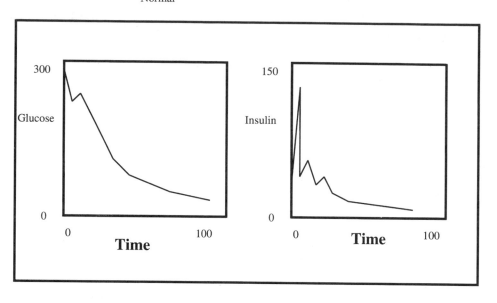

Diabetes

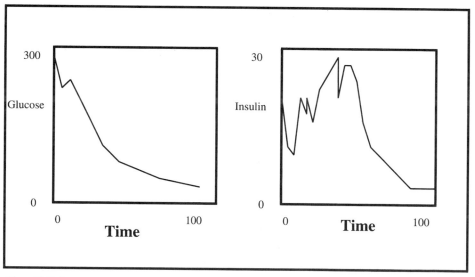

Fig. 8.14   Normal (top) versus abnormal (diabetes) (bottom) response to a bolus injection of glucose.

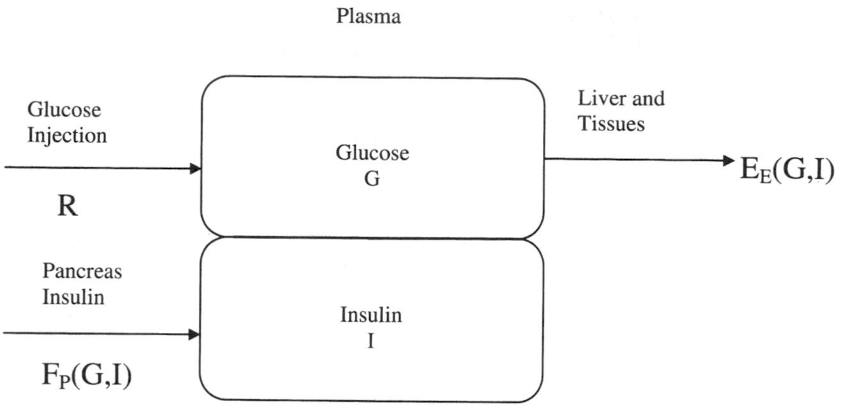

Fig. 8.15 Schematic of a modified compartmental analysis for glucose distribution.

These coupled differential equations can be solved by Laplace transforms with the initial conditions on g and p both zero (by their definitions):

$$s\tilde{g} = -K_1\tilde{g} - K_2\tilde{p} - \tilde{R} \tag{8.64a}$$

$$s\tilde{p} = K_3\tilde{g} - K_4\tilde{p} \tag{8.64b}$$

where s is the Laplace variable and the $\sim$ refers to the transformed function. Solving for $\tilde{p}$ and $\tilde{g}$ gives

$$\tilde{p} = \frac{1}{K_2}\left[\tilde{R} - (s + K_1)\tilde{g}\right] \tag{8.65a}$$

$$\tilde{g} = \frac{(s + K_4)\tilde{R}}{[s^2 + (K_1 + K_4)s + (K_1K_4 + K_2K_3)]} \tag{8.65b}$$

By specifying $R(t)$ as a bolus injection, i.e., the Dirac delta function $\delta(t)$, then $\tilde{R} = 1.0$ and Eq. (8.65b) becomes

$$\tilde{g} = \frac{s + K_4}{(s + \alpha - j\omega_0)(s + \alpha - j\omega_0)} \tag{8.66}$$

where

$$\alpha = \frac{(K_1 + K_4)}{2} \quad \text{and} \quad \omega_0^2 = (K_1K_4 + K_2K_3) - \alpha^2 \tag{8.67}$$

This inverts to

$$g(t) = \exp(-\alpha t)[\cos(\omega_0 t) + \beta\sin(\omega_0 t)] \quad \text{with} \quad \beta = (K_4 - \alpha)/\omega_0 \tag{8.68}$$

This relatively simple model does have its usefulness even though it reduces data poorly. It can help in classifying normal versus diabetic behavior. For example, the parameter ω_N, the system natural frequency as defined from ω_0, i.e., $\omega_N^2 = \omega_0^2 + \alpha^2$, is used to obtain a descriptive period, $2\pi/\omega_N$, which tends to be less than 4 h for normal behavior and is longer for most diabetics, regardless of the severity of the condition.

This model is also useful in the planning or design of new experiments. The results may lead to extensions of the model and permit a wider class of phenomena to be described. Numerous tracer experiments demonstrate that this linearized model is oversimplified. It is not isomorphic with living individuals; however, it is of sufficient similarity to simulate various types of normal and abnormal responses in several types of glucose tolerance tests. This model has proved adequate for a surprising number of situations. The fact that it fails with respect to some fine details should not be disturbing. It does, however, raise the following question: What minimum number of extensions/modifications must be made to the model to broaden its range of validity? To probe this further is beyond the scope of this text. The answers can be sought in more advanced coverage of this topic in texts listed under Suggested Reading.

Example Problem 8.6

A chemostat, with a residence time of 1.5 min, is operating with an inlet substrate (S) composition of 0.75 mol/liter. The kinetics are second order in substrate concentration with a rate constant $k = 0.667$ liter/(mole, min). Find the change in exit concentration of S as a function of time for a step change in inlet concentration to 0.9 mol/liter.

Solution

Since a chemostat is a well-mixed bioreactor, its dynamics are governed by an equation similar to Eq. (8.4). The difference is that the rate term is now nonlinear, $r = kS^2$; therefore,

$$\frac{dS}{dt} = \left(\frac{1}{\theta}\right)(S_i - S) - kS^2$$

which is awkward because of the S^2 term. The equation is linearized using a Taylor series expansion about a steady-state value for the rate ($\bar{r}$). Using deviation variables defined in terms of the steady-state value for the substrate ($\bar{S}$) as $C = S - \bar{S}$, the system becomes

$$r = kS^2 \cong k\bar{S}^2 + 2k\bar{S}(S - \bar{S})$$

and

$$\frac{dC}{dt} = \frac{1}{\theta}[(\bar{S} + C_i) - (\bar{S} + C_i)] - k\bar{S}^2 - 2k\bar{S}C, \quad \text{with} \quad C_i = S_i - \bar{S}$$

Note that at steady state, $S = \bar{S}$ and thus $C = 0$. This equation then becomes $k\theta\bar{S}^2 + \bar{S} - S_i = 0$, from which a value of $\bar{S}$ is obtained. Furthermore, this steady-state relationship can be used to simplify the system dynamic equation to give

$$\frac{dC}{dt} = \frac{1}{\theta}(C_i - C) - k_a C$$

where $k_a = 2k\bar{S}$.

Taking the Laplace transform, with $C(t = 0)$ gives

$$s\hat{C} = \frac{1}{\theta}(\hat{C}_i - \hat{C}) - k_a\hat{C} \quad \text{or} \quad \frac{\hat{C}}{\hat{C}_i} = 1/[(s + k_a)\theta + 1] = K/(Ts + 1)$$

where

$$K = 1/(1 + k_a\theta) \quad \text{and} \quad T = \theta/(1 + k_a\theta).$$

For a step change in $C_i = \beta$, i.e., $\hat{C}_i = \dfrac{\beta}{s}$,

$$\hat{C} = \frac{\beta K}{s(Ts + 1)} = \frac{\beta K}{s} - \frac{\beta KT}{(Ts + 1)} = \frac{\beta K}{s} - \frac{\beta K}{(s + 1/T)}$$

Inverting back to the time domain gives

$$C(t) = S(t) - \bar{S} = \beta K[1 - \exp(-t/T)]$$

To compute values for $S(t)$ $\bar{S}$ must be obtained from the steady-state equation which becomes

$\bar{S}^2 + \bar{S} - 0.75 = 0$ and thus $\bar{S} = 0.5$, $k_a = 2k\bar{S} = 2(0.667)(0.5) = 0.667$ min^{-1}
$K - 1/(1 + (0.667)(1.5)) = 0.5$, $T = 1.5/(1 + (0.667)(1.5)) = 0.75$, $\beta = 0.9 - 0.75 - 0.15$

Therefore,

$$S(t) = 0.5 + 0.075[1 - \exp(-1.333t)]$$

A new steady state is computed, i.e., $\bar{S} = 0.575$. This compares favorably with the value 0.5724 obtained from the nonlinear analysis. ∎

8.5 CONVECTIVE TRANSPORT BETWEEN PHYSIOLOGICAL COMPARTMENTS

The majority of models previously described generally lump the body into one or more compartments without explicitly drawing fluid streams as entering or leaving a compartment. Substance streams enter only by absorption and exit only by biochemical reactions (e.g., metabolism) and elimination. These processes are all characterized in a kinetic fashion, even though, as in urinary elimination, an actual convective flow exists. This works well for modeling the body as a whole since no fluid flows are considered to cross the boundaries. However, when modeling a local region, the significant inlet and exit fluid flows of blood (and to a lesser degree lymph flow or other extracellular fluids) across the boundaries must be explicitly involved. In this section, a modified compartmental analysis will be used to consider the transfer of substances between compartments by bulk fluid flow, i.e., by convective transport. In the first analysis of the dialysis patient, Eqs. (8.36) and (8.37) did consider a convective flow in the general sense. However, in that case the body was considered as one compartment and only mixing occurred within the compartment. There was no other form of transport into or out of the body since urea production was initially neglected. In the second analysis that used two compartments for the body, blood flow was not used explicitly (only in determining the clearance rate constant) since the process time constants were widely spaced as stated in the assumptions for that model. The analysis for the following systems would be inadequate if convective transport were ignored.

The simplest local (or flow) model is a two-compartment model that assumes equilibrium exists between the compartments. Here the main compartment is considered to consist of blood and tissue subregions as depicted in Fig. 8.16. A mass balance for a solute within the control volume (main compartment, blood, and tissue) is

$$V_T > \frac{dC_T}{dt} + V_B \frac{dC_B}{dt} = Q(C_{Bi} - C_B) \tag{8.69}$$

where Q is the volumetric flow of blood, C_{Bi} and C_B are the inlet and exit concentrations of solute in the blood (note that C_B exit $= C_B$ in the well-mixed blood compartment), C_T is the concentration in the tissue, and V_T and V_B are the respective tissue and blood region volumes.

The assumption of equilibrium provides a relationship between C_B and C_T, i.e., $C_B = K_{eq} C_T$, and in many regions of the body $V_B \ll V_T$. Therefore, V_T will be used as the total (main) compartment volume, V, and the term $V_B \dfrac{dC_B}{dt}$ in Eq. (8.69) will be ignored. This now becomes

$$\left(\frac{V}{K_{eq}}\right) \frac{dC_B}{dt} = Q(C_{Bi} - C_B) \tag{8.70}$$

If an instantaneous increase in solute concentration from zero to some steady value C_{Bi}^0 at $t = 0$ and $C_B(0) = 0$ occurs, then

$$C_B = C_{Bi}^0 \left[1 - \exp\left(\frac{-K_{eq}Q}{V} t\right) \right] \tag{8.71}$$

and C_T is obtained from the equilibrium statement.

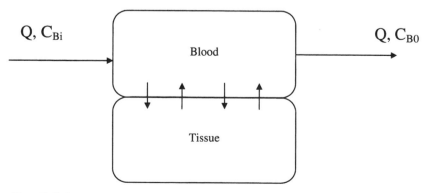

Fig. 8.16   Schematic of a two-compartment flow model for convective transport.

It is worthwhile to compare this result to that obtained from the type of model that was used previously, i.e., representing all mass transport processes in chemical–kinetic fashion (the absorption, distribution, and elimination as in Fig. 8.17):

$$\frac{dD}{dt} = V\frac{dC_D}{dt} = k_0 A - k_1 D$$

or
(8.72)

$$\frac{dC_D}{dt} = k_0\frac{A}{V} - k_1 C_D$$

For a step change in A from zero to A_0, and $C_D(0) = 0$

$$C_D = \left(\frac{k_0 A_0}{V k_1}\right)[1 - \exp(- K_1 t)]$$
(8.73)

Since the total volume of the compartment is used in both cases, the following correspondence between these two models can be observed:

$$C_D = C_B; \quad C_{Bi}^0 = \left(\frac{k_0 A_0}{k_1 V}\right); \quad k_1 = \frac{Q K_{eq}}{V}$$
(8.74)

Either model could be used to represent a physical system, depending on which seems to be most appropriate for the given situation or the circumstances of the necessary analyses.

Only a slight modification is needed to consider finite mass transfer resistances which show this approach to equilibrium. To characterize intercompartmental transfer, how far the system is from equilibrium must be known since the transport rate is linearly proportional to this. This is written as $(C_B - K_{eq}C_T)$, which is the difference between the actual blood concentration and the value it would have if the system was at equilibrium, i.e., $K_{eg}C_T = (C_B)_{eq}$. Furthermore, the actual transport rate is dependent on a mass transfer coefficient K (related to molecular diffusivity) and the interfacial area between the two compartments, S, which in this case for blood and tissue is the total area of capillary walls. The solute balances are

$$V_T\frac{dC_T}{dt} = KS(C_B - K_{eq}C_T)$$
(8.75a)

$$V_B\frac{dC_B}{dt} = Q(C_{Bi} - C_B) - KS(C_B - K_{eq}C_T)$$
(8.75b)

Fig. 8.17 Schematic for kinetics model comparison with a flow model.

For a step change as before,

$$C_B(t) = C_{Bi}^0[\kappa \exp(-\alpha t) + (1 - \kappa)\exp(-\beta t)] \tag{8.76}$$

where κ, α, and β are related to K, K_{eq}, S, V_B, V_T, and Q. The exact relationships are given in an exercise at the end of this chapter. What is of importance at this point is to observe the functional form of the solution, i.e., biexponential. This is useful for generalizing to other multicompartmental representations. It must be stressed, however, that a given functional form does not imply the type of physical system. As will be demonstrated later, the biexponential type of solution as obtained previously (for a two subregion compartment with finite mass transfer resistance) is the form obtained from models consisting of two parallel compartments, each having two subregions in which equilibrium exists. Agreement of experimental data with a biexponential expression implies either physical system could be applicable. Consequently, physiological reasoning must also be employed to decide which is most appropriate. Two situations are considered in this chapter to emphasize the nature of compartments in parallel modeling. The first involves blood flow to white and gray matter in the brain and the second is the clearance of inert gases from humans, which is presented in the context of an example problem and as an exercise at the end of the chapter.

The clearance of a radiolabeled substance from brain tissue provides an example for demonstrating parallel compartments and appropriate analysis procedures. The white matter of the brain (mainly bundles of nerve axons) is one compartment and the second is the cluster of nerve cells, i.e., the gray matter.

An aliquot of saline solution containing a dissolved radioactive compound is injected rapidly into one of the carotid arteries leading to the brain. It reaches the brain "instantly" and is transported into the various brain tissues according to the different blood perfusion routes. Total gamma-ray activity in the brain cavity at any time is monitored with conventional technology by placing a detector on the skull. The data obtained, counts per second versus time, are then compared with model predictions for parameter estimation and model validation.

The model equations are developed for each major compartment [white (1) and gray (2)] assuming each subregion is well mixed and equilibrium exists as discussed previously. Note that $C_{B1i} = C_{B2i} = 0$ for all time (after the injection period):

$$\left(\frac{V_1}{K_{eq1}}\right)\frac{dC_{B1}}{dt} = -Q_1 C_{B1} \tag{8.77a}$$

$$\left(\frac{V_2}{K_{eq2}}\right)\frac{dC_{B2}}{dt} = -Q_2 C_{B2} \tag{8.77b}$$

and of course $C_{T1} = C_{B1}/K_{eq1}$ and $C_{T2} = C_{B2}/K_{eq2}$.

The solution obtained with initial conditions $C_{B1}(0) = C_{B1}^0$ and $C_{B2}(0) = C_{B2}^0$ is

$$C_{B1}(t) = C_{B1}^0 \exp\left(\frac{-Q_1 K_{eq1}}{V_1}t\right) \tag{8.78a}$$

$$C_{B2}(t) = C_{B2}^0 \exp\left(\frac{-Q_2 K_{eq2}}{V_2}t\right) \tag{8.78b}$$

The blood flow to each parallel compartment is not equal and is given in terms of fractions (yet to be determined) of the total flow, Q, i.e., $Q_1 = f_1 Q$ and $Q_2 = f_2 Q$. The data collected reflect the mixed value (for the output of each compartment) and are represented by

$$C_B(t) = f_1 C_{B1} + f_2 C_{B2} = C_{B1}^0 f_1 \exp(-k_1 t) + C_{B2}^0 f_2 \exp(-k_2 t) \qquad (8.79)$$

where $k_1 = K_{eg1} Q_1 / V_1$ and $k_2 = K_{eq2} Q_2 / V_2$. This overall blood concentration is of biexponential form and is consistent with earlier statements. If k_1 and k_2 differ by an order of magnitude, the data analysis is simplified since one will dominate at short times and the other for long times. Each will give rise to a linear region on a semilog plot of counts per second versus time, where the individual k values can be obtained from the slopes and the intercepts will be equal to $f_1 C_{B1}^0$ and $f_2 C_{B2}^0$. It is reasonable to assume that the initial concentration in the blood is uniform throughout the brain, i.e., $C_{B1}^0 = C_{B2}^0 = C_B^0$. The values for f_1 and f_2 are therefore readily obtained.

Data has been obtained for radioactive krypton (^{85}Kr) from experiments as described previously. A plot of counts per second versus time showed the exponential decay typical of solutes being washed out of a tissue region. Data analysis yields a rate constant for the white matter of $k_1 = 0.286$ min^{-1} and for the gray matter of $k_2 = 1.67$ min^{-1}. The data further suggested that $f_1 = 0.2$ and $f_2 = 0.8$, indicating that gray matter is perfused with blood to a much greater extent.

Example Problem 8.7

N_2 depletion from the body occurs during a surgical procedure when the patient is administered a high O_2 content gas. This happens gradually as the N_2 that is dissolved in the body tissue under normal atmospheric conditions is transported to the O_2-rich stream. Develop a model to predict the rate of N_2 removal as a function of time and to obtain the fraction remaining in the body. Compare the amount removed during a 30-min operation to that for a 3-h operation.

Solution

The compartments in parallel modeling approach provides an adequate representation for the clearance of all inert gases (N_2, He, H_2, A, Kr, and Xe) from the human body. Data have shown that these inert gases are cleared at similar rates, even though their relative diffusion rates through tissues are quite different. These data support the conclusion that the clearance is blood flow limited. Therefore, when modeling any body region as two subregions, i.e., blood and tissue, the assumption of equilibrium between the subregions is justified. Noting that the volume of the tissue compartment, V_T, is much greater than that for the blood compartment, V_B (in each body region), and that the subregions are well mixed, an equation similar to Eq. (8.69) can be written for each body region (e.g., liver, thyroid, and kidneys).

$$V_T \frac{dC_T}{dt} = Q(C_{Bi} - C_B)$$

Assuming that the elimination of any inert gas in the lungs is efficient, then the blood returning to any body region will have an inert gas concentration near zero, i.e., take $C_{Bi} = 0$. Furthermore, taking $C_B = K_{eq} C_T$ gives

$$\frac{dC_T}{dt} = \left(\frac{-QK_{eq}}{Vt} \right) C_T$$

and given the initial condition, $C_T(0) = C_T^0$ the solution is

$$\frac{C_T(t)}{C_T^0} = \frac{A(t)}{A_0} = \exp\left[-\left(\frac{QK_{eq}}{VT} \right) t \right]$$

where $A(t)$ is the amount of gas present in this particular body region (e.g., ml of gas @ STP). For a set of n regions in parallel, the total amount, A_{tot}, would therefore be

$$A_{tot} = \sum_{i=1}^{n} A_i^0 \exp\left[-\left(\frac{Q_i(K_{eq})_i}{V_{Ti}} \right) t \right]$$

To illustrate for N_2, select a two-region representation of the entire body, i.e., fatty tissue and nonfatty tissue. At atmospheric pressure these tissues contain (@STP) approximately 0.05 ml N_2/ml of fatty tissue and 0.01 ml N_2/ml of nonfatty tissue, or $A_f^0 = 500$ ml and $A_{nf}^0 = 364$ ml, with corresponding values of

$$k = \left(\frac{QK_{eq}}{V_T} \right)_i$$

as $k_f = 0.0085$ min^{-1} and $k_{nf} = 0.098$ min^{-1}. Thus,

$$A_{tot} = 500\exp(-0.0085t) + 364\exp(-0.098t)$$

At $t = 30$ min, $A_{tot} = 406.7$ ml N_2 and the fraction removed $f_r = 1 - (406.7/864) = 0.47$. At $t = 180$ min, $A_{tot} = 108.3$ ml N_2 and $f_r = 0.875$.   ■

EXERCISES

1. Obtain an analytic solution to Eq. (8.3) using deviation variables and Laplace transforms. You will need to verify that as $t \to \infty$ (i.e., steady state) you obtain Eq. (8.5). Solve this same system by "conventional" methods and compare the techniques.

2. Using your result from exercise 1 (in the Laplace domain) compare the solution to the system described by Eq. (8.4) (also in the Laplace domain) after you take the limit as n (the number of compartments) goes to ∞. As suggested in the text, be able to draw the same conclusions when Eqs. (8.10) and (8.12) are compared.

3. Show that $E(t) = A_0 - A(t) - B(t)$ for the system given by Eqs. (8.13)–(8.15) with the initial conditions that $A(0) = A_0$ and $B(0) = E(0) = 0$. Note: You need not have the solution form for $A(t)$ or $B(t)$ to be able to show how $E(t)$ depends on them.

4. (a) Explain how you could obtain the model parameters, k_0 and k_1, with Eqs. (8.16)–(8.20), both analytically and graphically. Use Fig. 8.5 as a "data" reference. You may want to consider techniques as described in Section 8.3.1 for a biexponential form. (b) Using the k values from Fig. 8.5 as obtained in (a) as a base case, do a series of plots similar to it for various sets of k values to verify graphically the comments made about $k_1/k_0 \ll 1$, obtained from Eq. (8.19) and (8.20).

5. (a) Using the results from Example Problem 8.4, determine an injection schedule to maintain a drug plasma concentration of between 70 and 20% of initial dosage. In other words, after the first injection $B_0 = 0.2$ g/liter, what dose level and how frequent must doses be to ensure that blood plasma levels never exceed 0.14 g/liter or drop below 0.04 g/liter. (b) For this case, which is easier to use — half-life or system time constant? Why?

6. Confirm the results/comments made pertaining to Eqs. (8.24)–(8.27) with respect to bolus injection versus intravenous infusions, etc.

7. Explain how you could recover the parameters K_2 and K_{eq} from Eq. (8.35) given a value for E_0 and a plot of r versus $[S]$ data.

8. (a) Using a model as determined from Fig. 8.7, find the extent of treatment, i.e., patient BUN, after 4 h on this device. (b) How long must the patient be on a device that only has an area of 0.8 m² and $K = 0.015$ cm/min?

9. (a) Using the "tablet" data given in Fig. 8.8 and assuming it is "solution" data, obtain system parameters as done in the example given for sustained release medication. (b) Using the parameters obtained in (a) determine a response curve for a tablet given an initial dose (A_o) of 150 mg/kg with pepsin test results that give $k_r = 0.33$ h^{-1}, $f_r = 0.9$, and $f_i = 0.1$.

10. Verify Eqs. (8.52)–(8.54) by repeating the analysis starting with Eqs. (8.46) and (8.47).

11. Describe quantitatively the distribution of methotrexate (MTX) in a laboratory mouse using the flow model and parameters as discussed in Bischoff *et al.* (1970). (a) Use their two-compartment model plus gut lumen to develop the appropriate system equations in the nomenclature used in this chapter. Use Lightfoot's (1974) text, pages 378 and 479, for comparison. (b) Do the computations as suggested by Lightfoot on page 479 in his text.

12. Complete the solution to Eqs.(8.75a), (8.75b) and (8.76). That is, determine the relationship between the constants in Eq (8.76) to the system parameters in Eqs. (8.75a) and (8.75b) stated following Eq. (8.76).

13. Consider the regulation of carbon dioxide in the human body. Derive the equations of the two-compartment (tissue and lung) flow model which can be used to describe the system response to positive and negative "steps" of CO_2 in the inspired air. The compartments are connected by arterial circulation from the lung to the tissue and by venous circulator from the tissue to the lung. The only input from the tissue is metabolic CO_2 due to work. The lung obviously has airflow into and out of it to provide transport of CO_2 out of the body (Fig. 8.18). Consider the following assumptions: (i) Respiration is functionally represented by CO_2 concentration alone

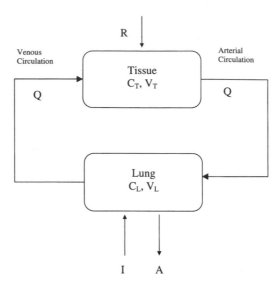

Fig. 8.18 Schematic for CO_2 clearance from the body.

within the range of the model (O_2 and pH are ignored); (ii) the rapid phasic changes in alveolar and blood gas concentrations with each respiratory cycle are ignored and alveolar volume is considered constant; (iii) cardiac output is constant (iv) CO_2 dissociation curves are linear and equal within the range of the model for arterial blood, venous blood, and tissue; (v) arterial partial pressure of CO_2 is constant and equal to alveolar partial pressure of CO_2; (vi) venous CO_2 partial pressure is equal to tissue CO_2 partial pressure; and (vii) circulation times are ignored. Comment on the validity of each assumption and its impact on model development and/or the need for experimental verification. You may find Cooney's (1976) Chapter 6, Lightfoot's (1974) Chapter 6 (page 387), or Ackerman and Gatewood's (1979) Chapter 3 to be helpful in addition to earlier sections of this chapter.

14. The history of infectious or communicable disease modeling dates to 1760 when D. Bernoulli studied the population dynamics of small pox with a mathematical model. In Bailey's (1975) text descriptions of the Kermack and McKendrick's compartmental model for the course of an acute epidemic in a closed population is presented. The concept is that this behavior is a function of the number of susceptibles (x) and the infective rate (βxy) between susceptibles and infectives (y). A third compartment must be considered, i.e., the number of immune individuals (z) which is updated as the infectives are removed from that state and become "immune" at the rate (γy). The total population (n) is the sum $x + y + z$. (a) Draw the compartmental structure (block diagram) showing the interconnections. (b) Write

the differential equations for the mass balance on each portion of the population. (c) Note you can remove some nonlinearities by taking ratios of the derivatives, for example, $dx/dt \div dz/dt$, and use of Taylor series expansions. Recall also that $dn/dt = 0$; therefore, only two of the dependent variables need to be solved. Select x and z. When developing a solution for z, retain only second-order terms from a Taylor series. If a relative removal rate $\rho = \gamma/\beta$ is defined and $y(0)$ and $z(0)$ are assumed to be negligible so that $x(0) = n$, find expression for $z(t)$ and $z(\infty)$, i.e., the total size of the epidemic.

SUGGESTED READING

Ackerman, E., Gatewood, L. C. (1979). *Mathematical Models in the Health Sciences: A Computer-Aided Approach*. Univ. of Minnesota Press, Minneapolis.

Bergman, R. N., Phillips, L. S., Cobelli, C. (1981). Physiologic evaluation of factors controlling glucose tolerance in man. *J. Clin. Invest.* **68**, 1456–1467.

Bird, R. B., Stewart, W. E., Lightfoot, E. N. (1960) Transport phenomena. Wiley, New York.

Bischoff, K. B., Brown, R. G. (1966) Drug distribution in mammals. *Chem. Eng. Prog. Symp. Ser* 62 (66).

Bischoff, K. B., Dedrick, R. L., and Zaharko, D. S. (1970). Preliminary model for methotrexate pharmaco-kinetics. *J. Pharm. Sci.* **59**, 149.

Blanch, H. W., Clark, D. S. (1996). *Biochemical Engineering*. Dekker, New York.

Cobelli, C., Saccomani, M. P. (1995). Compartmental models of physiologic systems. In *The Biomedical Engineering Handbook*. J. D. Bronzino, Ed. CRC Press, Boca Raton, FL.

Cooney, D. O. (1976). *Biomedical Engineering Principles: An Introduction to Fluid, Heat and Mass Transport Processes*. Dekker, New York.

Fisher, R. J. (1993). Flow Reactors. In *Magill's Survey of Science: Applied Science*. F. N. Magill, Ed. Salem Press, Pasadena, CA.

Fung, L. K., Shin M., Tyler. B., Brem, H., Saltzman, W. M. (1996). Chemotheropeutic drugs released from polymers: Distribution of 1,3-bis(2-chloroethyl)-1-nitrosourea in the rat brain. *Pharmacol. Res.* **13** (5).

Gibaldi, M. (1984). *Biopharmaceutics and Clinical Pharmacokinetics*. 3rd ed. Lea & Febiger, Philadelphia.

Glantz, S. A. (1979). *Mathematics for Biomedical Applications*. Univ. of California Press, Berkeley.

Hoppe, W., Lohmann, W., Marki, H., Ziegler, H. Eds. (1983). *Biophysics*, Springer-Verlag, New York.

Jacquez, J. A. (1985). *Compartmental Analysis in Biology and Medicine*. 2nd ed. Univ. of Michigan Press, Ann Arbor.

Jacquez, J. A., Simon, C. P. (1993). Qualitative theory of compartmental systems. *SIAM Rev.* **35**, 43.

Jain, R. K. (1994). Transport phenomena in tumors. In *Advances in Chemical Engineering: Volume 19*. Academic Press, New York.

Lauffenburger, D. A., Linderman, J. J. (1993). *Receptors: Model for Binding, Trafficking and Signaling*. Oxford Univ. Press, New York.

Lightfoot, E. N. (1974). *Transport Phenomena and Living Systems*. Wiley, New York.

Martin, A. (1993). *Physical Pharmacy*. 4th ed. Williams & Wilkins, Baltimore.

Shuler, M. L., and Kargi, F. (1998). *Bioprocess Engineeering: Basic Concepts.* 2nd ed. Prentice-Hall, Englewood Cliffs, NJ.

Shuler, M. L., Ghanem, A., Quick, D., Wang, M. C., and Miller, P. (1996). A self-regulating cell culture analog device to mimic animal and human toxicological responses. *Biotech. Bioeng.,* **52,** 45–60.

Stockton, E., Fisher, R. J. (1998). Designing biomimetic membranes for ion transport. *Proc. IEEE/NEBC,* **24,** 49–52.

Thoresen, K., Fisher, R. J. (1995). Use of supported liquid membranes as biomimetics of active transport processes. *Biomimetics 3* (1) 31–66.

9 BIOMECHANICS

Chapter Contents

At the conclusion of this chapter, the reader will be able to:

- Understand the application of engineering kinematic relations to biomechanical problems
- Understand the application of engineering kinetic relations to biomechanical problems
- Understand the application of engineering mechanics of materials to orthopedic structures
- Use MATLAB to write and solve biomechanical static and dynamic equations
- Use Simulink to study viscoelastic properties of biological tissues
- Understand how kinematic equations of motion are used in clinical analysis of human gait
- Understand how kinetic equations of motion are used in clinical analysis of human gait
- Explain how biomechanics applied to human gait is used to quantify pathological conditions, to suggest surgical and clinical treatments, and to quantify their effectiveness

9.1 INTRODUCTION

Biomechanics combines engineering and the life sciences by applying principles from classical mechanics to the study of living systems. This relatively new field covers a broad range of topics, including strength of biological materials, biofluid mechanics in the cardiovascular and respiratory systems, material properties and interactions of medical implants and the body, heat and mass transfer into biological tissues (e.g., tumors), biocontrol systems regulating metabolism or voluntary motion, and kinematics and kinetics applied to study human gait. The great breadth of the field of biomechanics arises from the complexities and variety of biological organisms and systems. This chapter focuses on orthopedic biomechanics. The first four sections serve as a broad overview of concepts from undergraduate mechanical engineering courses applied to orthopedic structures. The last section shows how these techniques are applied in the analysis of human gait data. The limited scope of this chapter seeks to present the human body as a complex machine, with the skeletal system and ligaments forming the framework and the muscles and tendons serving as the motors and cables. The main mechanical features of these biological components are presented in Section 9.5 with references so that the interested reader may continue to delve into the increasingly specialized subfields as desired.

Human gait biomechanics may be viewed as a structure (skeleton) composed of levers (bones) with pivots (joints) that move as the result of net forces produced by pairs of agonist and antagonist muscles. Consequently, the strength of the structure and the action of muscles will be of fundamental importance. Muscle contraction research has a long history, as chronicled in the book Machina Carnis by Needham (1971). Recently, histories of biomechanics research have been written by Fung (1993) and Nigg (1994). For a more comprehensive history of medicine see Singer and Underwood's (1962) book. The following is an overview of the highlights, especially those related to gait:

Galen of Pergamon (129–199): Published extensively in medicine, including *De Motu Musculorum* (*On the Movements of Muscles*). He realized that motion requires muscle contraction.

Leonardo da Vinci (1452–1519): Made the first accurate descriptions of ball-and-socket joints, such as the shoulder and hip, calling the latter the "polo dell'omo" (pole of man). His drawings depicted mechanical force acting along the line of muscle filaments.

Andreas Vesalius (1514–1564): Published *De Humani Corporis Fabrica* (*The Fabric of the Human Body*). Based on human cadaver dissections, his work led to a more accurate anatomical description of human musculature than Galen's and demonstrated that motion results from the contraction of muscles that shorten and thicken.

Galileo Galilei (1564–1642): Studied medicine and physics, integrated measurement and observation in science, and concluded that mathematics is an essential tool of science. His analyses included the biomechanics of jumping and the gait analysis of horses and insects, as well as dimensional analysis of animal bones.

Santorio Santorio (1561–1636): Used Galileo's method of measurement and analysis and found that the human body changes weight with time. This observation led to the study of metabolism and, thereby, ushered in the scientific study of medicine.

Giovanni Borelli (1608–1679): Mathematician who studied body dynamics, muscle contraction, animal movement, and motion of the heart and intestines. He published *De Motu Animalium* (*On the Motion of Animals*) in 1680.

Jan Swammerdam (1637–1680): Introduced the nerve–muscle preparation, stimulating muscle contraction by pinching the attached nerve in the frog leg. He also showed that muscles contract with little change in volume, refuting the previous beliefs that muscles contract when "animal spirits" fill them, causing bulging.

Robert Hooke (1635–1703): Devised Hooke's law, relating the stress and elongation of elastic materials, and used the term "cell" in biology.

Isaac Newton (1642–1727): He did no biomechanics work, but he invented the calculus, the classical laws of motion, and the constitutive equation for viscous fluid, all of which are fundamental to biomechanics.

Nicholas André (1658–1742): Coined the term "orthopaedics" at the age of 80 and believed that muscular imbalances cause skeletal deformities.

Leonard Euler (1707–1783): Generalized Newton's laws of motion to continuum representations that are used extensively to describe rigid body motion and studied pulse waves in arteries.

Thomas Young (1773–1829): Studied vibrations and voice and wave theory of light and vision, and devised Young's modulus of elasticity.

Ernst Weber (1795–1878) and Eduard Weber (1806–1871): Published "Die Mechanik der meschlichen Gerwerkzeuge" (*On the Mechanics of the Human Gait Tools*) in 1836 and pioneering the scientific study of human gait.

Hermann von Helmholtz (1821–1894): Studied an immense array of topics, including optics, acoustics, thermodynamics, electrodynamics, physiology, and medicine, including opthalmoscopy, fluid mechanics, nerve conduction speed, and the heat of muscle contraction.

Etienne Marey (1830–1904): Analyzed the motion of horses, birds, insects, fish, and humans. His inventions included force plates to measure ground reaction forces and the "Chronophotographe a pellicule," or motion picture camera.

Wilhelm Braune and Otto Fischer (research conducted from 1895 to 1904): Published "Der Gang des Menschen" (*The Human Gait*), containing the mathematical analysis of human gait and introducing methods still in use. They invented "cyclography" (now called interrupted-light photography with active markers), pioneered the use of multiple cameras to reconstruct 3-D motion data, and applied Newtonian mechanics to estimate joint forces and limb accelerations.

9.2   BASIC MECHANICS

This section reviews some of the main points from any standard introductory mechanics (statics and dynamics) course. Good references abound, e.g., *Engineering Mechanics* by Merriam and Kraige (1997). A review of vector mathematics is followed by matrix coordinate transformations, a topic new to some students. Euler's equations of motion (see Section 9.2.5) may also be new material. For both topics, *Principles of Dynamics* by Greenwood (1965) provides a comprehensive reference.

9.2.1   Vector Mathematics

Forces are either written in terms of **scalar components** and **unit vectors** or in **polar form** with **magnitude** and **direction**. Figure 9.1 shows that the 2-D vector **F** is composed of the **i** component, F_x, in the x direction, and the **j** component, F_y, in the y direction, or

$$\mathbf{F} = F_x\mathbf{i} + F_y\mathbf{j} \tag{9.1}$$

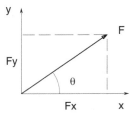

Fig. 9.1   2-D representation of vector **F**.

as in $20\mathbf{i} + 40\mathbf{j}$ lb. In this chapter vectors are set in bold type. This same vector may be written in polar form in terms of the vector's magnitude $|F|$, also called the **norm**, and the vector's angle of orientation, θ,

$$|F| = \sqrt{F_x^2 + F_y^2} \tag{9.2}$$

$$\theta = \arctan\frac{F_y}{F_x} \tag{9.3}$$

yielding $|F| = 44.7$ lb and $\theta = 63.4°$. Vectors are similarly represented in three dimensions in terms of their $\mathbf{i}$, $\mathbf{j}$, and $\mathbf{k}$ components:

$$\mathbf{F} = F_x\mathbf{i} + F_y\mathbf{j} + F_z\mathbf{k} \tag{9.4}$$

Often, a vector's magnitude and two points along its line of action are known. Consider the 3-D vector in Fig. 9.2. $\mathbf{F}$ has magnitude of 10 lb, and its line of action passes from the origin (0,0,0) to the point (2,6,4). $\mathbf{F}$ is written as the product of the magnitude $|F|$ and a unit vector that points along its line of action:

$$\mathbf{F} = 10 \text{ lb}\left(\frac{2\mathbf{i} + 6\mathbf{j} + 4\mathbf{k}}{\sqrt{2^2 + 6^2 + 4^2}}\right)$$

$$\mathbf{F} = 2.67\mathbf{i} + 8.02\mathbf{j} + 5.34\mathbf{k} \text{ lb}$$

$$|F| = \sqrt{2.67^2 + 8.02^2 + 5.34^2}$$

$$- 10 \text{ lb}$$

The quantity in parentheses is the unit vector of $\mathbf{F}$, or

$$\mathbf{e_F} = \left(\frac{2\mathbf{i} + 6\mathbf{j} + 4\mathbf{k}}{\sqrt{2^2 + 6^2 + 4^2}}\right) = 0.267\mathbf{i} + 0.802\mathbf{j} + 0.534\mathbf{k}$$

The vector $\mathbf{F}$ in Fig. 9.2 may also be defined in 3-D space in terms of the angles between its line of action and each coordinate axis. Consider the angles θ_x,

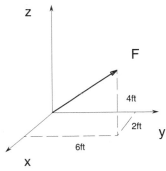

Fig. 9.2 3-D vector defined by its magnitude and line of action.

θ_y, and θ_z that are measured from the positive x, y, and z axes, respectively, to **F**. Then

$$\cos \theta_x = \frac{F_x}{|F|} \tag{9.5}$$

$$\cos \theta_y = \frac{F_y}{|F|} \tag{9.6}$$

$$\cos \theta_z = \frac{F_z}{|F|} \tag{9.7}$$

These ratios are termed the **direction cosines** of **F**. The unit vector $\mathbf{e_F}$ is equivalent to

$$\mathbf{e_F} = \cos \theta_x \mathbf{i} + \cos \theta_y \mathbf{j} + \cos \theta_z \mathbf{k} \tag{9.8}$$

or, in general

$$\mathbf{e_F} = \left(\frac{F_x \mathbf{i} + F_y \mathbf{j} + F_z \mathbf{k}}{\sqrt{F_x^2 + F_y^2 + F_z^2}} \right) \tag{9.9}$$

The angles θ_x, θ_y, and θ_z in this example are consequently

$$\theta_x = \arccos \frac{2.67}{10} = 74.5°$$

$$\theta_y = \arccos \frac{8.02}{10} = 36.7°$$

$$\theta_z = \arccos \frac{5.34}{10} = 57.7°$$

Vectors are added by summing their components:

$$\mathbf{A} = A_x \mathbf{i} + A_y \mathbf{j} + A_z \mathbf{k}$$

$$\mathbf{B} = B_x \mathbf{i} + B_y \mathbf{j} + B_z \mathbf{k}$$

$$\mathbf{C} = \mathbf{A} + \mathbf{B} = (A_x + B_x)\mathbf{i} + (A_y + B_y)\mathbf{j} + (A_z + B_z)\mathbf{k}$$

or in general

$$\mathbf{R} = \sum F_x \mathbf{i} + \sum F_y \mathbf{j} + \sum F_z \mathbf{k} \tag{9.10}$$

Vectors are subtracted similarly by subtracting vector components.

Vector multiplication consists of two distinct operations, the **dot** and **cross** products. The dot, or scalar, product of vectors **A** and **B** produces a scalar via

$$\mathbf{A} \cdot \mathbf{B} = AB \cos \theta \tag{9.11}$$

where θ is the angle between the vectors. For an orthogonal coordinate system, where all axes are 90° apart,

$$\mathbf{i} \cdot \mathbf{i} = \mathbf{j} \cdot \mathbf{j} = \mathbf{k} \cdot \mathbf{k} = 1$$
$$\mathbf{i} \cdot \mathbf{j} = \mathbf{j} \cdot \mathbf{k} = \mathbf{k} \cdot \mathbf{i} = \cdots = 0 \tag{9.12}$$

For example,

$$A = 3i + 2j + k \text{ lb}$$

$$B = -2i + 3j + 10k \text{ lb}$$

$$A \cdot B = 3(-2) + 2(3) + 1(10) = 10 \text{ lb}$$

Note that the dot product is commutative, i.e., $A \cdot B \equiv B \cdot A$.

The physical interpretation of the dot product $A \cdot B$ is the projection of A onto B, or, equivalently, the projection of B onto A. For example, **work** is defined as the force that acts in the same direction as the motion of a body. Figure 9.3 (left) shows a force vector F dotted with a direction of motion vector d. The work W done by F is given by $F \cdot d = Fd \cos \theta$. Dotting F with d yields the component of F acting in the same direction as d.

The **moment** of a force about a point or axis is a measure of its tendency to cause rotation. The cross, or vector, product of two vectors yields a new vector that points along the axis of rotation. For example, Figure 9.3 (right) shows a vector F acting in the x–y plane at a distance from the body's coordinate center O. The vector r points from O to the line of action of F. The cross product $r \times F$ is a vector that points in the z direction along the body's axis of rotation. If F and r have k components, their cross product will have additional components of rotation about the x and y axes. The moment M resulting from crossing r into F is written as

$$M = M_x i + M_y j + M_z k \qquad (9.13)$$

where M_x, M_y, and M_z cause rotation of the body about the x, y, and z axes, respectively.

Cross products may be taken by crossing each vector component term by term, for example,

$$A \times B = 3(-2)i \times i + 3(3)i \times j + 3(10)i \times k$$

$$+ 2(-2)j \times i + 2(3)j \times j + 2(10)j \times k$$

$$+ 1(-2)k \times i + 1(3)k \times j + 1(10)k \times k$$

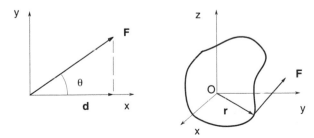

Fig. 9.3 (Left) The dot, or scalar, product of vectors **F** and **d** is equivalent to the projection of **F** onto **d**. (Right) The cross, or vector, product of vectors **r** and **F** is a vector that points along the axis of rotation.

The magnitude $|A \times B| = AB \sin \theta$, where θ is the angle between **A** and **B**. Consequently, for an orthogonal coordinate system all like terms equal zero, and $i \times j = k$, $j \times k = i$, $k \times i = j$, $i \times k = -j$, and so on. The previous example yields

$$A \times B = 9k - 30j + 4k + 20i - 2j - 3i$$
$$= 17i - 32j + 13k$$

Note that the cross product is not commutative, i.e., $A \times B \neq B \times A$.

Cross products of vectors are commonly computed using matrices. The previous example $A \times B$ is given by the matrix

$$A \times B = \begin{vmatrix} i & j & k \\ A_x & A_y & A_z \\ B_x & B_y & B_z \end{vmatrix}$$

$$= \begin{vmatrix} i & j & k \\ 3 & 2 & 1 \\ -2 & 3 & 10 \end{vmatrix} \tag{9.14}$$

$$= i(20 - 3) - j(30 + 2) + k(9 + 4)$$
$$= 17i - 32j + 13k$$

Example Problem 9.1

The vector **F** in Fig. 9.4 has a magnitude of 10 kN and points along the dashed line as shown. (a) Write **F** as a vector. (b) What is the component of **F** in the x–z plane? (c) What moment does **F** generate about the origin (0,0,0)?

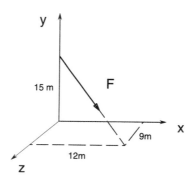

Fig. 9.4 Force vector **F** has a magnitude of 10 kN (Example Problem 9.1).

Solution

This example problem is solved using MATLAB. The ≫ prompt denotes input and the percent sign (%) precedes comments (ignored by MATLAB). Lines that begin without the ≫ prompt are MATLAB output. Some spaces in the following output below were omitted to conserve space.

```
≫ %(a) First write the direction vector d that points along F
≫ % as a 1D array:
≫ d = [12 −15 9]

d =   12   −15   9
≫ % Now write the unit vector of F, giving its direction:
≫ unit_vector = d/norm(d)

unit_vector =   0.5657   −0.7071   0.4243

≫ % F consists of the magnitude 10 kN times this unit vector
≫ F = 10*unit_vector

F =   5.6569   −7.0711   4.2426

≫ % Or, more directly
≫ F = 10*(d/norm(d))

F =   5.6569   −7.0711   4.2426

≫ % (b) First write the vector r_xz that points in the xz plane:
≫ r_xz = [12 0 9]
r_xz =   12   0   9

≫ % The dot product is given by the sum of all the term-by-term
≫ % multiplications of elements of vectors F and r_xz
≫ F_dot_r_xz = sum(F.*r_xz)
≫ % or simply, dot (F,r_xz)

F_dot_r_xz =   106.0660

≫ % (c) Cross F with a vector that points from the origin to F.
≫ % The cross product is given by the cross function
≫ r_xz_cross_F = cross(r_xz,F)

r_xz_cross_F =   63.6396   0   −84.8528
```

```
≫ % Note that the cross product is not commutative
≫ cross(F,r_xz)
ans =   -63.6396   0   84.8528

≫ % Vectors are added and subtracted in MATLAB using the + and -
≫ % operations, respectively.
```
■

9.2.2 Coordinate Transformations

3-D Direction Cosines

When studying the kinematics of human motion, it is often necessary to transform body or body segment coordinates from one coordinate system to another. For example, coordinates corresponding to a coordinate system determined by markers on the body (a moving coordinate system) must be translated to coordinates with respect to the fixed laboratory (inertial coordinate system). These 3-D transformations use direction cosines that are computed as follows.

Consider the vector **A** measured in terms of the uppercase coordinate system XYZ, shown in Fig. 9.5 in terms of the unit vectors **I**, **J**, **K**:

$$\mathbf{A} = A_x\mathbf{I} + A_y\mathbf{J} + A_z\mathbf{K} \tag{9.15}$$

The unit vectors **I**, **J**, **K** can be written in terms of **i**, **j**, **k** in the *xyz* system:

$$\mathbf{I} = \cos\theta_{xX}\mathbf{i} + \cos\theta_{yX}\mathbf{j} + \cos\theta_{zX}\mathbf{k} \tag{9.16}$$

$$\mathbf{J} = \cos\theta_{xY}\mathbf{i} + \cos\theta_{yY}\mathbf{j} + \cos\theta_{zY}\mathbf{k} \tag{9.17}$$

$$\mathbf{K} = \cos\theta_{xZ}\mathbf{i} + \cos\theta_{yZ}\mathbf{j} + \cos\theta_{zZ}\mathbf{k} \tag{9.18}$$

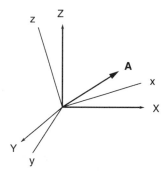

Fig. 9.5 Vector **A**, measured with respect to coordinate system *XYZ*, is related to coordinate system *xyz* via the nine direction cosines of Eq. (9.20).

where θ_{xX} is the angle between **i** and **I**, and similarly for the other angles. Substituting Eqs. (9.16)–(9.18) into Eq. (9.15) gives

$$\mathbf{A} = A_x \left[\cos\theta_{xX}\mathbf{i} + \cos\theta_{yX}\mathbf{j} + \cos\theta_{zX}\mathbf{k}\right]$$
$$+ A_y \left[\cos\theta_{xY}\mathbf{i} + \cos\theta_{yY}\mathbf{j} + \cos\theta_{zY}\mathbf{k}\right] \qquad (9.19)$$
$$+ A_z \left[\cos\theta_{xZ}\mathbf{i} + \cos\theta_{yZ}\mathbf{j} + \cos\theta_{zZ}\mathbf{k}\right]$$

or

$$\mathbf{A} = (A_x \cos\theta_{xX} + A_y \cos\theta_{xY} + A_z \cos\theta_{xZ})\,\mathbf{i}$$
$$+ (A_x \cos\theta_{yX} + A_y \cos\theta_{yY} + A_z \cos\theta_{yZ})\,\mathbf{j} \qquad (9.20)$$
$$+ (A_x \cos\theta_{zX} + A_y \cos\theta_{zY} + A_z \cos\theta_{zZ})\,\mathbf{k}$$

Consequently, **A** may be represented in terms of **I, J, K** or **i, j, k**.

Euler Angles

The coordinates of a body in one orthogonal coordinate system may be related to another orthogonal coordinate system via Euler angle transformation matrices. For example, one coordinate system might correspond to markers placed on the patient's pelvis, whereas the other coordinate system might correspond to the patient's thigh. The two coordinate systems are related by a series of rotations about each original axis in turn. Figure 9.6 shows the xyz coordinate axes with a y–x–z rotation sequence. First, xyz is rotated about the y axis (top), transforming the **ijk** unit vectors into the **i'j'k'** unit vectors, via the equations

$$\mathbf{i}' = \cos\theta_y\mathbf{i} - \sin\theta_y\mathbf{k} \qquad (9.21)$$

$$\mathbf{j}' = \mathbf{j} \qquad (9.22)$$

$$\mathbf{k}' = \sin\theta_y\mathbf{i} + \cos\theta_y\mathbf{k} \qquad (9.23)$$

This new primed coordinate system is then rotated about the x axis (Fig. 9.6, middle), giving the double-primed system:

$$\mathbf{i}'' = \mathbf{i}' \qquad (9.24)$$

$$\mathbf{j}'' = \cos\theta_x\mathbf{j}' + \sin\theta_x\mathbf{k}' \qquad (9.25)$$

$$\mathbf{k}'' = -\sin\theta_x\mathbf{j}' + \cos\theta_x\mathbf{k}' \qquad (9.26)$$

Finally, the double-primed system is rotated about the z axis, giving the triple-primed system:

$$\mathbf{i}''' = \cos\theta_z\mathbf{i}'' + \sin\theta_z\mathbf{j}'' \qquad (9.27)$$

$$\mathbf{j}''' = -\sin\theta_z\mathbf{i}'' + \cos 0_z\mathbf{j}'' \qquad (9.28)$$

$$\mathbf{k}''' = \mathbf{k}'' \qquad (9.29)$$

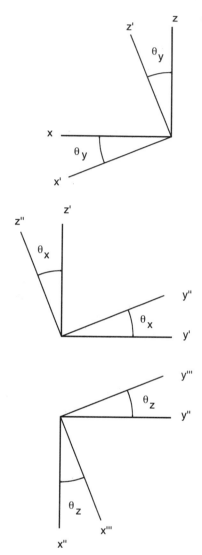

Fig. 9.6 The unprimed coordinate system *xyz* undergoes three rotations — about the *y* axis (top), about the *x* axis (middle), and about the *z* axis (bottom) — yielding the new triple-primed coordinate system *x'''y'''z'''*.

The three rotations may be written in matrix form to directly translate **ijk** into **i'''j'''k'''**:

$$\begin{bmatrix} i''' \\ j''' \\ k''' \end{bmatrix} = \begin{bmatrix} \cos\theta_z & \sin\theta_z & 0 \\ -\sin\theta_z & \cos\theta_z & 0 \\ 0 & 0 & 1 \end{bmatrix} \begin{bmatrix} 1 & 0 & 0 \\ 0 & \cos\theta_x & \sin\theta_x \\ 0 & -\sin\theta_x & \cos\theta_x \end{bmatrix} \begin{bmatrix} \cos\theta_y & 0 & -\sin\theta_y \\ 0 & 1 & 0 \\ \sin\theta_y & 0 & \cos\theta_y \end{bmatrix} \begin{bmatrix} i \\ j \\ k \end{bmatrix} \qquad (9.30)$$

$$= \begin{bmatrix} \cos\theta_z & \sin\theta_z\cos\theta_x & \sin\theta_z\sin\theta_x \\ -\sin\theta_z & \cos\theta_z\cos\theta_x & \cos\theta_z\sin\theta_x \\ 0 & -\sin\theta_x & \cos\theta_x \end{bmatrix} \begin{bmatrix} \cos\theta_y & 0 & -\sin\theta_y \\ 0 & 1 & 0 \\ \sin\theta_y & 0 & \cos\theta_y \end{bmatrix} \begin{bmatrix} i \\ j \\ k \end{bmatrix}$$

$$\begin{bmatrix} i''' \\ j''' \\ k''' \end{bmatrix} = \begin{bmatrix} \cos\theta_z\cos\theta_y + \sin\theta_z\sin\theta_x\sin\theta_y & \sin\theta_z\cos\theta_x & -\cos\theta_z\sin\theta_y + \sin\theta_z\sin\theta_x\cos\theta_y \\ -\sin\theta_z\cos\theta_y + \cos\theta_z\sin\theta_x\sin\theta_y & \cos\theta_z\cos\theta_x & \sin\theta_z\sin\theta_y + \cos\theta_z\sin\theta_x\cos\theta_y \\ \cos\theta_x\sin\theta_y & -\sin\theta_x & \cos\theta_x\cos\theta_y \end{bmatrix} \begin{bmatrix} i \\ j \\ k \end{bmatrix} \qquad (9.31)$$

If the angles of coordinate system rotation (θ_x, θ_y, θ_z) are known, coordinates in the xyz system can be transformed into the $x'''y'''z'''$ system. Alternatively, if both the unprimed and triple-primed coordinates are known, the angles may be computed as follows:

$$\mathbf{k'''} \cdot \mathbf{j} = -\sin\theta_x \rightarrow \theta_x = -\arcsin(\mathbf{k'''} \cdot \mathbf{j}) \qquad (9.32)$$

$$\mathbf{k'''} \cdot \mathbf{i} = \cos\theta_x \sin\theta_y \rightarrow \theta_y = \arcsin\left[\frac{\mathbf{k'''} \cdot \mathbf{i}}{\cos\theta_x}\right] \qquad (9.33)$$

$$\mathbf{i'''} \cdot \mathbf{j} = \sin\theta_z \cos\theta_x \rightarrow \theta_z = \arcsin\left[\frac{\mathbf{i'''} \cdot \mathbf{j}}{\cos\theta_x}\right] \qquad (9.34)$$

Example Problem 9.2

Write the Euler angle transformation matrices for the y–x–z rotation sequence using the MATLAB symbolic math toolbox.

Solution

```
% eulerangles.m
%
% Euler angles for y-x-z rotation sequence
% using MATLAB symbolic math toolbox
%
% x, y and z are thetax, thetay and thetaz, respectively
%
syms x y z

A = [ cos(y),  0,  -sin(y);
       0,  1,        0;
      sin(y),  0,   cos(y)]
```

$$B = \begin{bmatrix} 1, & 0, & 0; \\ 0, & \cos(x), & \sin(x); \\ 0, & -\sin(x), & \cos(x) \end{bmatrix}$$

$$C = \begin{bmatrix} \cos(z), & \sin(z), & 0; \\ -\sin(z), & \cos(z), & 0; \\ 0, & 0 & 1 \end{bmatrix}$$

$$D = C*B*A$$

The resulting transformation matrix from the previous m-file is

$$D =$$

$$\begin{bmatrix} \cos(z)*\cos(y) + \sin(z)*\sin(x)*\sin(y), & \sin(z)*\cos(x), & -\cos(z)*\sin(y) + \sin(z)*\sin(x)*\cos(y) \\ -\sin(z)*\cos(y) + \cos(z)*\sin(x)*\sin(y), & \cos(z)*\cos(x), & \sin(z)*\sin(y) + \cos(z)*\sin(x)*\cos(y) \\ \cos(x)*\sin(y), & -\sin(x), & \cos(x)*\cos(y) \end{bmatrix}$$

which is the same as Eq. (9.31). ■

The Euler transformation matrices are used differently depending on the available data. For example, if the body coordinates in both the fixed (unprimed) and body (triple-primed) systems are known, the body angles θ_x, θ_y, and θ_z can be computed (e.g., Eqs. 9.32–9.34 for a y–x–z rotation sequence). Alternatively, the body's initial position and the angles θ_x, θ_y, and θ_z may be used to compute the body's final position.

Example Problem 9.3

An aircraft undergoes 30 degrees of pitch (θ_x), then 20 degrees of roll (θ_y), and finally 10 degrees of yaw (θ_z). Write a MATLAB function that computes the Euler angle transformation matrix for this series of angular rotations.

Solution

Since computers use radians for trigonometric calculations, first write two simple functions to compute cosines and sines in degrees:

```
function y = cosd(x)
%COSD(X) cosines of the elements of X measured in degrees.
y = cos(pi*x/180);
```

```
function y = sind(x)
%SIND(X) sines of the elements of X measured in degrees.
y = sin(pi*x/180);
```

Next write the x–y–z rotation sequence transformation matrix:

```
function D = eulangle(thetax, thetay, thetaz)
%EULANGLE matrix of rotations by Euler's angles.
%   EULANGLE(thetax, thetay, thetaz) yields the matrix of
%   rotation of a system of coordinates by Euler's
%   angles thetax, thetay and thetaz, measured in degrees.

A = [1    0                0
     0    cosd(thetax)     sind(thetax)
     0    -sind(thetax)    cosd(thetax)];

B = [cosd(thetay)    0    -sind(thetay)
     0               1    0
     sind(thetay)    0    cosd(thetay)];

C = [cosd(thetaz)     sind(thetaz)    0
     -sind(thetaz)    cosd(thetaz)    0
     0                0               1];

% for x-y-z rotation sequence
D = C*B*A;
```

Now use this function to compute the numerical transformation matrix:

```
≫ eulangle(30, 20,10)
ans =
    0.9254     0.3188    -0.2049
   -0.1632     0.8232     0.5438
    0.3420    -0.4698     0.8138
```

This matrix can be used to convert any point in the initial coordinate system (pre-maneuver) to its position after the roll, pitch, and yaw maneuvers have been executed. ∎

9.2.3 Static Equilibrium

Newton's equations of motion applied to a structure in static equilibrium reduce to the following vector equations:

$$\sum \mathbf{F} = 0 \tag{9.35}$$

$$\sum \mathbf{M} = 0 \tag{9.36}$$

These equations are applied to biological systems in the same manner as standard mechanical structures. Analysis begins with a drawing of the free-body diagram of the body segment(s) of interest with all externally applied loads and reaction forces at the supports. Orthopedic joints can be modeled with appropriate ideal joints (e.g., hinge and ball-and-socket) as discussed in Chapter 2 (see Fig. 2.33).

Example Problem 9.4

Figure 9.7 (top) shows a Russell's traction rig used to apply an axial, tensile force to a fractured femur for immobilization. (a) What magnitude weight w must be suspended from the free end of the cable to maintain the leg in static equilibrium? (b) Compute the average tensile force applied to the thigh under these conditions.

Solution

The free-body diagram for this system is shown in the bottom of Fig. 9.7. If the pulleys are assumed frictionless and of small radius, the cable tension T is constant throughout. Using Eq. 9.35,

$$\mathbf{F_1} + \mathbf{F_2} + \mathbf{F_3} + \mathbf{F}_{femur} - mg\mathbf{j} = 0$$

Writing each force in vector form,

$$\mathbf{F_1} = -F_1\mathbf{i} = -T\mathbf{i}$$

$$\mathbf{F_2} = (-F_2 \cos 30°)\mathbf{i} + (F_2 \sin 30°)\mathbf{j}$$

$$= (-T \cos 30°)\mathbf{i} + (T \sin 30°)\mathbf{j}$$

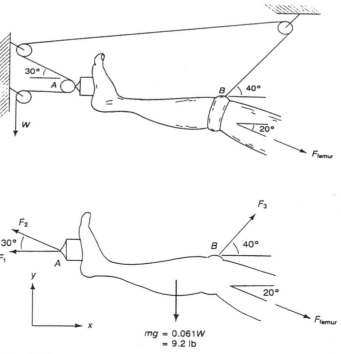

Fig. 9.7 (Top) Russell's traction mechanism for clinically loading lower extremity limbs. (Bottom) Free-body diagram of the leg in traction (adapted from Davis, 1986, Figs. 6.14 and 6.15, pp. 206–207).

$$\mathbf{F}_3 = (F_3 \cos 40°)\mathbf{i} + (F_3 \sin 40°)\mathbf{j}$$
$$= (T \cos 40°)\mathbf{i} + (T \sin 40°)\mathbf{j}$$
$$\mathbf{F}_{\text{femur}} = (F_{\text{femur}} \cos 20°)\mathbf{i} - (F_{\text{femur}} \sin 20°)\mathbf{j}$$

Using Table 9.1, and neglecting the weight of the thigh, the weight of the foot and leg is 0.061 multiplied by total body weight, yielding

$$mg\mathbf{j} = (0.061)(150\mathbf{j} \text{ lb}) = 9.2\mathbf{j} \text{ lb}$$

Summing the **i** components gives

$$-T - T \cos 30° + T \cos 40° + F_{\text{femur}} \cos 20° = 0$$

Summing the **j** components gives

$$T \sin 30° + T \sin 40° - F_{\text{femur}} \sin 20° - 9.2 \text{ lb} = 0$$

The last two expressions may be solved simultaneously, giving T, which is equal to the required externally applied weight and the axial tensile force F_{femur}

$$T = 12.4 \text{ lb}$$
$$F_{\text{femur}} = 14.5 \text{ lb}$$ ∎

Example Problem 9.5

A 160-lb person is holding a 10-lb weight in his or her palm with the elbow fixed at 90° flexion (Fig. 9.8, top). (a) What force must the biceps generate to hold the forearm in static equilibrium? (b) What force(s) does the forearm exert on the humerus?

Solution

Figure 9.8 (bottom) shows the free-body diagram of this system. Due to the increased number of unknowns compared to the previous example, both Eqs. (9.35) and (9.36) will be used. Summing moments about the elbow at point O, the equilibrium equation $\Sigma\mathbf{M} = 0$ can be written as

$$-\mathbf{r}_{OE} \times (-\mathbf{F}_A) + \mathbf{r}_{OB} \times (-10 \text{ lb})\mathbf{j} + \mathbf{r}_{OP} \times (-3.5 \text{ lb})\mathbf{j} = 0$$
$$(-2 \text{ in.})\mathbf{i} \times (-F_A)\mathbf{j} + (12 \text{ in.})\mathbf{i} \times (-10 \text{ lb})\mathbf{j} + (9.25 \text{ in.})\mathbf{i} \times (-3.5 \text{ lb})\mathbf{j} = 0$$
$$(2 \text{ in.}) F_A\mathbf{k} - (120 \text{ lb in.})\mathbf{k} - (32.4 \text{ lb in.})\mathbf{k} = 0$$

Solving this last expression for the one unknown, F_A, the vertical force at the elbow is

$$F_A = 76.2 \text{ lb}$$

To find the unknown horizontal force at the elbow, F_C, and the unknown force the biceps must generate, F_B, the other equation of equilibrium, $\Sigma\mathbf{F} = 0$, is used:

$$F_C\mathbf{i} - F_A\mathbf{j} + (-F_B \cos 75°\mathbf{i} + F_B \sin 75°\mathbf{j}) - 10\text{lb}\mathbf{j} - 3.5\text{lb}\mathbf{j} = 0$$

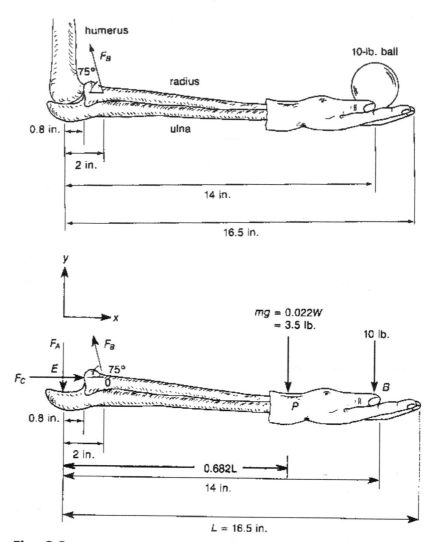

Fig. 9.8 (Top) The forearm held statically fixed in 90° flexion while holding a 10-lb weight at the hand. (Bottom) Free-body diagram of the forearm system (adapted from Davis, 1986, Figs. 6.16 and 6.17, pp. 208–209).

Summing the **i** and **j** components gives

$$F_C - F_B \cos(75°) = 0$$
$$-F_A + F_B \sin(75°) - 10 \text{ lb} - 3.5 \text{ lb} = 0$$

Solving these last two equations simultaneously, and using $F_A = 76.2$ lb, gives the force of the biceps muscle, F_B, and the horizontal elbow force, F_C:

$$F_B = 92.9 \text{ lb}$$
$$F_C = 24.1 \text{ lb}$$

Example Problem 9.6

The force plate depicted in Fig. 9.9 has four sensors, one at each corner, that read the vertical forces F_1–F_4. If the plate is square, with side of length ℓ, and forces F_1–F_4 are known, write two expressions that will give the x and y locations of the resultant force R.

Solution

The resultant magnitude R can be computed from the sum of forces in the z direction:

$$\sum F_z = 0$$
$$F_1 + F_2 + F_3 + F_4 - R = 0$$
$$R = F_1 + F_2 + F_3 + F_4$$

The force plate remains horizontal; hence, the sum of the moments about the x and y axes must be zero. Taking moments about the x axis,

$$\sum M_x = 0$$
$$F_2\ell + F_3\ell - Ry = 0$$
$$y = \frac{(F_2 + F_3)\ell}{R}$$

Similarly, summing moments about the y axis,

$$\sum M_y = 0$$
$$F_1\ell + F_2\ell - Rx = 0$$
$$x = \frac{(F_1 + F_2)\ell}{R}$$

The coordinates x and y locate the resultant R.

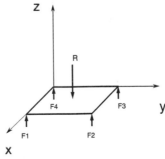

Fig. 9.9 A square force plate with sides of length ℓ is loaded with resultant force R and detects the vertical forces at each corner, F_1–F_4.

9.2.4   Anthropomorphic Mass Moments of Inertia

The resistance of a body (or a body segment such as a thigh in gait analysis) to rotation is quantified by the body or body segment's moment of inertia I:

$$I = \int_m r^2 dm \qquad (9.37)$$

where m is the body mass and r is the the moment arm to the axis of rotation. The incremental mass dm can be written as ρdV. For a body with constant density ρ the moment of inertia can be found by integrating over the body's volume V:

$$I = \rho \int_V r^2 dV \qquad (9.38)$$

This general expression can be written in terms of rotation about the x, y, and z axes:

$$I_{xx} = \int_V (y^2 + z^2)\rho dV$$

$$I_{yy} = \int_V (x^2 + z^2)\rho dV$$

$$I_{zz} = \int_V (x^2 + y^2)\rho dV \qquad (9.39)$$

The **radius of gyration**, r_g, is the moment arm between the axis of rotation and a single point where all the body's mass is concentrated. Consequently, a body segment may be treated as a point mass with moment of inertia,

$$I = mr_g^2 \qquad (9.40)$$

where m is the body segment mass. The moment of inertia with respect to a parallel axis I is related to the moment of inertia with respect to the body's center of mass I_{cm} via the **parallel axis theorem**:

$$I = I_{cm} + md^2 \qquad (9.41)$$

where d is the perpendicular distance between the two parallel axes. Anthropomorphic data for various body segments is listed in Table 9.1.

Example Problem 9.7

A 150-lb person has a thigh length of 17 in. Find the moment of inertia of this body segment with respect to its center of mass in SI units.

Solution

$$m_{body} = (150 \text{ lb})(0.454 \text{ kg/lb}) = 68.1 \text{ kg}$$

Using Table 9.1,

$$m_{thigh} = (0.100)(68.1 \text{ kg}) = 6.81 \text{ kg}$$

Using the given thigh segment length and the proximal segment data from Table 9.1,

$$\ell_{thigh} = 17 \text{ in.} = 0.432 \text{ m}$$

$$d = (0.432 \text{ m})(0.433) = 0.187 \text{ m}$$

TABLE 9.1 Anthropomorphic Data[a]

Segment	Definition	Segment weight/ body weight	Center mass/ segment length		Radius gyration/ segment length	
			Proximal	Distal	Proximal	Distal
Hand	Wrist axis/knuckle II middle finger	0.006	0.506	0.494	0.587	0.577
Forearm	Elbow axis/ulnar styloid	0.016	0.430	0.570	0.526	0.647
Upper arm	Glenohumeral axis/elbow axis	0.028	0.436	0.564	0.542	0.645
Forearm and hand	Elbow axis/ulnar styloid	0.022	0.682	0.318	0.827	0.565
Total arm	Glenohumeral joint/ulnar styloid	0.050	0.530	0.470	0.645	0.596
Foot	Lateral malleolus/head metatarsal II	0.0145	0.50	0.50	0.690	0.690
Leg	Femoral condyles/medial malleolus	0.0465	0.433	0.567	0.528	0.643
Thigh	Greater trochanter/femoral condyles	0.100	0.433	0.567	0.540	0.653
Foot and leg	Femoral condyles/medial malleolus	0.061	0.606	0.394	0.735	0.572
Total leg	Greater trochanter/medial malleolus	0.161	0.447	0.553	0.560	0.650
Head and neck	C7-T1 and 1st rib/ear canal	0.081	1.000		1.116	
Shoulder mass	Sternoclavicular joint/glenohumeral axis		0.712	0.288		
Thorax	C7-T1/T12-L1 and diaphragm	0.216	0.82	0.18		
Abdomen	T12-L1/L4-L5	0.139	0.44	0.56		
Pelvis	L4-L5/greater trochanter	0.142	0.105	0.895		
Thorax and abdomen	C7-T1/L4-L5	0.355	0.63	0.37		
Abdomen and pelvis	T12-L1/greater trochanter	0.281	0.27	0.73		
Trunk	Greater trochanter/glenuhumeral joint	0.497	0.50	0.50		
Trunk, head, neck	Greater trochanter/glenohumeral joint	0.578	0.66	0.34	0.830	0.607
Head, arm, trunk	Greater trochanter/glenohumeral joint	0.678	0.626	0.374	0.798	0.621

[a] Adapted from Winter (1990, Table 3.1, pp. 56-57).

The moment of inertia of the thigh with respect to the hip is

$$I_{thigh/hip} = mr_g^2 = (6.81 \text{ kg})[(0.54)(0.432 \text{ m})]^2 = 0.371 \text{ kg m}^2$$

The thigh moment of inertia with respect to the hip is related to that with respect to the thigh's center of mass:

$$I_{thigh/hip} = I_{thigh/cm} + md^2$$

so

$$I_{thigh/cm} = I_{thigh/hip} - md^2$$

$$I_{thigh/cm} = 0.371 \text{ kg m}^2 - (6.81 \text{ kg})(0.187 \text{ m})^2 = 0.133 \text{ kg m}^2 \qquad \blacksquare$$

9.2.5   Equations of Motion

Vector equations of motion are used to describe the translational and rotational kinetics of bodies.

Newton's Equations of Motion

Newton's second law relates the net force **F** and the resulting translational motion as

$$\mathbf{F} = m\ddot{\mathbf{r}}_c \qquad (9.42)$$

where $\ddot{\mathbf{r}}_c$ is the acceleration of the body's center of mass for translation. For rotation

$$\mathbf{M} = \dot{\mathbf{H}} \qquad (9.43)$$

where **H** is the body's angular momentum. Hence, the rate of change of a body's angular momentum is equal to the net moment **M** acting on the body. These two vector equations of motion are typically written as a set of six x, y, and z component equations.

Euler's Equations of Motion

Newton's equations of motion describe the motion of the center of mass of a body. More generally, Euler's equations of motion describe the motion of a rigid body with respect to its center of mass. For the special case in which the xyz coordinate axes are chosen to coincide with the body's principal axes,

$$\sum M_x = I_{xx}\alpha_x + (I_{zz} - I_{yy})\omega_y\omega_z \qquad (9.44)$$

$$\sum M_y = I_{yy}\alpha_y + (I_{xx} - I_{zz})\omega_z\omega_x \qquad (9.45)$$

$$\sum M_z = I_{zz}\alpha_z + (I_{yy} - I_{xx})\omega_x\omega_y \qquad (9.46)$$

M_i is the net moment, I_{ii} is the body's moment of inertia with respect to the principal axes, and α_i and ω_i are the body's angular acceleration and angular velocity, respectively. Euler's equations require angular measurements in radians. Their derivation is outside the scope of this chapter, but may be found in any intermediate dynamics book. Equations (9.44)–(9.46) will be used in Section 9.6 to compute body segment moments.

9.3 MECHANICS OF MATERIALS

Just as kinematic and kinetic relations may be applied to biological bodies to describe their motion and its associated forces, concepts from mechanics of materials may be used to quantify tissue deformation, to study distributed orthopedic forces, and to predict the performance of orthopedic implants and prostheses and of surgical corrections. Since this topic is very broad, some representative concepts will be illustrated with the following examples.

A bone plate is a flat segment of stainless steel used to screw two failed sections of bone together. The bone plate in Fig. 9.10 has a rectangular cross section, A, measuring 4.17×12 mm and made of 316L stainless steel. An applied axial load, F, of 500 N produces axial stress, σ, (force/area):

$$\sigma = \frac{F}{A} = 10 \text{ MPa} \tag{9.47}$$

The maximum shear stress, τ_{max}, occurs at a 45° angle to the applied load:

$$\tau_{max} = \frac{F_{45°}}{A_{45°}} \tag{9.48}$$

$$= \frac{(500 \text{ N}) \cos 45°}{\left[\dfrac{(0.00417 \text{ m})(0.012 \text{ m})}{\cos 45°} \right]} = 5 \text{ MPa}$$

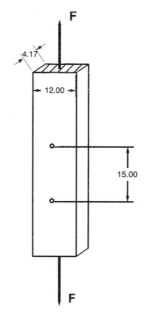

Fig. 9.10 Bone plate used to fix bone fractures, with applied axial load. Dimensions are in millimeters. Example adapted from Burstein and Wright (1994, pp. 104–108).

which is 0.5 σ, as expected. Prior to loading, two points were punched 15 mm apart on the long axis of the plate, as shown. After the 500 N load is applied, these marks are an additional 0.00075 mm apart. The plate's strain, $\in$, relates the change in length, $\Delta\ell$, to the original length, ℓ:

$$\in = \frac{\Delta\ell}{\ell} \tag{9.49}$$

$$= \frac{0.00075 \text{ mm}}{15 \text{ mm}} = 5 \times 10^{-5} = 50 \text{ μ (microstrain)}$$

The elastic modulus, E, relates stress and strain and is a measure of a material's resistance to distortion by a tensile or compressive load:

$$E = \frac{\sigma}{\in} \tag{9.50}$$

For linearly elastic (Hookean) materials, E is a constant, and a plot of σ as a function of $\in$ is a straight line with slope E. For the bone plate,

$$E = \frac{10 \times 10^6 \text{ Pa}}{50 \times 10^{-6}} = 200 \text{ GPa}$$

Materials such as metals and plastics display linearly elastic properties only in limited ranges of applied loads. Biomaterials have even more complex elastic properties. Figure 9.11 shows tensile stress–strain curves measured from longitudinal and transverse sections of bone. Taking the longitudinal curve first, from 0 to 7000 μ bone behaves as a purely elastic solid, with $E \approx 12$ GPa. At approximately 85 MPa, the stress–strain curve becomes nonlinear, yielding into the plastic region of defor-

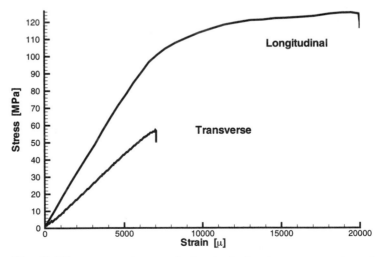

Fig. 9.11 Stress–strain curves for longitudinal and transverse sections of bone (adapted from Burstein and Wright, 1994, Fig. 4.12, p. 116).

mation. This sample ultimately fails at 120 MPa. Table 9.2 shows elastic moduli, yield stresses, and ultimate stresses for some common orthopedic materials, both natural and implant.

Figure 9.11 also shows that the elastic properties of bone differ depending on whether the sample is cut in the longitudinal or transverse direction, i.e., bone is anisotropic. Bone is much weaker and less stiff in the transverse compared to the longitudinal direction, as is illustrated by the difference in the yield and ultimate stresses and the slopes of the stress–strain curves for the two samples.

Figure 9.12 shows that the elastic properties of bone also vary depending on whether the load is being applied or removed, displaying hysteresis. From a thermodynamic view, the energy stored in the bone during loading is not equal to the energy released during unloading. This energy difference becomes greater as the maximum load increases (curves A to B to C). The "missing" energy is dissipated as heat due to internal friction and damage to the material at high loads.

The anisotropic nature of bone is sufficient that its ultimate stress in compression is 200 MPa while in tension it is only 140 MPa and in torsion 75 MPa. For torsional loading the **shear modulus** or **modulus of rigidity**, denoted G, relates the shear stress to the shear strain. The modulus of rigidity is related to the elastic modulus via Poisson's ratio, ν, where

$$\nu = \frac{\epsilon_{transverse}}{\epsilon_{longitudinal}} \tag{9.51}$$

Typically, $\nu \approx 0.3$, meaning that longitudinal deformation is three times greater than transverse deformation. E, G, and ν are related by

$$G = \frac{E}{2(1 + \nu)} \tag{9.52}$$

One additional complexity of predicting biomaterial failure is the complexity of physiological loading. For example, bone is much stronger in compression than in tension. This property is demonstrated in "boot-top" fractures in skiing. Since the foot is fixed, the skier's momentum causes a clockwise moment over the ski boot top

TABLE 9.2 Representative Yield and Ultimate Stresses and Elastic Moduli (E) for Some Common Orthopedic Materials[a]

Material	σ_{yield} (MPa)	$\sigma_{ultimate}$ (MPa)	E (GPa)
Stainless steel	700	850	180
Cobalt alloy	490	700	200
Titanium alloy	1100	1250	110
Bone	85	120	18
PMMA (fixative)		35	5
UHMWPE(bearing)	14	27	1
Patellar ligament		58	

[a] Data from Burstein and Wright (1994, Table 4.2, p. 122).

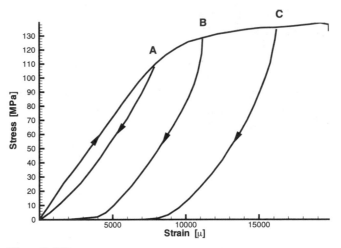

Fig. 9.12 Bone shows hysteresis and shifting of stress–strain curves with repeated loading and unloading (adapted from Burstein and Wright, 1994, Fig. 4.15, p. 119).

and produces three-point bending of the tibia. In this bending mode the anterior tibia undergoes compression, whereas the posterior is in tension and potentially in failure. Contraction of the triceps surae muscle produces high compressive stress at the posterior side, reducing the amount of bone tension. The following example shows how topics from statics and mechanics of materials may be applied to biomechanical problems.

Example Problem 9.8

Figure 9.13 (left) shows a nail plate used to fix an intertrochanteric fracture. The hip applies an external force of 400 N, as shown. The nail plate is rectangular stainless steel with cross-sectional dimensions of 5 × 10 mm, and the 135° angle is 6 cm from the 400 N load. What forces, moments, stresses, and strains will develop in this orthopedic structure?

Solution

As for any statics problem, the first task is constructing a free-body diagram, including all applied forces and moments and all reaction forces and moments that develop at the supports. If the nail plate is well fixed with screws along its vertical section (outside of bone) and is a friction fit into the trochanteric head, one reasonable description of fixity is a cantilever beam of length 0.06 m with a combined loading, as depicted in Fig. 9.13 (right, top). The applied 400 N load consists of axial and transverse components:

$$F_x = 400 \text{ N} \cos 20° = 376 \text{ N}$$
$$F_y = 400 \text{ N} \sin 20° = 137 \text{ N}$$

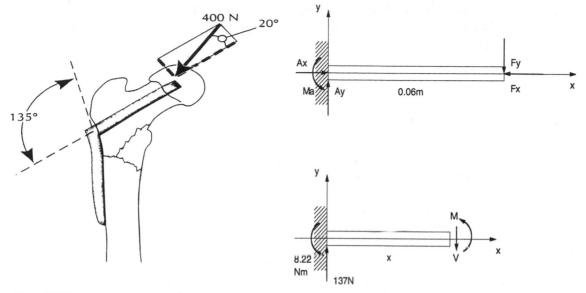

Fig. 9.13 (Left) An intertrochanteric nail plate used in bone repair (adapted from Burstein and Wright, 1994, Fig. 5.5, p. 141). (Right) The free-body diagram of the upper section of this device; (lower right) the free-body diagram of a section of this beam cut at a distance x from the left-hand support.

The axial load produces compressive normal stress:

$$\sigma_x = -F_x/A$$

$$= -\frac{376 \text{ N}}{(0.005 \text{ m})(0.01 \text{ m})} = -7.52 \text{ MPa}$$

The maximum shear stress due to the axial load is

$$\tau_{max} = \sigma_x/2 = -3.76 \text{ MPa}$$

and occurs at 45° from the long axis. The axial strain can be computed using the elastic modulus for stainless steel:

$$E = \frac{\sigma}{\in} = \frac{F/A}{\Delta\ell/\ell}$$

$$\in = \frac{F}{EA}$$

$$= \frac{376 \text{ N}}{180 \times 10^9 \text{ Pa}(0.005 \text{ m})(0.01 \text{ m})} = 4.18 \times 10^{-5}$$

From this strain the axial deformation can be computed:

$$\Delta\ell_{axial} = \in\ell = 2.51 \times 10^{-6} \text{ m}$$

which is negligible.

The transverse load causes the cantilever section to bend. The equations describing beam bending can be found in any mechanics of materials text. Beam tip deflection δy is equal to

$$\delta y = \frac{Fx^2}{6EI}(3L - x) \tag{9.53}$$

where x is the axial distance along the beam, L is the total beam length, and I is the beam's cross-sectional area moment of inertia. For this example the cross section is rectangular, so $I = 1/12\ bh^3$, with b and h denoting the beam's cross-sectional width and height, respectively, and maximum deflection will occur at $x = L$:

$$I = -\frac{1}{12}bh^3 \tag{9.54}$$

$$= 1.042 \times 10^{-10}\ \text{m}^4$$

$$\delta y_{\text{max}} = \frac{FL^3}{3EI} \tag{9.55}$$

$$= \frac{137\ \text{N}(0.06\ \text{m})^3}{3(180 \times 10^9\ \text{N/m}^2)(1.042 \times 10^{-10}\ \text{m}^4)}$$

$$= 5.26 \times 10^{-4}\ \text{m} = 0.526\ \text{mm}$$

which is also negligible.

Computation of maximum bending and shear stresses requires maximum shear force V and bending moment M. Starting by static analysis of the entire freebody:

$$\sum F_x : A_X = F_x = 376\ \text{N}$$

$$\sum F_y : A_y - 137\ \text{N} = 0$$

$$\sum M_A : M_a - 137\ \text{N}(0.06\ \text{m}) = 0$$

Solving these equations gives $A_x = 0$, $A_y = 137\,\text{N}$, and $M_a = 8.22\ \text{N m}$. Taking a cut at any point x to the right of A and isolating the left-hand section gives the freebody in Fig. 9.13 (right, bottom). Applying the equations of static equilibrium to this isolated section yields

$$\sum F_y :\ 137\ \text{N} - V = 0$$

$$V(x) = 137\ \text{N}$$

$$\sum M_A :\ 8.22\ \text{N m} - (137\ \text{N})(x\ \text{m}) + M = 0$$

$$M(x) = (137\ \text{N m})\ x - 8.22\ \text{N m}$$

These last two equations can be plotted easily using MATLAB, giving the shear stress and bending moment diagrams shown in Fig. 9.14.

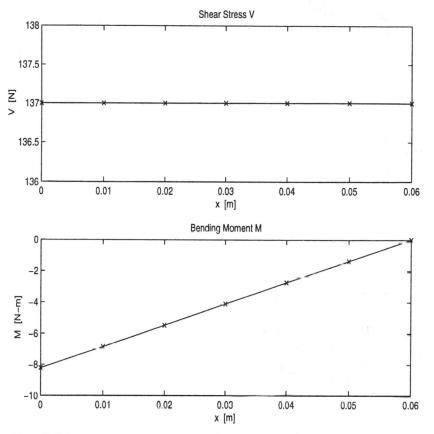

Fig. 9.14 Shear stress V (top) and bending moment M (bottom) computed for the nail plate in Fig. 9.13 as functions of the distance x along the plate.

```
% Use MATLAB to plot shear stress and bending moment diagrams
x = [0:0.01:0.06];
V = x.*0 + 137;
M = 137.*x - 8.22;

figure
subplot(2,1,1), plot(x,V,x,V,'x'), . . .
xlabel('x  [m]'), . . .
ylabel('V  [N]'), . . .
title('Shear Stress V'), . . .
subplot(2,1,2), plot(x,M,x,M,'x'), . . .
xlabel('x  [m]'), . . .
ylabel('M  [N-m]'), . . .
title('Bending Moment M')
```

The maximum bending and shear stresses follow as

$$\sigma_{b\text{max}} = \frac{M_{\text{max}}c}{I} \qquad (9.56)$$

where c, the distance to the beam's neutral axis, is $h/2$:

$$\sigma_{b\text{max}} = \frac{-8.22 \text{ Nm}[0.5(5 \times 10^{-3} \text{ m})]}{1.042 \times 10^{-10} \text{ m}^4} = -197 \text{ MPa}$$

$$\tau_{b\text{max}} = \frac{V_{\text{max}}h^2}{8I} \qquad (9.57)$$

$$= \frac{137 \text{ N}(5 \times 10^{-3} \text{ m})^2}{8(1.042 \times 10^{-10} \text{ m}^4)} = 4.11 \text{MPa}$$

All these stresses are well below σ_{yield} for stainless steel (700 MPa).   ■

9.4   VISCOELASTIC PROPERTIES

The Hookean elastic solid is a valid description of materials only within a narrow loading range. For example, an ideal spring that relates force and elongation by a spring constant k is invalid in nonlinear low-load and high-load regions. Furthermore, if this spring is coupled to a mass and set into motion, the resulting perfect harmonic oscillator will vibrate forever, which experience shows is not valid. Missing is a description of the system's viscous or damping properties. In this case, energy is dissipated as heat in the spring and air friction on the moving system.

Similarly, biomaterials all display viscoelastic properties. Different models of viscoelasticity have been developed to characterize materials with simple constitutive equations. For example, Fig. 9.15 shows three such models that consist of a series ideal spring and dashpot, a parallel spring and dashpot, and a series spring and dashpot with a parallel spring. Each body contains a dashpot, which generates force in proportion to the derivative of its elongation. Consequently, the resulting models exhibit stress and strain properties that vary in time.

The dynamic response of each model can be quantified by applying step changes in force F and noting the models' resulting change in length, or position x, denoted the **creep** response. The converse experiment applies a step change in x and measures the resulting change in F, denoted **stress relaxation**. Creep and stress relaxation tests for each dynamic model can be easily carried out using the Simulink program. Figure 9.16 shows a purely elastic material subjected to a step change in applied force F. The material's subsequent position x follows the change in force directly. This material exhibits no creep. Figure 9.17 shows the purely elastic material subjected to a step change in position x. Again, the material responds immediately with a step change in F, i.e., no stress relaxation is observed.

James Clerk Maxwell (1831–1879) used a series combination of ideal spring and dashpot to describe the viscoelastic properties of air. Figure 9.18 shows the Maxwell

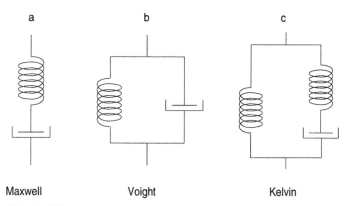

Fig. 9.15 Three simple viscoelastic models: (a) the Maxwell model, (b) the Voight model, and (c) the Kelvin body or standard linear solid model.

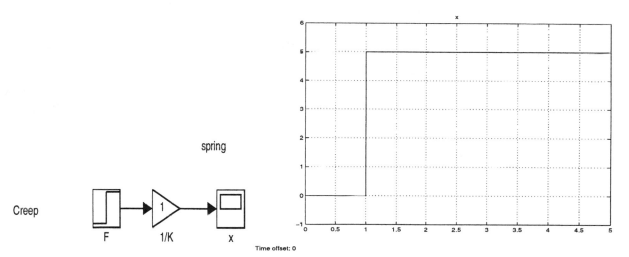

Fig. 9.16 (Left) Simulink model of the creep test for a purely elastic material (simple spring); (Right) The elastic creep response to a step increase in applied force F.

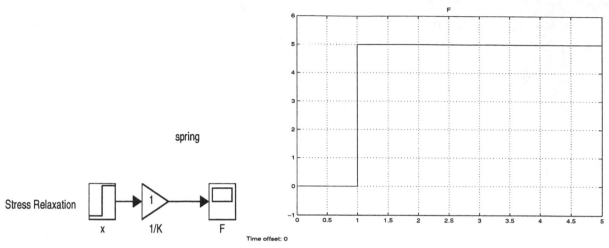

Fig. 9.17 (Left) Simulink model and (Right) stress relaxation test of the purely elastic model (simple spring).

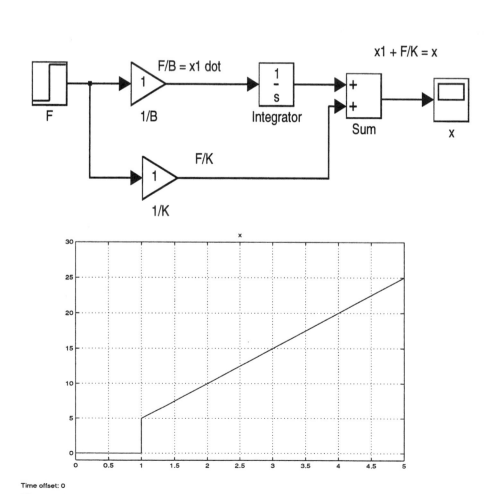

Time offset: 0

Fig. 9.18 Creep of the Maxwell viscoelastic model (series spring and dashpot).

viscoelastic model subjected to a step change in applied force, and Figure 9.19 shows the Maxwell model's stress relaxation response. The latter exhibits an initial high stress followed by stress relaxation back to the initial stress level. The creep response, however, shows that this model is not bounded in displacement since a perfect dashpot may be extended forever.

Woldemar Voight (1850–1919) used the parallel combination of an ideal spring and dashpot in his work with crystallography. Figure 9.20 shows the creep test of the Voight viscoelastic model. Figure 9.21 shows that this model is unbounded in force. When a step change in length is applied, force goes to infinity since the dashpot cannot immediately respond to the length change.

William Thompson (Lord Kelvin, 1824–1907) used the three-element viscoelastic model (see Fig. 9.15c) to describe the mechanical properties of different solids in the form of a torsional pendulum. Figure 9.22 shows the three-element Kelvin

Stress Relaxation

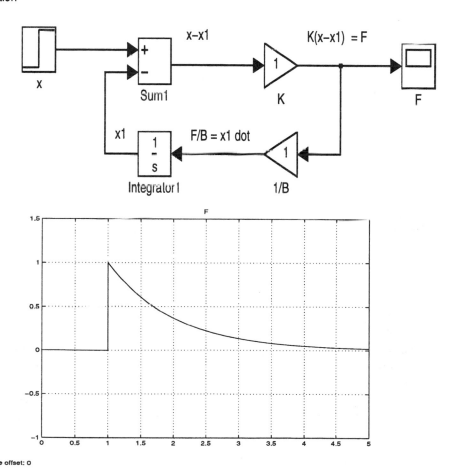

Time offset: 0

Fig. 9.19 Stress relaxation of the Maxwell viscoelastic model (series spring and dashpot).

Creep

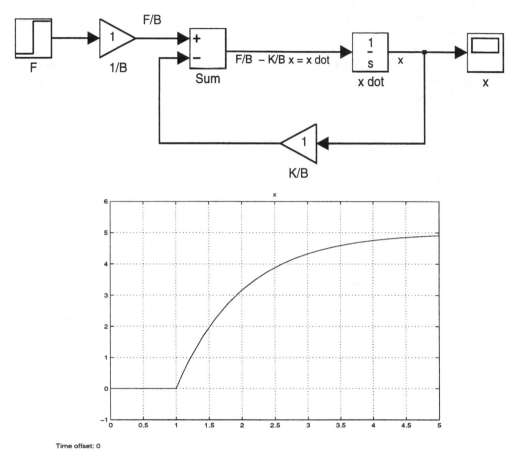

Fig. 9.20 Creep of the Voight viscoelastic model (parallel spring and dashpot).

model's creep response. This model has an initial rapid jump in position with subsequent slow creep. Figure 9.23 shows the Kelvin model stress relaxation test. Initially, the material is very stiff with subsequent stress decay to a nonzero steady-state level that is due to the extension of the dashpot. The three-element Kelvin model is the simplest lumped viscoelastic model that is bounded both in extension and force.

The three-element viscoelastic model describes the basic features of stress relaxation and creep. Biological materials often exhibit more complex viscoelastic properties. For example, plotting hysteresis as a function of frequency of strain gives discrete curves for the lumped viscoelastic models. Biological tissues demonstrate broad, distributed hysteresis properties. One solution is to describe biomaterials with a distributed network of three-element models. A second method is to use the

Stress Relaxation

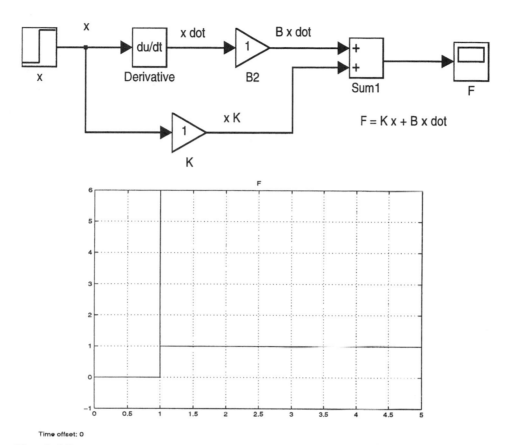

Fig. 9.21 Stress relaxation of the Voight viscoelastic model (parallel spring and dashpot).

generalized viscoelastic model of Westerhof and Noordergraaf (1970) to describe the viscoelastic wall properties of blood vessels. Making the elastic modulus (mathematically) complex yields a model that includes the frequency-dependent elastic modulus, stress relaxation, creep, and hysteresis exhibited by arteries. Furthermore, the Voight and Maxwell models emerge as special (limited) cases of this general approach.

9.5 CARTILAGE, LIGAMENT, TENDON, AND MUSCLE

The articulating surfaces of bones are covered with articular cartilage, a biomaterial composed mainly of collagen. Isolated collagen fibers have high tensile strength that is comparable to nylon (50–100 MPa) and an elastic modulus of approximately

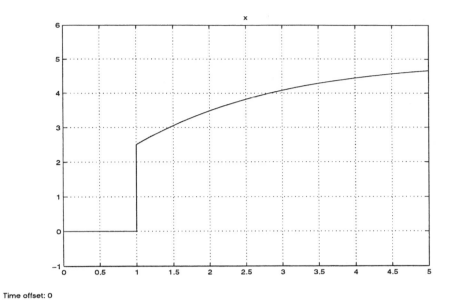

Time offset: 0

Fig. 9.22 Creep of the Kelvin three-element viscoelastic model.

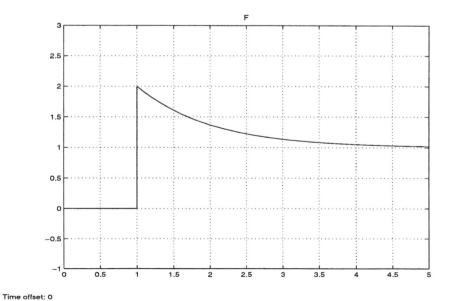

Time offset: 0

Fig. 9.23 Stress relaxation of the Kelvin viscoelastic model.

1 GPa. Elastin is a protein found in vertebrates and is particularly important in blood vessels and the lungs. Elastin is the most linearly elastic biosolid known, with an elastic modulus of approximately 0.6 MPa. It gives skin and connective tissue their elasticity. Collagen is the main structural material of hard and soft tissues in animals.

9.5.1 Cartilage

Cartilage serves as the bearing surfaces of joints. It is porous and its complex mechanical properties arise from the motion of fluid in and out of the tissue when subjected to joint loading. Consequently, articular cartilage is strongly viscoelastic with stress relaxation times in compression on the order of 1–5 s. Cartilage is anisotropic and displays hysteresis during cyclical loading. Ultimate compressive stress of cartilage is on the order of 5 MPa.

9.5.2 Ligaments and Tendons

Ligaments join bones together and consequently serve as part of the skeletal framework. Tendons join muscles to bones and transmit forces generated by contracting muscles to cause movement of the jointed limbs. Tendons and ligaments primarily transmit tension; hence, they are composed mainly of parallel bundles of collagen fibers and have similar mechanical properties. Human tendon has an ultimate stress of 50–100 MPa and exhibits very nonlinear stress–strain curves. The middle stress–strain range is linear with an elastic modulus of approximately 1 or 2 GPa. Both tendons and ligaments exhibit hysteresis, viscoelastic creep, and stress relaxation. These materials may also be "preconditioned," whereby initial tensile loading can affect subsequent load-deformation curves. The material properties shift due to changes in the internal tissue structure with repeated loading.

9.5.3 Muscle Mechanics

Chapter 2 introduced muscle as an active, excitable tissue that generates force by forming crossbridge bonds between the interdigitating actin and myosin myofilaments. The quantitative description of muscle contraction has evolved into two separate foci — lumped descriptions based on A. V. Hill's contractile element and crossbridge models based on A. F. Huxley's description of a single sarcomere. The earliest quantitative descriptions of muscle are lumped whole muscle models with the simplest mechanical description being a purely elastic spring. Potential energy is stored when the spring is stretched, and shortening occurs when it is released. The idea of muscle elastance can be traced back to Ernst Weber (1846), who considered muscle as an elastic material that changes state during activation via conversion of chemical energy. Subsequently, investigators retained the elastic description but ignored metabolic alteration of muscle stiffness. A purely elastic model of muscle can be refuted on thermodynamic grounds since the potential energy stored during stretching is less than the sum of the energy released during shortening as work and heat. Still, efforts to describe muscle by a combination of traditional springs and dashpots continued. In 1922, Hill coupled the spring with a viscous medium, thereby reintroducing viscoelastic muscle descriptions that can be traced back to the 1840s.

Quick stretch and release experiments show that muscle's viscoelastic properties are strongly time dependent. In general, the faster a change in muscle length occurs, the more severely the contractile force is disturbed. Muscle contraction clearly arises from a more sophisticated mechanism than a damped elastic spring. In 1935, Fenn and Marsh added a series elastic element to Hill's damped elastic model and concluded that "muscle cannot properly be treated as a simple mechanical system." Subsequently, Hill embodied the empirical hyperbolic relation between load and initial velocity of shortening for skeletal muscle as a model building block, denoted the contractile element. Hill's previous viscoelastic model considered muscle to possess a fixed amount of potential energy whose rate of release is controlled by viscosity. Energy is now thought to be controlled by some undefined internal mechanism rather than by friction. This new feature of muscle dynamics varying with load was a step in the right direction; however, subsequent models, including heart studies, built models based essentially on this hyperbolic curve that was measured for tetanized skeletal muscle. This approach can be criticized on two grounds: (i) Embodiment of the contractile element by a single force-velocity relation sets a single, fixed relation between muscle energetics and force, and (ii) it yields no information on the contractile mechanism behind this relation. Failure of the contractile element to describe a particular loading condition led investigators to add passive springs and dashpots liberally with the number of elements reaching at least nine by the late 1960s. Distributed models of muscle contraction to date have been conservative in design and have depended fundamentally on the Hill contractile element. Recent models are limited to tetanized, isometric contractions or to isometric twitch contractions.

A second, independent focus of muscle contraction research works at the ultrastructural level with the sliding filament theory serving as the most widely accepted contraction mechanism. Muscle force generation is viewed as the result of crossbridge bonds formed between thick and thin filaments at the expense of biochemical energy. The details of bond formation and detachment are under considerable debate, with the mechanism for relaxation particularly uncertain. Prior to actual observation of crossbridges, A. F. Huxley (1957) devised the crossbridge model based on structural and energetic assumptions. Bonds between myofilaments are controlled via rate constants f and g that dictate attachment and detachment, respectively. One major shortcoming of this idea was the inability to describe transients resulting from rapid changes in muscle length or load, similar to the creep and stress relaxation tests previously discussed.

Subsequent models adopt increasingly complex bond attachment and detachment rate functions and are often limited in scope to description of a single pair of myofilaments. Each tends to focus on description of a single type of experiment, e.g., quick release. No model has been shown to broadly describe all types of contractile loading conditions. Crossbridge models have tended to rely on increasingly complex bond attachment and detachment rate functions. This trend has reversed the issue of describing complex muscle dynamics from the underlying (simpler) crossbridges to adopting complex crossbridge dynamics to describe a particular experiment.

Alternatively, Palladino and Noordergraaf (1998) recently proposed a large-scale, distributed muscle model that manifests both contraction and relaxation as the result of fundamental mechanical properties of crossbridge bonds. As such, muscle's complex contractile properties emerge from the underlying ultrastructure dynamics, i.e., function follows from structure. Bonds between myofilaments, which are biomaterials, are described as viscoelastic material. The initial stimulus for contraction is electrical. Electrical propagation through cardiac muscle occurs at finite speed, implying spatial asynchrony of stimulation. Furthermore, Ca^{2+} release from the sarcoplasmic reticulum depends on diffusion for availability at the myosin heads. These effects, as well as nonuniformity of structure, strongly suggest that contraction is asynchronous throughout the muscle. Recognition of muscle's distributed properties by abandoning the assumption of perfect synchrony in contraction and consideration of myofilament mass allow for small movements of thick with respect to thin filaments. Such movements lead to bond detachment and heat production. Gross movement, e.g., muscle shortening, exacerbates this process. Quick transients in muscle length or applied load have particularly strong effects and have been observed experimentally. Muscle relaxation is thereby viewed as a consequence of muscle's distributed properties.

The new distributed muscle model is built from the following main features: Sarcomeres consist of overlapping thick and thin filaments connected by crossbridge bonds which form during activation and detach during relaxation. Figure 9.24 shows a schematic of a muscle fiber (cell) composed of a string of series sarcomeres. Crossbridge bonds are each described as three-element viscoelastic solids and myofilaments as masses. Force is generated due to viscoelastic crossbridge bonds that form and are stretched between the interdigitating matrix of myofilaments. The number of bonds formed depends on the degree of overlap between thick and thin filaments and is dictated spatially and temporally due to finite electrical and chemical activation rates. Asynchrony in bond formation and unequal numbers of bonds formed in each half sarcomere, as well as mechanical disturbances such as muscle shortening and imposed length transients, cause small movements of the myofilaments. Since myofilament masses are taken into account, these movements take the form of damped vibrations with a spectrum of frequencies due to the distributed system properties. When the stress in a bond goes to zero, the bond detaches. Consequently, myofilament motion and bond stress relaxation lead to bond detachment and produce relaxation without assumption of bond detachment rate functions. In essence, relaxation results from inherent system instability. Although the model is built from linear, time-invariant components (springs, dashpots, and masses), the highly dynamic structure of the model causes its mechanical properties to be highly nonlinear and time varying, as is found in muscle fibers and strips.

Sensitivity of the model to mechanical disturbances is consistent with experimental evidence from muscle force traces, aequorin measurements of free calcium ion, and high speed X-ray diffraction studies that all suggest enhanced bond detachment. The model is also consistent with sarcomere length feedback studies in which reduced internal motion delays muscle relaxation.

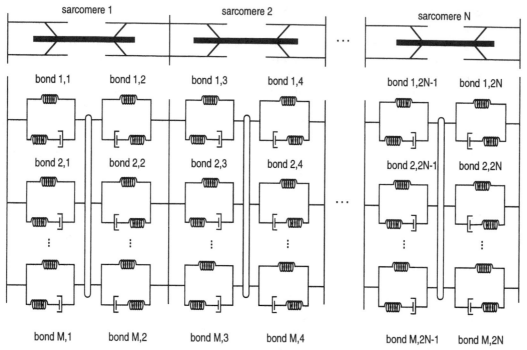

Fig. 9.24 Schematic diagram of a muscle fiber built from a distributed network of N sarcomeres. Each sarcomere has M parallel pairs of crossbridge bonds (adapted from Palladino and Noordergraaf, 1998, Fig. 3.4, p. 44).

This model proposes a structural mechanism for the origin of muscle's complex mechanical properties and predicts new features of the contractile mechanism, e.g., a mechanism for muscle relaxation and prediction of muscle heat generation. This new approach computes muscle's complex mechanical properties from physical description of muscle anatomical structure, thereby linking subcellular structure to organ-level function.

This section describes some of the high points of biological tissues' mechanical properties. More comprehensive references include Fung's *Biomechanics: Mechanical Properties of Living Tissues* (1993), Nigg and Herzog's *Biomechanics of the Musculo-Skeletal System* (1994), and Mow and Hayes' *Basic Orthopaedic Biomechanics* (1997).

9.6 CLINICAL GAIT ANALYSIS

Clinical gait analysis involves the measurement of the parameters that characterize a patient's gait pattern, the interpretation of the collected and processed data, and the recommendation of treatment alternatives. It is a highly collaborative process that requires the cooperation of the patient and the expertise of a multidisciplinary team that typically includes a physician, a physical therapist or kinesiologist, and an engineer or technician. The engineer is presented with a number of challenges with respect to clinical gait analysis. The fundamental objective in data collection is to monitor the patient's movements accurately and with sufficient precision for clinical use without altering the patient's typical performance. While measurement devices for clinical gait analysis are established to some degree, i.e., commercially available, the protocols for the use of the equipment are still under development. The validity of these protocols and associated models and the care with which they are applied ultimately dictate the meaning and quality of the resulting data provided for interpretation. This is one area in which engineers in collaboration with their clinical partners can have a significant impact on the clinical gait analysis process.

Generally, data collection for clinical gait analysis involves the placement of markers or targets on the surface of the patient's skin. These external markers then either emit or reflect infrared light to an array of video cameras that surround the measurement volume. The instantaneous location of each of these markers can then be determined stereometrically based on the images obtained simultaneously from two or more cameras. Other aspects of gait can be monitored as well, including ground reactions via force platforms embedded in the walkway and muscle activity via electromyography with either surface or intramuscular or fine wire electrodes, depending on the location of the particular muscle. Increasingly, metabolic energy, e.g., oxygen consumption or carbon dioxide production, is also assessed to reflect a global measure of gait efficiency.

In keeping with the other material presented in this chapter, this section will focus on the biomechanical aspects of clinical gait analysis and includes an outline of the computation of segmental and joint kinematics and joint kinetics and a brief illustration of how the data are interpreted.

9.6.1 The Clinical Gait Model

The gait model is the algorithm that connects the data collected during walking trials and the information required for clinical interpretation. Its design is predicated on a clear understanding of the needs of the clinical interpretation team, e.g., the specific aspects of gait dynamics of interest. To meet these clinical specifications, the gait model development is constrained both by the technical limitations of the measurement system and by the broad goal of developing protocols that may be appropriate for a wide range of patient populations that vary in age, gait abnormality, walking ability, etc. An acceptable model must be sufficiently general to be used for many different types of patients (e.g., adults and children with varying physical and cognitive involvement), be sufficiently sophisticated to allow detailed biomechanical questions to be addressed, and incorporate data-collection protocols that can be executed by the clinical staff.

9.6.2 Kinematic Data Analysis

The placement of three noncolinear markers or points of reference on a body segment allows the tracking of the spatial orientation of that segment through space. These markers form a plane from which a segmentally fixed coordinate system may be derived. Any three markers will allow the segment motion to be monitored, but unless these markers are referenced to the subject's anatomy, such kinematic quantification is of limited clinical value. Markers must be placed either directly over palpable bony landmarks on the segment or at convenient (i.e., visible to the measurement cameras) locations on the segment that are referenced to the underlying bone(s). An examination of the pelvic and thigh segments illustrates these two alternatives.

Motion of the Pelvis

For the pelvis, markers placed over the right and left anterior–superior–iliac–spine (ASIS) and either the right or left posterior–superior–iliac–spine (PSIS) will allow for the computation of an anatomically referenced segmental coordinate system, as described in the following problem.

Example Problem 9.9

Given the following three dimensional locations, in meters, for a set of pelvic markers expressed relative to an inertially fixed, laboratory-based coordinate system (x_{lab}, y_{lab}, z_{lab}) (Fig. 9.25),

$$\text{Right ASIS: } \mathbf{RASIS} = -0.850\mathbf{i} - 0.802\mathbf{j} + 0.652\mathbf{k}$$

$$\text{Left ASIS: } \mathbf{LASIS} = -0.831\mathbf{i} - 0.651\mathbf{j} + 0.652\mathbf{k}$$

$$\mathbf{PSIS} = -1.015\mathbf{i} - 0.704\mathbf{j} + 0.686\mathbf{k}$$

compute an anatomical coordinate system for the pelvis.

Solution

1. Subtract vector **RASIS** from vector **LASIS** to find

$$\mathbf{r}_1 = 0.0190\mathbf{i} + 0.1510\mathbf{j} + 0.0000\mathbf{k}$$

and the associated unit vector and y axis for the pelvis of

$$\mathbf{e}_{r1} = \mathbf{e}_{pay} = 0.1248\mathbf{i} + 0.9922\mathbf{j} + 0.0000\mathbf{k}$$

2. Subtract vector **RASIS** from vector **PSIS** to find

$$\mathbf{r}_2 = -0.1650\mathbf{i} + 0.0980\mathbf{j} + 0.0340\mathbf{k}$$

3. Take the cross product $\mathbf{e}_{pay} \times \mathbf{r}_2$ to yield

$$\mathbf{r}_3 = 0.0337\mathbf{i} - 0.0042\mathbf{j} + 0.1759\mathbf{k}$$

with the associated unit vector that defines a z axis for the pelvis, $\mathbf{e}_{paz}$:

$$\mathbf{e}_{r3} = \mathbf{e}_{paz} = 0.1882\mathbf{i} - 0.0237\mathbf{j} + 0.9818\mathbf{k}$$

4. Take the cross product $\mathbf{e}_{pay} \times \mathbf{e}_{paz}$ to yield

$$\mathbf{e}_{pax} = 0.9742\mathbf{i} - 0.1226\mathbf{j} - 0.1897\mathbf{k}$$

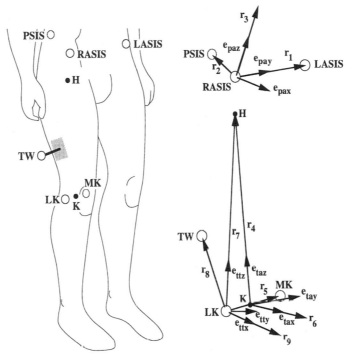

Fig. 9.25 Kinematic markers used to define pelvis and thigh coordinate systems. For the pelvis, PSIS denotes posterior–superior iliac–spine, H, is hip center, and RASIS and LASIS denote right and left anterior–superior iliac spine markers, respectively. For the thigh, TW is thigh wand, K is knee center, and MK and LK are medial and lateral knee (femoral condyle) markers, respectively.

By monitoring the motion of the three pelvic markers, the instantaneous orientation of an anatomically referenced segmental coordinate system for the pelvis, $\{e_{pa}\}$, composed of vectors e_{pax}, e_{pay}, and e_{paz}, can be determined. The absolute angular displacement of this coordinate system can then be computed via Euler angles as pelvic tilt, obliquity, and rotation. An example of these angle computations is presented later in this section. ■

Motion of the Thigh

The thigh presents a more significant challenge than the pelvis since three bony landmarks are not readily available as reference points during gait. A model based on markers placed over the medial and lateral femoral condyles and the greater trochanter is appealing but plagued with difficulties. A marker placed over the medial femoral condyle is not always feasible during gait, e.g., with patients whose knees make contact while walking. A marker placed over the greater trochanter is often described in the literature but should not be used as a reference because of its significant movement relative to the underlying greater trochanter during gait (skin motion artifact).

In general, the approach used to quantify thigh motion (and the shank and foot) is to place additional anatomical markers on the segment(s) during a static subject calibration process so that the relationship between these static anatomical markers (that are removed before gait data collection) and the motion markers (that remain on the patient during gait data collection) may be calculated. It is assumed that this mathematical relationship remains constant during gait, i.e., the instrumented body segments are assumed to be rigid.

Example Problem 9.10

Given the following marker coordinate data that have been acquired while the patient stands quietly,

$$\text{Lateral femoral condyle marker } \mathbf{LK} = -0.881\mathbf{i} - 0.858\mathbf{j} + 0.325\mathbf{k}$$

$$\text{Medial femoral condyle marker } \mathbf{MK} = -0.855\mathbf{i} - 0.767\mathbf{j} + 0.318\mathbf{k}$$

compute an anatomical reference system for the thigh.

Solution

The location of the knee center of rotation may be approximated as the midpoint between the medial and lateral femoral condyle markers, giving

$$\text{Knee center location } \mathbf{K} = -0.868\mathbf{i} - 0.812\mathbf{j} + 0.321\mathbf{k}$$

The location of the center of the head of the femur, referred to as the hip center, is commonly used in this calculation by approximating its location based on patient anthropometry and a statistical model of pelvic geometry. In this case, it is located at approximately

$$\text{Hip center location } \mathbf{H} = -0.906\mathbf{i} - 0.763\mathbf{j} + 0.593\mathbf{k}$$

An anatomical coordinate system for the thigh may then be computed as follows:

1. Subtract the vector $\mathbf{K}$ from $\mathbf{H}$, giving

$$\mathbf{r_4} = -0.038\mathbf{i} + 0.049\mathbf{j} + 0.272\mathbf{k}$$

and then determine the longitudinal axis of the thigh via the unit vector:

$$\mathbf{e_{r4}} = \mathbf{e_{taz}} = -0.137\mathbf{i} + 0.175\mathbf{j} + 0.975\mathbf{k}$$

2. Subtract $\mathbf{LK}$ from $\mathbf{MK}$:

$$\mathbf{r_5} = 0.026\mathbf{i} + 0.091\mathbf{j} - 0.007\mathbf{k}$$

3. Form the cross product $\mathbf{r_5} \times \mathbf{e_{taz}}$ to yield

$$\mathbf{r_6} = 0.090\mathbf{i} - 0.024\mathbf{j} + 0.017\mathbf{k}$$

and the associated unit vector, the fore–aft coordinate axis for the thigh:

$$\mathbf{e_{r6}} = \mathbf{e_{tax}} = 0.949\mathbf{i} - 0.258\mathbf{j} + 0.180\mathbf{k}$$

4. Determine the medial–lateral axis of the thigh, e_{tay}, from the cross product $e_{taz} \times e_{tax}$:

$$e_{tay} = 0.284i + 0.950j - 0.131k$$

This defines an anatomical reference fixed to the thigh, $\{e_{ta}\}$, composed of vectors e_{tax}, e_{tay}, and e_{taz}, but its basis includes an external marker (medial femoral condyle, MK) that must be removed before the walking trials. This dilemma is resolved by placing another marker on the surface of the thigh such that it forms a plane with the hip center and lateral knee marker. These three reference points can then be used to compute a "technical" coordinate system for the thigh to which the knee center location may be referenced. ∎

Example Problem 9.11

Continuing Example Problem 9.10 and given the coordinates of another marker placed on the thigh, but not anatomically aligned,

$$\text{Thigh wand marker } TW = -0.890i - 0.937j + 0.478k$$

compute a technical coordinate system for the thigh.

Solution

A technical coordinate system for the thigh can be computed as follows:

1. Subtract vector **LK** from **H** to form

$$r_7 = -0.025i + 0.094j + 0.268k$$

and the associated unit vector

$$e_{r7} = e_{ttz} = -0.088i + 0.330j + 0.940k$$

2. Subtract **LK** from **TW** to compute

$$r_8 = -0.009i - 0.079j + 0.153k$$

3. Calculate the cross product $r_7 \times r_8$ to yield

$$r_9 = 0.036i + 0.001j + 0.003k$$

with the associated unit vector

$$e_{r9} = e_{ttx} = 0.996i + 0.040j + 0.079k$$

4. Compute the cross product $e_{ttz} \times e_{ttx}$ to yield

$$e_{tty} = -0.012i + 0.943j + 0.333k$$

To reiterate, the result of this subject calibration for the thigh is an anatomical co-ordinate system $\{e_{ta}\}$ and a technical coordinate system $\{e_{tt}\}$ that can be mathematically related if the segment is assumed rigid. That is, the anatomical coordinate system may now be recreated from the technical coordinate system by using

- the three Euler angles relating $\{e_{ta}\}$ to $\{e_{tt}\}$, or
- the nine direction cosines associated with $\{e_{ta}\}$ relative to $\{e_{tt}\}$, or

■ the location of the anatomical points required to reconstruct $\{e_{ta}\}$ expressed in terms of $\{e_{tt}\}$ ■

Application to the Knee

Continuing Example Problem 9.11 with the origin of $\{e_{tt}\}$ at

$$\text{Lateral knee marker } \mathbf{LK} = -0.881\mathbf{i} - 0.858\mathbf{j} + 0.325\mathbf{k}$$

the location of the required anatomical reference (knee center) required to compute $\{e_{ta}\}$ in terms of the $\{e_{tt}\}$ coordinate directions can be expressed as

$$\text{Knee center } \mathbf{K_{tt}} = 0.015\mathbf{i} + 0.044\mathbf{j} + 0.011\mathbf{k}$$

In each frame of motion data for the walking trial, the technical coordinate system of the thigh may be computed from the locations of the hip center (in laboratory coordinates), the thigh wand marker and the lateral knee marker. After transforming the locations of the knee anatomical centers (stored in terms of $\{e_{tt}\}$) back into the laboratory frame of reference, the anatomical coordinate system of the thigh may be computed from the locations of the hip center, the lateral knee marker and the knee center.

Tracking the anatomical coordinate system for each segment allows for the determination of either the absolute angular orientation (or attitude) of each segment in space or the angular position of one segment relative to another. In the example presented previously, the three pelvic angles that define the position of the pelvic coordinate system $\{e_{pa}\}$ composed of e_{pax}, e_{pay}, and e_{paz}, relative to the laboratory (inertially fixed) coordinate system, can be computed from the Euler angles. For example, using Eqs. (9.32)–(9.34), with a y–x–z rotation sequence, and where the proximal, unprimed coordinate system is the laboratory and the distal, triple-primed coordinate system is the pelvis, gives the angles of pelvic tilt, obliquity, and rotation as follows. Equation (9.32) can be solved for θ_x, pelvic obliquity:

$$\theta_x = -\arcsin(\mathbf{k'''} \cdot \mathbf{j})$$

$$= -\arcsin(e_{paz} \cdot \mathbf{j})$$

$$= -\arcsin(-0.0237) = 1.35°$$

Similarly, Eqs. (9.33) and (9.34) are solved for pelvic tilt, θ_y, and pelvic internal–external rotation, θ_z, respectively:

Pelvic tilt θ_y	11°
Pelvic obliquity θ_x	1°
Pelvic (internal–external) rotation θ_z	−7°

As another example, the three hip angles that define the position of the thigh anatomical coordinate system $\{e_{ta}\}$ relative to $\{e_{pa}\}$ are similarly computed, yielding

Hip flexion–extension θ_y	20°
Hip abduction–adduction θ_x	−9°
Hip internal–external rotation θ_z	−8°

In this hip example, the unprimed coordinate system is the pelvis and the triple-primed system is the thigh. This process may be repeated for other body segments, such as the shank (or lower leg), foot, trunk, arms, and head, with the availability of properly defined anatomical coordinate systems.

9.6.3 Kinetic Data Analysis

The marker displacement or motion data provide an opportunity to appreciate segment and joint kinematics. Kinematic data can be combined with ground reaction data, i.e., forces and torque, and their points of application, referred to as the centers of pressure. Combined with estimates of segment mass and mass moments of inertia, the net joint reactions, i.e., joint forces and moments, may then be computed. To illustrate the computation of the reactions at the ankle (Fig. 9.26), consider the following patient with mass of 25.2 kg. Data for one instant in the gait cycle are as follows:

	Symbol	Units	x_{lab}	y_{lab}	z_{lab}
Ankle center location	A	m	0.357	0.823	0.056
Toe marker location	T	m	0.421	0.819	0.051
Center of pressure location	CP	m	0.422	0.816	0.000
Ground reaction force vector	F_g	N	3.94	−15.21	242.36
Ground reaction torque vector	T_g	N m	0.000	0.000	0.995
Foot anatomical coordinate system	e_{fax}		0.977	−0.0624	−0.202
	e_{fay}		0.0815	0.993	0.0877
	e_{faz}		0.195	−0.102	0.975
Foot linear acceleration vector	a_{foot}	m/s^2	2.09	−0.357	−0.266
Foot angular velocity vector	ω_{foot}	rad/s	0.0420	2.22	−0.585
Foot angular acceleration vector	α_{foot}	rad/s^2	−0.937	8.85	−5.16
Ankle angular velocity vector	ω_{ankle}	rad/s	−0.000759	1.47	0.0106

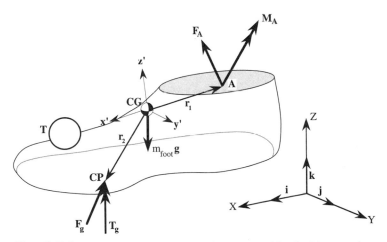

Fig. 9.26 Ankle A and toe T marker data are combined with ground reaction force data **F$_g$** and segment mass and mass moment of inertia estimates to compute the net joint forces and moments.

The mass of the foot, m_{foot}, may be estimated to be 1.45% of the body mass, or 0.365 kg. The location of the center of gravity is approximated as 50% of the foot length and is located, in this case, at

$$\text{Center of gravity } \mathbf{CG} = 0.389\mathbf{i} + 0.821\mathbf{j} + 0.054\mathbf{k}$$

With a foot length of 0.064 m and a centroidal radius of gyration per segment length of 0.475, the centroidal mass moment of inertia may be estimated as

$$I_{x'x'} = 3.31 \times 10^{-5} \text{ kg m}^2$$
$$I_{y'y'} = 3.41 \times 10^{-4} \text{ kg m}^2$$
$$I_{z'z'} = 3.41 \times 10^{-4} \text{ kg m}^2$$

where x', y', and z' approximate the principal axes of the foot.

The unknown ankle reaction force, $\mathbf{F_A}$, may then be found by using Newton's second law, or $\Sigma\mathbf{F} = m\mathbf{a}$:

$$\mathbf{F_g} + \mathbf{F_A} - m_{foot}\, g\mathbf{k} = m_{foot}\mathbf{a_{foot}}$$
$$\mathbf{F_A} = m_{foot}\mathbf{a_{foot}} - \mathbf{F_g} + m_{foot}\, g\mathbf{k}$$
$$= -3.18\mathbf{i} + 15.08\mathbf{j} - 238.87\mathbf{k}$$

Euler's equations of motion (Eqs. 9.44–9.46), in which x', y', and z' are fixed to the foot and approximate its principal axes, $\omega_{x'}$, $\omega_{y'}$, and $\omega_{z'}$ are the components of $\boldsymbol{\omega}_{foot}$ in foot coordinates, and α'_x, α'_y, and α'_z are the components of $\boldsymbol{\alpha}_{foot}$ in foot coordinates, are all used to compute the unknown ankle moment reaction, $\mathbf{M_A}$, as follows:

1. All vectors required for the computation are transformed into the foot coordinate system. The direction cosines, Eqs. (9.16)–(9.18), are used where the lab coordinates (x_{lab}, y_{lab}, z_{lab}) correspond to the $\mathbf{I}$, $\mathbf{J}$, $\mathbf{K}$ axes and the foot anatomical coordinates correspond to the $\mathbf{i}$, $\mathbf{j}$, $\mathbf{k}$ axes. The coefficients for e_{fax}, e_{fay}, and e_{faz} given previously represent a 3×3 matrix of direction cosines, $[e_{fa}]$, The lab coordinates (unprimed) are related to the foot coordinates (primed) via the matrix relation

$$[\mathbf{A}] = [e_{fa}][\mathbf{A'}] \qquad (9.58)$$

Consequently a vector $[\mathbf{A}]$ expressed relative to lab coordinates may be transformed into foot coordinates by transposing the matrix $[e_{fa}]$,

$$[\mathbf{A'}] = [e_{fa}]^T[\mathbf{A}] \qquad (9.59)$$

These vector operations are easily performed using MATLAB:

$$\mathbf{r_1} = -0.032\mathbf{i} + 0.002\mathbf{j} + 0.003\mathbf{k}$$
$$= -0.032\mathbf{i'} - 0.004\mathbf{k'} \text{ m}$$
$$\mathbf{r_2} = 0.033\mathbf{i} - 0.005\mathbf{j} - 0.054\mathbf{k}$$
$$= 0.0435\mathbf{i'} - 0.007\mathbf{j'} - 0.0457\mathbf{k'} \text{ m}$$

$$\mathbf{F_g} = 3.94\mathbf{i} - 15.21\mathbf{j} + 242.36\mathbf{k}$$
$$= -44.16\mathbf{i'} + 6.47\mathbf{j'} + 238.62\mathbf{k'}\,\text{N}$$
$$\mathbf{T_g} = 0.995\mathbf{k}$$
$$= -0.201\mathbf{i'} + 0.0873\mathbf{j'} + 0.970\mathbf{k'}\,\text{N m}$$
$$\mathbf{F_A} = -3.18\mathbf{i} + 15.1\mathbf{j} - 239\mathbf{k}$$
$$= 44.2\mathbf{i'} - 6.23\mathbf{j'} - 235\mathbf{k'}\,\text{N}$$
$$\boldsymbol{\omega}_{\text{foot}} = 0.0420\mathbf{i} + 2.22\mathbf{j} - 0.585\mathbf{k}$$
$$= 0.021\mathbf{i'} + 2.16\mathbf{j'} - 0.789\mathbf{k'}\,\text{rad/s}$$
$$\boldsymbol{\alpha}_{\text{foot}} = -0.937\mathbf{i} + 8.85\mathbf{j} - 5.16\mathbf{k}$$
$$= -0.425\mathbf{i'} + 8.26\mathbf{j'} - 6.116\mathbf{k'}\,\text{rad/s}^2$$

2. The cross products are computed for the terms on the left side of Euler's equations (Eqs. 9.44–9.46):

$$\mathbf{r_1} \times \mathbf{F_A} = 0.0625\mathbf{i'} - 7.68\mathbf{j'} + 0.215\mathbf{k'}\,\text{N m}$$
$$\mathbf{r_2} \times \mathbf{F_g} = -1.38\mathbf{i'} - 8.35\mathbf{j'} - 0.0283\mathbf{k'}\,\text{N m}$$

3. The values for the inertial terms on the right-hand sides of the equations are calculated:

$$I_{x'x'}\alpha_{x'} + (I_{z'z'} - I_{y'y'})\omega_{y'}\omega_{z'} = -1.4 \times 10^{-5}$$
$$I_{y'y'}\alpha_{y'} + (I_{x'x'} - I_{z'z'})\omega_{z'}\omega_{x'} = 2.8 \times 10^{-3}$$
$$I_{z'z'}\alpha_{z'} + (I_{y'y'} - I_{x'x'})\omega_{x'}\omega_{y'} = -2.1 \times 10^{-3}$$

4. All values are then substituted into back Euler's equations, Eqns. (9.44)–(9.46), to yield $\mathbf{M_A}$ in terms of foot (primed) coordinates

$$\mathbf{M_A} = [\mathbf{I}]\boldsymbol{\alpha}_{\text{foot}} + \boldsymbol{\omega}_{\text{foot}} \times [\mathbf{I}]\boldsymbol{\omega}_{\text{foot}} - \mathbf{r_1} \times \mathbf{F_A} - \mathbf{r_2} \times \mathbf{F_g} - \mathbf{T_g}$$
$$= 1.50\mathbf{i'} + 15.9\mathbf{j'} - 1.16\mathbf{k'}$$

that is, in turn, transformed back into lab (unprimed) coordinates, i.e. Eqn. (9.59):

$$\mathbf{M_A} = 2.54\mathbf{i} + 15.9\mathbf{j} - 0.037\mathbf{k}\,\text{N m}$$

By combining the ankle moment with the ankle angular velocity, the instantaneous ankle power may be computed as

$$\mathbf{M_A} \cdot \boldsymbol{\omega}_{\text{ankle}} = (2.54\mathbf{i} + 15.9\mathbf{j} - 0.037\mathbf{k}\,\text{N m}) \cdot (-0.000759\mathbf{i} + 1.47\mathbf{j} + 0.0106\mathbf{k}\,\text{rad/s})$$
$$= 23.3\,\text{Watts}$$

or

$$\mathbf{M_{A'}} \cdot \boldsymbol{\omega}_{\text{ankle}} = (1.50\mathbf{i'} + 15.9\mathbf{j'} - 1.16\mathbf{k'}\,\text{N m}) \cdot (-0.0946\mathbf{i'} + 1.46\mathbf{j'} + -0.140\mathbf{k'}\,\text{rad/s})$$
$$= 23.3\,\text{Watts}$$

which is thought to be a quantitative measure of the contribution by the ankle to propulsion.

9.6.4 Clinical Gait Interpretation

The information and data provided for treatment decision making in clinical gait analysis include not only the quantitative variables described previously, i.e., 3-D kinematics (such as angular displacement of the torso, pelvis, hip, knee, and ankle/foot) and 3-D kinetics (such as moments and power of the hip, knee, and ankle), but also

- Clinical examination measures
- Biplanar video recordings of the patient walking stride and temporal gait data, such as step length and walking speed
- Electromyographic (EMG) recordings of selected lower extremity muscles

Generally, the interpretation of gait data involves the identification of abnormalities, the determination of the causes of the apparent deviations, and the recommendation of treatment alternatives. As each additional piece of data is incorporated, a coherent "picture" of the patient's walking ability is developed by correlating corroborating data sets and resolving apparent contradictions in the information. Experience allows the team to distinguish a gait anomaly that presents the difficulty for the patient from a gait compensatory mechanism that aids the patient in circumventing the gait impediment to some degree.

To illustrate aspects of this process, consider the data presented in Figs. 9.27–9.29 that were measured from a 9-year-old girl with cerebral palsy spastic diplegia. Cerebral palsy is a nonprogressive neuromuscular disorder that is caused by an injury to the brain during or shortly after birth. The neural motor cortex is most often affected. In the ambulatory patient, this results in reduced control of the muscles required for balance and locomotion, causing overactivity, inappropriately timed activity, and muscle spasticity. Treatment options include physical therapy, bracing (orthoses), spasmolytic medications such as botulinum toxin and Baclofen, and orthopedic surgery and neurosurgery.

The sagittal plane kinematics for the left side of this patient (Fig. 9.27) indicate significant involvement of the hip and knee. Her knee is effectively "locked" in an excessively flexed position throughout stance phase (0–60% of the gait cycle) when her foot is contacting the floor. The knee's motion in the swing phase (60–100%) is also limited, with the magnitude and timing of peak knee flexion in swing reduced and delayed. The range of motion of her hip during gait is less than normal, failing to reach full extension at the end of stance phase. The motion of her pelvis is significantly greater than normal, tilting anteriorly in early stance coincident with extension of the hip and tilting posteriorly in swing coincident with flexion of the hip.

The deviations noted in these data illustrate neuromuscular problems commonly seen in this patient population. Inappropriate hamstring tightness, observed during the clinical examination, and inappropriate muscle activity during stance, seen in Fig. 9.28, prevent the knee from properly extending. This crouched knee position also impedes normal extension of the hip in stance due to hip extensor weakness (also observed during the clinical examination). Hip extension is required in stance to allow the thigh to rotate under the advancing pelvis and upper body. To com-

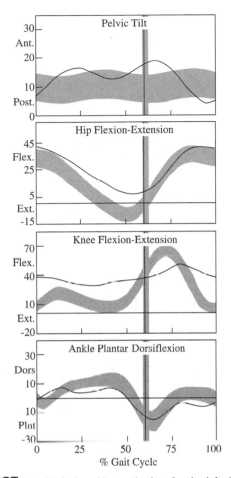

Fig. 9.27 Sagittal plane kinematic data for the left side of a 9-year-old patient with cerebral palsy spastic diplegia (solid curves). Shaded bands indicate ±1 SD about the performance of children with normal ambulation. Stance phase is 0–60% of the gait cycle and swing phase is 60–100%, as indicated by the vertical solid lines.

pensate for her reduced ability to extend the hip, she rotates her pelvis anteriorly in early stance to help move the thigh through its arc of motion. The biphasic pattern of the pelvic curve indicates that this is a bilateral issue to some degree.

The limited knee flexion in swing combines with the plantar flexed ankle position to result in foot clearance problems during swing phase. The inappropriate activity of the rectus femoris muscle (Fig. 9.28) in midswing suggests that spasticity of that muscle, a knee extensor (also observed during clinical examination), impedes knee flexion. Moreover, the inappropriate activity of the ankle plantar flexor, primarily the gastrocnemius muscle, in late swing suggests that they are overpowering the pretibial muscles, primarily the anterior tibialis muscle, resulting in plantar flexion of the ankle or "foot drop."

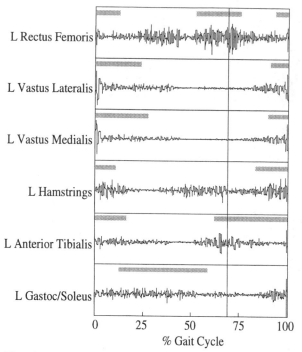

Fig. 9.28 EMG data for the same cerebral palsy patient as in Fig. 9.27. Plotted are EMG activity signals for each of six left lower extremity muscles, each plotted as a function of percentage of gait cycle. Gray bars represent mean normal muscle activation timing.

The sagittal joint kinetics for this patient (Fig. 9.29) demonstrate asymmetrical involvement of the right and left sides. Of special note is that her right knee and hip are compensating for some of the dysfunction observed on the left side. Specifically, the progressively increasing right knee flexion beginning at midstance (Fig. 9.29, first row, center) and continuing into swing aids her contralateral limb in forward advancement during swing, i.e., her pelvis can rock posteriorly along with a flexing hip to advance the thigh. One potentially adverse consequence of this adaptation is the elevated knee extensor moment in late stance that increases patella–femoral loading with indeterminate effects over time. The asymmetrical power production at the hip also illustrates clearly that the right lower extremity, in particular the muscles that cross the hip, provides the propulsion for gait with significant power generation early in stance to pull the body forward and elevate its center of gravity. Moreover, the impressive hip power generation, with respect to both magnitude and timing, at toe-off accelerates the stance limb into swing and facilitates knee flexion despite the elevated knee extensor moment magnitude. This is important to appreciate given the bilateral spastic response of the plantar flexor muscles, as evidenced by the premature ankle power generation and the presentation of a spastic stretch reflex in the clinical examination. This girl uses her hip musculature, right more

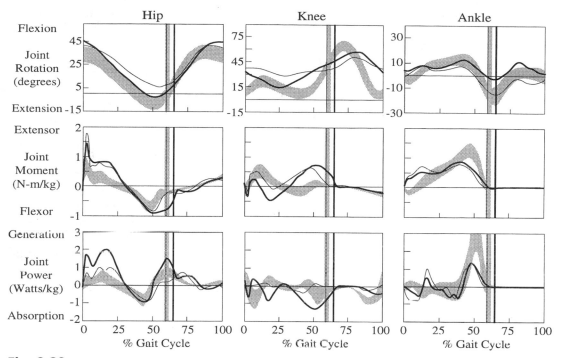

Fig. 9.29 Sagittal joint kinetic data for the same cerebral palsy patient as in Fig. 9.27. For the hip, knee, and ankle, joint rotation, joint moment, and joint power are plotted as functions of percentage of gait cycle. Dark and light solid curves denote right and left sides, respectively. Bands indicate ±1 SD of the normal population.

than left, to a much greater degree than her ankle plantar flexors to propel herself forward during gait.

This cursory case examination illustrates the process whereby differences from normal gait are recognized and the associated biomechanical etiology is explored. Some of the effects on gait of neuromuscular pathology in the sagittal plane have been considered in this discussion. Clinical gait analysis can also document and elucidate gait abnormalities associated with static bony rotational deformities. It is also useful in areas of clinical research by documenting treatment efficacy associated with bracing, surgery, etc. It should be noted, however, that although engineers and applied physicists have been involved in this work for well over 100 years, there remains significant opportunity for improvement in the biomechanical protocols and analytical tools used in clinical gait analysis — there remains much to learn.

Exercises

1. Write all the vector expressions of Eqs. (9.1)–(9.14) using MATLAB.
2. Repeat Example Problem 9.2 using an x–y–z rotation sequence.

3. Write the free-body diagrams for each of the three orientations of the humerus in Fig. 2.36. For a particular load and fixed position, write and solve the equations of static equilibrium.

4. Solve Example Problem 9.5 for any angle of orientation of the forearm. Using MATLAB, plot the required biceps muscle force F_B for static equilibrium as a function of this angle. How much does this force vary?

5. Considering the previous problem, explain why Nautilus weight machines at the gym use asymmetric pulleys.

6. The force plate in Fig. 9.9 is 75×75 cm^2. At a particular instant of the gait cycle each transducer reads as follows: $F_1 = 150$ N, $F_2 = 220$ N, $F_3 = 240$ N, and $F_4 = 160$ N. Compute the resultant force and its location.

7. For your own body, compute the mass moment of inertia of each body segment in Table 9.1 with respect to its center of mass.

8. Repeat Example Problem 9.8 using a titanium rod with a circular cross-sectional diameter of 10 mm.

9. Write the Simulink models of the three-element Kelvin viscoelastic description and perform the creep and stress relaxation tests, the results of which appear in Figs. 9.22 and 9.23.

10. Use the three-element Kelvin model to describe the stress relaxation of a biomaterial of your choice. Using a stress response curve from the literature, compute the model spring constants K_1 and K_2 and the viscous damping coefficient β.

11. Write and solve the kinematic equations defining an anatomically referenced coordinate system for the pelvis, $\{e_{pa}\}$, using MATLAB.

12. Using the kinematic data of Section 9.6.3, compute the instantaneous ankle power of the 25.2 kg patient using MATLAB.

13. Given the pelvis and thigh anatomical coordinate systems defined in Example Problems 9.9 and 9.10, compute the pelvis tilt, obliquity, and rotation angles using an x–y–z rotation sequence.

14. Repeat Problem 13 (using an x–y–z rotation sequence) solving for hip flexion–extension, hip abduction–adduction, and internal–external hip rotation. Hint: $\{e_{ta}\}$ is the triple-primed coordinate system and $\{e_{pa}\}$ is the unprimed coordinate system in this case.

15. Using MATLAB, determine the effect that a 1-cm perturbation in each coordinate direction would have on ankle moment amplitude.

SUGGESTED READING

Allard, P., Stokes, I. A. F., and Blanchi, J. P. (Eds.) (1995). *Three-Dimensional Analysis of Human Movement*. Human Kinetics, Champagne, IL.

Burstein, A. H., and Wright, T. M. (1994). *Fundamentals of Orthopaedic Biomechanics*. Williams & Wilkins, Baltimore.

Cappozzo, A. (1984). Gait analysis methodology. *Human Movement Sci.* **3**, 27–50.

Davis, R. B. D., III (1986). Musculoskeletal biomechanics: Fundamental measurements and analysis. In *Biomedical Engineering and Instrumentation* (J. D. Bronzino, Ed.). PWS Engineering, Boston.

Davis, R. B., Ōunpuu, S., Tyburski, D. J., and Gage, J. R. (1991). A gait analysis data collection and reduction technique. *Human Movement Sci.* **10**, 575–587.

Davis, R., and DeLuca, P. (1996). Clinical gait analysis: Current methods and future directions. In *Human Motion Analysis: Current Applications and Future Directions* (G. Harris and P. Smith, (Eds.), pp. 17–42. IEEE Press, Piscataway, NJ.

Fung, Y. C. (1993). *Biomechanics: Mechanical Properties of Living Tissues,* 2nd ed., pp. 1–10. Springer-Verlag, New York.

Gage, J. R. (1991). *Gait Analysis in Cerebral Palsy.* MacKeith, London.

Greenwood, D. T. (1965). *Principles of Dynamics.* Prentice-Hall, Englewood Cliffs, NJ.

Harris, G. F., and Smith, P. A. (Eds.) (1996). *Human Motion Analysis.* IEEE Press, Piscataway, NJ.

Meriam, J. L., and Kraige, L. G. (1997). *Engineering Mechanics,* 4th ed. Wiley, New York.

Mow, V. C., and Hayes, W. C. (1997). *Basic Orthopaedic Biomechanics,* 2nd ed, Lippencott–Raven, Philadelphia.

Needham, D. M. (1971). *Machina Carnis. The Biochemistry of Muscular Contraction in Its Historical Development.* Cambridge Univ. Press, Cambridge, UK.

Nigg, B. M. (1994). In *Biomechanics of the Musculo-Skeletal System* (Chapter 1: Introduction) (B. M. Nigg and W. Herzog, Eds.), pp. 3–35. Wiley, New York.

Nordin, M., and Frankel, V. H. (1989). *Basic Biomechanics of the Musculoskeletal System,* 2nd ed. Lea & Febiger, Philadelphia.

Palladino, J. L., and Noordergraaf, A. (1998). Muscle contraction mechanics from ultrastructural dynamics. In *Analysis and Assessment of Cardiovascular Function* (G.M. Drzewiecki and J. K-J. Li, Eds.), pp. 33–57. Springer-Verlag, New York.

Perry, J. (1992). *Normal and Pathological Function.* Slack, Thorofare, NJ.

Roark, R. J. (1989). *Formulas for Stress and Strain,* 6th ed. McGraw-Hill, New York.

Singer, C., and Underwood, E. A. (1962). *A Short History of Medicine,* 2nd ed. Oxford Univ. Press, New York.

Westerhof, N., and Noordergraaf, A. (1970). Arterial viscoelasticity: A generalized model. *J. Biomech.* **3**, 357–379.

Winter, D. A. (1990). *Biomechanics and Motor Control of Human Movement.* Wiley, New York.

10 CARDIOVASCULAR MECHANICS

Chapter Contents

At the conclusion of this chapter, the reader will be able to:

- Define a fluid, both physically and mathematically, with special emphasis on characteristic parameters that distinguish one type of fluid from another
- Quantify physiologic mechanisms that produce flow in cardiovascular channels, with special emphasis on the human heart and the cardiac cycle
- Develop field equations that account for the conservation of mass, energy, and momentum in fluid dynamic systems
- Apply the field equations to formulate and solve problems in biofluid mechanics, with special emphasis on flow through the cardiovascular system
- Identify some basic physiologic mechanisms that are responsible for controlling flow through the human cardiovascular system

10.1   INTRODUCTION

Mechanics is the engineering science that is concerned with studying, defining, and mathematically quantifying various "interactions" that take place among various "things" in our universe. Our awareness that such interactions are taking place is embedded in the concept of a **force,** which may be defined as the physically perceptible manifestation of an interaction taking place between two things or among several things. For purposes of engineering analysis, the things are classified according to the criteria defined later as being solid, fluid, or some combination of both (such as viscoelastic). Thus, more scientifically, mechanics may be defined as the engineering science that is concerned with studying, defining, and mathematically quantifying the behavior of solids, fluids, and viscoelastic materials that are being subjected to distributed systems of forces. The "behavior," in turn, is further classified according to whether (dynamics) or not (statics) the things move in response to being acted on by forces and also whether (deformation and/or flow) or not (rigid-body mechanics) the things become distorted under the action of such forces. If the things do move, one may study the motion itself — without regard for the forces that actually caused it — in a branch of mechanics called kinematics, or one may study the motion and also take into account the forces that are causing it in a branch of mechanics called kinetics. With this brief introduction to the subject of mechanics, we examine more specifically what we mean by a "fluid" and how this definition relates to some fundamental principles that govern the behavior of physiologic fluids, most notably, blood.

10.2   DEFINITION OF A FLUID AND BASIC PRINCIPLES OF BIOFLUID MECHANICS

The distinction between solid behavior and fluid behavior is illustrated in Fig. 10.1. A mass, m, of material rests on a cross-hatched surface but is not free to move along the surface because the bottom of m is firmly secured to the cross-hatched region

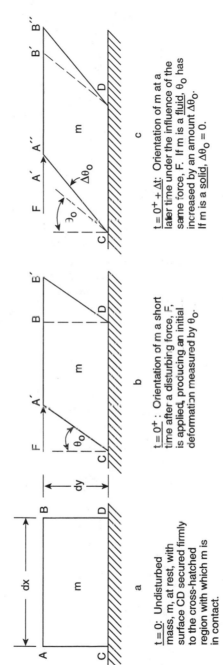

Fig. 10.1 Definition of solid and fluid behavior based on the response of a material to a shearing load, *F*.

along the contact area labeled *CD*. In Fig. 10.1a *m* is shown at time $t = 0$, lying at rest on the cross-hatched surface and assuming the shape of a rectangular parallelepiped, with side *AC* of length *dy* being perpendicular to side *CD* of length *dx*. Shortly thereafter, at time $t = 0^+$, *m* is subjected to a disturbing force, *F*, applied in a direction parallel to the cross-hatched surface, at the point *A* shown in Fig. 10.1b. In response to the applied shearing load, *F*, suppose the faces *CA* and *DB* of mass *m* deform uniformly to the new orientations, *CA'* and *DB'* (recall that *C* and *D* cannot move because they are secured rigidly to the cross-hatched surface). Then, one may take the angle, θ_0, between *AC* (undeformed orientation) and *A'C* (deformed orientation) to be, in some sense, a measure of the shearing deformation (distortion) of *m* under the influence of *F*, with *dy*, if infinitesimally small enough, assumed to remain a straight line (or a plane surface) throughout the deformation process.

Now suppose that as long as *F* is maintained at some constant value for all subsequent time, *m* equilibrates to the deformed orientation *CA'B'D*, with angle θ_0 (the amount of deformation) remaining finite and constant for all subsequent time. Then, the material behavior of *m* is said to be that of a solid. Moreover, if $\theta_0 \to 0$ when *F* is removed, i.e., *m* assumes exactly its original configuration *CABD* as $F \to 0$, then the solid behavior is termed perfectly elastic; in addition, again under the assumption that *dx* and *dy* are small enough to always remain essentially straight lines, if θ_0 is directly and linearly proportional to *F*, i.e.,

$$F = k\theta_0 \tag{10.1}$$

then *m* is said to be a linearly elastic (or "Hookean") solid. The constant of proportionality, *k*, is a property of the material defining in some sense an "elasticity modulus" or "linear spring constant," which is a measure of the material's ability to resist being deformed. The process of finding *k* (which depends on the specific atomic and molecular characteristics of the material of which *m* is "constituted") and of mathematically relating disturbance, *F* (or some equivalent measure of stress), to response, θ_0 (or some equivalent measure of strain), in the sense of Eq. 10.1 or some corresponding nonlinear form of it is called **constitutive modeling**, which we address further in Section 10.2. Finally, if θ_0 does not go to zero when *F* is removed, i.e., the material retains a permanent "set" after the disturbance is withdrawn, then its behavior is termed plastic.

We now go one step further and suppose that *m* does not equilibrate to a fixed θ_0 under the influence of a persistent *F* but rather the material continues to deform to a new strain, $\theta_0 + \Delta\theta_0$ at a later time, $t = 0^+ + \Delta t$, as is illustrated in Fig. 10.1c. Indeed, suppose *m* deforms continuously (a condition called flow) as long as *F* persists and regardless of how small *F* may be. In other words, suppose *m* shows no resistance to being deformed continuously under the influence of the very slightest disturbance. In this case, the material behavior is said to be that of a fluid, and its response to a disturbance is measured more appropriately by the rate of deformation, i.e., by the quantity, $\lim_{\Delta t \to 0} \dfrac{\Delta\theta_0}{\Delta t} = \dfrac{d\theta_0}{dt}$, rather than by the amount of deformation, θ_0, since θ_0, varying continuously with time, does not mean anything for a fluid. Note that strain rate, $d\theta_0/dt$, is actually a measure of the angular velocity, ω, of the line AC so that, for a linear, perfectly elastic, so-called "Newtonian" fluid, Eq. (10.1) would take the corresponding form:

$$F = \kappa^*\omega \tag{10.2}$$

with κ^* defining a "viscous modulus" or "linear shear coefficient."

It is important to realize that behavior classified as being "solid" or "fluid" is not absolute for any given material. For example, a material such as steel may show solid behavior when it is in one thermodynamic state, such as room temperature and pressure, and fluid behavior when it is in another thermodynamic state, such as at 3500°F. Moreover, a material such as glass may behave like a solid under impact loading at room temperature but may show fluid behavior at the same temperature when subjected to a slower varying, continuous disturbance, such as the pull of gravity on a vertically mounted windowpane. In the latter case, the material gradually "creeps" (flows) over a period of years [κ^* is very high, making ω small for a given F in Eq. (10.2)]. Thus, solid or fluid behavior may depend both on the thermodynamic state of the material and on the time rate at which it is being deformed. Rate-dependent material behavior, which is exhibited by virtually all physiologic tissues, is called **viscoelastic** (literally, "fluid-solid"). Material behavior may be further constrained by time and space boundary conditions that are imposed on both the disturbing force systems applied and the consequent material response, as we have already seen (i.e., mass m was secured firmly to the surface with which it was in contact, a boundary condition known in fluid mechanics as the no-slip condition along a surface such as is shown in Fig. 10.1).

Example Problem 10.1

Starting with Eq. (10.2) derive Newton's law of viscosity:

$$\tau = \mu\frac{dv}{dy}$$

where τ is shear stress, dv/dy is shear rate, and μ is the dynamic coefficient of viscosity of the mass m, which is a measure of its rate resistance to flow.

Solution

Let the mass m in Fig. 10.1 have depth dz extending in the third dimension (into the paper and perpendicular to it). Then define the shearing stress, τ, corresponding to the applied force, F, as

$$\tau = \frac{F}{dxdz}$$

where $dxdz$ represents the area parallel to the cross-hatched surface, over which F is distributed. Furthermore, under the influence of τ, let surface $dxdz$ located at dy start to deform continuously, at a rate corresponding to $d\theta_0/dt$. Along the direction AB, let ds represent the displacement of any point, such as A, and note that $ds = (dy)(d\theta)$ for infinitesimally small movements. Then, $ds/dt = dy\,(d\theta/dt) =$ the velocity, dv, of any point along AB; therefore, $d\theta/dt \equiv \omega = dv/dy \equiv$ the shear rate of the flow. Equation (10.2) thus becomes

$$\tau(dxdz) = \kappa^* \frac{dv}{dy}$$

and if we define the quantity $\kappa^*/dxdz$ to be the dynamic coefficient of viscosity of the material m, we arrive at Newton's law of viscosity as given previously.   ■

Having defined in a formal sense what we mean by a fluid, and by constitutive modeling, we now examine in more detail the process of developing constitutive models for biological fluids, most notably blood.

10.3   CONSTITUTIVE MODELING OF PHYSIOLOGIC FLUIDS: BLOOD

Recall that constitutive modeling of material behavior refers to the process of describing mathematically the response, which may take the form of a strain for solids or a strain rate for fluids, of the material to an imposed disturbance, which is usually in the form of an externally applied force or force per unit area, i.e., stress. The word constitutive establishes the fact that the response of the material being modeled is intimately dependent on the attributes of the substances of which the material is constituted; such attributes manifest themselves as measurable material properties, for example, the dynamic coefficient of viscosity, μ, defined for a linear Newtonian fluid. Thus, constitutive equations in their most general mathematical form are nonlinear, time-dependent, stress–strain (for a solid) or stress–strain–rate (for a fluid) relationships that contain as many specific material properties as are necessary to quantify how the particular material being modeled responds the way it does to the particular disturbance being imposed. Each material property accounts for a corresponding physical attribute and therefore the more material parameters a constitutive model contains (within reasonable limits of mathematical tractability), the more accurately it defines the actual physics of the behavior of the material and the more degrees of freedom it gives the model in quantifying such behavior. Clearly, a one-parameter model, such as Newton's linear law of viscosity, is limited to defining only the very simplest of fluids, such as air and water. The model breaks down when defining a complicated fluid such as blood.

10.3.1   Blood: Hematology

Accounting for about $8 \pm 1\%$ of total body weight and averaging 5200 ml, blood is a complex, heterogeneous suspension of formed elements — the blood cells or hematocytes [from the Greek *haima* ("blood") and *kytos* ("cell")] — suspended in a continuous, straw-colored fluid called plasma [from the Greek *plássein* ("to form, or mold"). Nominally, the composite fluid has a mass density of 1.057 ± 0.007 g/cm^3, and it is three to six times as viscous as water. The hematocytes (Table 10.1)

TABLE 10.1 Hematocytes

Cell type[a]	Number of cells/mm³ blood[a]	Corpuscular diameter (μm)[a]	Corpuscular surface area (μm²)[a]	Corpuscular volume (μm³)[a]	Mass density (g/cm³)[a]	% Water[a]	% Protein[a]	% Extractives[a,b]
Erythrocytes (red blood cells; 26 trillion total cells]	4.2–5.4 × 10⁶ ♀ / 4.6–6.2 × 10⁶ ♂ (5 × 10⁶)	6–9 (7.5) Thickness, 1.84–2.84 "Neck," 0.81–1.44	120–163 (140)	80–100 (90)	1.089–1.100 (1.098)	64–68 (66)	29–35 (32)	1.6–2.8 (2)
Leukocytes (WBCs); [white blood cells, 16–65 billion total cells]	4000–11000 (7500)	6–10	300–625	160–450	1.055–1.085	52–60 (56)	30–36 (33)	4–18 (11)
Granulocytes (polymorphonuclear phagocytes; 56–76% of all WBCs]								
Neutrophils 55–70% WBCs (65%)	2–6 × 10³ (4875)	8–8.6 (8.3)	422–511 (467)	268–333 (300)	1.075–1.085 (1.080)	—	—	—
Eosinophils 1–4% WBCs (3%)	45–480 (225)	8–9 (8.5)	422–560 (491)	268–382 (321)	1.075–1.085 (1.080)	—	—	—
Basophils 0–1.5% WBCs (1%)	0–113 (75)	7.7–8.5 (8.1)	391–500 (445)	239–321 (278)	1.075–1.085 (1.080)	—	—	—
Agranulocytes (23–43% WBCs)								
Lymphocytes 20–35% WBCs (25%)	1000–4800 (1875)	6.75–7.34 (7.06)	300–372 (336)	161–207 (184)	1.055–1.070 (1.063)	—	—	—
Monocytes 3–8% WBCs (6%)	100–800 (450)	9–9.5 (9.25)	534–624 (579)	382–449 (414)	1.055–1.070 (1.063)	—	—	—
Thrombocytes (platelets; 2.675 × 10⁵/mm³)	1.4(♂), 2.14(♀) – 5 × 10⁵	2–4 (3) Thickness, 0.9–1.3	16–35 (25) (700 billion–2 trillion total cells)	5–10 (7.5)	1.04–1.06 (1.05)	60–68 (64)	32–40 (36)	Negligible

[a] Normal physiologic range with "typical" value in parentheses.

[b] Extractives include mostly minerals (ash), carbohydrates, and fats (lipids).

Reprinted (with modifications) with permission, from: Schneck, Daniel J., "An Outline of Cardiovascular Structure and Function," in *The Biomedical Engineering Handbook*, edited by Joseph D. Bronzino. Copyright CRC Press, Boca Raton, Florida. © 1995. Visit the CRC Press Home Page at: http://www.crcpress.com

include three basic types of cell: red blood cells (RBCs) [erythrocytes, from the Greek *erythros* ("red"), totaling nearly 95% of the formed elements], white blood cells (WBCs) [leukocytes, from the Greek *leukos* ("white"), averaging <0.15% of all hematocytes], and platelets [thrombocytes, from the Greek *thrombos* ("lump"), on the order of 5% of all blood cells]. Hematocytes are all derived in the active ("red") bone marrow (about 1500 g) of adults from undifferentiated stem cells called hemocytoblasts, and all reach ultimate maturity in the bone marrow, the spleen, and some specialized lymphatic tissue (hence the designation lymphocytes for some types of WBCs) via a process called hematocytopoiesis. It only takes 9 ounces (about 255 g) of active bone marrow to manufacture 20 billion RBCs, and in the course of the average life span of any given individual, bone marrow will manufacture more than three-quarters of a ton of RBCs, each of which takes 6 days to complete and survives about 120 days in the circulation. Just prior to reaching full maturity, RBCs lose their nucleus, which gives them their characteristic bicon-cave disk shape [although as many as 250 billion normoblasts (immature, young, still nucleated RBCs) and young but nonnucleated reticulocytes may be found at any time circulating along with the 26 trillion fully mature erythrocytes]. These oxygen-carrying cells are removed from the circulation by the spleen, the largest collection of reticuloendothelial cells in the body. The phagocytic cells of this large (12 to 13 cm long × 7 cm wide × 3 to 4 cm deep) lymphatic organ can also store and dis-charge blood cells into the vascular system as necessary. The term hematocrit refers to the percentage by volume of RBCs in blood, which can normally range from 30 to 54% for adults (average, 45%).

The water contained within hematocytes accounts for approximately 6% of the body's total intracellular water, even though the cells account for approximately 25% of the total number of cells in the body. The primary function of erythrocytes is to aid in the transport of blood gases — about 30–34% (by weight) of each cell consisting of the oxygen and carbon dioxide-carrying protein, hemoglobin ($64,000 \leq MW \leq 68,000$), and a small portion of the cell containing the enzyme carbonic anhydrase that catalyzes the reversible breakdown of carbonic acid (from carbon dioxide and water) into the bicarbonate and hydrogen ions. The primary function of leukocytes is to endow the human body with the ability to identify and dispose of foreign substances (such as infectious organisms) that do not belong: Agranulocytes (lymphocytes and monocytes) essentially identify and granulocytes [neutrophils, basophils, and eosinophils (also known as acidophils)] dispose of the foreign substances. The primary function of platelets is to participate in the blood clotting process.

Example Problem 10.2

Consider a red blood cell of maximum dimension 7.5 μm and a volume of 90 μm³. What would be the hematocrit of blood if each cell were completely free to spin around itself in any direction without interference from a neighboring cell? Assume a face-centered cubic packing factor.

Solution

Consider a face-centered cubic (FCC) configuration composed of cubes having side length b, such that the length of a diagonal of one of the cube faces is $b\sqrt{2}$. In an FCC packing arrangement, geometry tells us that this same diagonal length $b\sqrt{2}$ is also equal to $4r$, where r is the radius of a sphere of diameter 7.5 μm — thus enclosing a physical space within which a red blood cell would be completely free to spin around itself in any direction without restraint. We therefore conclude that $b\sqrt{2} = 4(7.5/2)$ or $b = 10.6066$ μm. In a 1-mm^3 volume, we could pack $1/b^3 = 1/(10.6066 \times 10^{-3})^3 = 838{,}052.5$ cubic cells of side 10.6 μm, each containing four RBCs of volume 90 μm^3. The volume of cells per mm^3 of whole blood would then be $(838{,}052.5)4(90 \times 10^{-9}) = 0.3017$, yielding a hematocrit of 30.2%.   ∎

Removal of all hematocytes from blood by centrifugation or other separating techniques leaves behind the aqueous (91% water by weight, 94.8% water by volume), saline (0.15 N) suspending medium called plasma, which has an average mass density of 1.035 ± 0.005 g/cm^3 and a viscosity one and one-half to twice that of water. Plasma is the liquid portion of blood, in which all the cellular elements (blood cells or corpuscles) are suspended, and it accounts for 17–20% of the body's total extracellular water. Approximately 6.5–8% by weight of plasma (7% average) consists of the plasma proteins, of which there are three major types (albumin, the globulins, and fibrinogen) and several of lesser prominence (Table 10.2). The primary functions of albumin (56% of total protein) are to help maintain the osmotic (oncotic) transmural pressure differential that ensures proper mass exchange between blood and interstitial fluid in the microcirculation and to serve as a transport carrier molecule for several hormones and other small biochemical constituents (such as some metal ions). The primary functions of the globulin class of proteins (38% of total protein) are to act as transport carrier molecules (mostly of the α and β class) for large biochemical substances, such as fats (lipoproteins, 6% of total protein) and certain carbohydrates (muco- and glycoproteins), and heavy metals (mineraloproteins) and to work together with leukocytes in the body's immune system. The latter function is primarily the responsibility of the γ class of immunoglobulins (IgGs), which have antibody activity and account for 86% of the 12–14% of total proteins that are immunoglobulins; the others are IgA (7%), IgD (2.33%), and IgM (4.67%). Glycoproteins, mineraloproteins, and transport carrier free globulins collectively make up the majority of the 18–20% of total proteins that are not lipoproteins or immunoglobulins but are still in the globulin class of protein. The primary function of fibrinogen (4% of total protein, and also known as clotting factor I) is to work with thrombocytes in the formation of a blood clot — a process also aided by one of the most abundant (0.25% of total protein) of the lesser proteins, prothrombin (also known as clotting factor II, MW ≃ 62,000).

Of the remaining approximately 2% (by weight) of plasma, just under half (0.95% or 983 mg/dl plasma) consists of minerals (inorganic ash), trace elements (e.g., protein-bound iodine, iron, sulfur, and silicon), and electrolytes — mostly the cations sodium, potassium, calcium (also known as clotting factor IV), and magnesium and the anions chlorine, bicarbonate, phosphate, and sulfate [the latter three helping as buffers to maintain the fluid at a slightly alkaline pH between 7.35 and 7.45 (average 7.4)]. What is left, about 1087 mg of material per deciliter of plasma, includes (i) mainly (0.76% by weight) three major types of emulsified fat, i.e., cholesterol (in a free and esterified form), phospholipid (a major ingredient of cell membranes), and triglyceride, with lesser amounts (0.04%) of the fat-soluble vita-

TABLE 10.2 Plasma

Constituent	Concentration range (mg/dl plasma)	Typical plasma value (mg/dl)	Molecular weight range	Typical value	Typical size (nm)
Total protein (7% by weight)	6000–8300	7245	21,000–1,200,000	—	—
Albumin (56% TP)	2800–5600	4057	66,500–69,000	69,000	15×4
α_1-Globulin (5.5% TP)	300–600	400	21,000–435,000	60,000	5–12
α_2-Globulin (7.5% TP)	400–900	542	100,000–725,000	200,000	50–500
β-Globulin (13% TP)	500–1230	942	90,000–1,200,000	100,000	18–50
γ-Globulin (12% TP)	500–1800	869	150,000–196,000	150,000	23×4
Fibrinogen (4% TP)	150–470	290	330,000–450,000	390,000	$(50–60) \times (3–8)$
Other (2% TP)	70–210	145	70,000–1,000,000	200,000	$(15–25) \times (2–6)$
Inorganic ash (0.95% by weight)	930–1140	983	20–100	—	— (radius)
Sodium	279–340	325	—	22.98977	0.102 (Na^+)
Potassium	13–21	17	—	39.09800	0.138 (K^+)
Calcium	8.4–11.0	10	—	40.08000	0.099 (Ca^{2+})
Magnesium	1.5–3.0	2	—	24.30500	0.072 (Mg^{2+})
Chloride	336–390	369	—	35.45300	0.181 (Cl^-)
Bicarbonate	110–240	175	—	61.01710	0.163 (HCO_3^-)
Phosphate	2.7–4.5	3.6	—	95.97926	0.210 ($HPO_4^=$)
Sulfate	0.5–1.5	1.0	—	96.05760	0.230 ($SO_4^=$)
Other	0–100	80.4	20–100	—	0.1–0.3
Lipids (fats; 0.80% by weight)	541–1000	828	44,000–3,200,000	= Lipoproteins, 6% TP	Up to 200 or more
Cholesterol (34% TL)	12–105 Free, 72–259 esterified, 84–364 total	59, 224, 283	386.67	Contained mainly in the 2.52% intermediate to low-density β-lipoproteins; Higher in women	
Phospholipid (35% TL)	150–331	292	690–1,010	Contained mainly in the 2.16% high- to very high-density α_1-lipoproteins	
Triglyceride (26% TL)	65–240	215	400–1,370	Contained mainly in the 1.32% very low-density α_2-lipoproteins and chylomicra	
Other (5% TL)	0–80	38	280–1,500	Fat-soluble vitamins; prostaglandins; fatty acids	
Extractives (0.25% by weight)	200–500	259	—	—	—
Glucose	60–120 fasting	90	—	180.1572	0.86 diameter
Urea	20–30	25	—	60.0554	0.36 diameter
Carbohydrate	60–105	83	180.16–342.3	—	0.74–0.108 diameter
Other	11–111	61	—	—	—

Reprinted (with modifications) with permission, from: Schneck, Daniel J., "An Outline of Cardiovascular Structure and Function," in *The Biomedical Engineering Handbook*, edited by Joseph D. Bronzino. Copyright CRC Press, Boca Raton, Florida. © 1995. Visit the CRC Press Home Page at: http://www.crcpress.com

mins (up to 0.015% of A, D, E, and K), free fatty acids, chylomicrons, and other lipids; and (ii) "extractives" (0.25% by weight), of which about two-thirds include glucose and other forms of carbohydrate, with the remainder (0.083%) consisting mostly of the water-soluble vitamins (up to 0.015% of B complex and C), a trace of certain enzymes, nonnitrogeneous and nitrogenous waste products of metabolism (up to 0.03%, including urea, creatine, and creatinine), clotting factors [thromboplastin (clotting factor III), proaccelerin (clotting factor V), proconvertin (clotting factor VII); antihemophilic clotting factor VIII, and as many as seven others], organic acids (up to 0.03%, including glucuronic acid, lactic acid, acetic acid, and formic acid), hormones (about 0.00014% endocrine secretions), a trace of antibodies (antigens), and many smaller amounts (totaling <0.008% by weight of plasma) of other biochemical constituents.

Removal from blood of all hematocytes and the protein fibrinogen (by allowing the fluid to completely clot before centrifuging) leaves behind a clear, slightly yellowish fluid called serum, which has a density of about $1.018 + 0.003$ g/cm^3 and a viscosity up to 1.5 times that of water, the study of which is called serology. Tables 10.1 and 10.2, together with the very brief summary presented previously, give the reader an immediate appreciation for why blood is often referred to as the "river of life." This river is made to flow through the vascular piping network by two central pumping stations arranged in series: the left and right sides of the human heart (discussed in Section 10.4). More to the point, however, the previous discussion makes abundantly clear the reasons why, from a biofluid mechanics point of view, blood would not be expected to obey the very simple, one-parameter, linearized law of viscosity as developed by Newton: Blood is a nonhomogeneous, anisotropic, polarized (ionic), composite fluid composed of a suspension of many asymmetric, relatively large, viscoelastic particles carried in a liquid that contains high-molecular-weight, asymmetric, polarized, long-chain polymers that behave in a complicated way under shear-type loading. Thus, blood exhibits non-Newtonian (nonlinear), time-dependent (viscoelastic) deformation (flow) characteristics that can only be modeled by higher order constitutive equations, such as the Power-Law paradigm.

10.3.2 A Power-Law Constitutive Model for Blood

The fact that blood exhibits non-Newtonian behavior was actually first recognized around the turn of the century. Also vaguely appreciated at this time were observations that the macroscopic rheological (deformation) properties of red blood cell suspensions are influenced at least by shear rate, hematocrit, degree of cellular interactions (aggregation), and the mechanical properties of the cells. Significant attempts to define such non-Newtonian behavior, however, did not appear until the 1960s, when variable-shear rotational viscometers were introduced. Since then, literally dozens of constitutive models have been proposed that attempt to relate shear stress to shear rate in the fluid: The most practical of these is an empirical power-law formulation that generalizes Newton's law of viscosity to the form:

$$\underline{T} = K\underline{S}^n + \underline{T}^*$$

(10.3)

where $\underline{T}$ is a generalized, three-dimensional, time-dependent stress dyadic (a second-order tensor comparable to one-dimensional shear stress, τ, in Newton's law of viscosity); $\underline{S}$ is a generalized, three-dimensional, time-dependent strain (for a solid) or rate of strain (for a fluid) dyadic (a second-order tensor comparable to the steady shear rate, dv/dy, in Newton's law of viscosity); K is a material-dependent constitutive parameter, which is most generally a complex (in the mathematical sense, i.e., containing a real and imaginary component) higher order tensor quantity which may have as many as 81 components, but which reduces to the real dynamic coefficient of viscosity for a Newtonian fluid; n is a material-dependent nonlinearity index (non-Newtonian parameter for a fluid), which cannot be negative but may be fractional; and $\underline{T}^*$ is a generalized, three-dimensional yield stress dyadic which defines so-called Bingham plastic fluids, wherein a minimum amount of force is necessary to initiate flow. Thus, Eq. (10.3) holds for $\underline{T} \geq \underline{T}^*$; and $\underline{S}$ (representing strain rate) $\equiv 0$ for $\underline{T} < \underline{T}^*$ (material behavior is that of a solid).

The nonlinearity parameter, n, is equal to zero for a perfectly rigid body — an idealization that presumes that a material does not deform at all when subjected to a disturbing force — or, more accurately, that any material deformations are negligibly small in comparison with other static and dynamic response characteristics of the material being examined. Values of n between 0 and 1 define pseudoplastic material behavior, illustrated in Fig. 10.2a for a material with (Bingham pseudoplastic) and without (simple pseudoplastic) a yield stress. Pseudoplastic behavior is exhibited by blood and other physiologic fluids, as well as hard tissue such as bone; it is characteristic of composite materials having internal microstructures that show less resistance to being sheared the more they are sheared; hence the decreasing slope of the stress–strain rate curves as $\underline{T}$ increases. When $n = 1$, Eq. (10.3) reduces to the linearized Hookean elastic, viscoelastic, or Newtonian forms already discussed — with (Bingham plastic) or without (perfectly elastic) a yield stress. Finally, when n is larger than unity, materials exhibit dilatant behavior, such as that illustrated in Fig. 10.2b for the human carotid artery. Dilatant behavior — with (Bingham dilatancy) or without (simple dilatancy) a yield stress — is typical of composite materials such as physiologic soft tissues (e.g., muscles, smooth connective tissue, cartilage, arteries, and veins), all of which show more resistance to being deformed the more they are deformed; hence the increasing slope of the stress–strain (or strain rate) curves as $\underline{T}$ increases. For the human body, n will generally lie between 0.5 and 5 for all of its constituent tissues. The mathematical definition of raising a tensor quantity to some arbitrary power is beyond the scope of this book, but we shall examine some physiologically-relevant simplified forms of equation (10.3) that have practical value.

Human blood shows a very slight Bingham plastic property attributable mainly to the presence of the protein fibrinogen (i.e., the Bingham plastic behavior disappears in the absence of fibrinogen), but appearing only in blood samples having

Fig. 10.2 Pseudoplastic (a) and dilatant (b) non-Newtonian material behavior with (Bingham plastic) and without a corresponding yield stress (illustrated in a only).

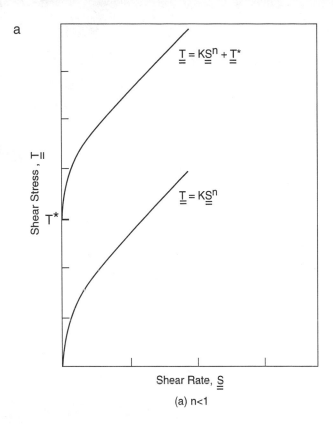

$$\underline{\underline{T}} = K\underline{\underline{S}}^n + \underline{\underline{T}}^*$$

$$\underline{\underline{T}} = K\underline{\underline{S}}^n$$

Shear Stress , $\underline{\underline{T}}$

T^*

Shear Rate, $\underline{\underline{S}}$

(a) n<1

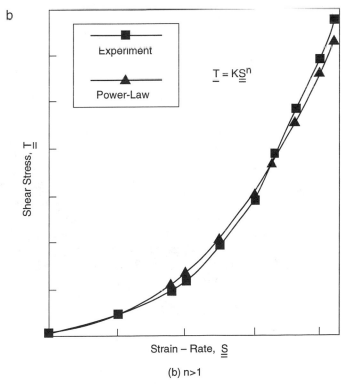

Experiment

Power-Law

$$\underline{T} = K\underline{\underline{S}}^n$$

Shear Stress, $\underline{\underline{T}}$

Strain – Rate, $\underline{\underline{S}}$

(b) n>1

hematocrit values exceeding 10%. An empirical formula that describes this relationship is given by

$$\sqrt{T^*} = (H - 0.1)(C_F + 0.5), \text{ for } C_F > 0$$
$$= 0 \text{ for } C_F = 0 \tag{10.4}$$

where H is the hematocrit ($>10\%$), expressed as a fraction; C_F is the plasma concentration of fibrinogen, expressed in grams per 100 ml (so-called "grams-percent"); and T^* is given in dynes (10^{-5} N) per square centimeter (10^{-4} m^2), which is equivalent to 0.1 Pa (N/m^2); thus, T^* is given in tenths of a Pascal.

Example Problem 10.3

Use Eq. (10.4) and Tables 10.1 and 10.2 to calculate a typical yield stress, in Pascals, for blood.

Solution

From Table 10.1, we find that, typically, the number of RBCs averages 5 million per cubic millimeter of whole blood, with each cell having a mean corpuscular volume of 90 μm^3. Thus,

$$H = 5,000,000 \frac{\text{cells}}{\text{mm}^3} \times 90 \times 10^{-9} \frac{\text{mm}^3 \text{ cell}}{\text{cell}} = 0.45$$

Furthermore, from Table 10.2 we find that, typically, plasma contains 290 mg/dl of fibrinogen, so

$$C_F = 290 \frac{\text{mg}}{\text{dl}} \times \frac{1 \text{ g}}{1000 \text{ mg}} = 0.290 \frac{\text{g}}{100 \text{ ml plasma}}$$

Finally, from Eq. (10.4),

$$T^* = (0.45 - 0.10)^2(0.290 + 0.5)^2 = 0.076 \frac{\text{dyn}}{\text{cm}^2} = 0.008 \text{ Pa} \qquad \blacksquare$$

The yield stress for blood is quite small ($\ll 0.025$ Pa), and Bingham plastic behavior is significant only when a fluid is being started up from rest (i.e., it is a static property which has little significance in dynamic situations such as those encountered in the human cardiovascular system). Thus, $\underline{T}^*$ in Eq. (10.3) is often neglected with impunity; not so, however for K and n. Both the non-Newtonian index, n, and the "consistency index," K, have a primary functional dependence on hematocrit, H, and on certain of the plasma proteins — most notably, the globulins [expressed for anticoagulated human blood as, essentially, total serum protein minus the albumin fraction (TPMA)]. This functional dependence takes the following form under time-independent circumstances:

$$K = C_1 e^{C_2 H} \tag{10.5}$$

and:

$$n = 1 - C_3 H \tag{10.6}$$

where C_1, C_2, and C_3 are species-specific coefficients that have been determined for several animal groups (see Tables 10.3 and 10.4). Note, in particular, the explicit influence of hematocrit, which also accounts for the actual number of RBCs per unit volume of blood (red cell count) and their approximate size [as measured by the mean corpuscular volume (MCV); see Tables 10.1, 10.4 and Example Problem 10.3]. That is, since hematocrit is defined as the product

$$H = (RBC) \times (MCV) \times 100 \text{ (in \%)} \tag{10.7a}$$

the influence of the number of cells per unit volume, and their approximate size, on the variables K and n, and hence on the rheological behavior of the fluid, are specifically defined by Eqs. (10.3)–(10.7a). What about additional effects that have to do with the degree of cellular aggregation, the mechanical properties of the cells, and the chemistry of the plasma? How are these accounted for in the previous equations?

The coefficients C_1, C_2, and C_3 do represent the influence of all of the previous factors on the stress–strain rate characteristics of blood. For example, the coefficient C_1 for human blood does not have the constant value listed in Table 10.3, but, rather, depends on plasma globulin concentration and on hematocrit, according to the relation

$$C_1 = C_1^* e^{C_4 \frac{\text{TPMA}}{H^2}} \tag{10.7b}$$

where TPMA is the total protein minus the albumin fraction (or, essentially, the plasma globulin concentration), $C_1^* = 0.00797 \text{ d Pa (s)}^n$, and $C_4 = 145.85 \text{ dl/g}$. Similary, an examination of values for C_2 and TPMA in Table 10.3 suggests that the two are directly correlated; C_2 increases directly with TPMA (the effect of hematocrit on C_2 is somewhat obscure), and both C_1 and C_2 show more sensitivity to the viscosity of blood plasma (i.e., to blood chemistry) than to any factors related to the blood cells (hematocytes). The latter may be illustrated more clearly by introducing for blood the concept of an "apparent viscosity," μ_a, defined in the Newtonian sense as the ratio of shear stress to shear rate (see Example Problem 10.1):

$$\mu_a \equiv \frac{\underline{\underline{T}}}{\underline{\underline{S}}} \tag{10.8}$$

Substituting the power-law equation (Eq. 10.3), with $\underline{\underline{T}}^*$ considered to be negligible, into Eq. (10.8) results in

$$\mu_a = K(\underline{\underline{S}})^{n-1} \tag{10.9}$$

and, utilizing eqs. (10.5) and (10.6):

$$\mu_a = C_1 e^{C_2 H}(\underline{\underline{S}})^{-C_3 H} \tag{10.10}$$

Now, for hematocrit, H, equal to zero, which reduces whole blood to blood plasma, Eq. (10.10) clearly shows that $C_1 = \mu_a$, the viscosity of plasma. The latter, in turn, is most sensitive to the highly labile and uttermost variable of all the plasma proteins, i.e., the globulins. However, as Eq. (10.7) also shows, this sensitivity is moderated for whole blood by the hematocrit of the fluid. The columns labeled C_1 and plasma μ_{rel} in Table 10.3 verify the direct correlation between these two vari-

TABLE 10.3 Species-Specific Coefficients, Viscous Behavior, and Some Biochemical Data for Seven Species of Animals

Animal species	$C_1 (\times 10^{-2})$	$C_2 (\times 10^{-2})$	K	Plasma (μ_{rel})	$\mu_{232.8}$	pH	Serum total protein	Albumin	TPMA	FIB
Human	1.4800	5.1200	0.1352	1.80	4.175	7.39	7.15	4.56	2.59	0.29
Rat	1.1286	6.5950	0.1904	1.70	4.807	7.35	7.30	3.20	4.10	0.17
Horse (equine)	1.1148	5.8390	0.1019	1.65	4.254	7.38	7.07	3.10	3.97	0.30
Sheep (ovine)	1.0804	4.9136	0.0552	1.60	3.379	7.44	6.35	3.78	2.57	0.28
Cow (bovine)	1.0610	5.4000	0.0920	1.55	6.348	7.38	7.11	3.90	3.21	0.55
Pig (swine)	1.0570	6.2650	0.1332	1.55	3.915	7.35	7.55	3.52	4.03	0.30
Dog (canine)	1.0500	5.4000	0.1259	1.50	4.074	7.36	6.57	3.22	3.35	0.28

Note. The coefficient C_1 is tabulated such that the consistency index and apparent viscosities will all be expressed in SI units of tenths of a Pascal second; C_2 is dimensionless. Apparent whole blood viscosity at a shear rate of 232.8 s^{-1} (in mPa-sec). μ_{rel}, Nondimensional plasma viscosity, relative to water at the same temperature. Serum total protein (TP), albumin (Alb), total protein minus albumin (globulins; TPMA), and fibrinogen (FIB) are all expressed in grams per deciliter. All tabulated quantities are typical values for the respective species indicated.

Reprinted (with modifications) with permission, from: Schneck, Daniel J., "On the Development of a Rheological Constitutive Equation for Whole Blood," in *Biofluid Mechanics* *3: *Proceedings of the Third Mid-Atlantic Conference on Biofluid Mechanics*, edited by Daniel J. Schneck and Carol L. Lucas, New York: New York University Press, 1990.

TABLE 10.4 Some Rheologic, Morphologic, and Physiologic Parameters for Seven Species of Animals

Animal species	C_3 ($\times 10^{-3}$)	H (%)	n	T(°C)	ESR (mm/h)	Hgb	RBC dia.	RBC count	MCV (μm^3)	Surface area (μm^2)
Rat	5.8952	42.85	0.7474	38.1	1.25	20.7	5.9	7.14	60.0	—
Pig (swine)	5.5550	40.45	0.7753	39	5.35	19.0	6.0	6.37	63.5	113
Human	4.9900	43.20	0.7844	37	2.36	15.7	7.5	5.08	90.0	140
Dog (canine)	4.500	46.00	0.7930	38.9	3.12	15.0	7.0	6.57	74.3	121
Horse (equine)	4.2308	37.89	0.8397	37.8	>18 ≤50	14.4	5.7	8.61	44.0	84
Sheep (ovine)	2.7144	33.20	0.9099	38	0.55	13.5	4.5	10.54	36.8	66
Cow (bovine)	1.7000	40.00	0.9320	38.5	1.17	11.0	5.5	7.84	51.0	110

Note. The coefficient C_3 is dimensionless, as is the non-Newtonian index, n. H, hematocrit; T, temperature; ESR, erythrocyte sedimentation rate; Hgb, hemoglobin (expressed in grams per deciliter of blood); RBC dia., red blood cell diameter [in μm (10^{-6} m)]; RBC count, red blood cell count (in millions per cubic millimeter of whole blood); MCV, mean corpuscular volume; Surface area, average red blood cell surface area; All tabulated quantities are typical values for the respective species indicated.

Reprinted (with modifications) with permission, from: Schneck, Daniel J., "On the Development of a Rheological Constitutive Equation for Whole Blood," in *Biofluid Mechanics *3: Proceedings of the Third Mid-Atlantic Conference on Biofluid Mechanics*, edited by Daniel J. Schneck and Carol L. Lucas, New York: New York University Press, 1990.

ables, whereas the column labeled $\mu_{232.8}$ (the apparent viscosity of whole blood having the values of H listed in Table 10.4 and measured at a shear rate of 232.8 reciprocal seconds) verifies that the effect is moderated by the hematocrit of the fluid.

Example Problem 10.4

Show that Eq. (10.9) is intrinsically linear and, on this basis, explain how one might experimentally obtain values for K and n for any given blood sample.

Solution

By intrinsically linear, we mean that an equation can be manipulated in such a way that it assumes a linear form. In the case of Eq. (10.9) we can take the natural log of both sides to get

$$\ln \mu_a = \ln K + (n - 1) \ln \underline{S}$$

which is a linear equation of the form $Z = m_0 X + b_0$, where Z is the dependent variable $= \ln \mu_a$, X is the independent variable $= \ln \underline{S}$, m_0 is the slope of the line $= (n - 1)$, and, b_0 is the z intercept $= \ln K$. Thus, using an appropriate laboratory viscometer, we can measure μ_a (apparent viscosity) vs $\underline{S}$ (shear rate) for any particular blood sample and, using a linear least squares regression analysis, calculate $n = m_0 + 1$ and $K = e^{b_0}$. The results of such a series of studies are illustrated in Fig. 10.3, in which Eq. (10.9) is plotted based on the data given in Tables 10.3 and 10.4. ■

Whereas the coefficients C_1 and C_2 depend both on hematocrit and on the chemistry of the plasma, the coefficient C_3 seems to be relatively independent of the plasma chemistry and appears to be remarkably constant for a given animal species. The value of C_3 does, however, represent a RBC form factor (hence its species specificity as discussed later) that depends on the degree of red cell aggregation [as measured by the erythrocyte sedimentation rate (See Example Problem 10.5) and Rouleaux formation]; the erythrocyte boundary geometry (specifically, the degree of cell biconcavity); erythrocyte size (surface area, RBC diameter, mean corpuscular volume, and so on; see Table 10.1); and red cell deformability (a measure of which is its hemoglobin content). Thus, we observe in Fig. 10.3 that bovine blood is the most nearly Newtonian ($n = 0.9320$) of all the animal species shown, probably due at least in part to the fact that bovine erythrocytes show little or no Rouleaux formation of chains at low (<50 reciprocal s) rates of shear. The word Rouleaux is a derivative of the French word for a "roll," as in a roll of coins; it refers to the tendency of red cells to aggregate — to attach themselves side by side to form long chains that resemble rolls of coins. Aggregation results from cellular attraction that derives from polarized groups in the surface membrane of the cells, and it is enhanced by cellular immobility and the presence of fibrinogen and the plasma globulins (TPMA). The presence and degree of Rouleaux formation are reflected in a rheologic parameter called the erythrocyte sedimentation rate (ESR).

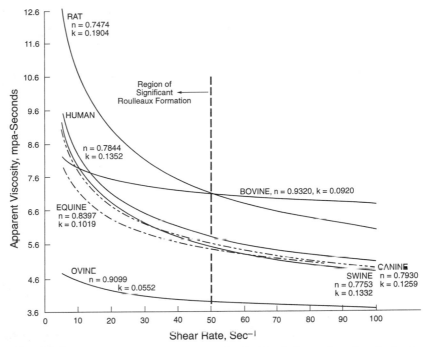

Fig. 10.3 Apparent viscosity vs shear rate for the blood of several species of animals, calculated for typical values of their respective non-Newtonian and consistency indices. Reprinted (with modifications) with permission, from: Schneck, Daniel J., "On the Development of a Rheological Constitutive Equation for Whole Blood," in *Biofluid Mechanics *3: Proceedings of the Third Mid-Atlantic Conference on Biofluid Mechanics*, edited by Daniel J. Schneck and Carol L. Lucas, New York: New York University Press, 1990.

Example Problem 10.5

Well-mixed, anticoagulated (prevented from clotting) blood placed carefully in a 2.5-mm diameter by 100-mm long cylindrical tube will eventually separate into a clear upper (plasma) layer and an opaque lower (erythrocyte) layer, with the two separated by a thin, buffy intermediate layer containing leukocytes and platelets. This precipitation occurs because the hematocytes, being denser than plasma, tend to settle out of solution when the fluid is allowed to stand at rest for a long period of time, for example, an hour or more. The rate at which this "settling out" occurs has some clinical significance — 0–9 mm/h being "normal" for a tube of the size given — and this rate is called the sedimentation rate (or sed-rate) for hematocytes.

A reasonable approximation for the sed-rate may be derived using Stoke's law for the drag force, F_D, acting on a falling sphere, i.e.,

$$F_D = 6\pi\mu v r_0$$

where μ is the dynamic viscosity of the fluid (plasma) through which a sphere (e.g., a RBC approximation) of outer radius r_0 falls (or "settles out") with speed (sedimentation rate) v.

Assuming a steady state (i.e., v = constant), derive the fact that

$$v = \frac{2}{9} r_0^2 \frac{\rho_s - \rho_f}{\mu} g_0$$

where ρ_s is the sphere (erythrocyte) mass density, ρ_f is the fluid (plasma) mass density, and g_0 is the acceleration of gravity.

Solution

The weight of a single RBC is given by Newton's second law of motion as $\rho_s(MCV)g_0$, where, for a sphere, $MCV = \frac{4}{3} \pi r_0^3$. The buoyant force acting on each cell is given by Archimedes' principle as the weight of displaced plasma volume, $\rho_f \frac{4}{3} \pi r_0^3 g_0$, and the drag force is given by Stoke's law as $6\pi\mu v r_0$. In the steady state (v = constant, so acceleration = 0) the net downward force on each cell must be zero, i.e., the sedimentation rate equilibrates to a value, v, defined by the equation

$$\frac{4}{3} \pi r_0^3 \rho_s g_0 - \frac{4}{3} \pi r_0^3 \rho_f g_0 - 6\pi\mu v r_0 = 0$$

from which $v = \frac{2}{9} r_0^2 \frac{\rho_s - \rho_f}{\mu} g_0$.   ∎

The sedimentation rate for women (4–7 mm/h) is usually higher than that for men (1–3 mm/h) and may climb as high as 100 mm/h or more in human disease conditions. Returning to our discussion of C_3, n, and the effect of Rouleaux formation on both sedimentation rate and Newtonian behavior, however, we note that the RBCs of sheep and goats also (as is true of bovine blood) do not show significant cellular aggregation (Rouleaux), and thus indeed one notes from the results reported in Table 10.4 and Fig. 10.3 that ovine (sheep) blood is second only to bovine blood in terms of its nearly Newtonian ($n = 0.9099$) behavior. Since the significant appearance of Rouleaux at shear rates below 50 s^{-1} is known to be a major cause of non-Newtonian behavior in this region of fluid deformation, one is led to infer that the coefficient C_3 in Eq. (10.6) is at least in some sense a measure of such red cell aggregation. However, this explanation must not tell the whole story because the blood of equines, which is the next most Newtonian ($n = 0.8397$), is known to exhibit considerable Rouleaux formation, whereas the blood of rats, which shows the highest degree of non-Newtonian behavior ($n = 0.7474$), is associated with a comparatively small tendency toward Rouleaux. Thus, one must look beyond red cell aggregation in trying to establish some physical significance to the coefficient C_3.

Taking the analysis one step further, there is also some suggestion that RBC geometry may also be correlated with the coefficient C_3. For example, it is well documented that goat and sheep RBCs ($C_3 = 2.7144 \times 10^{-3}$) show little biconcavity, whereas cat and horse erythrocytes ($C_3 = 4.2308 \times 10^{-3}$) are more moderately biconcave compared to dog ($C_3 = 4.5 \times 10^{-3}$) and human ($C_3 = 4.99 \times 10^{-3}$) erythrocytes, which are markedly biconcave. One therefore observes an increase in C_3 with the degree of concavity of the red cell surface curvature or a higher degree of non-Newtonian behavior (decreasing n) as the cell shape increasingly deviates from being a perfect disk or sphere to become instead more tortuous in shape. Moreover, sheep, horse, and bovine erythrocytes are among the smallest in physical size (see RBC diameter, MCV, and surface area columns of Table 10.4) of all the animal cells considered thus far; human, dog, pig, and rat erythrocytes are among the largest erythrocytes. Also, smaller RBCs tend to give the fluid a high degree of Newtonian behavior (smaller C_3, higher n) than do larger red cells, and C_3 is indeed also related to the size and geometry of the cell.

Two last points: First, One cannot help but notice in Table 10.4 that C_3 correlates directly with the cellular hemoglobin concentration. That is, C_3 (dimensionless) and Hgb (expressed as a concentration in grams per deciliter) both increase and decrease together. As C_3 decreases (n increases, fluid becomes more Newtonian) from rat blood to swine, human, canine, equine, ovine, and bovine species, respectively, so does the average hemoglobin concentration — from a high of 20.7 g/dl to a low of 11.0 g/dl on average. Now, it is known that hemoglobin content is the major determinant of intracellular erythrocyte viscosity, leading one to presume that the coefficient C_3 is somehow related to the latter. Second, an increase in shear rate will deform softer erythrocytes, thus lowering the suspension apparent viscosity, which defines precisely pseudoplastic behavior, so that one would expect that the softer the cells (all other factors being held constant), the more non-Newtonian (n decreasing, C_3 increasing) the fluid and vice versa. That is, the harder the cells, the more Newtonian (but also the more viscous) the fluid because increasing the shear rate has no effect on the viscosity of hardened red cell suspensions, even though the latter have a higher viscosity than do comparable suspensions of softer cells having the same basic size, geometry, and number (hematocrit) immersed in plasma solutions having the same chemical composition. Thus, observe, for example, in Fig. 10.3, that bovine blood is the most Newtonian ($n = 0.9320$) but also that it is the most viscous at high rates of shear (above 50 s^{-1}), where cellular aggregation is minimal.

In summary, the species-specific non-Newtonian parameters C_1, C_2, and C_3 define, for a given sample of blood,

1. the chemistry of the plasma, specifically its globular protein composition (apparent viscosity shows only a very weak dependence on plasma lipids, albumin concentration, and other chemical constituents);
2. the chemistry of the red blood cell, specifically its hemoglobin concentration (which accounts for about one-third of the cell contents, the remaining two-thirds being almost entirely water);
3. the number (hematocrit), size (MCV, RBC diameter, and surface area), and degree of cellular aggregation (Rouleaux) of red blood corpuscles;
4. the erythrocyte shape, geometry (degree of cell biconcavity), and deformability; and
5. the mass density of RBCs, plasma, and the composite whole blood.

The coefficient C_3, a form factor that remains virtually constant for any given animal species (characterized by its own unique RBCs), shows the greatest variability from species to species of any of the three coefficients defined previously. By contrast, the coefficient C_1, a consistency parameter most closely related to the apparent viscosity of blood plasma, shows the least variation from species to species of all the parameters thus far defined. Also, as revealed in Table 10.3, there does appear to be a direct relationship between decreasing values of C_1 and decreasing values of the nondimensional plasma viscosity, μ_{rel}, measured relative to that of water at the same temperature. Keeping the temperature constant is an important consideration: Studies have shown that in the 30–40°C range, the consistency index, K, of human blood changes about 2.5% per degree centigrade change in temperature (the relationship being essentially inverse), whereas the non-Newtonian index, n, remains virtually constant. Likewise, the alkalinity or acidity of blood, as measured by its pH (see Table 10.3), is another important consideration. A marked increase in blood viscosity has generally been reported as blood becomes more acidic (i.e., as pH decreases). While this trend is not obvious from the data reported in Table 10.3, we do note that the least viscous blood (sheep, $\mu_{232.8} = 3.38$) is associated with the highest pH (7.44), whereas the blood which is *most* viscous (rat, $\mu_{232.8} = 4.81$) is associated with the lowest pH (7.35). Finally, we discuss the role of WBCs and platelets.

Since WBCs and platelets make up only about 5% of all hematocytes — which, in turn, account for only about 45% by volume of whole blood — it is not surprising that neither leukocytes nor thrombocytes play a significant role in the rheological behavior of the fluid, at least when the latter is examined in the larger blood vessels. In the microcirculation, however (i.e., those blood vessels in which the diameter of the tube lumen becomes comparable to the size of the cells) WBCs in particular, though constituting the smallest population of formed elements in the fluid, nevertheless exert a significant hemodynamic influence. This influence results from both the fact that WBCs can be very large (see Table 10.1) and that they can be very mobile, migrating in and out of small blood vessels under their own propulsion systems. Similarly, the large number of very small platelets that exist in blood can, nevertheless, have a significant rheological effect when these hematocytes form large clusters of masses as they participate in the blood clotting process. Blood clotting and the mechanics of the microcirculation are beyond the scope of this chapter, which addresses mainly the biofluid mechanics of larger blood vessels, so we shall now discuss the time-dependent behavior of blood.

Suppose that in Eq. (10.8) the shear stress, $\underline{T}$, which in general is a second-order tensor (i.e., a dyad), is also a time-dependent quantity which can be expressed as the simple periodic function,

$$\tau = Be^{i\omega_0 t} \tag{10.11}$$

where τ is shear stress (reduced to two-dimensions) t is time, and $i = \sqrt{-1}$. The modulus, B, represents the amplitude of a periodic cosine/sine wave that has a circular frequency $\omega_0 = 2\pi f_0$. Furthermore, let the generalized strain rate dyad, $\underline{S}$, be written as

$$\frac{dv}{dy} = D(t)e^{i(\omega_0 t + \phi)} \tag{10.12}$$

where $D(t)$ is the time-dependent amplitude of the periodic response function, dv/dy, to the driving function, τ. Note that dv/dy may be out of phase with τ by an amount ϕ. Thus, for example, in Fig. 10.4 the time-dependent behavior of a pseu-

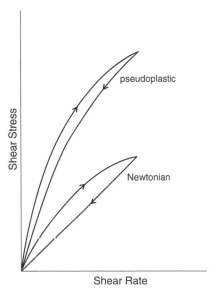

Fig. 10.4 Hysteresis loops for a thixotropic fluid such as blood.

doplastic fluid is demonstrated which, when initially loaded, appears to be quite viscous (a large τ is required to generate a given dv/dy) and very non-Newtonian. However, when unloaded, the fluid appears to be less viscous (smaller stress required to generate the same shear rate at any given point) and more Newtonian. Also, when loaded again, it shows an even lower apparent viscosity. Under cyclic loading, there is a progressive structural breakdown of the internal fluid architecture such that with each cycle of shear the fluid increasingly tends toward a Newtonian limit.

The fluid behavior illustrated in Fig. 10.4 is typical for blood and is called thixotropic behavior, the opposite of which is rheopectic behavior. The area enclosed by any two successive loading–unloading curves is called a hysteresis loop, the area of which approaches zero as the number of loading–unloading cycles approaches infinity. We can quantify this behavior by first substituting Eqs. (10.11) and (10.12) into Eq. (10.8) to get

$$\mu_a = \frac{Be^{i\omega_0 t}}{D(t)e^{i(\omega_0 t + \phi)}} = \mu_e(\omega_0, t)e^{-i\phi(\omega_0, t)} \qquad (10.13)$$

where $\mu_e(\omega_0, t)$ is the effective coefficient of viscosity of blood, which, like ϕ, can depend on both time and ω_0. Expanding Eq. (10.13), we can write

$$\mu_a = \mu_e \cos \phi - i\mu_e \sin \phi = \mu_V - i\mu_E \qquad (10.14)$$

where μ_V is the real ($\mu_e \cos \phi$), energy-dissipating part ("visco-") of the complex apparent coefficient of viscosity, μ_a, and μ_E is the imaginary ($\mu_e \sin \phi$), energy-storing part ("-elastic") of μ_a. Then,

$$\tan \phi = \frac{\mu_E}{\mu_V}$$

and

$$\mu_e = \sqrt{\mu_V^2 + \mu_E^2}$$

The energy-storing part of μ_a corresponds to both the ability of the elastic membrane of RBCs to store energy and the erythrocytes to aggregate reversibly. The energy-dissipating part of μ_a corresponds to the viscous properties of blood plasma and to the viscous properties of the intracellular hemoglobin fluid of RBCs. The energy-storing coefficient of viscosity is an order of magnitude smaller than is the energy-dissipating coefficient of viscosity for blood, and although there is more energy dissipation at higher frequencies (more frictional heat under increased cyclic loading), and red cells also become stiffer at higher rates of cyclic disturbance, these two effects are more than offset by the other consequences of high-rate periodicities in fluid-driving forces. Thus, the lack of time to aggregate, and to recover from being deformed at high oscillatory frequencies, actually cause both μ_V and μ_E for blood to decrease significantly with increasing ω_0. Moreover, both viscoelastic coefficients tend to an asymptotic limit as $t \rightarrow \infty$, this tendency being reasonably exponential in nature. That is, the structural breakup with increasing shear and structural recovery with decreasing shear, at any given ω_0, reaches a steady state after a large number of cycles have elapsed ($t \rightarrow \infty$), with the steady state being a decreasing function of ω_0. The reader is referred to Suggested Reading for more details. Next, we examine the pumping system that drives blood to within a few micrometers of each and every one of the approximately 100 trillion cells in the human body — the heart.

10.4 GENERATION OF FLOW IN THE CARDIOVASCULAR SYSTEM: THE HUMAN HEART (CARDIOLOGY) AND THE CARDIAC CYCLE

Barely the size of the clenched fist of the individual in whom it resides — an inverted, conically shaped, hollow muscular organ measuring 12 to 13 cm from base (top) to apex (bottom) and 7 or 8 cm at its widest point (about the size of a 3 × 5-in. index card), and weighing just under three-fourths of a pound (about 0.474% of the individual's body weight or, on average, approximately 325 g) — the human heart occupies a small region between the third and sixth ribs in the central portion of the thoracic cavity of the body. It rests on the diaphragm, between the lower part of the two lungs, with its base-to-apex axis leaning mostly toward the left side of the body and slightly forward. The heart is divided by a tough muscular wall, the interatrial/interventricular septum, into a somewhat crescent-shaped right side and cylindrically shaped lift side (Fig. 10.5), each a self-contained pumping station but both connected in series. The left heart (pump) drives oxygen-rich blood through the aortic semilunar outlet valve into the systemic circulation, which carries the fluid to within a differential neighborhood of each cell in the body, from which it returns to the right side of the heart low in oxygen and rich in carbon dioxide. The right heart (pump) then drives this oxygen-poor blood through the pulmonary semilunar (pulmonic) outlet valve into the pulmonary circulation, which carries the fluid to the lungs, where its oxygen supply is replenished and its carbon dioxide content is purged before it returns to the left side of the heart to begin the cycle all over again.

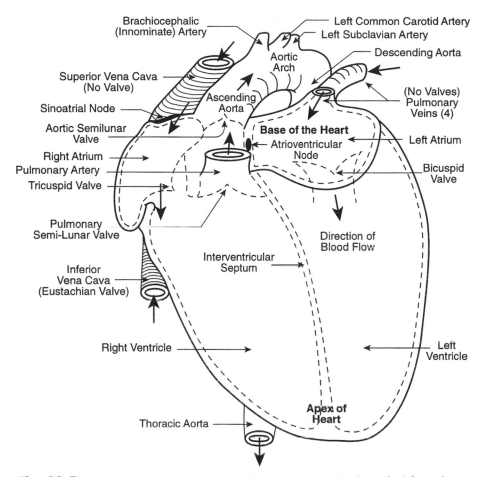

Fig. 10.5 Anterior view of the human heart, showing the four chambers, the inlet and outlet valves, the inlet and outlet major blood vessels, the wall separating the right side from the left side, and the two cardiac pacing centers — the sinoatrial node and the atrioventricular node. Large arrows show the direction of flow through the heart chambers, the valves, and the major vessels. Reprinted (with modifications) with permission, from: Schneck, Daniel J., "An Outline of Cardiovascular Structure and Function," in *The Biomedical Engineering Handbook,* edited by Joseph D. Bronzino. Copyright CRC Press, Boca Raton, Florida. © 1995. Visit the CRC Press Home Page at: http://www.crcpress.com

Because of the anatomical proximity of the heart to the lungs, the right heart does not have to work very hard to drive blood through the pulmonary circulation, and so it functions as a low-pressure ($p \leq 40$ mmHg gauge) pump compared with the left heart, which does most of its work at a high pressure (up to 140 mmHg gauge or more) to drive blood through the entire systemic circulation to the furthest extremes of the organism. The complete journey out the left ventricle, through the systemic circulation into the right atrium, and then out the right ventricle, through the pulmonary circulation and back into the left atrium, takes from as little as 18–24 s to as much as 1 min in normal circumstances. This is called the circulation time.

Each cardiac (heart) pump is further divided into two chambers: a smaller upper receiving chamber, or atrium (auricle), separated by a one-way valve from a lower discharging chamber, or ventricle (literally, "little stomach"), which is about twice the size of its corresponding atrium. In order of size, the spherically shaped left atrium is the smallest chamber, holding about 45 ml of blood (at rest), operating at pressures on the order of 0–25 mmHg gauge, and having about a 3-mm wall thickness. The pouch-shaped right atrium is next (63 ml fluid, 0–10 mmHg gauge, 2-mm wall thickness), followed by the conical/cylindrically shaped left ventricle (100 ml blood, up to 140 mmHg gauge, variable wall thickness up to 12 mm), and the crescent-shaped right ventricle (about 130 ml fluid, up to 40 mmHg gauge, and a wall thickness on the order of one-third that of the left ventricle, up to about 4 mm). The heart chambers collectively have a capacity of approximately 325–350 ml (just over one-third quart of fluid), or about 6.5% of the total blood volume in a typical individual, but these values are nominal since the organ alternately fills and expands, contracts, and then empties as it generates a cardiac output.

10.4.1   The Cardiac Cycle

Like the lungs, which alternately inflate to take in a fresh, oxygen-rich supply of air and deflate to expire the waste products of metabolism (carbon dioxide and water), the human heart cyclicly fills (diastole) and empties (systole) to expel a bolus of fluid with each "stroke." There is thus no specific beginning or end to this cyclic process and so, for descriptive and analytical purposes, it is arbitrarily divided into four phases, the first of which, the filling phase, starts just after the heart has finished its most recent contraction and is beginning to reload in preparation for its next contraction. The beginning of the filling phase is set to be equal to time, $t = 0$, and we move forward from there to examine four variables that define the events of the systemic cardiac cycle: the volume (in ml) of the left ventricle; the pressure (in mmHg) in the left atrium, the left ventricle, and the root of the aorta just as this main blood vessel of the systemic circulation leaves the heart; the electrical voltage (electrocardiogram) that traces the propagation of excitation signals through the musculature of the heart [the myocardium, from the Greek *mys* ("muscle") and *kardia* ("heart")], expressed in millivolts; and the speed (in meters per-second) at which these excitation signals course through the said myocardial musculature. All four variables are plotted in Fig. 10.6, which we now describe as we discuss the four phases of the cardiac cycle. At the beginning of the filling phase ($t = 0$) the volume of blood in the left ventricle is approximately 70 ml (note that the heart does not empty itself completely with each stroke) and there is little or no measurable electrical activity in the myocardium (the heart has just completed contracting and is "resting" passively). The aortic root pressure is approximately 100 mmHg, significantly greater than the left ventricular pressure of approximately 2 or 3 mmHg, so the 2.25-cm-diameter aortic semilunar outlet valve is closed (as is the 2.4-cm-diameter pulmonary or pulmonic outlet valve leading from the right ventricle into the pulmonary artery); however, the 3.1-cm-diameter left ventricular inlet bicuspid (or mitral) valve is open because the left atrial pressure is slightly higher than the left ventricular pressure, the situation that prevails at $t = 0$. The heart is filling and continues to fill passively for approximately 200 or so milliseconds (during which time

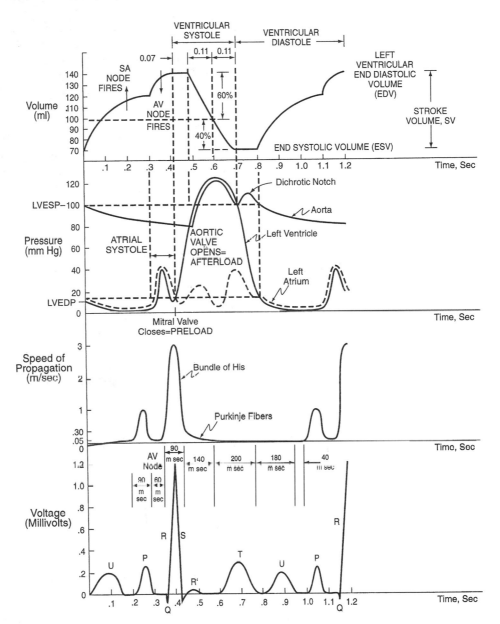

Fig. 10.6 The events of the systemic cardiac cycle, recorded as the time course of left ventricular volume; left ventricular, left atrial, and aortic root pressure; the electrical activity of the ventricular musculature; and the speed of propagation of ventricular waves of depolarization.

the 3.8-cm-diameter tricuspid valve from right atrium to right ventricle is also open).

About 200 ms into a typical cardiac cycle of 800 ms (approximately one-quarter of the way into the period of an average 75 beat per minute heart rate) the pacemaker of the heart, its sinoatrial (SA) node (located, as shown in Fig. 10.5, where the superior vena cava enters the right atrium), fires, initiating a wave of electrical depolarization through the musculature of both atria. This wave of depolarization spreads through the atrial musculature at a speed of about 1 m/s and is recorded in the electrocardiogram as a "P-wave" of approximately 0.20–0.30 (to as much as 0.40) mV amplitude and 90 ms period. However, the firing of the SA node does not show up immediately in the volume or pressure graphs because there is a 90- to 100-ms delay (the so-called latent period) from the time the node actually fires to the time the atrial musculature actually contracts. The delay results both from the time that it takes for the signal to permeate the musculature and from the anatomic and physiologic architectural changes that must occur in order to activate a muscle from its resting state to a contractile state. Thus, after the SA node fires, the left ventricle continues to fill passively for the next 90–100 ms, and the pressure curves continue their unremarkable, inactive trends.

At about $t = 300$ ms, and lasting for about 110 ms, the two atria contract, suddenly injecting a final bolus of fluid into their respective ventricles — bringing the left ventricle to its so-called end-diastolic-volume (EDV) of about 140 ml of fluid. This active filling phase is called atrial systole and is accompanied by a corresponding jump in left atrial pressure to a maximum of approximately 40–45 mmHg. Since the latter is still less than the aortic root pressure of approximately 82–85 mmHg during this same period of time, the aortic valve remains closed during atrial systole; however, the mitral valve remains open (which is why the left ventricular pressure curve parallels the activity of the left atrial pressure curve) because the pressure in the contracting atria during this time is always greater than the pressure in the relaxed ventricles. The latter situation is guaranteed by delaying, for about 60 ms, the propagating wave of depolarization from reaching the ventricular musculature. The delay is accomplished by the atrioventricular (AV) node, a small mass of slowly conducting tissue located in the right atrial wall of the heart near its juncture with the interventricular septum (see Fig. 10.5) which ensures that the two ventricles will not be attempting to empty (i.e., will not themselves be contracting) at the same time that the contracting atria are attempting to fill them. The AV node delay shows up on the electrocardiogram graph as a 60-ms isolectric pause at 0 mV corresponding to a near-zero depolarization wave speed, following which the propagating signal leaves the AV node to excite the ventricular musculature.

Excitation of the ventricular musculature starts about half-way through atrial systole, at about $t = 350$ ms, as the wave of depolarization leaves the AV node and heads down the interventicular septum toward the apex of the heart. This event is demarcated on the ECG graph by the 10-ms, short, downward-sloping "Q wave," following which the upward-sloping, 40-ms "R wave" and downward-sloping, 40-ms "S wave" trace the propagating signal through the ventricular muscle bundle of His to the terminal Purkinje fibers. The entire so-called QRS complex lasts about 90 ms, reaching a sharp peak voltage of about 1.2 mV (1.0–1.5 mV range); again, there is a delay from the time the signal leaves the AV node to the time that the ventricles actually begin contracting. This time, however, the delay is only 60 ms, rather than the previous 100 ms, because, among other things, the depolarization wave travels three times faster (up to 3 m/s) through the ventricular musculature than it did through the atrial musculature. Thus, at about $t = 410$ ms, approximately two-thirds of the way through the QRS complex and just past the middle of the entire cardiac cycle, ventricular systole begins and the ventricles start to contract.

Since the two atria are now relaxed (atrial systole having been completed by the time ventricular systole begins), almost immediately at the start of myocardial contraction the pressure in the left ventricle becomes higher than that in the left atrium — the two respective pressure curves cross over one another at about $p = 12$–14 mmHg — and the mitral (inlet) valve snaps shut at what is termed the cardiac preload pressure. This is opposed to the so-called left ventricular end-diastolic pressure (LVEDP), which is the slightly lower (about 10 mmHg) pressure in the left ventricle at the start of ventricular systole. Once the mitral valve on the left side of the heart (and the corresponding tricuspid valve on the right side of the heart) closes, no blood can enter the ventricle. However, as long as the aortic root pressure (and the corresponding pulmonary artery root pressure) is higher than the left ventricular pressure, the aortic semilunar (and, correspondingly, the pulmonary semilunar) outlet valve also remains closed, so blood cannot leave the ventricle either. Thus, phase two occurs during the next 70 ms of the cardiac cycle, the period of isovolumetric contraction, during which the elastic energy (strain energy) of myocardial muscular contraction is being converted into fluid pressure energy (potential energy) in a process being carried out at constant volume. Isovolumetric contraction lasts until the ventricular pressure exceeds the aortic root pressure, at which time the two respective pressure curves cross over one another at about $p = 80$–82 mmHg, and the aortic semilunar (outlet) valve opens at the cardiac afterload pressure. Also known as the systemic diastolic blood pressure, the cardiac afterload pressure opens the aortic valve with such force that blood literally "squirts" out of the left ventricle, producing a fluid jet with an audible "first heart sound" that heralds the start of phase three of the cardiac cycle — the period of ejection.

The period of ejection lasts about 220 ms and is broken down further into a 110-ms "rapid ejection phase," during which time the heart will expel 60% (approximately 42 ml of fluid) of the total amount of blood that will eventually leave the heart, and a 110-ms "slow ejection phase," during which time will be expelled the remaining 40% of blood (approximately 28 ml of fluid). Thus, the electrically induced vigorous contraction of cardiac muscle normally allows the ventricles to empty about half of their contained volume with each heartbeat: the 70 ml ejected is called the stroke volume (SV) and the 70 ml left behind is called the cardiac reserve volume or the end-systolic volume (ESV). More generally, the stroke volume (the volume of blood expelled from the heart during each systolic interval) will be equal to the difference between the actual EDV and the actual ESV, and the ratio of SV to EDV is called the cardiac ejection fraction or ejection ratio (0.5–0.75 = normal; 0.4–0.5 signifies mild cardiac damage; 0.25–0.40 implies moderate heart damage; and <0.25 warns of severe damage to the heart's pumping ability). If the stroke volume is multiplied by the number of systolic intervals (usually approximately 75) per minute, or heart rate (HR), one obtains the total cardiac output (CO):

$$CO = HR \times (EDV - ESV) \tag{10.15}$$

Since during the ejection phase the aortic valve is open, the aortic pressure curve "tracks" the left ventricular pressure curve, whereas since the mitral valve is closed the left atrial pressure curve shows no remarkable behavior, except for some mild undulations induced by the chamber's proximity to the contracting myocardium. Similarly, the electrical propagating activity of the myocardial musculature has ceased (speed of propagation → 0) as it now goes through the process of repolarizing (T and U waves on the ECG tracing) in preparation for the next cardiac cycle. The maximum pressure reached during the ejection phase is approximately 120–125 mmHg (to as much as 140 mmHg); this is termed the systemic systolic blood pressure and much of this systolic blood pressure (potential energy) is converted into corresponding fluid kinetic energy as the blood squirts out of the heart at peak speeds reaching 1.5 m/s or more. Some of the pressure energy is also converted into tissue elastic energy because the sudden expulsion of a bolus of fluid from the heart causes the flexible walls of the root of the aorta to expand momentarily. The difference between the peak systolic blood pressure of approximately 120 mmHg and the systemic diastolic blood pressure of approximately 80 mmHg is called the pulse pressure; their ratio, 120/80, is how clinical blood pressure is normally reported.

Having completed contracting, the ventricles now relax, but the distended elastic walls of the aorta (and pulmonary artery) "recoil," producing a sudden jump in the aortic pressure curve called the dichrotic notch. This sudden rise in aortic blood pressure causes the ventricular pressure curve and the aortic root pressure curve to cross once again as the aortic semilunar valve snaps shut; the event occurs at about $p = 100$ mmHg [called the left ventricular end-systolic pressure (LVESP)], $t = 700$ ms, and it is accompanied by an audible characteristic "second heart sound." With the closing of the aortic valve, no blood can leave the ventricle; however, as long as the ventricular pressure still exceeds the left atrial pressure, the one-way mitral valve also remains closed, so no blood can enter the ventricle. Thus, for approximately the next 100 ms, phase four, isovolumetric relaxation, is completed, during which time blood pressure energy is converted back into myocardial strain energy as the heart relaxes after having contracted. Isovolumetric relaxation continues until the venous return to the atrium, coupled with the relaxing ventricle, causes the ventricular pressure curve and the atrial pressure curve to cross once again at $t = 800$ ms and $p = 14$–16 mmHg — at which point the mitral valve opens and blood once again starts to enter the ventricle, with a characteristic "third heart sound" that heralds the start of the next filling phase. The cycle begins again, starting at $t = 0$. A "fourth heart sound" will accompany each atrial systole.

In summary, we note that a typical 800-ms cardiac cycle consists of four phases: (i) a 410-ms filling phase (which includes a 300-ms, slow, passive part followed by a 110-ms, fast, active atrial systole); (ii) a 70-ms period of isovolumetric contraction; (iii) a 220-ms emptying phase (half of which accounts for 60% of the ejected fluid, with the remaining 40% being ejected more slowly during the second half); and (iv) a 100-ms period of isovolumetric relaxation. Two phases, isovolumetric relaxation + filling = 510 ms or about 63.75% (just over 5/8), of the cardiac cycle are designated to be ventricular diastole, during which time the heart is relaxed and not working. The remaining two phases, isovolumetric contraction + emptying = 290 ms or about 36.25% (just under 3/8), of the cardiac cycle are designated to be ventricular systole, during which time the heart musculature is shortening as the organ does work to generate a stroke volume. Thus, contrary to some popular beliefs that the heart works continuously, nonstop throughout the entire life span, the myocardial musculature is actually resting nearly two-thirds of the time! This observation is not intended to denigrate the "work ethic" of this vital organ but merely to point out that its longevity can be traced to its ability to "pace" itself and to protect itself from suffering the long-term consequences of continuous fatigue loading. Bearing this in mind, we next examine some of the salient features related to the energy required to generate a cardiac output.

10.4.2 Cardiac Mechanics

We mentioned earlier that an empirical power-law formulation gives very practical results in modeling the constitutive behavior of physiologic tissue; indeed, this is true not only for biological fluids, such as blood, but also for biological soft tissues, such as cardiac muscle. Thus, if we apply an equation such as Eq. (10.3) to experimental data obtained for the passive stress–strain behavior of cardiac muscle tissue in healthy human adults, we find that, typically, such muscle tissue exhibits dilatant characteristics, with, n, a non-Hookean index for soft tissues, lying between 1.4 and 1.6 for cardiac muscle, and K, a nonlinear resiliency modulus for soft tissues lying between 1.5×10^5 and 3.0×10^5 N/m^2 for cardiac muscle. If we let σ be the myocardial tissue normal stress (force per unit area) component in a given direction, and ε be the corresponding strain, then we may write, approximately, for cardiac tissue:

$$\sigma = K\varepsilon^n \tag{10.16}$$

where K and n are as given previously. Then we may use Eq. (10.16) to relate myocardial wall tension to corresponding intraventricular blood pressure for various states of cardiac contraction (ε).

Example Problem 10.6

Suppose, as an approximation, that the ventricular chamber is modeled by an ellipsoid-like prolate spheroid, with principal radii of curvature R_1 and R_2 in two mutually perpendicular cross-sectional planes, respectively, where both R_1 and R_2 vary with time as the heart goes through each cardiac cycle (see accompanying diagram on next page). For any given time, t, let the internal pressure in the ventricle be p; finally, assume that the prolate spheroid has a uniform, thin wall of thickness h, such that the normal wall stress, σ, can be considered to be also uniformly distributed throughout the wall cross section. Under these conditions, derive an equation that relates wall tension, T, per unit length of wall surface area (i.e., $T = \sigma h$) to ventricular pressure, p, for any given values of R_1 and R_2.

Solution

The accompanying figure is a free body diagram of a differential piece of the ventricular wall, subtending the angle $d\theta$ in the cross-sectional plane of the ellipsoid where the instantaneous radius of curvature is R_1, and $d\phi$ in the orthogonal cross-sectional plane of the ellipsoid, where the instantaneous radius of curvature is R_2. The wall stresses normal to the surfaces of the differential piece are designated σ in the circumferential directions (these are so-called "hoop stresses") and are assumed to be uniformly distributed in all directions and across the wall thickness, h. The internal surface of the wall section is subjected to a uniformly distributed pressure, p, and the external surface is assumed (for simplicity) to be free of stress or, certainly, to be subjected to an external pressure distribution that is far less than p.

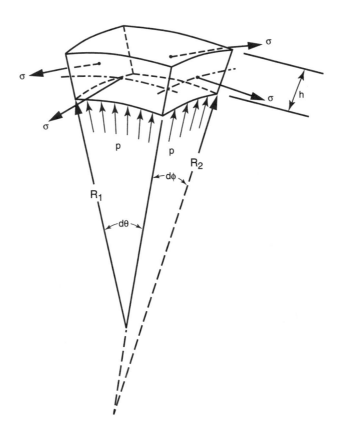

If one now examines a balance of forces in a direction along an axis perpendicular to the internal and external surfaces of our differential volume, and passing through the centroid of said volume, with the outer normal being taken to define the positive direction, then one obtains along this axis, and per unit length of wall surface,

$$\Sigma F_{\substack{\text{normal to} \\ \text{ellipsoid} \\ \text{surface}}} = -2\sigma h(R_2 d\phi) \sin \frac{d\theta}{2} - 2\sigma h(R_1 d\theta) \sin \frac{d\phi}{2}$$

$$+ \, p \cos d\theta (R_1 d\theta R_2 d\phi) = \rho_w (R_1 d\theta R_2 d\phi) h a_n$$

where $\sigma h = T = $ wall tension per unit length of wall surface, ρ_w is the mass density of the ventricular wall, and a_n is the acceleration of the differential piece of the ventricular wall in a direction along the axis under consideration. In the limit, for $d\theta$ small $\left(\sin\dfrac{d\theta}{2}\approx\dfrac{d\theta}{2},\ \cos\theta\approx 1\right)$, $d\phi$ small $\left(\sin\dfrac{d\phi}{2}\approx\dfrac{d\phi}{2}\right)$, and the differential volume element approaching a point ($R_1 d\theta R_2 d\phi h$ approaching zero faster than do the lower-order quantities $R_2 d\phi h$, $R_1 d\theta h$, or $R_1 d\theta R_2 d\phi$, which are surface elements, compared with a volume element), one can write

$$-TR_2 d\phi d\theta - TR_1 d\theta d\phi + pR_1 R_2 d\theta d\phi = 0$$

from which

$$T = p\frac{R_1 R_2}{R_1 + R_2} = \sigma h$$

$$p = \sigma\left[h\left(\frac{1}{R_1} + \frac{1}{R_2}\right)\right]$$

which is known as LaPlace's equation for thin-walled chambers. Experiments have shown that the nondimensional ratio, $h\,(1/R_1 + 1/R_2)$ measured at various locations in the ventricular chambers, is approximately constant, being 0.055 for the right ventricle and 0.360 for the left ventricle. The ratio of the two $0.360/0.055 = 6.5454$ is about the same as the ratio of peak systolic pressures in the two ventricles, which suggests also (from the equation derived previously) that the wall stresses must be about the same in both ventricles. ∎

Under extremely relaxed conditions (such as sleeping) it might take as little as 1–1.5 W of stroke power for the heart to generate the required cardiac output to sustain life. During peak load requirements, the estimated 10^{11} myocytes (muscle cells) of the cardiac musculature can actually generate an order of magnitude more — up to 12 W of stroke power for brief periods of time (such as running a 100-yard dash). In between, sedentary work requires about 4 W of cardiac stroke power, whereas climbing a flight of stairs might require as much as 7 W.

It has been suggested that the cardiac output (in ml/min) is proportional to the weight, W_0 (in kg) of an individual according to the equation:

$$CO = 224 W_0^{3/4} \tag{10.17}$$

and the "normal" heart rate obeys very closely the relation:

$$HR = 229 W_0^{-1/4} \tag{10.18}$$

For a typical 68.7-kg individual (blood volume = 5200 ml), Eqs. (10.15), (10.17), and (10.18) yield: CO = 5345 ml/min, HR = 80 beats/min (cardiac cycle period = 754 ms), and SV = $\dfrac{CO}{HR} = \dfrac{224 W_0^{3/4}}{229 W_0^{-1/4}} = 0.978\,W_0 = 67.2$ ml/beat, which are very reasonable values for routine conditions related to the activities of daily living. Furthermore, assuming this individual lives about 75 years, his or her heart will have cycled over 3.1536 billion times, pumping a total of 0.2107 billion liters of blood (55.665 million gallons or 8134 quarts per day). All this pumping activity will have been accomplished at an extremely low cardiac cycle efficiency (from an engineering point of view) — on the order of only 5% (or even lower, depending on the reference energy input to which the cardiac output power is compared) — because the human heart sacrifices efficiency in favor of effectiveness and endurance.

The blood pressure generated in the ventricles of the heart during the systolic phase of the cardiac cycle is the driving force that propels blood through a major highway network which brings this river of life to within a differential neighborhood of any given cell in the body. Not only does this major highway network — the vascular system — allow for the transport of mass throughout various regions of the body but also, acting as a countercurrent heat exchanger, the vascular system transports energy, giving it a thermoregulatory function as well. That is, when one envisions the anatomical configuration of the vascular network of pipes and channels through which the blood flows as it courses through the body and compares this anatomy with, for example, the radiator of an automobile, the potential for thermoregulation afforded by the cardiovascular system becomes immediately obvious. Like engine coolant coursing through the coils of a radiator, blood in the physiologic system can absorb the heat generated by the complex biochemical reactions of life; also, like engine coolant, blood can carry (convect) heat to the 3000-in.^{2} surface area (skin) of the organism or to the 300 million alveoli of the lungs to be expelled to the environment. Indeed, various estimates suggest that as much as 95% of the heat generated by the body at any given time flows through vascular convection to the cutaneous circulation supplying the outer layers of the skin and to the pulmonary circulation coursing through the lungs, where 60% of it is lost to the environment by radiation, 25% by evaporation (through sweating and latent heat loss), 12% by convection (and sensible heat loss), and 3% by conduction. Conversely, this thermoregulatory role can be reversed when heat is to be conserved by the body. Again, in a manner similar to the thermostat in the radiator of an automobile (which keeps the engine coolant away from the heat-dissipating surface area when the engine is cold), vascular smooth muscle tissue acting at the arteriolar level (the control points of the cardiovascular system) can completely shut off (thus totally bypassing) or significantly reduce flow to the periphery when this becomes necessary to maintain core body temperature at 37°C. Thus, keeping in mind that flow through the vascular system has both a mass-flux function and a thermoregulatory function, we next examine the mechanics of such flow, starting with the basic fluid dynamic field equations that govern the fluid behavior.

10.5 FLUID DYNAMIC FIELD EQUATIONS: CONSERVATION OF MASS, ENERGY, AND MOMENTUM

In our introductory remarks to this chapter, we defined mechanics as the engineering science that describes the behavior of materials subjected to distributed systems of forces. Such behavior may be quantified by defining for a finite quantity, M, of any such material, some intensive property, η, per unit mass of material — η being most appropriately mass, energy, and momentum — to the extent that changes in these variables result from disturbances (forces) imposed on an equilibrated control system of mass, M. Thus, it can be shown that, for any arbitrary open control region, V, in space, bounded by a control surface, A, defined by the outer normal unit vector $\hat{\lambda}$, the following generic relationship (sometimes known as the generalized Reynolds transport equation) holds in the absence of any sources or sinks for η:

$$\underbrace{\left\{ \frac{\partial}{\partial t} \iiint_V \eta \rho dV + \iint_A \eta \rho (\vec{v} \cdot \hat{\lambda}) dA \right\}}_{\text{For control volume, } V} = \underbrace{\left\{ \frac{d(\eta M)}{dt} \right\}}_{\substack{\text{For control} \\ \text{system, } M}} \tag{10.19}$$

where $\vec{v}$ is a characteristic transport velocity for the quantity (ηM) and ρ represents mass density. From the generic Reynolds transport equation (Eq. 10.19), we may write the following specific transport equations — so-called field equations — for the mechanics of solids, fluids, or viscoelastic materials:

1. For the transport of mass: $\eta = M/M = 1$, $dM/dt = 0$ (by definition since M is a fixed control system of finite mass), and Eq. (10.19) reduces to

$$\frac{\partial}{\partial t} \int_V \rho dV + \int_A \rho (\vec{v} \cdot \hat{\lambda}) dA = 0, \tag{10.20}$$

which is known as the equation for conservation of mass or the continuity equation.

2. For the transport of linear momentum: $\eta = M\vec{v}/M = \vec{v}$, and, from Newton's second law of motion, $\frac{d}{dt}(M\vec{v}) = \Sigma \vec{F}_{\text{external}} - \vec{F}_{\text{net}}$, so Eq. (10.19) reduces to

$$\frac{\partial}{\partial t} \int_V \rho \vec{v} dV + \int_A \rho \vec{v} (\vec{v} \cdot \hat{\lambda}) dA = \vec{F}_{\text{net}} \tag{10.21}$$

which is known as the equation for conservation of linear momentum.

3. For the transport of energy: $\eta = E/M$ = internal energy per unit mass (by definition since M is defined to be a fixed quantity of mass, which makes it a closed system in the thermodynamic sense — capable of exchanging only energy, not mass, with its environment and thus experiencing only changes in internal energy) $= u = c_v T_a$, where c_v is the specific heat of the material at constant volume and T_a is its absolute temperature. Thus, from the first law of thermodynamics, which stipulates that the change in total internal energy, dE, of any given system is equal to the heat, dQ, added to the system minus the work, dW, done by the system, we may write Eq. (10.19) as

$$\frac{\partial}{\partial t} \int_V \rho u dV + \int_A \rho u (\vec{v} \cdot \hat{\lambda}) dA = \frac{dQ}{dt} - \frac{dW}{dt} \tag{10.22}$$

which is known as the equation for conservation of energy.

Note that Eq. (10.20)–(10.22), which are actually five in number because the vector Eq. (10.21) can actually be written as three orthogonal scalar component equations, still contain (most generally) eight unknown quantities: three velocity components, three components of force, mass density, and temperature. Thus, additional constitutive and entropic relationships, such as the perfect gas law, Newton's law of dynamic viscosity, and the second law of thermodynamics, need to be introduced if one is to get unique solutions to well-posed problems. We may eliminate one unknown quantity, mass density, ρ, by assuming (as is appropriate for blood) that the material is incompressible, which also uncouples the continuity and momentum equations from the energy equation and eliminates this variable entirely from Eq. (10.20), reducing it to

$$\frac{\partial V}{\partial t} + \int_A (\vec{v} \cdot \hat{\lambda}) dA = 0 \qquad (10.23)$$

If we further assume (for now) that the flow is steady ($\partial/\partial t = 0$) and confined to cylindrical channels with impermeable walls that deviate only slightly from being perfectly parallel (i.e., the flow is essentially one-dimensional) then between any two sections, 1 and 2 of the channel, Eq. (10.23) stipulates that

$$A_1 v_1 = A_2 v_2 = q \qquad (10.24)$$

where A_1 is the cross-sectional area of the channel at point 1, where the velocity normal to A_1 is $-v_1$ (into the control volume); A_2 is the cross-sectional area of the channel at point 2, where the velocity normal to A_2 is v_2 (out of the control volume); and q is the volumetric discharge through the control volume.

Example Problem 10.7

At the root of the aorta, just as it leaves the left ventricle, the blood vessel has an internal diameter approximately equal to 2.815 cm, through which flows blood at a mean velocity corresponding to a cardiac output of 5345 ml/min. What is the mean velocity of blood flow at this point and how does it compare to a typical maximum velocity at the peak of systole? To what corresponding flow rate does the maximum velocity correspond?

Solution

The cross-sectional area of the root of the aorta is $\dfrac{\pi D_0^2}{4} = \dfrac{\pi (2.815)^2}{4} = 6.224 \text{ cm}^2$.
The volumetric discharge out of the heart is $q = 5345$ ml/min $= 89.083 \text{ cm}^3/\text{s}$. Thus, the mean velocity of blood flow at this point is

$$\frac{89.083 \text{ cm}^3}{6.224 \text{ cm}^2 \text{s}} = 14.31 \text{ cm/s} = 0.1431 \text{ m/s}$$

which is on the order of only 10% of the maximum velocity at the peak of systole — the velocity of which corresponds to a momentary, instantaneous volumetric flow rate of

$$6.224 \text{ cm}^2 \times \frac{14.31 \text{ cm}}{(0.1) \text{ s}} = 890.833 \text{ cm}^3/\text{s} = 53,450 \text{ ml/min} \qquad \blacksquare$$

In keeping with the assumptions imposed thus far, the energy equation (Eq. 10.22) applied to a steady-state situation becomes

$$\int_A \rho u (\vec{v} \cdot \hat{\lambda}) dA = \frac{dQ}{dt} - \frac{dW}{dt} \qquad (10.25)$$

Disregarding for now any body forces that derive from subnuclear, nuclear, electrical, magnetic, or free surface tension effects, we note that the internal energy of a pure thermodynamic system (fixed quantity of mass) consists of its energy due to motion (i.e., kinetic energy per unit mass, $\frac{1}{2}v^2$) plus its energy due to state (i.e., intrinsic energy per unit mass, e_i, due to intermolecular forces that derive from spacing and other configurational, interactive effects and that are dependent on state variables such as thermodynamic pressure, mass density, or absolute temperature) plus its energy due to position (i.e., gravitational potential energy per unit mass, $g_0 z$, due to relative height, z, above the surface of the earth). Moreover, we note further that the work done by a thermodynamic system consists basically of a reversable energy-conserving portion — pdV, due to pressure forces acting on moving boundaries (sometimes called "flow work") — plus an irreversible, energy-dissipating portion, designated W_S due to shearing forces and commonly called "shaft work" because of one particular example which involves the torque applied to a rotating shaft to keep it spinning. Finally, if we temporarily ignore any heat transfer, dQ, between the thermodynamic system and its environment, Eq. (10.25) reduces further to

$$\int_A \rho \left[\frac{1}{2} v^2 + e_i + g_0 z \right] (\vec{v} \cdot \hat{\lambda}) dA = - \left\{ \frac{d}{dt} [pdV + W_S] \right\} \qquad (10.26)$$

For steady-state pressure not a function of time, the divergence theorem tell us that $\frac{d}{dt}(pdV) = p\frac{dV}{dt}$ may be written as $p(\vec{v} \cdot \hat{\lambda})dA$, such that Eq. (10.26) becomes

$$\int_A \rho \left[\frac{p}{\rho} + \frac{v^2}{2} + g_0 z + e_i \right] (\vec{v} \cdot \hat{\lambda}) dA = - \frac{dW_S}{dt} \qquad (10.27)$$

Once again, for incompressible flow confined to cylindrical channels with impermeable, nearly parallel, rigid walls, we may integrate Eq. (10.27) between any two points in the channel to get

$$\left(\frac{p_2}{\rho_2} + \frac{v_2^2}{2} + g_0 z_2 + e_{i_2} \right) \rho_2 v_2 A_2$$

$$- \left(\frac{p_1}{\rho_1} + \frac{v_1^2}{2} + g_0 z_1 + e_{i_1} \right) \rho_1 v_1 A_1 = - \int_A \frac{dW_S}{dt} \qquad (10.28)$$

and since $\rho_1 = \rho_2 = \rho$ for incompressible flow and $A_1 v_1 = A_2 v_2 = q$ from the continuity equation (Eq. 10.24) it is convenient to divide Eq. (10.28) by ρq and, to simply, let $\frac{1}{\rho q} \int_A \frac{dW_S}{dt} =$ "Losses from 1 to 2," per unit mass of fluid flowing, to get

$$\frac{p_1}{\rho} + \frac{v_1^2}{2} + g_0 z_1 + e_{i_1} = \frac{p_2}{\rho} + \frac{v_2^2}{2} + g_0 z_2 + e_{i_2} + losses_{1-2} \qquad (10.29)$$

One simplified form of Eq. (10.29), where losses and intrinsic energy effects are neglected, is the Bernoulli equation applied along a fluid streamline, which energy equation asserts that potential and kinetic energies are conserved, i.e., that

$$\frac{p}{\rho} + \frac{v^2}{2} + g_0 z = \text{constant} \tag{10.30}$$

It would be beneficial for the student to review all the assumptions that are inherent in the Bernoulli equation!

Example Problem 10.8

Suppose the heart (in particular, the left ventricle, where there is no appreciable kinetic energy) is taken to be a reference point ($z = 0$) for the measurement of hydrostatic pressure, p. At $t = 0$ in the cardiac cycle, calculate p in the ankle of an individual standing erect if the joint is located 1.2 m below the level of the heart. Give your answer both in Pascals and in units of mmHg.

Solution

Equation (10.30) can be applied to get

$$\frac{p_{\text{ventricle}}}{\rho_{\text{blood}}} + \frac{v^2_{\text{ventricle}}}{2} + g_0 z_{\text{ventricle}} = \frac{p_{\text{ankle}}}{\rho_{\text{blood}}} + \frac{v^2_{\text{ankle}}}{2} + g_0 z_{\text{ankle}}$$

At $t = 0$: From Fig. 10.6, $p_{\text{ventricle}} = 2\text{–}3$ mmHg ~ 0

$v^2_{\text{ventricle}} \sim 0$ since there is no appreciable kinetic energy in the ventricle

$z_{\text{ventricle}} = 0$ since it is a reference point

$v^2_{\text{ankle}} \sim 0$ if we assume that the blood at this point comes momentarily to rest as it reverses direction from a downward flow away from the heart to an upward flow back toward the heart

Then $p_{\text{ankle}} = -\rho_{\text{blood}} g_0 z_{\text{ankle}}$, which is known as Pascal's law for hydrostatic ($v \sim 0$) pressure; $\rho_{\text{blood}} = 1.057 \text{g/cm}^3 = 1.057 \text{dyn-s}^2/\text{cm}^4 = 1.057 \times 10^{-5}$ N-s^2/10^{-8}m^4;

$z_{\text{ankle}} = -1.2$ m and $g_0 = 9.806$ m/s^2

$p_{\text{ankle}} = -(1.057 \times 10^3)(9.806)(-1.2)$

$= 12{,}438$ N/m$^2 = 12{,}438$ Pa

The hydrostatic pressure corresponding to 1 mmHg is 133.3616 Pa, so 12,438 Pa converts to approximately 93 mmHg, which is quite close to the ankle blood pressures actually measured.   ■

As mentioned earlier, the losses in Eq. (10.29) are due mainly to the effects of fluid shear, which can be accounted for by the momentum equation (Eq. 10.21). For steady, equilibrated flow, the momentum equation reduces to $\vec{F}_{\text{net}} = 0$ on any given fluid volume element. Thus, consider again the volume element between two sections, 1 and 2, of a cylindrical channel with impermeable, quasi-parallel walls. Let this volume element have an arbitrary radius, r, less than the nominal lumen radius R of the channel, and a length, l, between axial locations 1 and 2. Let p_1 be the pressure along the cross-sectional area A_1 at the upstream location 1 of l, p_2 be the pressure along the cross-sectional area A_2 at the downstream location 2 of l, and τ be the shearing stress acting along the side elements of the cylindrical volume, with τ being directed parallel to the walls of the channel and opposite to the direction of steady fluid flow, which is assumed to be from 1 toward 2. Then, for steady, equilibrated laminar flow in this channel, conservation of momentum requires that

$$p_1 A_1 - p_2 A_2 - \tau(2\pi r l) = \vec{F}_{net} = 0 \tag{10.31}$$

where body forces are not considered to be acting on the free body under consideration. Since $A_1 = \pi r^2 \cong A_2$ under the assumption of nearly parallel walls, Eq. (10.31) becomes

$$\tau = \frac{r}{2l}(p_1 - p_2) \tag{10.32}$$

which indicates that the axial pressure gradient, $(p_1 - p_2)/l \rightarrow \partial p/\partial x$ for l infinitesimally small, is necessary to overcome fluid shear, τ, acting at the arbitrary radial location, r. The latter, in turn, has propagated radially inward from the wall surface, where friction impedes and arrests fluid motion (the so-called "no-slip" boundary condition), according to a constitutive equation such as Eq. (10.3) for power-law fluids. In particular, for the fully developed (velocity not a function of the axial coordinate, x), axially symmetric [velocity independent of the azimuthal (angular) coordinate in the plane of the flow channel cross-section], steady (velocity not a function of time) flow of a power-law fluid exhibiting negligible Bingham plastic behavior ($T^* = 0$), Eq. (10.3) reduces to the simple form

$$\tau = K\left(\frac{dv}{dr}\right)^n \tag{10.33}$$

if the fluid is also considered to be essentially homogeneous, isotropic, incompressible, and in a state of thermodynamic equilibrium. Note that dv/dr is a negative quantity in Eq. (10.33) if r is measured radially outward from the axis of the channel, since then v decreases with increasing r toward the walls of the flow tube. Taking this fact into consideration, substituting Eq. (10.33) into Eq. (10.32) and rearranging yields

$$-dv = \left\{\frac{1}{2Kl}(p_1 - p_2)\right\}^{1/n} r^{1/n} dr$$

which can be integrated to get

$$-v = \left\{\frac{1}{2Kl}(p_1 - p_2)\right\}^{1/n}\left[\frac{n}{1+n}\right] r^{(1+n)/n} + \text{constant}$$

We can determine the constant of integration by imposing the no-slip boundary condition, which requires that $v = 0$ at the wall, $r = R$. Thus,

$$\text{constant} = -\left\{\frac{1}{2Kl}(p_1 - p_2)\right\}^{1/n}\left[\frac{n}{1 + n}\right]R^{(1+n)/n}$$

so that

$$v = \left\{\frac{1}{2Kl}(p_1 - p_2)\right\}^{1/n}\left[\frac{n}{1 + n}\right]R^{(1+n)/n}\left\{1 - \left[\frac{r}{R}\right]^{(1+n)/n}\right\} \tag{10.34}$$

The volumetric flow rate, q, through the channel can be obtained by examining a differential annulus of fluid located at r, of thickness dr, through which the flow rate is $dq = v(r)dA = (2\pi rdr)v(r)$, and integrating dq across the channel cross section using Eq. (10.34). The result of such an integration is

$$q = \int dq = \left\{\frac{\Delta pR}{2Kl}\right\}^{1/n}\frac{\pi R^3 n}{3n + 1} \tag{10.35}$$

from which one may determine the mean velocity of flow, q/A (where $A = \pi R^2$), to be

$$v_{\text{mean}} = \left\{\frac{\Delta pR}{2Kl}\right\}^{1/n}\frac{nR}{3n + 1} \tag{10.36}$$

where $\Delta p = p_1 - p_2 = $ the pressure drop across the fluid volume element of length l. Finally, substituting Eq. (10.36) into Eq. (10.34), and again doing some rearranging, one can write

$$\left(\frac{v}{v_{\text{mean}}}\right) = \frac{3n + 1}{n + 1}\left[1 - \left(\frac{r}{R}\right)^{(n+1)/n}\right] \tag{10.37}$$

which nondimensionalized equation can be plotted to give velocity profiles (v/v_{mean}) as a function of (r/R), with n acting as a parameter.

For $n = 1$, Eq. (10.37) reduces to the classic, parabolic velocity profiles that characterize laminar Poiseuille flow of a Newtonian fluid, i.e.,

$$\frac{v}{v_{\text{mean}}} = 2\left[1 - \left(\frac{r}{R}\right)^2\right] \tag{10.38}$$

where $v_{\text{max}} = 2v_{\text{mean}}$ along the centerline ($r = 0$) of the channel, and $v = 0$ at the wall surface ($r = R$). At the one extreme, $n = 0$, Eq. (10.37) becomes $v/v_{\text{mean}} = 1$ or $v = v_{\text{mean}} = $ constant, since $\frac{r}{R}$, a quantity having a value always less than 1 (except at the wall, where $r/R = 1$ and $v = 0$ for any value of n), goes to zero when raised to an infinite power (i.e., $(n + 1)/n = 1 + (1/n) \rightarrow \infty$ as $n \rightarrow 0$). Thus, $n = 0$ represents the behavior of an inviscid fluid (as mentioned earlier, an idealized rigid body that does not deform at all when subjected to a disturbing force but simply moves with some uniform velocity v_{mean}), having a flat velocity profile that experiences a sudden discontinuity ($dv/dr \rightarrow \infty$ at $r = R$) at the wall surface (again, an idealization). At the other extreme, $n = \infty$, $1/n \to 0$, so Eq. (10.37) reduces to the straight line:

$$\frac{v}{v_{\text{mean}}} = 3\left(1 - \frac{r}{R}\right) \tag{10.39}$$

with $v_{\text{max}} = 3v_{\text{mean}}$ at $r = 0$ and, again, $v = 0$ at $r = R$. Note, however, that this velocity profile is also an idealization in that now there is a discontinuity at the centerline of the channel, where the conical volume represented by the three-dimensional representation of Eq. (10.39) comes to a sharp point at which there is again an infinite change in slope. Thus, in real life, $0 < n < \infty$, and for human blood, $0.67 < n < 1.00$, as already discussed.

Example Problem 10.9

All other factors being equal, compare the volumetric flow rate in a given channel containing a pseudoplastic ($n < 1$) fluid to that in the same channel containing a Newtonian ($n = 1$) fluid when both fluids are driven by the same pressure gradient and explain, especially in the case of blood flow, why there is a difference between the two flows.

Solution

From Eq. (10.35), for $n = 1$,

$$q = \frac{\Delta p R^4 \pi}{8\mu l}$$

where $K \to \mu$ for a Newtonian fluid, and so q is directly proportional to Δp, all other factors being held constant. Moreover, for a pseudoplastic fluid such as blood, for which n is typically 0.7844 (see Table 10.4), Eq. (10.35) yields

$$q = \frac{\Delta p^{1.275} R^{4.275} \pi}{10.344(Kl)^{1/n}}$$

Thus, $q_{\text{non-Newtonian}}/q_{\text{Newtonian}} = 0.77 \dfrac{\mu}{K(Kl)^{0.275}} > 1$, for any values of the variables involved, or, $q_{\text{non-Newtonian}} > q_{\text{Newtonian}}$ since now q is proportional to $\Delta p^{1.275}$ which is bigger than Δp, all other parameters being equal. In fact, the limit of Eq. (10.35) as $n \to 0$ is infinite because $\{\Delta p R/2Kl > 1\}^{1/n}$ goes to infinity much faster than $(\pi R^3)/(3 + (1/n))$ goes to zero as $n \to 0$. The pseudoplastic flow is therefore higher than the Newtonian flow for the same pressure gradient in the same tube. The reason that this is so has to do with the fact that pseudoplastic velocity profiles are blunter than Newtonian velocity profiles because with decreasing n, the curves progressively go from parabolas ($n = 1$) to uniformly flat, inviscid ($n = 0$) straight lines. Thus, pseudoplastic fluids appear to be less viscous than corresponding Newtonian fluids and so yield higher flow rates for the same driving pressures. In the case of blood flow, a blunting of the velocity profiles (so-called "plug flow") is due to the axial accumulation of RBCs (due partly to a magnus effect), a streamline RBC alignment (with the development of a consequent "plasma skimming layer") also known as a shear effect, a sigma effect, and a consequent so-called Fahraeus–Lindqvist effect — all of which will be discussed in the next section. Suffice it to observe further, for now, that the dependence of blood flow on the fourth (and higher) power of the tube radius makes small vessels highly resistant to flow and also gives the body a convenient means for regulating blood pressure by controlling the internal diameter of downstream resistance vessels (discussed later). In fact, when internal radii drop below 1 mm, blood basically ceases to act as a liquid and must be examined as a composite, microcontinuum, with anisotropic, nonhomogeneous, multifluid viscoelastic properties. ■

The generic Eqs. (10.19)–(10.22) are valid for any continuum (solid, fluid, or viscoelastic) within which the quantity η (mass, energy, or momentum) is not being created (source) or destroyed (sink). More specific forms of these equations will apply to different situations, such as those defined by Eqs. (10.23)–(10.39), depending on the nature of the material involved (as quantified by its corresponding constitutive equation) and the time and space boundary conditions that constrain any given physical process. We have already discussed the constitutive properties of blood and the characteristics of the pumping mechanisms that propel it through the vascular system. We now address some of the boundary conditions that constrain flow through the vascular system in order to see how the equations developed in this section may be applied to the study of blood flow (hemodynamics) in arteries and veins. As usual, we will begin with a description of the anatomy and physiology of blood vessels as they relate to the function of these cardiovascular flow channels.

10.6 HEMODYNAMICS IN VASCULAR CHANNELS: ARTERIAL (TIME DEPENDENT) AND VENOUS (STEADY)

The vascular system is divided by a microscopic capillary network into an upstream, high-pressure, efferent arterial side (Table 10.5) consisting of relatively thick-walled, viscoelastic tubes that carry blood away from the heart, and a downstream, low-pressure, afferent venous side (Table 10.6) consisting of correspondingly thinner (but having a larger caliber) elastic conduits that return blood back to the heart. Except for their difference in thickness, the walls of the largest arteries and veins consist of the same three distinct, well-defined, and well-developed layers. From innermost to outermost, these layers are the thinnest tunica intima — a continuous lining (the vascular endothelium) consisting of (i) a single layer of simple, squamous (thin, sheet-like) endothelial cells "glued" together by a polysaccharide (sugar) intercellular matrix, surrounded by (ii) a thin layer of subendothelial connective tissue, interlaced with a number of circularly arranged elastic fibers to form the subendothelium, and separated from the next adjacent wall layer by (iii) a thick elastic band called the internal elastic lamina; the thickest tunica media — composed of (i) numerous circularly arranged elastic fibers, especially prevalent in the largest blood vessels on the arterial side (allowing them to expand during systole and to recoil passively during diastole), a significant amount of (ii) smooth muscle cells arranged

TABLE 10.5 Arterial System[a]

Blood vessel type	Systemic typical number	Internal diameter range	Length range[b]	Wall thickness	Systemic volume (ml)	Pulmonary typical number	Pulmonary volume (ml)
Aorta	1	1.0–3.0 cm	30–65 cm	2–3 mm	156	—	—
Pulmonary artery	—	2.5–3.1 cm	6–9 cm	2–3 mm	—	1	52
Wall morphology: Complete tunica adventitia, external elastic lamina, tunica media, internal elastic lamina, tunica intima, subendothelium, endothelium, and vasa vasorum vascular supply							
Main branches	32	5 mm–2.25 cm	3.3–6 cm	≈2 mm	83.2	6	41.6
Along with the aorta and pulmonary artery, the largest, most well-developed of all blood vessels							
Large arteries	288	4.0–5.0 mm	1.4–2.8 cm	≈1 mm	104	64	23.5
A well-developed tunica adventitia and vasa vasorum, although wall layers are gradually thinning							
Medium arteries	1,152	2.5–4.0 mm	1.0–2.2 cm	≈0.75 mm	117	144	7.3
Small arteries	3,456	1.0–2.5 mm	0.6–1.7 cm	≈0.50 mm	104	432	5.7
Tributaries	20,736	0.5–1.0 mm	0.3–1.3 mm	≈0.25 mm	91	5,184	7.3
Well-developed tunica media and external elastic lamina, but tunica adventitia virtually nonexistent							
Small rami	82,944	250–500 μm	0.2–0.8 cm	≈125 μm	57.2	11,664	2.3
Terminal branches	497,664	100–250 cm	1.0–6.0 mm	≈60 μm	52	139,968	3.0
A well-developed endothelium, subendothelium, and internal elastic lamina, plus about two or three 15-μm-thick concentric layers forming a very thin tunica media; no external elastic lamina							
Arterioles	18,579,456	25–100 μm	0.2–3.8 mm	≈20–30 μm	52	4,094,064	2.3
Wall morphology: More than one smooth muscle layer (with nerve association in the outermost muscle layer), a well-developed internal elastic lamina; gradually thinning in 25 to 50 μm vessels to a single layer of smooth muscle tissue, connective tissue, and scant supporting tissue							
Metarterioles	238,878,720	10–25 μm	0.1–1.8 mm	≈5–15 μm	41.6	157,306,536	4.0
Well-developed subendothelium; discontinuous contractile muscle elements; one layer of connective tissue							
Capillaries	16,124,431,360	3.5–10 μm	0.5–1.1 mm	0.5–1 μm	260	3,218,406,696	104
Simple endothelial tubes devoid of smooth muscle tissue; one cell-layer-thick walls							

[a] Values are approximate for a 68.7-kg individual having a total blood volume of 5200 ml.
[b] Average uninterrupted distance between branch origins (except aorta and pulmonary artery, which are total length).

Reprinted (with modifications) with permission, from: Schneck, Daniel J., "An Outline of Cardiovascular Structure and Function," in *The Biomedical Engineering Handbook*, edited by Joseph D. Bronzino. Copyright CRC Press, Boca Raton, Florida. © 1995. Visit the CRC Press Home Page at: http://www.crcpress.com

TABLE 10.6 Venous System

Blood vessel type	Systemic typical number	Internal diameter range	Length range	Wall thickness	Systemic volume (ml)	Pulmonary typical number	Pulmonary volume (ml)
Postcapillary venules	4,408,161,734	8–30 μm	0.1–0.6 mm	1.0–5.0 μm	166.7	306,110,016	10.4
	Wall consists of thin endothelium, exhibiting occasional pericytes (pericapillary connective tissue cells) which increase in number as the vessel lumen gradually increases						
Collecting venules	160,444,500	30–50 μm	0.1–0.8 mm	5.0–10 μm	161.3	8,503,056	1.2
	One complete layer of pericytes; one complete layer of veil cells (veil-like cells forming a thin membrane); occasional primitive smooth muscle tissue fibers that increase in number with vessel size						
Muscular venules	32,088,900	50–100 μm	0.2–1.0 mm	10–25 μm	141.8	3,779,136	3.7
	Relatively thick wall of smooth muscle tissue						
Small collecting veins	10,241,508	100–200 μm	0.5–3.2 mm	≃30 μm	329.6	419,904	6.7
	Prominent tunica media of continuous layers of smooth muscle cells						
Terminal branches	496,900	200–600 μm	1.0–6.0 mm	30–150 μm	206.6	34,992	5.2
	A well-developed endothelium, subendothelium, and internal elastic lamina; well-developed tunica media but fewer elastic fibers than corresponding arteries and much thinner walls						
Small veins	19,968	600 μm–1.1 mm	2.0–9.0 mm	≃0.25 mm	63.5	17,280	44.9
Medium veins	512	1–5 mm	1–2 cm	≃0.50 mm	67.0	144	22.0
Large veins	256	5–9 mm	1.4–3.7 cm	≃0.75 mm	476.1	48	29.5
	Well-developed wall layers comparable to large arteries, but about 25% thinner						
Main branches	224	9.0 mm–2.0 cm	2.0–10 cm	≃1.00 mm	1,538.1	16	39.4
	Along with the vena cava and pulmonary veins, the largest, most well-developed of all blood vessels						
Vena cava	1	2.0–3.5 cm	20–50 cm	≃1.50 mm	125.3	—	—
Pulmonary veins	—	1.7–2.5 cm	5–8 cm	≃1.50 mm	—	4	52

Wall morphology: Essentially the same as comparable major arteries but a much thinner tunica intima, a much thinner tunica media, and a somewhat thicker tunica adventitia; contains vasa vasorum

Total systemic blood volume: 4394 ml − 84.5% of total blood volume; 19.5% in arteries (~3:2 large:small), 5.9% in capillaries, 74.6% in veins (~3:1 large:small); 63.0% of volume is in vessels >1 mm internal diameter

Total pulmonary blood volume: 468 ml − 9.0% of total blood volume; 31.8% in arteries, 22.2% in capillaries, 46% in veins; 58.3% of volume is in vessels >1 mm internal diameter; remainder of blood is in heart

Reprinted (with modifications) with permission, from: Schneck, Daniel J., "An Outline of Cardiovascular Structure and Function," in *The Biomedical Engineering Handbook*, edited by Joseph D. Bronzino. Copyright CRC Press, Boca Raton, Florida. © 1995. Visit the CRC Press Home Page at: http://www.crcpress.com

in spiraling layers around the vessel wall, especially prevalent in medium-size arteries and arterioles (allowing them to function as control points for blood distribution), and (iii) some interlacing collagenous connective tissue, elastic fibers, and intercellular mucopolysaccharide substance (extractives), all separated from the next adjacent wall layer by (iv) another thick elastic band called the external elastic lamina; and the medium-sized tunica adventitia — an outer vascular sheath consisting entirely of connective tissue.

The largest blood vessels, such as the aorta, the pulmonary artery, and the pulmonary veins, have such thick walls that they require a separate network of tiny blood vessels (the vasa vasorum) just to service the vascular tissue. As one moves toward the capillaries from the arterial side (Table 10.5), the vascular wall keeps thinning, as if it were shedding 15-μm-thick, onion peel-like concentric layers. Also, while the percentage of water in the vessel wall stays relatively constant at 78% (by weight), the ratio of elastin to collagen decreases (actually reverses) from 3 : 2 in large arteries (9% elastin and 6% collagen by weight) to 1 : 2 in small tributaries (5% elastin and 10% collagen), and the amount of smooth muscle tissue increases from 7.5% by weight (most of it water and protein) in large arteries to 15% in small tributaries. By the time one reaches the capillaries, one encounters single-cell-thick endothelial tubes, devoid of any smooth muscle tissue, elastin, or collagen, downstream of which the vascular wall gradually "reassembles itself," layer by layer, as it directs blood back to the heart through the venous system (Table 10.6).

10.6.1 Anatomy and Physiology of the Human Vascular System: Angiology

Blood vessel structure is directly related to function. The thick-walled large arteries and main distributing branches that direct blood to a major systemic location are designed to withstand the pulsating 80–130 mmHg blood pressures and high-speed flows that they must endure. The smaller elastic conducting vessels that direct blood specifically through the vascular bed unique to the systemic location involved need only operate under steadier blood pressures in the range 70–90 mmHg, but they must be thin enough to penetrate and course through organs and tissues without unduly disturbing the anatomical integrity of the mass involved. Controlling arterioles operate at much lower fluid speeds, and at blood pressures between 50 and 70 mmHg, but they are heavily endowed with smooth muscle tissue (hence they are referred to as muscular vessels) so that they may be actively shut down when flow to the capillary bed they service is to be restricted (for any reason); the smallest capillary resistance vessels (which operate at blood pressures on the order of 10–50 mmHg) are designed to optimize conditions for transport to occur between blood and the surrounding interstitial fluid. Traveling back up the venous side, one encounters relatively steady blood pressures continuously decreasing from approximately 30 mmHg all the way down to near zero, so these vessels can be thin-walled without significant consequence. However, the gravity-impeded, low blood pressure, slower and steady (time-independent) flow, thin walls, and larger caliber that characterize the venous system causes blood to tend to pool in veins, allowing them to act like reservoirs; hence, they are referred to as capacitance vessels. It is not surprising that at any given instant, one normally finds about two-thirds of the total human blood volume residing in the venous system (up to 75% on the systemic side; down to 46% on the pulmonary side), with the remaining one-third being divided among the heart (6.5%), the microcirculation (7% combined in systemic and pulmonary capillaries), and the arterial system (19.5–20%). Table 10.7 and Fig. 10.7 provide a more specific breakdown of the distribution of blood by volume (i.e., the quantity of blood that the given vascular bed can hold as a result of its nominal geometric dimensions, size, and caliber), by flow (i.e., the quantity of blood streaming through the given vascular bed, as a typical percentage of cardiac output), and by pressure (in mmHg) as one travels among the various organs, tissues, and blood vessels of a typical 68.7-kg individual. Note, in particular (Fig. 10.7), the differences from a fluid dynamic point of view between flow on the arterial side of the vascular system (upstream of capillaries) and that on the venous side of the vascular system (downstream of capillaries) — the former being high-speed, high-pressure, pulsatile (unsteady), gravity-aided flow through thick-walled, viscoelastic, valve-free channels having a cumulative cross-sectional area that increases as the fluid moves away from the heart (unstable diverging-flow) and in which turbulence and flow disturbances ("anomalies") are fairly common — of a complex fluid that is even chemically significantly different from that in the latter case, which is experiencing low-speed, low-pressure, steady, gravity-impeded flow through thin-walled, linearly elastic, valve-endowed (to prevent backflow) channels having a cumulative cross-sectional area that decreases as the fluid moves back toward the heart (stable, converging flow) and in which turbulence and flow anomalies are much less common.

Example Problem 10.10

The coefficient of friction (or drag coefficient) for steady, laminar flow in a pipe is defined by

$$C_f = \frac{\tau_w}{\frac{1}{2}\rho_{blood}v_{mean}^2}$$

where τ_w is the shear stress at the wall. Moreover, if C_f is also given by $16/Re$ for steady, laminar flow, where Re = the **Reynolds number** (ratio of fluid convected inertial effects to viscous effects) of the flow, use Eq. (10.35) and the definition of C_f to derive a generalized Reynolds number for a power-law fluid.

Solution

From the definition of $C_f = \dfrac{\tau_w}{1/2(\rho v^2)} = \dfrac{16}{Re}$,

$$Re = \frac{8\rho_{blood}v_{mean}^2}{\tau_w}$$

Now, from Eqs. (10.24) and (10.35),

$$\frac{q}{A} = v_{mean} = \frac{n\pi R^3}{3n+1}\left[\frac{\Delta p R}{2Kl}\right]^{1/n}\frac{1}{\pi R^2}$$

TABLE 10.7 Blood Distribution by Volume and Flow

Physiologic structure	% of total blood volume	Typical value (ml)	% of total blood flow	Typical value (ml/min)
Aorta	3.00	156.00	(100.0)	5345 at the aortic valve
All vascular vasa vasorum	0.20	10.40	Negligible	
	5.0% loss to coronary circulation			5078 in the aortic arch
	13% loss to cerebral circulation (brain) + 15% loss to circulation of the head, neck, and arms			3581 in the thorax
	3% loss to bronchial circulation (lungs) + 2.0% loss to circulation of the thoracic cavity			3314 in the abdomen
	20.5% loss to splanchnic circulation (GI tract) + 6.0% loss to hepatic circulation (liver) + 21.5% loss to renal circulation (kidney) + 2.0% loss to circulation of the abdominal cavity			636 at end to all lower extremities
	11.9% (636 ml/min) loss to circulation of the legs and lower body			
Arteries, large distributing	3.60	187.20	—	—
Brain: Cerebral circulation	1.50	78.00	13.0	694.85
Gastrointestinal tract:				
Splanchnic circulation	9.10	473.20	20.6	1101.070
Spleen: Splenic circulation	(1.50)	(78.00)	(3.67)	(196.338)
Pancreas	(0.60)	(31.20)	(1.05)	(55.933)
Stomach	(2.00)	(104.0)	(1.33)	(71.153)
Intestines	(5.00)	(260.0)	(14.55)	(777.646)
Liver: Portal circulation	—	—	(20.60)	(1101.070)
(fed from splanchnic circulation, 80% from mesenteric stomach and intestines, 20% from pancreas and spleen)				
+ hepatic circulation	7.00	364.00	6.0	320.700
Heart chambers	6.50	338.00	—	—
Heart wall: Coronary circulation	1.50	78.00	5.00	267.250
Kidneys: Renal circulation	1.00	52.00	21.50	1149.175

(continued)

TABLE 10.7 Blood Distribution by Volume and Flow (*Continued*)

Physiologic structure	% of total blood volume	Typical value (ml)	% of total blood flow	Typical value (ml/min)
Lungs				
Pulmonary circulation	9.00	468.00	(97.0)	(5184.65) at the pulmonic valve
(5345 ml/min drains back into the left atrium)				
Bronchial circulation	1.00	52.00	3.00	160.35
(Bronchial circulation feeds back into pulmonary circulation)				
Lymphatic system (nodes)	0.29	15.08	~0.1%	5.345
Muscle, skeletal (inactive)	14.00	728.00	15.00	801.750
Skeleton	7.00	364.00	5.60	299.320
Active bone marrow	(4.00)	(208.00)	(3.36)	(179.592)
Inactive bone marrow	(1.00)	(52.00)	(0.56)	(29.932)
Skeleton minus marrow	(2.00)	(104.00)	(1.68)	(89.796)
Skin: Integumentary circulation	5.00	260.00	8.00	427.600
Adipose tissue	6.00	312.00	(6.61)	(353.305)
Thyroid gland	0.07	3.64	1.8%	96.210
Testes	0.04	2.08	Negligible	Negligible
Veins, large collecting	17.79	925.08	—	—
Vena cava	2.41	125.32	(97.0)	(5184.65) at the right atrium)
All other tissues	4.00	208.00	0.4%	21.380
Totals	100.00	5200.00	100.00	5345.000

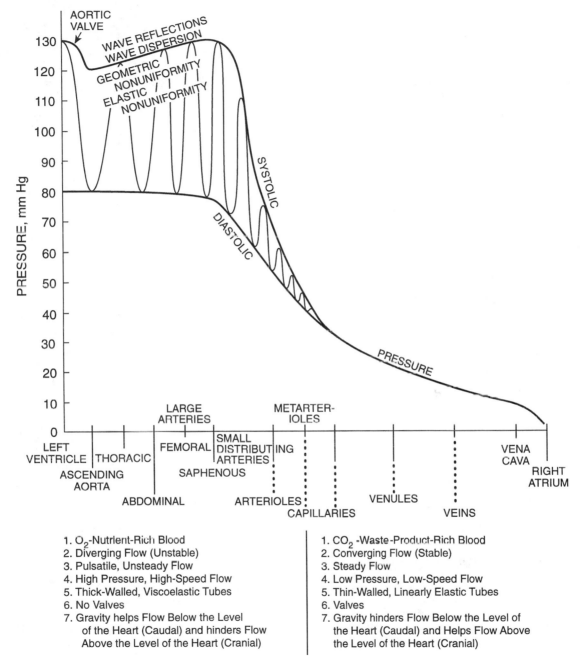

1. O₂-Nutrient-Rich Blood
2. Diverging Flow (Unstable)
3. Pulsatile, Unsteady Flow
4. High Pressure, High-Speed Flow
5. Thick-Walled, Viscoelastic Tubes
6. No Valves
7. Gravity helps Flow Below the Level of the Heart (Caudal) and hinders Flow Above the Level of the Heart (Cranial)

1. CO₂-Waste-Product-Rich Blood
2. Converging Flow (Stable)
3. Steady Flow
4. Low Pressure, Low-Speed Flow
5. Thin-Walled, Linearly Elastic Tubes
6. Valves
7. Gravity hinders Flow Below the Level of the Heart (Caudal) and Helps Flow Above the Level of the Heart (Cranial)

Fig. 10.7 Pressure variations in the systemic circulation. Note especially the drop in systolic blood pressure across the aortic valve, and the increase in pulse pressure (systolic-diastolic) — due to a combination of wave reflections, wave dispersion (a Fourier effect), geometric and elastic nonuniformities — as one moves initially away from the heart, following which the pulse pressure completely attenuates by the time one reaches the capillaries.

and from Eq. (10.32)

$$\tau_w = \frac{R}{2l}\Delta p = v_{mean}^n \left\{\frac{nR}{3n+1}\right\}^{-n} K$$

Substituting the previous results into the generalized definition of C_f, and algebraically simplifying, gives the final result:

$$Re = \frac{\rho_{blood} v_{mean}^{2-n} R^n}{\left\{\dfrac{3n+1}{n}\right\}^n \dfrac{K}{8}}$$

Note: For $n = 1$ (Newtonian fluid), we get the more standard definition of the Reynolds number $= \dfrac{\rho v_{mean} R(2)}{\mu}$, where $2R$ is the diameter of the tube, and $K \to \mu$ for $n = 1$, from which the fact that $C_f = 16/Re$ was originally derived for laminar flow. ∎

Table 10.8 provides a sampling of the many major arteries and veins of the systemic circulation. This vascular system arises from the left ventricle via the largest artery of the body — the aorta, which has four major divisions: the ascending aorta, the aortic arch, the thoracic descending aorta, and the abdominal descending aorta. At its origin, which is the aortic semilunar valve, the aorta has a nominal internal diameter that may be as large as 3 cm, but the blood vessel tapers considerably, down to slightly less that 1 cm by the time it terminates at the iliac bifurcation in the inguinal (groin) region of the body. A reasonable approximation for the rate of aortic tapering is given by the equation:

$$D(x) = 2.815\sqrt{e^{-0.10x/2.815}} \tag{10.40}$$

where $D(x)$ is the diameter of the blood vessel at any location, x, measured from the aortic semilunar valve ($x = 0$) along the centerline of the aorta up to the iliac bifurcation (approximately $x = 65$ cm). Along its path, the aorta gives off a total of 58 branches — 2 from the ascending aorta ($0 \le x \le 4$ cm, $2.622 \le D(x) \le 2.815$ cm), 3 from the aortic arch ($4 \le x \le 12$ cm, $2.275 \le D(x) \le 2.622$ cm), 31 from the thoracic aorta ($12 \le x \le 43$ cm, $1.312 \le D(x) \le 2.275$ cm), and 22 from the abdominal aorta ($43 \le x \le 65$ cm, $0.887 \le D(x) \le 1.312$ cm) — before reaching the area of the groin, at which point the blood vessel bifurcates into the right and left common iliac arteries (Table 10.8).

The 5 arterial branches of the ascending aorta and aortic arch give rise to about 107 total major offshoots in the first few generations of branching. Those 31 branches of the thoracic aorta eventually divide themselves into approximately 85 major downstream offshoots, and the 22 branches of the abdominal aorta spawn another 70 daughters so that the 58 branches of the main aortic trunk soon become 262 smaller limbs of the arterial tree. These 262 blood vessels continue to divide further along the vascular tree (Table 10.5) so that the total number of major arteries in the body ultimately reaches 5×10^3. These approximately 5000 arteries give rise to 21,000 smaller tributaries, which spawn 83,000 ramifications (rami), which di-

TABLE 10.8 Major Arteries and Veins of the Systemic Circulation

The ascending aorta: Branches 1 and 2: Coronary circulation

Major arteries of the head and neck:

Branch 3: Brachiocephalic trunk (innominate artery, 1.24 cm internal diameter $\times$ 3–5 cm L)

Branch 4: Left common carotid artery, 6–7.5 mm D $\times$ 21 cm L)

Branch 5: Left subclavian Artery, 8–8.5 mm D $\times$ 10 cm L

Right common carotid artery from innominate; 0.74 cm D $\times$ 18 cm L

Right subclavian artery from the innominate; 0.8 cm D $\times$ 6.8 cm L

Left internal carotid: 0.36 cm D $\times$ 5.9 cm L + 0.26 cm D $\times$ 5.9 cm L + 0.16 cm D $\times$ 5.9 cm L; tapers along its length

Left external carotid: 0.30 cm D $\times$ 11.8 cm L; right internal and external carotids, same sizes (approximately) as corresponding left arteries

Cerebral circulation

Major arteries of the torso

Thoracic aorta, right intercostobronchial trunk (3 mm D), aortic branch 6 to right bronchial artery

Branch 7 to intercostal space 3

Branch 8: Left, cranial bronchial artery

Branch 9: Left, caudal bronchial artery

Bronchial circulation

Branches 10–11: Left and right posterior intercostal arteries to space 4

Branches 12–13: Left and right pericardial arteries to dorsal pericardium

Branches 14–15: Left and right posterior intercostal arteries to space 5

Branches 16–17: Left and right superior esophageal arteries

Branches 18–19: Left and right posterior intercostal arteries to space 6

Branches 20–21: Left and right inferior esophageal arteries

Branches 22–27: Left and right posterior intercostal arteries to spaces 7–9 of thoracic vertebrae

Branches 28–29: Left and right mediastinal arteries to mediastinum

Branches 30–33: Left and right posterior intercostal arteries to spaces 10 and 11 of thoracic vertebrae

Branches 34–35: Left and right subcostal arteries

Branch 36: Superior phrenic artery

Abdominal aorta, Branches 37–38: Left and right inferior phrenic arteries

Branch 39: Celiac axis trunk to splanchnic circulation

Branches 40–41: Left and right suprarenal arteries to adrenal glands

Branch 42: Superior mesenteric artery

Branches 43–44: Right and left renal arteries (kidney)

Branches 45–46: Left and right lumbar arteries to space 1

Branches 47–48: Left and right gonadal (Genital) arteries

Branches 49–50: Left and right lumbar arteries to space 2

Branch 51: Inferior mesenteric artery

Branches 52–55: Left and right lumbar arteries to spaces 3 and 4

Branch 56: Middle sacral artery

Branches 57–58: Left and right common iliac arteries to lower limbs

Blood vessels of the upper limbs

Axillary (armpit) artery and vein

Brachial (arm) artery and vein

Ulnar (forearm) artery and vein

Radial (forearm) artery and vein

Deep palmar arch (hand) arteries

Superficial palmar venous network

Superficial palmar arch arteries

Common palmar digital arteries of the fingers

Median cubital and basilic veins

Blood vessels of the lower limbs

Femoral (thigh) arteries and veins

Popliteal (knee) arteries and veins

Anterior tibial (shank) artery and vein

Posterior tibial arteries and veins

Profunda femoris artery

Dorsalis pedis artery of the foot

Dorsal venous arch

Small and great saphenous veins

Major veins of the torso

Testicular or ovarian veins

Glomerular and tubular renal veins

Renal vein and iliac veins

Portal and hepatic (liver) veins

Pelvic and perineal veins

Superior and inferior vena cava

Major veins of the head and neck

Right and left brachiocephalic veins

Right and left subclavian veins

Four jugular veins

Note. D, diameter; L, length.

vide into 500,000 terminal branches, which distribute themselves into approximately 1.9×10^7 arterioles and 2.4×10^8 metarterioles. The metarterioles finally empty into a huge capillary network of approximately 10–20 billion (1 to 2×10^{10}) tiny blood vessels of average diameter 3.5 μm (nutritive capillaries) to 10 μm (operant capillaries) and length 1.1 mm or less. With over 10^{10} blood vessels of this size supplying about 10^{14} cells of average size on the order of 10 μm, it is possible for any given cell to be within a cell size of the nearest capillary.

Capillaries recombine to form 4.4×10^9 venous capillaries and postcapillary venules as blood begins its journey back to the heart. The latter coalesce into approximately 1.6×10^8 collecting venules, which merge to form 32 million muscular venules, which become consolidated into about a third as many small collecting veins and about 500,000 terminal venous branches, corresponding to the same number of terminal branches on the arterial side (Table 10.6). The terminal branches soon join to form 20,000 small veins, about 512 larger medium veins, 256 large veins, and 224 even larger main venous branches — all of which eventually join to form two large vena cava: the superior vena cava collecting gravity-aided blood flow from regions above the level of the heart and the inferior vena cava collecting gravity-impeded blood flow from regions below the level of the heart. Both vena cava become one as they unite just prior to emptying into the right atrium of the heart. Counting duplicate pairs (e.g., two arms, two legs, two eyes, and two ears), even smaller rami, and a significant network of potential spaces, i.e., anastomoses that provide bypass connections between any two vascular spaces, various estimates suggest that if all of the blood vessels of this system were laid end to end next to one another, they would stretch around the world approximately 2.5 times, totaling perhaps as much as 60,000–100,000 miles of vascular pathways (the study of which is called angiology).

From an engineering point of view, for purposes of fluid dynamic analysis, it may be postulated that, on average, the number of branches, N, in any generation, j, of blood vessels contained in a particular vascular network, may be written in power-law form as

$$N = B_c^j \qquad (10.41)$$

where B_c defines the branching configuration, i.e., $B_c = 2$ identifies a bifurcating network wherein the parent branch of each upstream generation divides at its termination to spawn two daughter branches in the next downstream generation; $B_c = 3$ represents a trifurcating network, and so on. Furthermore, the mean diameter, d_j, and characteristic length, l_j, of a typical blood vessel in generation j are related to the diameter d_0 and length l_0 of the main parent vessel ($j = 0$) from which the vascular network was derived by the equations

$$d_j = d_0(D_R)^j \quad \text{and} \quad l_j = l_0(L_R)^j \qquad (10.42)$$

respectively, where D_R is the branching diameter ratio and L_R is the branching length ratio. Anatomic data from the literature, for at least the pulmonary, renal, coronary, bronchial, and cerebral circulations, suggest that in general, the branching diameter ratio, D_R, shows a remarkable consistency, between 0.600 (for the coronary circulation) and 0.616 (for the renal circulation). The total number of generations from parent vessel to terminal capillaries is also consistent, being 12 for the bronchial circulation and 17 for the pulmonary circulation, but 13 for the other three organs (kidney, heart, and brain) thus far examined. The branching length ratio, L_R, shows somewhat wider variability, between two-thirds for the pulmonary circulation and three-fourths for the coronary circulation; however, this range is also not extraordinarily large. Going one step further, it is observed that vascular branching configurations typically start out binary ($B_c = 2$) near the parent vessel (i.e., for the first two or three generations) and then become tertiary ($B_c = 3$) for several generations beyond that. They subsequently become increasingly more complex (increasing B_c) as the progeny approach the microvasculature but simplify again as the vascular networks reconverge en route back through the venous system to the heart. Regional blood volume estimates and capillary transport areas calculated from these approximations are reasonably close to those reported in the literature for the corresponding organs, as are ratios of venous diameters to arterial diameters. For the purposes of engineering design calculations, it thus appears that the morphometry (anatomical measurement) of the human vascular system may be reasonably modeled by simple power-law formulations, which, as we have already illustrated, have widespread applicability for approximating physiologic structure and function.

Example Problem 10.11

Consider a decimal branching system, i.e., one for which $B_c = 10$ in Eq. (10.41), such that in each generation, $j = 0,1,2,\cdots$, the number of "daughter" vessels increases in number by a factor of 10 over the previous generation. Assume this situation prevails because each blood vessel in each generation gives off eight symmetrical side branches before bifurcating (branches 9 and 10) at its termination (much like what happens when the abdominal aorta reaches the inguinal region) and assume (for simplicity) that all blood vessels in the same generation have the same diameter and the same characteristic length.

(a) Derive an equation for the ratio of diameters, d_{j+1}/d_j, between any two successive generations of blood vessels that would result in the total cross-sectional area of all daughter branches, being taken collectively, equalling exactly twice that of the corresponding area in the immediately preceding generation. That is, for what value of (d_{j+1}/d_j) will the total cross-sectional area of the systemic circulation double, every time the number of vessels increases by an order of magnitude (i.e., by 10)? Compare your result with Murray's law, a modified form of which asserts that $d_0^3 = d_1^3 + d_2^3 + \cdots + d_j^3$ for a parent ("0") giving off j daughters.

(b) From your results to part (a), generalize to get an expression for d_j/d_0, i.e., the diameter of the vessels in the jth generation relative to the original parent vessel, which, in the case of the systemic circulation, is the aorta. How do your results compare with values for D_R quoted earlier for the systemic circulation? Suppose, instead, that we have a branching network that gives off four daughter branches for each parent and obeys Murray's law with an exponent 2.7 (as has been suggested for the systemic circulation) rather than 3, which exponent corresponds to a geometry that offers the least possible resistance to the flow of blood at least through a bifurcation. Do the results improve?

Solution

(a) From Eq. (10.41), the number of branches, N, in the jth generation of a decimal ($B_c = 10$) branching system is $N = 10^j$ so that the total cross-sectional area of all N branches in the jth generation is $(\pi d_j^2/4)10^j$. Similarly, the total cross-sectional area of all N branches in the $(j + 1)$th generation is $(\pi d_{j+1}^2/4)10^{j+1}$. If we want the latter area to be twice the former, then $(\pi d_{j+1}^2/4)10^j 10 = 2(\pi d_j^2/4)10^j$, from which

$$\frac{d_{j+1}}{d_j} = \frac{\sqrt{5}}{5} = 0.4472$$

Now, according to Murray's law, $d_j^3 = d_1^3 + d_2^3 + \cdots + d_{j+1}^3$, so for $d_1 = d_2 = d_3 = \cdots = d_{j+1}$, and, for 10 daughter branches in generation $j + 1$ for each parent in generation j, we have

$$\frac{d_{j+1}}{d_j} = \frac{1}{\sqrt[3]{10}} = 0.4642$$

which is less than 4% higher than 0.4472 — not a bad match!

(b) From the equation derived in (a), $d_1 = (\sqrt{5}/5)d_0$, $d_2 = (\sqrt{5}/5)d_1 = (\sqrt{5}/5)^2 d_0 \cdots$; extrapolating to the jth generation, we have

$$d_j = \left(\frac{\sqrt{5}}{5}\right)^j d_0 \quad \text{or} \quad D_R = \frac{\sqrt{5}}{5} = 0.4472$$

(see Eq. 10.42). This value is 72–75% of the 0.60–0.62 range quoted earlier for the systemic circulation. On the other hand, for a "Murray network" with exponent 2.7, giving off four equal branches per generation,

$$d_j^{2.7} = 4d_{j+1}^{2.7}$$

from which

$$\frac{d_{j+1}}{d_j} = D_R = 0.6$$

Thus, this result is more realistic as a model of the human circulation. ■

Having quantified the major features of the human vascular system as these features relate to both the establishment of boundary conditions that constrain the movement of blood through arteries and veins and the maximization via optimal branching configurations of the ratio of transport surface area to the limited volume within which this area is contained, we will return to this system as we examine some additional hemodynamic considerations.

10.6.2 Fluid Dynamics in Vascular Channels

Recall from Section 10.3.2 that the single most significant reason why blood exhibits pseudoplastic behavior is that it consists of a suspension of many (nearly half the volume of blood) asymmetric (biconcave and disc shaped), relatively large (almost visible to the naked eye), deformable RBCs that interact with one another in several shear-dependent ways (e.g., Rouleaux formation) as they are propelled both through and with their plasma suspending fluid. Among the consequences of this physical milieu are that (i) straightforward mathematical integrations that take us from equations such as (10.32) and (10.33) to (10.34) and (10.35), under the assumption that we are dealing with differential (infinitesimally small), arbitrary regions that are statistically representative of any discretionary portion of a uniformly distributed homogeneous continuum, cannot be performed so "casually" or precariously, and (ii) secondary effects, such as RBC spin and axial alignment, significantly alter actual flow patterns compared with those that are predicted from analytical formulations which do not take into consideration the composite material nature of blood. For example, we use Eq. (10.35), for the simple case of $n = 1$, to illustrate what would happen if dr were not infinitesimally small, i.e., if we were dealing instead with a fluid that was flowing in J concentric layers of annular laminae, each having a finite layer thickness, δ (e.g., equal to the typical size of a RBC) which is the same for each annulus so that $R =$ lumen radius $= J\delta$. Suppose further that at some arbitrary location, r, in the blood vessel, we have encapsulated I annuli of thickness δ so that $r = I\delta$, $I = 0$ (along the centerline of the tube), 1, 2, 3, $\cdots J$ (at the wall of the tube, where $r = R$), and $dr/dI = \delta$. Then, we may write

$$dq = (2\pi r\, dr)v(r) = 2\pi \left\{ d\left[\frac{v(r)r^2}{2} \right] - \frac{r^2}{2}dv(r) \right\}$$

from the definition of the derivative of a product such as $(r^2v(r))/2$. The perfect differential, $d[(r^2v(r))/2]$, examined between limits $r = 0$ and $r = R$, cancels out because of the no slip, rigid-wall assumption, i.e., $v(R) = 0$, and the fact that $r = 0$ at the centerline of the tube. Moreover, writing $dv(r) = \frac{dv(r)}{dr}dr$, $\frac{dv(r)}{dr} = -\frac{\tau}{\mu}$ (for a Newtonian fluid), and $\tau = \Delta pr/2l$ (see Eq. 10.32), we have $dq = 2\pi\left(-\frac{r^2}{2}\right)\left(-\frac{\Delta pr}{2\mu l}\right)dr$. Finally, since $r = I\delta$, $dr = (dr/dI)dI = \delta dI$, and $dI = 1$ (i.e., the difference in integer values between any two adjacent concentric lamellae is unity), the continuous integral for q reduces to the finite sum

$$q = \sum_{I=0}^{I=J} \frac{\Delta p\pi\delta^4 I^3}{2\mu l} \tag{10.43}$$

Now, $\sum_{I=0}^{I=J} I^3 = \dfrac{J^2(J+1)^2}{4}$ and, since $J = \dfrac{R}{\delta}$, Eq. (10.43), after some algebraic rearranging, becomes

$$q = \frac{\Delta p \pi R^4}{8l \dfrac{\mu}{(1 + \delta/R)^2}} \qquad (10.44)$$

Observe that for $\delta \rightarrow 0$, or $\delta/R \ll 1$ (i.e., for tubes having radii much larger than the size of a typical RBC), Eq. (10.44) reduces to Eq. (10.35) (for $n = 1$) obtained by straightforward integration techniques (see Example Problem 10.9). On the other hand, as we approach the microcirculation, where $\delta \cong R$, or $\delta/R \rightarrow$ unit order, Eq. (10.44) predicts a volumetric flow rate which is four times higher than that predicted by Eq. (10.35) with $n = 1$. In other words, this so-called sigma effect — due to the form of Eq. (10.43) — once again illustrates the fact that purely analytic, closed-form integral solutions that assume a homogeneous distribution of material often are on the low side when they are used to calculate the volumetric flow rate of suspensions containing relatively large-size particles. This can also be shown by defining, using Eq. (10.44), an apparent viscosity given by

$$\mu_{apparent} = \frac{\mu}{\left(1 + \dfrac{\delta}{R}\right)^2} \qquad (10.45)$$

so that Eq. (10.44) may be written in an equivalent Newtonian form as

$$q = \frac{\Delta p R^4 \pi}{8l \mu_{apparent}}$$

except that now $\mu_{apparent}$ decreases with tube radius [as opposed to fluid shear rate; see Eq. (10.9) and Fig. 10.3] as the latter becomes of the same order as δ (again, a pseudoplastic phenomenon). The decrease in fluid apparent viscosity moving from the macrocirculation into the microcirculation is named after the investigators who first observed this phenomenon during the late 1920s, hence the name Fahraeus–Lindqvist effect.

Example Problem 10.12

Compare the pressure drop, Δp, required to drive blood through a typical arteriole, assuming the fluid is a homogenous continuum, to that required to drive a suspension of particles, where δ is of the order of magnitude of a typical RBC, through the same vessel at the same flow rate. Use Tables 10.1 and 10.5.

Solution

From Example Problem 10.9 and Eq. (10.44),

$$q_{\substack{homogenous \\ continuum}} = \frac{(\Delta p)_h R^4 \pi}{8l\mu} = q_{\substack{suspension \\ of\ particles}} = \frac{(\Delta p)_s \pi R^4}{8l\mu \left(1/\left(1 + \dfrac{\delta}{R}\right)^2\right)}$$

so

$$\frac{(\Delta p)_s}{(\Delta p)_h} = \frac{1}{\left(1 + \dfrac{\delta}{R}\right)^2} = \frac{1}{\left(1 + \dfrac{7.5}{31.25}\right)^2} = 0.65$$

so it takes 35% less of a pressure drop to drive a suspension of particles than it does to drive a homogeneous solution through the same small tube at the same flow rate because, in the former case, the concentric annular flow of lamellae having a finite thickness requires shearing less individual streamlines than does the continuously distributed parabolic Poiseuille flow of a homogeneous fluid. ■

A second consequence of the relatively large size of RBCs is that they cut across several streamlines of plasma fluid at the same time, as whole blood flows through vascular channels. Those streamlines closer to the centerline of the channel represent fluid velocities that are faster than the ones associated with fluid streamlines closer to the wall of the blood vessel (Eq. 10.37). This gradient in velocity across any given RBC starts the cell spinning in such a way that the side of the cell facing the blood vessel centerline moves in a direction downstream (with the flow) while the side of the cell facing the wall of the vascular channel moves in a direction upstream (against the flow) thus exacerbating even more the radial differential in velocity across the cell. Now, according to the energy equation (see Eq. 10.30), all other factors being essentially equal, whenever there is a drop in velocity (loss in kinetic energy) there is a corresponding increase in pressure (conversion to potential energy) and vice versa. Thus, there results from the velocity gradient across each cell a corresponding radial pressure gradient that drives the cell in, toward the centerline of the tube and away from the wall surface. This so-called axial drift (or accumulation) of RBCs leads to the existence of a cell-free zone of pure plasma fluid called the plasma skimming layer near the wall of a blood vessel. The physical process that produces the axial drift of RBCs is called the magnus effect (basically a conservation of energy phenomenon), and it is quite similar to the effect that causes a baseball to travel along a tortuous path (i.e., to "curve") when released with the proper "spin" by an adept pitcher.

The axial accumulation of RBCs leaves behind a lower viscosity plasma skimming layer near the wall of the blood vessel, where frictional effects are most prominent. Thus, the axial pressure gradient does not have to work as hard to drive the fluid through the vascular channel, and so the whole blood appears to be less viscous (a pseudoplastic effect). Moreover, in regions of vascular branching, more cell-free plasma fluid along the boundaries of the flow tends to enter the downstream branches than does the cell-rich whole blood contained in the central core of the flow. Thus, one notices a decreasing hematocrit as one moves through the vascular system from larger blood vessels to smaller ones; this is a blood thinning effect which also helps to explain why blood appears to be less viscous in smaller blood vessels than in larger ones (and hence requires a lower pressure gradient to drive it). Finally, it should be noted that the axial drift of RBCs does not actually take them all the way into the centerline of the channel. Other considerations related to conservation of mass, linear, and angular momentum reveal that the cells will drift inward only to some equilibrated radial location, which depends on the hematocrit of the fluid and the size of the tube. Using a simple fluid dynamic model which assumes the steady flow of blood through a circular tube of radius R, and which divides the tube cross section into a cell-rich core zone of whole blood viscosity μ_b occupying the region $(R - \delta)$ of the flow and a cell-free plasma skimming zone of thickness δ near the wall of the tube, where the viscosity, μ_p, is that of pure plasma, it can be shown by an analysis very similar to that from which Eq. (10.45) was derived that

$$\mu_{apparent} = \frac{\mu_p}{1 - (1 - (\delta/R))^4 \ (1 - (\mu_p/\mu_b))} \tag{10.46}$$

Example Problem 10.13

Consider the steady flow of blood in a 100-mm-diameter tube. If the apparent viscosity of the fluid is measured to be 3.3 mPa-s, the viscosity of plasma is 1.2 mPa-s, and the viscosity of packed RBCs, is 6.5 mPa-s, approximately what thickness can be attributed to the plasma skimming layer?

Solution

From Eq. (10.46), with $\mu_p = 1.2$, $\mu_b = 6.5$, $\mu_{apparent} = 3.3$, and $R = 50$ mm, we calculate δ to be approximately 3 mm, which is on the order of 6% of the tube radius or 400 times the mean diameter (7.5 μm) of a typical RBC.   ◼

A third consequence of the relatively large size and number of RBCs derives from the asymmetry of the erythrocyte contour, i.e., their donut-shaped surface. After they start to spin and drift toward the vascular channel centerline, the cells eventually equilibrate at some characteristic radial location; however, as they do so, they align with the flow streamlines in such a way that they offer the least profile drag in the direction of flow. Thus, as blood is progressively sheared from an initial state in which the RBCs are originally oriented randomly with respect to the flow streamlines, causing much cell-to-cell interference and a large profile-drag resistance to motion, to an eventual equilibrated state in which the RBCs are neatly aligned with the flow streamlines, offering a minimal resistance to motion, the apparent viscosity of the fluid seems to decrease (a pseudoplastic effect). Because of the mechanism that produces this axial alignment (as opposed to axial accumulation) of RBCs, the pseudoplastic behavior caused by a progressive decrease in profile drag and cellular interaction is referred to as a shear effect. Together with the magnus effect (axial accumulation of erythrocytes), the sigma effect (annular laminated flow of a nonhomogeneous suspension of particles), and the Fahraeus–Lindqvist effect (blood thinning in small vessels), the shear or profile drag effect helps to explain the physics of why blood exhibits non-Newtonian fluid dynamic behavior. Next, we examine some other features of hemodynamics, especially as they relate to the time dependence of the flow and to the elasticity of the vascular channels.

10.6.3   Pulsatile Flow in Elastic Channels

Figure 10.8 illustrates a typical graph of blood instantaneous axial velocity (averaged across the tube cross section) vs time ($t = 0 =$ start of ventricular ejection) for a typical large artery (such as the aorta) in the systemic circulation. The velocity–

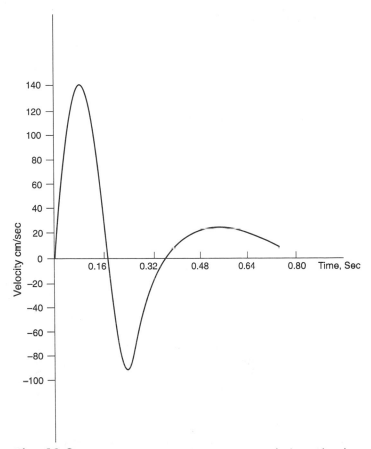

Fig. 10.8 Typical space-averaged instantaneous velocity vs time in a large systemic artery, where $t = 0$ = the start of ventricular ejection.

time behavior results from the fluctuations in driving pressure that accompany the cardiac cycle. In normal arteries, the fluid accelerates rapidly (over about 75–80 ms) to a maximum speed, v_{max}, of approximately 0.4–1.4 m/s (acceleration $\sim$ 0.5–2 g_0) during early systole. Toward the end of systole, vascular recoil (see discussion of the dichrotic notch), coupled with pressure pulse waves being reflected back upstream from downstream branch points and other geometric discontinuities, produces a short period ($\sim$150 ms) of flow reversal, during which antegrade movement the fluid reaches minimum speeds, v_{min}, on the order of 0.2–0.9 m/s (backwards). The flow then recovers during diastole to drain away from the heart (forward flow) at a much lower speed. Cardiac cycle velocity fluctuations are sometimes quantified by parameters such as the pulsatility index (PI)

$$\text{PI} = \frac{v_{max} - v_{min}}{(v_{max})_{\substack{\text{time averaged over} \\ \text{several cardiac cycles}}}} \tag{10.47}$$

or the resistive index (RI)

$$RI = \frac{v_{max} - v_{min}}{(v_{max})_{instantaneous}} \tag{10.48}$$

or the amplitude ratio, $|v_{max}/v_{min}|$. As was the case for pseudoplastic flow vs Newtonian flow, so too is the flow rate greater for pulsatile flow than it is for nonpulsatile, steady flow (given the same fluid, the same mean driving pressure, and the same vascular channels) provided the oscillation frequency is not too great.

In a more rigorous engineering sense, the effects of pulsatility are embedded in a frequency-dependent nondimensional parameter called the Strouhal number, St, defined by

$$St = \frac{f_0 l}{v_{mean}}$$

In the same sense that the previously defined Newtonian Reynolds number (see Example Problem 10.10) is a measure of the ratio of fluid convected inertial effects (as measured by the quantity $2Rv_{mean}$) to fluid viscous effects (as measured by the kinematic viscosity, $\mu/\rho = v$), the Strouhal number is a measure of the ratio of fluid unsteady inertial effects (as measured by the frequency parameter, $f_0 l$) to fluid convected inertial effects (as measured by the kinematic mean velocity, v_{mean}). Thus, if we multiply the Strouhal number by the Reynolds number, and scale this product by the geometric ratio, (R/l), and the factor π, we get $(2\pi R^2 f_0)/v$, which is a measure of unsteady fluid inertial effects (as measured by the kinematic transport coefficient, $\omega_0 R^2$, where $\omega_0 = 2\pi f_0$ is the circular radian frequency of the periodic motion) to fluid viscous effects (as measured again by the kinematic viscosity, v). The quantity $\omega_0 R^2/v$ is sometimes called an unsteady Renolds number, the square root of which, i.e., $R\sqrt{\omega_0/v} = \alpha$, defines a so-called Womersley parameter or kinetic Reynolds number that characterizes the unsteady features of periodic flows. The kinetic Reynolds number actually arises naturally in the argument of the complex Bessel functions obtained when one solves Eq. (10.21) for the periodic (cyclicly time-dependent) fully developed flow through rigid circular channels.

Example Problem 10.14

Calculate the kinetic Reynolds number at the root of the aorta for a 75 beats per minute heart rate.

Solution

From Table 10.3, for a typical sample of human blood, we may take the apparent viscosity at the root of the aorta to be $\mu_{232.8} = 4.175$ mPa-s $= 0.004175$ N-s/m^2. The density, ρ, of blood is approximately 1.057 g/cm^3 = 1057 N-s^2/m^4 and the radius of the aorta at its root is, from Eq. (10.40) with $x = 0$, 2.815 cm/2 = 0.02815/2 m. Finally, for a pulse rate of 75 beats per minute = 1.25 cycles per second,

$$\alpha = R\sqrt{\frac{\omega_0}{v}} = \frac{0.02815}{2}\sqrt{\frac{2\pi(1.25)}{.004175/1057}} = 19.85 \sim 20 \qquad \blacksquare$$

As mentioned previously, the kinetic Reynolds number arises naturally when one solves the unsteady momentum equation for the flow of blood through rigid circular channels. If the vessels through which the fluid flows are not rigid, another important variable appears due to the coupling of the fluid dynamic equations with the elastic wall momentum equations at the blood/blood vessel boundary interface. This additional parameter relates the speed, c, at which the pressure pulse initiated in the aorta travels through the vascular system, to the elasticity of the blood vessel walls as measured by their respective modulus of elasticity, E_m, through the simple Moens–Korteweg approximation for thin-walled cylindrical tubes of wall thickness h and internal radius R [i.e., tubes that obey LaPlace's law (see the equation derived in Example Problem 10.6, with $R_1 = R$ and $R_2 \rightarrow \infty$)]. Thus,

$$c = \sqrt{\frac{E_m h}{2R\rho_{\text{blood}}}} \tag{10.49}$$

Example Problem 10.15

Calculate the speed of propagation of the arterial pulse wave through the femoral artery if $E_m = 3 \times 10^6$ dyn/cm^2 and $h = 0.05$ cm.

Solution

For the femoral artery, $2R = 0.48$ cm; therefore, $h/2R = 0.05/0.48 = 0.10 \ll 1$, so the femoral artery may be considered a thin-walled elastic tube in which Eq. (10.49) holds. Thus,

$$c = \sqrt{\frac{3 \times 10^6 \text{ g/cm/s}^2 (0.05 \text{ cm})}{0.48 \text{ cm } (1.057 \text{ g/cm}^3)}}$$

$$= 543.735 \text{ cm/s}$$

or $c = 5.44$ m/s, which is very close to values actually measured in the human femoral artery. (Note: 10^5 dyn = 1 N). ∎

We have already seen that, for a given mean pressure gradient, pseudoplastic flow is higher than Newtonian flow (see Example Problems 10.9 and 10.12) and that pulsatile flow is greater than steady flow. We now point out that, for the same driving pressures, the average fluid velocity through a tube having perfectly elastic, incompressible walls (i.e., walls that suffer no volume change when they are stretched) is approximately 10% higher than it is for a comparable rigid-walled tube. This increase in flow is due, at least in part, to the ability of elastic walls to store energy as the vessel is pressurized and to return that energy to the fluid by "recoiling" and thus helping to drive the fluid along in the channel (see discussion of the dischrotic notch). Therefore, at least three properties of the cardiovascular system — the pseudoplasticity of the fluid, the pulsatile nature of the driving pump, and the elasticity of the vascular channels — coupled with the aid of gravity all help to optimize its performance. Indeed, one of the fundamental features of all physiologic function is the extent to which it is highly optimized; a second basic feature is the extent to which physiologic function is carefully controlled.

10.7   GENERAL ASPECTS OF CONTROL OF CARDIOVASCULAR FUNCTION

In a global sense, one can think of the human cardiovascular system — using an electrical analog — as a voltage energy source (the heart), two storage capacitors (a large elastic venous system and a smaller arterial system), and a dissipative resistor (the microcirculation taken as a whole). Blood flow and the dynamics of the system represent electrical inductance (inertia), and useful engineering approximations can be derived from such a simple lumped-parameter model. As previously discussed, the cardiovascular system is designed to bring blood to within a capillary size of all the more than 10^{14} cells of the body; however, which cells receive blood at any given time, how much blood they get, the composition of the fluid coursing by them, and related physiological considerations are all matters that are not left to chance. Blood flows through organs and tissues either to nourish and sanitize them (so-called nutritive pathways) or to be processed in some sense (operant pathways) [e.g., to be oxygenated (pulmonary circulation), stocked with nutrients (splanchnic circulation), dialyzed (renal circulation), cooled (cutaneous circulation), or filtered of dilapidated RBCs (splenic circulation)]. Thus, any given vascular network normally receives blood according to the metabolic needs of the region it perfuses and/or the function of that region as a blood treatment plant and/or thermoregulatory pathway. However, it is not feasible to expect that the physiologic transport system can be "all things to all cells, all of the time," especially when resources are scarce and/or time is a factor. Thus, the distribution of blood is further prioritized according to three basic criteria: (i) How essential the perfused region is to the maintenance of life (e.g., we can survive without an arm, leg, stomach, or even a large portion of our small intestine, but not without a brain, heart, and at least one functioning kidney and lung); (ii) how essential the perfused region is in allowing the organism to respond to a life-threatening situation (e.g., digesting a meal is among the least of the body's concerns in a "flight-or-fight" circumstance); and (iii) how well the perfused region can function and survive on a decreased supply of blood (e.g., some tissues, such as striated skeletal and smooth muscle, have significant anaerobic capability; others, such as several forms of connective tissue, can function quite effectively at a significantly decreased metabolic rate when necessary; some organs, such as the liver, are larger than they really need to be, and some anatomical structures, such as the eyes, ears, and limbs, have duplicate pairs, giving them a built-in redundancy).

Within this generalized prioritization scheme, control of cardiovascular function is accomplished by mechanisms that are based either on the inherent physicochemical attributes of the tissues and organs (intrinsic control) or on responses that can be attributed to the effects on cardiovascular tissues of other organ systems in the body (most notably, the autonomic nervous system and the endocrine system; extrinsic control). For example, the accumulation of wastes and depletion of oxygen and nutrients that accompany the increased rate of metabolism in an active tissue both lead to an intrinsic relaxation of local precapillary sphincters (rings of muscle), with a consequent widening of corresponding capillary entrances, which reduces the local resistance to flow and thereby allows more blood to perfuse the active region.

On the other hand, the extrinsic innervation by the autonomic nervous system of smooth muscle tissue in the walls of arterioles allows the central nervous system to completely shut down the flow to entire vascular beds (such as the cutaneous circulation) when this becomes necessary (such as during exposure to extremely cold environments).

In addition to prioritizing and controlling the distribution of blood, physiologic regulation of cardiovascular function is directed mainly at four other variables: cardiac output, blood pressure, blood volume, and blood composition. From Eq. (10.15), cardiac output can be increased by increasing the heart rate (a chronotropic effect), increasing the end-diastolic volume (allowing the heart to fill longer by delaying the onset of systole), decreasing the end-systolic volume (an inotropic effect), or doing all three of these at once. Indeed, under the extrinsic influence of the sympathetic nervous system and the adrenal glands, HR can triple (to approximately 240 beats per minute if necessary), EDV can increase by as much as 50% (to approximately 200 or more ml of blood), and ESV can decrease a comparable amount (the so-called cardiac reserve) to about 30–35 ml or less. The combined result of all three effects can lead to more than a sevenfold increase in cardiac output from the normal 5–5.5 liters/min to as much as 40 or 41 or more liters/min for very brief periods of strenous exertion.

The control of blood pressure is accomplished mainly by adjusting at the arteriolar level the downstream resistance to flow-an increased resistance leading to a rise in arterial backpressure, and vice versa. This effect is conveniently quantified by a fluid dynamic analog to the famous Ohm's law in electromagnetic theory, with voltage drop in the latter being equated to fluid pressure drop, Δp, electric current corresponding to volumetric flow [cardiac output (CO)], and electric resistance being associated with an analogous vascular "peripheral resistance" (PR). Thus,

$$\Delta p = (CO)(PR) \tag{10.50}$$

Normally, the total systemic peripheral resistance is 15–20 mmHg/liter/min of flow, but it can increase significantly under the influence of the vasomotor center located in the medulla of the brain, which controls arteriolar muscle tone.

The control of blood volume is accomplished mainly through the excretory function of the kidney. For example, antidiuretic hormone secreted by the pituitary gland acts to prevent renal fluid loss (excretion via urination) and thus increases plasma volume, whereas perceived extracellular fluid overloads such as those that result from the peripheral vasoconstriction response to cold stress lead to a sympathetic/adrenergic receptor-induced renal diuresis (urination) that tends to decrease plasma volume, if not checked, to sometimes dangerously low dehydration levels. Blood composition is also maintained primarily through the activity of endocrine hormones and enzymes that enhance or repress specific biochemical pathways. Since these pathways are too numerous to itemize here, suffice it to say that in the body's quest for homeostasis and stability, virtually nothing is left to chance, and every biochemical end can be arrived at through a number of alternative means. In a broader sense, as the organism strives to maintain life, it coordinates a variety of different functions, and central to its ability to do this is the role played by the cardiovascular system in transporting mass, energy, and momentum and the biofluid mechanical principles that govern such transport.

EXERCISES

1. Using both Newton's Law of Viscosity and his Second Law of Motion ($F = ma$), show, formally, that the dynamic coefficient of viscosity, μ, represents, physically, the transport of linear momentum in a direction normal to the direction of flow, per unit area in that normal direction — which is to say, μ is a measure of the ease with which a disturbance in one direction (F, along x) propagates laterally in a perpendicular direction adjacent to it (y). In other words, "viscosity" is a measure of how well fluid layers "drag" one another along in the direction of flow, and hence, how easy it is to shear a fluid. The greater is the viscosity of a fluid, the more difficult it is to deform it at a given rate, and the more rigid (solid-like) is its behavior. Show, further, that κ^* is, in fact, precisely equal to the lateral transport of momentum in a flowing stream.

2. In a laboratory experiment, the following data are obtained for an unknown sample of fluid maintained at constant temperature:

dv/dy (rad/s)	0	11.64	23.28	46.56	116.40	232.80	116.40	46.56
τ (N/m^2)	0.008	0.073	0.130	0.232	0.500	0.890	0.40	0.085

Plot stress vs strain rate, and classify the material.

3. What would be the hematocrit of blood if the cells were packed into a honeycomb arrangement (like a beehive), each cell enclosed in its own hexagonal regular polygon of side length 4.50 μm and thickness 2.25 μm and all polygons packed into a tight matrix?

On the basis of the answers to Example Problem 10.2 and Exercise 3 and the earlier discussion of the normal range for adult hematocrits, can you reconcile these numbers in view of the physiologic role of erythrocytes? What would happen if the hematocrit were too low (e.g., <30%), a condition called anemia? What would happen if the hematocrit were too high (e.g., >60%), a condition called polycythemia? In what sense, then, is a hematocrit range from 40 to 50% "optimum"?

4. A fluid is defined to be a substance that
   (a) always reconfigures itself to completely fill the volume within which it is contained;
   (b) is virtually incompressible;
   (c) cannot be subjected to shear forces;
   (d) cannot remain at rest under the action of any shear force, no matter how small;
   (e) has the same shear stress at any point within its domain, regardless of its motion.

5. Plot Eq. (10.4) as $\sqrt{T^*}$ versus C_F for $0.150 \leq C_F \leq 0.470$ g% (see Table 10.2), with hematocrit acting as a parameter. Choose $H = 30, 35, 40, 45, 50, 55,$ and 60% to get a family of seven curves. From your results, and the form of Eq. (10.4), comment on the role that hematocrit plays in affecting the variation of T^* with C_F, and give a reasonable explanation for why this should be the case.

6. Substitute Eq. (10.7b) into Eq. (10.10) and then plot μ_a (in mPa-s) vs $\underline{\underline{S}}$ (in reciprocal seconds for a simple two-dimensional viscometric configuration) for the shear rate range, $25 \leq \underline{\underline{S}} \leq 250$ s^{-1}, and for $C_1^* = 0.00797$ (dPa-s)^n, $C_2 = 0.0512$ (dimensionless), $C_3 = 0.00499$ (dimensionless), and $C_4 = 145.85$ dl/g. First, obtain a family of curves for $H = 35\%$ (expressed as a percentage, not a decimal as in the aforementioned equations) and TPMA acting as a parameter, taking on values 1.5, 2.0, 2.5, 3.0, 3.5, and 4.0 g%. Next, obtain a family of graphs for the same parametric values of TPMA, but for increasing hematocrit values of 40, 45, 50, 55, and 60%. From your results, comment on the effects of TPMA (the plasma globulins) on μ_a (the apparent viscosity of whole blood) for a given hematocrit, H, and on how this effect is mediated by H. Can you explain why?

7. Use the results from Example Problem 10.5 to compute the terminal velocity (sedimentation rate), v, in units of mm/h, for a sphere (RBC) of volume 90 μm^3 and mass density 1.098 g/cm^3 falling through a fluid (plasma) of mass density 1.035 g/cm^3 and viscosity 1.8 mPa-s. Compare your answer with the value of ESR listed for human blood in Table 10.4 and comment on the validity of the approximation for calculating the ESR.

8. For a typical volumetric flow rate (see Table 10.7) across a typical aortic semilunar valve, calculate the mean velocity of flow (in cm/s) through such a valve. Then, considering blood to be a Newtonian ($n = 1$) fluid having a viscosity $\mu_{232.8}$ (see Table 10.3) and a typical mass density, calculate a typical Newtonian Reynolds number for flow across the aortic valve.

9. In a thermodynamic sense the cardiac stroke power may be approximated by the product of mean ventricular blood pressure during systole, times stroke volume, divided by the ventricular ejection time during the cardiac emptying phase. If the maximum systolic ventricular blood pressure at rest is measured to be 120 mmHg, and the minimum ventricular blood pressure at the very start of systolic ejection is measured to be 80 mmHg, calculate the cardiac stroke power — in both watts and horsepower — required to generate 70 ml of stroke volume in 220 ms of ejection time.

10. Use Tables 10.3 and 10.4, together with whatever additional data you need from Exercise 8, to calculate a corresponding non-Newtonian Reynolds number for flow across the human aortic semilunar valve. How does the non-Newtonian Reynolds number compare with the Newtonian Reynolds number calculated in Exercise 8?

11. The quantity,

$$\frac{p}{T} = \frac{R_1 + R_2}{R_1 R_2}$$

derived earlier for a prolate spheroid is called the mechanical advantage of the heart, i.e., the ratio of how much internal pressure, p, can be generated in the chambers of the heart by the tension, T, generated in its walls by the ventricular musculature. For $R_1 = 5$ cm, plot the mechanical advantage of the heart vs the ratio, $R_1 : R_2$, and comment on what happens for $R_1 \gg R_2$ and for $R_1 \ll R_2$.

12. In a pathologic condition known as coarctation of the aorta, the lumen (internal space) of the blood vessel becomes narrowed due to an external compression of its walls (a squeezing action). If such aortic coarctation reduces its root diameter by one-third, calculate the effect that this reduction has on both the mean velocity of flow through the constricted area and the corresponding volumetric discharge through this region.

13. Consider a pilot executing a vertical acrobatic "loop" of radius 230 m. If the speed of the aircraft at the bottom of the path is 134 m/s, calculate the centripetal acceleration, in "g"s, to which the pilot is exposed as he or she passes through this point in the loop. If a major artery in the head of this pilot is approximately 50 cm above the level of his or her heart, within which the mean blood pressure at this point in the loop is 100 mmHg, use Pascal's law to determine whether or not this pilot will suffer a blackout (resulting from a momentary loss of blood flow to his or her brain) during this maneuver. On this basis, what would you predict to be the limit of human tolerance to "positive-g" acceleration without blackout, and what could this pilot do (quantitatively, in this particular maneuver) to stay within that limit?

14. Using the results of Example Problem 10.7, calculate the mean kinetic energy attributable to the stroke volume of 70 ml ejected by the left ventricle in 220 ms of ejection time, assuming blood has a mass density of 1.057 g/ml. Calculate the efficiency of the conversion of cardiac stroke power into fluid kinetic energy (converted into corresponding power; in watts). Comment on why this efficiency is so low (on the order of only one-tenth of a percent!).

15. Plot Eq. (10.37) with n acting as a parameter taking on the values 0, 0.7844, 1.0, 1.5, and infinity, to get a family of curves with $0 \leq r/R \leq 1.00$ as the ordinate (vertical axis) and v/v_{mean} as the abscissa (horizontal axis). As $n \to \infty$ (see Eq. 10.39), comment on what happens to the volumetric flow rate, q (see Eq. 10.35).

16. (a) From the information contained in Table 10.5 for the arterial system (use 2.815 cm for a typical aortic internal diameter), and Table 10.6 for the venous system (use 3.000 cm for a typical internal diameter of the vena cava), prepare a bar graph on semilog paper. Plot the area ratio (A/A_0) as the ordinate, where A is the total cross-sectional area of all blood vessels contained in a given systemic vascular generation (assuming they all have the same internal diameter which is the mean of the range quoted in the tables), and A_0 is the cross-sectional area of the aorta — vs. the corresponding blood vessels that comprise that particular vascular generation as the abscissa, (i.e., where you are located in the systemic vascular system as you move from aorta to vena cava).

(b) For a cardiac output (CO) of 5345 ml/min, together with the areas you calculated in (a), and the principal of conservation of mass, i.e.,

$$CO = Av$$

calculate and plot the corresponding mean velocity, v, that one is likely to measure in a typical blood vessel of a given generation of arteries and veins as one moves progressively through the systemic circulation from end to end.

(c) Taking the typical values for K and n from Tables 10.3 and 10.4 for human blood and its mass density to be 1.057 g/cm^3, calculate typical non-Newtonian Reynolds numbers based on tube diameter and mean flow velocity that characterize different regions of the systemic circulation.

(d) Evaluate the results you obtained in terms of the following fluid dynamic considerations: Where is flow in the cardiovascular system most likely to be laminar? Where are you most likely going to encounter at least periods of turbulent and "disturbed" flow patterns? What is good about a diverging flow (A increasing)? What is bad about it? What is good about a converging flow (A decreasing)? What is bad about it? What can you surmise about the presence or absence of turbulence in the systemic circulation: Is it there? Is it not? If it is, what is good about it? Bad? If it is not, is this good? Bad? Finally, why might you not be surprised to find valves in veins but not in arteries?

17. Many investigators believe that transport between blood and the surrounding interstitial fluid actually takes place in three regions of the microcirculation: the metarterioles, the capillaries, and the postcapillary venules (where, some say, most of the transport actually occurs). For purposes of approximation, let all such blood vessels of the systemic circulation have a mean length of 0.6 mm, with respective diameters of 19, 6, and 9 μm (see Tables 10.5 and 10.6), and let the pulmonary vessels have lengths 0.3, 0.85, and 0.6 mm and average internal diameters of 10, 7, and 8.5 μm, respectively for the metarterioles, capillaries, and postcapillary venules. Based on these numbers, calculate the total microcirculatory surface area available for transport between blood and the surrounding interstitial fluid in the pulmonary and systemic circulations and see how your results compare with published estimates of 60–80 m^2 for the pulmonary circulation and 500–700 m^2 for the systemic circulation. Also compare your answers with the total surface area of the RBCs contained in the pulmonary microcirculation (Tables 10.1 and 10.7 also may be useful for this part) and with the total surface area of the RBCs contained in the systemic microcirculation assuming a typical hematocrit of 45%.

18. (a) If the nominal diameter of the pulmonary artery is 3.00 cm and a typical pulmonary capillary has a bore size of 7 μm, verify from Eq. (10.42) that, for $D_R = 0.6114$, the total number of generations from parent vessel to terminal capillaries in the pulmonary circulation is about 17.

(b) Using the typical number of capillaries shown in Table 10.5 for the pulmonary circulation, using Eq. (10.41), calculate an average B_c for this vascular system, i.e., on average, what type of branching network exists if your results to (a) are valid?

(c) Suppose we think of the systemic circulation also as a vascular network that starts with a parent vessel of nominal diameter 2.815 cm (see Eq. 10.40) and proceeds through 17 generations to a typical capillary of bore size 6.568 μm. What would D_R be for this system? From Table 10.5 and Eq. (10.41), what would be the corresponding value of B_c? Do these results sound realistic? (Hint: They are probably not what you might expect).

(d) If we impose the requirement (more realistic) that the systemic circulation should be at least a binary (or more) branching network, i.e., $B_c \geq 2$, how many generations of branching would be a more likely estimate for the systemic vascular network if the number of capillaries given in Table 10.5 is about accurate and $B_c = 2$?

(e) For $d_0 = 2.815$ cm, N calculated from Eq. (10.41), and your result from (d), what would be the diameter of a typical capillary in this branching network? How does this compare with the estimate given in (c)?

(f) With $d_0 = 2.815$ cm, d_j = the value calculated in (e), j = the value calculated in (d), and Eq. (10.42), what is the value of D_R for this model of the systemic circulation? Does this value seem reasonable? If so, why? If not, why not? Any insights or suggestions?

19. Using Eq. (10.45), with δ of order of 7.5 μm, plot the nondimensional ratio, $[\mu_{apparent}/\mu]$ (ordinate) vs R (abscissa) in millimeters, and using Table 10.5 define the various regions of the systemic arterial system along the abscissa. Where does the Fahraeus–Lindqvist effect start to become very significant (e.g., $\mu_{apparent} \leq 0.9\mu$)?

20. Consider the steady flow of blood in a 0.10-mm-diameter tube which bifurcates (splits) into two symmetric, equal daughter branches of 0.05-mm diameter each. The total volume of blood removed from a 1-cm upstream section of the parent branch (just ahead of the bifurcation), together with that collected from a 1-cm section of each daughter branch (just downstream of the bifurcation), collectively yields a composite average hematocrit for the total blood volume of approximately 45%.

(a) If the hematocrit of the blood contained in each daughter branch is 40%, what is the hematocrit of the blood flowing through the main branch?

(b) Assuming that all the cells in each branch are packed tightly in a circular cylindrical configuration at the center of the tube (i.e., that the fluid is experiencing "plug flow" with a blunt velocity profile), what fraction of the total cross-sectional area is composed of a cell-free layer of fluid in each branch?

(c) From your answers to (a) and (b), what would you conclude about the apparent viscosity of blood in the daughter branches vs that in the main or parent branch?

21. The human femoral artery is a 0.48-cm-diameter blood vessel that branches off of the external iliac artery to supply blood to the genital organs, the wall of the abdomen, the muscles of the thigh, and, through downstream branches, the lower extremities of the body. Using Eq. (10.45), with $\mu = 4.175$ mPa-S and $\delta = 7.5$ μm, calculate the kinetic Reynolds number, α, for a typical femoral artery and a 75-beats per minute pulse rate. Then, using Tables 10.3 and 10.4 and Eq. (10.9), calculate an average shear rate for the femoral artery. Finally, if this average shear rate can be approximated by twice the mean flow velocity in the blood vessel divided by its radius, calculate the mean velocity of flow (in cm/s) and the corresponding volumetric flowthrough, q (in ml/min) through a typical femoral artery. Compare your mean flow velocity calculation with measured peak systolic common femoral velocity measurements of 0.9–1.4 m/s and reverse diastolic velocities in the 0.3–0.5 m/s range.

22. Blood flows at a constant velocity through an inclined artery. Along one stretch of blood vessel there is a 10-cm vertical drop due to the incline. The blood pressure at the entrance to this stretch is 50 mmHg. Assuming the total fluid energy remains unchanged, calculate the blood pressure at the exit from this stretch of blood vessel.

23. Equation (10.30) suggests that, for arbitrary z, p should increase if v decreases and vice versa. However, from Fig. 10.7, as one moves through the cardiovascular system from the heart out to the capillaries, blood slows down (losing kinetic energy), yet this drop in velocity is not accompanied by a corresponding rise in blood pressure. Instead, blood pressure also decreases, so the fluid loses both kinetic and potential energy. Explain why this happens.

24. Volume flow rate through a blood vessel is regulated primarily by _____ and _____ .

25. According to Eq. (10.49), an increase in fluid density decreases the speed of propagation of the arterial pulse wave through a blood vessel. Explain.

26. Based on your answer to Exercise 25, explain why the speed of propagation of a wave through less dense air is lower than it is through more dense water?

SUGGESTED READING

Astarita, G., and Marrucci, G. (1974). *Principles of Non-Newtonian Fluid Mechanics*. McGraw-Hill, New York.

Bronzino, J. D. (1995). *Biomedical Engineering Handbook*. CRC Press, Boca Raton, FL.

Caro, C. G., Pedley, T. J., Schroter, R. C., and Seed, W. A. (1978). *The Mechanics of the Circulation*. Oxford Univ. Press, New York.

Chandran, K. B. (1992). *Cardiovascular Biomechanics*. New York Univ. Press, New York.

Chmiel, H., and Walitza, E. (1980). *On the Rheology of Blood and Synovial Fluids*, Research Studies Press, New York.

Dawson, T. H. (1991). *Engineering Design of the Cardiovascular System of Mammals*. Prentice Hall, Englewood Cliffs, NJ.

Fung, Y. C. (1984) *Biodynamics: Circulation*. Springer-Verlag, New York.

Li, J. K.-J. (1987). *Arterial System Dynamics*. New York Univ. Press, New York.

Mazumdar, J. N. (1992). *Biofluid Mechanics*. World Scientific, Singapore.

Noordergraaf, A. (1978). *Circulatory System Dynamics*. Academic Press, New York.

Schneck, D. J. (1978). *Biofluid Mechanics, Volume 1*. Virginia Polytechnic Institute & State University, Blacksburg.

Schneck, D. J. (1980). *Biofluid Mechanics, Volume 2*. Plenum Press, New York.

Schneck, D. J. (1990). *Engineering Principles of Physiologic Function*. New York Univ. Press, New York.

Schneck, D. J. (1992). *Mechanics of Muscle*, 2nd ed. New York Univ. Press, New York.

Schneck, D. J., and Lucas, C. L. (1990). *Biofluid Mechanics, Volume 3*. New York Univ. Press, New York.

Schroeder, M. (1991). *Fractals, Chaos, Power Laws*. Freeman, New York.

Shepherd, J. T. Vanhoutte, P. M. (1979). *The Human Cardiovascular System*: *Facts and Concepts*. Raven Press, New York.

Streeter, V. L., and Wylie, E. B. (1985). *Fluid Mechanics*, 8th ed. McGraw-Hill, New York.

Welkowitz, W. (1987). *Engineering Hemodynamics*: *Application to Cardiac Assist Devices*, 2nd ed. New York Univ. Press, New York.

Wolf, S., and Werthessen, N. T. (1979). *Dynamics of Arterial Flow*. Plenum Press, New York.

11 BIOMATERIALS

Chapter Contents

At the conclusion of this chapter, the reader will be able to:

- Define different classes of materials

- List failure modes of different materials

- List mechanical tests and determine which are appropriate for a given application

- Understand the considerations for selection of appropriate materials for specific applications

- Describe the biological responses to implanted materials

- Define inappropriate biological responses

- Determine correct biocompatibility testing protocols

11.1 INTRODUCTION

The definition of "biomaterial" has been difficult to formulate. One current acceptable definition is that a biomaterial is any material, natural or man-made, that comprises whole or part of a living structure or a biomedical device which performs, augments, or replaces a natural function. Thus, biomaterials may be inert materials or they may be from a biological source. Since medical devices are composed of one or many materials, it is important to understand the nature of each material in the device because, if one material is inappropriate, the entire device may fail from either adverse biological reactions or material failure.

Another important caveat to include in thinking about the definition of biomaterials is that it is difficult to improve on normal tissues or organs. Thus, the biomaterial or device may perform, augment, or replace a function that has been lost through disease or injury, but it will not restore the defective part to the original function of the living structure. In the public view, there has been much disappointment regarding medical devices since the devices have not performed miracles. For example, patients who receive a total knee replacement for an arthritic knee will usually have relief of pain and increased motion in the knee compared to the diseased knee. However, the knee will not be restored to normal function, and it may not last the life of the patient. The terms device and material may be used interchangeably in this chapter, and the reader should be alert to the subtleties involved.

The materials used in medical devices come from a variety of sources. In general, many of these materials were developed for other uses, and very few were actually manufactured purely for medical purposes. The amount used in medical devices is minuscule compared to that used in other applications such as food packaging, automobiles, or cookware. Thus, the material science involved in the field of biomaterials involves the same discipline as material science for any material. Once the fundamentals are known, then the modification for medical use can be considered.

11.2 MECHANICAL PROPERTIES AND MECHANICAL TESTING

Some basic terminology regarding the mechanical properties of materials is necessary for a discussion of materials and their interactions with biological tissues. The most common way to determine mechanical properties is to pull a specimen apart and measure the force and deformation. Materials are also tested by crushing them in compression or by bending them. Standardized test protocols have been developed to facilitate comparison of data generated from different laboratories. The vast majority of those used in the biomaterials field are from the American Society for Testing and Materials (ASTM). For example, tensile testing of metals can be done according to ASTM E8, ASTM D412 is for rubber materials, and ASTM D638 is for tensile testing of rigid plastics. These methods describe specimen shapes and dimensions, conditions for testing, and methods for calculating and reporting the results.

Tensile testing according to ASTM E8 is done with a "dog bone"-shaped specimen that has its large ends held in some sort of a grip while its narrow midsection is the "test" section. The midportion is marked as the "gage length" where deformation is measured. A mechanical test machine uses rotating screws or hydraulics to stretch the specimen. **Force** is measured in Newtons (N), and how much the specimen stretches, **deformation**, is measured in millimeters. Since specimens of different dimensions can be tested, measurements must be normalized to be independent of size. **Stress** σ (N/m^2 or Pascals), is calculated as force divided by the original cross-sectional area, whereas **strain**, ε (%), is calculated as change in length divided by the original length:

$$\sigma \ (\text{N/m}^2) = \text{force/cross-sectional area} \tag{11.1}$$

$$\varepsilon \ (\%) = [(\text{deformed length} - \text{original length})/\text{original length}] \times 100\% \tag{11.2}$$

A stress–strain curve can be generated from these data (Fig. 11.1), and there are a number of material properties that can be calculated. Region A in Fig. 11.1 is known as the elastic portion of the curve. If a small stress is applied to a metal, such as up to point 1, it will deform elastically. This means that, like a rubber elastic band, it will return to its original length when the stress is removed. The slope of the elastic portion of the stress–strain curve is a measure of the stiffness of the material and is called the **elastic modulus** (E) or **Young's modulus**:

$$E = \sigma/\varepsilon \quad \text{initial slope} = \text{stress/strain} \tag{11.3}$$

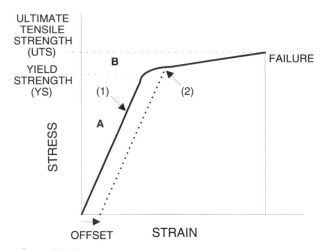

Fig. 11.1 Typical stress–strain curve for a metal that stretches and deforms (yields) before breaking. Stress is measured in N/m^2 (Pa), whereas strain is measured as a percentage of the original length. The minimum stress that results in permanent deformation of the material is called the yield strength (YS). The ultimate strength (UTS) is the maximum stress that is tolerated by the material before rupturing. The stress at which failure occurs is called the failure strength. Region A represents the elastic region since the strain increases in direct proportion to the applied stress. If a small stress is applied, e.g., to point 1, the material will return to its original length when the stress is removed. Region B represents the plastic region in which changes in strain are no longer proportional to changes in stress. Stresses in this region result in permanent deformation of the material. If a stress is applied which results in the strain at point 2, the material will follow the dotted line back to the baseline when the stress is removed and will be permanently deformed by the amount indicated by the offset.

As the applied stress is increased, a point is reached at which the metal begins to deform permanently, the yield point (YS). If at point 2 in Fig. 11.1 the stress is now released, the stress–strain recording will come down the dotted line parallel to the elastic region. The permanent amount of deformation is now shown as the offset yield. Since it may be difficult to determine the yield point for a material, an offset yield point is often used in place of the original yield point. For metals, yield is typically defined as 0.2%, whereas a 2% offset is often used for plastics. If the metal is loaded again, the recording will follow the dotted line starting at the offset yield, reaching the upper curve, and continuing to show a gradual increase in stress with increasing strain. This is known as the plastic region of the curve. The peak stress that is attained is called the tensile or ultimate tensile strength (UTS). Eventually, the metal will break at the failure or fracture strength (FS), and the percentage of elongation or compression to failure can be determined.

Example Problem 11.1

A 7-mm cube of bone was subjected to a compression loading test in which it was compressed in increments of approximately 0.05 mm. The force required to produce each amount of deformation was measured, and a table of values was generated. Plot a stress–strain curve for this test. Determine the elastic modulus and the ultimate tensile strength of the bone.

Deformation (mm)	Force (N)
0.00	0
0.10	67.9
0.15	267.6
0.20	640.2
0.26	990.2
0.31	1265.1
0.36	1259.9
0.41	1190.9
0.46	1080.8
0.51	968.6
0.56	814.2

Solution

First, determine the cross-sectional area of the cube in meters (0.007×0.007 m = 49×10^{-6} m). Use this value to determine the stress at each measuring point where stress (σ) equals force divided by cross-sectional area. For example, the stress when the cube was compressed by 0.10 mm was 67.9 N/0.000049 m^2 = 1.39 MPa (1 N/m^2 = 1 Pa). Next, determine the strain at each measuring point. The deformed length when the cube was compressed by 0.10 mm was $7 - 0.10$ mm = 6.9 mm, so the strain was $[(6.9 - 7.0 \text{ mm})/7.0 \text{ mm}] \times 100\% = 1.43\%$. This is the same value that would be obtained by dividing the amount of deformation by the original length. The minus sign is ignored since it merely indicates that the sample was subjected to compression rather than to tension.

Strain (%)	Stress (MPa)
0.00	0
1.43	1.39
2.14	5.46
2.86	13.07
3.71	20.21
4.43	25.82
5.14	25.71
5.86	24.30
6.57	22.06
7.29	19.77
8.00	16.61

The resulting stress–strain curve is shown in Fig. 11.2. Linear regression was used to determine the line shown in Fig. 11.2. The elastic modulus, i.e., the slope of the line, was 8.4, and the UTS was 25.82 MPa.

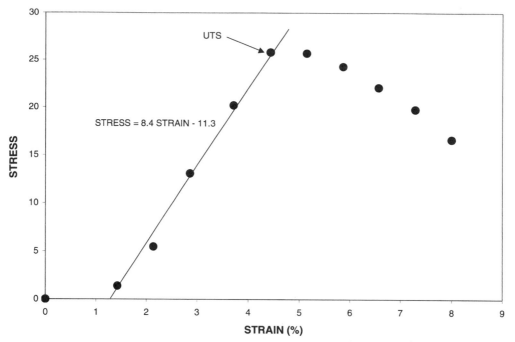

Fig. 11.2 The stress–strain curve for the bone data from Example Problem 11.1. Linear regression analysis was used to find the line which best fit the data for strains of 1.43–4.43%, i.e., the linear portion of the curve. The slope of the line, 8.4, represents the elastic modulus of the bone. The ultimate tensile strength (UTS) of the material was 25.82 MPa. ◼

Several other terms are applied to the test results. The slope of the elastic portion, the elastic modulus, is often called **stiffness.** If the metal stretches a great deal before failure it is said to be **ductile** (Fig. 11.3). If the material does not deform or yield much before failure, it is said to be **brittle.** The area under the curve has units of energy and is called **toughness.** Although not directly available from a stress–strain curve, the strength of a material can be related to its **hardness.** Stronger materials are typically harder. Hardness is tested by measuring the indentation caused by a sharp object that is dropped onto the surface with a known force. Hardness is perhaps the most important property when considering a material's wear resistance.

An additional property that is not depicted in Fig. 11.3 is the **fatigue strength** or **endurance limit** of a material. If the material tested as in Fig. 11.1 is loaded to point 2 and unloaded, it would become permanently deformed. If this were repeated several times, like bending a paper clip back and forth, the material would eventually break. If, however, the metal is loaded to point 1 and then unloaded, it is not deformed or broken. If it is loaded again to point 1 and unloaded, it still will not break. For some metals, there is a stress level below which the part can theoretically be loaded and unloaded an infinite number of times without failure. In reality, a fatigue limit is defined at a specified number of cycles, such as 10^6 or 10^7. Clearly, fatigue strength is a critical property in the design of load-bearing devices such as to-

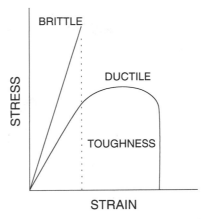

Fig. 11.3 Brittle materials reach failure with only a small amount of deformation (strain), whereas ductile materials stretch or compress a great deal before failure. The area under the stress–strain curve is called toughness and is equal to the integral from ε_0 to ε_f σ $d\varepsilon$.

tal hips, which are loaded on average a million times a year, or heart valves, which are loaded 40 million times a year.

Table 11.1 shows the properties of common materials. In general, metals are strong and tough, whereas ceramics are more rigid or stiffer but brittle. Polymers have a lower modulus so they are softer and weaker than metals and ceramics but often have much higher elongations to failure. Polymers may not be tougher because they are not as strong. Bone is stiffer and stronger than collagen. Metals are typically 10 times stiffer than bone, whereas polymers are much less stiff than bone.

Example Problem 11.2

Why would the following uses not be a good idea?
   a. A metal tube blood vessel replacement
   b. A silicone rubber for ligament or tendon replacement
   c. A silicone rubber pin for fixing a bone fracture
   d. A ceramic skin graft

Solution

(a) A metal tube is too rigid or stiff to match the blood vessel. The blood vessel moves and stretches with movement and each heartbeat and would get torn apart at the connection (anastamosis). (b) Silicone rubber is too stretchy for a ligament replacement. A joint with a torn ligament, e.g., a knee, would be much too lax if a silicone rubber ligament replacement was used. In addition, silicone is not strong enough. (c) Silicone is not stiff enough for keeping a fracture lined up. Its modulus is too low. (d) A ceramic is too stiff for skin. It is also brittle, and a thin skin graft would probably crack or shatter. ∎

TABLE 11.1 Mechanical Properties of Some Materials[a]

Material	Yield (Mpa)	Utimate Tensile Strength (Mpa)	Deform (%)	Modulus (GPa)
Metals				
High-strength carbon steel	1600	2000	7	206
F138, annealed	170	480	40	200
F138, cold worked	690	860	12	200
F138, wire	—	1035	15	200
F75, cast	450	655	8	200
F799, forged	827	1172	12	200
F562, hot forged	1000	1200	10	200
F136, Ti64	795	860	10	105
Gold		2–300	30	97
Aluminum, 2024-T4	303	414	35	73
Polymers				
PEEK		93	50	3.6
PMMA cast		45–75	1.3	2–3
Acetal (POM)		65	40	3.1
UHMWPE		30	200	0.5
Silicone rubber		7	800	0.03
Ceramics				
Alumina		400	0.1	380
Zirconia, Mg partially stabilized		634		200
Zirconia, yttria stabilized		900		200
Carbons and Composites				
LTI pyrolytic carbon + 5–12% Si		600	2.0	30
PAN AS4 fiber		3980	1.65	240
PEEK, 61% C fiber, long		2130	1.4	125
PEEK, 61% C fiber, ±45		300	17.2	47
PEEK, 30% C fiber, chopped		208	1.3	17
Biologic Tissues				
Hydroxyapatite (HA) mineral		100	0.001	114–130
Bone (cortical)		80–150	1.5	18–20
Collagen		50		1.2

[a] Literature values or minimum values from standards.

11.3 GENERAL CLASSIFICATION OF MATERIALS USED IN MEDICAL DEVICES

11.3.1 Metals and Metal Alloys

Metals are composed of atoms held together by metallic bonds with each atom surrounded by a "cloud" of valence electrons with high mobility. This mobility gives metals good conductivity of electricity and heat. The concept of a phase is important to understanding alloys. A **phase** is a singular form (composition or structure) of a material. Ice, water, and steam are three phases of H_2O. When metals solidify from a melt, they undergo a phase transformation from liquid to solid. During solidification, atoms take on an orderly array known as a crystal structure, much like salt crystals forming as salty water evaporates or rock candy forms on a string.

The nature of the relationship between atoms is described by what is called a primitive cell, or Bravais lattice. For example, atoms may arrange themselves with atoms at eight corners of a cube and with one in the middle called body-centered cubic (BCC) as shown in Fig. 11.4A. Figure 11.4B shows the open circles arranged on the faces of the cube and is called face-centered cubic (FCC). Some metals, such as iron, cobalt, and titanium, will undergo allotropic transformations which means that they change crystal structure with changes in temperature. Thus, for example, after iron undergoes the phase transformation from liquid to solid BCC, it will undergo additional phase transformtions to FCC and then back to BCC during cooling.

As molten metal solidifies in a mold, solidification will begin on the surface of the mold. If the mold is very hot, there will be only a few locations where the solid will begin to form (nucleate) and grow. If the mold is cold, there will be many nucleation sites. At each site, atoms will be laid down on the solid in an orderly crystallographic manner. Solidification will continue until two growth areas meet and form a boundary. Each of these growth sites is called a crystal or grain, and each boundary is a crystal or grain boundary.

Metallic alloys are mixtures of several elements in a solid solution. For elements of similar atomic charge, diameter, and crystal structure, there is no limit to the solubility of one element in another so they solidify as a single phase. For example, copper and nickel are fully soluble in each other.

When atoms are of different sizes or electronegativity, they will have limited solubility; for example, lead and tin that are used for electrical solder. At intermediate temperatures, they have limited solubility, but at room temperature the solder exists as a two-phase alloy. More significantly, addition of one to the other lowers the melting temperature to a minimum that is referred to as the eutectic temperature. This means the liquid transforms directly to solid at a defined temperature without any intermediate solid–liquid mixtures. For lead–tin, the eutectic temperature occurs at 61.9% tin, which is the standard 60–40 solder. By using other ratios, higher melting temperatures can be achieved for multiple soldering jobs.

Small differences in the diameter of the two (or more) elements in a single-phase alloy or two-phase alloy provide strengthening. The presence of large atoms in a lattice of smaller atoms will produce a localized strain in the lattice, and it will be under localized compression. Similarly, a few small atoms in a lattice of larger atoms will be under localized tension. These localized strains increase the strength of the metal by a mechanism known as solid solution strengthening.

Elements with markedly different properties or crystal structures will have limited solubility. For example, carbon is much smaller than iron. In small quantities, carbon is soluble in iron, but at higher concentrations it will precipitate out as a second phase, such as graphite, or form a carbide. A number of alloy systems utilize the precipitation of second phases as a strengthening mechanism known as precipitation hardening.

Carbon will also influence the crystal structure of iron. At room temperature, iron has a BCC crystal structure and is known as α-ferrite. With heating, it undergoes a phase transformation to FCC and is known as γ-austenite. With further heat-

a

b

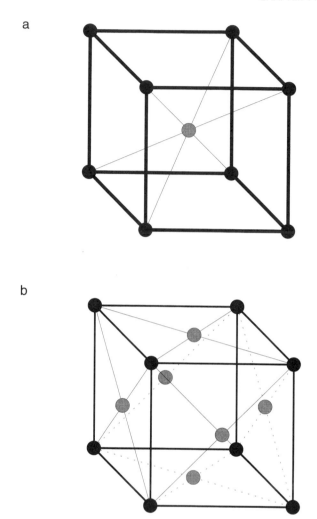

Fig. 11.4 Schematic diagrams of atoms in crystallographic arrays of body-centered cubic (a) and face-centered cubic (b).

ing, it goes back to BCC (δ-ferrite) before it melts. Since the spaces, or interstices, between atoms are larger in FCC than in BCC, the smaller carbon atoms fit better in the FCC and thus have a higher solubility in FCC. This has several implications. At low concentrations, carbon will increase the stability of the FCC. In other words, the presence of carbon will lower the temperature at which the BCC α-ferrite converts to the FCC γ-austenite and will increase the temperature at which the FCC converts back to BCC δ-ferrite. Carbon also provides interstitial solid solution strengthening of iron.

Point or line defects can occur in lattice structures. When viewed as a two-dimensional grid, there will be an occasional line that ends at a dislocation (Fig. 11.5). Above the dislocation, the atoms are in compression, whereas they are in ten-

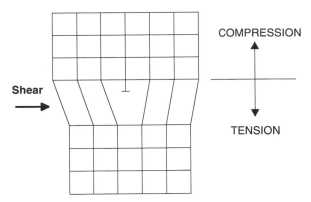

Fig. 11.5 Regions of tension (below) and compression (above) around an edge dislocation in a simple cubic lattice. Application of shear force as indicated would move the dislocation one position at a time to the right.

sion below. By pushing against the side of the dislocation line with a shear force, the position of the dislocation can move one line or plane at a time. The presence of dislocations greatly reduces the force necessary to deform a metal.

When a metal solidifies, there are few dislocations. If the metal is then mechanically worked, as in pounding with a hammer or bending (like bending a paper clip), dislocations move around, and their number greatly increases. This is the mechanism of plastic deformation of metals. The increased number of dislocations, each with its localized stress field, makes it more difficult to do more plastic deformation. The dislocations get in each other's way. Thus, the metal gets stronger and harder by the mechanism of cold working (work hardening). If the deformation results in too many dislocations in a particular location, they will coalesce into a crack, and the metal will begin to break. The strength of a metal can therefore be controlled by controlling the grain size. By keeping the grain size small, few dislocations can accumulate within a grain, and thus the metal can deform more before it begins to break.

After extensive cold working, the alloy can be heated to anneal or stress relieve it. This results in the formation of new crystals with few dislocations. Controlling the temperature and time can result in a soft, fine-grained metal that can be subsequently cold worked. This is the process used to produce small pieces of metal from a large ingot. Metal can be rolled to form a heavily cold-worked but thinner plate. Before it breaks, the metal is annealed to stress relieve it, and then it is rolled again. Similarly, bars and wires are formed by repeated drawing and annealing. The differences due to annealing and cold working can be seen in the data for F138 stainless steel in Table 11.1. Similarly, cast F75 is typically large grained. Minor compositional differences and forging result in a fine-grained heavily cold-worked and much stronger F799.

All these concepts are utilized in designing the alloy systems used for medical devices. Consider stainless steel as an example of how an alloy "can be designed."

Steel is an alloy of iron, carbon, and other elements. Corrosion resistance is the main concern for stainless steel, and the precipitation of carbon to form a two-phase alloy is undesirable since the two phases can lead to mixed metal or galvanic corrosion. For stainless steel, it is important to prevent the precipitation of carbon. One way to avoid this is to keep the concentration of carbon low, typically in the 0.08–0.03% range. It is also important to have the iron in the FCC form since the carbon solubility is higher in the FCC than in the BCC form.

A minimum of 12% chromium is added to make steel "stainless." Since chromium is BCC, its addition will stabilize the BCC form of iron. Carbon has a great affinity for chromium and forms chromium carbides with a typical composition of $Cr_{23}C_6$ that results in a two-phase and carbon precipitation problem. In the region surrounding the carbide, the chromium concentration will be depleted (it all moved into the carbide), and thus the corrosion resistance of the steel surrounding the carbide will be reduced.

Nickel is added to the chromium–iron–carbon steel to solve this apparent dilemma. Nickel is FCC so its addition will stabilize the FCC form of iron and keep the carbon in solution. This is stainless steel. Stainless-steel knives, forks, and spoons are typically "18–8": 18% chromium and 8% nickel. To make surgical stainless steel according to ASTM F138, 17–19% chromium, 13–15.5% nickel, and 2 or 3% molybdenum are added for improved corrosion resistance. There are two grades: high (<0.08%) and low (<0.03%) carbon. The result is a homogeneous, single-phase, corrosion-resistant stainless-steel alloy. While stainless steel has good corrosion resistance, the options for strengthening mechanisms have been limited to only cold working. As is shown in Table 11.1, the mechanical properties of ASTM F138 surgical stainless steel in the annealed condition can be greatly increased by cold working and are even higher for heavily drawn wire. While yield and ultimate strength increase, the ductility (elongation) decreases. Cold working makes stainless steel stronger but more brittle.

Example Problem 11.3

Cold working increases the strength of stainless steel, but plates and screws used for fracture fixation are not made with the same degree of cold working. Why?

Solution

The strongest material is desired for both plate and screw, but there is a difference in the way the devices are used. The surgeon will often bend and twist a metallic plate so it fits the contour of the bone. If the plate were heavily cold worked, it might exceed its plastic limit during bending and break. Therefore, a moderate degree of cold working is used for plates. By contrast, the screws can be heavily cold worked to give them strength.   ∎

These basic concepts are utilized in the design and selection of other metals used for biomedical applications. The metallurgy is more complicated but the concepts are the same. The two other major surgical implant alloys are made from cobalt or titanium, and both undergo allotropic transformations. Both have more inherent corrosion resistance than iron so it is not critical for them to be single-phase alloys.

Cobalt alloys typically have chromium and cobalt carbides for precipitation strengthening. Titanium is typically used as a two-phase alloy with a variety of structures and morphologies. Typical properties and chemical compositions are listed in Tables 11.1 and 11.2.

11.3.2 Ceramics and Glasses

Unlike metals, in which atoms are loosely bound and able to move, ceramics are composed of atoms that are ionocovalently bound into compound forms. This atomic immobility means that ceramics do not conduct heat or electricity. Two very obvious properties that are different from metals are those of melting point and brittleness. Ceramics have very high melting points, generally above 1000°C, and are brittle. This means that it is very expensive to melt ceramics and impossible to cold work them. Some ceramics also form inorganic glasses. Glasses are different from crystalline materials in that they do not have a regularly arranged repeatable structure. Crystals, like ice, will melt at very distinct temperatures, whereas glasses will get softer and runnier over a range of temperatures.

Many of the more common ceramics are metal oxides such as alumina, sapphire (Al_2O_3), and silica (SiO_2). Ceramics also include many nonoxide compounds, such as carbides, nitrides, sulfides, and silicides. Some are single-element materials such as carbon (graphite and diamond) and silicon, which is used for integrated circuits. In some ceramics, many different elements form a single compound such as the structural ceramic $Si_3Al_3O_3N_5$ (Sialon), the mineral phase of bone $Ca_{10}(PO_4)_6(OH)_2$ (hydroxyapatite), or the superconductor series ceramics (e.g., $YBa_2Cu_3O_{7-\delta}$ and $Bi_2Sr_2CaCu_2O_8$).

The atomic planes in ceramic materials cannot slide one position at a time by the mechanism of dislocation. For a metal, the atoms and bonds surrounding the dislocation are all similar (Fig. 11.5). In contrast, ceramics have directional bonds that prevent shifting one position at a time. Thus, they are brittle and do not bend. Instead, they crack and break. If a metal is hit, it will absorb the energy by bending or deforming. A ceramic that is hit will convert the strain energy to surface free energy, i.e., it will form a crack. Ceramics are typically strong and hard, which gives them excellent wear resistance. Grain boundary impurities often lead to failure and grain excavation in situations of wear or abrasion. One method of toughening ceramics is to engineer the grains and grain boundaries so that the crack tends to go through the boundary and is deflected by the grain. Another approach is to develop com-

TABLE 11.2 Weight Percent Chemical Composition of Some Surgical Alloys

F138, wrought stainless steel: 17–19 Cr, 13–15.5 Ni, 2–3 Mo, <2 Mn, <0.08 or <0.03 C
F75, cast cobalt–chromium–molybdenum alloy: 27–30 Cr, <1.0 Ni, 5–7 Mo, <1 Mn
F799, wrought Co–Cr–Mo alloy: 26–30 Cr, <1.0 Ni, 5–7 Mo, <1.0 Mn, <1.5 Fe, <1.5 C
F136, titanium 6AI-4V alloy: 5.5–6.5 Al, 3.5–4.5 V, <0.015 N, <0.13 O, <0.08 C

pounds, such as yttria and magnesia partially stabilized zirconia, in which a phase transformation is induced by stress and crack propagation is prevented.

This wide variety of compounds also yields a wide variety of properties. One way to describe the differences among ceramics is to distinguish them by bond types. A material such as diamond or silicon carbide has a very directional, covalent bond. These bonds are very strong and very dependent on where each atom is located. The bonds in ionic materials such as halite (NaCl) are weaker and less directional. Transition metal compounds, such as tungsten carbide and titanium nitride, exhibit a combination of metallic and covalent bonding and, hence, exhibit the properties of metals and ceramics. They work well as hard wear-resistant coatings for metal drill bits and bearing surfaces. A stainless-steel and "gold" watchband probably has titanium nitride coating as the gold.

Ceramics begin as a powder that has been specially processed to be of high purity for most biomedical applications. Appropriate amounts of raw material powders are mixed together into a batch. The batched powder is then processed and formed into a shape that is similar to the final part. This unfired ceramic is called a **green body.** The green body is then raised to a temperature at which the powder coalesces into a single structure in a process called sintering. In sintering, the powder coalesces into a hard material in which the particles have fused together to form grains. Grains are crystalline and have amorphous boundaries that often result in the inclusion of higher levels of any impurities. The sintered parts are then machined into their final shape by grinding and polishing methods. Glasses, because they soften as opposed to melt at a single temperature, are usually heated until soft and then poured and made into a final shape while still hot.

The formation of carbon materials is quite different. Many are formed by pyrolysis or burning of an organic material. Graphite is made by putting coke in a furnace and subjecting it to very high temperatures for several days to burn off all the noncarbon elements. Graphite makes good electric motor brushes or pencil leads, but it is too soft for most biomedical applications so it is coated. One method produces low-temperature isotropic LTI pyrolytic carbon. For example, to make a heart valve, like the floating disc in Fig. 11.6, preformed graphite discs are placed in a col-

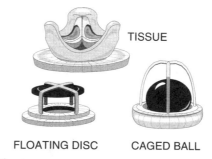

TISSUE

FLOATING DISC   CAGED BALL

Fig. 11.6   Three types of artificial valves that are used to replace diseased or malformed human valves. The tissue valve generally contains valve leaflets from pigs, whereas the other two types are composed entirely of manufactured materials.

umn through which methane or propane are pumped to levitate the discs in a "flu-idized bed." An alloy of carbon and silicon can be made using methyltrichlorosilane (CH_3SiCl_3). The column is heated to 1500°C to pyrolyze the gas. The result is the formation of small crystallites of carbon that are deposited on the graphite discs. The 1500°C is considered a low temperature. The crystallites are random in orientation, and the coating has isotropic properties, hence the name LTI carbon. This produces a very strong and hard coating as shown in Table 11.1. Carbon fibers can be made from polymeric fibers such as polyacrylonitrile (PAN). Fibers are subjected to a multistage process of heating to burn off the noncarbon atoms and create long chains of the hexagonal crystalline carbon structure. Temperatures in the final stages are 1500–2000°C depending on the process. The result is very high strength fibers as shown in Table 11.1.

The biomedical applications of ceramics continue to grow. Their corrosion resistance and good wear properties and electrical insulation offer major advantages, whereas their brittleness with catastrophic failure is a disadvantage. Careful engineering and design are currently being used to investigate new ceramics.

11.3.3 Polymers

As the name implies, polymers are made up of many "mers" or basic building blocks. They are formed in long, sometimes branching chains, with a backbone of carbon or silicon atoms that are held together with covalent bonds. Perhaps the simplest and most common one is polyethylene (PE), which is made from many ethylene molecules.

$$[-\underset{\underset{\displaystyle H}{|}}{\overset{\overset{\displaystyle H}{|}}{C}}-\underset{\underset{\displaystyle H}{|}}{\overset{\overset{\displaystyle H}{|}}{C}}-]_n$$

The number of mers (n) will determine many of the properties of the polymer that is formed. Low-molecular forms of polyethylene are used for soda pop bottles, whereas ultrahigh-molecular-weight polyethylene (UHMWPE) is used for the bottom of snow skis and the plastic-bearing component of total joint replacements.

One very important consideration for biomedical applications is the nature of polymerization. PE polymers are formed by addition reactions. Heat and a catalyst are used to add ethylene mers to the string. In contrast, nylon is formed by a condensation reaction, in which two different molecules are combined and then strung together as a chain. Combining the two molecules releases a by-product, in this case water, which does not have any adverse biological effects. By comparison, the polymerization of silicone adhesive "bathtub caulk" releases acetic acid, and the area smells like vinegar. Other condensation reactions release by-products with potentially adverse biological effects. The issue is not only what is released during polymerization but also what may be trapped in the material that may slowly leach out over time and expose the patient to potential toxic components.

In the structure PE discussed previously, the polymer is represented as a linear chain with hydrogen atoms on all the side binding sites. In many systems, one of the hydrogens is replaced with one or more ligands. For example, replacing one hydrogen with a chlorine atom produces polyvinyl chloride (PVC). Replacing all four with fluorine atoms results in polytetraflouroethylene (PTFE) or Teflon. In some cases, the polymer may not be linear but may have side branches. Low-density polyethylene may have a branched structure with side chains of PE coming off at a number of the hydrogen sites that were shown in the PE structure. It is also possible to tie chains together by cross-linking. The process of vulcanization, which was developed by Goodyear, used sulfur to form a bridge between chains of natural rubber and created a three-dimensional structure.

Polymers can be classified as thermoplastic or thermosetting. A thermoplastic polymer has a linear or branched structure. As a solid it is like a bowl of spaghetti since the chains can slide over one other. With heating, the chains can slide more easily, and the polymer melts or flows. Thus, thermoplastic polymers can be heated, melted, molded, and recycled. Differences in properties can be achieved with the addition of different ligands. PVC is more rigid than PE because the chlorine atoms are larger and tend to prevent the sliding of one molecule over another. Polymethylmethacrylate (PMMA), as shown in Table 11.1, is stronger, stiffer, and much more brittle than UHMWPE. In this case, two of the four hydrogen atoms are replaced, one with a methyl group (CH_3) and the other with an acrylic group ($COOCH_3$). These large side groups make sliding much more difficult, hence the increase in strength and modulus. They also make it difficult for the molecules to orientate in an orderly, crystalline pattern. As a result of this amorphous structure, PMMA (Plexiglas or Lucite) is optically transparent.

In contrast, a thermosetting polymer is composed of chains that are cross-linked. They do not melt with heating but degrade. The term thermoset implies that there is a chemical reaction, often involving heat, which results in setting in a three-dimensional cross-linked structure. A common example is "5-minute epoxy." When the two parts are mixed, the catalyst causes setting and cross-linking of the epoxy. Once set, it cannot be heated and reused. The amount of cross-linking affects the mechanical properties. Few cross-links are used in rubber gloves. Adding more sulfur and cross-linking produces a car tire. Even more cross-links are added to make the hard casing of a car battery.

Mixing two-part polymer systems does not necessarily imply a thermosetting reaction. Dental acrylic and bone cement are two-part systems used for making dentures or fixing total joint replacements. The powder is ground up polymer (PMMA) with a chemical initiator, benzyol peroxide (BP). The liquid is monomeric and is a chemical that acts as an activator. Mixing them softens the powder and causes activation of the BP. The BP then causes the polymerization of the monomer, resulting in solidification of the powder–liquid in 5–10 min. There is no cross-linking so the solid could be heated and molded after solidification.

Another term used to describe the mechanical properties of polymers is elastomer. Elastomers can be either thermoplastic or thermosetting. Elastomer implies that the polymer will return to its original shape after deformation like a rubber band. Molecules with zigzag backbone structures can straighten out with stress and then spring back. Cross-linked thermosets can stretch the bonds, straighten out the structure, and then spring back. Note the very high elongation (% deformation) for silicone elastomer (Table 11.1).

Many devices are made of plastics. A **plastic** is a polymer with any number of additives. Colorants are added to make them look nice, and antioxidants may be added to increase stability. Plasticizers may be added to increase softness. These are low-molecular-weight substances that act as a lubricant so the molecules can slide over each other, much like butter or tomato sauce in a bowl of spaghetti. An early problem with PVC blood bags involved the addition of a plasticizer that tended to leak or leach out of the plastic and had toxic effects in the blood of the recipients.

11.3.4 Composites

Composite materials are produced by mixing or joining two or more materials to give a combination of properties of the individual components. Typically, the property of each material is not changed but contributes to the property of the composite. Based on the shapes of the parts, composites can be placed in three categories: particulate (concrete), fiber (fiberglass), and laminar (plywood). Many properties can be determined by the rule of mixtures, which states that the property of the mixture is equal to the sum of the volume fractions of each constituent times that property for that constituent. For example, density of a polymer–fiber composite is equal to the sum of the fraction of polymer matrix times its density plus the fraction of fiber times its density.

For example, consider composites of carbon fiber in a polyetheretherketone (PEEK) matrix. The tensile strength for PEEK is 93 MPa, and its modulus is 3.6 GPa (Table 11.1). For carbon fibers made of PAN the values are 3980 and 240, respectively. A composite made by adding 30% chopped fiber to PEEK increases the strength (208) and modulus (17), but the ductility is reduced from 50% to only 1.3%. Long-fiber composites produce even higher strength and modulus. A composite made from sheets of 61% carbon fibers and tested in a direction longitudinal to the fibers has a strength of 2130 and modulus of 125. By using a 45/45 cross-ply, the properties drop to 300 and 47, but there is a substantial increase in elongation. Thus, adjusting the nature and amount of the fibers can produce a wide range of mechanical properties.

Many biological materials are also composites. Bone, for example, is a composite of mineral hydroxyapatite (HA) and collagen fibers. As shown in Table 11.1, HA is strong, rigid, and brittle. Collagen has a much lower modulus, and though it is difficult to discern from the literature it has much more elongation than HA. The composite bone is a strong and more flexible material.

Since bone growth and structure are responsive to stress, the use of some rigid metallic devices has been associated with alterations in bone healing and structure. Composite technology has been used to tailor make materials to match bone. The 30% chopped-carbon PEEK has a modulus close to that of bone and has been considered for use as a moderately flexible device to hold a fracture together during healing. The long-fiber composites are stiffer than bone but not as stiff as stainless steel or cobalt alloys. Thus, they have been considered for high-strength devices to replace a bone or joint section. The nature and amount of the composite constituents can be adjusted to produce the optimum properties.

11.4 DEGRADATION OF MATERIALS

11.4.1 Metallic Corrosion

There are a number of mechanisms by which metals can corrode, and corrosion resistance is one of the most important properties of metals used for implants. The ones that are of most significance to implant applications in aqueous saline solutions are galvanic (or mixed-metal) corrosion, crevice corrosion, and fretting corrosion.

Galvanic (mixed-metal) corrosion results when two dissimilar metals in electrical contact are immersed in an electrolyte. There are four essential components that must exist for a galvanic reaction to occur: an anode, a cathode, an electrolyte, and an external electrical conductor. The *in vivo* environment contains electrolytes. A patient with two total hip replacements (THRs) made of different alloys is not subject to mixed-metal corrosion since there is no electrical connection. However, a THR made of two alloys or a fracture plate of one metal fixed with screws of another metal may be susceptible to mixed-metal corrosion.

When two dissimilar metals are connected in an electrochemical cell, one will act as an anode while the other will be the cathode. Metal oxidation will occur at the anode, as shown in Eq. (11.4). The reaction at the cathode will depend on the pH of the environment (Eqs. 11.5 and 11.6). The direction of the reaction can be determined by examining the electromotive force (EMF) series, a short listing of which is shown in Table 11.3. These potentials represent half-cell potentials of metals in equilibrium with 1-M solutions of their ionic species. The potential for hydrogen is

TABLE 11.3 Electromotive Force Series[a]

$K = K^+ + e^-$	-2.93	Active (more anodic)
$Na = Na^+ + e^-$	-2.71	
$Al = Al^{+3} + 3e^-$	-1.66	
$Ti = Ti^{+2} + 2e^-$	-1.63	
$Zn = Zn^{+2} + 2e^-$	-0.76	
$Cr = Cr^{+3} + 3e^-$	-0.74	
$Fe = Fe^{+2} + 2e^-$	-0.44	
$Co = Co^{+2} + 2e^-$	-0.28	
$Ni = Ni^{+2} + 2e^-$	-0.25	
$Sn = Sn^{+2} + 2e^-$	-0.14	
$H_2 = 2H^+ + 2e^-$	0.000	
$Cu = Cu^{+2} + 2e^-$	0.34	
$Ag = Ag^+ + e^-$	0.80	
$Pt = Pt^{+2} + 2e^-$	1.20	
$Au = Au^{+3} + 3e^-$	1.50	Noble (more cathodic)

[a] Standard reduction potentials (E^0 V) in aqueous solution at 25°C.

defined as zero. As shown, the standard potential for iron is -0.44 V. If iron is connected to copper with an EMF of 0.34 V, the potential difference is 0.78 V. Since iron is the anode, iron oxidation will occur according to the reaction shown in Eq. (11.4). The reaction at the copper cathode will depend on the pH of the solutions as shown in Eqs. (11.5) and (11.6):

$$\text{Anodic reaction} \quad Fe \rightarrow Fe^{+2} + 2e^- \tag{11.4}$$

$$\text{Cathodic reaction (acidic solution)} \quad 2H^+ + 2e^- \rightarrow H_2 \tag{11.5}$$

$$\text{Cathodic reaction (neutral or basic)} \quad O_2 + 2H_2O + 4e^- \rightarrow 4OH^- \tag{11.6}$$

Because the free energy per mole of any dissolved species depends on its concentration, the free energy change and electrode potential of any cell depends on the composition of the electrolyte. Thus, the direction and rate of the reactions also depend on the concentration of the solutions. Increasing the concentration of Fe^{+2} in the environment will shift the potential in the positive or Noble direction. As the Fe^{+2} concentration increases, the potential difference between the iron and copper will become less as the iron becomes more cathodic. Similarly, the concentration of oxygen at the cathode will affect the EMF of the cell. Increasing O_2 will make it more noble, whereas decreasing O_2 will make it more anodic. In fact, crevice corrosion is initiated by changes in oxygen concentration as discussed later.

Galvanic cells occur not only with different alloys but also with differences within an alloy. Carbides, grain boundaries, and different phases within an alloy also present differences in EMF and thus the possibility for localized galvanic cells. Cold working also increases the free energy of metal and thus its susceptibility to corrosion. Bending a plate or pounding on a nail head causes localized cold working and makes that area anodic to the rest to the piece.

Galvanic corrosion can also be utilized to prevent corrosion by cathodically polarizing the part to be protected. Steel ships are protected from rusting by the attachment of blocks of zinc. The zinc blocks ("zincs") serve as a sacrificial anode and protect the steel hull. Metal pumps and other metallic components on ships are also protected with zincs. A power supply can be attached to a part, such as a steel underground pipeline, to make the pipe cathodic to a replaceable anode. This protects the pipeline.

Electrode size also has an effect on galvanic reaction rates. The classic example is the difference between galvanized and tin-plated iron. As Table 11.3 shows, zinc is anodic to iron. Thus, galvanizing results in coating the iron with an anodic material. When the zinc is scratched and the iron is exposed, the small size of the iron cathode limits the reaction, and there is minimal corrosion. In contrast, tin is cathodic to iron. When a tin plate is scratched, the small iron anode is coupled with a large cathode. Anodic corrosion and removal of iron from the scratch results in an increased area of the exposed iron and thus an increase in corrosion rate. This self-accelerating corrosion can be prevented by coating the tin cathode with a nonconductive material such as paint or varnish.

Crevice corrosion can occur in a confined space that is exposed to a chloride solution. The space can be in the form of a gasket-type connection between a metal and a nonmetal or between two pieces of metal bolted or clamped together. Crevice corrosion involves a number of steps that lead to the development of a concentration cell, and it may take 6 months to 2 years to develop. Crevice corrosion has been observed in some implanted devices where metals were in contact, such as in some THR devices, screws and plates used in fracture fixation, and some orthodontic appliances. The initial stage is uniform corrosion within the crevice and on the surfaces outside the crevice. Anodic and cathodic reactions occur everywhere with metal oxidation at the anode and reduction of oxygen and OH^- production at the cathode. After a time, the oxygen within the crevice becomes depleted because of the restricted convection of the large oxygen molecule. The cathodic oxygen reduction reaction ceases within the crevice, but the oxidation of the metal within the crevice continues. Metal oxidation within the crevice releases electrons which are conducted through the metal and consumed by the reduction reaction on the free surfaces outside the crevice. This creates an excess positive charge within the crevice that is balanced by an influx of negatively charged chloride ions. Metal chlorides hydrolyze in water and dissociate into an insoluble metal hydroxide and a free acid (Eq. 11.7). This results in an ever-increasing acid concentration in the crevice and a self-accelerating reaction:

$$M^+Cl^- + H_2O \rightarrow MOH + H^+Cl^- \tag{11.7}$$

The surgical alloys in use today owe their corrosion resistance to the formation of stable, passive oxide films — a process called **passivation**. Titanium, which appears as an active metal on the EMF series in Table 11.3, forms a tenacious oxide that prevents further corrosion. Stainless steels and cobalt alloys form chromium oxide films. As indicated in Table 11.3, they are active, but in the environment in which this oxide film is formed, they become passive or noble.

To be self-passivating, stainless steels must contain at least 12% chromium. However, carbon has a strong affinity for chromium, and chromium carbides form with the average stoichiometry of $Cr_{23}C_6$. The formation of a carbide results from the migration of chromium atoms from the bulk stainless-steel alloy into the carbide. The result is that the carbide has high chromium content, whereas the alloy surrounding the carbide is depleted in chromium. If the chromium content is depleted and drops below 12% Cr, then there is insufficient Cr for effective repassivation, and the stainless steel becomes susceptible to corrosion. As a safety factor, surgical stainless steel contains 17–19% chromium, and the carbon content in surgical alloys is kept low at <0.08 or $<0.03\%$.

The problem of carbide formation is especially important with welded stainless-steel parts. If steel is heated to the "sensitizing range" of 425–870°C, the chromium can diffuse in the solid and form carbides. At temperatures above 870°C, the carbon is soluble in the FCC lattice. Below 425°C, the mobility is too low for carbide formation. If the peak temperature in the metal away from the weld is in the sensitizing range, carbides can form. This is known as weld decay or corrosion of the sensitized metal on each side of the weld. By heat treating after welding, the carbides can be redissolved, and the metal quickly quenched to avoid reformation.

With the oxide film intact, surgical alloys are passive and noble. If the film is damaged, as with scratching or fretting, the exposed metal is active. Reformation or repassivation results in restoration of the passive condition. Fretting corrosion involves continuous disruption of the film and the associated oxidation of exposed metal. Devices that undergo crevice corrosion are also examples in which fretting corrosion has accelerated crevice corrosion.

Example Problem 11.4

Old cars in the northern United States are often rusted on the bottom of their doors and on their trunk lids, and the tailgates of old pickup trucks are also often rusted. Name and discuss five reasons for this. Would bolting on a zinc help?

Solution

(i) The edges and bottoms of doors and lids are formed by bending the metal back on itself. This causes cold working at the bend, which makes it anodic to the rest of the metal. (ii) The crimps are then spot-welded closed. This creates an area of different microstructure, which leads to a galvanic situation. (iii) The roads are salted in the winter, and the salt water spray gets caught in the crimp. This is the electrolyte. Sand may also get in the crimped space and help maintain a moist environment. (iv) Car manufacturers put a decorative strip of chromium-plated steel along the bottom of the lid. The chromium is cathodic to the steel and provides another source of galvanic corrosion. (v) There are potholes in the roads. These cause bouncing of the lid against the frame. This can chip the paint and expose the unprotected metal, or it may be a cause of fretting corrosion.

Bolting on a zinc would not help except for corrosion of metal in the same electrolyte pool. A zinc in the crevice would help the crevice. If the car fell into the ocean, then the zinc would protect the whole car. ∎

Example Problem 11.5

Your grandmother has a stainless-steel total hip. Now she needs the other hip replaced, and the doctor wants to use one made of a cobalt chromium alloy. Is that a problem for corrosion?

Solution

No. One of the four essential elements for galvanic corrosion is missing. There is an anode, the stainless steel; there is a cathode, the cobalt alloy; and there is an electrolyte, the saltwater of the body. However, there is no electrical connection so there is no problem. If she were to fall and fracture her pelvis, and the break was repaired with an external fixator, then there might be an electrical connection and a problem. However, these alloys are so corrosion resistant and similar electrochemically that there is probably no need to worry. ∎

11.4.2 Polymer Degradation

Degradation of polymeric materials can result in alterations in the properties of the material as well as release of degradation products that may cause adverse biological responses in the surrounding tissues. There are a number of ways that polymers can degrade, but the mechanisms are beyond the scope of this text. However, discussion of what changes can occur can provide insight into the biological performance of polymers.

Chemical degradation starts with breaking the long polymer chain into smaller fragments, or chain scission. There are a number of mechanisms by which this can occur, such as hydrolysis, in which H_2O splits to H^+ and OH^- attacks a bond. This can split the chain and the H^+ and OH^- can attach to the ends of the fragments forming end caps. Some enzymes and ionizing radiation are also capable of attacking polymers. Once split, the smaller fragments may react as free radicals to form end caps, react with other chains to form side branches, or react with other chains to form cross-links. Thus, the polymer may be reduced in molecular weight, increasing its solubility and potential for further breakdown, or it may become harder and more brittle due to cross-linking.

Properties of polymers may degrade over time by physical interaction with the environment. Low-molecular-weight substances, especially lipids, may adsorb into the material and cause discoloration. These low-molecular-weight substances may act as plasticizers and soften the material or get trapped in the matrix of macro-molecules and make the material more brittle. Such adsorption may be accelerated by mechanical forces such as cyclic loading or flexing. These mechanical forces can cause a sort of pumping action with space between molecules opening and closing with each cycle and thus entrapping the low-molecular-weight substances that enter the spaces. Similarly, low-molecular-weight substances added to the material during manufacture, such as plasticizers, may diffuse out over time. This can result in changes in the mechanical or other physical properties of the device. Such leaching may also stimulate an adverse biological reaction.

11.4.3   Wear

Relative motion between parts can cause mechanical damage and release of small particles due to wear. There are several mechanisms that are applicable to biomedical applications. The surface of an implanted material is not perfectly smooth on a microscopic scale but rather has small protrusions or asparities on the surface. Contact is not across the entire surface but is localized to the asparities. Thus, a relatively low contact pressure for the entire surface can actually result in very high local pressures on the asparities. Such localized contact pressures can result in fusion or adhesion between asparities of two surfaces. Subsequent movement tears the asparities off and produces damage due to adhesive wear. These asparities then break off as particulate debris.

When one surface is much harder than the other, abrasive wear may occur where the harder plows into the softer. This is especially likely in three-body wear that occurs when small particles get trapped between the wear surfaces. This is analogous to sandpaper made with hard abrasive particles. Three-body wear of the metal component in metal–polymer joints can lead to a confounding situation of polymer wear. The hard particles plow into the polymer and cause wear, but they also scratch the metal surface. Since the metals are ductile, scratching will create a trough with metal pushed up on each side of the trough like a plow moving through dirt. These ridges on each side of the scratches probably contribute to a major portion of the subsequent abrasive wear of the polymeric component. In contrast, ceramic materials are not ductile. They may get scratched, resulting in troughs, but they will not plastically deform to form ridges. This may be one reason for the much lower wear rates observed with ceramic–polymer total joints compared to metal–polymer combinations.

Wear can occur due to fatigue in situations of point loading or with large sliding distances of a smaller component over a larger surface. Consider sliding your fingertip back and forth across the palm of your hand. The fingertip is constantly loaded, but the palm tissue is loaded and then unloaded as the finger passes by. This is analogous to the loading of a tibial component in a total knee replacement. Now think of sticking your finger in a bowl of gelatin. Pushing down and up pushes the gelatin aside, back and forth. There is a shear stress under your finger due to the compressive force of pushing. The same is true for the tissue in your palm or the polymeric bearing of the total knee. Repeated loading and unloading causes cyclic shear stress under the contact point. Eventually, this may lead to the formation of fatigue cracks parallel to the joint surface. These flake off and produce particulate debris due to fatigue wear.

11.5 BIOLOGICAL EFFECTS

11.5.1 Cells Involved in Responses to Biomaterials

It is important to know the names and functions of some of the cells that circulate in the blood and are active in responses to biomaterials. The blood cells are usually identified by taking a small drop of blood and smearing it on a glass slide (Fig. 11.7). The smeared blood is then dried, fixed, and stained. An additional sample of blood is used to determine cell numbers. There are several distinctive types of cells, and typical numbers exist for the concentration of each cell type in a sample of blood.

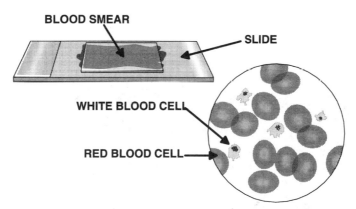

Fig. 11.7 A small droplet of blood is placed on a slide and then smeared with a coverslip in order to count the number of different types of blood cells that are present in a given volume of blood. In an actual sample, the white blood cells would vary from 8 to 12 μm in diameter, whereas the red blood cells would be approximately 8 μm in diameter.

The red blood cells (RBCs) are biconcave discs that have no nucleus (Fig. 11.8). They carry the hemoglobin that contains iron and is responsible for oxygen transport. They are 8 μm in diameter and are present at 5 million/mm^3. The white blood cells (leukocytes) have a nucleus and are present at about 8500/mm^3. Leukocytes are involved in the body's immune response. An increase in the white cell count occurs with disease, especially infection.

There are several types of white cells. The first type is the granulocyte. These cells are about 8 μm in diameter, have discrete bodies called granules in their cytoplasm, and have multilobed nuclear material. The nuclear material makes the cell appear to be multinucleated when observed under the microscope so these cells are called polymorphonuclear (many forms of nuclear material). There are three major types of granulocytes that are shown in Fig. 11.8.

The predominant type of granulocyte is the neutrophil, which accounts for 65–70% of the white cells. This cell type contains acid and basic material in the granules so that the balance is neutral. Since this is the predominant type of granulocyte in the blood and tissue, it is usually referred to as a polymorphonuclear leukocyte (PMN) or poly. This is not the correct terminology, but it is common usage. Neutrophils provide a protective mechanism since their function is to engulf and destroy foreign material (phagocytosis). These cells are active phagocytes and are important in the body's defense against bacteria and other foreign substances. Another type of granulocyte, the eosinophil, contains acid material in the granules that stain pink-red in the conventional bloodstains. This cell type is associated with allergic responses and usually comprises about 3% of the white cells in blood. The third type of granulocyte contains basic material in the granules, stains blue, and is called a basophil. This cell type accounts for <1% of the white cells in blood. The basophil is difficult to preserve so its importance in the host response to biomaterials is not well understood and may be underestimated.

The second group of white blood cells is distinctly different from the PMNs and

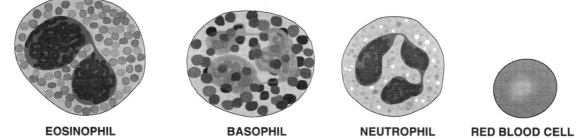

EOSINOPHIL   **BASOPHIL**   **NEUTROPHIL**   **RED BLOOD CELL**

Fig. 11.8   The three large cells on the left are types of granular leukocytes, or granulocytes. Eosinophils make up 2–4% of the circulating white blood cells and are about 12 μm in diameter. Their granules stain darkly with the red dye, eosin, and with other acid dyes. Basophils are 8–10 μm in diameter and have granules that stain darkly with basic dyes. They make up <1% of the circulating white blood cells. Neutrophils are the most numerous and represent 50–70% of the circulating white blood cells. They are similar in size to eosinophils but have chemically neutral granules that are difficult to stain. Each mature neutrophil has a multilobed nucleus that looks like beads on a string. Red blood cells are about 8 μm in diameter and lack a nucleus when they are mature.

has a single nucleus (mononuclear) with characteristic features. In this group is the monocyte, which accounts for up to 10% of the white blood cells, is about 10 μm in size, and has a large nucleus that has an indentation that makes it look like a kidney bean. This cell is also a phagocytic cell and has the capability of engulfing and destroying foreign substances. When a monocyte leaves the blood and goes into the tissue, it changes its appearance and differentiates into a macrophage (histiocyte). The macrophage is a large cell, up to about 20 μm in size, and is actively phagocytic (Fig. 11.9). When a macrophage is overwhelmed in its attempt to phagocytize material, it coalesces with other cells and forms a multinucleated giant cell. There are several types of giant cells. The one of concern to biomaterials is called the foreign body giant cell. There is often evidence of foreign material, or wear debris, associated with these cells. The presence of the foreign body giant cell indicates that the phagocytic system is incapable of ridding the body of the foreign matter. Tissue destruction often occurs and results in loss of normal tissue, loss of function of the organ, or significant bone resorption, depending on the site of inflammation.

The final group of white cells are also mononuclear and are called lymphocytes. These cells comprise about 20% of the white blood cells in blood and are usually 10–12 μm in size. These cells are responsible for our immune system that protects us from disease. The cells are active protein factories and have a large nucleus that almost completely fills the cell. The small amount of cytoplasm contains the factory components for making and secreting proteins. This area of immunology is important for understanding the body's response to biomaterials. It is a complex, interactive system, and the student is encouraged to pursue this topic independently from this course.

Scattered throughout the blood cells are fragments called platelets that are

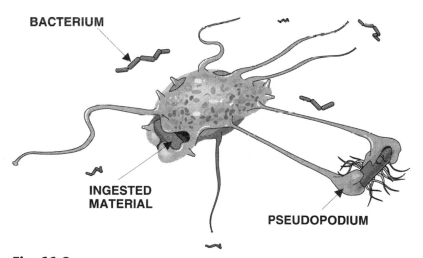

BACTERIUM

INGESTED MATERIAL

PSEUDOPODIUM

Fig. 11.9 In this figure, a macrophage is ingesting foreign material and has pseudopodia that are preparing to help bring another foreign body to the cell. In a real macrophage, the pseudopodia would flow out and around the bacterium or tissue debris and would engulf it without forming such long, thin connections to the cell.

1–3 μm in size. They have no nucleus but stain purple in the blood smear and are important in the clotting mechanisms that form a plug to prevent excessive bleeding when tissue is injured. They are present at about $150,000/mm^3$.

11.5.2   Inflammation

The human body is dependent on skin to protect it from the onslaught of microorganisms and some toxic substances. Whenever the skin is broken, a pathway is created for entry of organisms and chemicals. The cells and tissues of the mammalian body initially respond in a predictable manner. Inflammation is the first response of any vascularized tissue (tissue with blood vessels) to tissue damage. There are four signs (cardinal signs) associated with the inflammatory response.

1. The site turned red (rubor). This was due to the presence of RBCs just under the skin. These came because of an alteration in the permeability of the blood vessel that allowed the RBCs to leave the vessel.
2. The site became swollen (tumor) due to the fluid that came with the blood leaving the blood vessel.
3. The site became warm (calor), which is also the result of the warm blood escaping into the cooler tissue. There may also have been some fever (pyrogenic response) from release of tissue substances.
4. The site became painful (dolor) due to impingement on and damage to local nerve networks.

The easiest way to remember these signs is to think about the last time that you had a splinter. Also, remember that this is a response of tissue with blood vessels and most of the response is due to changes in these blood vessels. These responses are rapid and are normal. Only when the response becomes prolonged is there a problem.

Injury and vascular damage also triggers the two branches of the blood coagulation system. The first involves activation of platelets, which stick to exposed surfaces or foreign objects and form a plug. The other involves a series of proteolytic reactions, starting with the activation of a protein circulating in the blood by exposure to damage to vascular tissue or to a foreign object. This sequence results in the final pathway of conversion of fibrinogen to fibrin and the formation of a clot. The reaction involves a complicated series of reactions and is often associated with regions of sluggish or stagnant blood flow. These clots are often red due to entrapment of RBCs. The coagulation reaction can be blocked with drugs such as heparin, which blocks several of the reactions, or by administration of drugs that inhibit the formation of several components in the liver.

When the tissue around the splinter is examined, a characteristic tissue or cellular response (histologic response) is seen. The presence of RBCs and fluid can be observed early. Then there is an influx of the polys. These cells appear in the first 24 h after injury and continue to appear with a maximum response in about 3–5 days. This early pattern is referred to as acute inflammation. If the source of the injury is removed, e.g., the chemical is washed away or the splinter removed, then the response resolves, and the tissue returns to a normal appearance.

If the source of injury is not removed and the response continues, then inflammation progresses through various stages to chronic inflammation. In the continued response, the polys are replaced by larger cells from the monocyte/macrophage line. These cells are also phagocytic and are intended to clean up remaining tissue debris and the offending substance. When they cannot do the job, they appear to coalesce into very large, multinucleated giant cells. This represents chronic inflammation and is usually associated with a large mass of tissue debris that can cause harm to the tissue and organ. In some circumstances, the response is organized into a mass with a characteristic cellular response, including multinucleated giant cells, and is called a granuloma. These can be very painful, harmful, and a cause for concern. Thus, the acute inflammatory response is a normal response to injury. However, if the response continues for more than a week and leads to chronic inflammation, then harm can result.

11.5.3 Wound Healing

Whenever there is an injury with destruction of tissue, there must be repair. The first step in the repair process is acute inflammation. Changes occur in the vascular system, and the polys go into the tissue to clean up the damaged tissue and the foreign material. The platelets get activated and form clots to prevent further leakage of blood into the site. The remaining debris is then phagocytized by the polys. Finally, actual tissue repair begins. In order for the tissue or organ to function, there must be a blood supply so small vessels begin to grow into the wound. Tissue cells, called fibroblasts, come in and start to synthesize collagen. This forms a structural network called granulation tissue, which is pearly and often translucent and twinkles. There are blood vessels in it so it looks pink. This very delicate tissue is easily wounded. If this occurs, there is more bleeding, and the repair process must start again.

There are basically three types of tissue to be considered for wound healing. One is called labile. This tissue is constantly repairing and regenerating. Any wound to labile tissue is likely to result in complete healing and a return to normal. These tissues include the bone marrow, the blood cells, most of the blood vessels, the epidermis (external skin), the gastrointestinal tract, and bone. The second type is called stable. Repair and regeneration of this tissue is likely to successfully restore function, but there may be some scarring. This includes liver and lung. The third type is called permanent, and damage to these cells, tissues, or organs results in permanent loss without replacement and with extensive scarring and loss of function. This includes the nervous system, most of the muscular system, and the heart.

11.5.4 Healing in the Presence of a Biomaterial

When a biomaterial is inserted into the body, all of the features described previously are involved. First, there is damage to the tissue during the insertion, and an acute inflammatory response occurs. This is followed by the wound healing response. If the material is there for a short period of time and then removed, wound healing proceeds normally. However, if the material remains in the body, then the response is altered. Normal tissue repair cannot take place since there is a foreign body in the way. The first reaction of the body is to try to eliminate it. The phagocytes come in, and the acute inflammatory response may progress to a chronic response. If it is not a degradable or phagocytizable material, then the reaction will follow one of two paths. The usual response is for the fibroblasts in the healing response to lay down layers of fibrous tissue around the material (fibrous capsule) to "wall it off" from the rest of the body. It is generally believed that the thinner the fibrous capsule, the more acceptable the material is to the host, i.e., it is biocompatible. A thicker capsule means the material is not biocompatible. The more unusual response is for the inflammatory response to continue as a chronic inflammatory response and to progress to giant cells and granulomas. This occurs when the material is not biocompatible, and the host reaction is still trying to neutralize it.

There can be problems associated with the host reaction to the material. Return to the splinter. In most cases, the splinter will be walled off with the fibrous capsule, but since the splinter is close to the skin, the response will be to form this capsule and move the splinter to the external layers where it eventually will be sloughed off from the skin. This happens to materials that are implanted just under the skin or through the skin (percutaneous). The fibrous layer forms and attempts are made to extrude the material. For this reason, it is difficult to maintain materials in the epidermis and dermis. Percutaneous devices develop a fibrous layer around them, and the seal between the epidermis and the material is lost. This enables the formation of a pathway from the external environment to the deeper tissue and removes the normal protective skin barrier so that infection becomes a problem.

The formation of the fibrous capsule is an indication that the material is biocompatible and will occur as an early step in healing. Fibrous capsule formation needs to be considered when predicting device function. For instance, if the device is to serve as a drug delivery system, the formation of the fibrous capsule may alter the permeability of the device and the diffusion of the drug so that the function of the device will not be anticipated from laboratory studies.

Formation of a fibrous capsule generally does not occur with porous or textured materials. If the interstices are large enough, local vascularized tissue will grow into the pores rather than form the capsule. Long-term percutaneous catheters have been coated with velour to facilitate ingrowth and anchorage. Vascular prostheses made of fabric demonstrate ingrowth of vascular-like tissue. Joint replacement prostheses are sometimes coated with metal beads or wire mesh to facilitate ingrowth of bone and biological anchorage. It is important to remember that acute inflammation and wound healing are a necessary part in the use of materials and devices that enter the body so that the effect of these responses on the function of the device must be considered.

11.5.5   Infection

Whenever there is damage to the skin or mucous membranes and bacteria can enter, there is the risk of infection. The presence of infection will prevent the resolution of inflammation, and a chronic inflammatory response will occur. Whenever there is a chronic inflammatory response, the wound healing response will not be completed, and excessive scarring will result.

Consider the impact of the presence of a foreign body (implanted device or splinter) on the infection. The skin or mucous membranes have been damaged due to insertion of the device. The combination of the injury and the presence of the foreign material will initiate the inflammatory response. The material is in the tissue, and wound healing will be altered. The presence of bacteria will only add to the problems. It is well recognized that the presence of a foreign body greatly increases the infection risk and markedly decreases the number of bacteria required to cause an infection from 10^6 to 10^2. Not only is there an increased risk of infection but also the infection will be difficult to cure. Once again consider a splinter. If the splinter site becomes infected, the only way to cure the infection is to remove the splinter. Unfortunately, the same is true for implanted devices. The only way to cure the infection is to remove the device. The consequences of this depend on the need for the device. Removal of sutures may cause little impairment to healing, whereas removal of the total artificial heart would result in death.

Issues of implant site infection remain an active area of investigation from many different viewpoints: microbiology, infectious diseases, material science, device design, surface coatings with antimicrobial agents to inhibit bacterial attachment and growth, and surgical technique. It is an important issue since implant site infections are a major source of health care expenditures with prolonged hospital care and, in some cases, the implantation of a new device.

Example Problem 11.6A

A laboratory facility that does biological testing for device manufacturers has examined the histological response to one of the polymeric materials that is used by your company in a device for implantation. They examined the tissue at the implantation site at 3 days and found RBCs, PMN leukocytes, and some fibroblasts. Among your responsibilities in your company is the interpretation of testing and presentation of reports on the suitability of the material for the device. What is your response to this report?

Solution

This is consistent with acute inflammation and normal wound healing and the 3 days of implantation is too short to determine the biocompatibility of the material. The testing lab is asked to do studies over a longer time period. ∎

Example Problem 11.6B

The laboratory subsequently submits a report that they examined the tissue after 21 days of implantation. The implantation site was hot and swollen, and the animal was lethargic and appeared to have a fever. The evaluation of the tissue revealed accumulations of large numbers of PMN leukocytes, some lymphocytes, a large accumulation of fluid, and some areas of pus. Again you are asked to interpret this. What is your response to this report?

Solution

Acute inflammation should have resolved by this time. This is an abnormal response that is indicative of acute inflammation that has not resolved and a wound that is not healing. The signs and symptoms in the test animal and the histology in the tissue are consistent with infection.   ■

Example Problem 11.6C

You ask the laboratory the following questions: (a) Did they culture the site to determine if an infection was present? and (b) Did they sterilize the material according to the instructions that you gave them?

Solution

The testing laboratory answers that the culture was done. They just received the report that stated that bacteria grew from the site. They also replied that they had not followed the sterilization instructions since polymers are difficult to sterilize and they did not have the specified equipment. Thus, they had implanted the material without sterilizing it.   ■

Example Problem 11.6D

What are your conclusions and what do you do with this test report and your material?

Solution

These results should be filed in your records with the note that this test needs to be repeated since the implant had not been sterilized and infection ensued as would be expected. You plan to supply sterile material for the next series of tests. The ability to sterilize is an important consideration in selection of materials for devices. Polymeric materials tend to be more difficult to sterilize since many undergo degradation with heat, harsh chemicals, or oxidative events. Suitable sterilization processes need to be described for materials and devices.   ■

11.5.6   Immune Response

The immune system is an important and complex protective system that reacts specifically and with memory. This is why vaccinations are important for common diseases such as polio and measles. The lymphocytes are the key element in the immune system. Although the immune system is an important defense mechanism, sometimes it can cause harm to the host. These reactions are called allergy or autoimmunity (if the immune system attacks the host tissue).

Allergic responses occur to foreign substances through a variety of mechanisms. One type of allergy is called atopic or type I. The most common example of this is hay fever, which is the result of the formation of specific types of antibodies to ragweed pollen. An inflammatory response occurs, and the nature of the reaction depends on the site, with itching at skin sites and runny nose and eyes from aerosol contact. This type of reaction must be avoided by minimizing use of materials with substances that stimulate the reaction. The current concern in this area is with latex materials, such as surgical gloves, which contain a protein that may cause this type of allergy.

A second type of reaction is called contact dermatitis (type IV). A common example of this is the skin reaction that occurs from contact with poison ivy. It is important in evaluating biomaterials to determine that stimulation of allergic responses similar to this does not occur, and it is generally the degradation products from the materials that are of concern. Care is taken to avoid the use of chemicals that are known to cause allergy or to minimize their release into the body. Some of the metals, such as nickel and chromium, are common causes of allergy when contacted as metal salts. Allergic reactions to metallic devices are a concern, but unless there is corrosion and release of metal ions, there will not be allergic responses. All the methods described earlier that are used to make metal alloys corrosion resistant are important in minimizing the chances of allergic reactions to implants.

The issue of autoimmunity is also important in selection of biomaterials and evaluation of responses. In autoimmune reactions, there is an immune reaction that causes damage to the host tissue. This occurs because the material altered the host tissue in some manner or the immune response to the biomaterial also reacted with the host tissue. The bioprosthetic heart valve in Fig. 11.6 that is a porcine (pig) valve is an example of this concern. The tissue is treated so that immune responses do not occur to it and to the host heart tissue.

11.5.7 Biological Effects and Biomaterials

It is important to keep these fundamental biological events in mind when evaluating materials for use in or on the body. Acute inflammation is a necessary response for tissue repair; however, a chronic response is harmful. If a material is not tolerated by the host or is toxic, then chronic inflammation will occur. Any material implanted into the body must be sterile; otherwise, an infection will occur. Sterilization methods may destroy materials or alter the materials' properties. This is a major issue in developing or adapting materials for use in the biological environment. Infection in the presence of an implanted material is of special concern because the only way to cure the infection is to remove the implant. Again, consider the splinter. An infected site will not heal while the splinter remains.

It is also critical to consider the immune system when evaluating a material for use. The material must not interfere with the functioning of the immune system (i.e., be toxic to the system). Although stimulation of the immune system by the material components may not be harmful, it is desirable that the material not stimulate the system or cause an allergic reaction.

The issue of wound healing in the presence of an implanted material is important and not thoroughly understood. The wound healing response described previously follows a standard pathway for repair. However, if there is a material in the wound, then the response is altered. Since the blood vessels cannot go through it, they go around it. Similarly, the fibroblasts cannot go through it so they also go around it. In the process of doing this, they lay down the fibrous capsule. This fibrous capsule is normal host tissue. The inflammatory response is not needed anymore so the rest of the tissue returns to normal. If the capsule is thin, the material is tolerated well. If the capsule is thick, then there is something irritating the site. If there is no capsule and a chronic inflammatory response occurs, then the material is not biocompatible.

11.6 IMPACT OF DEGRADATION MATERIALS ON THE BIOLOGICAL SYSTEM

As was discussed previously, there can be degradation, wear, or corrosion of materials. Degradation will have important implications for device performance and biological responses.

11.6.1 Degradation

Some materials are designed to be biodegradable or resorbable. This means that a simple chemical process that can occur in the body will cause the material to undergo changes. The most common process is hydrolysis in which the chemical processes in the body introduce water into a large molecule and cause it to break into increasingly smaller pieces. This is common in sutures that are used to close wounds. After a period of time, they lose mass and strength and eventually disappear. This concept is being used to make materials and devices that slowly degrade after their function has been fulfilled. Often when a device has fulfilled its function, a surgical procedure is undertaken to remove the device. One example of this is the removal of plates and screws from bone fracture sites that have healed. Removal is often important to maintain the bone structure. If a degradable material could be used, then this extra surgery would not be needed.

However, there are special precautions that must be taken when using degradable materials. The material must maintain strength and other necessary mechanical properties for the specified duration, and the degradation products must not be toxic to the host tissue. For example, many of the suture materials produce acid when they degrade. The small size of the suture releases a small amount of acid which is easily neutralized by body chemistry, but larger pieces pose a problem due to tissue damage from the acid.

11.6.2 Corrosion

Some materials will undergo degradation through the corrosion mechanisms that were described previously. Corrosion can occur in the body and result in the release of metal ions into the tissue. The distribution and elimination of these ions must be considered in determining the effect on the host. Many of the elements used in metal alloys are essential trace elements for human metabolism. Nickel, chromium, cobalt, and iron are all necessary for metabolic processes, and there are normal mechanisms for maintaining proper concentrations. Thus, only in extreme excess will they impact on the host. Some, such as nickel and cobalt ions, are rapidly excreted in the kidney, whereas others, such as chromium with a valence of 6+, are stored in the tissues. Thus, the toxicity issues include tissue accumulation as well as total concentration from the material.

11.6.3 Wear

As described previously, wear is often a complication when hard materials rub against softer components. Wear may result in the production of small pieces (particles) that can stimulate a biological reaction. Polys and macrophages will attempt to engulf and degrade these foreign particles. If they succeed, then there will be no additional response. However, if they cannot succeed, then chronic inflammation will result with foreign body giant cells, granulomas, and the release of various soluble mediators. The generation of wear particles is currently a hot topic in studies on orthopedic devices such as total joint replacements. Metal, ceramic, and polymeric particles may be released.

Most of the particles are about the size of bacteria (a few micrometers) and are easily engulfed by the phagocytes. However, degradation may be another issue. In general, the metallic particles will undergo corrosion in the acid environment of the cell and be released and eliminated as ions. The ceramics are less prone to corrosion or dissolution so the degradation of any particles from wear may be slow. The polymers are usually very resistant to further degradation. They tend to be the major cause of prolonged cellular response with tissue and organ damage (such as loss of bone) from extensive, but ineffective, phagocytosis. Not only do the frustrated phagocytes send out signals (cytokines) to recruit more cells but also they die in their effort to destroy particles, thus releasing all their intracellular enzymes, acids, and superoxides.

11.7 BIOCOMPATIBILITY TESTING

A number of factors need to be considered in the selection of materials to be used to make devices. One of these is biocompatibility. Biocompatibility is difficult to define, but the following definition is usually used: Biocompatibility is the ability of a material to perform with an appropriate host response in a specific application. Will this material stimulate the appropriate biological response for the intended use? The material to be tested must be in the identical form to that when used, i.e., it must be sterilized in the appropriate manner and any surface or bulk modifications completed. The tests may be done *in vitro* using tissue or cell cultures or may involve injection of substances or implantation into an animal.

In some tests, the material will be used directly. In other tests the material will be extracted in liquid form first and the extract will be tested. The extract used will vary depending on the test methods to be used or on the nature of the material. Generally, two solvents, one polar and one nonpolar, are used. Water or saline and vegetable oil or serum are the usual choices. The intention in these methods is to extract susbstances that would normally be extracted in the biological environment. In some cases, harsher extractions are used to more completely degrade the material and determine the content. These extracts may include alcohols, acids, bases, or even enzymes.

The tests are designed to test for cytotoxicity (damage to cells), stimulation of the immune response (especially allergy), irritation to tissues, provocation of chronic inflammation, effects on blood and blood components, and effects on genetic factors including mutations and tumor formation. The selection of tests to be done and the methods used are the subject of much investigation and deliberation. In general, guidelines for the selection of tests and the methods for performing these tests may be obtained from standards writing groups such as the ASTM Committee F04 on Medical and Surgical Materials and Devices and the International Standards Organization Technical Committee 194. Regulatory agencies, such as the Food and Drug Administration in the United States, will give the developer and manufacturer of devices guidelines about what testing is advisable.

Biocompatibility testing is desired on all new materials or old materials used in new applications. The selected tests should be carefully considered and matched to the ultimate use of the device. For example, testing for a new type of wound dressing involves very different tests than those for a total heart replacement. Biocompatibility testing is an inexact science, and some materials pass the testing but ultimately do not perform as expected in human patients. Mechanical performance, immune responses, and infection do not have appropriate biocompatibility tests so device performance is evaluated in clinical trials on patients. Despite the limitations, biocompatibility evaluation remains a necessary screening test before a device can be approved.

Example Problem 11.7A

You have developed a new heart valve and are now undertaking biological testing. You first examine the stability of the device by soaking it in saline (saltwater) for 48 h. Is this appropriate?

Solution

It is a good start. However, since this device is to be used in contact with blood, it would be better to test the stability using blood. ■

Example Problem 11.7B

The results of the stability testing in blood reveal that the valve occluder mechanism has swollen considerably and its shape has changed. What are the possible biological consequences of this?

Solution

There are many severe biological consequences of this that include the following: (i) The occluder will not seat correctly; thus, the valve will not close correctly or open correctly and will not function in a satisfactory manner, and (ii) the flow of blood will be altered, and there may be areas of stagnation. There may be platelet damage and red cell damage. This may lead to clots and to hemolysis and severe consequences. ■

Example Problem 11.7C

The design of the device is such that the struts are a soft polymeric material and the occluder mechanism is a stiff, hard, highly reinforced polymeric material. What are the consequences of having the swollen, misshapen hard material constantly rub and impinge against the soft polymeric struts?

Solution

There will be wear and degradation of the polymers (mostly the soft material) with the release of particles into the tissue and the bloodstream. This will result in local and systemic biological responses to the particles by means of chronic inflammatory responses. The degradation of the device will continue with more reduction in the mechanical match between the struts, the occluder, and the seating ring and with biological responses that will alter the local tissue. The seating ring will eventually detach. ∎

11.8   BIOMATERIALS AND DEVICE DESIGN CRITERIA

The developments of two types of prosthetic devices, the artificial heart valve and the total joint replacement, have demonstrated the multifaceted subject of biomaterials and their interaction with the biological system. Each type of device has its own set of problems based on its intended function and implantation site.

11.8.1   Prosthetic Heart Valves

The two mechanical heart valves shown in Fig. 11.6 have four essential components: an occluder, such as a disc or ball; a seating ring against which the occluder sits when the valve is closed; a capture mechanism, such as a cage, that constrains the occluder when the valve is open; and a sewing ring that permits attachment of the valve to the heart. The occluder bounces back and forth from the seating ring to the capture mechanism with each heartbeat. The more promptly it moves, the more efficient the opening and closing. Thus, weight, or mass, of the occluder and wear resistance of the occluder material are critical features. Early materials that were used included lightweight plastics or hollow metal balls. Silicone rubber proved very effective as a ball. The advent of the low-profile disc valves for the mitral position set a new material constraint, namely stiffness. Silicone rubber was too soft for the disc designs. Polyoxymethylene or poly acetal is stronger and stiffer than silicone rubber and therefore was used in early disc designs. However, it had a problem with wear. The discs were supposed to be free to spin in the cage so as to distribute the wear evenly around the edge of the disc. However, as the disc moved up and down in the cage, wear tracks developed on the edge, preventing spinning and leading to the development of deep wear tracks and valve failure. Due to its high-fatigue strength and wear resistance, pyrolytic carbon was selected as one of the prime materials for the occluder.

The capture mechanism and seating ring required strength and stiffness to maintain their shape. Furthermore, they had to be made of a material that could be sprung open for insertion of the occluders. Metal has typically been used, either as machined parts or as separate parts that are welded together. Early designs used the cobalt alloys due to their strength and corrosion resistance. Recently, titanium alloys have been used, due in part because they are lighter than cobalt alloys. Concerns of allergic reactions to cobalt alloy cages have also been expressed over the years.

Both the occluder and the containment system are in direct contact with blood. Contact with foreign materials can cause blood to clot and can lead to platelet attachment and clotting. If the design results in eddy flow or stagnation, the clotting factors can accumulate and lead to thrombus formation. Therefore, both the material selection and device design must consider problems of blood contact and blood flow. Since these problems have not been solved, patients with mechanical heart valves receive medication to reduce their tendency to form blood clots.

In contrast to mechanical valves, bioprosthetic tissue valves do not have the problem of thrombus formation, and patients do not need to be anticoagulated for the rest of their lives. The use of foreign tissue for the valves does require special treatment to minimize immune reactions. The major problem with tissue valves is that they calcify over a period of 5–10 years. This accumulation of calcium in the leaflets leads to stiffening and valvular incompetence and eventually to mechanical failure of the valve.

Both mechanical and bioprosthetic valves need to be attached to the heart. This involves the use of many sutures to sew the valve in place. Thus, the sewing ring must either have many holes for stitching or be made of a fabric or cloth that can be stitched to the muscle. Being in contact with muscle and connective tissue, the sewing ring also has to provide some measure of suppleness at the interface between the metal seating ring and the pumping heart. This is a situation in which material selection can take advantage of host reactions. The use of a porous material, such as a fabric made of polyester, for the sewing ring permits ingrowth of connective tissue into the interstices of the fabric. This provides a biological anchorage of the valve and an effective seal between tissue and the implant.

11.8.2 Total Joint Replacements

The evolution of the modern-day total joint replacements (TJR), or arthroplasties, such as the THR and total knee replacement depicted in Fig. 11.10, dates back to the 1940s and demonstrates a wide range of material and biological issues. Early artificial hip joints involved replacement of the head of the femur and thus were hemiarthroplasties. Balls made of metal or plastic were held in place with a nail placed

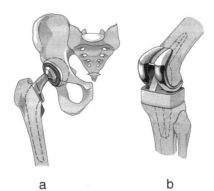

a   b

Fig. 11.10 Artificial hip (a) and knee (b) joints are held in place by a special cement [polymethylmethacrylate (PMMA)] and by bone ingrowth. Special problems occur at the interfaces due to the different elastic moduli of the materials (110 GPa for the titanium implant, 2.2 GPa for PMMA, and 20 GPa for bone).

in the neck of the femur. The plastic devices failed rapidly due to severe wear and the subsequent severe chronic inflammatory reaction to the wear particles. Cast cobalt chromium molybdenum alloy (F75 in Table 11.3) had sufficient wear resistance and is still in use today as femoral head and neck replacements. Hemiarthroplasty does not solve the problem of breakdown of the acetabular cartilage. The success of the F75 hemiarthroplasty led to the development of the THR in the mid-1950s with metal acetabular cups articulating with the metal femoral head and neck replacements. Of all the alloys, only the cast F75 had sufficient wear and corrosion resistance to function as this metal–metal ball-and-socket bearing. These were metal devices with screws to anchor the acetabular component and a large metal stem inserted down the femoral medullary canal to anchor the femoral component. However, some of these failed early due to wear, breakdown of the metal prosthesis–bone interface, or seizing of the metal–metal bearing. In many respects, this set the stage for the next 40 years of total joint development.

The search for the ideal TJR has focused on two issues: strength of materials and interface phenomena. Throughout the history of total hips, there have been concerns about mechanical failure due to fatigue fracture of metallic components. Processing and design improvements have resulted in virtual elimination of this concern. Along the way, there have been modifications that actually resulted in a significant number of failures. Technology has developed alloys of steel, cobalt, and titanium with sufficient strength for most applications, but the interface phenomena have proved to be the most challenging.

A TJR consists of at least two parts and has at least three interfaces: the bearing or articulation, the interfaces between component and acetabular bone, and the femoral stem–bone interface. As shown in Fig. 11.10, some devices are "fixed" with PMMA "bone cement." This adds an additional interface — between the cement and the prosthesis. In his pioneering work on the development of the THR, the late Sir John Charnley strove to develop a "low-friction arthroplasty" to solve the friction problems with the metal–metal total hips. He also wanted a way to fix the device in the bone which would accommodate for differences in anatomy of his patients. This led to the development of bone cement.

For low friction, Sir John chose a polymeric acetabular cup and a stainless-steel femoral component. His search for a low-friction plastic led to a fluorocarbon material similar to Teflon. Unfortunately, this material had very poor wear resistance and shredded *in vivo*. The particles were the size of bacteria and were phagocytized. Since the particles were chemically inert, the macrophages and giant cells could not digest them and died trying. The result was a severe chronic inflammatory response and bone resorption. Sir John had to remove all the cups within a matter of 1 or 2 years. His next choice for a plastic was UHMWPE. With minor variations in processing and sterilization, UHMWPE is still the polymeric material of choice today.

For the component fixation, Sir John sought a leuting agent, a space filler that could provide a forgiving bed for the prosthesis. At the suggestion of Dr. Dennis Smith, a dentist from Toronto, Sir John used dental acrylic as a space filler. This is a two-part material used for making the pink gum portion of dentures. Powdered polymer with an activator is mixed with liquid monomer. The monomer is polymerized, in effect gluing the powder particles together to create a solid mass within

10–15 min. After removing the old joint tissue and carving the medullary tissue, the cement is mixed and inserted into the medullary space, and the prosthesis is inserted into the cement. Ten minutes later it is fixed. Thus, the first hips were fixed with pink dental cement.

This method created an additional interface: bone–cement. The cement acts as a space filler, not an adhesive, and is also a cushion. The metallic components of stainless steel or cobalt have an elastic modulus of 200 GPa, whereas Titanium is about 110 GPa. These are very stiff compared to bone with a modulus of about 20 GPa. Bone cement with a lower modulus of 2 or 3 GPa acts as a cushion to distribute the stress more evenly than could be achieved with direct bone–metal contact.

Bone cement also has its problems. During polymerization, it gives off heat, and the cement mass gets hot. It also tends to release some monomer, which has potential pharmacologic effects. Over time, the cement can crack, break, and become ground up. If the PMMA particles are small relative to the macrophages, they are phagocytized. If they are large, they are encapsulated. If there are many particles, the result is a severe inflammatory response that leads to bone resorption, device loosening, and failure. This led to the phenomena of "cement disease" and the search for an alternative to the use of bone cement.

As was mentioned with regard to biological fixation of heart valves with a textured sewing ring, it is generally true that the encapsulation of foreign objects does not occur if the device is porous or has a porous coating with sufficient pore size for the local, vascularized tissue to grow into the pores. For bone ingrowth, the pore size needs to be in the range of 100 μm for "biological fixation." As a solution to cement disease, the stems of TJRs were given porous coatings with small beads, a wire mesh, or micromachined technologies. These techniques solved the problem of cement disease but opened the door for new problems. Metals are more rigid than bone. Therefore, fixing a metal device to bone will deprive the bone of mechanical deformation with normal loading. Since bone formation and resorption is in response to stress, alteration in loading, as with shielding the bone from stress with a big metallic THR stem, can lead to bone resorption and loosening. On the UHMWPE side, the use of a metal backing for bony ingrowth creates another interface: the back side of metal–plastic. Currently, the debate is still active: biological fixation versus improved cementing techniques.

The issue of biomechanics and stress shielding has also led to the development of alternative materials. The early use of stainless steel and cobalt alloy stems was followed by the development of more flexible titanium stems. Recently, the use of carbon fiber-reinforced composites has been considered. Heat-resistant tiles for the space shuttle are made of carbon fiber-reinforced carbon. This is a high-strength material but has poor wear resistance. With an LTI pyrolytic carbon coating, it became a candidate material, but cost and quality control problems terminated this development. Carbon fiber-reinforced PEEK is currently being evaluated as a candidate material for THR stems. As indicated in Table 11.1, there are a number of different fiber orientations that can be utilized to produce a variety of mechanical properties. These are all being considered.

Finally, consider the issue of articulation. Many improvements have been made in surface finish of the metal and in processing of the UHMWPE. However, the polymer does wear, and wear particles do cause inflammation. Ceramics are harder than metals and can take a finer finish. Thus, ceramic (alumina or zirconia)–UHMWPE devices are being developed. Ceramics on ceramic bearings are also being studied. Metallurgy has also advanced and the metal–metal hip joints are being revisited. The early devices were made of cast F75 alloy, but modern devices are made of fine-grained wrought F799 or similar cobalt chromium molybdenum alloys that have much better wear characteristics.

The question of which interface problem is driving technology and device development represents a multifaceted problem. Currently, it is the wear at the articulation that is receiving much of the attention. The bottom line has not changed. The biological response is the key. If particles are produced, either from the cement or from the articulation, the inflammatory response may lead to loosening. If the device is too strong and stiff, then the bone will resorb. If the device is too big, it will not fit and will be of no value. Anatomy, physiology, material science, and biomechanics must all be considered.

EXERCISES

1. The following table provides the output data from a compression loading test of a cancellous bone specimen that was 8 mm long, 5 mm wide, and 6 mm thick. Plot a stress–strain curve. Determine the UTS and the elastic modulus.

Displacement (mm)	Force (N)
0.00	0
0.05	21
0.10	49
0.15	118
0.20	190
0.25	301
0.30	399
0.35	503
0.40	602
0.45	700
0.50	810
0.55	960
0.60	1090
0.65	1200
0.70	1298
0.75	1390
0.80	1487
0.85	1523
0.90	1490
0.95	1350

2. The bottom lip of the trunk lid of a car often shows evidence of severe corrosion. Why?
3. Your neighbor covered the bottom lip of the trunk lid with a plastic strip in order to protect it. Is this a good idea? Why?

4. A company has asked your advice on a product they are considering making. It is a polymer used to fill defects in the face. It is to be injected as a liquid, and then it will polymerize in the tissue to form a soft mass. What questions do you want to ask about this polymer before considering it for this use?

5. A company has developed a new thermoplastic polymer for heat-seal packaging. They claim that, after use, the packaging can be heated and reused as heat-seal packaging. Is this likely to work?

6. Because of concern about nickel toxicity, a medical device manufacturer is making a stainless steel without nickel. What are the properties of this material likely to be?

7. A new material is needed that is hard, wear resistant, and corrosion resistant. What type of material is most likely to meet these criteria?

8. Now that the material in question 6 has been developed, it is proposed for use as a plate to hold broken bones together. During use, large forces will be applied occasionally. Will this material be successful in this new application?

9. Mechanical testing of materials is done in tension, compression, shear, and rotation. List a biomedical application for which tension testing is most critical. List a biomedical application for which compression testing is most critical. List a biomedical application for which testing in shear is most critical. List a biomedical application for which testing in rotation is most critical.

10. A company has asked your advice. They have synthesized a new polymer to make fracture fixation devices to hold pieces of bone together for healing. Although the modulus is acceptable, the ultimate strength is too low. What is the definition of modulus? What advice can you give them to increase the ultimate strength? What will this do to the modulus?

11. Figure 11.6 shows a caged-ball heart valve replacement. What material characteristics would you want for the ball? What class of material would best fit these characteristics? What material characteristics would you chose for the cage? What match of material would you want between the cage and the ball?

12. Figure 11.10A shows a total hip replacement. What material characteristics are necessary for this device?

13. A new material has been surgically implanted into a rabbit to test for biocompatibility. The site of implantation is examined at 5 days. The site is slightly red and swollen. There are numerous polys at the site. What concern does this raise about the biocompatibility of the material?

14. The material in question 13 has been left in place for 30 days. Examination of the site reveals the presence of macrophages and giant cells. What concern does this raise about its biocompatibility?

15. A biosensor for calcium has been implanted into muscle. The signals are monitored for several days successfully. However, beginning at Day 10, the signal is weaker, and the biosensor is not detecting much calcium. The sig-

nal continues to get weaker. Why might the sensitivity of the biosensor decrease after implantation into tissue?

16. Stainless-steel sutures (small-diameter wire) are often used to close skin wounds in dogs. Why would the veterinarian choose stainless steel?

17. Stainless-steel sutures are not used to close skin wounds in humans. Why?

18. When you touch your amalgam fillings with a fork, you feel a tinge of pain. Why?

19. The orthodontist uses many materials for braces and wires and hooks. One doctor decides to make the wire out of nickel and the brackets the wire goes into out of copper. What will happen?

20. Infection is a major concern with the use of implanted devices. Why is this and what are the consequences of infection at the site of an implant.

21. Biocompatibility testing is required for all medical devices. However, the nature of the testing differs depending on the site of use of the device. Why do devices that contact only intact skin have different requirements than devices that are implanted?

22. For each of the following devices list the material characteristics desirable and what class of material would probably suit the application best. Some may be multicomponent and need more than one material. Chapter 2, which discusses anatomy and physiology, and a medical dictionary may be helpful.

    Contact lens
    Intraocular lens
    Wound dressing
    Surgical glove
    Urinary catheter
    Syringe
    Needle
    Screw for fracture fixation
    Rod for fracture fixation
    Plate and screw for fracture fixation of jaw
    Plate and screw for fracture fixation of tibia
    Denture
    Complete artificial tooth (base into jaw and exposed tooth)
    Cap for tooth
    Orthodontic wire
    A–V shunt (used to drain fluid from the brain in hydrocephalic children or in brain injury)
    Pacemaker electrode
    Biosensor for glucose implanted into cardiovascular system
    Suture to stitch a wound closed
    Finger joint replacement
    Hip replacement
    Total heart replacement
    Heart valve replacement
    Replacement of part of a blood vessel (vascular graft)

Replacement of a piece of an entire blood vessel
Bag for respirator for a ventilatory assist device
Filling in tooth

SUGGESTED READING

American Society for Testing and Materials, Annual Book of Standards, 100 Bar Harbor Drive, West Conshohocken, PA 19428-2959; *http://www.astm.org.*
D412, Test Methods for Rubber Properties in Tension.
D638, Test Method for Tensile Properties of Plastics.
E8, Test Methods for Tension Testing of Metallic Materials.
F75, Specifications for Cobalt-28 Chromium-6 Molybdenum Casting Alloy and Cast Products for Surgical Implants (UNS R30075).
F136, Wrought Titanium-6 Aluminum, -4 Vanadium ELI (Extra Low Interstitial) Alloy (R56401) for Surgical Implants.
F138, Specification for Stainless Steel Bar and Wire for Surgical Implants (Special Quality).
F562, Specification for Wrought Cobalt-35 Nickel-20 Chromium-10 Molybdenum Alloy for Surgical Implant Applications.
F748, Practice for Selecting Generic Biological Test Methods for Materials and Devices.
Askeland, D. F. (1985). *The Science and Engineering of Materials.* PWS Engineering, Boston.
Marek, M. I. (1987). Thermodynamics of aqueous corrosion. In *Corrosion,* Metals Handbook Vol. 13, pp. 18–27, Metals Park, OH.
Park, J. B., and Lakes, R. S. (1992). *Biomaterials — An Introduction, 2nd ed.* Plenum, New York.
Ratner, B. D., Hoffman, A. S., Schoen, F. J., and Lemons, J. E. (1996). *Biomaterials Science.* Academic Press, San Diego.
Vincent, J. (1990). *Structural Biomaterials, Rev. ed.* Princeton Univ. Press, Princeton, NJ.
Williams, D. F., and Roaf, R. (1973). *Implants in Surgery.* Saunders, London.

12 TISSUE ENGINEERING

Chapter Contents

At the conclusion of this chapter, the reader will be:

- Familiar with the growing area of cellular therapies
- Able to understand the cellular dynamics underlying tissue function
- Able to qualitatively describe the importance of stem cells in tissue function
- Able to quantitatively describe cellular fate processes
- Able to perform order of magnitude estimation of cellular communication processes
- Knowledgable about the parameters that characterize the tissue microenvironment and how to approach mimicking them in vitro.
- Able to describe the issues fundamental to scale-up
- Familiar with many of the issues that one encounters when delivering cellular therapies to patients

12.1 CELLULAR THERAPIES

Cellular therapies promise to become major therapeutic modalities of the next century. In order to implement these therapies successfully, a number of important scientific and engineering challenges have to be met, including basic cell and developmental biology issues, bioengineering analysis and design, and clinical implementation.

Cellular therapies represent clinical applications of rapidly advancing knowledge in cell and molecular biology. Bioengineering, biochemical engineering, and biomaterial science are all needed for the implementation of cellular therapies. Challenges revolve around the design of tissues and tissue-like conditions so that clinically meaningful numbers of cells can be produced. Bioengineering, the production of therapeutic doses of primary cells, promises to have an important impact on the delivery of health care in the next century.

For a historical comparison, the discovery of penicillin alone was not enough to affect the delivery of health care. Mass production methods of clinical-grade material had to be developed. The development of large-scale production of antibiotics arguably represents the most significant contribution of engineering to the delivery of health care. In a similar fashion, the provision of primary human cells in therapeutically meaningful numbers will enable the routine use of cellular therapies in the next century. Tissue engineering is focused on the manipulation and design of human tissues to meet these clinical needs.

Several basic issues of cell biology need to be understood and quantitatively described. These include the key cellular fate processes of cell differentiation, division (mitosis), migration (motion), and death (apoptosis) that underlie tissue function. The role of tissue-specific stem cells in organ function, genesis, and repair is critical, and the basics of stem cell biology will be discussed. Unlike pure cell cultures, the complex cell cultures that reconstitute tissue function involve many cell types. Each cell type has a particular function, but the communication and cell–cell interactions among them must be re-established in culture to accurately reconstitute tissue function. The characteristics of the tissue microenvironment will be discussed. The role of the extracellular matrix, direct cell–cell contact, and soluble signals in cellular communication will be described.

Many bioengineering challenges must be met. The function, choice, manufacturing, and treatment of biomaterials for cell growth and for device construction is important. Fluid mechanics and mass transfer play an important role in normal tissue function and therefore become critical issues in cell culture device design and operation. Systems analysis of metabolism, cell communication, and the cellular fate processes plays a key role. A properly designed *ex vivo* tissue culture process must appropriately balance the rates of biological and physicochemical processes in order for tissues to function. This balancing in turn requires the formulation of dimensionless groups that describe ratios of characteristic time constants for the processes to be compared. Many of these dimensionless property ratios will be new, involving the ratio of a physical time constant to a biological time constant.

The implementation of cellular therapies in the clinic requires the recognition and resolution of several difficult issues, including tissue harvest, cell processing and isolation, safety testing, cell activation/differentiation, assay and medium development, storage and stability, and quality assurance and quality control issues. These challenges will be described but not analyzed in detail. Manufacturing of cell therapy products is and will continue to be fundamentally challenging.

12.1.1   Human Cells as Therapeutic Agents

Cellular therapies use human cells as therapeutic agents to affect a pathological condition. It is important to note that cellular therapies are not new. Blood transfusion, basically of red blood cells, into anemic patients to restore adequate oxygen transport has been practiced for decades with great therapeutic success and benefit. Similarly, platelets have been successfully transfused into patients who have blood clotting problems. Bone marrow transplantation (BMT) has been practiced for almost two decades, with tens of thousands of cancer patients undergoing high-dose chemo- and radiotherapies and receiving BMT annually. Recently, the use of transplantation of mobilized peripheral blood stem cells has increased rapidly and is expanding the clinical use of hematopoietic stem cell transplants. These are all applications of cell therapies associated with the blood cells and blood cell generation (hematopoiesis, *hemato* "blood" and *poiesis* "generation of"). Numerous patients are currently receiving therapeutic benefit from cellular therapies.

Transplants can be carried out in an allogeneic setting (donor to patient) or in an autologous setting (patient to patient) (There are also xenogenic transplants, from one species to another, and syngenetic transplants between identical twins). Historically, the allogeneic setting is well known, which is of course also the case with traditional whole organ transplantation. However, with the advent of cell culture and manipulation procedures, the autologous transplantation setting is becoming more common and will undoubtedly represent the delivery of many cellular therapies in the future. There are, however, attempts under way in several laboratories to make "universal donor" cell sources and cell lines. Such cell sources should have minimal or no immunogenicity problems (analogy: type O, Rh− blood).

The ability to reconstitute tissues *ex vivo* (tissue engineering) and produce cells in clinically meaningful numbers has a wide spectrum of applications. Table 12.1 summarizes the incidence of organ and tissue deficiencies and the number of procedures performed annually to treat these deficiencies in the United States. The aggregate cost of these procedures was estimated at a staggering $400 billion per year. The potential socioeconomic impact of cellular therapies is therefore substantial. Effective treatment and/or cure of these deficiencies will greatly improve the quality of life of the affected individuals, and make them more productive with the concomitant economic impact on society.

The concept of directly engineering tissues was pioneered by Y. C. Fung in 1985. The first symposium was organized by Richard Skalak and Fred Fox in 1988, and since then the field of tissue engineering has grown significantly. A number of collections of articles have appeared since then, and the journal, *Tissue Engineering*, is now published. Newer cellular therapies include various immunotherapies using T cells, chondrocytes for cartilage repair, liver and kidney cells for extracor-

TABLE 12.1 Incidence of Organ and Tissue Deficiencies, or the Number of Surgical Procedures Related to These Deficiencies in the United States[a]

Indicator	Procedure or Patients per Year
Skin	
Burns[b]	2,150,000
Pressure sores	150,000
Venous stasis ulcers	500,000
Diabetic ulcers	600,000
Neuromuscular disorders	200,000
Spinal cord and nerves	40,000
Bone	
Joint replacement	558,200
Bone graft	275,000
Internal fixation	480,000
Facial reconstruction	30,000
Cartilage	
Patella resurfacing	216,000
Chondromalacia patellae	103,400
Meniscal repair	250,000
Arthritis (knee)	149,900
Arthritis (hip)	219,300
Fingers and small joints	179,000
Osteochondritis dissecans	14,500
Tendon repair	33,000
Ligament repair	90,000
Blood Vessels	
Heart	754,000
Large and small vessels	606,000
Liver	
Metabolic disorders	5,000
Liver cirrhosis	175,000
Liver cancer	25,000
Pancreas (diabetes)	728,000
Intestine	100,000
Kidney	600,000
Bladder	57,200
Ureter	30,000
Urethra	51,900
Hernia	290,000
Breast	261,000
Blood Transfusions	18,000,000
Dental	10,000,000

[a] From Langer and Vacanti (1993).
[b] Approximately 150,000 of these individuals are hospitalized and 10,000 die annually.

poreal support devices, β-islet cells for diabetes, skin for ulcers and burns, and genetically modified myocytes for treatment of muscular dystrophy. The challenges faced with each tissue are different. A few examples are provided for illustrative purposes.

Bone Marrow Transplanatation

Bone marrow is the body's most prolific organ. It produces approximately 400 billion myeloid cells daily, all of which originate from a small number of pluripotent stem cells (Fig. 12.1). The bone marrow is composed of 500–1000 billion cells and, therefore, regenerates as many cells as it is composed of every 2 or 3 days. This cell production rate represents normal hematopoietic function. Individuals under hematopoietic stress, such as systemic infection or sickle cell anemia, will have blood cell production rates that exceed the basal level. The prolific nature of bone marrow makes it especially susceptible to damage from radio- and chemotherapies. Bone marrow damage limits the extent of these therapies, and some regimens are fully myoablative. Without any hematopoietic support, patients that receive myoablative dose regimens will die due to hematopoietic failure.

BMT was developed to overcome this problem. In an autologous setting, the bone marrow is harvested from the patient prior to radio- and chemotherapies. It is cryopreserved during the time period that the patient undergoes treatment. After a few half-lives of the chemotherapeutic drugs have passed, the bone marrow is rapidly thawed and returned to the patient. The bone marrow cells are simply put into circulation, and the bone marrow stem cells "home in" on the marrow cavity

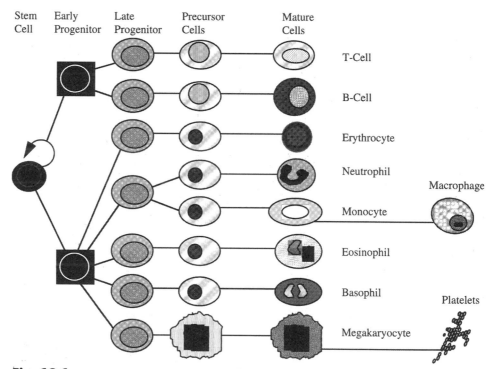

Fig. 12.1   Hematopoietic cell production. The production fluxes through the lineages can be estimated based on the known steady-state concentration of cells in circulation, the total volume of blood, and the half-lives of the cells. Note that the 400 billion cells produced per day arise from a small number of stem cells (from Koller and Palsson 1993).

and reconstitute bone marrow function. In other words, the hematopoietic tissue is rebuilt *in vivo* by these stem cells. This process takes on the order of weeks, during which time the patient is immunocompromised.

Autologous BMT as a form of cellular therapy simply involves removing the cells from the patient and storing them temporarily outside the patient's body. There are several advantages to growing the harvested cells and newer therapies and treatments are being developed based on *ex vivo* culture of hematopoietic cells. Newer methods to harvest bone marrow stem cells have been developed recently. These methods rely on using cytokines or cytotoxic agents to "mobilize" the stem cells into circulation. The hematopoietic stem and progenitor cells are then collected from the circulation using leukophoresis.

Myoablative regimens are used in allogeneic settings. In the case of leukemia, this not only removes the bone marrow but also hopefully the disease. The donor's cells home in on the marrow and repopulate the bone cavity, just as in the autologous setting. The primary difficulty with allogeneic transplants is high mortality (10–15%) primarily due to graft versus host disease, in which the transplanted cells recognize the recipient's tissues as foreign. Overcoming this rejection problem would significantly advance the use of allogeneic transplantation.

BMT is a well-developed and accepted cellular therapy for a number of indications, including allogeneic transplants for diseases such as leukemia and autologous transplants for diseases such as lymphoma and breast and testicular cancer. Significant growth has occurred in the use of BMT since the mid-1980s, and approximately 20,000 patients were treated worldwide in 1995 — about evenly split between autologous and allogenic transplants.

Skin

Skin is the body's third most prolific tissue. It basically consists of two layers: the dermis, whose main cellular component is fibroblasts, and the epidermis, whose main cellular component is keratinocytes that are at various stages of differentiation (see Fig. 12.8). Both cell types grow very well in culture, and *ex vivo* cultivation is not the limiting factor with this tissue. Interestingly, transplanted dermal fibroblasts have proven to be surprisingly nonimmunogenic.

Victims of burns and diabetic ulcers have severe problems with skin healing. To treat these problems, skin can be cultured *ex vivo* and applied to the affected areas. There are several ways of culturing and introducing skin to the patients. Technologically, this cell therapy is relatively well developed. Several commercial concerns exist that deliver human skin equivalents for clinical purposes.

Pancreas/β-Islet Cells

The pancreas in insulin-dependent diabetics has lost its ability to produce and secrete insulin. Cadaverous pancreas can be used as a source of islets of Langerhans which contain the insulin-secreting β-islet cells. These cells can be injected into the liver portal vein. The islets then lodge in the liver and secrete the life-sustaining insulin. Unfortunately, it is not possible to grow β-islet cells in culture without losing their essential properties, and thus this procedure is constrained by the severely limited supply of tissue.

This cellular therapy is basically an allogeneic transplantation procedure. Cells from the cadaver are used to treat the diabetic. The graft will eventually be rejected; therefore, the treatment is only temporary. The duration of the graft can be prolonged by immunosuppressing the recipient or by using a well-matched donor.

The immune rejection problem is an important concern in cellular therapies and is discussed later. However, one way to overcome rejection is the physical separation of the donor's cells from the immune system by a method that allows exchange between the graft and the host across a semipermeable membrane (Fig. 12.2).

Cartilage and Chondrocytes

Cartilage is an unusual tissue. It is avascular, alymphatic, and aneural. It consists mostly of extracellular matrix, in which chondrocytes are dispersed at low densities — on the order of 1 million per cm^3 compared to a cell density of a few hundred million cells per cm^3 found in most tissues. Chondrocytes can be cultured *ex vivo* to increase their cell numbers by approximately 10-fold. Deep cartilage defects in the knee can be treated by autologous cell transplantation. In this case, a biopsy is collected from the patient's knee outside the affected area (Fig. 12.3). The chondrocytes are liberated from the matrix by an enzymatic treatment and allowed to grow in a two-stage cell culture process. The cells are then harvested from the culture and introduced into the affected area in the knee joint. The transplanted chondrocytes are able to heal the defect.

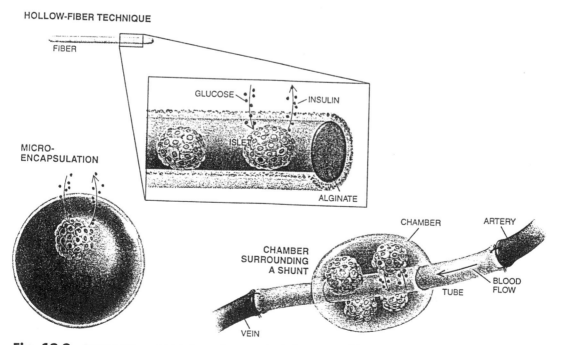

Fig. 12.2   Encapsulation of islets in semiporous plastic is one promising way to protect them from attack by the immune system (from Lacey, 1995).

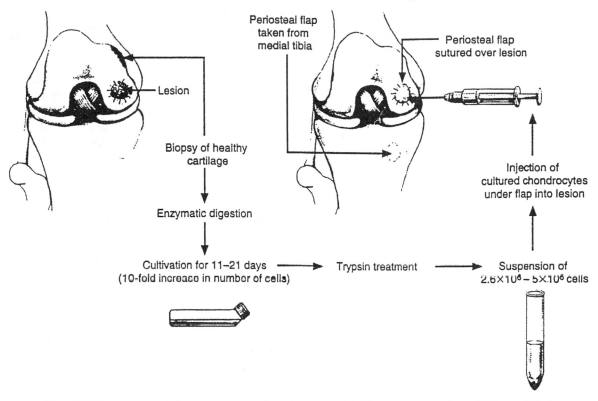

Fig. 12.3 Diagram of chondrocyte transplantation in the right femoral condyle (from Brittberg, 1994).

This procedure is relatively simple in principle, but there are still challenges with clinical implementation. Numerous patients are afflicted with knee problems which effectively leave them immobile. Over 200,000 patients are candidates for this type of cellular therapy annually in the United States.

12.1.2 Fundamental Questions

Having illustrated the range and potential of cell therapies, how do these therapies become clinical practice? Before proceeding into the detailed technical and scientific issues, the cell numbers required for cellular therapies and the growth potential of human cells will be reviewed. These numbers help set overall goals.

Cell Numbers *in Vivo*: Orders of Magnitude

The cell densities in human tissues are on the order of 1.0–3.0 billion cells per milliliter. The volume of a 70-kg human is about 70,000 ml. Therefore, the human body consists of about 100 trillion (trillion = 10^{12}) cells. A typical organ is about 100–500 ml in size and, therefore a organ contains about 100–1500 billion (10^9) cells. Organs are composed of functional subunits, as illustrated in Section 12.7.

Their typical linear dimensions are approximately 100 μm. The cell number in a cube that is 100 μm on each side is estimated to be about 500–1000. These cell numbers are summarized in Table 12.2.

Therefore, a typical organ will have a few hundred million functional subunits. This number will be determined by the capability of each subunit and the overall physiological need for its particular function. For example, the number of nephrons in the kidney is determined by the maximal clearance need of toxic by-products and the clearance capability that each nephron possesses.

Several important conclusions can be derived from these numbers. The fundamental functional subunit of most tissues contains only a few hundred cells, and this is a mixed cell population. As outlined in Section 12.6, most organs have accessory cells that can be as much as 30% of the total cell number. Furthermore as illustrated later, the tissue type-specific cells may be present at many stages of differentiation.

The nature of the tissue microenvironment and the dynamics of the cellular, communication, and metabolic processes that take place need to be understood in order to reconstitute tissue function accurately. To generate a therapeutic dose of cells requires a large number of microenvironments. These microenvironments must be relatively similar in order to have all the functional subunits perform in a similar fashion. Therefore, the design of cell culture devices and processes must be such that uniformity in the supporting factors, such as nutrient, oxygen, and growth factor concentration, must be reasonably homogeneous down to about 100 μm. Below this size scale, nonuniformity would be expected and in fact needed for proper functioning of the fundamental units of tissue function.

These considerations influence the approach to cell transplantation. Currently, it

TABLE 12.2   Cell Numbers in Tissue Biology and Tissue Engineering: Orders of Magnitude

Cell numbers *in vivo*	
Whole body	10^{14}
Human organ	10^{9}–10^{11}
Functional subunit	10^{2}–10^{3}
Cell production *in vivo*	
Theoretical maximum from a single cell (Hayflick limit)	$2^{30-50} < 10^{15}$
Myeloid blood cells produced over a lifetime	10^{16}
Small intestine epithelial cells produced over a lifetime	5×10^{14}
Cell production *ex vivo*	
Requirements for a typical cellular therapy	10^{7}–10^{9}
Expansion potential[a] of human tissues	
Hematopoietic cells	
Mononuclear cells	10-fold
CD34 enriched	100-fold
Two or three antigen enrichment	10^{6}- to 10^{7}-fold
T cells	10^{3}- to 10^{4}-fold
Chondrocytes	10- to 20-fold
Muscle, dermal fibroblasts	$>10^{6}$-fold

[a] Expansion potential refers to the number of cells that can be generated from a single cell in culture.

is unlikely that fully functioning organs will be produced *ex vivo* for transplantation purposes. Integration of such *ex vivo* grown tissues into the recipient may prove problematic. The alternative is to produce stem and progenitor cells that in sufficient number and properly placed will proliferate *in vivo* to regenerate organ function. BMT, for instance, relies on this approach.

What are clinically meaningful numbers of cells? The cells required for currently practiced and experimental cell therapy protocols seem to fall into the range of a few tens of millions to a few billion (Table 12.2). Since tissue-like cultures require densities higher than 10 million cells per milliliter, the size of the cell culture devices for cell therapies should be in the range of 10 to a few hundred milliliters in volume.

What are the fundamental limitations to the production of primary cells? The number of cell divisions that a primary cell undergoes is subject to the so-called Hayflick limit. This limitation will be discussed in greater detail in Section 12.3, but primary human cells can undergo about 30–50 doublings in culture, depending on the age of the cell donor. Therefore, a single cell can theoretically produce 10^{10}–10^{15} cells in culture. Given the requirements for cell therapies, the Hayflick limit does not present a stifling limitation. Fibroblasts and skeletal myocytes grow extensively in culture, but the same is not true for other cells, such as liver and β-islet cells. The challenge in the latter case is to develop methods to get the cells to grow.

How rapidly do primary cells grow in culture? Growth rates of primary cells vary greatly in culture. Hematopoietic progenitors have been individually clocked at 11 or 12-h doubling times, which represents the minimum cycle time for adult human cells. Dermal foreskin fibroblasts grow with doubling times of 15 h, a fairly rapid rate which may be partially attributable to their early ontological age. Adult chondrocytes grow slowly in culture with doubling times of about 24–48 h.

How are these cells currently produced? In most cases, the production methods currently used are primitive by cell culture engineering standards. They are mostly produced in bags (T cells) and in T flasks (chondrocytes). Special chambers have been developed for the production of skin and bone marrow.

12.2 TISSUE DYNAMICS

Tissues are composed of many different cell types of various developmental origins (Fig. 12.4). The dynamic behavior of cells and their interactions determine overall tissue formation, state, and function. The activities of individual cells are often substantial. However, the time scales of cellular processes that determine tissue function are relatively long, and therefore their importance tends to be overlooked.

Overall, the following three dynamic states of tissues are of concern:

1. *Tissue histogenesis:* The normal steady-state function of tissue. Some tissues produce cells (bone marrow and skin) as their main function, whereas others produce a secreted product (glands). Some tissues primarily carry out mass

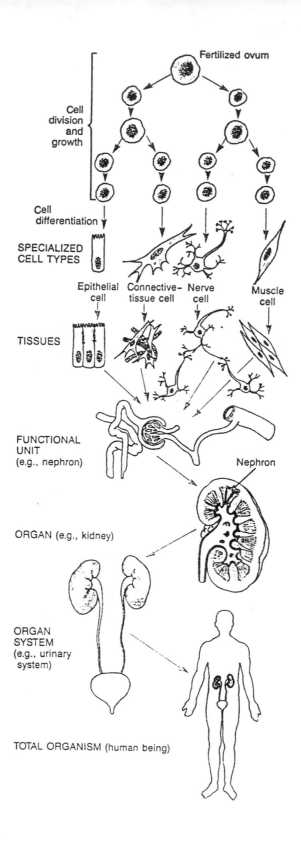

Fertilized ovum

Cell division and growth

Cell differentiation

SPECIALIZED CELL TYPES

Epithelial cell   Connective-tissue cell   Nerve cell   Muscle cell

TISSUES

FUNCTIONAL UNIT (e.g., nephron)

Nephron

ORGAN (e.g., kidney)

ORGAN SYSTEM (e.g., urinary system)

TOTAL ORGANISM (human being)

transfer operations (lungs and kidneys), whereas others are biochemical "refineries" (liver).

2. *Tissue formation:* The formation of tissue is the field of developmental biology. The way tissues form is of key concern in cellular therapies, and the younger the cell source, the more organogeneic potential it possesses.

3. *Tissue repair.* Wounded tissue displays a healing process that may be of concern in cell therapies and tissue engineering. A biopsied piece of tissue is expected to initially display a healing-type response after being placed in culture.

All of these are dynamic processes that involve interplay among many different cell types. The cells communicate and coordinate their efforts through the principal cellular fate processes (Fig. 12.5). The biology and dynamics of these processes are discussed in detail in Section 12.4.

12.2.1 Tissue Histogenesis

All tissues are dynamic. For instance, tissue dynamics can be illustrated by comparing the cell numbers that some organs produce over a lifetime to the total number of cells in the human body. As stated previously, the human bone marrow produces about 400 billion myeloid cells daily in a homeostatic state. Over a 70-year lifetime, the cell production from bone marrow accumulates to a staggering 10^{16} cells. This cell number is several hundred times greater than the total number of cells that are in the body at any given time. Similarly, the intestinal epithelium, the body's second most prolific tissue, produces about 5×10^{14} cells over a lifetime — 10 times the number of total number of cells in the human body.

Tissues have their own characteristic turnover rates (Table 12.3). Bone marrow is the most proliferative tissue in the body, followed by the lining of the small intestine and then by the epidermis. The turnover rate of these two tissues is on the order of a few days. Other tissues are much slower in their cell turnover times. The liver turns over in about 1 year, whereas fully formed neural tissue is believed to be stationary in cell number.

The cellular fate processes that underlie the dynamic states of tissue function are classified into the following categories:

1. *Cell replication* — an increase in cell number
2. *Cell differentiation* — changes in gene expression and the acquisition of a particular function
3. *Cell motility* — the motion of a cell into a particular niche or location
4. *Cell apoptosis* — the controlled death of a cell, distinguished from necrotic death,
5. *Cell adhesion* — the physical binding of a cell to its immediate environment, which may be a neighboring cell, extracellular matrix, or an artificial surface

Fig. 12.4 Levels of cellular organization in tissues and the diverse developmental origins of cells found in tissues (from Vander *et al.,* 1994).

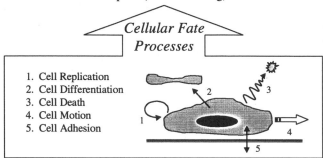

- Tissue Function (homeostasis)
- Tissue Formation (developmental biology)
- Tissue Repair (wound healing)

Cellular Fate
Processes

1. Cell Replication
2. Cell Differentiation
3. Cell Death
4. Cell Motion
5. Cell Adhesion

Fig. 12.5   Tissue dynamics. The three dynamic states of tissues and the underlying cellular fate processes.

These processes are illustrated in Fig. 12.5. What is known about each one of these processes will be briefly described in Section 12.4, with particular emphasis on quantitative and dynamic descriptions.

Tissue-specific stem cells provide the source of cells that undergo the cellular fate processes and produce the dynamic states of tissues. The function and fate of tissues is therefore significantly influenced by the behavior of stem cells. Stem cells are therefore the fundamental cellular material from which cells are built. The most dramatic demonstration of this fact is the reconstitution of multilineage hematopoietic cells following a stem cell transplant, which in mice has been demonstrated to occur with only a handful of cells.

TABLE 12.3   Cell Renewal Rates in Tissues

Tissue	Species	Turnover Time (days)
Erythropoiesis	Rat	2.5
Myelopoiesis	Rat	1.4
Hematopoiesis	Human	2.5
Small intestinal epithelium	Human	4–6
	Rat	1–2
Epidermis	Human	7–100
Coreneal epithelium	Human	7
Lymphatic cells	Rat (thymus)	7
	Rat (spleen)	15
Epithelial cells	Rat (vagina)	3.9
	Human (cervix)	5.7
Spermatogonia	Human	74
Renal interstitial cells	Mouse	165
Hepatic cells	Rat	400–500

12.2.2 Histogenesis in Highly Prolific Tissues

Many tissues produce cells at a high rate. Prolific tissues have been found to have stem cells that are the source of all the basic types of parenchymal cells in these tissues. These tissues display a highly organized pattern of stem cell replication and differentiation to mature progeny.

Bone Marrow and Blood Cell Formation

Hematopoiesis was the first tissue function for which a stem cell model was established (see Fig. 12.1), and now there is a growing number of tissues that have been shown to harbor tissue-specific stem cells. A general model for the interaction of the three fate processes of differentiation, replication, and apoptosis in the production of mature cells from tissue-specific stem cells is shown in Fig. 12.6. Tissue-specific stem cells form preprogenitor cells that replicate slowly. Further divisions make them progenitor cells (with colony-forming potential in semisolid media). Progenitor cells divide at a maximal cycling rate.

Apoptosis is very active at the progenitor stage, and in fact most of the hematopoietic growth factors that act at this stage are believed to be survival factors (i.e., antiapoptotic) that permit the progenitors to pass this stage of differentiation to become precursor cells. The precursors replicate slowly and mature into fully morphologically developed cells. These cells then take on their mature cell function. They leave the bone marrow and enter circulation where they perform their mature parenchymal cell function.

The mature cells eventually die and have to be replaced. The rate of death of ma-

	Stem Cell	Early Progenitor	Late Progenitor	Precursor Cell	Mature Cell
Cell Number	Potential 2^{30}-2^{50} per cell				Need About 10^{16} total over lifetime
Cell Cycling	Very slow ($t_d \sim 1/6$wks.)	Slow ($t_d \sim 60$-100hrs.)	Very rapid ($t_d \sim 12$ hrs.)	Slow	Zero (can be activated in special cases)
Apoptosis	Inactive	Inactive	Very Active (1:5000 survives)	Slow	Inactive (can be induced)
Motility	Zero (except during homing)	Zero	Low	Higher	Function of Physiological State
Regulation	Cell-Cell Contact	Cell-Cell Contact	Soluble Growth Factors	Soluble Growth Factors	Soluble Growth Factors

Fig. 12.6 Model for cell production in prolific tissues. This model was derived from decades-long research in hematology. The columns represents increasingly differentiated cells, and the rows indicate the cellular fate processes and other events that cells undergo at different states of differentiation (t_d denotes doubling time).

ture cells (not necessarily by apoptosis) determines the need for the cell production rate in the tissue and, ultimately, the number of stem cell commitments that are needed. The specific hematopoietic lineage cell production is given in Fig. 12.1.

The Villi in the Small Intestine

The lining of the small intestine is composed of villi, shown schematically in Fig. 12.7, that absorb nutrients. The intestinal epithelial cell layer is highly dynamic. Its cellular content turns over approximately every 5 days, and is thus the body's second most prolific tissue.

In between the villi are tube-shaped epithelial infoldings. An infolding is known as a cryptus. All intestinal epithelial cell production takes place in the crypt. Once the cells are mature, they are at the outer edge of the crypt. They then move, over a period of about 5 days, from the base of villus to the top, where they die and slough off. The villi thus represents a tube or a cylinder with a slowly moving layer of mature intestinal epithelial cells that emerge from the crypt and fall off once they reach

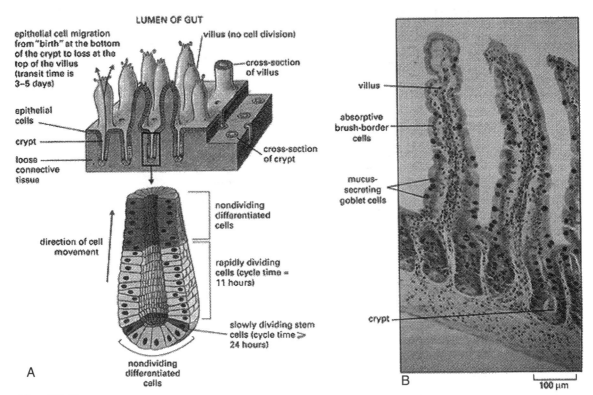

Fig. 12.7   Villi in the small intestine. (A) Rows of villi of epithelial intestinal cells (the diameter of a villi is about 80 μm). (B) A schematic showing the villi and the crypt indicating the mitotic state of the cells in various locations (from Alberts *et al.*, 1994).

the outer edge of the villi. During this passage the cells carry out their organ-specific function as mature parenchymal cells. They function in the absorption and digestion of nutrients that come from the lumen of the gut.

Toward the bottom of the crypt is a ring of slowly dividing tissue-specific stem cells. The number of stem cells per crypt is about 20. After division, the daughter cell moves up the crypt where it becomes a rapidly cycling progenitor cell, with a cycling time on the order of 12 h. The cells that are produced move up the crypt and differentiate. Once they leave the crypt, they are mature and enter the base of the villi.

Skin

Human skin has two principal cell layers separated by a basal lamina (Fig. 12.8). Collagen VII is an important component of this basal lamina. A connective tissue layer, the dermis, is under the basal lamina and is composed primarily of fibroblasts. On the outside of the basal lamina is the epidermis. It is a cell layer composed of differentiating keratinocytes, with the least differentiated ones located at the basal lamina.

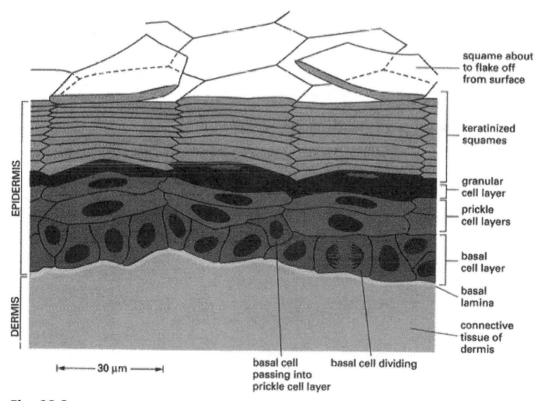

Fig. 12.8 The cellular arrangement and differentiation in skin. The cross section of skin and the cellular arrangement in the epidermis and the differentiation stages that the cells undergo (from Alberts *et al.,* 1994).

Thin skin has a squames columnar (each column is about 30 µm in diameter) organization. The cells located at the basal lamina then undergo differentiation that divides the epidermis into well-defined cell layers of increased differentiation in the direction away from the basal lamina. Only the cells on the basal lamina are cycling, while cells in the outer layer are differentiating. It is believed that approximately, 1 in 10–12 basal cells is an epidermal stem cell that is responsible for the cell production in the squames column above it. As the cells leave the basal lamina, they stop cycling and undergo differentiation to pricle cells and then to granular cells, which develop into keratinized squames that eventually flake off.

The net proliferative rate of skin depends on the region of the body. The turnover of skin is on the order of a few weeks. It is estimated that about half of dust found in houses originates from human skin. The skin is the body's third most prolific tissue.

12.2.3   Tissue Formation

The homeostatic function of many tissues involves cell production as outlined previously. Organogenesis, or tissue formation, involves the growth and formation of tissue from immature tissue-specific stem cells. These tissue-specific stem cells are formed in early embryogenesis. Our knowledge about the early embryonic events is growing. The basic events that lead to the formation of the basic body plan and the commitment to various organs can in many cases be traced in detail. Understanding this tissue-formation process is important to the tissue engineer.

Developmental Biology of Hematopoiesis

During vertebrate ontogeny, hematopoiesis sequentially occupies the yolk sac, fetal liver, spleen, and bone marrow. Variations exist among vertebrate species. The earliest identification of hematopoietic cells is their assignment to the progeny of the C4 blastomere in the 32-cell embryo. The blastula grows to about 1000 cells (10 doublings) and assumes a spherical shape. Then, the blastula undergoes gastrulation — not unlike pushing a finger into an inflated balloon. The point of invagination is the endoderm (the vegetal pole) that eventually forms the gut. After gastrulation, the ectoderm is brought into apposition with the endoderm and a third germ layer is formed between the two — the mesoderm. This middle layer is called the marginal zone and is formed via cell–cell interactions and soluble growth factor action. Several tissue-specific stem cells originate from the mesoderm, including hematopoietic tissue, mesenchymal tissue, muscle, kidney, and notochord.

Blood cells originate from the ventral mesoderm. Some hematopoietic cells migrate into the yolk sac to form blood islands — mostly erythroid cells ("primitive" hematopoiesis). Intraembryonic hematopoiesis originates from the aortic region in the embryo and leads to "definitive" hematopoiesis. It appears that the embryonic origin of hematopoietic cells is from bipotent cells that give rise to both the vasculature (the endothelium) and the hematopoietic cells. Hematopoietic stem cells are then found in the liver in the fetus. Near birth, the hematopoietic stem cells migrate from the liver into the bone marrow, where they reside during postnatal life. Interestingly, the umbilical cord blood contains hematopoietic stem cells capable of engrafting pedriatic, juvenile, and small adult patients.

This developmental process illustrates the asymmetric nature of stem cell division during development and increasing restriction in developmental potential. Furthermore, the migration of stem cells during development is important. Understanding the regulatory and dynamic characteristics of the stem cell fate processes is very important to tissue engineering.

12.2.4 Tissue Repair

When tissue is injured, a healing response is induced. The wound healing process is composed of a coordinated series of cellular events. These events vary with ontological age. Fetal wound healing proceeds rapidly and leads to the restoration of scarless tissue. In contrast, postnatal healing is slower and often leads to scarring, which generally permits satisfactory tissue restoration, while not always fully restoring normal tissue structure. Some pathological states resemble wound healing. A variety of fibrotic diseases involve similar processes to tissue repair and subsequent scarring.

The Sequence of Events That Underlie Wound Healing

Immediately following injury, control of bleeding starts with the rapid adhesion of circulating platelets to the site of damage. Within seconds, the platelets are activated, secrete contents from their storage granules, spread, and recruit more platelets to the thrombus that has started to develop. Within minutes of injury, hemorrhaging is contained through the constriction of surrounding blood vessels.

The next phase of the wound healing process involves the release of agents from the platelets at the injured site that cause vasodilatation and increased permeability of neighboring blood vessels. The clotting cascade is initiated and results in the cleavage of fibrinogen by thrombin to form a fibrin plug. The fibrin plug, along with fibronectin, holds the tissue together and forms a provisional matrix. This matrix plays a role in the early recruitment of inflammatory cells and later in the migration of fibroblasts and other accessory cells.

Inflammatory cells now migrate into the injured site. Neutrophils migrate from circulating blood and arrive early on the scene. As the neutrophils degranulate and die, the number of macrophages at the site increase. All tissues have resident macrophages and their number at the injury site is enhanced by macrophages migrating from circulation. They act in concert with the neutrophils to phagocytose cellular debris, combat any invading microorganisms, and provide the source of chemoattractants and mitogens. These factors induce the migration of endothelial cells and fibroblasts to the wound site and stimulate their subsequent proliferation. If the infiltration of macrophages into the wound site is prevented, the healing process is severely impaired.

At this time, so-called granulation tissue has formed. It is composed of a dense population of fibroblasts, macrophages, and developing vasculature that is embedded in a matrix that is composed mainly of fibronectin, collagen, and hyaluronic acid. The invading fibroblasts begin to produce collagen, mostly types I and III. The collagen increases the tensile strength of the wound. Myofibroblasts contract at this time, shrinking the size of the wound by pulling the wound margins together.

The matrix undergoes remodeling, which involves the coordinated synthesis and degradation of connective tissue protein. Remodeling leads to a change in the composition of the matrix with time. For instance, collagen type III is abundant early on but gives way to collagen type I with time. The balance of these processes determine scar formation. Although the wound appears healed at this time, chemical and structural changes continue to occur within the wound site. The final step of the wound healing process is the resolution of the scar. The formation and degradation of matrix components returns to its normal state in a process that may take many months. Finally, the composition of the matrix and the spatial location of the cells return to close to those of the original state.

Understanding the would healing process is important to the tissue engineer since the placement of disaggregated tissues in *ex vivo* culture induces responses reminiscent of the wound healing process.

12.2.5   Organization of Tissues into Functional Subunits

The body has 11 major organ systems (Table 12.4) with the muscular and skeletal systems often considered together as the musculoskeletal system. These organ systems carry out major physiological functions, such as respiration, digestion, and mechanical motion. Each one of these organs systems in turn is composed of organs. The major organs that participate in digestion are shown in Fig. 12.9. Each organ system and organ have homeostatic functions that can be defined based on their physiological requirements. These can be thought of as "spec sheets." An example is given in Table 12.5.

There are a significant number of important conclusions that can be arrived at using simple order of magnitude analysis of the information found in such spec sheets. Insightful and judicious order of magnitude analysis will continue to be the primary mode of analysis of tissue function in the future. Detailed analysis and calculations cannot substitute for focused and well-justified experimental work.

Organs function at a basal rate but have the ability to respond to stress. The total circulation rate and organ distribution under strenuous exercise differs significantly from that at rest. Similarly, under hematopoietic stress, such as infection or in patients with sickle cell anemia, the basal blood cell production rate can significantly exceed the basal rate given in Fig. 12.1.

Organs, in turn, are composed of functional subunits. These subunits include the alveoli in the lung and the nephron in the kidney (Fig. 12.10). These functional units are comprised of a mixture of different kinds of cells that together constitute tissue function. Separating the functional subunits into their individual cell cohorts leads to the loss of tissue specific function, but specific cell properties can be studied with such purified preparations.

TABLE 12.4 The Major Organ Systems of the Body

Circulatory	Heart, blood vessels, blood (some classifications also include lymphatic vessels and lymph in this system)	Transport of blood throughout the body's tissues
Respiratory	Nose, pharynx, larynx, trachea, bronchi, lungs	Exchange of carbon dioxide and oxygen; regulation of hydrogen–ion concentration
Digestive	Mouth, pharynx, esophagus, stomach, intestines, salivary glands, pancreas, liver, gallbladder	Digestion and absorption of organic nutrients, salts, and water
Urinary	Kidneys, ureters, bladder, urethra	Regulation of plasma composition through controlled excretion of salts, water, and organic wastes
Musculoskeletal	Cartilage, bone, ligaments, tendons, joints, skeletal muscle	Support, protection, and movement of the body; production of blood cells
Immune	Spleen, thymus, and other lymphoid tissues	Defense against foreign invaders; return of extracellular fluid to blood; formation of white blood cells
Nervous	Brain, spinal cord, peripheral nerves and ganglia, special sense organs	Regulation and coordination of many activities in the body; detection of changes in the internal and external environments; states of consciousness; learning; cognition
Endocrine	All glands secreting hormones: Pancreas, testes, ovaries, hypothalamus, kidneys, pitutitary, thyroid, parathyroid, adrenal, intestinal, thymus, heart, pineal	Regulation and coordination of many activities in the body
Reproductive	Male: Tetses, penis, and associated ducts and glands	Production of sperm; transfer of sperm to female
	Female: Ovaries, uterine tubes, uterus, vagina, mammary glands	Production of eggs; provision of a nutritive environment for the developing embryo and fetus; nutrition of the infant
Integumentary	Skin	Protection against injury and dehydration; defense against foreign invaders; regulation of temperature

A fundamental statement follows from this observation: Tissue function is a property of cell–cell interactions. The size of the functional subunits is on the order of 100 μm, whereas the size scale of a cell is 10 μm. Each organ is then composed of tens to hundreds of millions of functional subunits. The sizing of organs represents an evolutionary challenge that is also faced by tissue engineering in scaling up the function of reconstituted tissues *ex vivo*.

12.2.6 The Challenges Facing the Tissue Engineer

The previous observations lead to the identification of the following fundamental challenges that face the tissue engineer in the implementation of cell therapies (Fig. 12.11):

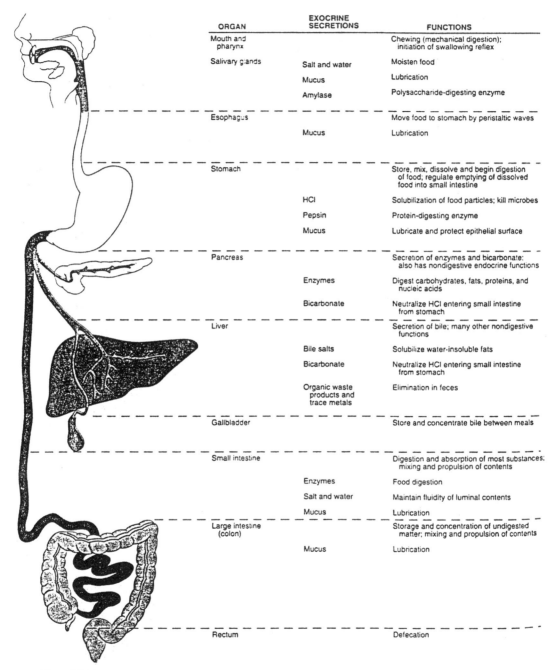

ORGAN	EXOCRINE SECRETIONS	FUNCTIONS
Mouth and pharynx		Chewing (mechanical digestion); initiation of swallowing reflex
Salivary glands	Salt and water	Moisten food
	Mucus	Lubrication
	Amylase	Polysaccharide-digesting enzyme
Esophagus		Move food to stomach by peristaltic waves
	Mucus	Lubrication
Stomach		Store, mix, dissolve and begin digestion of food; regulate emptying of dissolved food into small intestine
	HCl	Solubilization of food particles; kill microbes
	Pepsin	Protein-digesting enzyme
	Mucus	Lubricate and protect epithelial surface
Pancreas		Secretion of enzymes and bicarbonate; also has nondigestive endocrine functions
	Enzymes	Digest carbohydrates, fats, proteins, and nucleic acids
	Bicarbonate	Neutralize HCl entering small intestine from stomach
Liver		Secretion of bile; many other nondigestive functions
	Bile salts	Solubilize water-insoluble fats
	Bicarbonate	Neutralize HCl entering small intestine from stomach
	Organic waste products and trace metals	Elimination in feces
Gallbladder		Store and concentrate bile between meals
Small intestine		Digestion and absorption of most substances; mixing and propulsion of contents
	Enzymes	Food digestion
	Salt and water	Maintain fluidity of luminal contents
	Mucus	Lubrication
Large intestine (colon)		Storage and concentration of undigested matter; mixing and propulsion of contents
	Mucus	Lubrication
Rectum		Defecation

Fig. 12.9 Functions and organization of the gastrointestinal organs (from Vander *et al.,* 1994).

TABLE 12.5 Standard American Male[a]

Age	30 years
Height	5 ft 8 in. or 1.86 m
Weight	150 lb or 68 kg
External surface	19.5 ft^2 or 1.8 m^2
Normal body temperature	37.0°C
Normal mean skin temperature	34°C
Heat capacity	0.86 cal/(g) (°C)
Capacities	
Body fat	10.2 kg or 15%
Subcutaneous fat layer	5 mm
Body fluids	ca.51 liters or 75%
Blood volume	5.0 liters (includes formed elements, primarily red cells, as well as plasma)
	Hematrocit = 0.43
Lungs	
Total lung capacity	6.0 liters
Vital capacity	4.2 liters
Tidal volume	500 ml
Dead space	150 ml
Mass transfer area	90 m^2
Mass and energy balances at rest	
Energy conversion rate	72 kcal/h or 1730 kcal/day [40 kcal/(m^2)(h)]
O_2 consumption	250 ml/min (respiratory quotient = 0.8)
CO_2 production	200 ml/min
Heart rate	65/min
Cardiac output	5.0 l/min (rest)
	3.0 + 8 M in general (M = O_2 consumption in liters per minute)
Systemic blood pressure	120/80 mmHg

[a] From Lightfoot (1974)

1. The proper reconstitution of the microenvironment for the development of tissue function
2. The scale-up to generate enough properly functioning microenvironments to be clinically meaningful
3. The automation of the operation of a system to a clinically meaningful scale
4. The implementation of the automated device in a clinical setting with all the cell-handling and preservation procedures that are required for the implementation of cellular therapies

The primary focus of this chapter is on the first two issues, although some of the challenges faced with the latter two are discussed in Section 12.8. The first two issues are illustrated in Fig. 12.12 from the point of view of every cell in the body. At the center of this diagram is a cell. The diagram represents the environment that every cell in the body senses. First, there is the microenvironment. It is characterized by neighboring cells, extracellular matrix, cyto/chemokine and hormone trafficking,

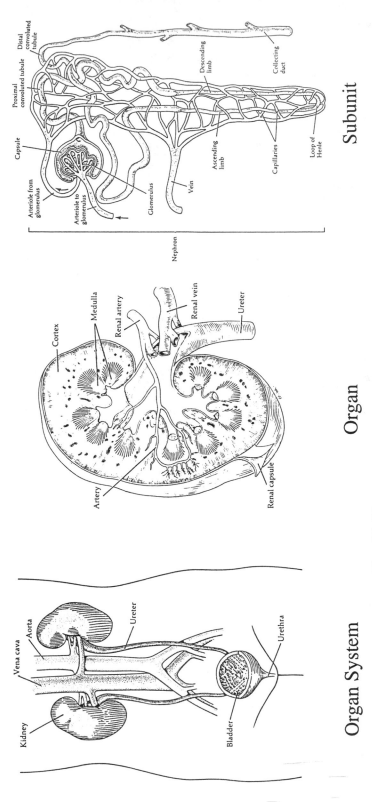

Organ System

10 cm

Vena cava
Aorta
Kidney
Ureter
Bladder
Urethra

Organ

1 cm

Cortex
Medulla
Renal artery
Renal vein
Ureter
Artery
Renal capsule

Subunit

100 μm

Distal convoluted tubule
Proximal convoluted tubule
Descending limb
Collecting duct
Capsule
Arteriole from glomerulus
Arteriole to glomerulus
Glomerulus
Vein
Ascending limb
Capillaries
Loop of Henle
Nephron

Fig. 12.10   An organ system, an organ, and a functional subunit.

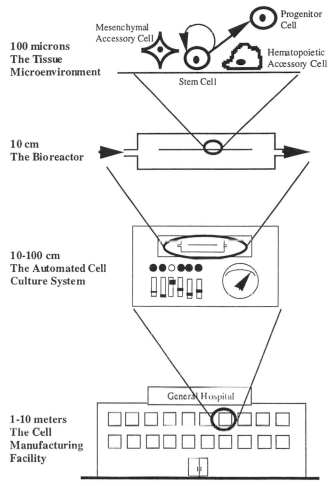

Fig. 12.11 The four principal size scales in tissue engineering and cellular therapies.

its geometry, the dynamics of respiration, the provision of nutrients, and the removal of metabolic by-products.

The microenvironment is thus very complex. To achieve proper reconstitution, these dynamic, chemical, and geometric variables must be accurately replicated. This task is difficult. A significant fraction of this discussion is devoted to developing quantitative methods to describe the microenvironment. These methods can then be used to develop an understanding of key problems, formulation of solution strategy, and analysis for its experimental implementation.

The microcirculation connects all the microenvironments in every tissue to their larger whole body environment. With few exceptions, essentially all metabolically active cells in the body are located within a few hundred micrometers from a capillary. The capillaries provide a perfusion environment that connects every cell (the

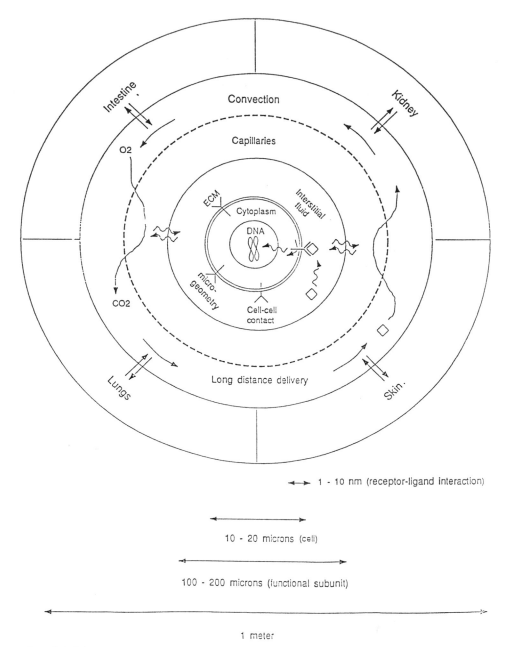

Fig. 12.12 A cell and its communication with other body parts (modified from Lightfoot, 1974).

cell at the center of the diagram) to a source of oxygen and sink for carbon dioxide (the lungs), a source of nutrients (the small intestine), the clearance of waste products (the kidney), and so forth. The engineering of these functions *ex vivo* is the domain of bioreactor design. Such culture devices have to appropriately simulate and provide respiratory, gastrointestinal, and renal functions. Furthermore, these cell culture devices have to respect the need for the formation of microenvironments and thus have to have perfusion characteristics that allow for uniformity down to the 100 μm length scale. These are stringent design requirements.

12.2.7 Estimating Tissue Function from "Spec Sheets"

Most of the useful analysis in tissue engineering is performed with approximate calculations that are based on physiological and cell biological data — a tissue spec sheet (e.g., Table 12.5). These calculations are useful for interpreting organ physiology and for providing a starting point for an experimental program. One example will be provided in each category.

The Respiratory Functions of Blood

Remarkably insightful calculations leading to interpretation of the physiological respiratory function of blood have been carried out. The basic functionalities and biological design challenges can be directly derived from tissue spec sheets.

Blood needs to deliver about 10 mM of O_2 per minute to the body. The gross circulation rate is about 5 liters per minute. Therefore, blood has to deliver to tissues about 2 mM oxygen per liter during each pass through the circulation. The pO_2 of blood leaving the lungs is about 90–100 mmHg, whereas pO_2 in venous blood at rest is about 35–40 mmHg. During strenuous exercise the venous pO_2 drops to about 27 mmHg. These facts indicate the basic requirements that circulating blood must meet to deliver adequate oxygen to tissues.

The solubility of oxygen in aqueous media is low. Its solubility is given by

$$[O_2] = \alpha_{O_2}\, pO_2$$

where the Henry's law coefficient is about 0.0013 mM/mmHg. The oxygen that can be delivered with a partial pressure change of 95–40 = 55 mmHg is thus about 0.07 mM, far below the required 2 mM (by a factor of about 30-fold). Therefore, the solubility or oxygen content of blood must be substantially increased and the concentration dependency of the partial pressure must be such that 2 mM is given up when the partial pressure changes from 95 to 40 mmHg. Furthermore, during strenuous exercise the oxygen demand doubles and 4 mM must be liberated for a partial pressure change from 95 to 27 mmHg.

The evolutionary solution is to put an oxygen-binding protein into circulating blood to increase the oxygen content of blood. To stay within the vascular bed such a protein would have to be 50–100 kDa in size. With a single binding site the required protein concentration for 10 mM oxygen is 500–1000 g/liter which is too concentrated from an osmolarity standpoint, and the viscosity of such a solution may be 10-fold that of circulating blood, which is clearly impractical. Furthermore, circulating proteases would lead to a short plasma half-life of such a protein.

Four sites per oxygen-carrying molecule would reduce the protein concentration to 2.3 m*M* and confining it to a red cell would solve both the viscosity and proteolysis problems. These indeed are the chief characteristics of hemoglobin. A more elaborate kinetic study of the binding characteristics of hemoglobin shows that positive cooperativity will give the desired oxygen transfer capabilities both at rest and under strenuous exercise (Fig. 12.13).

Perfusion Rates in Human Bone Marrow Cultures

The question of how often the medium should be replenished is important in designing cell culture conditions. Normally, the *in vivo* situation provides a good starting point for experimental optimization. Thus, a dynamic similarity analysis is in order.

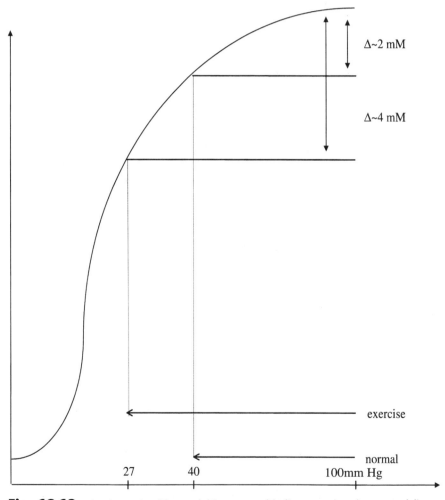

Fig. 12.13   A schematic of hemoglobin–oxygen binding curves and oxygen delivery. The change in oxygen concentration during passage through tissues (i.e., oxygen delivery) is shown as a function of the concentration of oxygen in blood (modified from Garby and Meldon, 1977).

The blood perfusion though bone marrow is about 0.08 ml/cm^3/min. Cellularity in marrow is about 500 million cells per cm^3. Therefore, the cell-specific perfusion rate is about 2.3 ml/10 million cells/day. Cultures of mononuclear cell populations from murine bone marrow were developed in the mid- to late 1970s. These cultures had long-term viability. Initial attempts to use the same culture protocols for human bone marrow cultures in the early 1980s were largely unsuccessful. The culture protocol called for medium exchange about once per week.

To perform a dynamic similarity analysis of perfusion rates, or medium exchange rates, between *in vivo* and *in vitro*, the per cell medium exchange rate in culture is calibrated to that calculated for the *in vivo* situation discussed previously. Cell cultures are typically started with cell densities on the order of a million cells per milliliter. Therefore, 10 million cells would be placed in 10 ml of culture medium which contains about 20% serum (vol/vol). A full daily medium exchange would hence correspond to replacing the serum at 2 ml/10 million cells/day, which is similar to the number calculated previously.

Experiments using this perfusion rate and the cell densities were performed in the late 1980s and led to the development of prolific cell cultures of human bone marrow. These cultures were subsequently scaled up to produce a clinically meaningful number of cells and are currently undergoing clinical trials. Thus, a simple similarity analysis of the *in vivo* and *in vitro* tissue dynamics led to the development of culture protocols that are of clinical significance. Such conclusions can be derived from tissue spec sheets.

These examples serve to illustrate the type of approximate calculations that will govern initial analysis in tissue engineering. Well-organized fact or spec sheets provide the basic data. Then, characteristic time constants, length constants, fluxes, rates, concentrations, etc. can be estimated. Then, the relative magnitudes of such characteristics serve as a basis for order of magnitude judgments.

12.3 STEM CELLS

A model for the histogenesis of proliferative tissues was presented in the previous section. This model relied on a stem cell as the source of all the mature cells in the tissue. This section gives a brief summary of stem cell biology.

12.3.1 Some Facts about Stem Cells

What are stem cells? Stem cells are cells that are undifferentiated and have the potential to differentiate and generate a large number of mature cells of one or more lineages. It is believed that they have unlimited replication potential, i.e., would self-renew. Self-renewal is a division without any commitment to differentiation. True stem cells can reconstitute and maintain organ function.

What evidence is there that stem cells exist? Lethally irradiated mice that otherwise die from complete hematopoietic failure can be rescued with as few as 20 selected stem cells. These animals reconstitute multilineage durable hematopoiesis as predicted by the stem cell model. Genetically marked mesenchymal stem cells, found in bone marrow, will in sublethally irradiated animals give rise to cells in multiple organs over a long time period. These investigations and many others unambiguously establish the presence of cells that have multi-lineage potential and persist over long periods of time *in vivo*. These cells are termed stem cells.

What do stem cells look like? Hematopoietic stem cells have been isolated. They are small spherical cells that are about 6–8 μm in diameter that have no particular morphological features. They have a high nuclear-to-cytoplasmic ratio. They appear to express receptors for most of the known hematopoietic growth factors. Furthermore, they appear to grow very slowly and may commit to growth and differentiation in a stochastic manner, with a first-order rate constant that is about 1–6 weeks.

Their cycling times have been measured by means of time-lapse videography. They commit to differentiation in culture. Thus, the first and second doubling take about 60 h, then the cycling rate speeds up to about 24 h cycling time. By the fifth and sixth doubling they are dividing at a maximal rate (12- to 14-h doubling time).

Which tissues have stem cells? As stated previously, tissues have characteristic turnover times. The more rapidly proliferating tissues, such as bone marrow, small intestinal epithelium, and skin, all appear to have stem cell systems. The same may be true for organs with slow turnover times, such as the liver and even β-islet cells in the pancreas.

What is the role of stem cells in tissue dynamics? Stem cell commitment initiates organ function, repair, and genesis. The organogenic processes, cell proliferation, cell death, cell motion, and cell differention then determine the fate of the committed cell. Stem cell depletion may lead to full or partial loss of organ function.

12.3.2 Examples of Stem Cell Systems

Two more stem cell systems are described here in addition to the three examples given in Section 12.2.2.

Mesenchymal Stem Cells

The existence of a cell population that has the potential to differentiate into a number of connective tissue cell types has been demonstrated. The developmental hierarchy is illustrated in Fig. 12.14. Interestingly, this cell population is present in the bone marrow. The existence of a multilineage potential mesenchymal stem cells (MSCs) is not fully proven because culture experiments from single cells have not shown the ability to produce multiple mature cell types. The mature cell types that can be derived from a purified population of punative MSC include osteocytes (bone), chondrocytes (cartilage), myoblasts (muscle), fibroblasts (tendon), and possibly adipocytes.

The creation of a transgenic mouse carrying an assayable collagen gene has recently been used to show that multiple tissue types can be obtained from a subpopulation that resides in bone marrow. The marrow from the transgenic mouse was transplanted into a sublethally radiated mouse and the presence of the collagen minigene was monitored over time in various tissues. The results show that there is a subpopulation present in bone marrow that has the capability to produce a num-

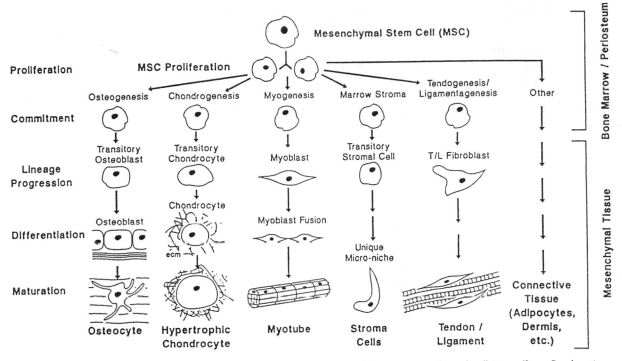

Fig. 12.14 The differentiation of mesenchymal stem cells into different mesenchymal cell types (from Bruder *et al.,* 1994).

ber of mature cell types in different organs. These observations support, but do not prove, the existence of a multipotent MSC.

The Liver as a Stem Cell System

Controversy has existed for years about the existence of a liver stem cell. Recent observations support that the liver is a stem cell system, albeit slowly replicating (Fig. 12.15). The turnover time of liver may be on the order of 1 year.

12.3.3 Models for Stem Cell Proliferative Behavior

It should be clear that the replication functions of stem cells are critical to tissue function and tissue engineering. How do stem cells divide and what happens when they divide? The following models describe the dynamic behavior of the stem cell population:

1. The clonal succession concept: Cellular systems are maintained by a reservoir of dormant cells which are available throughout a lifetime and which can be challenged one after another to enter a complex process of cell proliferation and differentiation. Once triggered, such a stem cell would give rise to a large clone of mature cells. Any one of these clones would have a limited life span because no further input is maintaining the cell production flux. After a long

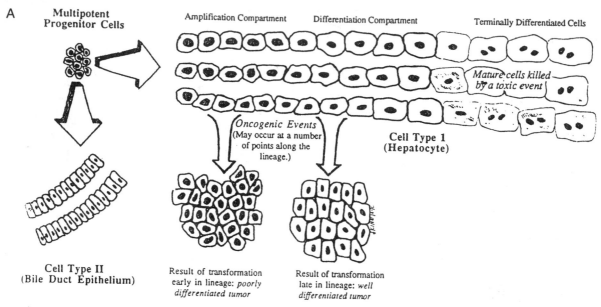

Fig. 12.15 (A) Lineage mechanisms affecting oncogenesis in liver (B) The cellular organization of the liver (from Sigal and Brill, 1992).

time, such a clone will "burn out" and a new stem cell clone will take over the cell production role.

2. Deterministic self-maintenance and self-renewal: This model relies on an assumption that stem cells can truly self-renew. Following a stem cell division, there is a 50% probability that one of the daughter cells maintains the stem cell characteristics while the other undergoes differentiation. The probability of self-renewal is regulated and may not be exactly 50% depending on the dynamic state of the tissue.

3. Stochastic model: This model considers that the progeny of a stem cell division can generate zero, one, or two stem cells as daughter cells (notice that the clonal succession model assumes zero and the deterministic model one). The assumption is that each of the three outcomes has a particular property.

12.3.4 Stem Cells and Tissue Engineering

Stem Cells Build Tissues

As just pointed out, stem cells are the source for cells in tissues. Stem cells thus build and maintain tissues *in vivo*.

Ex Vivo Growth and Manipulation of Stem Cells

The ability to grow stem cells *ex vivo* — that is, to have them divide without differentiation — would theoretically provide an inexhaustible source of stem cells.

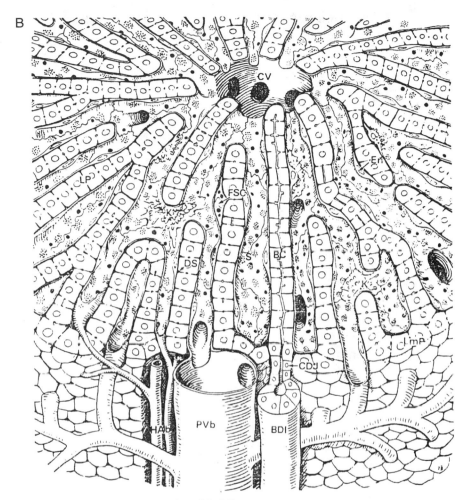

Fig. 12.15 (*Continued*)

However, stem cells appear to age and this capability may never be realized. The most likely way to success is to tap the potential that early stem cells provide, i.e., from fetal liver and cord blood. This capability has not been unambiguously established.

Isolation of Stem Cells for Scientific and Clinical Purposes

Several methods have been developed to isolate stem cells or enrich the stem cell content of a cell population. These include affinity columns, magnetic columns, and counterflow elutriation. The difficulty with positively identifying the presence of stem cells lies with the assay needed, namely, the reconstitution of an ablated animal. Stem cells are rare. These are believed to be about 1 in 100,000 hematopoietic stem cells in human bone marrow and 1 in 10,000 in mouse bone marrow.

Engraftment Potential of Cultured Stem Cells

Isolating stem cells and culturing them for transplantation purposes has a number of attractive attributes from a clinical standpoint. Unfortunately, the accumulated data obtained to date suggest that this is not possible. The stem cells differentiate and lose their engraftment potential quickly in culture. The only potential way to prevent differentiation is to keep the proper number of accessory cells present that function in a physiological fashion. The role of accessory cells and the importance of cell–cell communication in tissue function will be discussed in the next section.

12.3.5   Stem Cell Aging

Telomerases, DNA Stability, and Natural Cell Senescence

When linear DNA is replicated, the lagging strand is synthesized discontinuously through the formation of the so-called Okazaki fragments. The last fragment cannot be initiated, and therefore the lagging strand will be shorter than the leading strand. Linear chromosomes have noncoding repeating sequences on their ends that are called telomeres. These telomeres can be rebuilt using an enzyme called telomerase. This enzyme is active in microorganisms, such as yeast, and has recently been discovered to be active in the majority of human cancer cells.

Normal human somatic cells lack this activity, and the telomeres are shortened by about 50–200 bp per replication. This shortening gives rise to the so-called mitotic clock. The length of the telomeres is about 9–11 kbp and when it reaches about 5–7 kbp the chromosomes become unstable and replication ceases (Fig. 12.16). This mechanism is believed to underlie the Hayflick limit.

Telomerase is a ribonucleoprotein DNA polymerase that elongates telomeres in eukaryotes. When expressed, telomerase maintains the telomere length in growing cells. The telomere hypothesis implicates short telomere length and telomerase activation as critical players in cellular immortalization and cancer.

Stem Cells and Telomeres

Telomerase activity is found in somatic hematopoietic cells at a low activity level. There is evidence that telomeres in immature hematopoietic cells do shorten with ontogeny and with increased cell doublings *in vitro*. The rate of telomere shortening in stem cells is finite but may be slower than that in other somatic cells. Numerical evaluation of the consequences of stem cell aging strongly suggests that there has to be some form of self-renewal of stem cells in adults.

12.4   THE CELLULAR FATE PROCESSES

The key processes that determine cellular fates in tissues were introduced in Section 12.2 (Fig. 12.5). The coordinated activity of the cellular fate processes determines the dynamic state of tissue function. There is an increasing amount of information available about these processes in genetic, biochemical, and kinetic terms. The dynamics considerations that arise from the interplay of the major cellular fate processes is introduced at the end of the chapter, and the associated bioengineering challenges are described.

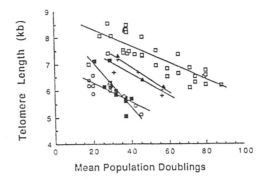

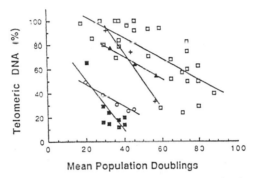

Fig. 12.16 Primary experimental data showing the shortening of telomere length with increasing cellular doubling in cell culture (from Harley et al., 1990).

12.4.1 Cell Differentiation

Describing Cellular Differentiation Biologically

Differentiation is a process by which a cell undergoes phenotypic changes to an overtly specialized cell type. The specialized cell type then carries out the physiological function for which it is designed. This process begins with a lineage and differentiation commitment and is followed by a coordinated series of gene expression events.

The term differentiation is derived from differential gene expression. Differentiation involves a change in the set of genes that are expressed in the cell, and this change is usually an irreversible change toward a particular functional state. This process involves a carefully orchestrated switching off and on of gene families. The final set of genes expressed are those that pertain to the function of the mature cell. In animals, the process of differentiation is irreversible.

Example Problem 12.1.

Production of red blood cells
How long does it take to produce a red blood cell?

Solution

Erythropoiesis replaces decaying mature red blood cells. About 200 billion need to be produced daily in a human adult. Many of the changes that a cell undergoes during the process of erythropoiesis are well known (Fig. 12.17). The cell size, the rates of RNA and DNA synthesis, and protein content all change in a progressive and coordinated fashion. The differentiation from a pronormoblast (earlier precursor

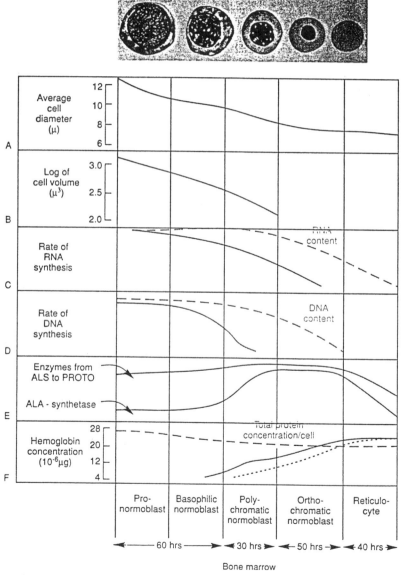

Fig. 12.17 Erythroid maturation sequence (from Granich and Levere, 1964).

stage) to a fully mature enucleated erythrocyte takes about 180 h or about 1 week. The replication activity is the highest at the preprogenitor and progenitor stage, but once the precursor stage is reached replication activity ceases sharply. ∎

Experimental Observations of Differentiation

The process of differentiation can be observed directly using fluorescent surface markers and/or light microscopy for morphological observation. Ultrastructural changes are also used to define the stage of differentiation.

A flow cytometer is often used to monitor the process of cellular differentiation. The basis for this approach is the fact that characteristic surface proteins are found on cells at different stages of differentiation. These surface markers can be used as binding sites for fluorescently conjugated monoclonal antibodies. The flow cytometer can be used to trace the expression of several surface markers.

Erythropoiesis can be traced based on expression of the transferrin receptor (CD71) and glycophorin-A. The latter is a erythroid-specific surface protein that is highly negatively charged and serves to prevent red cell aggregation in dense red cell suspensions. The transferring receptor plays a critical role during the stages in which iron is sequestered in hemoglobin. The measurement of this process is shown in Fig. 12.18.

Describing the Kinetics of Cell Differentiation

The process of differentiation is a slow one, often taking days or weeks to complete. The kinetics of this complex process can be described mathematically using two different approaches.

Compartmental Models. The traditional approach to describing cell growth and differentiation is to use compartmental models. The differentiation process involves a series of changes in cell phenotype and morphology typically becoming more pronounced at the latter stages of the process:

$$X_0 \rightarrow X_1 \rightarrow X_2 \rightarrow \cdots X_i \rightarrow \cdots \rightarrow X_n \rightarrow \text{turnover} \qquad (12.1)$$

where n can be as high as 16–18.

In the use of compartmental models in the past, the transition from one stage to the next was assumed to represent cell division. Thus, these models couple differentiation with cell replication. Mathematically, this model is described by a set of ordinary differential equations as

$$dX_i/dt = 2k_{i-1}X_{i-1} - k_iX_i \qquad (12.2)$$

The transition rate from one stage to the next is proportional to the number of cells present at that stage. This transition rate can clearly be a function of growth factor concentration and a number of other variables.

Differentiation as a Continuous Process. An alternative view is to consider the differentiation process to be a continuous process. Once the commitment to differentiation has been made, the differentiation process proceeds at a fixed rate.

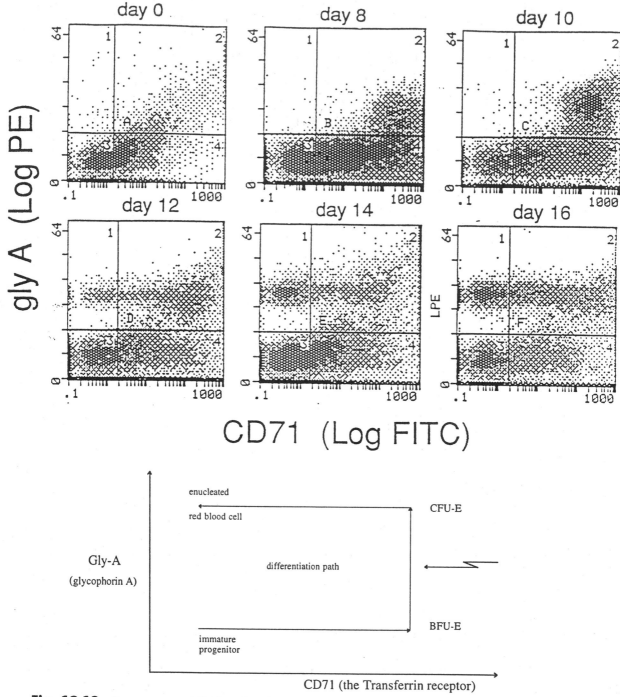

Fig. 12.18 Two-parameter definition of erythropoietic differentiation. Glycophorin A is found on erythroid cells post the blast-forming unit-erythroid (BFU-E) stage, whereas transferrin (CD71) is expressed at the progenitor stage [BFU-E and colony-forming unit-erythroid, (CFU-E)]. By measuring the two simultaneously using a flow cytometer, this differentiation process can be traced as a U-shaped path on a bivariate dot plot (from Rogers *et al.,* 1995).

This viewpoint leads to a mathematical description in the form of first-order partial differential equations:

$$dX/dt + \delta \, dX/da = (\mu(a) - \alpha(a))X \qquad (12.3)$$

where δ is the rate of differentiation and a is a parameter that measures the differentiation state of the cell. μ and α are the growth and death rates, respectively, and vary between zero and unity. Both μ and α can be a function of a.

12.4.2 Cell Motion

The Biological Roles of Cell Migration

Cell migration plays an important role in all physiological functions of tissues and also some pathological processes. Cell migration is important during organogenesis and embryonic development. It plays a role in the tissue repair response in both wound healing and angiogenesis. The immune system relies on cell migration, and pathological situations such as cancer metastasis are characterized by cell motility. Cell migration represents an integrated molecular process.

Animal cells exhibit dynamic surface extensions when they migrate or change shape. Such extensions, called lamellipodia and filopodia, are capable of dynamic formation and retraction. Local actin polymerization at the plasma membrane is a significant process in the generation of these structures. These extensions in turn rely on a complex underlying process of multiprotein interactions. In neurites, filopodia are believed to play a role in the progression of elongation by aiding the assembly of microtubules that are a significant component of these cells. The filopodia in neurites extend from the lamellipodial region and act as radial sensors. Filopodia on neurites have been found to be crucial to growth cone navigation. The filopodia on neurite growth cones have also been found to carry receptors for certain cell adhesion molecules. Mature leukocytes can extend cytoplasmic extensions. Recently, structures termed uropods have been found on T lymphocytes. These cytoplasmic projections form during lymphocyte–endothelial interaction. There is a redistribution of adhesion molecules, including ICAM-1 and -3, CD43, and CD44, to this structure. T cells have been found to use the uropods to contact and communicate directly with other T cells. Uropod development was promoted by physiologic factors such as chemokines. Cytoplasmic extensions, therefore, can perform a spectrum of functions in different cells that are related to migration and communication.

Describing Cell Motion Kinetically

Whole Populations. The motion of whole, nonreplicating cell populations is described with

$$dX/dx = J \qquad (12.4)$$

where J is the flux vector of the system boundary (cells/distance/time in a two-dimensional system), X is the cell number, and x is the flux dimension. The cellular fluxes are then related to cellular concentration and chemokine concentrations with

$$J = \text{random motility} + \text{chemokinesis} + \text{chemotaxis}$$

$$= \sigma \, dX/dx - (X/2)(d\sigma/da)(da/dx) + \chi X da/dx \qquad (12.5)$$

where σ is the random motility coefficient, χ is the chemotactic coefficient, and a is the concentration of a chemoattractant. The first term is similar to a diffusion term in mass transfer and is a measure of the dispersion of the cell population. The chemokinesis term is a measure of the changes in cell speed with concentration and is normally negligible.

If $\chi = 0$ and chemokinesis is negligible, then the motion of the cells is like random walk and is formally analogous to the process of molecular diffusion. The biased movement of the cells directly attributable to concentration gradients is given by the chemotaxis term, which is akin to convective mass transfer.

Individual Cells. As just described, motile cells appear to undergo a random walk process that mathematically can be described in a fashion similar to the diffusion process. Unlike the diffusion process, the motion of each moving entity can be directly determined (Fig. 12.19). Thus, the motility characteristics of migrating cells can be measured on an individual cell basis.

Such a description is given in terms of the cell speed (s), the persistence time (p), which is the length of time that the cell moves without changing its direction, and the orientation bias (θ) that is due to action of chemoattractants. The random motility coefficient is related to these parameters as

$$\sigma = s^2 p \qquad (12.6)$$

Typical values for these parameters are given in Table 12.6.

12.4.3   Cell Replication

Describing the Cell Cycle Biologically

The process of cell division is increasingly well known and understood in terms of molecular mechanisms (Fig. 12.20). The cycle is driven by a series of regulatory protein known as cyclin-dependent kinases. The cyclins exist in phosphorylated and dephosporylated states, basically a binary system. The network goes through a well-orchestrated series of switches, during which the cyclins are turned off and on. When in an on state, they serve to drive the biochemical processes that are needed during that part of the cell cycle. The sequence of events that describes the mammalian cell cycle has been the subject of many reviews. It should be noted that there are several important decisions associated with moving "in and out of cycle," namely, to move a cell from a quiescent state (the so-called G_0) to a cycling state and vice versa.

The cell cycle is such that the time duration of the S, G_2, and M phases is relatively fixed. Once a cell determines it needs to and can divide, it initiates DNA syn-

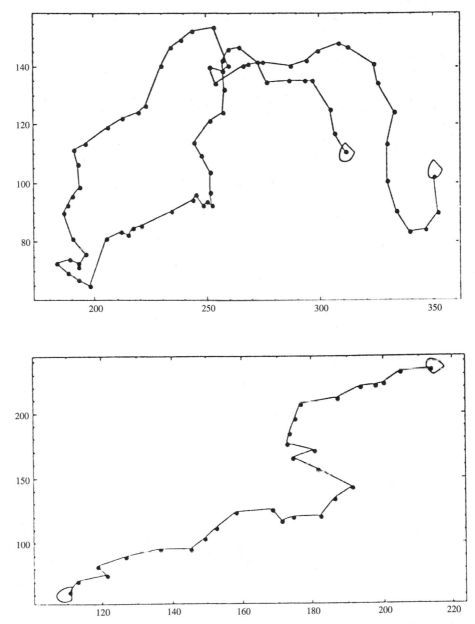

Fig. 12.19 Experimental paths of individual neutrophil leukocytes undergoing random motility in uniform environments (from Lauffenburger and Linderman, 1993).

thesis and subsequent cell division. This process has the overall characteristics of a zero-order kinetic process. Once initiated, it proceeds at a certain rate. The minimal cycling time of human cells is about 12 h. Progenitor cells can cycle at this rate.

TABLE 12.6   Random Motion — Measured Cell Speeds and Persistence Times

Cell Type	Speed	Persistence Time
Rabbit neutrophils	20 μm/min	4 min
Rat alveolar macrophages	2 μm/min	30 min
Mouse fibroblasts	30 μm/h	1h
Human microvessel endothelial cells	25–30 μm/h	4–5 h

Describing the Cell Cycle Dynamics

The dynamics of growth can be described in different ways.

Exponential Growth. If growth is unconstrained, the rate of formation of new cells is simply proportional to the number of cells present:

$$dX/dt = \mu X \ X(t = 0) \ = X_o \Rightarrow X(t) = X_o\exp(\mu t) \tag{12.7}$$

and exponential growth results. The growth rate μ is equal to $\ln(2)/t_d$, where t_d is the doubling time.

Age–time Structured Descriptions. If the phases of the cell cycle are to be described, then the cell cycle status of the cell needs to be incorporated in the dynamic description. This leads to first-order partial differential equations in time and cell cycle status:

$$dX/dt + \nu \ dX/da = \alpha(a)x \tag{12.8}$$

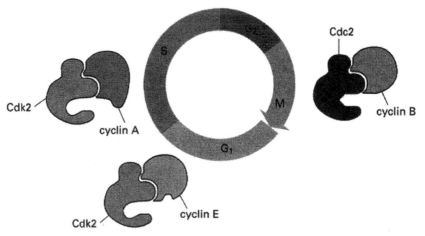

Fig. 12.20   Schematic representation of the eukariotic cell cycle (G_1–S–G_2–M) and the presence of cycling-dependent kinases.

where v is the rate at which the cell moves through the cell cycle, a is a variable that describes the cell cycle status ($a = 0$, newborn cell; $a = 1$, cell completing mitosis), and α is the death rate of the cell that can be cell cycle dependent. This population balance equation can be solved under the appropriate initial and boundary conditions.

Molecular Mechanisms. The cascade of cyclin-dependent kinases that constitute the molecular mechanism for cell cycling has been unraveled. Based on this knowledge, it is possible to describe the cell cycle in terms of the underlying molecular determinants. Such models have many components and need to be solved using a computer.

12.4.4 Interacting Cellular Fate Processes Determine Overall Tissue Dynamics

The differentiation process involves a series of changes in cell phenotype and morphology that typically become more pronounced at the latter stages of the process. The key event is milieu-dependent differentiation (or differential gene expression), which basically is the organogenic process to yield mature cells of a certain type and functionality. Similarly, embryonic induction can be described as a series of such events. Branching in this process will be described later. This process is schematically presented in Eq. (12.1).

This progression of changes is typically coupled to fundamental "driving forces," e.g., cell cycling (mitosis) and cell death (apoptosis). Thus, the basic cellular processes can be accounted for in a population balance at each stage of differentiation:

Changes in cell number

$$= \text{exit by input} - \text{differentiation} - \text{exit by apoptosis} + \text{entry by cell division}$$

or in mathematical terms:

$$dX/dt = I - \delta X - \alpha X + \mu X$$
$$= I - (\delta + \alpha - \mu)X \qquad (12.9)$$

where δ, α, and μ are the rates of differentiation, apoptosis, and replication, respectively. This equation can be rewritten as

$$dX/dt + k\,X = I \qquad (12.10)$$

where $k = \delta + \alpha - \mu$ and $t = 1/k$. The parameter k is the reciprocal of the time constant t that characterizes the dynamics of changes in the number of cells of type X. It is therefore evident that the ratios of μ/δ and α/μ are the key dimensionless groups that determine the overall cell production in tissue.

12.5   CELLULAR COMMUNICATIONS

12.5.1   How Do Cells Communicate?

Cells in tissues communicate with one another in three principal ways (Fig. 12.21):

1. They secrete soluble signals, known as cyto- and chemokines.
2. They touch each other and communicate via direct cell–cell contact.
3. They make proteins that alter the chemical microenvironment (extracellular matrix).

All these means of cellular communication differ in terms of their characteristic time and length scales and in terms of their specificity. Consequently, each is suitable to convey a particular type of a message.

12.5.2   Soluble Growth Factors

Growth factors are small proteins that are on the order of 15–20 kDa in size (1 Da is the weight of the hydrogen atom). They are relatively chemically stable and have long half-lives unless they are specifically degraded. Initially, growth factors were discovered as active factors that originated in biological fluids. For instance, erythropoietin was first isolated from urine and the colony-stimulating factors from conditioned medium from particular cell lines. The protein could be subsequently purified and characterized. With the advent of DNA technology, growth factors can

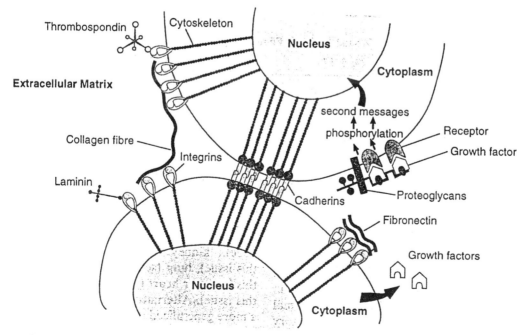

Fig. 12.21   Cell and extracellular matrix (ECM) protein interactions (from Mutsaers *et al.,* 1997).

be cloned directly as ligands for known receptors. This procedure was used to isolate thrombopoietin as the *c-mpl* ligand and the stem cell factor as the c-*kit* ligand. Growth factors are produced by a signaling cell and secreted to reach a target cells.

Example Problem 12.2

What is the maximal secretion rates of protein?

Solution

The maximal secretion rate of a protein from a single gene in a cell can be estimated based on the maximal rates of transcription and translation. Such estimates have been carried out for the production of immunoglobulins (MW = 150 kDa) whose genes are under the control of strong promoters. The estimate shows that the maximal secretion rate is approximately 2000–8000 protein molecules per cell per second, which corresponds to about 1 pg per cell per hour. This estimate compares favorably with measured maximal secretion rates. Since growth factors tend to be about one-tenth the size of immunoglobulin, a correspondingly higher secretion rate in terms of molecules per cell per time would be expected, although the total mass would stay the same. The secretion rates of protein from cells are expected to be some fraction of this maximum rate since the cell is making a large number of proteins at any given time. ■

Growth factors bind to their receptors, whcih are found in cellular membranes, with high affinities. Their binding constants are as low as 10–100 pM. The binding of a growth factor to a receptor is described as

$$G + R \leftrightarrow G:R \text{ and } [G][R]/[G:R] = K_a \qquad (12.11)$$

where $[G]$, $[R]$, and $[G:R]$ are the concentrations of the growth factor receptor and the bound complex, respectively, and K_a is the binding constant. Since the total number of receptors (R_{tot}) in the system is constant we have the mass conservation quantity:

$$R_{tot} = [R] + [G:R] \qquad (12.12)$$

and therefore Eq. (12.12) can be written as

$$R_{tot} = [G:R] K_a /[G] + [G:R] = [G:R](1 + K_a/[G]) \qquad (12.13)$$

or

$$([G:R]/R_{tot}) = [G]/([G] + K_a) \qquad (12.14)$$

Receptor occupancy rates ($[G:R]/R_{tot}$) need to be approximately 0.25–0.5 to reach a significant stimulation (Fig. 12.22), and therefore growth factor concentrations as low as 10 p M are sufficient to generate cellular response.

In many cases, the receptor:ligand complex is internalized, and a typical time constant for internalization is 15–30 min. The absolute values for growth factor uptake rates have been measured. For instance, interleukin-3 and the stem cell factor are consumed by immature hematopoietic cells at rates of about 10 and 100 ng per million cells per day. Furthermore, it has been shown that 10,000–70,000 growth

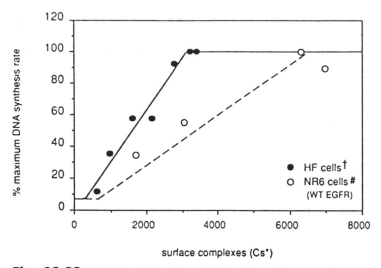

Fig. 12.22   Relationship between steady-state surface complexes and mitogenic response to EGF, for NR6 cells (○) and human fibroblasts (●) (data from Knauer *et al.*, 1984).

factor molecules need to be consumed to stimulate cell division in complex cell cultures.

Example Problem 12.3

How far can soluble signals propagate and how long does it take?

Solution

The maximum signaling distance can be estimated from a simple diffusion model that describes the secretion from a sphere. Under steady-state conditions, it can be shown that the concentration of a secreted molecule as a function of the distance from the cell is described as

$$c/K_a = \alpha\, R/r, \text{ where } \alpha = (R^2/D)/(K_a R/F) \tag{12.15}$$

where R is the radius of the cell, F is the secretion rate, and D is the diffusion coefficient of the growth factor. The distance that a signal reaches when $c = K_a$ is estimated by

$$r_{\text{critical}}/R = \alpha \tag{12.16}$$

Thus, the maximal secretion rate is given by the ratio of two time constants α, the time constant for diffusion away from the cell (R^2/D), and the secretion time constant ($K_a R/F$). Since it takes an infinite amount of time to reach a steady state, a more reasonable estimate of the signal propagation distance is the time at which the signal reaches one-half of its ultimate value. This leads to a time constant estimate for the signaling process of about 20 min and a maximal distance of about 200 μm.

This distance is shortened proportionally to the secretion rate (F) and inversely proportionally to the affinity (K_a), both of which are limited. ∎

The binding of a growth factor to its receptor triggers a complex signal transduction process (Fig. 12.23). Typically the receptor complex changes in such a way that its intracellular components take on catalytic activities. These activities are in many cases due to the so-called Janu kinases (JAKs). These kinases then operate on the STATs (signal transducers and activators of transcription) that transmit the signal into the nucleus. The kinetics of this process are complex and detailed analyses are becoming available for the epidermal growth factor signal transduction process.

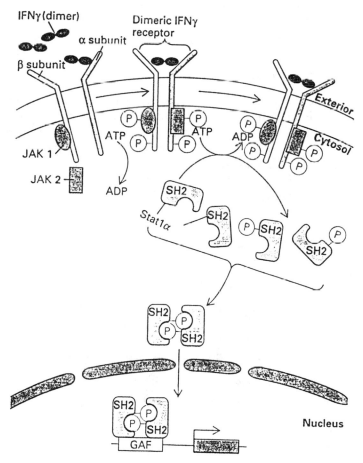

Fig. 12.23 A schematic representation of the interferon gamma signal transduction pathway (from Lodish *et al.*, 1995).

12.5.3   Direct Cell–Cell Contact

Cells are equipped with proteins on their surface that are called the cell adhesion molecules. These include the cadhedrins (adhesion belts and desmosome) and connexins (gap junctions). These molecules are involved in direct cell-to-cell contact.

Some of these are known as the cell junction molecules since junctions are formed between the adjacent cells allowing for direct cytoplasmic communication. Such junctions are typically approximately 1.5 nm in diameter and allow molecules below about 1 kDa to pass between cells.

A growing body of literature indicates how fluid mechanical shear forces influence cell and tissue function. Tissue function has a significant mechanical dimension. At the cellular level, the mechanical role of the cytoskeleton is becoming better understood. Signals may be delivered to the nucleus by cellular stretching in a way that is similar to the method in which growth factor binding to a receptor delivers signals. The integrins thus can perform as "mechanical transducers" of important signals. Furthermore, cells do know their location within a tissue and long-distance information must be transmitted between cells via weak mechanical interactions. The mechanical characteristics of the cellular microenvironment are thus of importance, as are the mechanical properties of the cells.

12.5.4.   Cell–Matrix Interactions

The extracellular matrix (ECM) comprises complex weaves, glues, struts, and gels that interconnect all the cells in a tissue and their cytoskeletal elements. The ECM is multifunctional and provides tissue with mechanical support and cells with a substrate on which to migrate as well as a place to locate signals for communications. The ECM thus has structural and informational functions. It is dynamic and is constantly being modified. For instance, ECM components are degraded by metalloproteases. In cardiac muscle about 3% of the matrix is turned over daily.

The ECM is composed of a large number of components that have varying structural and regulatory functions. On the cell surface, there are a number of adhesion and ECM receptor molecules that facilitate cell–ECM communications. These signals contain instructions for migration, replication, differentiation, and apoptosis. The nature of these signals is governed by the composition of the ECM, which in turn can be altered by the cells found in the tissue. Thus, all the cellular fate processes can be directed by the ECM, and it provides a means for cells to communicate. The signals in the ECM are more stable and can be made more specific and stronger than those delivered by diffusible growth factors. The components of the ECM and their functions are summarized in Table 12.7.

Example Problem 12.4

How many receptor sites are needed for various cellular functions?

Solution

The RGD (arginine–glycine–aspartic acid) tripeptide binding sequence has been immobilized on a cell growth surface at varying densities. Cell attachment, spreading, and growth were examined as a function of the surface density of RGD binding sites

TABLE 12.7 Components of the Extracellular Matrix[a]

Component	Function	Location
Collagens	Tissue architecture, tensile strength Cell–matrix interactions Matrix–matrix interactions	Ubiquitously distributed
Elastin	Tissue architecture and elasticity	Tissues requiring elasticity, e.g., lung, blood vessels, heart, skin
Proteoglycans	Cell–matrix interactions Matrix–matrix interactions Cell proliferation Binding and storage of growth factors	Ubiquitously distributed
Hyaluronan	Cell–matrix interactions Matrix–matrix interactions Cell proliferation Cell migration	Ubiquitously distributed
Laminin	Basement membrane component Cell migration	Basement membranes
Epiligrin	Basement membrane component (epithelium)	Basement membranes
Entactin (nidogen)	Basement membrane component	Basement membranes
Fibronectin	Tissue architecture Cell–matrix interactions Matrix–matrix interactions Cell proliferation Cell migration Opsonin	Ubiquitously distributed
Vitronectin	Cell–matrix interactions Matrix–matrix interactions Hemostasis	Blood Sites of wound formation
Fibrinogen	Cell proliferation Cell migration Hemostasis	Blood Sites of wound formation
Fibrillin	Microfibrillar component of elastic fibers	Tissues requiring elasticity, e.g., lung, blood vessels, heart, skin
Tenascin	Modulates cell–matrix interaction Antiadhesive Antiproliferative	Transiently expressed associated with remodeling matrix
SPARC[a] (osteonectin)	Modulates cell–matrix interaction Antiadhesive Antiproliferative	Transiently expressed associated with remodeing matrix
Thrombospodin	Modulates cell–matrix interaction	Platelet α granules
Adhesion molecules	Cell surface proteins mediating cell adhesion to matrix or adjacent cells Mediators of transmembrane signals	Ubiquitously distributed
von Willebrand factor	Mediates platelet adhesion Carrier for procoagulant factor VIII	Plasma protein Subendothelium

[a] Mutsaers, S., Bishop, J., McGrouther, G., and Laurent, G., "Mechanisms of Tissue Repair, from Wound Healing to Fibrosis," Int. J. of Biochem. Cell Biol. Vol. 29 No. 1 (p. 5–17) (1997).
[b] SPARC, secreted protein acidic and rich in cysteine.

for fibroblasts. The results showed that an average receptor spacing of 440 nm was sufficient for cell attachment and spreading and 160 nm for focal point adhesion formation.   ∎

Many tissue engineering efforts to date have been aimed at artificially constructing an ECM. These matrices have taken the form of polymeric materials, which in some cases are bioresorbable to allow the cells to replace this artificial ECM as they establish themselves and reconstruct tissue function. The properties of this matrix are hard to design since most of the functionalities are unknown and since it carries two-way communication between cells. The full spectrum of ECM functionalities can only be provided by the cells (Table 12.8).

12.6   THE TISSUE MICROENVIRONMENT

To this point, the overall states of tissue function, and the fact that this behavior is based on interacting cellular fate processes have been discussed. How these cellular fate processes are quantitatively described and the fact that there are complex genetic circuits that drive these cellular fate processes have been shown.

Synthesis is the next topic. How can tissue function be built, reconstructed, and modified? This task is one of tissue engineering. To set the stage, some fundamentals can be stated axiomatically:

TABLE 12.8   Adhesion Molecules with the Potential to Regulate cell–ECM Interactions

Adhesion molecule	Ligand
Integrins	
$\alpha_1\beta_1$	Collagen (I, IV, VI), laminin
$\alpha_2\beta_1$	Collagen (I–IV, VI), laminin
$\alpha_3\beta_1$	Collagen (I), laminin, fibronectin, entactin, epiligrin
$\alpha_4\beta_1$	Fibronectin$_{ALT}$, VCAM-1, thrombospondin
$\alpha_5\beta_1$	Fibronectin, thrombospondin
$\alpha_6\beta_1$	Laminin
$\alpha_V\beta_1$	Fibronectin
$\alpha_L\beta_2$	ICAM-1, ICAM-2, ICAM-3
$\alpha_M\beta_2$	ICAM-1, iC3b, fibrinogen, factor X, denatured protein
$\alpha_x\beta_2$	Fibrinogen, iC3b, denatured protein
$\alpha_V\beta_3$	Vitronectin, fibrinogen, fibronectin, thrombospondin
$\alpha_V\beta_5$	Vitronectin
$\alpha_V\beta_6$	Fibronectin
$\alpha_0\beta_4$	Laminin
$\alpha_4\beta_7$	Fibronectin$_{ALT}$ VCAM-1, MAdCAM-1
$\alpha_{1th}\beta_3$	Fibrinogen, fibronectin, vitronectin, vWF
LRI[h]	Fibrinogen, fibronectin, vitronectin, vWF, collagen (IV), entactin

- The developmental program and the wound healing response require the systematic and regulated unfolding of the information on the DNA through coordinated execution of "genetic subroutines and programs." Participating cells require detailed information about the activities of their neighbors. Proper cellular communications are of key concern.
- Upon completion of organogenesis or wound healing, the function of fully formed organs is strongly dependent on the coordinated function of multiple cell types. Tissues function is based on multicellular aggregates, so-called functional subunits of tissues.
- The microenvironment has a major effect on the function of an individual cell. The characteristic length scale of the microenvironment is 100 μm.
- The microenvironment is characterized by (i) neighboring cells: cell–cell contact, soluble growth factors, etc.; (ii) the chemical environment: the extracellular matrix and the dynamics of the nutritional environment; and (iii) the local geometry.

12.6.1 Cellular Function *in Vivo:* Tissue Microenvironment and Communication with Other Organs

An important requirement for successful tissue function is a physiologically acceptable environment in which the cells will express the desired tissue function. Most likely, but not necessarily, this environment should recreate or mimic the physiological *in vivo* situation. How can a physiologically acceptable tissue microenvironment be understood and recreated?

The communication of every cell with its immediate environment and other tissues can be illustrated using a topological representation of the organization of the body. The DNA, in the center of the diagram, contains the information that the tissue engineer wishes to express and manage. This cell is in a microenvironment that has important spatio temporal characteristics. Examples of tissue microenvironments are shown in Fig. 12.24 and some are discussed in Section 12.2.2.

Signals to the nucleus are delivered at the cell membrane and transmitted through the cytoplasm by a variety of signal transduction mechanisms. Some signals are delivered by soluble growth factors after binding to the appropriate receptors. These growth factors may originate from the circulating blood or from neighboring cells. Nutrients, metabolic waste products, and respiratory gases traffic in and out of the microenvironment in a highly dynamic manner. The microenvironment is also characterized by its cellular composition, the ECM, and local geometry and these components also provide the cell with important signals. The size scale of the microenvironment is approximately 100 μm. Within this domain cells function in a coordinated manner. If this arrangement is disrupted, cells that are unable to provide tissue function are obtained.

Example Problem 12.5

What do accessory cells do?

Liver

a

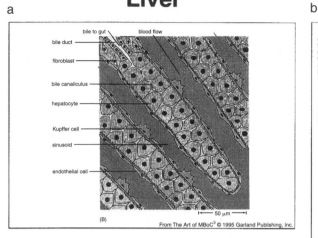

(B)

From The Art of MBoC³ © 1995 Garland Publishing, Inc.

Intestinal Epithelium

b

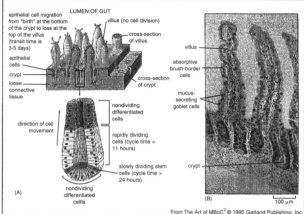

From The Art of MBoC³ © 1995 Garland Publishing, Inc.

Skin

c

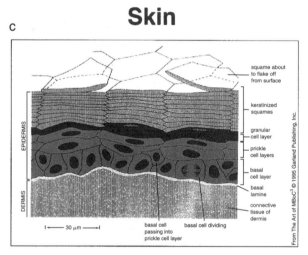

Fig. 12.24   *In vivo* tissue microenvironments. (Parts a, b, c, e, f [(A)] from Alberts *et al.,* 1994, Part d from Rhodin, 1974, Part f [(B)] from Krstic, 1985, Part f [(C)] from Fawcett, 1986).

Solution

The hepatic stellate cell is a mesenchymal cell that is located in the space of Dissie that is found between the sheets of hepatocytes and the hepatic sinusoidal endothelium (Fig. 12.25). In the normal histogenic state of liver, hepatic stellate cells are found in close proximity to a basement membrane-like matrix. This matrix consists of collagen type IV, laminin, and heparan sulfate proteoglycans. If the liver is injured, hepatic stellate cells are activated, and they begin to make matrix protein, mostly collagen types I and III. As a result of hepatic stellate cell activation, the phenotype of both the hepatocytes (lose their brush border) and endothelium (lose their fenestration) changes. If this condition persists, liver fibrosis results. Thus, a lack of

Bone Marrow

d

Bone

e

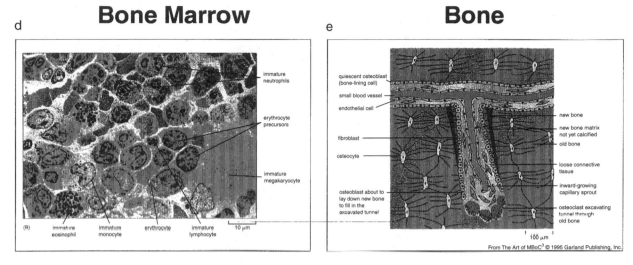

Mammary Gland

f

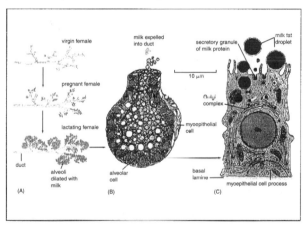

Fig. 12.24 *(Continued)*

coordination in cellular function in the tissue microenvironment can lead to loss of tissue function. ■

Tissues are perfused by the microcirculation and they are homogeneous from this standpoint down to a length scale of about 100 μm. The microcirculation then interfaces with the long-distance convective transport system that connects all the tissues in the body and thus all cells to a nutritional supply, exchange of respiratory gases, removal of toxic products, etc.

All tissues are composed of functional subunits, which are the irreducible cellular arrangements that generate tissue function. Separating the cells in this microen-

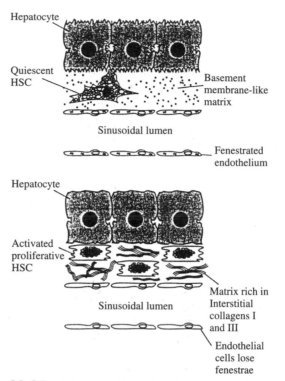

Fig. 12.25 Schematic of hepatic stellate cell activation and the process of liver fibrosis (from Iredale, 1997).

vironment leads to cellular function, but all of them together form tissue function. These microenvironments have cross-species similarities. The number of microenvironments per tissue is then determined by the overall requirements on the particular organ. The kidney in an elephant must be much larger than that in a mouse, although the fundamental unit, the nephron, may be very similar.

From a tissue engineering standpoint, these features lead to problems in decomposition, i.e., designing the macro- and the microenvironments. The challenges that face the tissue engineer can be broken down into two basic categories:

- The microenvironment (its chemical, geometric, cell architectural, and diffusional characteristics)
- Interactions with other tissues (source of nutrients, exchange of respiratory gases, removal of waste products, and delivery of soluble protein such as growth factors)

A device that mimics the tissue function must provide the function for the two outermost layers and locally produce an acceptable environment, represented by the remaining layers. Although much experience was gained during the 1980s in bioreactor design for cultures of pure continuous cell lines for protein production, tissue bioreactors clearly represent a series of new challenges. This topic will be addressed later. In this section, the discussion will focus on the variables that influence the microenvironment: its cellularity, its chemistry, its dynamics, and its geometry.

12.6.2 Cellularity

The packing density of cells is on the order of a billion cells per cm^3. Tissues are typically operating at one-third to one-half of packing density, leaving typical cell densities in tissues of approximately 100–500 million cells/cm^3. Since the master length scale is about 100 μm, the order of magnitude of the number of cells found in a tissue microenvironment can be estimated. A 100–μm cube at 500 million cells/cm^3 contains about 500 cells. Simple multicellular organisms, such as *Caenorhabditis elegans* a much studied small worm, have about 1000 cells and provide an interesting comparison.

The cellularity of the tissue microenvironment varies among tissues. At the low end there is cartilage. The function of chondrocytes in cartilage is to maintain the extracellular matrix. Cartilage is avascular, alymphatic, and aneural. Thus, many of the cell types found in other tissues are not in cartilage. The cellularity of cartilage is about 1 million cells/cm^3 or about one cell per cubic 100 μm. Thus, the microenvironment is simply one cell that maintains its surrounding ECM.

Many cell types are found in tissue microenvironments (Table 12.9). In addition to the parenchymal cells (the tissue type cells, e.g., hepatocytes in liver), a variety of accessory cells are found in all tissues. These are as follows:

- Mesenchymal cells (such as fibroblasts and smooth muscle cells) are present in all tissues. These cells are of connective tissue type.
- Monocytes are present in all tissues taking on very different morphologies (Fig. 12.26). Monocytes can differentiate into macrophages that, once

TABLE 12.9 Cells That Contribute to the Tissue Microenvironment

Stromal cells: derivates of a common precursor cell
   Messenchyme
   Fibroblasts
   Myofibroblasts
   Osteogenic/chondrogenic cells
   Adipocytes
Stromal-associated cells: histogenically distinct from stromal cells, permanent residents of a tissue
   Endothelial cells
   Macrophages
Transient cells: cells that migrate into a tissue for host defense either prior to or following an
   inflammatory stimulus
   Blymphocytes/plasma cells
   Cytotoxic T cells and natural killer (NK) cells
   Granulocytes
Parenchymal cells: cells that occupy most of the tissue volume, express functions that are definitive for
   the tissue, and interact with all other cell types to facilitate the expression of differentiated function

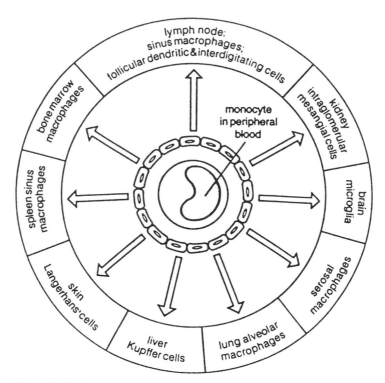

Fig. 12.26   Distribution of macrophages and their presence in different tissues. (from Hoffbrand and Pettit, 1988).

activated, produce a variety of cyto- and chemokines that influence the behavior of neighboring cells.

■ Endothelial cells are associated with the vasculature found in all tissues. These cells play a major role in the trafficking of cells in and out of tissue and may play a major role in determining tissue metabolism.

■ Lymphocytes and neutrophils have a transient presence in tissues, typically as a part of a host defense response or other cleanup functions.

These accessory cells are typically about 30% of the cellularity of tissue and the parenchymal cells make up the balance. An example is provided in Table 12.10, which lists the cellularity of the liver.

Example Problem 12.6

What happens if accessory cells are removed?

Solution

The role of accessory cell (called stroma) function in bone marrow cultures has been systematically studied. Since the immature cells can be isolated based on known antigens, the relative composition of the key parenchymal cells and the accessory

TABLE 12.10 Cells Contributing to the Hepatic Microenvironment

Cell type	Size (μm)	Relative percentage of total cells
Stroma		
Kupffer cells	12–16	8
Vascular endothelia	11–12	9
Biliary endothelia	10–12	5
Fat-storing cells	14–18	3
Fibroblasts	11–14	7
Pit cells	11–15	1–2 (variable)
Parenchymal cells		
Mononuclear (type I)	17–22	35
Binuclear (type II)	20–27	27
Acidophilic (type III)	25–32	5

cells can be varied. Such an experiment amounts to a titration of the accessory cell activity. The results from such an experiment are shown in Fig. 12.27. All cell production indices (total cells, progenitor cells, and preprogenitor cells) decline sharply as the accessory cells are removed. Supplying preformed irradiated stroma restores the production of total and progenitor cells but not the preprogenitors. This result is consistent with the expectation that specific accessory cell–parenchymal cells interactions are important for immature cells. ∎

12.6.3 Dynamics

The microenvironment is highly dynamic and displays a multitude of time constants.

Oxygenation

Generally, mammalian cells do not consume oxygen rapidly compared to microorganisms, but their uptake rate is large compared to the amount of oxygen available at any given time in blood or in culture media (Table 12.11). At 37°C, air saturated aqueous media contain only 0.21 mM of 210 μmol oxygen per liter. Mammalian cells consume oxygen at a rate in the range of 0.05–0.5 μmol/10^6 cells/h. With tissue cellularities of 500 million cells/cm^3 these oxygen uptake rates call for volumetric oxygen delivery rates of 25–250 μmol oxygen/cm^3/h. This rate needs to be balanced with the volumetric perfusion rates of tissues and the oxygen concentration in blood.

Metabolically active tissues and cell cultures, even at relatively low cell densities, quickly deplete the available oxygen. For instance, at cell densities of 10^6 cells/ml, oxygen is depleted in about 0.4–4 h. Oxygen thus must be continually supplied. To date, primary cell types (e.g., hepatocytes, keratinocytes, chondrocytes, and hematopoietic cells) have been grown *ex vivo* for the purpose of cell therapy. The effects of oxygen on hepatocytes has been systematically investigated. The reported

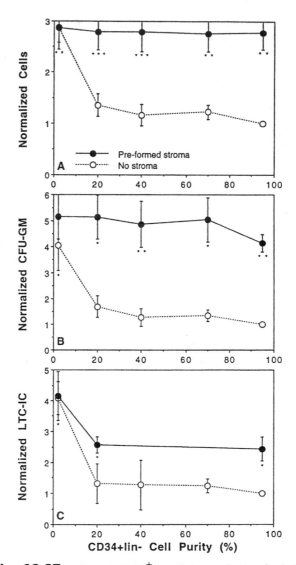

Fig. 12.27 Effect of CD34$^+$lin cell (a population of primitive hematopoietic cells) purity on culture output. With increasing purity the performance on a per cell basis drops due to loss of accessory cell function. CFU-GM, colony-forming units granulocyte/macrophage, LTC-IC, long-term culture-initiating cells (from Koller *et al.*, 1995a).

specific oxygen uptake rate (OUR) for hepatocytes is approximately 1.0 μmol/10^6 cells/h, which is relatively high for mammalian cells. Conversely, a much lower oxygen consumption rate of about 0.02 μmol/10^6 cells/h has been reported for rat bone marrow cells.

In addition to the supply requirements, the concentration of oxygen close to the

TABLE 12.11 Measured Oxygen-Demand Rates of Human Cells in Culture

Human	μmol $O_2/10^6$ cells/h
HeLa	0.1–0.0047
HLM (liver)	0.37
LIR (liver)	0.30
AM-57 (amnion)	0.045–0.13
Skin fibroblast	0.064
Detroit 6 (bone marrow)	0.43
Conjunctiva	0.28
Leukemia MCN	0.22
Lymphoblastoid (namalioa)	0.053
Lung	0.24
Intestine	0.40
Diploid embryo WI-38	0.15
MAF-E	0.38
FS-4	0.05

cells must be within a specific range. Oxygenation affects a variety of physiological processes, ranging from cell attachment and spreading to growth and differentiation. An insufficient concentration retards growth, whereas an excess concentration may be inhibitory or even toxic. For instance, several studies have shown that the formation of hematopoietic cell colonies in colony assays is significantly enhanced by using oxygen concentrations that are 5% of saturation relative to air, and an optimal oxygen concentration for bioreactor bone marrow cell culture has been shown to exist. The oxygen uptake rate of cells is thus an important parameter in the design of primary cell cultures.

Metabolism and Cell Signaling

Typically, there is no transport limitation for major nutrients, although cells can respond to their local concentrations. The reason for this is because their concentrations can be much higher than that of oxygen, especially for the nutrients consumed at high rates. Typical uptake rates of glucose are approximately 0.2 μmol/million cells/h, whereas the consumption rates of amino acids are in the range of 0.1–0.5 μmol/billion cells/h. The transport and uptake rates of growth factors have more serious transport limitations. The expected diffusional response times are given in Fig. 12.28.

Perfusion

The circulatory system provides blood flow to organs that is then distributed into the microenvironments. Overall, the perfusion rates in a human are about 5 liters/min/70 kg, or about 0.07 ml/cm^3/min. With 500 million cells/cm^3, this is equivalent to 0.14 μl/million cells/min. These numbers represent a whole body average. There are differences in the perfusion rates of different organs that typically correlate with their metabolic activity.

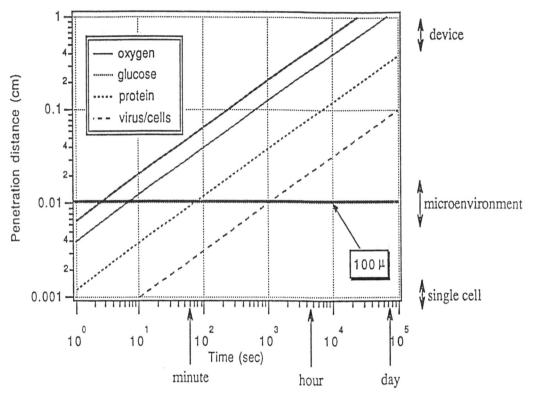

Fig. 12.28 The diffusional penetration lengths as a function of time for several classes of biomolecules.

Cell Motion

As described earlier, cells are motile and move at different rates. Neutrophils can move several cell diameters in a minute, whereas adherent cell types, such as keratinocytes, fibroblasts and endothelial cells, move a cell diameter per hour. These motilities represent rapid processes compared to cell replication and differentiation. Neutrophils have to be able to respond rapidly to invading microorganisms, whereas the adherent cell types move in response to dynamic tissue needs.

Matrix Turnover

The ECM is constantly being turned over. In the heart, about 3% of the ECM is turned over on a daily basis.

12.6.4 Size and Geometry

Geometry

The geometric shapes of microenvironments vary (Fig. 12.24) and so do their dimensionalities. Many microenvironments are effectively curved 2-D surfaces. The cellular arrangement in bone marrow has been shown to have a fractal geometry with effective dimensionality of 2.7, whereas the brain is a 3-D structure.

What Determines the Size of the Microenvironment?

The answer to this question is not clear. At the lower limit is the size of a single cell (about 10 μm). A cell aggregate must be some multiple of this distance. The factors determining the upper limit are not clear, However, estimates of effective growth factor signal propagation distances and experimental evidence of oxygen penetration distances suggest that the dynamics of cell communication and cell metabolic rates are important in determining the size scale of the microenvironment. These distances are determined by the process of diffusion. In both cases, the estimated length scale is about 100–200 μm. The stability issues associated with coordinating cellular functions become more important with an increased number of cells and may represent a limitation in practice.

12.7 SCALING UP

12.7.1 Fundamental Concept

As discussed previously, tissue dynamics is composed of an intricate interplay between the cellular fate processes of cell replication, differentiation, and apoptosis. They are properly balanced under *in vivo* conditions. The dynamics of the *in vivo* conditions are a balance of these biological dynamics and the constraining physicochemical processes. The basic concept of design in tissue reconstruction is to engineer a proper balance between the biological and physicochemical rates so that normal tissue function can occur.

12.7.2 Key Design Challenges

Within this philosophical framework many of the engineering issues associated with successful reconstitution of tissues can be delineated. This short survey is not intended to be a complete enumeration of the important issues of tissue engineering, nor is it intended to be a comprehensive representation of published literature. Its main goal is to provide an engineering perspective of tissue engineering and help define the productive and critical role that engineering needs to play in the *ex vivo* reconstruction of human tissues.

Important design challenges have arisen regarding the following issues:

- Oxygenation, i.e., providing adequate flux of oxygen at physiological concentrations
- Provision and removal of cyto- and chemokines
- Physiological perfusion rates and uniformity in distribution
- Biomaterials — functional, structural, toxicity, and manufacturing characteristics

These four issues will be discussed briefly. Other issues associated with the clinical implementation of cellular therapies include the design of disposable devices, optimization of medium composition, initial cell distribution, meeting the Food and Drug Administration requirements, and operation in a clinical setting. These cannot be detailed herein, but some of these issues are addressed later.

12.7.3 Time Scales of Mass Transfer

The importance of mass transfer in tissue and cellular function is often overlooked. The limitations imposed by molecular diffusion become clear if the average displacement distance with time is plotted for diffusion coefficients that are typical for biological entities of interest in tissue function (Fig. 12.28). The diffusional penetration lengths over physiological time scales are surprisingly short and constrain the *in vivo* design and architecture of organs. The same constraints are faced in the construction of an *ex vivo* device, and high mass transfer rates into cell beds at physiological cell densities may be difficult to achieve.

The biochemical characteristics of the microenvironment are critical to obtaining proper tissue function. Much information exists about the biochemical require-

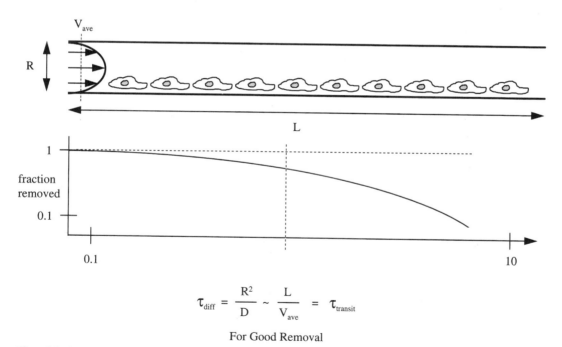

$$\tau_{diff} = \frac{R^2}{D} \sim \frac{L}{V_{ave}} = \tau_{transit}$$

For Good Removal

Fig. 12.29 A schematic of the concentration of growth factors in a liquid that is flowing across a bed of cells that consume it. (Bottom) The fractional removal of the growth factor as a function of the relative time constants of lateral diffusion and transit across the cell bed. The ratio of the two is the Graetz number (Gz). If the diffusional response time is slower than that of transit (Gz > 1), then there is insufficient time for the diffusing growth factor molecule to make good contact with the cell bed. Most of the growth factor leaves the system in the exit stream. Conversely, if the diffusion time is much shorter than that of transit, there will be ample time for the growth factor to make it to the cell bed. If it is rapidly consumed there, then negligible amounts will leave the system.

ments for the growth of continuous cell lines. For continuous cell lines, these issues revolve around the provision of nutrients and the removal of waste products. In cultures of primary cells, the nutrients may have other roles and directly influence the physiological performance of the culture. For instance, recently it has been shown that proline and oxygen levels play an important role in hepatocyte cultures.

In most cases, oxygen delivery is likely to be an important consideration. Too much oxygen will be inhibitory or toxic, whereas too little may alter metabolism. Some tissues, such as liver, kidney, and brain, have high oxygen requirements, whereas others require less. Controlling oxygen at the desired concentration levels, at the desired uniformity, and at the fluxes needed at high cell densities is likely to prove to be a challenge. Furthermore, the oxygen and nutritional requirements may vary among cell types in a reconstituted tissue that is composed of multiple cell types. These requirements further complicate nutrient delivery. Thus, defining, designing, and controlling the biochemical characteristics of the microenvironment may prove difficult, especially given the constraints imposed by diffusion and any requirements for a particular microgeometry.

Example Problem 12.7

How are specific oxygen uptake rates measured?

Solution

Most data on specific oxygen uptake rates are obtained with cells in suspension using standard respirometers. However, to obtain accurate and representative data for primary cells, they need to be adherent. This challenge has led to the design of a novel *in situ* respirometer (Fig. 12.30). This respirometer has been used to measure the oxygen uptake rates of hepatocytes in culture (OUR of 1.0 µmol/million cells/h). A similar device has been used to measure the OUR in bone marrow cultures, giving results of 0.03–0.04 µmol/million cells/h which are similar to the *in vivo* uptake rates. ∎

Example Problem 12.8

Oxygen can be delivered *in situ* over an oxygenation membrane as illustrated in Fig. 12.31

Solution

At slow flow rates compared to the lateral diffusion rate of oxygen, the oxygen in the incoming stream is quickly depleted, and the bulk of the cell bed is oxygenated via diffusion from the gas phase. This leads to oxygen delivery that can be controlled independently of all other operating variables and that is uniform. The gas phase composition can be used to control the oxygen delivery. ∎

12.7.4 Fluid Flow and Uniformity

The size scale of the microenvironment is 100 µm. A device that carries tens of millions of microenvironments must thus be uniform in delivery and removal of gases, nutrients, and growth factors. Achieving such uniformity is difficult. This difficulty

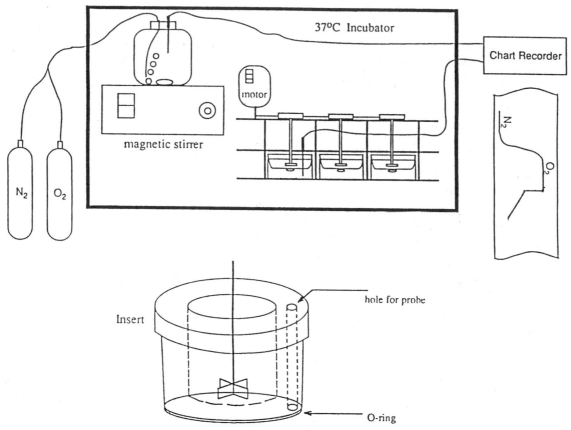

Fig. 12.30 Schematic diagram of the apparatus for measuring oxygen uptake rate (from Peng and Palsson, 1996a).

arises from the fact that fluid has a no slip condition as it flows past a solid surface. Thus, there will always be slower flowing regions close to any side walls in a bioreactor. These slow flowing regions under conditions of axial Graetz numbers of unity lead to mass transfer boundary layers that extend beyond the hydrodynamic boundary layer.

The origin of this problem is illustrated in Fig. 12.32. Fluid is flowing down a thin slit, representing an on-end view of the chamber shown in Fig. 12.29. Thin slits with high aspect ratios will satisfy the Hele–Shaw flow approximations, in which the flow is essentially the same over the entire width of the slit except close to the edges. The width of the slow flowing regions is on the order of the depth of the slit (R). Thus, if the aspect ratio is 10, fluid will flow the same way over about 90% of the width of the slit. In the remaining 10%, the slow flow close to the wall can create microenvironments with a different property than in the rest of the cell bed. Such nonuniformity can lead to differences in cell growth rates and to migration of cells toward the wall. Such nonuniformity in growth close to walls has been reported. This problem can be overcome by using radial flow.

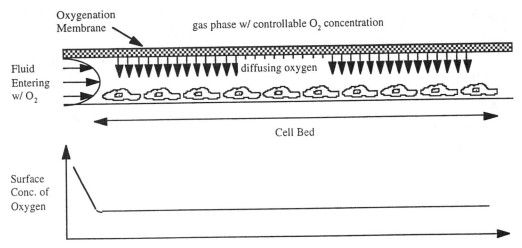

Fig. 12.31 Oxygen delivery across a membrane that is placed on top of a fluid that is flowing across a cell bed. If the fluid transit time is much slower than the diffusional time for oxygen, then oxygen is delivered primarily via diffusion. This leads to a small entrance effect where the oxygen in the incoming stream is consumed while the oxygen concentration over the rest of the cell bed is relatively constant.

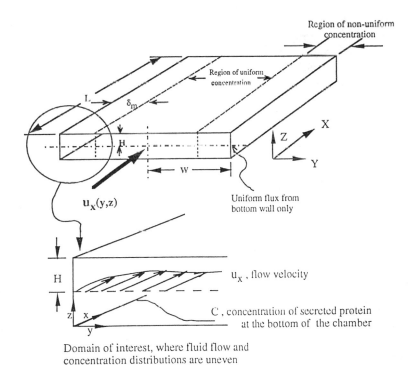

Fig. 12.32 Coordinate system for a rectangular chamber with production of biological factors secreted by cells lodged on the bottom wall. The fluid is slowed down close to the side wall creating a different concentration than that found in the middle of the slit. This leads to a very different microenvironment for cell growth and development of tissue function at the wall than elsewhere in the chamber (from Peng and Palsson, 1996b).

The example shown is for one type of a bioreactor for cell culturing. Other configurations will have similar difficulties in achieving acceptable uniformity in conditions, and careful mass and momentum transfer analyses need to be performed to guide detailed design.

12.7.5 Biomaterials

Biomaterials for tissue engineering present several challenging problems. There appear to be three length scales of primary interest. The shortest is at the biochemical level where concerns include the specific chemical features of the ECM and interactions with cellular receptors (e.g., see Section 12.5.3). Intact ECM components can be used to coat support material to ensure appropriate interactions between the cells and their immediate environment. More sophisticated treatments involve the synthesis of specific binding sequences from ECM protein and presenting them in various ways to the cells. Particular cellular arrangements can be obtained by micropatterning such materials. A combination of material manufacturing, biochemistry, and genetics is required to address these issues.

The next size scale of interest is the 100-μm size scale — the size of a typical organ microenvironment. Many organs have highly specific local geometries that may have to be engineered in an *ex vivo* system. Hence, a particular microgeometry with particular mechanical properties may have to be produced. Clearly, challenging material manufacturing issues arise. Furthermore, the support matrix may have to be biodegradable after transplantation and the degradation products nontoxic. Lactic and glycolic acid-based polymers are promising materials in this regard. If little restructuring of implants occurs following grafting, then the geometry of the support matrix over larger size scales may be important.

The largest size scale is that of the bioreactor. Bioreactors in tissue engineering are likely to be small with dimensions of approximately 10 cm. The materials issues that arise here are primarily those of biocompatibility. Although manufacturing technology exists for tissue culture plastic, it is likely that additional issues will arise. The tissue culture plastic that is commercially available is designed to promote adhesion, binding, and spreading of continuous cell lines. Although such features may be desirable for continuous cell lines, they may not be so for various primary cells.

12.8 DELIVERING CELLULAR THERAPIES IN A CLINICAL SETTING

In the previous section, the design challenges that the tissue engineer faces regarding scaling up microenvironments to produce cell numbers that are of clinical significance were surveyed. These problems are only a part of the challenges that must be met in implementing cellular therapies. In this section, some of the other problems that need to be solved will be discussed.

12.8.1 Donor-to-Donor Variability

The genetic variability in the human population is substantial. The outcome from an *ex vivo* growth procedure can vary significantly between donors. The standard deviation can exceed 50% of the mean value for key performance indices. Thus, even if the cell growth process, the production of materials, and the formulation of medium are essentially identical, a large variation in outcomes will be observed. Many experiments (10–12) need to be performed to obtain satisfactory answers.

This variability is due to intrinsic biological factors. Some regularization in performance can be achieved by using a full complement of accessory cells. Although interdonor variability is considerable, the behavior of the same tissue source is internally consistent. An example is shown in Fig. 12.33 in which the relative uptake rates of growth factors are shown for many donor samples. The uptake rates of growth factors can be highly correlated within many donor samples. The quantitative nature of the correlation changes from donor to donor, making it difficult to develop a correlation that would represent a large donor population.

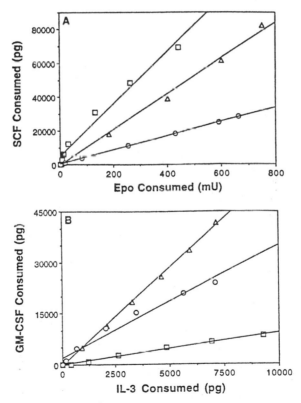

Fig. 12.33 The uptake of growth factors, i.e., stem cell factor (SCF) and erythropoietin (Epo), is highly correlated in cultures of human bone marrow. The correlation is strong for a single tissue sample. The slopes of the curves vary significantly between donors (from Koller *et al.,* 1995b).

12.8.2 Strongly Interacting Variables

Primary cell cultures are sensitive to many of the variables that can be controlled. Many, if not most, of these variables interact strongly. Thus, any change in a single cell culture variable will change the optimal value for all others. Thus, a statistical experimental design procedure that allows the fastest way to search many experimental variables should be employed. An example of such a 2-D search is shown in Fig. 12.34, in which the optimal progenitor production performance of human cell cultures is shown as a function of the inoculum density and the medium flow rate.

Figure 12.34 shows the optimal performance of the cultures as measured by two different objectives. The top panel shows the optimal number of progenitors produced per unit cell growth area. Optimizing this objective would lead to the smallest cell culture device possible for a specified total number of cells that is needed. The bottom panel shows the optimal expansion of progenitor cells, i.e., the output number relative to the input. This objective would be used in situations in which the starting material is limited and the maximum number of additional progenitors is desired. Note that the two objectives are found under different conditions. Thus, it is critical to clearly delineate the objective of the optimization condition from the outset. Sometimes this choice is difficult.

12.8.3 Immune Rejection

Allogeneic transplants have immune rejection problems. A variety of situations are encountered. Dermal fibroblasts, used for skin ulcers, for unknown reasons are effectively nonimmunogenic. This fact makes it possible to develop many grafts from a single source and transplant them into many patients. Conversely, β-islets face certain rejection (see Fig. 12.2). Lastly, in an allogeneic bone marrow transplantation, the graft may reject the immunocompromised host. The graft vs host disease is the main cause for the mortality resulting from allogeneic bone marrow transplants.

The cellular and molecular basis for the immune response is becoming better understood. The rejection problem is a dynamic process that relies on the interaction between subsets of $CD4^+$ cells ($CD4^+$ is a surface antigen on certain T cells), Th1, and Th2 that differ in their cytokine secretion characteristics and effector functions. The components of the underlying regulatory network are becoming known. Quantitative and systemic analysis of this system is likely to lead to rational strategies for manipulating immune responses for prophylaxis and therapy.

12.8.4 Tissue Procurement

The source of the starting material for a growth process is of critical importance. The source of dermal fibroblasts used for skin replacement is human foreskin obtained from circumcision. This source is prolific and can be used to generate a large number of grafts. Since the source is the same, the biological variability (see Section 12.8.1) in the growth process is greatly diminished. The costs associated with tissue procurement as reflected in the final product are minimal. Conversely, an adult autologous source may be expensive to obtain and will display highly variable performance.

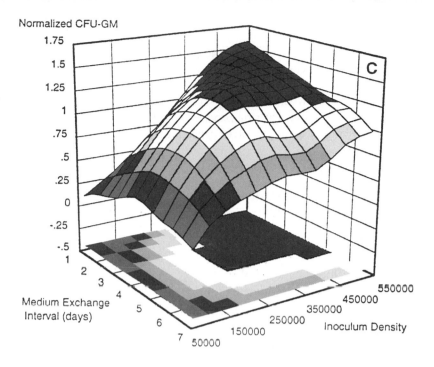

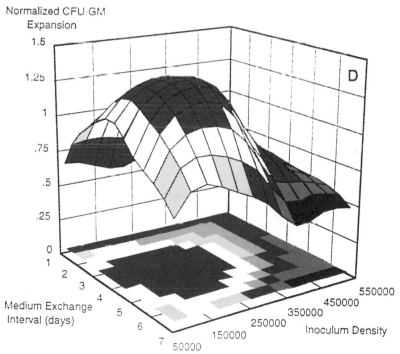

Fig. 12.34 The performance of a bone marrow culture system over a wide range of inoculum densities and medium exchange intervals. Note that the variables interact. The performance (on thez axis) in the top panel is the total number of progenitor cells (CFU-GM) per unit area, whereas in the bottom panel the expansion ratio is depicted (number out divided by number in). Note that the two measures are optimized under vastly different conditions (from Koller *et al.*, 1995c).

Ultimately, the most desirable starting material for cellular therapies is a tissue-specific stem cell whose rate of self-renewal and commitment to differentiation can be controlled. Furthermore, if such a source could be made nonimmunogenic, most of the problems associated with tissue procurement would be solved. The variability in the tissue manufacturing process would be reduced and would make any quality assurance and quality control procedures easier.

12.8.5 Cryopreservation

The scientific basis for cryopreservation is rooted in basic biophysics, chemistry, and engineering. Current cryopreservation procedures are clinically accepted for many tissues, including bone marrow, blood cells, cornea, germ cells, and vascular tissue. Recent experience has shown that, in general, the same procedures cannot be applied to human cells that have been grown *ex vivo*. New procedures need to be invented, developed, and implemented. Cryopreservation used in existing cellular therapies that rely on *ex vivo* manipulation calls for freezing the primary tissue prior to the desired manipulation.

12.8.6 Innoculum Composition and Purging

Some cell populations that are to be transplanted may contain subpopulations of unwanted cells. The primary example of this is the contamination of autologously harvested hematopoietic cell populations with the patient's tumor cells. Ideally, any such contamination should be removed prior to transplantation. Similarly, many biopsies are contaminated with accessory cells that may grow faster than the desired parenchymal cells. Fibroblasts are a difficult contaminant to eliminate in many biopsies and they often show superior growth characteristics in culture. For this reason, it is difficult to develop primary cell lines from many tumor types.

12.9   CONCLUSIONS

Many new cellular therapies are being developed that create challenges for engineering tissue function. To successfully understand tissue function, it must be possible to quantitatively describe the underlying cellular fate processes and manipulate them. Design can be accomplished by designing the physicochemical rate processes so that they match the requirements of the cellular processes that underlie tissue function. Tissue engineering is still in an embryonic stage, but the use of order of magnitude and dimensional analysis is proving to be valuable in designing and reconstituting tissue function. The ability to engineer tissues will undoubtedly markedly improve in the future.

12.10   GLOSSARY

Apoptosis: A cellular process that leads to cell death. This process is initiated by the cell itself.

Cellular therapies: The use of grafted or transfused primary human cells into a patient to affect a pathological condition.

Chondrocytes: Cells found in cartilage.

Colony-forming assay: Assay carried out in semisolid medium under growth factor stimulation. Progenitor cells divide, and progeny are held in place so that a microscopically identifiable colony results after 2 weeks.

Differentiation: The irreversible progression of a cell or cell population to a more mature state.

Engraftment: The attainment of a safe number of circulating mature blood cells after a BMT.

Extracellular matrix: A matrix of large protein molecules.

Flow cytometry: Technique for cell analysis using fluorescently conjugated monoclonal antibodies which identify certain cell types. More sophisticated instruments are capable of sorting cells into different populations as they are analyzed.

Functional subunits: The irreducible unit in organs that gives tissue function, i.e., alveoli in lung and nephron in kidney.

Graft versus host disease: The immunologic response of transplanted cells against the tissue of their new host. This response is often a severe consequence of allogenic BMT and can lead to death (acute GVHD) or long-term disability (chronic GVHD).

Hematopoiesis: The regulated production of mature blood cells through a scheme of multilineage proliferation and differentiation.

Lineage: Refers to cells at all stages of differentiation leading to a particular mature cell type.

Long-term culture-initiating cell (LTC-IC): Cell that is measured by a 7- to 12-week *in vitro* assay. LTC-ICs are thought to be very primitive, and the population contains stem cells. However, the population is heterogeneous so not every LTC-IC is a stem cell.

Mesenchymal cells: Immature cells of connective-type tissue, such as fibroblasts, osteoblasts (bone), chondrocytes (cartilage), and adipocytes (fat).

Microenvironment: Refers to the environment surrounding a given cell *in vivo*.

Mitosis: The cellular process that leads to cell division.

Mononuclear cell: Refers to the cell population obtained after density centrifugation of whole bone marrow. This population excludes cells without a nucleus (erythrocytes) and polymorphonuclear cells (granulocytes).

Myoablatron: The death of all myeloid cells (red, white, and platelets), as occurs in a patient undergoing high-dose chemotherapy.

Parenchymal cells: The essential and distinctive cells of a particular organ (i.e., hepatocytes in the liver or mytocytes in muscle).

Progenitor cells: Cells that are intermediate in the development pathway — more mature than stem cells but not yet mature cells. This is the cell type measured in the colony-forming assay.

Self-renewal: Generation of a daughter cell with identical characteristics as the original cell. Most often used to refer to stem cell division, which results in the formation of new stem cells.

Stem cells: Cells with unlimited proliferative and lineage potential.

Stromal cells: Heterogeneous mixture of support or accessory cells of the bone marrow. Also refers to the adherent layer which forms in bone marrow cultures.

EXERCISES

1. Given the following data, assess whether human hematopoietic stem cells can truly self-renew *in vivo:*

   - About 400 billion mature hematopoietic cells are produced daily
   - Best estimate of the Hayflick *limit* is 44–50
   - About 1:5000 progenitors do not apoptose
   - About 50–1000 mature cells are made per progenitor (6–10 doublings)
   - The entire differentiation pathway may be 17–20 doublings (soft fact)
   - *In vitro* about 10–30 million cells max can be made from a highly purified single hematopoietic stem/preprogenitor cell *in vitro*

   Present the order of magnitude calculations in constructing your decision. Also perform parameter sensitivity analysis of each of the parameters which govern the hematopoietic process. How important are the parameters which govern telemorase activity in determining the total number of mature progeny produced over a person's lifetime.

2. At $1 pM$ concentration, how many molecules are found in a volume of liquid that is equal to the volume of one cell (use a radius of 5 μm)

3. Use the continuum approach (Eq. 12.3) to show that in a steady state the number of cells produced during a differentiation process that involves replication but no apoptosis ($\alpha = 0$) is

$$X_{out}/X_{in} = e^{\mu/\delta}$$

   and is thus primarily a function of the ratio δ/μ. $a = 0$ is the completely undifferentiated state and $a = 1$ is the completely differentiated state. What is X_{out} if μ and δ are the same orders of magnitude and if δ is 10 times slower. Which scenario is a more reasonable possibility in a physiological situation? [Note that if the rates are comparable only two mature progeny will be produced. On the other hand, if the differentiation rate is 10 times slower than the replication rate (probably close to many physiologic situations, i.e., 20-h doubling time, and 200 h = 8 days differentiation time) then about 1 million cells will be produced. Thus, the overall dynamic state tissue is strongly dependent on the relative rates of the cellular fate processes.]

4. Kinetics of differentiation/continuous model: (i) Derive the first-order PDE that describes the population balance; (ii) make time dimensionless relative to the rate of differentiation; (iii) describe the two resulting dimensionless

groups (call the dimensionless group for apotosis A and the one for the cell cycle B); (iv) solve the equation in steady state for $A = 0$; (v) solve the equation(s) where A is nonzero for a portion of the differentiation process, i.e., between a_1 and a_2; and (vi) solve the transient equation for $A = 0$.

5. Consider two cells on a flat surface. One cell secretes a chemokine to which the other responds. Show that the steady-state concentration profile of chemokine emanating from the first cell is

$$C(r) = R^2F/D \times 1/r$$

where R is the radius of the cell, F is the secretion rate (molecules/area time), and D is the diffusion coefficient of the chemokine. The distance from the cell surface is r. Use the cell flux equation to calculate the time it would take for the responding cell to migrate to the signaling cell if there is no random motion ($\mu = 0$) given the following values:

$$\chi(C) = 20 \text{ cm}^2/\text{s} \cdot M$$

$$D = 10^{-6} \text{ cm}^2/\text{s}$$

$$R = 5 \text{ } \mu\text{m}$$

$$\text{Production rate} = 5000 \text{ molecules/cell/s}$$

Hint: Show that the constitutive equation for J reduces to $v = \chi dC/dr$, where v is the velocity of the cell.

6. The flux (F) of a molecule present at concentration C through a circular hole of diameter d on a surface that is adjacent to a fluid that it is diffusing in is given by

$$F = 4DdC$$

The total that can be transferred is the per pore capacity times the number of pores formed. Calculate the flux allowed though each pore if the diffusion coefficient is 10^{-6} cm^2/s the concentration is 1 mM, and the pore diameter is 1.5 nm. Discuss your results and try to estimate how many pores are needed to reach meaningful cell-to-cell communications. With the per pore flux estimated previously, derive the time constant for transfer of a metabolite from a particular cell to a neighboring cell. Assume that these are two epithelial cells whose geometry can be approximated as a box and that the two adjacent boxes are connected with transfer occurring through n pores.

7. If the cellularity in cartilage is about 1 million cells/cm^3, estimate the average distance between the cells. Discuss the characteristics of this microenvironment.

8. Use a one-dimensional analysis of the diffusion of oxygen into a layer of adherent cells to show that the maximum oxygen delivery per unit area (N_{ox}^{max}) in Example Problem 12.8 is given by

$$N_{ox}^{max} = DC^*/R$$

where C^* is the saturation concentration of oxygen and R is the thickness of the liquid layer.

9. Consider a neuron growth cone that is being influenced by a chemoattractant produced by a target cell. The geometry of the model system is shown in the figure. The target cell secretes a chemoattractant at a rate, P_r, which diffuses into a 3-D volume, with a diffusivity D. The governing equation for mass transport for a spherical source is

$$\frac{\partial C}{\partial t} = D \frac{1}{r^2} \frac{\partial}{\partial r}\left(r^2 \frac{\partial C}{\partial R}\right)$$

a. What are the boundary conditions for the system?

b. Considering a steady state, derive the concentration profile as a function of r. It has been found that the growth cone senses a target cell when the concentration difference across the growth cone is higher than 2%. It is believed that growth cones develop filopodia that extend radially out of the cones as a means to enhance their chemosensing ability. Let β_1 be the angle that a filopodia makes with the center radius line, R, and β_2 the angle made by a filopodia extending in the diametrically opposite direction. Filopodia can extend radially from the growth cone surface except where the cone connects to the axon as defined by α.

c. What are the appropriate limits of β_1 and β_2?

d. What are r_1 and r_2 as a function of β_1?

e. What is the percentage change in concentration, $\Delta C = |C_1 - C_2|/C_0$, across the effective growth cone radius, e.g., $R_{gc} + l$, as a function of r_1 and r_2 and consequently β_1?

f. What is β_1^{max}, for which ΔC is maximum. Plot the gradient change for the entire range of β_1 for filopodial lengths of 1, 5, 8 μm compared to no filopodia.

g. The limit for chemosensing ability is a concentration difference of 2%. For a target cell 500 μm away, what is the effect of filopodial length on enhancing chemosensing ability? Calculate for β_1^{max}

$$C_0 = \frac{PrR_t^2}{Dr_0}, \quad Pr = \frac{S_r}{4\pi R_t^2}$$

Where $D = 10^{-6}$ cm²/s, $S_r = 5000$ molecules/cell/s; $R_0 = R - R_{gc}$, $R_t = 20$ μm, $R_{gc} = 2.5$ μm, and $\alpha = 30°$.

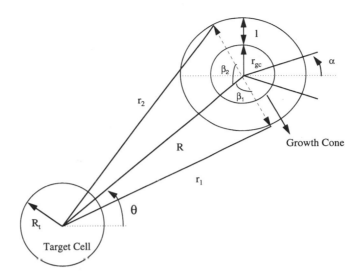

10. Use a compartmental model to calculate the number of mature cells produced from a single cell in a particular compartment. Use a doubling time of 24h $(\ln(2)/\mu)$ and a mature cell half-life of 8 h $(\ln(2)/k_D)$. Assume that self-renewal can only take place in the first compartment with a probability of 0.5. Use a total of 10 compartments and calculate the number of cells as a function of time with the initial conditions.

$$X_i(0) = 0 \text{ except } X_j(0) = 1$$

and vary j between 1 and 8. Plot all curves on the same plot. Discuss the implications of your results for transplantation.

11. Kinetics of differentiation/feedback control in a compartmental model:
    a. Consider a six-stage differentiation process $(N = 6)$, in which the last population, X_6, produces a cytokine, G, at a per cell rate of q_G. This cytokine has a half-life $t_{0.5}$ $(= 2 \text{ h})$ and influences the growth rate of the stem cells, i.e., $u_1 = fn([G])$, where $[G]$ is the concentration of the growth factor G. Extend the base set of differential equations to describe the dynamics of $[G]$.
    b. Incorporate into the equations $u_1([G]) = u/(1 + K[G]$, where K is the binding constant for the growth factor. What does the function $f([G])$ describe physiologically.
    c. Make the equations dimensionless using the growth rate as the scaling factor for time, and K for the cell concentration.
    d. Describe the meaning of the dimensionless groups and estimate their numerical values.
    e. obtain the numerical values for a simulation starting from a single stem cell: examine the effect of varying the numerical values of the parameters:

f. obtain the numerical values for a simulation starting from the steady-state solution and perturb the value of X_3 by 20%. Discuss your results.

SUGGESTED READING

Alberts, B., Bray, D., Lewis, J., Raff, M., Roberts, K., and Watson, J. D. (1994). *Molecular Biology of the Cell*, 3rd ed. Garland, New York.

Brittberg, M. (1994). Treatment of deep cartilage defects in the knee with autologous chondrocyte transplantation. *N. Engl. J. Med.* **331**, 889–95.

Bruder, S. P., Fink, D. J., and Caplan, A. (1994). Mesenchymal stem cells in bone development, bone repair, and skeletal regeneration therapy. *Cell. Biochem.* **56**, 283–294.

Fawcett, D. W. (1986). *A Textbook of Histology*, 11th ed. Saunders, Philadelphia.

Garby, L., and Meldon, J. (1977). *The Repiratory Functions of Blood.* Plenum, New York.

Granick, S., and Levere, R. (1964). Heme synthesis in erythroidal cells. In *Progress in Hematology* (C. Moore and E. Brown, Eds.), p. 1. Grune & Stratton, Orlando, FL.

Harley, C. B., Futcher, A. B., and Greider C. W. (1990). Telomeres shorten during aging of human fibroblasts. *Nature* **345**, 458–460.

Hoffbrand, A., and Pettit, J. (1988). *Clinical hematology.* In *Clinical Hematology* (A. Hoffbrand and J. Pettit, Eds.), p. 21. Gower Medical, London.

Iredale, J. P. (1997). Tissue inhibitors of metalloproteinases in liver fibrosis. *Int. J. Biochem. and Cell Biol.* **29**, 43–54.

Knauer, D. J., Wiley, H. S., and Cunningham, D. D. (1984). Relationship between epidermal growth factor receptor occupancy and mitogenic response: Quantative analysis using a steady state model system. *J. Biol. Chem.* **259**(9), 5623–5631.

Koller, M. R., and Palsson, B. O. (1993). Tissue engineering: Reconstitution of human hematopoiesis *ex vivo*. *Biotechnol. Bioeng.* **42**, 909–930.

Koller, M. R., Palsson, M. A., Manchel, I., and Palsson, B. O. (1995). Long-term culture-initiating cell expansion is dependent on prequent medium exchange combined with stromal and other accessory cell effects. *Blood* **86**, 1784–1793.

Koller, M. R., Bradley, M. S., and Palsson, B. O. (1995b). Growth factor consumption and production in perfusion cultures of human bone marrow correlates with specific cell production. *Exp. Hematol.*, **23**, 1275–1283.

Koller, M. R., Manchel, I., Palsson, M. A., Maher, R. J., and Palsson, B. O. (1995c). Different measures of *ex vivo* hematopoiesis culture performance is optimized under vastly different conditions. *Biotechnol. Bioeng.* **50**, 505–513.

Krstic, R. (1985). english edition, *General Histology of the Mammal*, Springer Verlag, NY, page 83, plate 40. Translated from the original German addition, Krstic, R. (1978) *Die Gewebes des Menschen und der Saugertiere*, Springer Verlag, Berlin.

Lacey, P. E., (1995, July). Treating diabetes with transplanted cells. *Sci. Am.*, 50–58.

Langer, R., and Vacanti, J. P. (1993). Tissue engineering. *Science* **260**, 920–926.

Lauffenburger, D. A., and Linderman, J. J. (1993). *Receptors: Models for Binding, Trafficking, and Signaling.* Oxford Univ. Press, New York.

Lightfoot, E. N. (1974). *Transport Phenomenon and Living Systems.* Wiley, New York.

Lodish, H., Baltimore, D., Berk, A., Zipursky, S. L., Matsudaira, P., and Darnell, J. (1995). *Molecular Cell Biology*, 3rd ed. Scientific American, New York.

Mutsaers, S. E., Bishop, J. E., McGrouther, G., and Laurent, G. J. (1997). Mechanisms of tissue repair: From wound healing to fibrosis. *Int. J. Biochem. Cell Biol.* **29**(1), 5–17.

Naughton, B. (1995). The importance of stroma cells. In *CRC Handbook on Biomedical Engineering* (J. D. Bronzino, Ed.), pp. 1710–1727. CRC Press, Boca Raton, FL.

Peng, C. A., and Palsson, B. O. (1996a). Determination of specific oxygen uptake rates in human hematopoietic cultures and implications for bioreactor design. *Ann. Biomed. Eng.* **24,** 373–381.

Peng, C., and Palsson, B. O. (1996b). Cell growth and differentiation on feeder layers is predicted to be influenced by bioreactor geometry. *Biotechnol. Bioeng.* **50,** 479–492.

Rhodin, J. A. G. (1974). *Histology: Text and Atlas,* Oxford University Press, Oxford.

Rogers, C. E., Bradley, M. S., Palsson, B. O., and Koller, M. R., (1995). Flow cytometric analysis of human bone marrow perfusion cultures: Erythroid development and relationship with burst-forming units-erythroid (BFU-E). *Exp. Hematol.* **24,** 597–604.

Sigal, S., and Brill, S. (1992). Invited review: The liver as a stem cell and lineage system. *Am. J. Physiol.* **263,** 139–148.

Vander, A. J., Sherman, J. H., and Luciano, D. S. (1994). *Human Physiology: The Mechanisms of Body Function,* 6th ed. McGraw–Hill, New York.

13   BIOTECHNOLOGY

Chapter Contents

At the conclusion of this chapter, the reader will be able to:

■ Define biotechnology and important terms that are used in biotechnology

■ Briefly describe important events in the history of biotechnology

■ Discuss the relationship between DNA, mRNA, and proteins

■ Briefly describe the basic techniques that are used in biotechnology

■ List and briefly describe the core technologies that are used in biotechnology

■ Briefly describe one therapeutic use of biotechnology

■ Briefly describe the Human Genome Project and its potential effects on health care

13.1   INTRODUCTION

The term biotechnology was recently created by combining the Greek words for life (*bios*) and systematic treatment (*technologia*), but the systematic treatment of biological systems by humans is not new. It is at least as old as the domestication of plants and animals that occurred 10,000 years ago. Humans have used microbes such as yeasts and bacteria to make food products such as wine, beer, cheese, bread, and yogurt for thousands of years. Within the past 100 years, microbes have been used to improve crop yields by inoculating the soil with nitrogen-fixing bacteria. Within the past five decades, microbes have been used to produce antibiotics, vaccines, and industrial enzymes. What is really "new" about biotechnology is the use of cells and biological molecules to solve problems and make useful products.

The National Research Council has defined biotechnology as techniques that use living organisms to make or modify products, improve plants or animals, and develop microorganisms for specific purposes. Biotechnology is a collection of technologies that are used to produce therapeutic drugs, to improve disease resistance and production in both plants and animals, to remediate toxic environmental sites, to produce enzymes and other chemical products, and to correct genetic defects in humans.

13.1.1   History

The roots of biotechnology date back to 1868 when Johann Friedrich Miescher isolated "nuclein" from the nuclear material of fish sperm and cells found in open wounds. He selected these cells because they were composed mostly of nuclear material, but the importance of his discovery was not recognized at the time. In 1928, Fred Griffith found that mice injected with heat-killed strains of bacteria that caused pneumonia would not develop the disease, whereas mice injected with harmless bacteria that had been allowed to grow in the presence of heat-killed pneumonia-causing bacteria died from pneumonia. He concluded that hereditary information had been transferred from the heat-killed pathogenic strain to the harmless strain and that this hereditary information transformed the harmless strain into disease-causing bacteria.

Although Miescher had isolated DNA (Table 13.1) from the nucleus and Griffith had shown that hereditary information could be transferred from one organism to another, they did not know whether DNA actually carried the hereditary information or whether it was carried by the complex proteins associated with DNA. Many scientists in the first half of the twentieth century favored proteins as the source of hereditary information. They reasoned that the informational blueprint specifying the complexity seen in organisms could only be encoded in very complex molecules. DNA seemed too simple for this task when compared to the complex chemical composition of proteins. It was not until 1944 that Oswald Avery and associates reported that the hereditary substance in the extracts used by Griffith was probably DNA and not proteins and that the material which transformed the bacteria from a nonpathogenic to a pathogenic form could be passed from one generation to the next. In 1952, Alfred Hershey and Martha Chase used radioactive isotopes of sulfur (found in proteins but not DNA) and phosphorus (found in DNA but not proteins) in an elegant experiment involving bacteria and bacteriophages (bacterial viruses) to confirm that hereditary information was passed in DNA and not proteins.

In 1953, Francis Crick and James Watson, with the help of X-ray crystallographers Rosalind Franklin and Maurice H. F. Wilkins, proposed a structural model for

TABLE 13.1 Important Abbreviations Used in Biotechnology

aRNA	Antisense RNA
BAC	Bacterial artificial chromosome
bp	Base pair (A–T, C–G)
cDNA	Complementary DNA formed from mRNA
DNA	Deoxyribonucleic acid
ELSI	Ethical, legal, and social issues
FISH	Fluorescence *in situ* hybridization
HGMIS	Human Genome Management Information System
HGP	Human Genome Project
kb	1,000 bp
Mb	1,000,000 bp
MAb, MCAb	Monoclonal antibody
mRNA	Messenger RNA
PAC	P1-derived artificial cloning system
PCR	Polymerase chain reaction
rDNA	Recombinant DNA
RFLF	Restriction fragment length polymorphism
RNA	Ribonucleic acid
STS	Sequence tagged sites
YAC	Yeast artificial chromosome

DNA that gave the macromolecule a helical, double-stranded form and predicted the mechanism by which hereditary information could be passed from a dividing cell to its offspring. The model proposed by Watson and Crick could not have been developed without contributions from Edwin Chargaff, a chemist who discovered that DNA molecules always have equal amounts of adenine and thymine and equal amounts of cytosine and guanine, and Linus Pauling, another chemist who described the rules that govern the formation of chemical bonds and published the details of a novel chemical structure called the α-helix. As shown in Fig. 13.1C, double-stranded DNA consists of two chains of paired nucleotides [adenine (A), guanine (G), cytosine (C), and thymine (T)] that wind into a double α-helix (see Fig. 2.3). Each chain of DNA nucleotides consists of several single nucleotides (Fig.

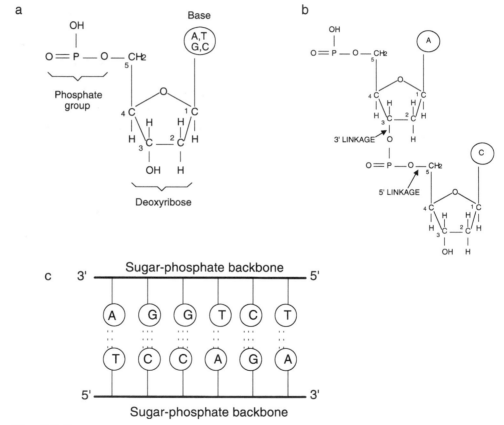

Fig. 13.1   (a) A single nucleotide of DNA consists of a phosphate group, a sugar (deoxyribose), and a base (A, adenine; T, thymine; C, cytosine; G, guanine). The carbon atoms of the deoxyribose are numbered according to chemical convention. (b) A chain of DNA nucleotides consists of several single nucleotides that are joined by phosphodiester bonds at their 3′ and 5′ carbons. (c) Hydrogen bonds (dashed lines) link complementary bases to form the double helix, which is shown as a flat molecule. One strand of DNA has sugar–phosphate linkages in the 3′ to 5′ direction, whereas the complementary strand has sugar–phosphate linkages in the 5′ to 3′ direction.

13.1A) that are joined by bonds at their 3' and 5' carbons (Fig. 13.1B). Each set of complementary nucleotides in a double-stranded DNA molecule (A–T, C–G) is called a base pair (bp). The base pairs in complementary strands of DNA are bonded together by hydrogen bonds. One strand is pointed in the 3' to 5' direction, whereas the complementary chain is pointed in the 5' to 3' direction (Fig. 13.1C).

Meselson and Stahl's description of DNA replication in *Escherichia coli* and the determination of the genetic code in 1961 followed the description of DNA's structure in 1958 by Nirenberg and Matthaei. The process of DNA replication that results in two identical strands of DNA during cell division was described in Chapter 2 (see Fig. 2.13). In active cells, stretches of DNA called genes can be used as a template to synthesize strands of messenger RNA (mRNA) that move into the cytoplasm and provide the genetic blueprint for protein biosynthesis. Information flows from DNA to RNA during a biosynthetic process called transcription (see Fig. 2.14) and from RNA to proteins through the complex mechanism called translation (see Fig. 13.2). Each gene provides the code for a single protein by means of mRNA or the code for ribosomal or transfer RNA (rRNA and tRNA, respectively).

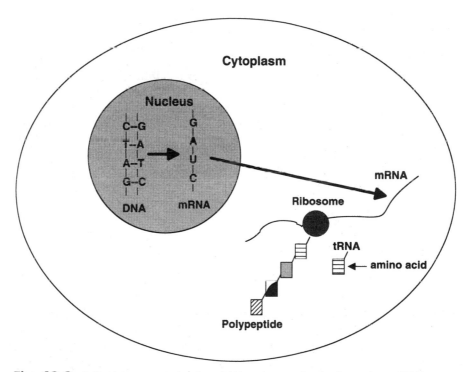

Fig. 13.2 Following transcription from DNA and processing in the nucleus, mRNA moves from the nucleus to the cytoplasm. In the cytoplasm, the mRNA joins with a ribosome to begin the process of translation. During translation, tRNA delivers amino acids to the growing polypeptide chain. Which amino acid is delivered depends on the three-base codon specified by the mRNA. Each codon is complementary to the anticodon of a specific tRNA. Each tRNA binds to a particular amino acid at a site that is opposite the location of the anticodon. For example, the codon CUG in mRNA is complementary to the anticodon GAC in the tRNA that carries leucine and will result in adding the amino acid leucine to the polypeptide chain.

Biotechnology in terms of the manipulation of the genetic content of cells became possible after the discovery and use of restriction endonucleases from bacteria in 1970 and the development of plasmid technology in 1973. Restriction endonucleases cleave DNA molecules at specific sites that contain particular short sequences of bases. These enzymes can be used to open plasmids (small, self-replicating pieces of circular DNA that are found in bacteria). Ligases, enzymes that join the ends of two DNA molecules together, are then used to insert new stretches of DNA into the open plasmids and to reclose them. With these techniques, DNA could be cut up into relatively small sequences of several hundred to several thousand bps that could be analyzed individually, and recombinant cells with DNA from other organisms could be created. The emergence of rDNA and hybridoma (which involves the fusion of antibody-producing cells with fast-dividing cancer cells) technologies helped start the biotechnology revolution.

Advances in biotechnology have progressed exponentially since the early 1970s. In 1975, E. M. Southern developed the first technique for detecting specific sequences of DNA. Other methods that enabled investigators to look at the sequences of bases in DNA were also developed in the 1970s. In the late 1970s, restriction fragment length polymorphisms (RFLPs) were used to study the mutation that causes sickle cell anemia. In the mid-1980s, scientists from the Cetus Corporation, a biotechnology company, described the polymerase chain reaction (PCR), which made it possible to produce useful amounts of DNA molecules of up to 1 million bps from very tiny starting amounts.

All of the hereditary material for a cell makes up its genome. Each bacteria cell has a single circular DNA molecule that contains about 1 million bps (10^6 bp = 1 Mb). *Escherichia coli,* which is often used in bacterial production systems, contains approximately 2000 genes, each of which is about 1200 bp long. The genome for *E. coli* is about 4×10^6 bp, whereas more complicated organisms, e.g., baker's yeast (*Saccharomyces cerevisiae*), have larger genomes (5800 genes in 13.5×10^6 bp). The haploid genomes for the fruit fly, mouse, and human are even larger — 1.65×10^8 (12,000 genes), 2.3×10^9, and 3.3×10^9 bp, respectively. The actual genome sizes for the fruit fly, mouse, and human are twice the given numbers because each cell is diploid and contains two sets of chromosomes (see Fig. 2.12).

PCR made efficient analysis of the human genome possible so the Human Genome Project (HGP) was begun in 1990. The ultimate goal of the HGP is to determine the base sequence of the human genome. While physical mapping of the 6.6×10^9 bp in the human genome is currently not possible, it is possible to identify and characterize all human genes since they represent approximately 3% of the total genome. As of 1994, about 5% of the estimated 80,000 genes had been mapped, including at least 770 genes that were associated with clinical abnormalities. These disease-causing genes were found by detecting abnormal sequences in affected individuals.

13.1.2 Proteins and Genes

Proteins consist of long chains of amino acids (Table 13.2), which are small organic compounds with an amino group ($-NH_2$), a carboxyl group ($-COOH$), and one or more atoms in a side group, designated R (Fig. 13.3). The R groups, which may

TABLE 13.2 Amino Acids in Proteins

Amino acid	Three-letter code	One-letter code
Alanine	Ala	A
Arginine	Arg	R
Asparagine	Asn	N
Aspartic acid	Asp	D
Cysteine	Cys	C
Glutamic acid	Glu	E
Glutamine	Gln	Q
Glycine	Gly	G
Histidine	His	H
Isoleucine	Ile	I
Leucine	Leu	L
Lysine	Lys	K
Methionine	Met	M
Phenylalanine	Phe	F
Proline	Pro	P
Serine	Ser	S
Threonine	Thr	T
Tryptophan	Trp	W
Tyrosine	Tyr	Y
Valine	Val	V

be uncharged or positively or negatively charged and which vary from a single hydrogen atom (glycine) to carbon chains (lysine) to complex ring structures (phenylalanine and tyrosine) and may include sulfur (cysteine and methionine), give each amino acid its distinctive properties. Chains that contain three or more amino acids joined by peptide bonds that link the amino group of one amino acid and the carboxyl group of another are called polypeptides.

Insulin (see Fig. 2.40; Fig. 13.4) was the first protein to be sequenced biochemically. It has a molecular weight of approximately 10 KDa (1 Da = 1/16 the mass of an oxygen atom, $\sim 1.65 \times 10^{-24}$ g) and consists of a single string of about 100 amino acids. A protein of this size is technically called a polypeptide, or peptide, and is considered too small and simple to be in the "protein" category. By contrast, antibodies, which the body uses to fight off infections, can be quite large and complex and often consist of two or more polypeptides that fold around each other to pro-

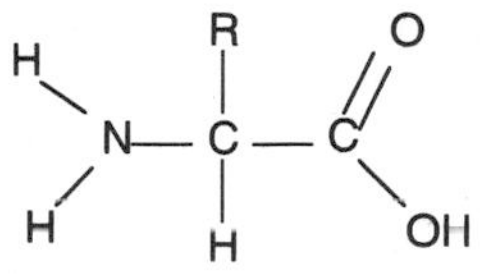

Fig. 13.3 The basic structure of an amino acid consists of an amino group ($-NH_2$), a carboxyl group ($-COOH$), and a side chain (R). The R group gives each amino acid its distinguishing physicochemical properties, e.g., charged, uncharged, acidic, and basic.

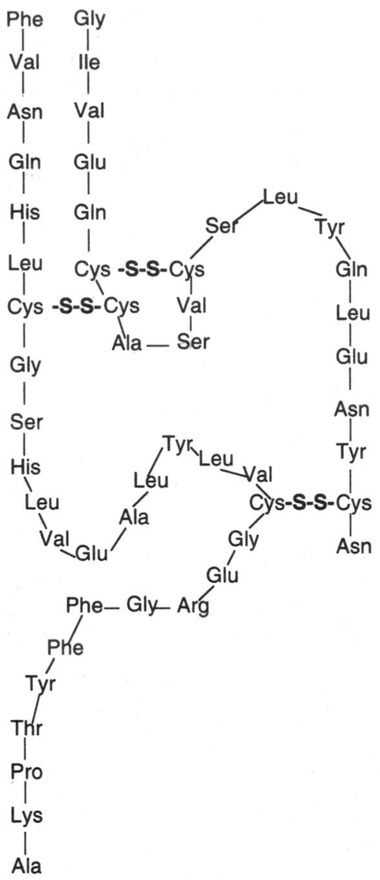

Phe   Gly
Val   Ile
Asn   Val
Gln   Glu
His   Gln   Leu
Leu   Ser   Tyr
Leu   Cys -S-S-Cys   Gln
Cys -S-S-Cys   Val   Leu
         Ala — Ser   Glu
Gly   Asn
Ser
His   Tyr Leu   Tyr
Leu   Leu   Val
     Ala   Cys-S-S-Cys
Val Glu   Gly   Asn
          Glu
Phe— Gly— Arg
Phe
Tyr
Thr
Pro
Lys
Ala

Fig. 13.4 Bovine insulin consists of two polypeptide chains that are joined by two disulfide bonds (-S–S-). Hydrogen bonds also exist between the chains and between segments of the same chain. The three-letter names stand for different amino acids (see Table 13.2).

duce a three-dimensional structure that then must be modified, through the addition of sugar residues, to achieve full activity. Hemoglobin, which transports oxygen in the blood, consists of four globin protein molecules. Each globin protein molecule is a polypeptide chain with an iron-containing heme group (Fig. 13.5). Myoglobin, a similar molecule that stores oxygen in muscles, is composed of one heme-containing polypeptide chain.

Proteins have several structural levels. The primary structure consists of the specific sequence of amino acids and gives rise to the protein's shape and function, i.e., whether it will act as an enzyme, a receptor on a cell's membrane, a channel through a cell's membrane, or an antibody. The protein's secondary structure, i.e., whether it is coiled into a helix or in sheet-like formations or both, depends on the hydrogen bonds that form between different amino acids in the chain. The R groups of the

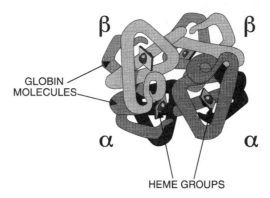

Fig. 13.5 Hemoglobin is a complex protein that consists of four globin polypeptide chains.

amino acids in the chain also interact and cause further folding of the protein that gives rise to its tertiary structure. Folding results in molecules with the hydrophobic parts in the center and the hydrophilic parts on the outside in aqueous solution. Quartenary structure occurs in molecules such as hemoglobin which are protein molecules that consist of two or more polypeptide chains that are held together by hydrogen bonds and other weak interactions.

Genes or DNA provide the blueprints for the cellular production of proteins. The structural portion of the gene consists of a linear sequence of nucleotide bases that specifies the linear sequence of amino acids in a peptide. A copy of the gene, a mRNA molecule, is made and moves to the cytoplasm where peptide synthesis occurs (Fig. 13.2). The ribosomes and associated tRNAs and translational regulatory proteins form the complex cellular machinery that can read the nucleotide message of the mRNA and produce the corresponding linear series of amino acids that are bonded to each other through peptide bonds to form a peptide (two amino acids) or polypeptide.

Genes are organized into sequence subsections that have different functions (Fig. 13.6). Some of these subsections directly affect transcription, whereas others affect translation. Of first importance in translation is the coding sequence, the actual DNA sequences that provide the blueprint (Table 13.3) for specifying the amino acid sequence of the encoded polypeptide. Upstream in the 5′ direction relative to the coding region are the regulatory sequences that are collectively known as the promoter. These sequences control the level of gene expression, the tissues and cell types in which the gene can be expressed, and the environmental conditions under which the gene can be active. The promoter region controls the expression of the gene by regulating transcription of the gene into a mRNA molecule, but the promoter itself is not transcribed. The coding region of DNA that is actually transcribed is used as a template to produce a mRNA with a complementary nucleotide sequence. Promoters and the coding region account for the majority of nucleotides in a gene.

There are also sequences within a gene that control posttranscriptional processing. The mRNA that is transcribed in bacteria (prokaryotes) is the molecule that is

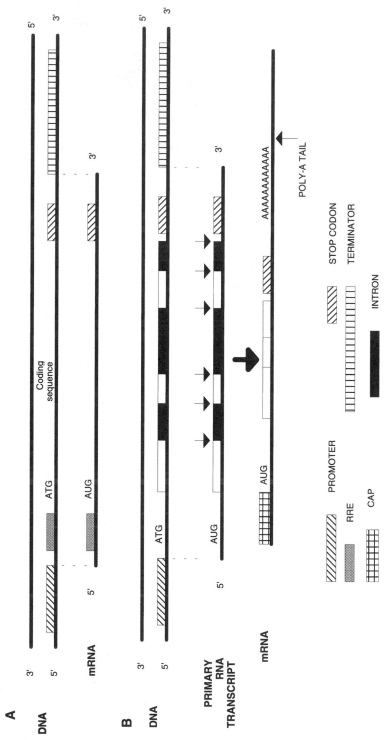

Fig. 13.6 (A) In prokaryotic cells (bacteria), a typical promoter sequence consists of 5'-TTGACA-17 bases-TATATT-3'. The ribosomal recognition element (RRE) often has a sequence of 5'-GAAGG-3' or 5'-AGGA-3' and is located 8–13 nucleotides upstream of the initiation codon (5'-ATG-3'). The terminator sequence stops transcription. The stop codon (UAG, UGA, or UAA in mRNA) stops translation. (B) In eukaryotic cells, the promoter region contains the sequence TATA, CAAT, or GC that is about 60 bases away from where transcription begins in mammals but only 30 bases away in yeast. Before the promoter sequence, eukaryotic DNA that is to be transcribed must have an enhancer sequence (not shown) that makes it possible for RNA polymerase to bind to the DNA. The primary RNA transcript, or pre-mRNA, is converted to mRNA by removal of the introns and addition of a cap (7-methyl guanosine) and a poly-A tail that consists of 100–200 adenines. An initiation region (not shown) precedes the initiation codon (AUG).

TABLE 13.3 The Genetic Code (mRNA)

First base	Second base				Third base
	A	**G**	**U**	**C**	
A	Lysine	Arginine	Isoleucine	Threonine	A
	Lysine	Arginine	Methionine[a]	Threonine	G
	Asparagine	Serine	Isoleucine	Threonine	U
	Asparagine	Serine	Isoleucine	Threonine	C
G	Glutamic acid	Glycine	Valine	Alanine	A
	Glutamic acid	Glycine	Valine	Alanine	G
	Aspartic acid	Glycine	Valine	Alanine	U
	Aspartic acid	Glycine	Valine	Alanine	C
U	Termination	Termination	Leucine	Serine	A
	Termination	Tryptophan	Leucine	Serine	G
	Tyrosine	Cysteine	Phenylalanine	Serine	U
	Tyrosine	Cysteine	Phenylalanine	Serine	C
C	Glutamine	Arginine	Leucine	Proline	A
	Glutamine	Arginine	Leucine	Proline	G
	Histidine	Arginine	Leucine	Proline	U
	Histidine	Arginine	Leucine	Proline	C

[a] Initiation when located at the leading end of the mRNA.

used as a template during translation (Fig. 13.6A). In higher organisms (eukaryotes, e.g., fungi, mammals, other animals, and plants), the mRNA is much longer and is actually called a primary RNA transcript or pre-mRNA. In eukaryotes, the coding sequence is not contiguous but comes in short sections (exons) that are separated by spacer regions (introns). During posttranscriptional processing, the introns are cut out of the pre-mRNA and the exons are spliced together to create the contiguous coding sequence that is used for translation (Fig. 13.6B). This greatly shortens the mRNA. Another modification of posttranscriptional processing involves the addition of a tail at the 3′ end of the shortened mRNA and a cap at the 5′ end. The tail, called a poly-A tail, consists of a sequence of 100–250 adenine nucleotides that are added to the mRNA, and the cap consists of 7-methyl guanosine. The cap appears to have a role in allowing the ribosome to bind to the mRNA so that protein translation can begin, and the poly-A tail plays a role in determining the number of times a given mRNA will be translated.

Once posttranscriptional processing is completed, the mRNA moves from the nucleus to the cytoplasm where other nucleotide sequences and the sequences for coding become important. Upstream of the coding region are initiation sequences that bind and align the protein synthesis machinery of the cells, the ribosomes, to the mRNA. The nucleotide sequence on the initiation region controls the efficiency and frequency of ribosome binding and thus controls the number of polypeptides that can be made from one mRNA molecule. These initiation sequences are not translated. The coding region starts with a special codon, AUG, which signals the first amino acid of the polypeptide, methionine. Due to posttranslational processing, the initial amino acid of the polypeptide that results from translation may not be the first amino acid in the active protein that is assembled at the appropriate subcellular location. There can be a leader sequence in the coding region that results in a short string of amino acids at the beginning of the peptide. This leader sequence acts as an address label that directs the protein to its proper subcellular location after translation, peptide folding, and protein assembly have occurred. The leader sequence is cut off the protein once it arrives at its destination. The termination sequence that causes detachment of the ribosome and termination of polypeptide synthesis is located at the end of the coding region. The termination sequence controls the efficiency of ribosome detachment, thereby controlling the rapidity by which ribosomes can move down the mRNA molecule. The termination sequence also controls the number of polypeptides that can be made from a single mRNA molecule. Finally, the length of the poly-A tail is directly proportional to how many times the mRNA is translated, i.e., the half-life of the mRNA. Messenger RNAs with short half-lives have short poly-A tails and do not produce as many copies of the polypeptide as do mRNAs with long half-lives.

13.2   BASIC TECHNIQUES

Much of the work that is done in biotechnology depends on knowing specific information about DNA. Restriction endonucleases from bacteria provided the first tool for cutting DNA into fragments with known ends and enabled scientists to form plasmid- and virus-based DNA libraries of entire genomes. Gel electrophoresis made it possible to sort fragments by size, whereas several techniques, including Southern blotting, allowed scientists to extract the DNA fragments from the gels. With PCR, thousands of DNA copies can be made from a single fragment in a short period of time. The development of DNA sequencing techniques enabled researchers to determine the exact sequence of nucleotides in DNA strands, and the discovery of reverse transcriptases allowed scientists to synthesize DNA molecules complementary to isolated mRNAs, the so-called cDNA molecule.

13.2.1   Restriction Endonucleases

Restriction endonucleases, also called restriction enzymes, are DNA-cutting proteins that defend bacterial cells from invading bacteriophages by cutting the bacteriophage's DNA into nonfunctional pieces. These enzymes recognize specific sites in DNA and cleave the molecule there. Many restriction enzymes recognize palindromic sequences, i.e., sequences that read the same in the opposite directions on the two DNA strands. For example, the restriction enzyme *Eco*RI (so named because it was purified from *E. coli*) recognizes GAATTC and cuts the DNA between G and A. Since the complementary strand will read CTTAAG and *Eco*RI will also cut it between G and A, the two pieces of DNA will have "sticky ends" that consist of AATTC and CTTAA (Fig. 13.7). *Hin*dIII cuts AAGCTT between AA, and *Bam*HI cuts GGATCC between GG. *Alu*I leaves blunt ends by cutting AGCT between the G and C. *Sma*I also leaves blunt ends when it cuts CCCGGG between C and G.

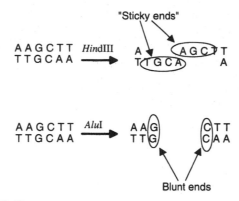

Fig. 13.7 Restriction enzymes are used to cut specific sequences of DNA. In this example, *Hind*III cuts the following sequence between the two As and leaves "sticky ends" that consist of single strands of DNA. *Alu*I leaves blunt ends because no lengths of single strands are formed by the cut through the DNA.

Example Problem 13.1

DNA with the following sequence is mixed with *Eco*RI and *Sma*I. What fragments will be produced?

```
ATTCGGAATTCTTCCCGGAAGGGCCCTTATATAGAAGCTTCCGGATTATT
TAAGCCTTAAGAAGGGCCTTCCCGGGAATATATCTTCGAAGGCCTAATAA
```

Solution

*Eco*RI recognizes GAATTC and cuts between G and A. *Sma*I recognizes CCCGGG and cuts between C and G. The following fragments will be produced:

```
ATTCGG            AATTCTTCCCGGAAGGG
TAAGCCTTAA           GAAGGGCCTTCCC
```

```
CCCTTATATAGAAGCTTCCGGATTATT
GGGAATATATCTTCGAAGGCCTAATAA
```

Restriction enzymes have made it possible for scientists to cut large molecules of DNA into smaller pieces that are easier to study since the same fragments are produced every time the same DNA is cut with the same restriction enzyme(s). Since some restriction enzymes produce fragments with sticky ends, rDNA can be made by attaching similar sticky ends from two different DNA molecules.

13.2.2 Gel Electrophoresis

The segments of DNA that result from cleavage by restriction enzymes can be separated by a process called gel electrophoresis (Fig. 13.8). Electrophoresis is a biochemical technique that is used to separate charged molecules in solution. For ease of handling and to allow separation by molecular size, the aqueous solution used to suspend DNA is gelled. The polysaccharide, agarose, is often used as the gelling agent. The synthetic polymer acrylamide is used as the gel for the smallest DNA pieces.

Gel electrophoresis is carried out by placing a thin slab of gel in a chamber filled with a buffer solution. The DNA samples are then placed in wells that have been made at one end of the gel. The chamber has electrodes so that an electric charge can be applied across the gel. The cathode (negative electrode) is at the end with the wells, and the anode (positive electrode) is at the far end of the gel. Since DNA is a negatively charged molecule due to its many phosphate groups, it will migrate through the gel from the cathode toward the anode. The rate at which these linear fragments of DNA migrate is inversely proportional to the $\log_{10}$ of the molecular weight of the fragment since the charge-to-mass ratio of DNA is independent of fragment length. In a given period of time, relatively short fragments will migrate farther than relatively long fragments. By including DNA standards (i.e., DNA that

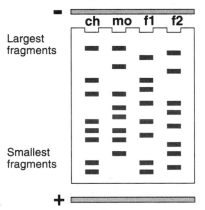

Fig. 13.8 Gel electrophoresis is used to separate fragments of DNA based on their molecular weights. The first well (ch) contains a DNA sample from a child. The sample was cleaved into fragments of various lengths by one or more restriction enzymes. The remaining wells (mo, f1, and f2) contain samples of DNA from the child's mother and two men who could have been the child's father. The samples from the mother and two possible fathers were exposed to the same restriction enzymes. An electric field was applied for a period of time long enough for the smallest fragments to migrate to the far end of the gel. The gel was stained to expose the DNA. The location of the fragments indicates that f2 was not the father of the child and f1 most likely was since four fragments from f1 match those of the child, whereas no fragments from f2 matched.

has been separated into several different fragments with known numbers of base pairs) in one well of the gel, it is possible to calculate the size of an unknown DNA fragment based on its migration rate. After electrophoresis, the agarose gels can be stained using ethidium bromide, a fluorescent dye that chelates into the DNA structure, to show the location of the DNA fragments. The fragments can be removed from the gel for additional study or to provide purified extracts.

Example Problem 13.2

The *E. coli* plasmid cloning vector pBR322 is used to move genes into *E. coli*. It has 4361 bps with the first T in the sequence ⋯ GAATTC ⋯ representing an *Eco*RI site designated as nucleotide number 1 (Fig. 13.9). The plasmid can also be cut by *Hind*III after nucleotide 29, by *Hinc*II after nucleotides 651 and 3905, and by *Bam*HI after nucleotide 375. What will be the lengths of the fragments that would be created when pBR322 is exposed to *Eco*RI, *Hind*III, *Hinc*II, and *Bam*HI? In what order will they appear if the mixture is subjected to gel electrophoresis?

Solution

*Eco*RI cuts the 4361-bp sequence between the G and A in GAATTC, i.e., between 4359 and 4360 since the first T is designated as the first nucleotide. The next cut will be done by *Hind*III after 29 so the first fragment (A) will be 31 bps long — 29 + 2 (4360 and 4361). *Bam*HI makes the next cut after 375 so this fragment (B) will be 375 − 29, or 346, bps in length. The next cut is made by *Hinc*II after 651 and will result in a fragment (C) that is 651 − 375, or 276, bps long. *Hinc*II also cuts after 3905. This will produce one 3254-bp fragment (D: 3905 − 651) and another (E) that is 454 bps in length (4359 − 3905). The final fragments will have 31 (A), 346 (B), 276 (C), 3254 (D), and 454 (E) bps. During gel electrophoresis, A will travel the longest and D will travel the shortest distance with D–E–B–C–A as the final sequence from top to bottom of the gel.

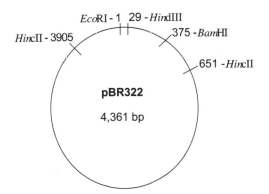

Fig. 13.9 The *E. coli* plasmid cloning vector, pBR322, is 4361 bp in length and can be cut by several different restriction enzymes. The locations of the restriction sites for *Eco*RI, *Hind*III, *Hinc*II, and *Bam*HI are shown.

13.2.3   Southern Blotting

In Southern blotting, DNA is cut into short pieces and then separated by gel electrophoresis. Following electrophoresis, the double-stranded DNA in the bands is denatured (unwound) into single-stranded fragments by soaking the gel in a solution of sodium hydroxide. The solution is then brought back to a neutral pH, and the entire gel is placed against a sheet of nitrocellulose paper and a stack of filter paper. The stack of filter paper acts as a wick to draw a salt solution through the gel. Since the salt solution brings the DNA strands with it as it moves through the gel and the filter paper, each DNA fragment ends up on the nitrocellulose paper in the same relative position that it occupied in the gel. Known sequences of DNA, labeled with radioactive nucleotides, are then placed in the solution that surrounds the nitrocellulose paper. The known sequences bind to complementary sequences of the DNA on the nitrocellulose paper. The locations of the radioactively labeled DNA sequences can be detected by autoradiography, a technique that involves putting the nitrocellulose paper on undeveloped photographic film. Areas that contain the radioactive label develop the adjacent film and indicate the location of the known sequences of DNA in the original gel.

13.2.4   Restriction Fragment Length Polymorphisms

Restriction fragment length polymorphisms provide a method that is used to cut DNA by one or more restriction enzymes and then investigate the lengths of the resulting fragments. During RFLP analysis of human DNA, for example, the DNA is cut into thousands of fragments which are separated into a continuous distribution of different sizes by means of gel electrophoresis. The DNA is then transferred by blotting to absorbent paper or a membrane where it sticks in the same smear of individual bands. The blotted DNA is next treated so that it unwinds and separates into individual strands. If the DNA blot is immersed in a solution containing a single-strand DNA molecule, or probe, whose sequence is complementary to one of the single-stranded DNA fragments on the blot, the probe will hybridize to form a double-stranded DNA and will stick to the blot. After unhybridized probe is removed, the probe DNA — blot DNA hybrids can be visualized by means of autoradiography if the original probe DNA was radioactively labeled or by fluorescence if the original probe was labeled with a chemiluminscent compound.

DNA from different individuals varies in the number and locations of sites that react with different restriction enzymes and with different probes. One individual can have restriction sites that do not appear in another individual. When DNA from different individuals is subjected to restriction enzymes, gel electrophoresis, and probes, different patterns emerge. This method can be used to establish the identity of individuals who have left samples of their DNA at a crime scene (e.g., in blood) to establish paternity, or to exonerate people who have been accused of a crime. For example, each child has some patterns that resemble those of the mother and some that resemble those of the father since half of the child's DNA came from the mother and half from the father. RFLPs can be used to indicate that a given male could not have been the father of a specific child or to indicate that another male could have been the father (Fig. 13.8). In other cases, the appearance of a particular band may be correlated with the expression of a particular disease but be absent in those individuals who do not have the disease. This method is being developed with the goal of creating each person's individual DNA fingerprint that can serve as identification and as a diagnostic tool to determine susceptibility to diseases and sensitivity to environmental toxins.

13.2.5 DNA Libraries and Gene Isolation

DNA libraries consist of a collection of DNA fragments that have been produced by restriction enzymes and inserted into plasmids or other cloning tools such as viruses. Viruses that are used for this purpose self-replicate after they have invaded a bacterial cell. The overall method has been designed to ensure that only a single virus invades a given bacterium. Lambda phage, for example, is a bacteriophage that has been widely used to make libraries in *E. coli*. Developing a plasmid or viral library of genes begins by using restriction enzymes to cut an entire set of DNA for an organism into gene-length pieces of about 10^5 bps each. The sticky ends of the pieces are then bound to the sticky ends of a plasmid or viral DNA that was cut with the same restriction enzyme. Each new plasmid or viral DNA that contains a gene-length piece of the organism of interest is used to infect a single bacterium, which then grows into an entire colony of bacteria that contain the same single gene-length piece of DNA from the original organism. Using this method, all of the nuclear DNA in a human can be represented by about 10^6 bacterial colonies.

Genes of interest are found by screening a library. However, some information about the gene or the protein it produces must be known in order to find the gene in the library. For example, if the protein sequence produced by the gene includes methionine–phenylalanine–lysine–tryptophan–asparagine, then the mRNA sequence that represents that genetic code will be 5′-AUG-UU(U,C)-AA(A,G)-UGG-AA(U,C)-3′. Either of the nucleotides in each set of parentheses will result in selection of the same amino acid sequence for the protein (Table 13.3). As is shown in this example and in Table 13.3, some tRNA molecules are constructed so that accurate base pairing is only required at the first two positions of the codon. In these cases, the third base can be one of two or more possibilities. This ability to tolerate a mismatch is called the wobble effect. DNA with a sequence of 3′-TAC-AAA-TTC-ACC-TTA-5′ is one of the possible sequences that would produce mRNA that would be translated into the amino acid sequence given previously. This, of course, ignores the possibility that the sequence could be interrupted by one or more introns. From this information, a probe tagged with radioactively labeled nucleotides could be constructed, e.g., 5′-ATG-TTT-AAG-TGG-AAT-3′, and used to locate the desired sequence in the DNA.

In order to use the probe to find the specific gene, all 10^6 different bacteria must be plated out on agar in petri dishes, with each dish containing about 5×10^4 different bacteria. The bacteria are grown overnight, and then a small amount of each growing colony is transferred to a nylon membrane by blotting. Thus, each membrane will contain a replica of the distribution of bacteria on a given petri dish. The bacteria on the membranes are then broken apart (lysed) and their DNA is denatured so that the strands will separate. The radioactive DNA probe is applied to all the membranes so that hybridization will occur. After excess probe is removed, the membranes are placed against an X-ray film for autoradiography. Dark spots will appear on those areas of the developed film which represent the location of the DNA–probe hybrids. The location of the bacterial colonies that contain the approrpriate DNA sequences can then be identified on the petri plates, and the colony is removed. The DNA sequence is reconfirmed several times before the bacterial colony is grown in large cultures from which enough DNA can be isolated to study the cloned gene's DNA sequence in detail.

Detailed gene analysis usually starts by determining the entire DNA sequence of the cloned gene. The same restriction enzyme that was used to insert DNA into the library cloning vector, either plasmid or virus, is used to cut the gene back out of the plasmid. This gene fragment is isolated, and several other restriction enzymes are then used to cut it into smaller DNA fragments. All the small fragments generated by each different restriction enzyme are individually sequenced. It is possible at this point to determine the entire sequence, which includes the coding sequence, the introns, and the flanking area that include the promoters, enhancers, etc. Once all the sequences of the fragments from several restriction enzymes are analyzed, overlapping sequences will be found that allow a complete DNA sequence of the entire gene to be assembled. New DNA sequencing equipment, developed as part of the HGP, has automated much of the work of DNA sequencing.

13.2.6   DNA Amplification

Another basic technique used in biotechnology involves DNA amplification. This can be done in a living system by means of cloning or in the laboratory by means of PCR. Restriction enzymes can be used to create fragments of DNA that will express a single gene and thus will make a single protein. Gel electrophoresis can then be used to separate the desired fragment from the others produced by the restriction enzymes. Once the fragment containing the desired gene has been identified and isolated, it can be introduced into chromosomal DNA of a bacterial or yeast cell. The new artificial chromosome will be replicated as the host reproduces itself as long as the artificial chromosome also contains the DNA sequences that initiate replication, mark the ends of the chromosomes, and are required for the chromosomes to separate when the cell divides.

PCR can be used to amplify DNA 100,000-fold in less than 3 h. Kary Mullis was awarded a Nobel prize for developing this enzymatic replication technique. The technique starts by assembling a solution of the DNA to be amplified, two short DNA sequences or primers (each of which is complementary to one or the other end of the DNA sequence to be replicated), a DNA polymerase to carry out the replication, and free nucleotides for DNA biosynthesis (Fig. 13.10). To start amplification, heat (usually 95°C) is used to separate the double-stranded DNA into single strands, a process called denaturation, which prepares the DNA for replication. Then the temperature is lowered quickly, allowing only the short primers to hybridize to the DNA to be amplified. These primers serve as the starting points for DNA replication (see Fig. 2.13). Since the quantity of available primers is excessive, the proba-

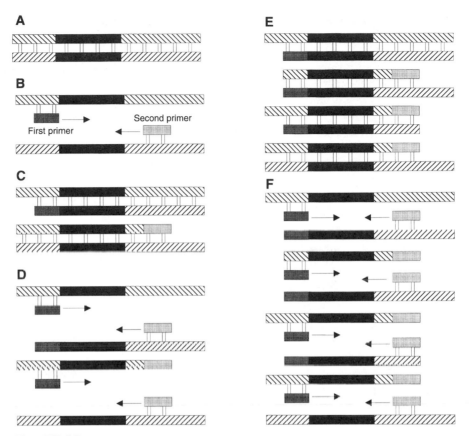

Fig. 13.10 The black area indicates the section that is of interest in the original DNA (A). PCR begins by heating the DNA to 95°C to break the weak hydrogen bonds (vertical lines) that hold the complementary strands together. When the solution is cooled to 50°C, primers in the solution attach to the single strands of DNA and form hybrid strands (B). Beginning with the primers, the remaining portions of the DNA strands are replicated in the presence of DNA polymerase and free nucleotides to produce two new strands of DNA (C). The cycle of denaturation, hybridization, and replication is repeated many times (D–F) to produce additional copies of the region of interest. The arrows in B, D, and F indicate the direction in which replication occurs. The number of copies of the region of interest doubled between A and C (2 to 4) and again between C and E (4 to 8). There will be 16 copies after F is completed.

bility of the two original DNA strands finding each other and reattaching before they attach to primers is infinitely small. The first type of primer molecule anneals with the DNA sequence on one end of the sample DNA, whereas the other type of primer molecule lines up with the DNA sequence at the other end on the complementary strand to produce hybrid double strands (Fig. 13.10B). At the lower temperature, once the primers hybridize to the DNA, the DNA polymerase commences DNA replication using the free nucleotides in the reaction mixture (Fig. 13.10C). After DNA polymerase begins at each primer and replicates the remainder of the DNA strand, the mixture contains two copies of the original DNA sample instead of just one. The cycles of denaturation, hybridization, and synthesis are repeated over and over. Each cycle results in doubling of the DNA sample of interest (Figs. 13.10C and 13.10E).

With the discovery of a DNA polymerase enzyme (*Taq* DNA polymerase) from a bacterium (*Thermus aquaticus*) that inhabits hot springs, it was possible to automate PCR since *Taq* DNA polymerase is not destroyed during the heating cycle. The first thermal cycler, Mr. Cycle, was invented by Cetus Instrument Systems. Many thermal cyclers are now available for PCR. PCR can produce enough DNA from a minute sample so that the product will be visible with gel electrophoresis and can be used for cloning. PCR can also be used as a diagnostic tool since the process depends on the specific base pairings of the primers, but special care must be taken to avoid false positives due to contamination from previous samples. Contamination concerns and problems are currently limiting the use of PCR in diagnostics.

13.2.7   DNA Sequencing

DNA sequencing is a technique that makes it possible to identify the order of nucleotides found in a segment of DNA. This method was initially developed by Frederick Sanger and uses chain terminators. These are chemicals, dideoxynucleotides, that specifically stop DNA polymerase from elongating a new strand of DNA. Chain terminators closely resemble nucleotides but lack the nucleotide 3′ hydroxyl group (Fig. 13.1A) that is essential for nucleotide addition during biosynthesis. DNA polymerase adds nucleotides to a growing DNA strand by forming a bond between the 5′ phosphate group of the new nucleotide and the 3′ OH group of the previous nucleotide (Fig. 13.1B). If the 3′ OH group is missing, DNA polymerase cannot add another nucleotide and the strand stops growing.

Sequencing begins with samples of the DNA molecule of interest. These are mixed with a primer and radioactive normal nucleotides. The master batch is then divided into four batches, each of which contains all four normal deoxynucleotides and a different dideoxynucleotide (ddA, ddT, ddC, and ddG). When DNA polymerase is added, new DNA strands are synthesized on the template molecules, all starting at the 3′ end of the primer. When the next nucleotide to be added is one for which there is a choice between a normal deoxynucleotide and the dideoxynucleotide, there is a random probability that the dideoxynucleotide will be used. The probability is determined by the relative concentration ratios of normal to dideoxynucleotides. If the dideoxynucleotide is added, DNA replication will be terminated at that particular site. For example, in the batch that contains ddT, every time an "A" occurs in the template sequence, some of the new molecules will have ddT added. Some strands will be terminated the first time an A is encountered, whereas others will be terminated at As in other locations. This will result in a set of new DNA molecules that have the new part of the strand terminated at the location of every A in the original template. In the other batches, synthesis will terminate each time a T, C, or G is reached in the template strand when ddA, ddG, or ddC is incorporated into the new strand.

After synthesis of the new strands has been completed, the DNA is denatured by heating and the ddA-, ddT-, ddC-, and ddG-reaction mixtures are loaded into separate lanes in a gel. Gel electrophoresis is used to separate the different strands by size. A photographic plate is exposed to the gel, and dark bands appear on the plate in each location that contains a fragment since the fragments were built with radioactive nucleotides. The ddT-reaction lane includes all the fragments that resulted when a ddT molecule terminated synthesis at the location of an A nucleotide in the template, whereas the ddC-reaction lane includes all those fragments that were terminated at a G in the template. The sequence of the DNA is read from the bottom of the film to the top by comparing the different lanes. The bottom-most site represents the second nucleotide in the original chain since this is the shortest fragment (one radioactive nucleotide and one dideoxynucleotide) and moves farther than any others in the gel (Fig. 13.11). The gel is then scanned for the next shortest fragment. If it appears in the lane with DNA from the ddC batch, then the third nucleotide in the DNA strand could be G. This procedure is used repeatedly up the gel, and the sequence is determined.

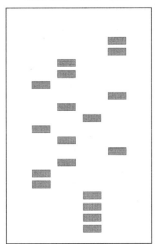

ddA ddT ddC ddG

Fig. 13.11 DNA sequencing uses dideoxynucleotides (ddA, ddT, ddC, and ddG) to stop synthesis of a template DNA strand at each nucleotide location. DNA synthesis stops when a thymine (T) is encountered in the template strand if ddA is added to the new strand instead of adenine (A). Thus, ddA indicates the locations of T on the DNA template, whereas ddT indicates the locations of A, ddC the locations of guanine (G), and ddG the locations of cytosine (C). Gel electrophoresis is used to separate the strands into fragments of different lengths. The shortest fragments travel fastest through the gel. The fragment that travels the farthest in the gel represents the first nucleotide in the sequence of the template DNA. In this example, the first 10 nucleotides in the DNA template were 3'-GGGGTTACAT-5'.

13.2.8   cDNA

Reverse transcriptases are important enzymes in biotechnology. These enzymes, made by RNA viruses (retroviruses), read the nucleotide sequence in RNA and synthesize complementary DNA (cDNA). Thus, they can be used to create a cDNA from any RNA, including mRNA. This is particularly useful in dealing with eukaryotic genes since eukaryotic mRNA represents the original DNA sequence after the introns have been removed (Fig. 13.6). Reverse transcriptase can convert mRNA into the coding sequence of a gene without the introns. The cDNA can then be inserted into bacteria for protein production or used as a probe to recover the original gene from a DNA library. Since prokaryotic cells cannot remove the introns from a primary RNA transcript to generate the mRNA needed for translation, the original eukaryotic gene could not be expressed in a prokaryotic cell if it still contained introns.

If the mRNA for a protein cannot be found, it may be possible to reconstruct the nucleotide sequence for a gene by reverse engineering the protein. Proteins consist of a series of amino acids, each of which is specified by one or more codons in the genetic code (Table 13.3). For example, one portion of cytochrome c, a protein component of electron transport chains in cells, has the following sequence of amino acids in a primate: threonine–asparagine–proline–lysine–lysine–isoleucine–proline–glycine–threonine–lysine–methionine–alanine. The possible mRNA sequences for this part of the polypeptide would be AC(A, G, U, C)-AA(U, C)-CC(A, G, U, C)-AA(A, G)-AA(A, G)-AU(A, U, C)-CC(A, G, U, C)-GG(A, G, U, C)-AC(A, G, U, C)-AA(A, G)-AUG-GC(A, G, U, C). Artifical DNA molecules can be synthesized using this information by coupling the DNA bases together in the correct order. Since transcription moves in the 3′ to 5′ direction on one strand of DNA and produces a 5′ to 3′ RNA, the bases must be put together in the appropriate order. In this case (using the first option in each parentheses to complete the codon), a correct sequence from RNA would be 5′-ACA-AAU-CCA-AAA-AAA-AUA-CCA-GGA-ACA-AAA-AUG-GCA-3′ and the corresponding sequence in cDNA would be 3′-TGT-TTA-GGT-TTT-TTT-TAT-GGT-CCT-TGT-TTT-TAC-CGT-5′. The cDNA sequence represents the coding sequence found in the gene.

Example Problem 13.3

Assume that the part of the cDNA for cytochrome c shown previously was put together in the opposite direction, i.e., 5′ to 3′. What would be the resulting sequence of amino acids that would be specified by mRNA?

Solution

DNA: 5′-TGT-TTA-GGT-TTT-TTT-TAT-GGT-CCT-TGT-TTT-TAC-CGT-3′

mRNA: 3′-ACA-AAU-CCA-AAA-AAA-AUA-CCA-GGA-ACA-AAA-AUG-GCA-5′

During protein translation, mRNA moves through the ribosome in the 5′ to 3′ direction so the codons would be read as ACG-GUA-AAA-ACA-AGG-ACC-AUA-AAA-AAA-ACC-UAA-ACA, which would translate into threonine–valine–lysine–threonine–arginine–threonine–isoleucine–lysine–lysine–threonine–termination–threonine. Since the code for termination now appears before the last codon, only part of a protein, which is not cytochrome c because the sequence of amino acids is different from that shown previously, would be produced. ■

13.3 OTHER CORE TECHNOLOGIES

The basic nucleic acid techniques that were described previously have many different applications in biotechnology. There are other, equally useful technologies derived from organisms. Monoclonal antibodies, for example, can be produced in special cell lines and used as diagnostics, therapeutics, drug delivery systems, and purification reagents. Bioprocessing technology is used to produce commercially important amounts of specific proteins, chemicals, and macromolecules from organisms such as bacteria or yeast. Animal cells and tissues can also be grown in the laboratory and used for testing pharmaceuticals, providing therapeutic materials, and studying molecular mechanisms. Bioanalytical sensors that are capable of detecting substances at extremely low concentrations have been developed using biotechnological techniques. Genetic engineering has provided the basic methods for correcting lethal defects in humans, whereas protein engineering technology is being used to improve existing proteins and create new ones. Antisense technology may provide another new therapeutic approach to inhibiting gene expression by blocking translation of mRNA into proteins.

13.3.1 Monoclonal Antibody Technology

When the human body is invaded by bacteria, viruses, or other infectious agents, it responds by producing antibodies that attach to specific chemical structures (the antigens) on the surface of the invaders. Antibodies are very specific — each type binds to and attacks only one particular antigen. These antigen–antibody complexes make it easier for the body's immune system to kill the infectious agent. Once the body has been exposed to a particular antigen, it has the ability to respond even more rapidly to a subsequent exposure because a "record" of the exposure is encoded permanently in rearranged DNA sequences in the family of antibody genes. This feature forms the basis for vaccination in which the body is exposed to killed or weakened bacteria or viruses that are unable to cause disease but still carry antigens that stimulate the body to produce antibodies. If virulent forms of the same infectious agent invade the body at a later date, the rearranged antibody genes formed as the permanent record of previous vaccination exposure to the antigens will produce antibody to attack the new, virulent invaders.

In addition to being used therapeutically to protect against disease, antibodies can also be used to diagnose a wide variety of illnesses and to detect unusual or abnormal substances in the blood. In the past, polyclonal antibodies were collected by injecting laboratory animals with a specific antigen and then harvesting the serum from clotted blood, the portion of the blood in which antibodies are found. This method produced antiserum with undesired substances and very small amounts of usable antibody. Monoclonal antibody technology makes it possible to produce large amounts of pure antibodies specific to individual antigens.

The initial step in the production of monoclonal antibodies involves injecting an antigen into a laboratory animal, usually a mouse. B lymphocytes, cells that produce antibodies to the given antigen, are then isolated from the animal's spleen and fused with tumor cells (usually a rapidly dividing white blood cell cancer) that have been grown in tissue culture. The resulting cells, called hybridomas, have the cancer cell's ability to grow continually and the B lymphocyte's ability to produce large amounts of pure antibody. Since the antibodies are produced by a single cell line, they are called monoclonal.

Monoclonal antibodies (MCAbs) are currently being used in many diagnostic and therapeutic applications. They are used in home pregnancy kits to detect the presence of a hormone that is produced by the placenta and in laboratories to diagnose diseases such as strep throat and gonorrhea. MCAbs can also be made that selectively bind to specific types of cancer cells. The MCAb can be tagged with a radioisotope to help identify metastatic cancer cells (those cells that have migrated from the primary tumor to another part of the body) or with a toxin that kills specific cancer cells. Since the MCAb will only attach to one specific antigen, e.g., a protein on the surface of a particular type of cancer cell, the toxin is delivered exclusively to cancer cells. This target-specific treatment results in much lower dosage requirements for the drug and minimal or no interactions with normal cells in the body. This could help prevent some of the undesirable side effects, e.g., nausea and allergic reactions, that are often produced by other chemotherapies because of the relatively high dosage required for nonspecific delivery to the entire body.

MCAbs are also used in basic research to identify and tag molecules of interest. They can be used to trace specific cells during embryonic development and to quantitate different polypeptides in abnormal cellular processes found in disease states. Several companies that specialize in the production of made-to-order monoclonal antibodies have recently been formed.

13.3.2   Bioprocessing Technology

In bioprocessing technology, living cells, generally one-celled microorganisms such as bacteria or yeast, are used to generate specific products. Fermentation is the oldest and most familiar example of bioprocessing technology. Originally, fermentation involved using the metabolic pathways in organisms to break glucose down into useful by-products, e.g., ethanol in beer and wine, carbon dioxide for leavening bread, lactic acid for making yogurt, and vinegar (acetic acid) for pickling foods. Microbial fermentation is now being used to produce antibiotics, amino acids, industrial enzymes, hormones, vitamins, pesticides, and many other products (Table 13.4).

13.3.3   Cell and Tissue Culture Technology

Both animal and plant cells and tissues can be grown in appropriate nutrients in laboratory containers and can be used to study the safety and efficacy of pharmaceuti-

TABLE 13.4 Microbial Producers of Industrial Products from Glucose[a]

Microbial source	Organic chemical	Industrial uses
Acetobacter	Acetic acid	Industrial solvent, rubber, plastics, food acidulant (vinegar)
	Tartaric acid	Acidulant, tanning, commercial esters for lacquers, printing
Aspergillus	Citric acid	Food, pharmaceuticals, cosmetics, detergents
	Gluconic acid	Pharmaceuticals, food, detergent
	Itaconic acid	Textiles, paper, manufacture, paint
	Malic acid	Acidulant
Clostridium	Acetone	Industrial solvent, intermediate for many organic chemicals
	Butanol	Industrial solvent, intermediate for many organic chemicals
	Isopropanol	Industrial solvent, cosmetic preparations, antifreeze, inks
Lactobacillus	Lactic acid	Food acidulant, fruit juice, soft drinks, dyeing, leather treatment, pharmaceuticals, plastic
Rhizopus	Fumaric acid	Intermediate for synthetic resins, dyeing, acidulant, antioxidant
Saccharomyces	Ethanol	Industrial solvent, fuel, beverages
	Glycerol	Solvent, sweetener, printing, cosmetics, soaps, antifreeze

[a] Adapted from Kreuzer and Massey (1996).

cal compounds, molecular mechanisms, and basic cell biochemistry and to provide therapeutic materials. Important issues in cellular and tissue engineering are discussed in Chapter 17.

13.3.4 Biosensor Technology

Within the context of biotechnology, a biosensor consists of a biological substance [e.g., a microbe, a single cell from a multicellular organism, or a cellular component (enzyme or antibody)] that is linked to a transducer and is used to detect substances at extremely low concentrations. Bioanalytical sensors are discussed in Chapter 10.

13.3.5 Genetic Engineering Technology

Genetic engineering technology, also known as recombinant DNA technology, involves joining genetic material from two different sources. This occurs naturally when crossing over occurs during meiosis and genetic materials from maternal and paternal chromosomes are recombined. It also occurs during fertilization when the egg and sperm fuse to form the zygote and when bacteria exchange genetic material through conjugation (which involves physical contact between two cells), transformation (which involves the uptake and incorporation of foreign DNA), and transduction (which involves the transfer of DNA from one bacterium to another by means of a bacteriophage). Increased genetic variation results from all these naturally occurring types of recombination.

The earliest examples of genetic manipulation involved selective breeding of domesticated organisms to alter their genetic makeup to better suit human needs. This method continues to be used today. Examples include the development of cows that produce large quantities of milk or corn that has fatter, sweeter kernels.

Genetic engineering involves molecular techniques that are used to join specific segments of DNA molecules together that do not occur that way naturally. As discussed earlier, restriction enzymes can be used to cut DNA in predictable ways, allowing isolation of valuable DNA sequences that encode recipient-specific desired characteristics. By using bacteria, viruses, or naked DNA, precise DNA sequences can then be moved into a target organism.

The genetic engineering strategy differs from the selective breeding strategy. In genetic engineering, a desired characteristic is moved into an organism by physical transfer of precisely defined DNA sequences with known functions, with the resulting transgenic organism expressing the new characteristic. In sexual breeding, the strategy is to move a desired characteristic into a breeding population through sexual recombinations of sets of genes that include the desirable one(s) and by using subsequent breeding cycles to rid the final population of other undesirable characteristics in the original set of genes. Selective breeding represents an iterative approach that relies on population statistics, whereas genetic engineering is theoretically specific since the transferred gene is well characterized. Selective breeding is limited to populations that can interbreed, i.e., members of the same species or some closely related genera, whereas genetic engineering allows genes to be transferred from any organism to any other. Genetic engineering has opened the door to using all the genetic diversity found in nature.

13.3.6 Protein Engineering Technology

Protein engineering, the process of changing a protein in a predictable and precise manner to bring about a change in the protein's function, is closely linked to genetic engineering since genes provide the blueprints for proteins. Thus, genetic engineering can be used to improve existing proteins or to create proteins that are not found in nature. In theory, it should be possible to create any protein by developing a gene that codes for it. Much protein engineering research to date has been directed at using physical protein data to develop computerized models that predict protein structure and function and molecular work to modify existing proteins, e.g., enzymes and antibodies.

Enzymes are catalytic protein molecules that make reactions that are already thermodynamically possible occur more readily. As catalysts, they bring about a reaction but are not permanently changed by the process. Thus, enzymes can be used repeatedly in a specific reaction until all the reactants have been converted to product or until some other equilibrium point has been reached. Most enzymes work best under conditions that are compatible with life. These conditions include neutral pH, mild temperature and pressure, and a water-based environment at relatively moderate osmotic potential. However, some enzymes, e.g., the *Taq* DNA polymerase, are active under extreme conditions. Much work is being done to isolate the genes that produce useful enzymes from organisms that live in extreme environments and to modify existing genes to produce enzymes that are stable under harsh conditions that are often found in nonbiological, commercial applications. Thermophilic endoglucanases, lipases, and proteases, for example, are enzymes that are used in hospital washing detergents since they can function at the high temperatures needed for sterilization.

Antibodies are also proteins that bind to specific molecules, but they do not commonly catalyze reactions. Another application of protein engineering has been in the area of abzyme development. Abzymes are antibodies with catalytic capabilities, i.e., antibodies that function as enzymes. These proteins can be custom designed to catalyze new reactions. For example, proteolytic abzymes could be designed to break down proteins at specific peptide bonds. This would be the equivalent of having a restriction enzyme for proteins and would make it possible to break apart existing proteins and create new ones with specific peptide sequences.

13.3.7   Antisense Technology

Antisense technology involves the inhibition of gene expression by blocking translation of mRNA into protein. This is achieved by having either an antisense DNA or antisense RNA bind to the mRNA. The antisense DNA and RNA are exactly complementary in sequence and opposite in polarity to the normal mRNA, i.e., 3′ to 5′ versus 5′ to 3′. These two characteristics allow the antisense sequence to bind with the mRNA to create a double-stranded RNA molecule or DNA–RNA hybrid. Such doublestranded nucleic acids cannot be translated into a protein and are quickly degraded in the cell cytoplasm, the site of translation.

This new technology could make it possible to develop therapies in which the antisense nucleic acid effectively cancels the proteins and the mRNA encoding them and prevents normal expression of specific proteins that are unique to the survival of an invading organism but are not required by the host. To do this, target RNA sequences in the invader must be identified. Also, ways must be developed to deliver the antisense nucleic acid to the infected cells via normal circulation. The antisense nucleic acid must also have a long enough half-life when inside the target cell so that it will be available to bind to the foreign mRNA. To achieve a sufficient half-life requires that the antisense nucleic acid be protected from cellular enzymes that break down RNA and DNA into free nucleotides. An additional challenge involves manufacturing and purifying nucleic acids at a low enough cost and in large enough quantities to make them therapeutically feasible.

With antisense technology, it is theoretically possible to suppress the expression of any gene. It could be used to decrease the amount of metabolites (e.g., cholesterol) in the body by preventing the production of one or more key enzymes in the metabolic, biosynthetic pathway. Since cancer and viral infections result from inappropriate expression of the body's own genes or from the expression of foreign genes, antisense technology could be used to treat these diseases. By blocking expression of genes that are unique to the disease, interference with processes in normal cells would be avoided and serious side effects would be eliminated. Clinical trials have been initiated for antisense treatment of human papillomavirus, which causes genital warts, and for an antisense molecule that is directed at an oncogene (a gene thought to be capable of causing the disease) for lung cancer. Antisense nucleic acids for a gene that promotes cell growth have also been used to stop the thickening of rat artery walls following balloon angioplasty surgery. This type of surgery is used to open clogged arteries in humans, but the artery wounding that sometimes occurs during the procedure can trigger cell growth in the artery walls so that the blood vessels reclose in about one-third of the cases.

13.4   MEDICAL APPLICATIONS

Several important medical applications have become possible due to developments in biotechnology, including the production of therapeutic proteins, the HGP, and tissue engineering, which is discussed in Chapter 17.

13.4.1   Production of Therapeutic Proteins

In the search for new drugs, pharmaceutical companies test chemicals or macromolecules for their ability to alter the activity of an enzyme or protein that is known to be associated with a particular disease. Knowing the mechanism of action of a drug is part of the information that is required to obtain Food and Drug Administration (FDA) approval of the drug for treating a disease. Drug development, whether small molecules such as antibiotics or polypeptides such as insulin, rests on fundamental knowledge of the molecular mechanisms of the disease state and how these mechanisms are affected by different compounds or macromolecules.

There are several types of therapeutic proteins. Some, such as insulin, function as regulatory hormones (see Fig. 2.40). In certain diabetic conditions, the pancreas produces insufficient insulin so glucose metabolism must be regulated by means of insulin injections. Another example is human growth hormone, a polypeptide that helps to regulate overall development. Lack of growth hormone causes dwarfism, whereas too much of the hormone results in another disease state called gigantism. Before biotechnology was developed, minute quantities of growth hormone were harvested from the pituitary glands of cadavers.

Studies have been performed that show the association of a specific peptide with a specific disease state. Therapeutic enzymes can be used to either block a disease or prevent complications. One therapeutic enzyme is the natural human protein, tissue plasminogen activator (t-PA), which is used to dissolve blood clots immediately after a stroke or heart attack. By dissolving the clot soon after its formation, blood flow can be restored to the downstream tissues, and the amount of permanent injury can be reduced. Streptokinase (fibrinolysin) is a bacterial enzyme that is reported to work as well as t-PA. Other therapeutic proteins that are under intense investigation are antibodies and the proteins that are associated with cancer. Biotechnology can also be used to produce protein vaccines.

The gene(s) that encodes the polypeptide(s) that forms a protein and a biological system that can transcribe the gene(s) and translate it into a polypeptide(s) are required to synthesize a protein. The biological system must be able to express the required genes, synthesize mRNA from the genes, process the mRNAs to prepare them for peptide synthesis, synthesize the peptides, process the peptides appropriately after translation, and deposit the protein to the appropriate subcellular compartment. How well the biological system performs these tasks and the complexity of the desired protein will to a large extent determine what biological system can be used to make the protein. Factors to consider include the biological source of the gene (e.g., bacteria, mammals, and plants), whether the protein is composed of one peptide or more than one peptide, and whether or not the functionality of the protein requires posttranslational processing.

There are two fundamentally different approaches used in the commercial synthesis of proteins. In one case, an organism already exists which is capable of producing the protein, whereas in the other case, a genetically engineered organism must be created. Medical, technological, economic, and regulatory constraints help determine which system is best for producing a particular protein product.

The first approach requires that an organism be identified that already makes the desired protein. This approach requires a sufficient and reliable source of the organism and the development of a purification scheme. The advantage of this method is that nothing needs to be known about the gene(s) that encodes the protein or how the organism stabilizes the protein. The disadvantage of this approach is that it is often not possible to find a sufficient and reliable source of the organism to produce commercially important amounts of protein.

The second approach requires that an efficient production organism be genetically engineered to produce the desired protein. This approach requires identification and isolation of the gene(s) that encodes the protein, a method for gene transfer to the production organism, and development of a purification scheme (Table 13.5). The advantage of this method is that once gene transfer has been accomplished, production of the organism is routine. The disadvantage of this approach is that a large amount of research must be done to achieve the protein production levels required commercially.

TABLE 13.5 Steps in Development of a Complete Bioprocess for Commercial Manufacture of a New Recombinant DNA-Derived Product

Step 1	The gene of interest is isolated from the chromosome where it is located.
Step 2	Restriction enzymes and DNA ligase are used to insert the desired gene into a plasmid.
Step 3	The recombinant plasmid is inserted into an appropriate organism, e.g., *E. coli*.
Step 4	The genetically engineered organism is allowed to reproduce in a small-scale culture (e.g., 250-ml shake flask) and is checked for gene expression. Conditions (e.g., medium composition, pH, temperature, and other environmental conditions) that allow optimal growth and productivity are determined.
Step 5	After the optimal culture conditions have been determined, the genetically engineered organisms are grown in a broth in bench-top bioreactors with capacities of 1 or 2 liters. Culture conditions can be more closely monitored in bioreactors than in shake flasks.
Step 6	The process is scaled up to a pilot-scale bioreactor (100–1000 liters) based on the results from the bench-top bioreactor studies.
Step 7	Following successful scale-up, the process is moved to an industrial-scale operation.
Step 8	Product recovery (downstream processing) occurs after the genetically engineered organism is grown in the industrial-scale operation and depends on the nature of the product and the broth in which it was produced.
Step 9	Packing and marketing occur after the desired product has been isolated in sufficient purity and quantity.

The exact DNA sequences in the desired gene(s) and the organization of these regions as well as the correct posttranscriptional and posttranslational processing and localization of the protein in the appropriate subcellular compartment are all critical to ultimate yields of foreign proteins in genetically engineered cells. Molecular genetics has yet to provide a comprehensive theory regarding how these sequences regulate transcription, posttranscriptional processing, translation, and, indirectly, posttranslational processing. Therefore, optimization for protein yield requires many empirical experiments, each transferring a somewhat modified sequence of the gene under study until a gene construct is found that gives optimal protein yield in the cell production system of choice.

Insulin provides a good example for comparing one type of protein production system to the other. Basic molecular research into the mechanisms that underlie juvenile-onset diabetes have shown that the disease is caused by insulin insufficiency. This information is the type that is required for FDA licensing. Lack of insulin in the bloodstream prompted the hypothesis that therapy for these patients might consist of insulin injections. To obtain sufficient insulin for therapeutic applications before the days of genetic engineering required identification of a biological source for insulin and efficient processing of this source to produce pure insulin polypeptide. The pharmaceutical industry found a source for insulin from pancreases available from the hog industry. Bioengineering research created the downstream processing procedures that were needed to produce pure insulin at feasible costs and quantities.

Many challenges are associated with protein production, whether from a natural or genetically engineered source. The natural source must be available in sufficient quantity to make commercialization feasible and it must be safe for use in humans. In the case of insulin, the hog industry provided a commercially viable source of the raw material. The next major hurdle is the downstream process for a polypeptide from the source tissue or cell to the final purified polypeptide. Downstream processing is a major cost in therapeutic protein development and can account for 50–80% of the total production costs.

The entire nature of the protein must be considered to create an appropriate downstream process. The purification scheme for each new protein is determine on a case by case basis. Several aspects of protein structure must be considered to determine the steps needed in purification, including (i) the molecular weight of the protein; (ii) the number of peptides; (iii) the presence or absence of small molecular cofactors; (iv) the stability of the protein as a function of temperature, pH, aqueous conditions, and oxidation reduction environment; (v) the isoelectric point of the protein; (vi) the hydrophobic or hydrophilic nature of the protein; and (vii) the presence and identity of small molecules that can bind to the protein. These factors also determine the range of possible protein manipulations that the purification procedure can have to prevent denaturation of the protein. Protein purification must also be done under specific "good manufacturing procedures" (GMPs) that are expensive and are required for FDA licensing.

Other issues arise after the protein is purified since therapeutic proteins are considered drugs and must be licensed for human use by the FDA. The therapeutic protein must be shown to be effective in the treatment of a specific disease, and its potential side effects must be documented using clinical trials that are expensive to conduct and that can take several years to run. If the therapeutic protein passes the regulatory requirements, the protein and the method by which it is produced are licensed for use in humans. This entire process usually takes 5–10 years.

The regulatory issues surrounding licensing have a major effect on quality control in production. GMPs must be strictly followed during production. Another quality control concern is the contamination of the protein preparation with allergens, pyrogens, endotoxins, and animal viruses. In the case of therapeutic proteins purified from animal systems, there is always a chance that a virus from the natural source production animal, a pathogenic virus of hogs in the case of insulin or from the animal cell line or transgenic animal in the genetic engineering case, will contaminate the protein preparation and cross over into humans when the protein is administered as a drug. Other contaminating substances of concern include possible allergens. Small amounts of allergens can cause large allergic responses that, in some cases, can be fatal. Contamination of the therapeutic protein with minute quantities of other proteins can result in allergic responses in some people.

The second approach for the production of a therapeutic protein involves the genetic engineering of an organism with a gene that encodes the desired protein. The strategy starts with the identification and cloning of the specific gene(s). The next decision involves the selection of the appropriate production organism and system. Many systems are currently used for biological production of proteins. The structure of the specified genes determines the system that is selected and, ultimately, the yields and costs of the manufactured proteins.

Genetically engineered protein production systems currently in use include (i) bacteria, e.g., *E. coli* and *Bacillus subtilis*; (ii) yeast and other fungi; (iii) mammalian cell cultures; (iv) insect cell cultures; (v) transgenic animals; (vi) plant cell cultures; and (vii) whole plants. Each system has advantages and disadvantages in terms of three criteria: efficacy of the final product, quality control, and cost.

Microbial cell cultures, including bacteria and yeast, are the oldest and best defined production systems. The two bacteria most commonly used are *E. coli* and *B. subtilis*. Several fungi are used, including *S. cerevisiae*, *Neuspora crassa*, and *Aspergillus nidulans*. The advantages of microorganisms for protein production include (i) fast growth rates with doubling times of minutes to hours; (ii) reproduction via cell division (clonal reproduction); (iii) growth in fluid media, (iv) simple, defined nutrient requirements; (v) high levels of protein production per unit time or per unit medium; and (vi) well-defined environmental requirements of temperature, oxygen level, pH, and osmotic potential. Commercial production methods center on the use of fermenters — large vessels that can contain thousands of liters of media in which the temperature, oxygen partial pressure, and pH are precisely controlled. Insulin is no longer produced from porcine pancreas. With the advent of genetic engineering of *E. coli*, recombinant insulin is made cheaply and of higher quality with reduced risk of contaminants using bacterial cells that are quickly grown to high titer in large 10,000- to 100,000-liter, air-lift fermenters.

Methods for transferring genes into bacteria and yeast are well established. Gene transfer is accomplished via direct DNA uptake, either by transformation of DNA into "competent" bacterial cells or via electroporation. During transformation, bacterial cells are treated with a specific salt solution and temperature regime that results in the cell membrane becoming permeable to the DNA that has been placed in the surrounding solution. The bacteria take up the DNA and incorporate it into their own DNA, allowing subsequent expression of the foreign gene, at a frequency of around 1×10^{-3}. Electroporation is a method in which a high-voltage electrical pulse is discharged through a solution containing suspended bacteria or yeast cells and the DNA of the gene to be transferred. During the electrical discharge (100–200 V for 1 or 2 ms), existing pores in the cell's membrane enlarge or small holes are created so that the DNA in the surrounding solution can enter the cell. The DNA is taken up and incorporated into the cell's genome. The frequency of gene transfer and incorporation following electroporation are generally higher than that resulting from transformation. In animal cells, DNA can also be transferred by means of microinjection, viral infection, and bombardment of cells with DNA-coated particles. In the last technique, minute gold particles are coated with the DNA that is to be transferred, and the particles are propelled at very high velocities at the target cells using high-pressure bursts of helium gas.

Although physical gene transfer is straightforward, high levels of gene expression depend on the specific gene construct and the source of the gene. Bacteria and fungi have very different promoter regions than do mammals, the source of most therapeutic proteins. Before a mammalian gene can be expressed effectively in a microbial system, the mammalian promoter must be replaced by one that is appropriate for expression in the microbial production species. Bacterial genes do not contain introns in their coding sequences. Therefore, bacteria lack the cellular posttranscriptional processing enzymes necessary to remove the introns from the mRNAs of higher organisms. Human genes, for example, cannot be properly translated in bacterial cells unless the DNA sequences are modified before gene transfer to remove the introns from the gene. This can be done by producing cDNA from mRNA and introducing the cDNA into the bacterial cells.

A more subtle problem affecting ultimate protein yields can arise due to codon usage. In the genetic code (Table 13.2), there are several triplet codons that specify the same amino acid. However, many organisms have a bias in codon usage and favor the use of one codon more than the others in the coding regions of genes. If the bias is large, a foreign gene will be incorrectly translated a significant proportion of the time or, in extreme cases, the mRNA will not be translated at all. In these cases, restructuring of the gene to more closely reflect the codon usage of the production organism would be required for the gene to be efficiently used by a particular organism. All these issues — promoters, other regulatory sequences, the presence and processing of introns, and codon usage — are some of the underlying reasons for the general rule that gene transfer between more similar species requires less modification to achieve high levels of expression than that between very dissimilar species.

Many modifications of the protein structure can occur during posttranslational processing of individual polypeptide strands into functional proteins that are fully folded, appropriately modified, and assembled (if there are several subunits). Cofactors (e.g., mineral ions and vitamins) can be added, and sugars can be covalently added to specific amino acids. These modifications are critical to the proper functioning of the protein. The addition of sugar residues to proteins, called glycosylation, is problematic with microbial production systems since glycosylation of proteins does not occur in bacteria, whereas yeast have a tendency to overly glycosylate proteins. In either case, microbial production systems would not be the best choice when production of a therapeutic protein requiring specific patterns of glycosylation is desired. Finally, the synthesis and assembly of multisubunit proteins, which require the coordinated regulation of production of the various subunits and similar subcellular localization for modifications and assembly, are often difficult to achieve with microbial systems, especially bacterial cells. In these cases, the engineer needs to examine alternative production systems.

Animal cell culture is one alternative if microbial systems are inadequate for the production of a stable therapeutic protein. Mammalian and insect cell cultures are available and are used commercially for the production of proteins. Animal cell cultures have several advantages when compared to microbial systems since animal cell cultures can produce proteins that microbial systems cannot. As a general rule, genes transferred from one organism to a similar organism (e.g., mammal to mammal as opposed to mammal to bacteria) will require little or no modification of the DNA sequence to obtain gene expression. This generality implies that it would be best to produce most therapeutic proteins in mammalian production systems because most genes encoding therapeutic proteins are mammalian (i.e., human) in origin. Unfortunately, mammalian culture systems are very expensive to run with protein production costs in the range of thousands of dollars per gram protein, commonly have low yields, and carry the risk of pathogen contamination and allergen problems. It is also difficult to produce commercially important amounts of some proteins (e.g., regulatory proteins) in mammalian systems because the large amounts of produced protein would be toxic to mammalian cell growth. To overcome the high cost of mammalian cell culture systems, genetically engineered pigs, sheep, and goats are being investigated as possible protein production systems.

The goal of therapeutic protein production is to make a final protein product that (i) shows high efficacy in application (i.e., works the way it should in the patient), (ii) complies with stringent regulations regarding quality control issues (contaminants, pathogens, etc.), and (iii) is made at the lowest possible cost. The issue of cost includes both the cost of producing the biological organisms and the costs associated with downstream processing to obtain purified protein. Different production systems require different equipment for organism production. This equipment can be quite expensive so there is an enormous economic incentive to try to achieve production of new products in systems with which a company has both experience and the necessary capital equipment. Because the costs of downstream processing are usually the largest proportion of overall protein production costs, any system that makes downstream processing cheaper has a decided advantage. Efficacy, quality, and cost should be considered in this order of importance. For each specific gene, there are different compromises to be made between protein efficacy/quality and production costs. The decisions that must be made by the production engineer when developing a genetically engineered therapeutic protein production system are shown in Fig. 13.12.

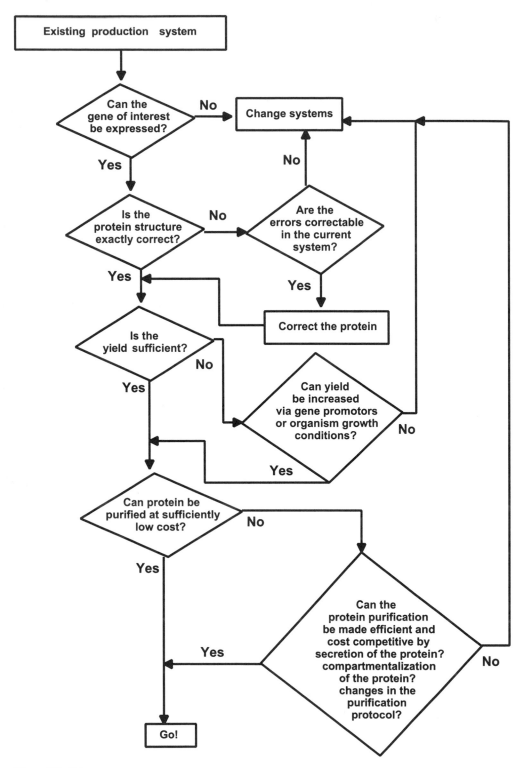

Fig. 13.12 Flowchart of the decisions to be made in the development of a production process for a new therapeutic protein from an existing production system.

13.4.2 Human Genome Project

The HGP is an international 15-year effort that began in 1990 and is being coordinated in the United States by the Department of Energy's (DOE) Human Genome Program and the National Human Genome Research Institute at the National Institutes of Health. The goal of the HGP is to discover all of the 60,000–100,000 human genes (the human genome) so that they can be made available for further study and to determine the complete sequence of the 3 billion DNA sequences in a haploid set of human chromosomes. Knowledge of the human genome will provide new strategies for diagnosing, treating, and perhaps even preventing human diseases as well as insights into embryonic development. Parallel efforts are under way to study additional organisms, e.g., *E. coli,* a yeast (*S. cerevisae*), a roundworm (*Caenorhabditis elegans*), the fruit fly (*Drosophila melanogaster*), and the laboratory mouse, since identifying genes that are common to many species will make it easier to map the human genome.

The first component of the HGP is the construction of genetic linkage maps of the 23 pairs of chromosomes, and the second component involves sequencing the DNA contained in all of the chromosomes and creating physical maps. The first component is well under way and should be accomplished by the Year 2000. The second component cannot be accomplished on a large scale until more efficient and low-cost sequencing technology has been developed. Current methodology is too slow and costly to enable scientists to sequence all the DNA in even a single human chromosome. As of 1997, approximately 2% of the human genome had been sequenced.

Many of today's most common diseases (e.g., heart disease, diabetes, immune system disorders, Alzheimer's disease, and many cancers) are probably the result of complex interactions between mutated genes and environmental factors, whereas other diseases (e.g., Huntington's disease, cystic fibrosis, and hemophilia) are known to result from gene mutations. Once the genes responsible for a given disease have been discovered, scientists can study how these genes interact with environmental factors such as diet, drugs, or pollutants. Once a gene has been identified, it will also be possible to determine what protein the gene encodes and to determine what that protein does and where it functions in the body. This information may lead to treatments such as correcting the abnormal gene, turning the faulty gene off, or replacing the faulty protein.

The HGP will also provide information about how genes or sets of genes are switched on and off throughout the life of a cell. During embryonic development, certain genes are turned on or off in order to produce all the different specialized tissues, e.g., muscle, nerve, and bone. At other times, the switching on or off of certain genes leads to apoptosis (programmed cell death) or to immortal cancer cells. For normal development to occur, genes must be switched on and off at the right time and in the correct sequence.

While the human population appears to be very diverse, the differences between any two people are the result of differences in only 2–10 million of the 3 billion bps in a haploid set of chromosomes. These genetic differences constitute <1% of the total DNA. Since the genetic differences between any two people are small, any person's DNA can be used as source material for the HGP. In some cases, such as people with genetic disorders, the differences are important and lead to information about the genetic basis of the disorder. Lymphocytes, one type of white blood cell, are generally used to provide DNA since these are easy to obtain from a blood sample. The lymphocytes are usually infected with a common virus (Epstein–Barr virus) that causes mononucleosis so that they are transformed into immortal cells that will grow and divide indefinitely. This method provides researchers with an unlimited amount of DNA for genome analysis.

Unraveling the secrets in the entire 2 m of DNA found in a human somatic cell is a huge project that is being undertaken one chromosome at a time. Different laboratories have been assigned different chromosomes for analysis. DOE funded sites are mapping chromosomes 2, 5, 11, X, 16, 19, and 21. In many cases, research groups are studying only portions of a single chromosome, and the results from the different laboratories will be combined to form a complete map of a chromosome. As of October 1997, almost 5800 genes had been mapped to particular chromosomes and 1400 genes had been identified whose locations had not been determined unequivocally. The Genome Database is the public repository for human genome mapping information and can be found at *http://gdbwww.gdb.org*.

The first goal of physical mapping of the human genome is to establish a marker every 100,000 bases across each chromosome. This will result in about 30,000 markers for the human genome. A physical map was published in July 1997 that featured about 8000 landmarks across the entire human genome. Previous efforts had produced detailed maps of only a few individual chromosomes.

Scientists at Lawrence Livermore National Laboratory have produced a physical map of chromosome 19 by using *Eco*RI restriction fragments that had been cloned in cosmids, synthetic cloning vectors that were modeled after bacteriophages. Overlapping fragments were used to produce contiguous blocks of clones, known as contigs. Since each fragment overlapped its neighbors, researchers were able to produce an ordered library of contigs that spanned an estimated 54 Mb (>95% of the chromosome outside of the centromere). Fluorescence *in situ* hybridization (FISH) has been used to map most of the contigs to visible chromosomal bands, and more than 200 of the cosmid contigs have been accurately ordered along the chromosome by a high-resolution FISH technique. In addition, the *Eco*RI restriction sites have been mapped on more than 45 Mb of the overall cosmid map, and more than 450 genes and genetic markers have been localized with nearly 300 of these incorporated into the ordered map. Of the 2000 genes that are likely to be found on chromosome 19, several have been identified. One of these is the gene responsible for the most common form of adult muscular dystrophy, and another is responsible for a form of dwarfism. Another gene that has been linked to congenital kidney disease has been localized to a single contig that spans 1 Mb.

A highly integrated map of chromosome 16 has been completed by the Los Alamos National Laboratory Center for Human Genome Studies. The Los Alamos effort used mouse chromosomes that contained fragments of human chromosome 16. Natural breakpoints were used to divide the chromosome into 78 fragments, each with lengths of about 1.1 Mb. Low-resolution contig maps were generated using 700 yeast artificial chromosome (YAC) clones and high-resolution contig maps were generated using 4000 cosmid clones. Sequence-tagged sites (STSs) that are short, unique stretches of DNA sequences have been used to tie the high- and low-resolution maps together. Genes that have been mapped to chromosome 16 are linked to blood disorders, another form of kidney disease, leukemia, and breast and prostate cancers.

While several hundred Mb have been sequenced and archieved in databases, most of these are from STSs on cloned fragments. Approximately 1% (30 Mb) of human DNA has been sequenced in longer stretches with the longest being about 685 kb. Eventually, many parts of the genome will have to be sequenced multiple times in order to reveal the differences that represent the different forms of various genes.

Efforts are also under way to develop new cloning vectors. YACs represent the classic tool for cloning large fragments of human DNA, but some portions of the human genome resist cloning in YACs. Bacterial artificial chromosomes, P1 bacteriophages, and P1-derived artificial cloning systems have also been developed. New approaches are needed to ensure that the entire genome can be represented in clone libraries without deletions, rearrangements, or insertions.

The HGP has provided many challenges in the areas of instrumentation and informatics. When the project began, it cost $2–10 to sequence a single bp, and a single researcher could produce 20–50 kb/year of contiguous, accurate sequence. From the beginning of the HGP, efforts have been made to reduce the cost of sequencing by developing new methods. One of the central goals for 1998 was to develop the capacity to sequence 50 Mb/year in long contiguous segments. If this is achieved, then a fully sequenced human genome should be produced by 2005.

One new sequencing method involves high-density nucleic acid arrays with more than 10,000 probes, different DNA fragments whose sequences are known, on a small chip (1.28 × 1.28 cm), a technique that was inspired by the semiconductor industry. In this method, fluorescently labeled target fragments of DNA are incubated with the chip. Regions in the target fragments that are complementary to the probes in the array form hybrids, with the extent of hybridization depending on the extent of the match to the probe. Color images can then be used to depict levels of fluorescence from bound target fragments. This method can be used to help sequence the genome, to look for mutations, and to detect specific pathogens.

A major development as a result of the HGP has involved the production of commercial robots that can perform repetitive tasks, such as the replication of large clone libraries, the pooling of libraries, and the development of arrays of clonal libraries for hybridization studies. High-speed, robotics-compatible, thermal cyclers have also been developed which have accelerated PCR amplifications. Another new instrument is used to rapidly sort human chromosomes for chromosome-specific libraries.

A less visible, but no less challenging, aspect of the HGP involves informatics. Information systems are needed for data acquisition, data analysis, and management and public dissemination of results. Laboratory data acquisition and management systems are used to construct genetic maps, physical maps, and DNA sequences and to analyze gene expression. Systems can also include software for controlling robots as well as databases for tracking biological materials, experimental procedures, and results. Data analysis systems are used to interpret map and sequence data. Oak

Ridge National Laboratory has developed the GRAIL system that provides a world-standard, gene identification tool. More than 180 Mb were analyzed with GRAIL in 1995.

The Human Genome Management Information System (HGMIS) was established by the DOE in 1989. HGMIS maintains web sites and publishes a newsletter that is used to disseminate information about the HGP and to keep scientists, policymakers, and the general public informed about the program's funded research. This reduces duplicative research efforts and fosters collaborations.

Additional ethical, legal, and social issues are related to the HGP. Issues of privacy and the fair use of genetic information must be resolved. Approximately 3–5% of the budget for the HGP has been devoted to studying these issues that surround the availability of genetic information. Some of the issues involve who will have access to genetic information and for what purposes. For example, will insurance companies be able to deny coverage to women who carry genes that predispose them to develop breast cancer? Will employers be able to withhold jobs from potential employees who carry genes for heart disease? In the future, we will be able to detect more congenital disorders using fetal genetic testing than we can today. Will this lead to decision making that involves selection of fetuses that carry desirable genes? The HGP has opened the door to solving many of life's mysteries and to curing many genetic disorders, but it has also opened the door to a plethora of ethical dilemmas.

EXERCISES

1. Search the literature and find one major breakthrough in biotechnology that occurred during the past year. Draw a time line that indicates the major breakthroughs that occurred in biotechnology between the late 1800s and the current year. Include the recent major breakthrough that you discovered in your search.

2. What fragments would be produced if the DNA shown in Example Problem 13.1 were cut with *Hind*III and *Alu*I? How many base pairs would each fragment contain? Which fragment would travel farthest if a mixture of the two was subjected to gel electrophoresis?

3. The *E. coli* plasmid cloning vector, pBR322, can be cut in many more sites than the few mentioned in Example Problem 13.2. What would be the lengths of the fragments that would result from cutting pBR322 with the following restriction enzymes: *Sgr*A I (409), *Eco*57I (3000 and 4048), and *Acc*I (651 and 2244)? Assume that the cuts are made after the locations given in parentheses. What would be the order of fragments on the gel if they were subjected to electrophoresis?

4. Assume that pBR322 was cut with the restriction enzymes used in Example Problem 13.2 and those used in Exercise 3. What would be the lengths of the resulting fragments and the order in which they would appear on the gel after electrophoresis?

5. Cutting a DNA standard by a series of restriction endonucleases resulted

in fragments that had the following numbers of base pairs: 23,130, 9416, 6557, 4361, 2322, 2027; 564, and 125. During gel electrophoresis, the fragments traveled 2.5, 2.7, 2.9, 3.2, 3.9, 4.1, 6.7, and 8.0 cm, respectively. Develop an equation that relates fragment size in bps to distance traveled for the standard DNA. What would be the sizes of fragments of an unknown DNA that traveled 3.4, 6.4, and 7.1 cm in the same gel?

6. What would be the sequence of nucleotides for the remaining eight nucleotides in the DNA shown in Fig. 13.11?

7. One part of normal hemoglobin includes a sequence of valine, histidine, leucine, threonine, proline, glutamic acid, and glutamic acid, whereas the same section in hemoglobin that results in sickle cell anemia includes a sequence of valine, histidine, leucine, threonine, proline, valine, and glutamic acid. Develop a sequence for cDNA that would represent the healthy gene. What change in the gene resulted in the hemoglobin that causes sickle cell anemia?

8. Using the World Wide Web (WWW) find a company that specializes in the production of monoclonal antibodies. Briefly describe the method that the company would use to produce a made-to-order MCAb.

9. Hemophilia is a sex-linked inherited disorder that is characterized by the inadequate production of clotting factors. Nearly all hemophilia A patients, those who cannot produce factor VIII, who were treated before 1985 with factor VIII made from human blood either carry HIV or have died from AIDS. The gene for factor VIII, a large complex glycoprotein (265 kDa), was cloned in 1984, and factor VIII is now synthesized in a mammalian kidney cell line (BHK). Explain why factor VIII must be grown in a mammalian cell line rather than in a bacteria such as *E. coli*. Why would it be advantageous to continue to obtain factor VIII from a genetically engineered cell line rather than from human blood even though the latter is now screened for and, most likely, free of HIV contamination?

10. Erythropoietin (EPO), produced in the kidneys, is a glycoprotein hormone (30–35 kDa) with an unusually high proportion of carbohydrate (40%). Since EPO stimulates the production of red blood cells in the bone marrow, kidney failure can result in anemia. Amgen, a biotechnology company, has produced a line of Chinese hamster ovary cells that express EPO and has marketed the product as EPOGEN. Briefly describe a recombinant product offered either by Amgen or by another biotechnology company that is used in the treatment of chronic hepatitis C virus.

11. Using the WWW, find and print a diagram of a human chromosome that has a gene which causes muscular dystrophy or breast cancer.

12. Briefly discuss one advance in health care technology that should result from the HGP.

13. Briefly describe a recent development in instrumentation that is used to sequence DNA.

14. Briefly describe a recent development in medical informatics that is directly related to the HGP.

SUGGESTED READING

Anonymous. Polymerase chain reaction (PCR), *http://www.gene.com/ae/AB/GG/polymerase.html*.

Anonymous. Polymerase chain reaction — From simple ideas, *http://www.gene.com/ae/AB/IE/PCR_From_Simple_Ideas.html*.

Anonymous. Tools of the trade, *http://www.ornl.gov/TechResources/Human_genome/tko/05_tools.html*.

Braham, R. (1997, January). Medical electronics. *IEEE Spectrum*, 99–102.

Butler, M. (1996). *Animal Cell Culture and Technology — The Basics*. IRL Oxford Univ. Press, Oxford.

Doran, P. M. (1995), *Bioprocess Engineering Principles*. Academic Press, London.

Gould, J. L., and Keeton, W. T. (1996), *Biological Science, 6th ed*. Norton, New York.

Kreuzer, H., and Massey, A. (1996). *Recombinant DNA and Biotechnology — A Guide for Students*. ASM, Washington, DC.

Lander, E. S., and Waterman, M. S. (Eds.) (1995). *Calculating the Secrets of Life*. National Academy Press, Washington, DC.

Larkin, P. (Ed.) (1994). *Genes at Work: Biotechnology*. CSIRO, East Melbourne, Victoria, Australia.

Lawrence, C. B. (Ed.) (1995). Managing data for the human genome project, *IEEE Eng. Med. Biol. Mag.* **14**(6), 688–761.

Le Doux, J. M., Morgan, J. R., and Yarmush, M. L. (1995). Antisense technology. In *The Biomedical Engineering Handbook* (J. D. Bronzino, Ed.), pp. 1472–1488. CRC Press, Boca Raton, FL.

Mardis, E. R. (1995). Technical improvements in high throughput genome sequencing. *IEEE Eng. Med. Biol. Mag.* **14**(6), 794–797.

Moses, V., and Moses, S. (1995). *Exploiting Biotechnology*. Harwood Academic Publishers, Reading, UK.

National Research Council (1992). *Putting Biotechnology to Work — Bioprocess Engineering*. National Academy of Sciences, Washington, DC.

Peters, P. (1993). *Biotechnology — A Guide to Genetic Engineering*. Brown, Dubuque, IA.

Ramabhadran, T. V. (1994). *Pharmaceutical Design and Development — A Molecular Biology Approach*. Ellis Horwood, New York.

Rudolph, F. B., and McIntire, L. V. (Eds.) (1996). *Biotechnology — Science, Engineering, and Ethical Challenges for the Twenty-First Century*. Joseph Henry Press, Washington, DC.

Russell, A. J., and Vierheller, C. (1995). Protein engineering. In *The Biomedical Engineering Handbook* (J. D. Bronzino, Ed.), pp. 1445–1450. CRC Press, Boca Raton, FL.

Smith, J. E. (1996). *Biotechnology, 3rd ed*. Cambridge Univ. Press, Cambridge, UK.

Starr, C. (1997). *BIOLOGY — Concepts and Applications, 3rd ed*. Wadsworth, Belmont, CA.

Sundaram, S., and Yarmush, D. M. (1995). Monoclonal antibodies and their engineered fragments. In *The Biomedical Engineering Handbook* (J. D. Bronzino, Ed.), pp. 1451–1471. CRC Press, Boca Raton, FL.

Wells, W. (1997). Extreme chemistry, *http://www.gene.com/ae/AB/IWT/1297xtremo.html*.

14 RADIATION IMAGING

Chapter Contents

At the conclusion of this chapter, the reader will be able to:

- Understand the fundamental principles of radioactivity

- Understand that ionizing radiation, as generally employed for medical imaging, is either externally produced and detected after it passes through the patient or introduced into the body, making the patient the source of radiation emissions

- Understand the operation of basic nuclear instrumentation and imaging devices such as scintillation counters and gamma cameras

- Become familiar with the basic concepts of computerized tomography systems.

14.1   INTRODUCTION

Throughout history, a persistent goal of medicine has been the development of procedures for determining the basic cause of a patient's distress. As a result, the search for tools capable of "looking into" the human organism with minimal harm to the patient has always been considered important. However, not until the later part of the twentieth century have devices capable of providing "images" of normal and diseased tissue within a patient's body been available. Today, modern imaging devices based on fundamental concepts in physical science (e.g., X-ray and nuclear physics and acoustics) and incorporating the latest innovations in computer technology and data processing techniques have not only proved to be extremely useful in patient care but also revolutionized health care.

Over a century ago (1895), the first Nobel laureate, physicist Wilhelm Conrad Roentgen, described a new type of radiation, X rays, that ultimately led to the birth of a new medical specialty, radiology, and the medical imaging industry. Initially, these radiographic imaging systems were very rudimentary, primarily providing images of broken bones or contrast-enhanced structures such as the urinary or gastrointestinal systems. However, since the 1970s, advances in imaging techniques and, in particular, the use of the computer have revolutionized the application of radiographic imaging techniques in medical diagnosis. The same excitement that surrounded Roentgen followed Allan Macleod Cormack and Godfrey Newbold Hounsfield. Cormack in the physics department at Tufts University and Hounsfield in the research laboratories of the English company, EMI Limited, worked independently of each other, but obviously under the spell of a common dream, and accomplished quite different things. Cormack elegantly demonstrated the mathematical rudiments of image reconstruction in a remarkable paper published in 1963. Less than a decade later, Hounsfield unveiled an incredible engineering achievement — the first commercial instrument capable of obtaining digital axial images with high-contrast resolution for medical purposes. In recognition of the significant advances made possible by the development of computerized, tomography, Cormack and Hounsfield shared the Noble prize in physiology and medicine in 1979. The excitement of their discovery has not yet subsided.

Ionizing radiation (i.e., radiation capable of producing ion pairs) as generally employed for medical imaging is (i) Radiation that is introduced into the body, thereby making the patient the "source" of radiation emissions, or (ii) externally produced radiation which passes through the patient and is detected by radiation-sensitive devices "behind" the patient.

During cellular or organ system function studies, gamma or X rays are emitted from radiopharmaceuticals and detected outside the patient, providing physiological (rate of decay or "washout") as well as anatomical (imaging) information. More than 90% of the diagnostic procedures in nuclear medicine use emission imaging techniques. On the other hand, externally produced radiation which passes through the patient is the basis of operation for X ray machines and computerized tomography. These devices and the procedures that use them are housed within departments of radiology. The purpose of this chapter is to present the fundamental principles of operation for each of these radiation imaging modalities.

14.2 EMISSION IMAGING SYSTEMS

Nuclear medicine is the branch of medicine that employs emission scanning for the purpose of helping physicians arrive at a proper diagnosis. An outgrowth of the atomic age and ushered in by the advances made in nuclear physics and technology, during World War II, nuclear medicine emerged as a powerful and effective approach for detecting and treating specific physiological abnormalities.

The field of nuclear medicine is a classic example of a medical discipline that has embraced and utilized the concepts developed in the physical sciences. Conceived as a "joint venture" between the clinician and the physical scientist, it has evolved into an interdisciplinary field of activity with its own body of knowledge and techniques. In the process, the domain of nuclear medicine has grown to include studies pertaining to

- The creation and proper utilization of radioactive tracers (or radiopharmaceuticals) that can be safely administered into the body
- The design and application of nuclear instrumentation devices and systems to detect and display the activity of these radioactive elements
- The determination of the relationship between the activity of the radioactive tracer and specific physiological processes

To better understand this radiation imaging modality, it is necessary to discuss radioactivity, its detection, and the instruments available to monitor the activity of radioactive materials.

14.2.1 Basic Concepts

In 1895, Roentgen opened a new realm of scientific inquiry when he announced the discovery of a new type of penetrating radiation, which he called X rays. X rays are a form of electromagnetic (EM) energy just like radio waves and light. The main difference between X rays and light or radio waves, however, is in their frequency or wavelength. Table 14.1 shows the EM radiation spectrum. X rays typically have a wavelength from 100 to 0.01 nm, which is much shorter than light or radio waves.

Spurred on by Roentgen's discovery, the French physicist Henri Becquerel inves-

TABLE 14.1 Electromagnetic Wave Spectrum

Energy (eV)	Frequency (Hz)		Wavelength (m)
4×10^{-11}	10^4		10^4
4×10^{-10}	10^5	AM radio waves	10^3
4×10^{-9}	10^6		10^2
4×10^{-8}	10^7	Short radio waves	10^1
		FM radio waves and TV	
4×10^{-7}	10^8		10^0
4×10^{-6}	10^9		10^{-1}
4×10^{-5}	10^{10}	Microwaves and radar	10^{-2}
4×10^{-4}	10^{11}		10^{-3}
4×10^{-3}	10^{12}	Infrared light	10^{-4}
4×10^{-2}	10^{13}		10^{-5}
4×10^{-1}	10^{14}	Visible light	10^{-6}
4×10^{0}	10^{15}	Ultraviolet light	10^{-7}
4×10^{1}	10^{16}		10^{-8}
4×10^{2}	10^{17}		10^{-9}
4×10^{3}	10^{18}	X−ray	10^{-10}
4×10^{4}	10^{19}		10^{-11}
4×10^{5}	10^{20}		10^{-12}
4×10^{4}	10^{21}	Gamma ray	10^{-13}
4×10^{7}	10^{22}	Cosmic ray	10^{-14}

tigated the possibility that known fluorescent or phosphorescent substances produced a type of radiation similar to the X rays discovered by Roentgen. In 1896, Becquerel announced that certain uranium salts also radiated, that is, emitted penetrating radiations, whether or not they were fluorescent. These results were starting and presented the world with a new and entirely unexpected property of matter.

Urged on by scientific curiosity, Becquerel convinced Marie Curie, one of the most promising young scientists at the École Polytechnique, to investigate exactly what it was in the uranium that caused the radiation he had observed. Madame Curie, who subsequently was to coin the term **radioactivity** to describe the emitting property of radioactive materials, and her husband Pierre, an established physicist by virtue of his studies of piezoelectricity, became engrossed in their search for the mysterious substance. In the course of their studies, the Curies discovered a substance far more radioactive than uranium. They called this substance radium and announced its discovery in 1898.

While the Curies were investigating radioactive substances, there was a tremendous flurry of activity among English scientists who were beginning to identify the constituent components of atoms. Since the turn of the nineteenth century, Dalton's chemical theory of atoms in which he postulated that all matter is composed of atoms that are indivisible reigned supreme as the accepted view of the internal composition of matter. However, in 1897, when J. J. Thompson identified the electron as a negatively charged particle having a much smaller mass than the lightest atom, a new concept of the basic elements comprising matter had to be formulated. Our current view of atoms was developed in 1911 by Ernest Rutherford, who demonstrated that the principal mass of an atom was concentrated in a dense positively charged nucleus surrounded by a cloud of negatively charged electrons. This idea became incorporated into the planetary concept of the atom, a nucleus surrounded by very light orbiting electrons, which was conceived by Niels Bohr in 1913.

With this insight into the elementary structure of atoms, Rutherford was the first to recognize that radioactive emissions involve the spontaneous disintegration of atoms. After the general acceptance of this basic concept, Rutherford was awarded a Nobel prize in chemistry in 1908, and the "mystery" began to unfold. By observing the behavior of these radioactive emissions in a magnetic field, the Curies discovered that there are three distinct types of active radiation emitted from radioactive material. The three, arbitrarily called alpha, beta, and gamma by Rutherford, are now known to be (i) alpha particles, which are positively charged and identical to the nucleus of the helium atom; (ii) beta particles, which are negatively charged electrons; and (iii) gamma rays, which are pure electromagnetic radiation with zero mass and charge.

14.2.2 Elementary Particles

By the 1930s, the research in this field had clearly identified three elementary particles: the electron, the proton, and the neutron, usually considered the building blocks of atoms. Consider the arrangement of these particles as shown in Fig. 14.1,

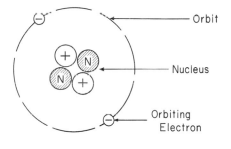

$\ominus$ = Electron: mass = 9.1×10^{-31} kilograms
charge = -1.6×10^{-19} Coulombs

$\oplus$ = Proton: mass = 1.6×10^{-27} kilograms
charge = $+1.6 \times 10^{-19}$ Coulombs

N = Neutron: mass = 1.6×10^{-27} kilograms
charge = 0

Fig. 14.1 Planetary view of atomic structure. The primary mass is in the nucleus, which contains protons and neutrons. The nucleus, which has a net positive charge, is surrounded by smaller orbiting electrons. In stable atoms, the net charge of electrons in orbit is equal and opposite to that of the nucleus. The atom illustrated is helium.

a common view of the atom. The atom includes a number of particles: (i) one or more electrons, each having a mass of about 9.1×10^{-31} kg and a negative electrical charge of 1.6×10^{-19} Cs; (ii) at least one proton with a mass of 1.6×10^{-27} kg, which is approximately 1800 times that of the electron; and (iii) perhaps neutrons, which have the same mass as protons but possess no charge.

Since the mass of an individual nuclear particle is very small and, when expressed in grams, involves the use of unwieldy negative exponents, the system of atomic mass units (amu) has been developed which uses carbon 12 ($_6^{12}$ C) as a reference atom. The arbitrary value of 12 mass units has been assigned to carbon-12. The masses of all other atoms are based on a unit which is 1/12 the mass of the carbon-12 atom. The mass of the lightest isotope of hydrogen is thus approximately 1 amu.

The mass in grams of an isotope that is numerically equal to its atomic mass is called a **gram-atomic mass**. A gram-atomic mass of a substance is also referred to as a mole or equivalent mass of the substance. Since gram-atomic masses have magnitudes proportional to the actual masses of the individual atoms, it follows that one mole of any substance contains a definite number of atoms. The number of atoms in one gram-atomic mass is given by Avogadro's number (N_A) and is equal to 6.023×10^{23}.

The mass of any atom in grams can thus be found by dividing the gram-atomic mass of the isotope by Avogadro's number. For example, the mass of an atom of carbon-12 is its gram-atomic mass (12 g) divided by Avogadro's number (N_A):

$$\text{mass of }^{12}\text{C} = \frac{12\text{g}}{6.023 \times 10^{23}}$$

Since ^{12}C is equal to 12 amu, the mass of 1 amu would be 1/12 of the mass of a ^{12}C atom:

$$1 \text{ amu} = \frac{1.99 \times 10^{-23}\text{g}/^{12}\text{C atoms}}{12 \text{ amu}/^{12}\text{C atom}}$$

$$1 \text{ amu} = 1.66 \times 10^{-24}\text{g}$$

The electrical charge carried by the electron is a fundamental property of matter, as is the mass of a particle. Since this is the smallest amount of electricity that can exist, it is usually expressed as a negative unit charge (-1). When expressing a unit charge in the metric system of units (meters, kilograms, and seconds), the value for the unit charge is 1.6×10^{-19} C. The charges carried by all other atomic particles are therefore some integral multiple of this value. Consequently, it is impossible for a particle to have, for example, a charge equal to 25 times that of the electron.

Neutrons and protons exist together in the nucleus and have been given the collective name of **nucleons**. The total number of nucleons in the nucleus of an element is called the **atomic mass**, or **mass number**, and is represented by the symbol (A), whereas the number of protons alone is referred to as the atomic number (Z). Different atoms having the same A are called isobars, whereas those with the same number of neutrons are called isotones.

The various combinations of neutrons and protons that may exist in nature are illustrated by examining the composition of the nuclei of the three types of hydrogen atom (hydrogen, deuterium, and tritium). Although hydrogen has a nucleus consisting of a single proton, the combination of one proton and one neutron exists as a single particle called a **demtron** and is the nucleus of the atom called **heavy hydrogen,** or deuterium. Extending this concept further, the combination of two protons and two neutrons forms a stable particle — the alpha particle, which in nature exists as the nucleus of the helium atom. Alpha radiation emitted from radioactive substances consists essentially of a stream of such particles. Tritium, the third atom of hydrogen, on the other hand, has a nucleus consisting of only one proton and two neutrons.

These three types of hydrogen are examples of atoms whose nuclei have the same number of protons [i.e., have the same atomic number (Z)] and at the same time may have a different number of neutrons and, thereby, a different atomic mass (A). Atoms exhibiting this characteristic were given the name **isotope** (from the Greek word meaning "same place"). The term **isotope** has been widely used to refer to any atom, particularly a radioactive one. However, current usage favors the word nuclide to refer to a particular combination of neutrons and protons. Thus, isotopes are nuclides that have the same atomic number. All elements with an atomic number $(Z) > 83$ and/or atomic mass $(A) > 209$ are radioactive, that is, they decay spontaneously into other elements, and this decay causes the emission of active particles.

Certain symbols have been developed in order to specifically designate each individual atom. In the process of reading the literature in this field, it is therefore necessary to become familiar with them. In the United States, it was the practice to place the atomic number as a subscript before and the atomic mass as a superscript after the chemical symbol of the atom $(_{53}I^{131})$. Since the chemical symbol itself also specifies the atomic number, one often omits it and simply writes I^{131}). In Europe, on the other hand, it was customarily written as a superscript prior to the chemical symbol (^{131}I). In an effort to achieve international standards, it was agreed in 1964 that the atomic mass should be placed as a superscript preceding the chemical symbol (^{131}I). When superscripts are not used, a more literal form of designation, such as cobalt 60, is commonly used. Referring to the helium atom, one simply writes ^{4}He, where 4 is equal to the atomic weight (because of the aggregation of protons and neutrons in the nucleus).

14.2.3 **Atomic Structure and Emissions**

As already mentioned, Bohr's model suggested the existence of an atomic structure that is analogous to the planetary system. In this system, electrons rotated in discrete orbits or shells around the nucleus, and the orbital diameters were determined by a quantum number (n) having integer values. These orbits were then represented by K, L, M, and N, corresponding to an increasing number of n, a nomenclature still in use today. This model was further refined in 1925 by Wolfgang Pauli in terms of quantum mechanical principles. Pauli's work on atomic structure explained various observed phenomena, including the estimates for binding energy of the electrons at various orbits of an atom. Pauli further observed that an atom can be defined by four quantum numbers: (i) n is the principal quantum number, which is an integer and scalar quantity; (ii) l is the angular momentum quantum number, a vector quantity which has integral values ranging from 0 to $n - 1$; (iii) m_1 is the magnetic quantum number with integral values ranging from -1 to 1; and (iv) m_s is the spin magnetic quantum number, which has the values of $1/2$ and $-1/2$. According to the **Pauli exclusion principle,** no two electrons in an atom can have the same set of quantum numbers.

In an electrically neutral atom, the number of orbital electrons exactly balances the number of positive charges in the nucleus. The chemical properties of an atom are determined by the orbital electrons since they are predominantly responsible for molecular bonding, light spectra, fluorescence, and phosphorescence. Electrons in the inner shells, on the other hand, are more tightly bound and may be removed from their orbits only by considerable energy such as by radiation interaction.

The amount of energy required to eject an electron from an orbit is equivalent to the binding energy for that shell, which is highest for the electrons at the innermost shell. The energy required to move an electron from an inner shell to an outer shell is equal to the difference in binding energies between the two shells. This energy requirement represents one of the natural characteristics of an element. When this characteristic energy is released as a photon, in the case of transition of an electron from an outer shell to an inner shell, it is known as a **characteristic X ray.** However, if instead of the emission of a photon, the energy is transferred to another orbital electron (called an Auger electron), which will be ejected from orbit. The probability for the yield of characteristic X ray in such a transition is known as **fluorescent yield.**

Protons and neutrons (also known as nucleons) experience a short-range nuclear force that is far greater than the electromagnetic force of repulsion between the protons. The movement of nucleons is often described by a shell model, analogous to orbital electrons. However, only a limited number of motions are allowed, and they are defined by a set of nuclear quantum numbers. The most stable arrangement is known as the **ground state.** The other two broad arrangements are (i) the **metastable state,** when the nucleus is unstable but has a relatively long lifetime before transforming into another state, and (ii) the **excited state,** when the nucleus is so unstable that it has only a transient existence before transforming into another state. Thus, an atomic nucleus may have separate existences at two energy levels, known as **isomers** (both have the same Z and same A). An unstable nucleus ultimately transforms itself to a more stable condition, either by absorbing or releasing energy (photons or particles) to a nucleus at ground state. This process is known as **radioactive transformation** or **decay.** As stated previously, naturally occuring heavier elements, having $Z > 83$, are all unstable.

Assessment of nuclear binding energy is important in determining the relative stability of a nuclide. This binding energy represents the minimum amount of energy necessary to overcome the nuclear force required to separate the individual nucleons. This can be assessed on the basis of mass–energy equivalence as represented by $E = mc^2$, where E, m, and c represent energy, mass, and speed of light, respectively. This has led to the common practice of referring to masses in terms of electron-volts (eV). The mass of an atom is always found to be less than the sum of the masses of the individual components (neutrons, protons, and electrons). This apparent loss of mass (Δm), often called mass defect or deficiency, is responsible for the binding energy of the nucleus and is equivalent to some change in energy (Δmc^2). As mentioned previously, the mass of a neutral ^{12}C atom has been accepted as 12.0 amu. The sum of the masses of the components of ^{12}C, however, is 12.10223 amu. The difference in masses (0.10223 amu) is equivalent to 95.23 MeV of binding energy for this nucleus, or 7.936 MeV/nucleon (obtained by dividing MeV by $A = 12$) for ^{12}C. For nuclei with atomic mass numbers > 11, the binding energy per nucleon ranges between 7.4 and 8.8 MeV. One atomic mass unit, therefore, is equal to $1.6605655 \times 10^{-24}$ g or, using the mass–energy relation, is equivalent to 931.502 MeV. The resting mass of an electron, on the other hand, is very small, i.e., only 0.511 MeV.

Example Problem 14.1

Verify that the energy released by 1 amu is 931 MeV.

Solution

The conversion can be expressed by Einstein's equation:

$$E = mc^2$$

where m is the mass in grams and c is the velocity of light in cm/s. Then, the energy equivalent (E) of 1 amu is given by

$$E = (1.66 \times 10^{-24}\text{g})(3 \times 10^{10}\text{cm/s})^2$$

$$E = 1.49 \times 10^{-3}\text{g/cm}^2\text{/s}^2$$

The unit, g/cm^2/s^2 is frequently encountered in physics and is termed the erg; thus,

$$E = 1.49 \times 10^{-3} \text{ ergs}$$

Another convenient unit of energy is the electron volt. An electron volt is defined as the amount of kinetic energy acquired by an electron when it is accelerated in an electric field produced by a potential difference of 1 V. Since the work done by a difference of potential, V, acting on a charge, e, is Ve and the charge on one electron is 1.6×10^{-19}C, it is possible to calculate the amount of energy in one electron volt as follows:

$$1 \text{ eV} = 1.6 \times 10^{-19} \text{ C} \times 1 \text{ V}$$

$$1 \text{ eV} = 1.6 \times 10^{-19}\text{J}$$

or

$$1 \text{ eV} = 1.6 \times 10^{-12} \text{ ergs}$$

Since the electron volt is a very small amount of energy, it is more commonly expressed in thousands of electron volts (keV) or millions of electron volts (MeV):

$$1 \text{ MeV} = 1,000,000 \text{ eV}$$

$$1 \text{ keV} = 1000 \text{ eV}$$

It is now possible to express the atomic mass unit in MeV:

$$1 \text{ amu} = 1.49 \times 10^{-3} \text{ ergs, and}$$

$$1 \text{ MeV} = 1.60 \times 10^{-6} \text{ ergs; therefore,}$$

$$1 \text{ amu} = 931 \text{ MeV}$$

Therefore, if 1 amu could be completely transformed into energy, 931 MeV would result. To gain a concept of the magnitude of the electron volt, 1 MeV is enough to lift only a milligram weight one millionth of a centimeter.   ■

An unstable nuclide, commonly known as a **radionuclide,** eventually transforms to a stable condition with the emission of ionizing radiation after a specific probability of life expectancy. In general, there are two classifications of radionuclides: natural and artificial. Naturally occurring radionuclides are those nuclides which emit radiation spontaneously and therefore require no additional energy from external sources. Artificial radionuclides, on the other hand, are essentially man-made and are produced by bombarding so-called stable nuclides with high-energy particles. Both types of radionuclides play an important role in emission scanning and nuclear medicine. The average life or half-life, the mode of transformation or decay, and the nature of emission (type and energy of the ionizing radiation) constitute the basic characteristics of a radionuclide.

All radioactive materials, whether they occur in nature or are artificially produced, decay by the same types of processes, i.e., they emit alpha, beta, and/or gamma radiations. As just discussed, the emission of alpha and beta particles involves the disintegration of one element, often called parent, into another, the daughter. The modes or phases of transformation in the process of a radionuclide passing from an unstable to stable condition may be divided into six different categories: (i) α decay/emission, (ii) β− (negatron) decay, (iii) β+ (positron) decay, (iv) electron capture (EC), (v) isomeric transition (IT), and (vi) fission.

Alpha Decay

Alpha particles are ionized helium atoms ($_2^4\text{He}$) moving at a high velocity. If a nucleus emits an alpha particle, it loses two protons and two neutrons, thereby reducing Z by 2 and A by 4. A number of naturally occurring heavy elements undergo such a decay. For example, a parent nucleus $_{92}^{238}\text{U}$ emits an alpha particle, thereby changing to a daughter nucleus $_{90}^{234}\text{Th}$ (thorium). In symbolic form this process may be written as follows:

$$_{92}^{238}\text{U} \rightarrow \, _{90}^{234}\text{Th} + \, _2^4\text{He}$$

Note the following facts about this reaction: (i) The atomic number (number of protons) on the left is the same as that on the right (92 = 90 + 2) since charge must be conserved, and (ii) the mass number (protons plus neutrons) on the left is the same as that on the right (238 = 234 + 4).

When one element changes into another, as in alpha decay, the process is called transmutation. In order for alpha emission to occur, the mass of the parent must be greater than the combined mass of the daughter and the alpha particle. In the decay process, this excess mass is converted into energy and appears in the form of kinetic energy in the daughter nucleus and the alpha particle. Most of the kinetic energy is carried away by the alpha particle because it is much less massive than the daughter nucleus. That is, because momentum must be conserved in the decay process, the lighter alpha particle recoils with a much higher velocity than the daughter nucleus. Generally, light particles carry off most of the energy in nuclear decays.

Example Problem 14.2

Radium, $^{226}_{88}$Ra, decays by alpha emission. What is the daughter element that is formed?

Solution

The decay can be written symbolically as

$$^{226}_{88}\text{Ra} \rightarrow X + {}^{4}_{2}\text{He}$$

where X is the unknown daughter element. Requiring that the mass numbers and atomic numbers balance on the two sides of the arrow, the daughter nucleus must have a mass number of 222 and an atomic number of 86:

$$^{226}_{88}\text{Ra} \rightarrow {}^{222}_{86}X + {}^{4}_{2}\text{He}$$

The periodic table shows that the nucleus with an atomic number of 86 is radon (Rn). ∎

Example Problem 14.3

In Example Problem 14.2, the $^{226}_{88}$Ra nucleus underwent alpha decay to $^{222}_{86}$Rn. Calculate the amount of energy liberated in this decay. Take the mass of $^{226}_{88}$Ra to be 226.025406 amu, that of $^{222}_{86}$Rn to be 222.017574 amu, and that of $^{4}_{2}$He to be 4.002603 amu.

Solution

After decay, the mass of the daughter, m_d, plus the mass of the alpha particle, m_α, is

$$m_d + m_\alpha = 222.017574 \text{ amu} + 4.002603 \text{ amu} = 222.020177 \text{ amu}$$

Thus, calling the mass of the parent nucleus M_p, the mass lost during decay is

$$\Delta_m = (M_p - (m_d + m_\alpha))$$
$$= 226.025406 \text{ amu} - 226.020177 \text{ amu} = 0.005229 \text{ amu}$$

Using the relationship, 1 amu = 931.5 MeV, the energy liberated is

$$E = (0.005229 \text{ amu})(931.50 \text{ MeV/amu}) = 4.87 \text{ MeV}$$ ∎

Negatron (β−, or β−, γ) Decay

When a radioactive nucleus undergoes beta decay, the daughter nucleus contains the same number of nucleons as the parent nucleus but the atomic number (Z) is increased by 1. A typical beta decay event is

$$\,_{6}^{14}C \rightarrow \,_{7}^{14}N + \,_{-1}^{0}e$$

The superscripts and subscripts on the carbon and nitrogen nuclei follow usual conventions, but those on the electron need explanation. The −1 indicates that the electron has a charge whose magnitude is equal to that of the proton but negative. The 0 used for the electron's mass number indicates that the mass of the electron is almost zero relative to that of carbon and nitrogen nuclei.

The emission of electrons from a nucleus is suprising because the nucleus is usually thought to be composed of protons and neutrons only. This apparent discrepancy can be explained by noting that the electron that is emitted is created in the nucleus by a process in which a neutron is transformed into a proton. This can be represented by the following expression:

$$\,_{0}^{1}n \rightarrow \,_{1}^{1}p + \,_{-1}^{0}e$$

Example Problem 14.4

Find the energy liberated in the beta decay of $\,_{6}^{14}C$ to $\,_{7}^{14}N$.

Solution

$\,_{6}^{14}C$ has a mass of 14.003242 amu and $\,_{7}^{14}N$ has a mass of 14.003074 amu. Here, the mass difference between the initial and final states is

$$\Delta_m = 14.003242 \text{ amu} - 14.003074 \text{ amu} = 0.000168 \text{ amu}$$

This corresponds to an energy release of

$$E = (0.000168 \text{ amu})(931.50 \text{ MeV/amu}) = 0.156 \text{ MeV} \quad \blacksquare$$

Only a small number of electrons have this kinetic energy. Most of the emitted electrons have kinetic energies less than this predicted value. If the daughter nucleus and the electron are not carrying away with this liberated energy, then the requirement that energy is conserved leads to the question, what accounts for the missing energy?

In 1930, Pauli proposed that a third particle must be present to carry away the "missing" energy and to conserve momentum. Enrico Remi later named this particle the **neutrino** (little neutral one) because it had to be electrically neutral and have little or no resting mass. Although it eluded detection for many years, the neutrino (symbol v) was finally detected experimentally in 1950. The neutrino has the following properties:

1. It has zero electric charge.
2. It has a resting mass smaller than that of the electron.
3. It interacts very weakly with matter and is therefore very difficult to detect.

Phosphorus 32 is a typical example of a pure $\beta-$ emitter (^{32}P is transformed to ^{32}S) that has been used for therapy.

Very often a nucleus that undergoes ratioactive decay is left in an excited energy state. The nucleus can then undergo a second decay to an even lower energy state by emitting one or more photons. The process is very similar to the emission of light by an atom. An atom emits radiation to release some extra energy when an electron "jumps" from a state of high energy to a state of lower energy. Likewise, the nucleus uses essentially the same method to release any extra energy it may have following a decay or some other nuclear event. In nuclear deexcitation, the jumps that release energy are made by protons or neutrons in the nucleus as they move from a higher energy level to a lower level. The photons emitted in such a deexcitation process are called gamma rays and have very high energy relative to the energy of visible light. Most of the radionuclides undergoing $\beta-$ decay also emit γ rays almost simultaneously. For example, iodine 131 emits several $\beta-$ and γ rays in this process.

The following sequence of events represents a typical situation in which gamma decay occurs:

$$\, ^{12}_{5}\text{B} \rightarrow \, ^{12}_{6}\text{C*} + \, ^{0}_{-1}e$$

and

$$\, ^{12}_{6}\text{C*} \rightarrow \, ^{12}_{6}\text{C} + \gamma$$

The first process represents a beta decay in which ^{12}B decays to ^{12}C*, where the asterisk indicates that the carbon nucleus is left in an excited state. The excited carbon nucleus then decays to a ground state by emitting a γ ray. Note that a γ ray emission does not result in any change in Z or A.

Positron ($\beta+$ or $\beta+$, γ) Decay

With the introduction of the neutrino, the beta decay process in its correct form can be written as

$$\, ^{14}_{6}\text{C} \rightarrow \, ^{14}_{7}\text{N} + \, ^{0}_{-1}e + \bar{v}$$

Where the bar in the symbol $\bar{v}$ indicates that this is an antineutrino. To explain an antineutrino, consider the following decay process:

$$\, ^{12}_{7}\text{N} \rightarrow \, ^{12}_{6}\text{C} + \, ^{0}_{1}e + v$$

When ^{12}N decays into ^{12}C, a particle is produced that is identical to the electron except that is has a positive charge of $+e$. This particle is called a **positron**. Because it is like the electron in all respects except charge, the positron is said to be antiparticle to the electron.

It should be noted that positrons ($\beta+$) are created in the nucleus, as if a proton was converted to a neutron and a positron. Positron decay produces a different element by decreasing Z by 1, with A being the same. For example, ^{11}C decays to the predominant stable isotope of boron (i.e., ^{11}C transforms to ^{11}B). Positrons (like negatrons) share their energy with neutrinos. Positron decay is also associated with γ-ray emissi.he positron, once emitted, however, is often annihilated as a result of a collision with an electron with 1 ns. This produces a pair of photons of 0.511 MeV that move in opposite directions. A minimum transition energy of 1.022 MeV is required for any positron decay.

Example Problem 14.5

^{12}N beta decays to an excited state of ^{12}C which subsequently decays to the ground state with emission of a 4.43 MeV γ ray. What is the maximum energy of the emitted beta particle?

Solution

The decay process for positive beta emission is

$$^{12}_{7}N \rightarrow ^{12}_{6}C^* + {^+e} + v \rightarrow ^{12}_{6}C + \gamma$$

Conservation of energy requires that a nucleus X will decay into a lighter nucleus X^1 with an emission of one or more particles (collectively designated as x) only if the mass of X is greater than the total mass $X^1 + x$. The excess mass energy is known as the Q value of the decay. In this process

$$Q = [m(^{12}_{7}N) - m(^{12}_{6}C^*) - 2\,m_e]c^2$$

To determine Q for this decay, the mass of the product nucleus ^{12}C in its excited state must be found. In the ground state, ^{12}C has a mass of 12.000000 amu so its mass in the excited state is

$$12.000000 \text{ amu} + \frac{4.43 \text{ MeV}}{931.5 \text{ MeV/amu}} = 12.004756 \text{ amu}$$

$Q = [12.018613 - 12.004756 \text{ amu} - 2 \times 0.000549 \text{ amu}]\, 931.5 \text{MeV/amu} - 11.89 \text{MeV}$

∎

Electron Capture or K Capture

An orbital electron, usually from the inner shell (K shell), may be captured by the nucleus (as if a proton captured an electron and converted itself to a neutron). An electron capture process produces a different element by decreasing Z by 1, with A being the same (similar to β+ decay). Indeed, some radionuclides have definite probabilities of undergoing either positron decay or electron capture (such as iron 52, which decays with about 42% EC and 58% β+ emission). Furthermore there could be an associated gamma emission with electron capture. For example, ^{51}Cr (chromium 51) transforms to ^{51}V (vanadium 51) with about 90% going directly to the ground state. The remaining 10% goes to an excited state of ^{51}V followed by transition to ground state with the emission of photons. An electron capture also causes a vacancy in the inner shell, which leads to the emission of a characteristic X ray or Auger electron. The absence of high-energy electrons (beta particles) in EC causes low-radiation absorbed dose to the tissue.

Isomeric Transition and Internal Conversion

Radionuclides at a metastable state emit only γ rays. The element remains the same with no change in A (isomeric transition). The atomic mass number of the iso-mer is therefore denoted by Am or mA. For example, ^{99}mTc (technetium 99m) decays to ^{99}Tc. However, there is a definite probability that instead of a photon coming out, the energy may be transferred to an inner orbital electron. This is known as **internal conversion**, and the internally converted electrons are close to monoenergetic beta particles. For example, barium 135m (decaying by IT) emits about 84% IC electrons. They also create a vacancy in the shell that subsequently leads to the emission of characteristic X rays and Auger electrons.

Nuclear Fission

Usually a heavy nuclide may break up into two nuclides that are more or less equal fragments. This may happen spontaneously, but it is more likely with the capture of a neutron. The uranium fission products mostly range between atomic numbers 42 and 56. Many medically useful radionuclides are produced as fission products, such as Xenon 133, which may be extracted by appropriate radiochemical procedures.

14.2.4 Radioactive Decay

The reduction of the number of atoms through disintegration of their nuclei is known as radioactive decay and is characteristic of all radioactive materials. Unaffected by changes in temperature, pressure, or chemical combination, the rate of the decay process remains constant with the same number of disintegrations occurring during each interval of time. Furthermore, this decay process is a random event. Consequently, every atom in a radioactive element has the same probability of disintegrating.

As the decay process continues, it is clear that fewer atoms will be available to disintegrate. This fraction of the remaining number of atoms that decay per unit of time is called the **decay constant** (λ). The half-life and decay constant are obviously related since the larger the value of the decay constant (λ), the faster the process of decay and, consequently, the shorter the half-life. In any event, the decay constant is an unchanging value throughout the decay process. Each radionuclide exhibits a distinctive disintegration process because of its inherent properties that is, its decay constant (λ) and half-life ($T_{\frac{1}{2}}$). All nuclides decay in the same manner but not at the same rate since this parameter is determined by the unique nature of the particular radioactive element in question.

Considering a number of atoms (N) of a specific type of radionuclide present at a time (t), the transformation rate can be defined by $-dN/dt$ (the minus sign denotes the decay/decrease), which would be proportional to the number of atoms, or

$$dN/dt = \lambda N \tag{14.1}$$

where λ is the decay constant.

Taking the initial number of atoms as N_0 ($N = N_0$ when $t = t_0$), integration gives

$$N = N_0 e^{-\lambda t} \tag{14.2}$$

where N is the number of radioactive nuclei present at time t, N_0 is the number present at time $t = 0$, and $e = 2.718$ is the base of the natural logarithm (Fig. 14.2).

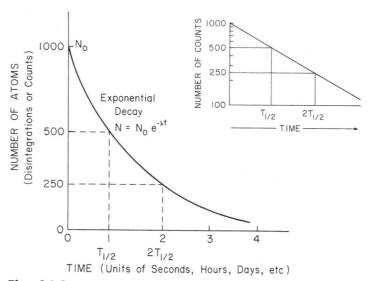

Fig. 14.2   The rate of decay of a radioactive material is exponential. The natural logarithm (see insert) of this decay process is therefore a straight line.

The unit of activity is the curie (Ci) and is defined as

$$1 \text{ Ci} \equiv 3.7 \times 10^{10} \text{ decays/s}$$

This number of decay events per second was selected as the original activity unit because it is the approximate activity of 1 g of radium. The SI unit of activity is the becquerel (Bq), where

$$1 \text{ Bq} = 1 \text{ decay/s}$$

Therefore, $1 \text{ Ci} = 3.7 \times 10^{10} \text{ Bq}$. The most commonly used units of activity are the millicurie (10^{-3} Ci) and the microcurie (10^{-6} Ci).

One of the most common terms encountered in any discussion of radioactive materials is half-life, $T_{\frac{1}{2}}$. This is because all radioactive substances follow the same general decay pattern. After a certain interval of time, half of the original number of nuclei in a sample will have decayed, and then in a second time interval equal to the first, half of those nuclei remaining will have decayed, and so on. The half-life is the time required for half of a given number of radioactive nuclei to decay. The half-life, or $T_{\frac{1}{2}}$ (the time corresponding to transformation of 50% of the nuclides when $N = N_0/2$), may be obtained by solving for $\lambda T_{\frac{1}{2}}$ in Eq. (14.2). Therefore,

$$T_{\frac{1}{2}} = \ln 2/\lambda = 0.693/\lambda \tag{14.3}$$

Example Problem 14.6

Derive the equation for $T_{\frac{1}{2}}$.

Solution

Start with $N = N_0 e^{-\lambda t}$. After a time interval equal to one half-life, $t = T_{\frac{1}{2}}$, the number of radioactive nuclei remaining is $N = N_0/2$. Therefore,

$$N_0/2 = N_0 e^{-\lambda t_{\frac{1}{2}}}$$

Dividing both sides by N_0 gives

$$1/2 = e^{-\lambda T_{\frac{1}{2}}}$$

Taking the natural logarithm of both sides of this equation eliminates the exponential factor on the right since $\ln e = 1$:

$$\ln 1/2 = -\lambda T_{\frac{1}{2}}$$

$$-0.693 = -\lambda T_{\frac{1}{2}}$$

$$T_{\frac{1}{2}} = 0.693 \qquad \blacksquare$$

After a second half-life, the number remaining is again reduced by one-half. Hence, after a time $2T_{\frac{1}{2}}$, the number remaining is $N_0/4$, and so forth. Half-lives range from about 10^{-22} s to 10^{21} years.

Example Problem 14.7

The half-life of the radioactive nucleus $^{226}_{88}\text{Ra}$ is 1.6×10^3 years. If a sample contains 3×10^{16} such nuclei, determine the activity.

Solution

First, calculate the decay constant, λ, using the fact that

$$T_{\frac{1}{2}} = 1.6 \times 10^3 \text{ years} = (1.6 \times 10^3 \text{ years}) (3.15 \times 10^7 \text{ s/year})$$

$$= 5.0 \times 10^{10}\text{s}$$

Therefore,

$$\lambda = \frac{0.693}{T_{\frac{1}{2}}} = \frac{0.693}{5.0 \times 10^{10}\text{s}} = 1.4 \times 10^{-11}\text{s}^{-1}$$

The activity, or decay rate, of the sample at $t = 0$ is calculated using the form $-(dN_0/dt = \lambda N_0 = R_0$, where R_0 is the decay rate at $t = 0$ and N_0 is the number of radioactive nuclei present at $t = 0$. Since $N_0 = 3 \times 10^{16}$,

$$R_0 = \lambda N_0 = (1.4 \times 10^{-11} \text{ s}^{-1})(3 \times 10^{16}) = 4.2 \times 10^5 \text{ decays/s}$$

Since 1 Ci$=3.7 \times 10^{10}$ decays/s, the activity, or decay rate, is

$$R_0 = 11.3 \text{ } \mu\text{Ci} \qquad \blacksquare$$

Progress in emission scanning and nuclear medicine has been linked to the availability of radionuclides that could be used in human subjects in appropriate chemical forms. The choice of a radionuclide depends on its physical characteristics in relation to diagnostic and therapeutic applications and the possibility of incorporating it into an appropriate chemical compound that is suitable for biomedical investigation.

A radiopharmaceutical is a specific chemical compound that is labeled with a radionuclide. This may be a simple inorganic salt or a complex organic molecule. It is well recognized that the chemical properties of an element reflect some of the possible biological behaviors, and the behavior of several elements within the same group of the periodic table appears similar. For example, strontium 85 has been used to represent calcium metabolism in bone. Table 14.2 summarizes some of the important radionuclides that are currently used in nuclear medicine.

14.2.5   Measurement of Radiation: Units

It should now be apparent that all radioactive substances decay and in the process emit various types of radiation (alpha, beta, and/or gamma) during this process. However, since they usually occur in various combinations, it is difficult to measure each type of radiation separately. As a result, measurement of radioactivity is usually accomplished by using one of the following techniques: (i) counting the number of disintegrations that occur per second in a radioactive material, (ii) noting how effective this radiation is in producing atoms (ions) possessing a net positive or negative charge, or (iii) measuring the energy absorbed by matter from the radiation that penetrates it. Using these fundamental concepts, three kinds of radiation units have been established: the curie (Ci), the roentgen (R), and the radiation-absorbed dose (rad).

Attention should be paid to the fact that the curie defines the number of disintegrations per unit time and not the nature of the radiation, which may be α, β, or γ rays. The curie is simply a measure of the activity of a radioactive source. The roentgen and the rad, on the other hand, are units based on the effect of the radiation on an irradiated object. Thus, while the curie defines a source, the roentgen and rad define the effect of the source on an object. One of the major effects of X-ray or

TABLE 14.2   Commonly Used Radionuclides in Nuclear Medicine

Radionuclide	Half-life	Transition	Production	Chemical form
Carbon 11	20.38 min	β^+	Cyclotron	3-N-methylspiperone
Flourine 18	109.77 min	β^+	Cyclotron	Flourodeoxyglucose
Phosphorus 31	14.29 days	β^-	Reactor	Phosphates
Chromium 51	27.704 days	EC	Reactor	Sodium chromate
Cobalt 57	270.9 days	EC	Cyclotron	Cyanocobalamin
Gallium 67	78.26 h	EC	Cyclotron	Citrate complex
Molybdenum 99	66.0 h	β^-	Reactor	Molybdate in column
Technetium 99m	6.02 h	IT	Generator	TcO_4 and complexes
Indium 111	2.83 days	EC	Cyclotron	DTPA and oxine
Iodine 123	13.2 h	EC	Cyclotron	Mainly iodide
Iodine 123	60.14 days	EC	Reactor	Diverse proteins
Iodine 131	8.04 days	β^-	Reactor	Diverse compounds
Xenon 133	5.245 days	β^-	Reactor	Gas
Thallium 201	3.044 days	EC	Cyclotron	Thallous chloride

gamma radiation is the ionization of atoms; that is, the creation of atoms possessing a net positive or net negative charge (ion pair). The roentgen is determined by observing the total number of ion pairs produced by X-ray or gamma radiation in 1 cm^3 of air at standard conditions (at 760 mmHg and 0°C). Since each ion pair has electrical charge, this can be related to electrical effects that can be detected by various instruments. One roentgen is defined as that amount of X-ray or gamma radiation that produces enough ion pairs to establish an electrical charge separation of 2.58×10^{-4} C per kilogram of air. Thus, the roentgen is a measure of radiation quantity, not intensity. The rad, on the other hand, is based on the total energy absorbed by the irradiated material. One rad means that 0.01 J (the unit of energy in the metric system) of energy is absorbed per kilogram of material.

Since human tissue is exposed to various types of radioactive materials in nuclear medicine, another unit of measure, the rem (roentgen equivalent man), is often used to specify the biological effect of radiation. Consequently, the rem is a unit of human biological dose that results from exposure of the biological preparation to one or many types of ionizing radiation.

These radiation units are used in various situations in nuclear medicine. The roentgen or more commonly its submultiple, the milliroentgen (mr), is used as a value for most survey meter readings. The rad is used as a unit to describe the amount of exposure received, for example, by an organ of interest following injection of a radiopharmaceutical. The rem is the unit used to express exposure values of some personnel-monitoring devices such as film badges.

The classical method of dosimetry, however, has been replaced by a modern method of radiation dose calculation through the effort of the Medical Internal Radiation Dose committee. The internationally accepted unit for radiation dose to tissue is the gray (Gy). One Gy is equivalent to the absorption of an energy of 1 J per kilogram of the tissue under consideration. Previously, the radiation absorbed dose was expressed as rad. One rad is equivalent to the absorption of an energy of 100 ergs per gram or 0.01 J per kilogram, which is equal to 0.01 Gy. Thus, an expression of 1 rad per millicurie is equivalent to 0.27027 milligray per megabecquerel (mGy/MBq), or 1 mGy/MBq is equivalent to 3.7 rad/mCi. Most of the computations on radiation dosimetry that are available for nuclear medicine procedures, however, have been expressed in terms of rads.

Exposure of living organisms, mammals in particular, to a high level of radiation induces pathological conditions and even death. This feature is utilized in therapy for malignant diseases in which the intention is to deliver a localized high-radiation dose to destroy the undesirable tissue. Exposure to lower levels of radiation may not show any apparent effect, but it increases the risk of cancer as a long-term effect. It also increases the probability of genetic defects if the gonads are exposed. This imposes the need for minimization of radiation exposure to the worker and specific guidelines for medical applications.

The United States Nuclear Regulatory Commission has adopted standards which limit maximum exposure for the general public to 0.5 rem per year. Limits for occupational exposure are 1.25 rem/3 months for the whole body and 18.75 rem/3 months for the extremities. Thus, one of the most important radiation safety procedures is to monitor both the personnel and the work area. Routine personnel monitoring is usually done with film badges and ring-type finger badges. Radiation survey and wipe tests are carried out at certain intervals and at times of incidental/accidental contamination. The common shielding material is lead, with its thickness depending on the energy of γ rays. In addition, radioactive wastes are stored at assigned areas with appropriate shielding. The procedures used for radioactive waste disposal depend on the nature and half-life of the waste. All require a temporary storage facility, and they are finally disposed of locally according to a prescribed set of procedures.

14.3   INSTRUMENTATION AND IMAGING DEVICES

From the discussion of radiation measurement, it should be clear that radiation energy can be measured only indirectly, that is, by measuring some effect caused by the radiation. Various indirect techniques are used to measure radioactivity, including

1. Photography: Film blackens when it is exposed to a specific type of radiation such as X- rays (which are the equivalent of γ rays and will be discussed later).
2. Ionization: The passage of radiation through a volume of gas established in the probe of a gas detector produces ion pairs. The function of this type of detector depends on the collection of these ion pairs in such a way that they may be counted. This technique has been most effective in measuring alpha radiation and least effective in measuring gamma radiation.
3. Luminescence: This involves the emission of light that is not due to incandescence. This technique is extremely useful since the flash of light produced by the bombardment of a certain type of material with penetrating-type radiation can be detected and processed. As a matter of fact, the fluorescent effect produced by ionizing radiation is the basis of the scintillation detector discussed later. This type of indirect detection scheme is excellent for observing the presence of all three types of radiation.

14.3.1   Scintillation Detectors

Since the majority of modern detector systems in nuclear medicine utilize probes based on the scintillation principle, this will be described in greater detail. Scintillation detectors may be used for all types of radiation, depending on the particular type of scintillator used and its configuration. Regardless of the application, however, the general technique is the same for all scintillation probes. Certain materials, for example, zinc sulfide and sodium iodide, emit a flash of light, or scintillation, when struck by ionizing radiation. The amount of light emitted, over a wide range, is proportional to the energy expended by the particle in this material. When the scintillator material is placed next to the sensitive surface of an electronic device called photomultiplier, the light from the scintillator is converted into a series of small electrical pulses in which the pulse heights are directly proportional to the energy of the incident γ ray. These electrical pulses can then be amplified and processed in such a way as to provide the operator with information regarding the amount and nature of the radioactivity striking the scintillation detector. Thus scintillators may be used for diagnostic purposes to determine the amount and/or distribution of radionuclides in one or more organs of a patient.

Figure 14.3 illustrates the basic scintillation detection system. It consists of (i) a detector, which usually includes the scintillation crystal, photomultiplier tubes, and preamplifier; (ii) signal processing equipment, such as the linear amplifier and the single-channel pulse analyzer; and (iii) data display units such as the scaler, scanner, and oscilloscope. Once the radioactive event is detected by the crystal and an appropriate pulse is generated by the photomultiplier circuitry, the resulting voltage pulses are still very small. To avoid any serious loss of information caused by distortion from unwanted signals (such as noise) and to provide a strong enough signal to be processed and displayed, the amplifier is used to increase the amplitude of the pulses by a constant factor. This process is called linear amplification.

In such a system, it should be apparent that because of the wide variation in energies of γ rays striking the scintillation crystal, the linear amplifier receives pulses having a wide variation in "pulse height." This is fine if detecting all the radiant energy under the probe is of interest. If, on the other hand, the operator is interested only in the activity of a specific radionuclide, additional processing is necessary. This is accomplished by the pulse height analyzer. Using this device, the operator can discriminate against all radiant energy other than the one of interest. In this way, the activity of a specific radionuclide is analyzed by allowing only those pulses related to it to be processed. Consequently, single-channel and multichannel pulse height analyzers are commonly found in nuclear medicine laboratories. These processed data must then be displayed in such a way as to provide information regarding the amount of radioactivity present and its location within the body. This information is very important in determining the status of the organs under investigation. Studies of the amount of radioactivity present as a function of time enable physicians to

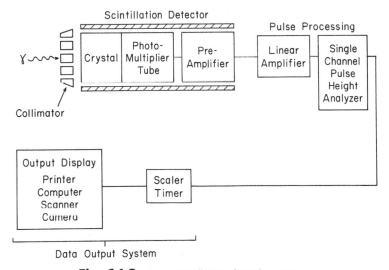

Fig. 14.3 Basic scintillation detection system.

ascertain whether the organ is functioning properly, whereas studies of the location of the radionuclide enable physicians to display the organ or abnormal tissue. Both types of studies are valuable.

In order to provide a measure of the scintillation events occurring during many nonimaging applications, it is often necessary to count these events and provide some means to display this information. The scaler is the most common type of electronic device found in nuclear medicine that accomplishes this task. A scaler is used to count the pulses produced by the detector system and processed by a variety of electronic pieces of equipment. Thus, the scaler is an electronic device that accepts signal pulses representing a range of energy levels (energy of the incident radiation) and counts them. The scaler is usually designed so that the operator has a choice between accepting a certain number of counts (preset count) or a predetermined period of time over which the counts can be accumulated (preset time). With preset time, the scaling device will count the number of events that occur during a set period of time and will then shut off automatically — thus the count is the variable. With preset count, the predetermined number of counts are accumulated, after which the scaler is automatically shut off. In this case, time is the variable.

Both types of data can be observed by the operator by viewing the front panel of the scaler. However, this information can also be supplied to other devices for analysis. Consider the case of operating the scaler in the present time mode. This information can be presented in a variety of ways. For example, it can be (i) displayed as counts per unit time (second, minute, etc.) for continual observation by the operator, (ii) supplied to a digital printer to provide a running tabulation of the counts as a function of time, or (iii) directed to a computer that can retain the counts in memory and perform a variety of calculations on the incoming data as they are being collected (if the process is slow enough) and at the same time provide the operator with a visual display (on the screen of an oscilloscope) in the form of a plot of radioactive decay. Thus, once the radiation is detected by the scintillation detector and subsequently measured by the scaler, the operator can be presented with information in a variety of formats that may be of clinical value.

Scintillation data output may also be processed through an imaging device. In general these devices take the pulse output from a detector and the electronic processing devices already described and place the pulse in some representation spatially, according to its point of origin in the radioactive source. Obviously, there is some distortion in this representation since the point of origin lies in a three-dimensional plane, whereas most common instruments are capable of only displaying that point of origin in two dimensions. However, despite this obvious shortcoming, this technique has proved to be of tremendous clinical value.

The most commonly used stationary detector system is the scintillation camera or gamma camera. This device views all parts of the radiation field continuously and is therefore capable of operating almost like a camera by building up an image quickly. Initially developed by Hal Anger during the late 1950s, the first Anger camera reflected the convergence of the disciplines of nuclear physics, electronics, optics, and data processing in a clinical setting. Its ultimate acceptance and further development has had a profound impact not only on the practice of clinical nuclear medicine but also on the entire diagnostic process. The initial concepts introduced by Anger became basic to the art of imaging specific physiological processes.

14.3.2 The Gamma Camera

The operation of the basic gamma camera is illustrated in Fig. 14.4. The detector of the gamma camera is placed over the organ to be scanned. A collimator is placed over the base of the scintillation crystal in order to localize the radiation from a given point in the organ and send it to an equivalent point on the detector. Since γ rays cannot be "bent," another technique must be used to selectively block those γ rays which, if allowed to continue on their straight-line path, would strike the detector at sites completely unrelated to their points of origin in the subject. This process of selective interference is accomplished by the collimator. To prevent unwanted off-axis γ rays from striking the crystal, collimators usually contain many narrow parallel apertures made of heavy-metal absorbers.

Consider the multihole collimator illustrated in Fig. 14.5A. In this case, the collimator consists of a flat lead plate through which narrow holes have been drilled. As can be seen, only a gamma event occurring directly under each hole will penetrate the collimator, and it will be represented at only one location on the face of the crystal. If this gamma event occurred much further away from the collimator, such as in region Y, then it would be represented by more than one location on the face of the crystal. When this occurs, the resolution, which may be defined as the ability of the detector to distinguish between two sources at various distances from the collimator, is greatly decreased. For the multihole collimator, then, the best resolution occurs when the area of interest is close to the collimator. Thus, as the subject is

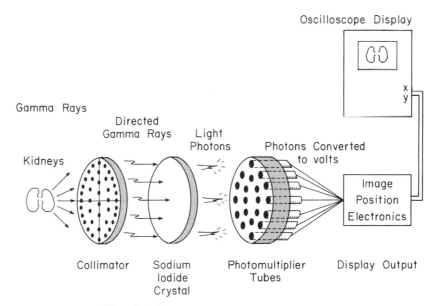

Fig. 14.4 Basic elements of gamma camera.

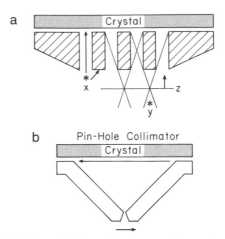

Fig. 14.5 (a) Multihole collimator used in conjunction with scintillation detector. Event x is represented at only one location in the crystal, whereas event y has multiple sites associated with its occurrence. (b) The pinhole collimator allows magnification and is used for viewing small organs at short range.

moved from point Z toward the detector, the resolution is improved. Obviously, when viewing an organ that lies beneath the surface of the skin, it is very important to closely approximate its distance from the probe relative to the degree of resolution required.

There are other types of collimators, for example, the pinhole collimator depicted in Fig. 14.5B. The pinhole collimator permits entry of only those rays aimed at its aperture. These γ rays enter the collimator and proceed in a straight line to the crystal where they are detected in inverted spatial correspondence to their source. When the source is located at a distance from the collimator equal to that of the pinhole to the crystal, then the source is represented on the crystal in exactly the same size as it exists. However, by proper positioning of the subject, it is possible to actually magnify or decrease the field of view of the detector. That is, as the source is moved closer to the aperture of the collimator, magnification occurs. A pinhole collimator, therefore, enlarges and inverts the image of the source located beneath it.

From this brief discussion, it can be seen that collimators in conjunction with a scintillation detector essentially "focus" the radioactivity that occurs at a particular point in space within the organ to a particular point on the surface of the crystal. The radiation passing through the collimator then impinges on or interacts with the scintillation crystal.

Currently, simple gamma camera system is the most important as well as the most basic instrument for diagnostic nuclear medicine. Interfacing with a digital computer makes the system even more versatile. Digital image-processing systems perform data collection, storage, and analyses. Data acquisition requires digitization of the image by an analog-to-digital converter, which divides a rectangular image area into small elements, or pixels, usually a 64 × 64, 128 × 128, or 256 × 256 matrix.

As a result, a particular region of interest can be selected to obtain certain quantitative information. Digital computation also improves dynamic studies since the regional rate of uptake and clearance pattern can easily be obtained from the serial images for any particular region, making it possible, for example, to study cardiac wall motion.

14.3.3 Positron imaging

The positrons emitted through transformation of a radionuclide can travel only a short distance in a tissue (a few millimeters) and then are annihilated. A pair of 511-keV photons that travel at 180° to one another is created. A pair of scintillation detectors can sense positron emission by measuring the two photons in coincidence. Annihilation coincidence detection provides a well-defined cylindrical path between the two detectors. Multihole collimators are unnecessary to define the position in a positron camera because electronic collimation accomplishes the task. Positron cameras have limited use; however, they have attained great importance in a highly modified form in positron-emitting transaxial tomography or positron emission tomography (PET scanning). Many small NaI(TI) detectors are arranged in an annular form so that annihilation photons in coincidence at 180° permit detection of position. Tomographic images are obtained by computer-assisted reconstruction techniques. Several large centers use positron emission tomography in conjunction with cyclotron-produced short-lived radiopharmaceuticals (mainly carbon 11, nitrogen 13, and fluorine 18) to provide structural as well as metabolic information.

14.4 RADIOGRAPHIC IMAGING SYSTEMS

This section discusses externally produced radiation which passes through the patient and is detected by radiation-sensitive devices behind the patient. These radiographic imaging systems rely on the differential attenuation of X rays to produce an image. Initial systems required sizable amounts of radiation to produce distinct images of tissues with good contrast (i.e., the ability to differentiate between body parts having minimally different density, such as fat and muscle). High-contrast differences, such as those between soft tissue and air (as in the lung) or bone and muscle, can be differentiated with a much smaller radiation dose. However, with the development of better film and other types of detectors, the radiation dosage has decreased for both high- and low-contrast resolution. Even more significant was the development of the computerized axial tomography (CAT) scanner in the 1970s. This device produces a cross-sectional view of a patient, instead of the traditional shadow graph recorded by conventional X-ray systems, by using computers to reconstruct the X-ray attenuation data. In the process, it provides the clinician with high-contrast resolved images of virtually any body part and can be reconstructed in any body plane. The basic concepts underlying routine X-ray imaging will be reviewed to aid in understanding the significance of this imaging modality.

14.4.1 Basic Concepts

All X-ray imaging systems consist of an X-ray source, a collimator, and an X-ray detector. Diagnostic medical X-ray systems utilize externally generated X-rays with energies of 20–150 keV. Since the turn of the twentieth century, conventional X-ray images have been obtained in the same fashion, that is, by using a broad-spectrum X-ray beam and photographic film. In general, X rays are produced by a cathode ray tube that generates a beam of X rays when excited by a high-voltage power supply. This beam is shaped by a collimator and passes through the patient, creating a latent image in the image plane. Depending on the type of radiographic system employed, this image is detected by X-ray film, an image intensifier, or a set of X-ray detectors. Using the standard film screen technique, X rays pass through the body, projecting an image of bones, organs, air spaces, and foreign bodies onto a sheet of film (Fig. 14.6). The "shadow graph" images obtained in this manner are the results of the variation in the intensity of the transmitted X-ray beam after it has passed through tissues and body fluids of different densities.

This technique has the advantages of offering high-resolution, high-contrast images with relatively small patient exposure and a permanent record of the image. On the other hand, its disadvantages include significant geometric distortion, inability to discern depth information, and incapability of providing real-time imagery. As a result, conventional radiography is the imaging method of choice for such tasks as dental, chest, and bone imagery. Since bone strongly absorbs X rays, fractures are readily discernible by the standard radiographic technique. When this procedure is used to project three-dimensional objects into a two-dimensional plane, however, difficulties are encountered. Structures represented on the film overlap, and it becomes difficult to distinguish between tissues that are similar in density. For this reason, conventional X-ray techniques are unable to obtain distinguishable/interpretable images of the brain, which consists primarily of soft tissue. In an effort to overcome this deficiency, attempts have been made to obtain shadow graphs from many different angles in which the internal organs appear in different relationships to one another and to introduce a medium (such as air or iodine solutions) that is either translucent or opaque to X rays. However, these efforts are usually time-consuming, sometimes difficult, sometimes dangerous, and often just not accurate enough.

Fig. 14.6 Overview of the three X-ray techniques used to obtain "medical images." (a) A conventional X-ray picture is made by having the X rays diverge from a source, pass through the body, and then fall on a sheet of photographic film. (b) A tomogram is made by having the X-ray source move in one direction during the exposure and the film in the other direction. In the projected image, only one plane in the body remains stationary with respect to the moving film. In the picture, all other planes in the body are blurred. (c) In computerized tomography, a reconstruction from projections is made by mounting the X-ray source and an X-ray detector on a yoke and moving them past the body. The yoke is also rotated through a series of angles around the body. Data recorded by the detector are processed by a special computer algorithm or program. The computer generates a picture on a cathode-ray screen.

a

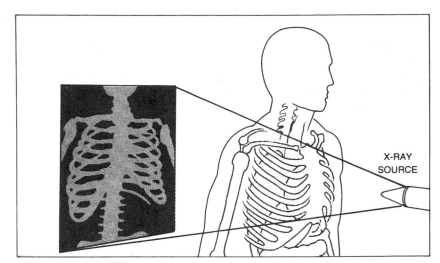

X-RAY
SOURCE

b

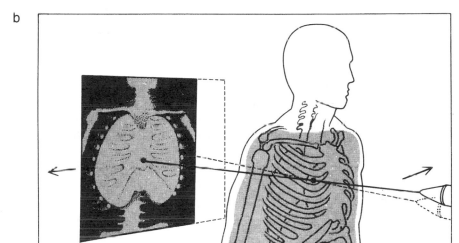

c

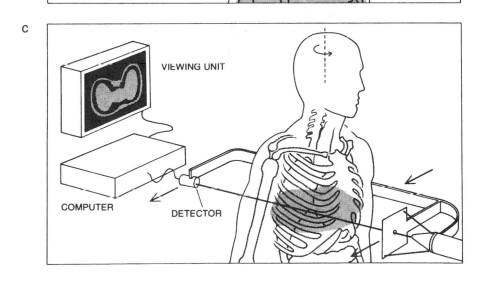

VIEWING UNIT

COMPUTER

DETECTOR

In the early 1920s, another X-ray technique was developed for visualizing three-dimensional structures. With this technique, known as plane tomography, the imaging of specific planes or cross sections within the body became possible. In plane tomography, the X-ray source is moved in one direction while the photographic film (which is placed on the other side of the body and picks up the X rays) is simultaneously moved in the other direction (Fig. 14.6b). The results of this procedure is that, while the X rays travel continuously changing paths through the body, each ray passes through the same point on the plane or cross section of interest throughout the exposure. Consequently, structures in the desired plane are brought sharply into focus and are displayed on film, whereas structures in all the other planes are obscured and show up only as a blur. Such an approach is clearly better than conventional methods in revealing the position and details of various structures and in providing three-dimensional information by such a two-dimensional presentation. There are, however, limitations in its use. First, it does not localize a single plane since there is some error in the depth perception that is obtained. Second, large contrasts in radiodensity are usually required in order to obtain high-quality images that are easy to interpret. In addition, X-ray doses for tomography are higher than those for routine radiographs, and because the exposures are longer patient motion may degrade the image content.

Computerized tomography (CT) represents a completely different approach. Consisting primarily of a scanning and detection system, a computer, and a display medium, it combines image-reconstruction techniques with X-ray absorption measurements in such a way as to facilitate the display of any internal organ in two-dimensional axial slices or by reconstruction in the z axis in three dimensions. The starting point is quite similar to that used in conventional radiography. A collimated beam of X rays is directed through the section of body being scanned to a detector that is located on the other side of the patient (Fig. 14.6c). With a narrowly collimated source and detector system, it is possible to send a narrow beam of X rays to a specific detection site. Some of the energy of the X rays is absorbed, whereas the remainder continues to the detector and is measured. In computerized tomography, the detector system usually consists of a crystal (such as cesium iodide or cadmium tungstate) that has the ability to scintillate or emit light photons when bombarded with X rays. The intensity of these light photons or "bundles of energy" is in turn measured by photodetectors and provides a measure of the energy absorbed (or transmitted) by the medium that is penetrated by the X-ray beam.

Since the X-ray source and detector system are usually mounted on a frame or "scanning gantry," they can be moved together across and around the object that is being visualized. In early designs, for example, X-ray absorption measurements were made and recorded at each rotational position traversed by the source and detector system. The result was the generation of an absorption profile for that angular position. To obtain another absorption profile, the scanning gantry holding the X-ray source and detector was then rotated through a small angle and an additional set of absorption or transmission measurements was recorded. Each X-ray profile or projection obtained in this fashion is basically one-dimensional. It is as wide as the body but only as thick as the cross section.

The exact number of these equally spaced positions determines the dimensions to be represented by the picture elements that constitute the display. For example, in order to generate a 160 × 160 picture matrix, absorption measurements from 160 equally spaced positions in each translation are required. It must be recalled, however, that each one-dimensional array constitutes one X-ray profile or projection. To obtain the next profile, the scanning unit is rotated a certain number of degrees around the patient and 160 more linear readings are taken at this new position. This process is repeated again and again until the unit has been rotated a full 180°. When all the projections have been collected, 160 × 180, or 28,800, individual X-ray intensity measurements are available to form a reconstruction of a cross section of the patient's head or body.

At this point, the advantages of the computer become evident. Each of the measurements obtained by the previously described procedure enters the resident computer and is stored in memory. Once all the absorption data have been obtained and located in the computer's memory, the software packages developed to analyze the data by means of image-reconstruction algorithms are called into action. These image-reconstruction techniques, which are based on known mathematical constructs developed for astronomy, were not used routinely until the advent of the computer because of the number of computations required for each reconstruction. Modern computer technology made it possible to fully exploit these reconstruction techniques.

To develop an image from the stored values of X-ray absorption, the computer initially establishes a grid consisting of many small squares for the cross section of interest, depending on the size of the desired display. The result of this process is analogous to the mesh of strings in a tennis racket. Since the cross section of the body has thickness, each of these squares represents a volume of tissue, a rectangular solid whose length is determined by the slice thickness and whose width is determined by the size of the matrix. Such a three-dimensional block of tissue is referred to as a voxel (or volume element) and is displayed in two dimensions as a pixel (or picture element) (Fig. 14.7).

During the scanning process, each voxel is irradiated by a narrow beam of X rays up to 180 times. Thus, the absorption caused by that voxel contributes to up to 180 absorption measurements, with each measurement part of a different projection. Since each voxel affects a unique set of absorption measurements to which it has contributed, the computer calculates the total absorption due to that voxel. Using the total absorption and the dimension of the voxel, the average absorption coefficient of the tissues in that voxel is determined precisely and displayed in a corresponding pixel as a shade of gray.

Example Problem 14.8

Demonstrate how X rays are attenuated as they pass through a cross-sectional area of a patient's body in multiple rays.

Solution

The cross-section of interest can be considered to be made up of a set of blocks of material (Fig. 14.8). Each block has an attenuating effect on the passage of the X-ray energy or photons, absorbing some of the incident energy passing through

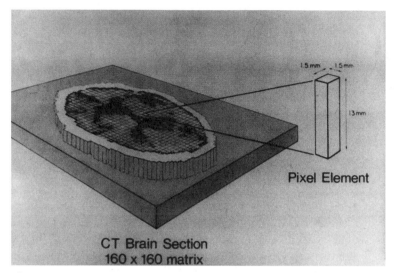

Fig. 14.7   Sketch of the cross-sectional image available using computerized tomography. Each image is divided into discrete three-dimensional sections of tissue referred to as voxels (or volume elements). On the computer monitor, the slice is viewed in two dimensions as pixels, or picture elements, in this case 1.5 × 1.5 mm.

it. In the line of blocks illustrated in Fig. 14.8, the first block absorbs a fraction A_1 of the incident photons, the second a fraction A_2, and so on so that the nth block absorbs a fraction A_n. The total fraction "A" absorbed through all the blocks is the product of all the fractions, whereas the logarithm of this total absorption fraction is defined as the measured absorption. Consider Fig. 14.8. The set of absorption measurements for an object that comprises only four blocks would be as follows:

$$\text{In position 1: } A(1) = A_1 \times A_3$$

$$\text{In position 2: } A(2) = A_2 \times A_4$$

$$\text{In position 3: } A(3) = A_1 \times A_2$$

$$\text{In position 4: } A(4) = A_3 \times A_4$$

In practice only the measured absorption factors $A(1)$, $A(2)$, $A(3)$, and $A(4)$ would be known. Therefore, problem is to compute A_1, A_2, A_3, and A_4 from the measured absorption values. The fact that the computation is possible can be seen from the previous four equations. There are four simultaneous equations and four unknowns, so a solution can be found. To reconstruct a cross section containing n rows of blocks and n columns, it is necessary to make at least n individual absorption measurements from at least n directions. For example, a display consisting of

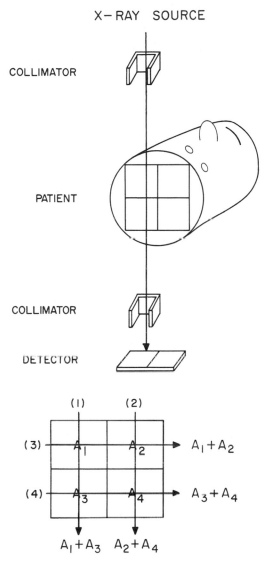

Fig. 14.8 Sketch of a manner in which the absorption coefficient values for each picture element are obtained.

320×320 picture elements requires a minimum of 320×320, or 102,400, independent absorption measurements. ∎

It was mentioned earlier that these absorption measurements are taken in the form of profiles. Imagine a plane parallel to the X-ray beam as defining the required slice. An absorption profile is created if the absorption of the emergent beam along a line perpendicular to the X-ray beam is plotted. This profile represents the total absorption along each of the X-ray beams. In general, the more profiles that are obtained, the better the contrast resolution of the resulting image.

From these individual measurements, a single two-dimensional plane can be reconstructed, and by simply (i.e., for the computer) stacking the appropriate sequence of such planes, it is possible to reconstruct a full three-dimensional picture. Image reconstruction of a three-dimensional object is therefore based primarily on a process of obtaining a cross section or two-dimensional image from many one-dimensional projections. The earliest method used to accomplish this was defined simply as "backprojection," which means that each of the measurement profiles was projected back over the area from which it was taken. Unfortunately, this simple approach was not totally successful because of blurring. To overcome this, iterative methods were introduced that successively modified the profile being backprojected until a satisfactory picture was obtained. Iteration is a good method but slow, and it requires several steps to modify the original profiles into a set of profiles that can be projected back to provide an unblurred picture of the original image. To speed up the process, mathematical techniques involving convolution or filtering were introduced that permit the original profile to be modified directly into the final one.

Whichever method of reconstruction is used, the final result of the computation is the same. In each case, a file, usually known as the picture file, is created in the computer memory. The picture file contains an absorption coefficient or density reading for each element of the final picture (e.g., 25,600 for 160×160 pixels or more than 100,000 for 320×320 pixels). The resultant absorption coefficients for each element can then be displayed as gray tones or color scales on a visual display. Most CT systems project an image onto the screen of a computer display terminal. Each element or pixel of the picture file has a value that represents the density (or more precisely the relative absorption coefficient) of a volume in the cross section of the body that is being examined.

The scale developed by Hounsfield that is shown in Fig. 14.9 demonstrates the values of absorption coefficients that range from air (-1000) at the bottom of the scale to bone at the top. This original scale covers approximately 1000 levels of absorption on either side of water, which is indicated as zero at the center. This is chosen for convenience since the absorption of water is close to that of tissue due to the high percentage of water that is found in tissue. In using this scale, the number 1000 is added to the absorption coefficient of each picture element. In the process, the absorption value of air becomes zero, whereas that of water becomes 1000. Having assigned each pixel to a particular value in this scale, the system is ready to display each element and create the reconstructed image.

Example Problem 14.9

Illustrate the process of backpropagation.

Solution

The process can be illustrated by means of a simple analog. In the following figure a parallel beam of X rays is directed past and through a cylinder-absorbing substance. As a result, a shadow of the cylinder is cast on the X-ray film. The density

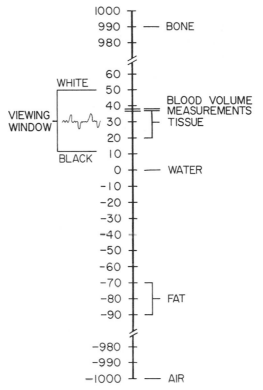

Fig. 14.9 Hounsfield scale for absorption coefficients.

of the exposed and developed film along the line AA can be regarded as a projection of the object. If a series of such radiographs are taken at equally spaced angles around the cylinder, these radiographs then constitute the set of projections from which the cross section has to be reconstructed.

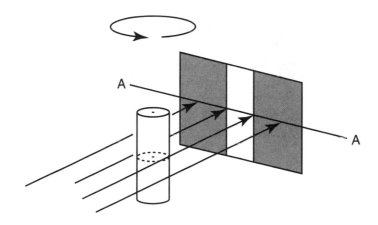

An approximate reconstruction can be reproduced by directing parallel beams of light through all the radiographs in turn from the position in which they were taken as shown in the following figure. The correct cross section can be reconstructed by backprojecting the original shadow as shown in the previous figure and subtracting the result of backprojecting two beams placed on either side of the original shadow as shown in the following.

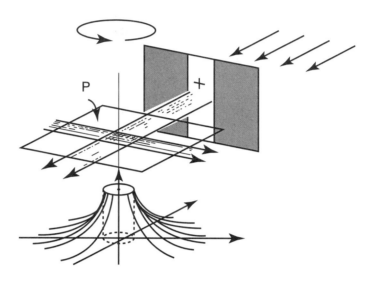

Mathematically, this is the equivalent of taking each transmission value in the projection and subtracting from it a quantity proportional to adjacent values. This process is called convolution and is actually used to modify projections. ■

In the display provided by any medical imaging device, factors affecting the resolution or accuracy of the image must be recognized and modified. Since the information provided by a CT scanner in essence reflects the distribution of X-ray absorption in a cross section of a patient, the smallest change in X-ray absorption that can be detected and projected in the image is defined as the density resolution. In addition, the smallest distance apart that two objects can be placed and still be seen as separate entities is defined as the spatial resolution. The density resolution reflects the sensitivity of the method to tissue change, whereas spatial resolution indicates the fineness of detail possible in the image produced by the CT scanner. These two characteristics of resolution are related to each other and to the radiation dose received by the patient. The clarity of the picture and hence the accuracy with which one can measure absorption values are often impaired by a low signal-to-noise ratio, which is caused by a reduction in the number of photons arriving at the detectors after penetrating the body. Fewer available photons, in turn, result in a reduction in the sharpness of the image since the resultant differences between the absorption coefficients of adjacent pixels are reduced. This is a situation that must be accepted as long as low-radiation doses are used. Despite this limitation, the reconstructed image can be enhanced by "display" units that constitute another major component of the CT system.

Display units usually have the capacity to display a given number (16, 32, etc.) of different colors or gray levels at one time. These different colors (or gray levels) can be used to represent a specific range of absorption coefficients. In this way, the entire range of absorption values in the picture can be represented by the entire range of available gray tones or colors (or both). To sharpen the resolution or detail of the display, the range of gray tones between black and white can be restricted to a very small part of the scale. This process of establishing a display "window" that can be raised or lowered actually enhances the imaging of specific organs. For example, if an image of heart tissue is desired, this window is raised above water density, but if an image of the detail in the lung is desired, it is lowered. The overall sensitivity can be further increased by reducing the window width. This permits ever-greater differentiation between absorption coefficients of specific organs and the surrounding tissue.

Other techniques are available to make the picture more easily interpretable. For instance, a portion of the picture can be enlarged or certain absorption values can be highlighted or made brighter. These methods do not add to the original information but enhance the image made possible by the data available in the computer's memory. Since a human observer can see an object only if there is sufficient contrast between the object and its surroundings, enhancement modifies the subjective features of an image to increase its impact on the observer, making it easier to locate and precisely measure obscure details.

There are many distinct advantages of CT scanning over conventional X-ray techniques. For example, CT provides three-dimensional information concerning the internal structure of the body by presenting it in the form of a series of slices (Fig. 14.10). Because of its contrast sensitivity, it can show small differences in soft tissue clearly, whereas conventional radiographs cannot. It accurately measures the X-ray absorption of various tissues, enabling the nature of these tissues to be studied, and yet the absorbed dose of X rays given to the patient by CT scanning is often less than the absorbed dose by a conventional X-ray technique such as tomography.

In comparing the CT approach with conventional tomography, which also images a slice through the body by blurring the image from the material on either side of the slice, it becomes apparent that in the tomographic approach only a thin plane of the beam produces useful information as the slice to be viewed. The remainder of the beam, which is by far the larger component, passes through material on either side of the slice, collecting unwanted information that produce artifacts in the picture. On the other hand, the X-ray beam in CT passes along the full length of the plane of the slice, thereby permitting measurements to be taken that are 100% relevant to only that slice. These measurements are not affected by the materials lying on either side of the section. As a result, the entire information potential of the X-ray beam is used to the fullest, and the image is more clearly defined.

The medical applications of such detailed pictures are considerable. Hard and soft tumors, cysts, blood clots, injured or dead tissue, and other abnormal morpho-

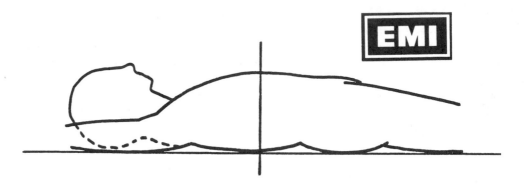

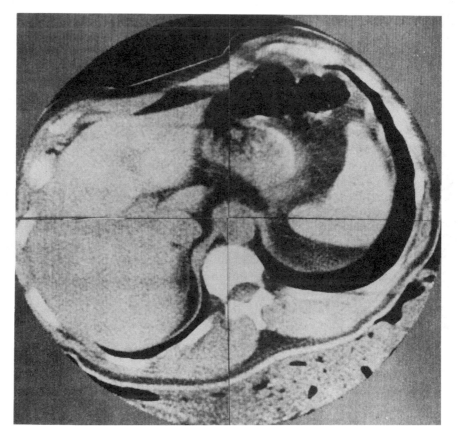

Fig. 14.10   Photography of a CT slice of the body.

logical conditions can be easily observed. The ability to distinguish between various types of soft tissue means that many diagnostic procedures can be simplified, thus eliminating the need for many methods that require hospitalization, are costly, and involve a potential risk to the patient due to their invasive nature. Since CT can accurately locate the areas of the body to be radiated and can monitor the actual progress of the treatment, it is effective not only in diagnosis but also in the field of radiotherapy.

14.4.2 CT Technology

Since its introduction, CT has undergone continuous refinement and development. In the process, CT scanning has developed into a technology of significant importance throughout the world. The manufacture of quality CT systems requires a strong base in X-ray technology, physics, crystallography, electronics, and data processing.

The evolution of CT scanners has resulted from improvements in (i) X-ray tube performance; (ii) scan times; (iii) image-reconstruction time; (iv) reduction and, in some cases, elimination of image artifacts; and (v) reduction in the amount of radiation exposure to patients. Achievements in these areas have focused attention on two key issues: image quality and its relationship to diagnostic value and scanner efficiency and its relationship to patient care and the economic impact it entails. In order to attain a better understanding of the operation and utility of current CT systems, it is important to understand some of the evolutionary steps.

CT scanners are usually integrated units consisting of three major elements: (i) the scanning gantry, which takes the readings in a suitable form and quantity for a picture to be reconstructed; (ii) the data handling unit, which converts these readings into intelligible picture information, displays this picture information in a visual format, and provides various manipulative aids to enhance the image and thereby assist the physician in forming a diagnosis; and (iii) a storage facility, which enables the information to be examined or reexamined at any time after the actual scan.

Scanning Gantry

The objective of the scanning system is to obtain enough information to reconstruct an image of the cross section of interest. All the information obtained must be relevant and accurate, and there must be enough independent readings to reconstruct a picture with sufficient spatial resolution and density discrimination to permit an accurate diagnosis. The operation of the scanning system is therefore extremely important.

CT scanners have undergone several major gantry design changes (Fig. 14.11). The earliest generation of gantries used a system known as "traverse and index." In this system, the tube and detector were mounted on a frame and a single beam of X rays traversed the slice linearly, providing absorption measurements along one profile. At the end of the traverse, the frame indexed through 1° and the traverse was repeated. This procedure continued until 180 single traverses were made and 180 profiles were measured. Using the traverse and index approach, the entire scanning procedure took approximately 4 or 5 min. The images it produced had excellent picture quality and hence high diagnostic utility. However, this approach had several disadvantages. First, it was relatively slow, resulting in relatively low patient output. Second, streak artifacts were common, and although they normally were not diagnostically confusing, the image quality did suffer. Finally, perhaps the biggest

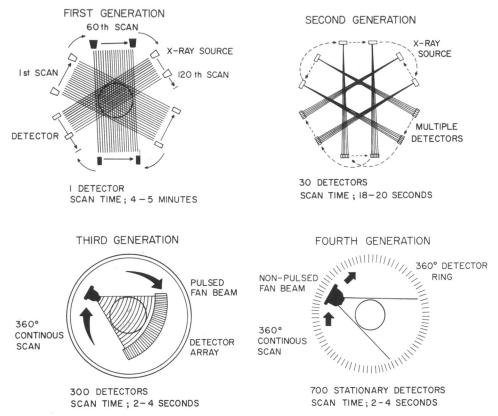

Fig. 14.11 Four generations of scanning gantry designs. With modern slip ring technology, third- or fourth-generation geometry allows spiral volumetric scanning using slice widths from 1 to 10 mm and pixel matrixes to 1024^2. Typically, a 50-cm volume can be imaged with a single breath hold.

drawback was that since the patient had to remain absolutely immobile during the scan, its use was restricted solely to brain scans.

In an attempt to solve these problems, a second generation of gantries evolved. The scanning procedures in this approach included a translation step as in the first generation, but incremented the scanning gantry 180° around the patient in 10° steps. In addition, in an effort to obtain more profiles with each traverse, a larger fan beam was used. By using a 10° fan beam, it was possible to take 10 profiles — at 1° intervals — with each traverse. By indexing through 10° before taking the next set of profiles, it was possible to obtain a full set of 180 profiles by making only 18 traverses. At the rate of approximately 1 s for each traverse, scanning systems of this type could operate in the range of 18–20 s. Even in this extremely reduced scan time, the scanning and detection system was able to obtain more than 300,000 precise absorption measurements during the complete (i.e., over 180°) scan. These values were then used to construct a picture of the slice under investigation. With this new generation of CT scanners, it became possible to obtain cross sections of any part of the body. However, it still was not fast enough to eliminate the streak artifacts, nor did it seem possible to further reduce scan time with this approach. In essence, the first and second generations had reached their limits. Scanning times could not be reduced further and still achieve an acceptable image quality. Nevertheless, this geometry had the potential to achieve the highest spatial and contrast resolutions of all the gantry designs.

A major redesign ensued and many of the scanning system components were improved. In the process, a third generation of gantries evolved that consisted of a pure rotational system. In these systems, the source-detector unit had a pulsing, highly collimated wide-angle (typically, 20–50°) fan beam, and a multiple detector array was rotated 360° around the patient. This single 360° smoothly rotating movement produced scan times as low as 500 ms, increased appreciably the reliability of the data because they were taken twice, and increased the quality of the reconstructed image. A by-product of this design was that the pulsing X-ray source could be synchronized with physiologic parameters, enabling rhythmic structures such as the heart to be more accurately imaged.

The main advantages of this type of system were its simplicity and its speed, but it had two major disadvantages. First, its fixed geometric system with a fan beam usually established for the largest patient was inefficient for smaller objects and in particular for head scanning. Second, it was particularly prone to circular artifacts. These "ring" artifacts were most obvious in early pictures from this type of machine. The magnitude of the difficulty in achieving stability can be appreciated when it is realized that an error of only one part in 10,000 can lead to an artifact that is diagnostically confusing.

Consequently, a fourth-generation gantry system consisting of a continuous ring (360°) of fixed detectors (usually between 600 and 4800) was developed. In this system, the X-ray source rotates as before, and the transmitted X rays are detected by the stationary detector ring. The fan beam is increased slightly so that the detectors on the leading and trailing edge of the fan can be continuously monitored and the data adjusted in case of shifts in detector performance. As a result, the image obtained using this approach is more reliable, and the ring artifact is usually eliminated. Fourth-generation scan times of approximately 500 ms or less with reconstruction times of 30 or less are now commonplace.

The development of high-voltage slip rings has enabled third and fourth-generation scanners to continuously rotate about the patient. By moving the patient couch at a uniform speed in or out of the scan plan, the X-ray beam will describe a helix or spiral path through the patient. Slice widths of 1–10 mm and pixel matrixes up to 1024^2 may be employed, and typically a 50-cm-long volume can be imaged in a single breath hold. Because of improvements in X-ray tube technology, spatial and contrast resolution are maintained. New display techniques allow real-time 3-D manipulations and virtual reality fly-through of tubular structures such as the aorta or large intestine.

Another gantry design, which might be called a fifth generation, utilizes a fixed detector array arranged in a 210° arc positioned above the horizontal plane of the patient and four target rings of 210° arc positioned below the horizontal plane and opposite to the detector array. There is no mechanical motion in the gantry except the movement of the couch into and out of the imaging plane (Fig. 14.12).

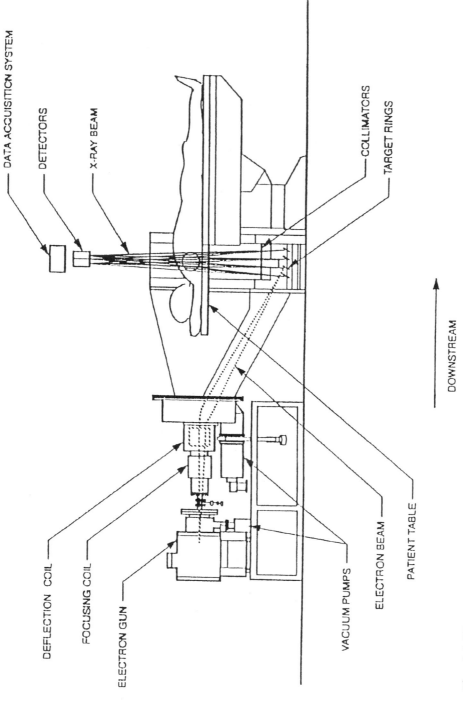

DOWNSTREAM

Fig. 14.12 Illustration of an ultrafast CT scanner. In this device, magnetic coils focus and steer the electron beam through the evacuated drift tube where they strike the target rings. The presence of the four target rings and multiple X-ray collimators allows four unique X-ray beams to pass through the patient. These beams are then interrupted by two contiguous detector arrays and form eight unique image slices (courtesy of Imatrol).

X rays are produced when a magnetic field causes a high-voltage focused election beam to sweep along the path of the anode target. Up to four 1.5-mm simultaneous slices are created in 50 or 100 ms and displayed in up to a 512^2 pixel matrix. This ultrafast CT can volumetrically examine the heart and evaluate the presence and severity of coronary calcific artherosclerosis. Although the ultrafast scanner can almost stop biologic motion, its contrast and spatial resolution are currently, less than those of third or fourth-generation mechanical scanners; therefore, its general applications are fewer.

X-Ray Tubes and Detectors

It must be emphasized that the whole purpose of the scanning procedure is to take thousands of accurate absorption measurements through the body. Taken at all angles through the cross section of interest, these measurements provide an enormous amount of information about the composition of the section of the body being scanned. Since these readings are taken by counting photons, the more photons counted, the better will be the quality of the information. With these factors in mind, it is essential that the radiation dose be sufficient to obtain good quality pictures. Since the total radiation dose is a function of the maximum power of the tube and its operating time, faster scanners have to use a significantly higher powered tube. These aspects of dose efficiency are very important in tube and detector design.

Currently, CT systems that operate in the region of 20 s use fixed-anode, oil-cooled tubes, which run continuously during the time scan, whereas those that operate in much less than 20 s have a rotating-anode, air- or water-cooled tube, which is often used in a pulse mode. The fixed anode tubes, being oil cooled, can be run continuously, and the sequence of measurements is obtained by electronic gating of the detectors. Faster scanners need tubes of substantially higher power, thereby requiring the rotating-anode type. There is a tendency to use these in a pulse mode largely for convenience because it simplifies the gating of the detectors. One of the problems encountered with fast machines, however, is the difficulty in providing enough dose within the scan time. For high-quality pictures, many of these scanners run at much slower rates than their maximum possible speed.

Of equal importance in the accurate detection and measurement of X-ray absorption is the detector itself. Theoretically, detectors should have the highest possible X-ray photon capture and conversion efficiency to minimize the radiation dose to the patient. Clearly, a photon which is not detected does not contribute to the generation of an image. The detection system of a CT scanner that is able to capture and convert a larger number of photons with very little noise will produce an accurate, high-quality image when aligned with suitable reconstruction algorithms. Consequently, the selection of a detector always represents a trade-off between image quality and low radiation dose.

Three types have been used: (i) scintillation crystals (such as sodium iodide, calcium fluoride, bismuth germinate, or cadmium tungstate) combined with photomultiplier tubes, (ii) gas ionization detectors containing xenon, and (iii) scintillation crystals with integral photodiodes. Recall that the principal factors in judging detector quality are how well the detector captures photons and its subsequent conversion efficiency, which results in optimal dose utilization. Systems employing scintillation crystals combined with photomultiplier tubes do not lend themselves to the dense packing necessary for maximal photon capture even though they offer a high conversion efficiency. The <50% dose utilization inherent in these detectors requires a higher dose to the patient to offset the resultant inferior-quality image. Although gas detectors are relatively inexpensive and compact and lend themselves for use in large arrays, they have a low conversion efficiency and are subject to drift. Consequently, use of xenon detectors necessitates increasing the radiation dose to the patient.

The advantage of scintillation crystals with integral photodiodes extends beyond their 98% conversion efficiency since they lend themselves quite readily to a densely packed configuration. The excellent image quality combined with high–dose utilization (i.e., approximately 80% of the applied photon energy) and detector stability ensure that solid-state detectors will continue to gain favor in future CT systems.

Data Handling Systems

In conjunction with the operation of the scanning and detection system, the data obtained must be processed rapidly to permit viewing a scan as quickly as possible. Data handling systems incorporating the latest advances in computer technology have also been developed to coincide with changes in the design of the scanning and detector components of CT scanners. For example, with the evolution of the third-generation scanners, the data handling unit had to digitize and store increasing amounts of data consisting of (i) positional information, such as how far the scanning frame was along its scan; (ii) reference information from the reference detector, which monitored the X-ray output; (iii) calibration information, which was obtained at the end of each scan; and (iv) the bulk of the readings, the actual absorption information obtained from the detectors. Electronic interfaces are now available that convert this information to digital form and systematically store it in the memory of the resident computer. The next step is the actual process of image reconstruction (Fig. 14.13).

Recall that the first stage of this computational process is to analyze all the raw data and convert them into a set of profiles, normally 180 or more, and then convert these profiles into information that can be displayed as a picture and used for diagnosis. This is the heart of the operation of a CT scanner, the element that makes CT totally different from conventional X-ray techniques and most other imaging techniques. The algorithms or computer programs for reconstructing tissue X-ray absorption properties from a series of X-ray absorption profiles fall into one of the following categories: simple backprojection (sometimes called summation), Fourier transforms, integral equations, and series expansions. The choice of an algorithm for a CT scanner depends on the algorithm's speed and accuracy.

Mathematical algorithms for taking the attenuation projection data and reconstructing an image can be classified into two categories, iterative and analytic. The iterative techniques [also known as the algebraic reconstruction technique (ART)], such as the one used by Hounsfield in the first-generation scanner, require an initial guess of the two-dimensional pattern of X-ray absorption. The attenuation projec-

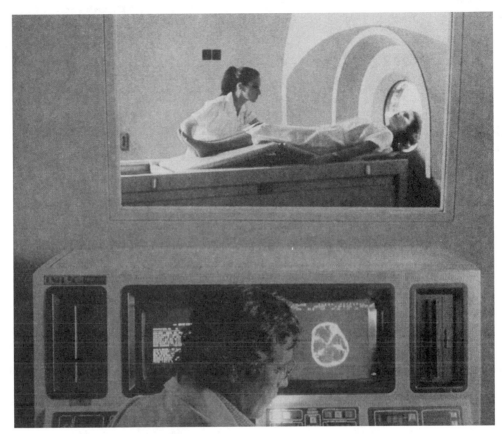

Fig. 14.13 Photograph of modern display terminal for CT scanner.

tion data predicted by this guess is then calculated, and the results are compared with the measured data. The difference between the measured data and predicted values is used in an iterative manner so that the initial guess is modified and the difference goes to zero. In general, many iterations are required for convergence, with the process usually halted when the difference between the calculated and the measured data is below a specified error limit. Many different versions of the ART were developed and used with first- and second-generation CAT scanners. Later-Generation scanners used analytic reconstruction techniques since the iterative methods were computationally slow and had convergence problems in the presence of noise.

Analytic techniques include the Fourier transform, back-projection, filtered back-projection, and convolution back-projection approaches. All of the analytic methods differ from the iterative methods in that the image is reconstructed directly from the attenuation projection data. Analytic techniques use the central section theorem (Fig. 14.14) and the two-dimensional Fourier transform. Given an image

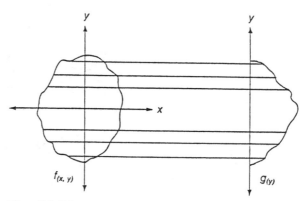

Fig. 14.14   Illustration of a projection via the central-section theorem.

$f(x, y)$, a single projection is taken along the x direction, forming a projection $g/(y)$ described by

$$g/(y) = \int_{-\infty}^{\infty} f(x, y)dx \tag{14.4}$$

This projection represents an array of line integrals as shown in Fig. 14.14. The two-dimensional Fourier transform of $f(x, y)$ is given by

$$F(u, v) = \int\int_{-\infty}^{\infty} f(x, y)\exp[-j2\pi(ux, vy)]dxdy \tag{14.5}$$

In the Fourier domain, along the line $u=0$, this transform becomes

$$F(0, v) = \int\int_{-\infty}^{\infty} f(x, y)\exp(-j2\pi vy)dxdy \tag{14.6}$$

which can be rewritten as

$$F(0, v) = \int\int_{-\infty}^{\infty} f(x, y)dx/\exp(-j2\pi vy)dy \tag{14.7}$$

or

$$F(0, v) = F_1[g, (y)]_1 \tag{14.8}$$

where $F_1[\ ]$ represents a one-dimensional Fourier transform. It can be shown that the transform of each projection forms a radial line in $F(u, v)$, and therefore $F(u, v)$ can be determined by taking projections at many angles and taking these transforms. When $F(u, v)$ is completely described, the reconstructed image can be found by taking the inverse Fourier transform to obtain $f(x, y)$.

The image reconstruction method used by most modern CT scanners is the filtered back-projection reconstruction method. In this method, attenuation projection data for a given ray or scan angle are convolved with a spatial filter function either in the Fourier domain or by direct spatial convolution. The filtered data are then back-projected along the same line, using the measured value for each point along the line, as shown in Fig. 14.15. The total back-projected image is made by summing the contributions from all scan angles. Depending on the filter technique, the image is obtained directly after back-projection summation or via the inverse Fourier transform of the back-projected image.

It should be obvious that compared to early head and body scanners, current models process information faster, use energy more efficiently, are more sensitive to differences in tissue density, have finer spatial resolution, and are less susceptible to artifact. CT scanners may be compared with one another by considering the following factors:

1. Gantry design, which affects scan speed, patient processing time, and cost-effectiveness.
2. Aperture size, which determines the maximum size of the patient along with the weight-carrying capacity of the couch.
3. Type of X-ray source, which affects the patient radiation dose and the overall life of the scanning device.
4. X-Ray fan beam angle and scan field, which affect resolution.
5. Slice thickness, as well as the number of pulses and the angular rotation of the source, which are important in determining resolution.
6. Number and types of detectors, which are critical parameters in image quality.
7. The type of minicomputer employed, which is important in assessing system capability and flexibility.
8. Type of data handling routines available with the system, which are important user and reliability considerations.
9. Storage capacity of the system, which is important in ascertaining the accessibility of the stored data.
10. Upgradeability and connectivity: Modern CT scanners should be capable of modular upgradeability and should communicate to any available network.

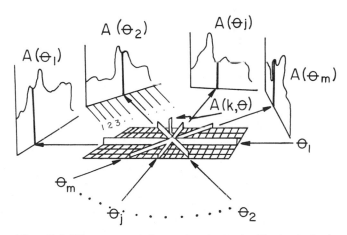

Fig. 14.15 Sketch of the results obtained with simple back-projection algorithm.

In reviewing the available CT systems, it is important to compare them not only in technical terms but also against what can be considered to be an ideal scanner. The ultimate objective of CT is to provide accurate diagnostic information that significantly improves patient care. In the ideal case, a CT scanner would provide an unambiguous diagnosis, thereby eliminating the necessity for further tests. Since financial considerations are also important and CT scanners require a high capital outlay and have considerable maintenance costs, the ideal system should process a large patient load quicky to provide high "patient throughput." A good scanner, therefore, must satisfy two major criteria: It must provide good diagnostic information and it must permit high patient throughput.

EXERCISES

1. Represent the decay process discussed in Example Problem 14.2 in symbolic form.
2. Compute the energy liberated when $^{238}_{92}U$ (amu = 238.050786) decays to $^{234}_{90}Th$(amu = 232.038054) via α emission.
3. Carbon 14, $^{14}_{6}C$ is a radioactive isotope of carbon that has a half-life of 5730 years. If an initial sample contained 1000 ^{14}C nuclei, how many would still be around after 22,920 years?
4. A 50-g sample of carbon is taken from the pelvis bone of a skeleton and is found to have a ^{14}C decay rate of 200 decays/min. It is known that carbon from a living organism has a decay rate of 15 decays/min $\times$ g and that ^{14}C has a half-life of 5730 years = 3.01×10^9 min. Find the age of the skeleton.
5. The half-life of a radioactive sample is 30 min. If you start with a sample containing 3×10^{16} nuclei, how many of these nuclei remain after 10 min?
6. Find the energy liberated in the beta decay of $^{14}_{6}C$ to $^{14}_{7}N$.
7. How long will it take for a sample of polonium of half-life 140 days to decay to one-tenth its original strength?
8. Suppose that you start with 10^{-3} g of a pure radioactive substance and 2 h later determine that only 0.25×10^{-3} g of the substance remains. What is the half-life of this substance?
9. The half-life of an isotope of phosphorus is 14 days. If a sample contains 3×10^{16} such nuclei, determine its activity.
10. How many radioactive atoms are present in a sample that has an activity of 0.2 μCi and a half-life of 8.1 days?
11. A freshly prepared sample of a certain radioactive isotope has an activity of 10 mCi. After 4 h, the activity is 8 mCi.
    (a) Find the decay constant and half-life of the isotope.
    (b) How many atoms of the isotope were contained in the freshly prepared sample?
    (c) What is the sample's activity 30 h after it is prepared?
12. Tritium has a half-life of 12.33 years. What percentage of the nuclei in a tritium sample will decay in 5 years?

13. For the following process $^{23}_{10}Ne_{13} \rightarrow ^{23}_{11}Ne_{12} + e + \bar{\nu}$, what is the maximum kinetic energy of the emitted electrons.

14. The half-life of ^{235}U is 7.04×10^8 years. A sample of rock which solidified with the earth 4.55×10^9 years ago contains N atoms of ^{235}U. How many ^{235}U atoms did the same rock have when it solidified?

15. Of the three basic types of radioaction — alpha, beta, and gamma — which have the greatest penetration into tissue? Explain the rationale behind your answer.

16. Discuss the principle of scintillation. How is it detected? How can it be used to generate an image?

17. One method of treating cancer of the thyroid is to insert a small radioactive source directly into the tumor. The radiation emitted by the source can destroy cancerous cells. Why do your suppose $^{131}_{53}I$ is used for this treatment?

18. Provide a clinical example of the use of instrumentation employed to detect the rate of "radioactive workout."

19. In a photomultiplier tube, assume that there are seven diodes with potentials 100, 200, 300, 400, 500, 600, and 700 V. The average energy required to free an electron from the dynode surface is 10 eV. For each incident electron released from rest, how many electrons are freed at the first dynode? At the last dynode?

20. Discuss the process of producing a tomograph. Provide figures to illustrate your answer.

21. Describe the operation of the iterative method to produce an image.

22. Provide an illustration of how the net absorption coefficient is obtained for each volume element.

SUGGESTED READING

Alpen, E. L. (1990). *Radiation Biophysics*. Prentice Hall, Englewood Cliffs, NJ.

Bronzino, J. D. (1982). *Computer Applications in Patient Care*. AddisonWesley, (Reading, MA.)

Cho, Z. H., Jones, J. P., and Singh, M. (1993) *Foundations of Medical Imaging*. Wiley, New York.

Croft, B. F., and Tsui, B. M. N. (1995). Nuclear medicine. In *The Biomedical Engineering Handbook*, pp. 1046–1076. CRC Press, Boca Raton, FL.

Cunningham, I. A., and Juoy, P. F. (1995) Computerized tomography. In *The Biomedical Engineering Handbook*, pp. 990–1005. CRC Press, Boca Raton, FL.

Hounsfield, G. N. (1973). Computerized transverse axial scanning (tomography) part 1, description of system. *Br. J. Radiol.* **46**, 1016–1022.

McCollough, C. H., and Morin, R. L. (1994). The technical design and performance of ultrafast computed tomography. *Cardiac Imagery* **32**, 521–536.

McCollough, C. H. (1992). Acceptance testing of a fifth generation scanning-electron-beam computed tomography scanner. *Med. Phys.* **19**, 846.

Serway, R. A., and Faughn, J. S. (1989). *College Physics*, 2nd ed. Saunders, Philadelphia.

Shroy, R. E., Jr., Van Lipel, M. S., and Yaffe, M. J. (1995), X-ray. In *The Biomedical Engineering Handbook*, pp. 953–989. CRC Press, Boca Raton, FL.

15 ULTRASOUND

Chapter Contents

At the conclusion of this chapter, the reader will be able to:

- Understand what ultrasound is and how an ultrasonic image is formed

- Understand how the Doppler principle can be used to measure blood flow

- Understand the advantages and disadvantages of ultrasound compared to other competing diagnostic imaging modalities

- Be aware of the potential biological effects of ultrasound and know how to use it prudently

- Know how the biological effects produced by ultrasound can be used to its advantage in therapeutic applications

15.1 INTRODUCTION

Ultrasonic imaging is the second most utilized diagnostic imaging modality today. Its potential as a diagnostic tool was known as early as the late 1940s when a few groups throughout the world started exploring diagnostic capabilities of ultrasound utilizing sonar technology developed for the Navy during World War II. Most notable in the United States was the team of John Wild, a physician, and John Reid, an engineer, who built a pulse-echo instrument demonstrating that ultrasound could be used to diagnose tumors in various tissues. In parallel with diagnostic efforts, the feasibility of using high-intensity ultrasound for therapeutic applications was also initiated by several investigators, including William Fry, Floyd Dunn, and their coworkers at the University of Illinois who used ultrasound to modify brain structures for the treatment of Parkinson's disease.

Although ultrasound has been applied to many medical problems since then, it did not become a well-accepted tool until the introduction in the 1970s of gray-scale ultrasonography in which signal compression was used to reduce the dynamic range of returned echo amplitude. Today, ultrasonic imaging is an integral part of all modern diagnostic imaging facilities. It not only compliments the more traditional approaches, e.g., X-ray, but also possesses unique characteristics which are advantageous in comparison to other competing modalities such as X-ray computed tomography, radionuclide emission tomography, and magnetic resonance imaging (MRI). The following are some of the unique characteristics and advantages of ultrasound:

1. Ultrasound is a form of nonionizing radiation and is considered safe to the best of current knowledge.
2. It is less expensive than imaging modalities of similar capabilities.

3. It produces images in real time, currently unattainable by any other methods.
4. It has a resolution in the millimeter range for the frequencies being clinically used today, which may be made better if the frequency is increased.
5. It can yield blood flow information by applying the Doppler principle.
6. It is portable and thus can be easily transported to the bedside of a patient.

For these reasons, ultrasound has been found to be a valuable diagnostic tool in such medical disciplines as cardiology, obstetrics, gynecology, surgery, pediatrics, radiology, and neurology. Because of heightened public concern with health care costs, which is quite evident in the recent precipitous drop in MRI equipment sales in the United States, the cost-effectiveness of an imaging tool is a crucial factor in planning diagnostic strategies. Diagnostic ultrasound is particularly attractive in this respect and has been anticipated by many, clinicians and manufacturers alike, to be the major modality of the future. Its future indeed appears to be very promising despite a few shortcomings. The disadvantages of diagnostic ultrasound include the following: (i) Organs containing gases and bony structures cannot be adequately imaged without introducing specialized procedures, (ii) only a limited window is available for ultrasonic examination of certain organs such as the heart and neonatal brain, (iii) it is operator skill dependent, and (iv) it is sometimes impossible to obtain good images from obese patients.

Although the pace of development in therapeutic ultrasound has not been as striking as that of diagnostic ultrasound, significant progress has been made in the past decades. These efforts have been primarily focused on developing better devices for hyperthermia, frequently in combination with chemotherapy or radiotherapy, for the treatment of tumors, and for tissue ablation. Ultrasonic therapeutic devices are commercially available, although they are not as well known as their diagnostic counterparts.

Despite the fact that ultrasound has been in existence for more than 40 years, technical advances in diagnostic and therapeutic ultrasound are still being made. In this chapter, fundamental concepts, recent important technical developments, and biological effects of ultrasound are discussed along with their clinical implications.

15.2 FUNDAMENTALS OF ACOUSTIC PROPAGATION

Ultrasound is a form of acoustic wave with frequencies higher than 20 kHz, the upper limit of the human audible range. Like sound, it needs a medium in which it can propagate. As an acoustic wave is launched into a medium, a mechanical disturbance of the medium is produced. If the source is sinusoidal, acoustic parameters such as pressure, medium velocity, medium displacement, density, temperature, and sound speed respond accordingly and vary as a function of time and space (Fig. 15.1). Figure 15.1A shows an acoustic parameter (e.g., the pressure which has a unit of Pascal or Newton/m^2 or medium displacement in a unit of m) at a fixed time as a function of space. The wavelength, λ, is defined as the distance between two adjacent points of the same amplitude in the wave. Figure 15.1B shows the

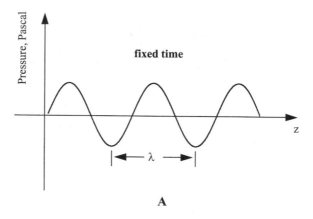

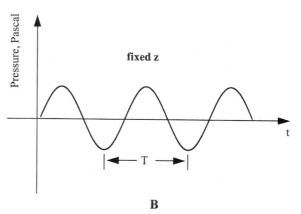

Fig. 15.1   (A) Sinusoidal acoustic wave propagation as a function of space at a fixed time. (B) Sinusoidal acoustic wave propagation as a function of time at a fixed distance *z*. Pressure is the acoustic parameter that is plotted here. The distance and time between two troughs or peaks are defined as the wavelength (λ) and period (*T*) of the wave, respectively.

pressure or displacement at a fixed point in space as a function of time, where T (the period) is the time required by the wave to travel one wavelength. Thus, by definition,

$$\lambda = cT \qquad\qquad (15.1)$$

where c is the sound speed in the medium with a unit of m/s. The sound speed, c, is the speed at which the sound energy travels and differs from the medium velocity, which is the velocity that the medium actually moves when perturbed. In most media, sound speed is much greater than the medium velocity. The frequency, f, of a wave is defined as the number of times per unit period of time a disturbance is repeated at a fixed point in space and is related to the period by

$$f = \frac{1}{T} \tag{15.2}$$

Frequency has a unit of Hertz or cycles per second. In medical ultrasound, the frequency typically used ranges from 1 to 30 MHz or million cycles per second.

15.2.1 Wave Equation

Acoustic wave propagation in an infinite medium is governed by a second-order differential equation called the wave equation:

$$\frac{\partial^2 U}{\partial z^2} = \frac{1}{c^2} \frac{\partial^2 U}{\partial t^2} \tag{15.3}$$

where U denotes medium displacement in the z direction that is produced by the wave and t denotes time. U in Eq. (15.3) can be replaced by any acoustic parameter mentioned previously. The solution of this equation has the form of $f(t \pm z/c)$, where the plus sign represents a wave propagating in the $-z$ direction and a minus sign indicates a wave propagating in the $+z$ direction. The sinusoidal solution representing a monochromatic plane wave traveling in the $+z$ direction as shown in Fig. 15.1 is given by

$$U = U_0 \, e^{j(\omega t - kz)} \tag{15.4}$$

where $\omega = 2\pi f$ is the angular frequency and $k = \omega/c$ is the wave number.

Equation (15.4) describes a longitudinal or compressional wave propagating in the z direction which has a displacement in the same direction as the direction of wave propagation, z. For the longitudinal wave, the sound speed is given by

$$c = \frac{1}{\sqrt{\rho G}} \tag{15.5}$$

where ρ and G respectively represent density in kg/cm^3 and compressibility, defined as the negative of the change in volume per unit change in pressure with a unit of cm^2/dyn or m^2/N. The sound speeds of a few materials are provided in Table 15.1. The sound speeds in soft tissues are similar and have been assumed to be a constant, 1540 m/s, in a majority of commercial scanners.

The displacement U in the wave equation can be replaced by a displacement in another direction, e.g., the x direction. The solution for the displacement in the x direction will be a function of z and t. This means that the displacement is perpendicular to the direction of wave propagation. This type of wave is called a shear or transverse wave. It should be noted that the shear wave sound speed is different from Eq. (15.5). Biological tissues except for bone can in general be treated as fluid-like media. As such, they cannot support shear wave propagation.

TABLE 15.1 Acoustic Properties of Biological Tissues and Relevant Materials

Material	Speed (m/s) at 20–25°C)	Acoustic impedance (MRayl)	Attenuation (coefficient) (neper/cm at 1 MHz)	Backscattering coefficient (cm^{-1}Sr^{-1} at 5 MHz)
Air	343	0.0004	1.38	—
Water	1480	1.48	0.00025	—
Fat	1450	1.38	0.06	—
Myocardium (perpendicular to fibers)	1550	1.62	0.35	8×10^{-4}
Blood	1550	1.61	0.02	2×10^{-5}
Liver	1570	1.65	0.11	5×10^{-3}
Skull bone	3360	6.00	1.30	—
Aluminum	6420	17.00	0.0021	—

Example Problem 15.1

In echocardiography, the frequencies range from 2.5 to 5 MHz. Calculate ultrasound wavelengths in cardiac tissue at these two frequencies.

Solution

Assume that the sound velocity in cardiac tissues is 1540 m/s and use Eqs. (15.1) and (15.2):

$$\lambda \text{ at } 5 \text{ MHz} = \frac{1540 \text{ m/s}}{5 \times 10^6 \text{ s}^{-1}} = 3.08 \times 10^{-4} \text{ m}$$

$$\lambda \text{ at } 2.5 \text{ MHz} = \frac{1540 \text{ m/s}}{2.5 \times 10^6 \text{ s}^{-1}} = 6.16 \times 10^{-4} \text{ m} \qquad \blacksquare$$

15.2.2 Characteristic Acoustic Impedance

The characteristic acoustic impedance of a medium is defined as the ratio of the pressure to medium velocity. These are analogous to the voltage and current in an electrical circuit. Thus, the acoustic impedance carries the same physical significance as the electrical impedance. For a fluid medium, the characteristic acoustic impedance, Z, is given by

$$Z = \rho c \qquad (15.6)$$

The acoustic impedance has a unit of kg/(m^{2}s) or Rayl. The acoustic impedance for water is approximately 1.48 MRayl. The acoustic impedances of tissues and other materials are given in Table 15.1.

15.2.3 **Intensity**

Analogous to the power in electrical circuits, the intensity of an acoustic wave, defined as the power carried by the wave per unit area propagating in a medium of acoustic impedance Z, is related to the pressure by

$$I(t) = p(t)u(t) = \frac{p^2(t)}{Z} \tag{15.7}$$

where $u = \partial U/\partial t$ is the medium velocity. For a sinusoidal wave, it can be easily shown that the average intensity $<I>$ is given by

$$<I> = \frac{1}{2}p_0 u_0 = \frac{1}{2}\frac{p_0^2}{Z} \tag{15.8}$$

where p_0 and u_0 are peak pressure and medium velocity. The unit for intensity is W/m^2 or W/cm^2. A conventional pulse-echo ultrasonic image is formed by launching an ultrasonic pulse into the body and then detecting and displaying the echoes that are reflected or scattered from these tissues. The pulse duration or the on time, τ, of the scanner is very short, typically <1 μs, but the pulse repetition period (PRP) is relatively long, typically >1 ms. This is shown in Fig. 15.2, in which the pressure waveform above and below a reference level, the ambient pressure, are called compressional and rarefactional, respectively. Consequently, the instantaneous intensity and pressure when the pulse is on can be quite high, sometimes reaching a few tens of W/cm^2 and a few MPa, but the time-averaged intensity over the whole cycle is very low, typically <100 mW/cm^2.

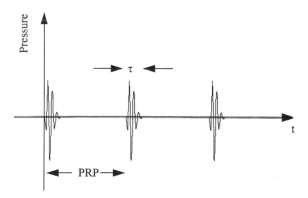

Fig. 15.2 A train of ultrasonic pulses. The duration of the pulse, typically defined as the duration between -3 dB points from the peak pressure, is denoted by τ, and the time between two pulses is the pulse repetition period (PRP). The pulse repetition frequency = 1/PRP. The horizontal axis represents ambient or static DC pressure. The waveform above the horizontal axis is called compressional pressure, and the waveform below the axis is called rarefactional pressure.

15.2.4   Attenuation, Reflection, and Scattering

Since the ultrasound wavelength illustrated in Example Problem 15.1 (0.3 mm at 5 MHz in tissues) is comparable to the dimension of many tissue structures, it interacts with the tissues in a very complex manner. As an ultrasonic wave penetrates a tissue, the ultrasonic energy is absorbed by the tissue, subsequently converted into heat, and diverted into other directions or scattered from the primary direction whenever the wave encounters a discontinuity in acoustic properties. For a plane wave, the sound pressure decreases as a function of the propagation distance, z, according to the following equation:

$$p(z) = p(0)e^{-\alpha z} \tag{15.9}$$

where $p(0)$ is the pressure at $z = 0$ and α is the attenuation coefficient in neper/m or neper/cm, which can also be expressed in dB/m or dB/cm. The value in neper/cm is multiplied by 8.68 to convert from an attenuation coefficient in neper/cm to one in dB/cm. The attenuation coefficient in tissues has been found to be linearly proportional to frequency in the range from 1 to 50 MHz or even higher as opposed to homogeneous media such as air and water in which the attenuation coefficient is proportional to frequency squared. As a first-order approximation, the attenuation coefficient in soft tissues is ~0.3 dB/cmMHz. The attenuation coefficients of a number of tissues and relevant materials are listed in Table 15.1. For n layers of tissues of thickness z_i and attenuation coefficient α_i, where $i = 1$ to n, the exponent in Eq. (15.9) should be replaced by $-(\alpha_1 z_1 + \cdots + \alpha_n z_n)$.

Example Problem 15.2

A plane ultrasonic wave with an initial temporal averaged intensity of 100 mW/cm^2 is being propagated into a heart. It has to traverse 1 cm of fat and 2 cm of heart muscle before reaching the heart valve to be imaged. Calculate the intensity at the heart valve for frequencies of 3 and 6 MHz.

Solution

Equation (15.8) shows that $<I>$ is proportional to the square of p. From Eq. (15.9) and by using the values for the attenuation coefficients for fat and heart muscle given in Table 15.1, which are 0.35 and 0.06 neper/cm at 1 MHz, the attenuation coefficients at 3 and 6 MHz can be found simply by multiplying the values at 1 MHz by 3 and 6, respectively. Thus, the intensities at the valve can be calculated using the modified version of Eq. (15.9) to account for the multiple-layered structure:

$$<I> \text{ at 6 MHz} = 100 \times e^{[(-0.06 \times 1) - (0.35 \times 2)] \times 6} = 1.05 \text{ mW/cm}^3$$

$$<I> \text{ at 3 MHz} = 100 \times e^{[(-0.06 \times 1) - (0.35 \times 2)] \times 3} = 10.2 \text{ mW/cm}^2 \qquad ■$$

The ultrasonic energy will also be lost due to a redistribution of the energy from the primary incident direction into other directions by the mechanisms of reflection or scattering. For an incident angle of θ_i and a plane wave, the pressure reflection coefficient Y and transmission coefficient Γ at a flat boundary between two media of different acoustic impedances, Z_1 and Z_2 (as illustrated in Fig. 15.3), are given by

$$Y = \frac{p_r}{p_i} = \frac{Z_2\cos\theta_i - Z_1\cos\theta_t}{Z_2\cos\theta_i + Z_1\cos\theta_t} \tag{15.10}$$

$$\Gamma = \frac{p_t}{p_i} = \frac{2Z_2\cos\theta_i}{Z_2\cos\theta_i + Z_1\cos\theta_t} \tag{15.11}$$

where p_r, p_t, and p_i are the reflected, transmitted, and incident pressures, respectively. The symbols θ_r and θ_t are angles of reflection and transmission and $\theta_i = \theta_r$. The angle of transmission or refraction is related to the angle of incidence by Snell's law as in optics:

$$\frac{\sin\theta_i}{\sin\theta_t} = \frac{c_1}{c_2} \tag{15.12}$$

where c_1 and c_2 refer to the velocities of sound in two different media with acoustic impedances of Z_1 and Z_2. For normal incidence, Eq. (15.10) reduces to

$$Y = \frac{Z_2 - Z_1}{Z_2 + Z_1} \tag{15.13}$$

From Eq. (15.13), it is apparent that the reflection at the interface of two media of very different acoustic impedances can be severe. In fact, almost 99% of the energy incident upon the interface between the lung, mostly consisting of air, and soft tissue is reflected. Thus, ultrasound is generally not useful for imaging anatomical structures that contain air, such as lung, and bones.

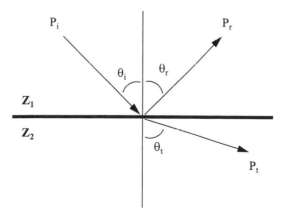

Fig. 15.3 A plane acoustic wave is incident upon a planar interface between two media of acoustic impedances Z_1 and Z_2. Here, p_i, p_r, and p_t are the incident, reflected, and transmitted pressure, respectively.

Example Problem 15.3

Calculate the reflection coefficients for pressure and intensity at a fat and bone interface and a bone and fat interface for normal incidence.

Solution

From Table 15.1, the acoustic impedances for fat and bone are found to be 1.38 and 6 MRayl. Using Eq. (15.13), the pressure reflection coefficient for a fat–bone boundary is

$$Y = \frac{6 - 1.38}{6 + 1.38} = 0.63$$

and the intensity reflection coefficient is $Y^2 = (0.63)^2 = 0.4$. For a bone–fat boundary, $Y = -0.63$, but the intensity reflection coefficient remains unchanged. The negative sign means that the reflected pressure is 180° out of phase relative to the incident pressure.   ∎

The ultrasonic scattering process in biological tissues is a very complicated phenomenon. Since echoes resulting from reflection or scattering in the tissues are used by pulse-echo ultrasonic scanners to form an image, a clear understanding of the scattering process is crucial for the development of ultrasonic imaging techniques. In this context, tissues have been treated both as dense distributions of discrete scatterers by some investigators and as continua with variations in density and compressibility by others.

Assume that an object is insonified by a plane ultrasound wave of intensity I_i as shown in Fig. 15.4. Part of the incident energy will be diverted from its original path. Let I_s denote the scattered intensity. The total power scattered by the object can be found by integrating I_s over the 4π solid angle. This is mathematically given by

$$P_s = \int_{4\pi} I_s d\Omega \tag{15.14}$$

where $d\Omega$ denotes the differential solid angle.

The term scattering cross section has been frequently used to quantitate the scattering strength of an object and is defined as the power scattered by the object per unit incident intensity. It is called a scattering cross section because it has a unit of

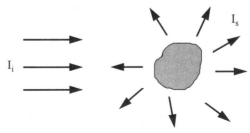

Fig. 15.4 An incident ultrasonic wave of intensity I_i is scattered by a volume of scatterers. I_s represents scattered intensity.

cm^2 or m^2. The scattering cross section of a scatterer of simple geometry, e.g., a sphere, for which the dimension is much smaller than the wavelength, can be solved analytically and is called Rayleigh scattering after Lord Rayleigh, who developed the theory. For a spherical scatterer with a radius, a, much smaller than the wavelength, λ, the scattering cross section is related to the compressibility, G_s, and density, ρ_s, of the scatterer by the following expression:

$$\sigma_s = \frac{4\pi k^4 a^6}{9}\left[\left|\frac{G_s - G}{G}\right|^2 + \frac{1}{3}\left|\frac{3\,\rho_s - 3\rho}{2\rho_\sigma + \rho}\right|^2\right] \tag{15.15}$$

where k is the wave number in the surrounding medium and equals $2\pi/\lambda$, and G and ρ are respectively the compressibility and the density of the surrounding medium.

Example Problem 15.4

The human red blood cell has the shape of a biconcave disk with a diameter of 8 μm and a thickness of 3 μm. For an ultrasound wave of 10 MHz with a wavelength of 154 μm, the dimensions of the red cell are all much smaller than the wave length, and the red cell can be approximated as a sphere with the same volume. Given the densities and compressibilities of the red cell and the plasma (G_s and $G = 34.1 \times 10^{-12}$ and 40.9×10^{-12} cm^2/dyn; ρ_s and $\rho = 1.09$ and 1.02 g/cm^3), calculate (a), the scattering cross section of a red cell at 10 MHz and (b) the power scattered by it if the incident intensity is 2 W/cm^2.

Solution

(a) The volume of a red cell, in this case a disk, $= \pi\,(4\,\mu m)^2\,(3\,\mu m) = 151\,\mu m^3$. The radius, a, of an equivalent sphere can be calculated from

$$\frac{4}{3}\pi a^3 = 151\,\mu m^3$$

Thus, $a = 3.3\,\mu$m. Substituting this value and other parameters into Eq. (15.15),

$$\sigma_s = \frac{4\pi\left(\dfrac{2\pi}{154 \times 10^{-6}}\right)^4(3.3 \times 10^{-6})^6}{9}\left[\left|\frac{34.1 - 40.9}{40.9}\right|^2 + \frac{1}{3}\left|\frac{3(1.09 - 1.02)}{2 \times (1.09) + 1.02}\right|^2\right]$$

$$= 1.5 \times 10^{-12}\,cm^2$$

The scattering indeed is quite small in the diagnostic ultrasound frequency range. (b) The power scattered by one red cell can be calculated from the definition of the scattering cross section:

$$P_s = 1.5 \times 10^{-12}\,cm^2 \times 2\,W/cm^2 = 3 \times 10^{-12}\,W \qquad\blacksquare$$

When there is more than one scatterer, the total scattered power can be obtained by summing all the contributions from individual scatterers. This is true, however, only if the volume concentration of the scatterers is very low, typically, <1% when independent scatterers can be assumed and secondary scattering can be ignored. For a dense distribution of scatterers such as biological tissues, e.g., human blood consisting of many red blood cells suspended in plasma ($5 \times 10^9/cm^3$), the red cells can no longer be considered independent, and the problem becomes very complicated. An alternative has been sought and a parameter called the scattering coefficient or volumetric scattering cross section, defined as the power scattered per unit incident intensity per unit volume of scatterers, has been used.

The scattering coefficient is a function of the ultrasound frequency and the sizes and acoustic properties of tissue components relative to the surrounding medium. Since the scattering strength of a scatterer or a scattering medium may be directionally dependent, parameters such as differential scattering cross section or differential scattering coefficient, defined as the power scattered in one solid angle at a certain direction relative to the incident direction per unit incident intensity per scatterer or per unit volume of scatterers, have been used as measures of scattering strength of a scatterer or scattering medium in a certain direction. The backscattering coefficients, which are differential scattering coefficients in the backward direction measured for several types of tissues, are listed in Table 15.1.

15.2.5   Doppler Effect

The Doppler effect describes a phenomenon in which an observer perceives a change in the frequency of the sound emitted by a source when the source or the observer is moving or both are moving. It can be shown that if the source and the receiver are both moving at a velocity, v, which is much smaller than the sound speed, c, in the medium and makes an angle of θ relative to the sound propagation direction, the Doppler frequency shift, f_d, is related to v by

$$f_d = \frac{2v\cos\theta}{c} f \tag{15.16}$$

where f is the frequency of the sound. This principle has been used extensively in medicine for measuring blood flow velocity in blood vessels. Figure 15.5 shows an ultrasound beam insonifying a blood vessel with blood flowing at a velocity of v. The angle between the ultrasound beam and the direction of the velocity, called the Doppler angle, is θ. The red blood cells in the blood act at first as observers and then as sources by scattering the ultrasound back to the ultrasound transducer that acts both as a source and as a detector. Therefore, Eq. (15.16) can be readily used to estimate the Doppler frequency shift which happens to be in the audible range for medical ultrasound frequencies. If extracted properly, it can be heard with a loudspeaker.

Example Problem 15.5

A 5-MHz ultrasound Doppler flowmeter is used to measure blood flow velocity in a human artery. The mean blood flow velocity in this artery is approximately 30 cm/s and the Doppler angle is 45°. Estimate the mean Doppler shift frequency.

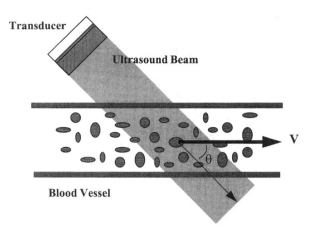

Fig. 15.5 An ultrasound beam indicated by the shaded area insonifies a blood vessel with blood flowing at a velocity v. The angle between the ultrasound beam and the direction in which the blood is flowing is θ.

Solution

Following Eq. (15.16) and assuming that c in blood is approximately 1540 m/s, which is much greater than $v = 30$ cm/s:

$$f_d = \frac{2 \times 30 \text{ cm/s} \times \cos(45°) \times 5 \times 10^6/\text{s}}{1.54 \times 10^5 \text{ cm/s}} = 1414 \text{ Hz} = 1.414 \text{kHz}$$

This is in the audible range.   ∎

15.3 DIAGNOSTIC ULTRASONIC IMAGING

As mentioned earlier, an ultrasonic image is constructed from echoes returned from tissues. There are various modes that have been utilized to display the image, namely, A-, B-, C-, and M-mode, whereas the B-mode is the most common. A, B, C, and M are the first letters for amplitude, brightness, constant depth, and motion, respectively. A-mode is simply a display of the echo amplitude versus the distance into the tissue, which is related to the ultrasonic speed, c, and the time of flight, t, by $z = ct/2$ since c is assumed to be a constant in soft tissues (1540 m/s). An A-mode display of a target is shown in Fig. 15.6A. The components of a system needed to obtain an A-line are shown Fig. 15.7. The center of the system is an energy-conversion device, the ultrasonic transducer, which converts the electric signal to ultrasound and vice versa.

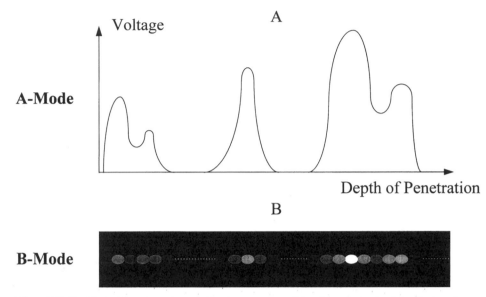

Fig. 15.6 (A) In A-mode, echo amplitude, represented by the voltage level, is displayed as a function of the depth of penetration. (B) In B-mode, the echo amplitude is represented by the gray or brightness level of the display. Each dot represents a pixel.

15.3.1 Transducers

The most critical component of an ultrasonic transducer is a piezoelectric element that is usually made from an artificial ferroelectric ceramic material such as lead zirconate titanate (PZT). A single-element transducer consisting of a circular disc, two

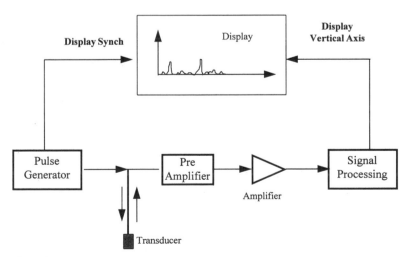

Fig. 15.7 Block diagram of an A-mode instrument. The display is synchronized by triggering pulses generated by the pulse generator.

matching layers, a lens, and a backing material is shown in Fig. 15.8. The disc is electroplated with gold or silver as electrodes. It outputs the most power at a resonance that is determined by its thickness. The thickness of the disc resonating at a frequency, *f*, should be one-half of the wavelength of sound in the piezoelectric material or one-half of sound speed in piezoelectric material divided by *f*. The matching layers serve the function of reducing the acoustic impedance mismatch between the piezoelectric materials, typically on the order of 30 MRayl or higher, and the loading medium, water or soft tissues, so that more energy can be transmitted into the loading medium. The lens behaves just like an optical lens and is used to focus the beam to a desired distance. The backing material performs the damping function for modifying the duration and shape of the excited acoustic pulse. This material should have an acoustic impedance similar to the piezoelectric material and be highly attenuative so that the energy entering the material is absorbed. Typically, broadband transducers, which produce short pulses and are required for imaging, have backing, whereas narrowband transducers that produce longer pulses and are used for Doppler measurements have no backing or are backed by air.

A transducer consisting of only one piezoelectric element is the simplest type used in biomedical ultrasonics. For imaging and hyperthermia, more complicated forms of transducers (e.g., one-dimensional (1-D) arrays and multidimensional arrays) have been used and will be discussed in more detail later. Although PZT is the material of choice in current commercial transducers, other materials such as piezoceramic composites and ferroelectric relaxor materials that have better properties are gaining importance.

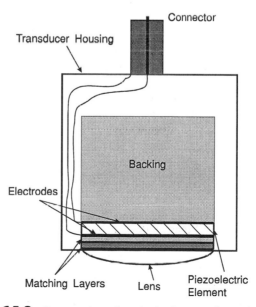

Fig. 15.8 Construction of a single-element ultrasonic transducer.

15.3.2   A-Mode and B-Mode Imaging

A block diagram for A-mode instruments is shown in Fig. 15.7. A high-energy pulser is used to excite the transducer. The returned echoes from the tissues are detected by the same transducer, amplified, and processed for display. Typical processing involves filtering and envelope detection. The type of information obtained by an A-mode instrument, called an A-line and depicted in Fig. 15.6A, can be displayed in an alternative format. In a B-mode display, the echo amplitude is represented by the intensity or gray level of the display. Figure 15.6B shows a B-mode display of the A-line shown in 15.6A. A majority of commercial scanners on the market today are B-mode scanners in which the beam position is also monitored. The positional information of the beam plus the amplitude of the returned echoes following amplification and envelope detection from the tissues are converted into a format compatible with a cathode ray tube monitor in a device which is called a scan converter and is almost invariably digital today. The amplitude of the echoes is normally logarithmically compressed to accentuate lower level signals. This compression process reduces the dynamic range of the returned echoes, which can be as high as 120 dB to as low as 40 dB. Modern B-mode scanners can acquire images fast enough to monitor the motion of an organ.

Depending on the mechanisms used to drive a transducer, the real-time scanners are classified into mechanical-sector and electronic-array scanners. Since electronic array systems generally produce images of better quality, modern ultrasonic scanners are almost exclusively array-based systems. In such systems, the array shown in Fig. 15.9A with a typical dimension of 1 cm × 3–10 cm consists of 128 or more small rectangular piezoelectric elements. The shaded areas indicate the array elements, and the kerf represents the space between elements that is usually filled with an acoustic insulating material. In a linear sequenced array, or simply linear array, a subaperture of 32 elements or more is fired simultaneously to form a beam. The beam is electronically focused and swept across the aperture surface as illustrated in Fig. 15.9B in which the beam, indicated by the solid line, is generated first and then another beam, indicated by the dashed line, is generated. Electronic dynamic focusing can be achieved by appropriately controlling the timing of the returned echoes to each individual element as shown in Fig. 15.10. In a linear phased array, the beam is also electronically steered, resulting in a pie-shaped image. Figures 15.11A–15.11C show, respectively, a photograph of a modern ultrasonic scanner, an image produced by a linear array, and an image produced by a linear phased array. Dynamic focusing can be applied to an extended region of the field of view, most frequently during reception.

A block diagram of a B-mode imaging system is shown in Fig. 15.12. The beam former can be either analog, consisting of a matrix of delay lines and amplifiers, or digital, consisting of an array of preamplifiers and analog-to-digital (A/D) converters. Several of the components in a B-mode scanner perform the same functions as those in the A-mode system. The signal processing chain performs functions such as time gain compensation (TGC), signal compression, demodulation, and filtering. TGC is used to compensate for the signal loss due to tissue attenuation and beam divergence as the beam penetrates the body. The operator can choose from various

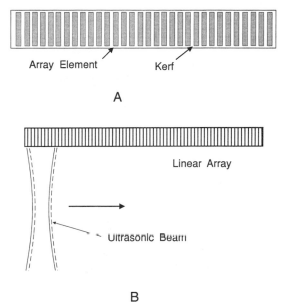

Array Element   Kerf

A

Linear Array

Ultrasonic Beam

B

Fig. 15.9 (A) A linear array with shaded areas indicating piezo-electric elements. The space between two elements is called a kerf. (B) In a linear sequenced array or linear array, the beam is swept from one end to the other electronically.

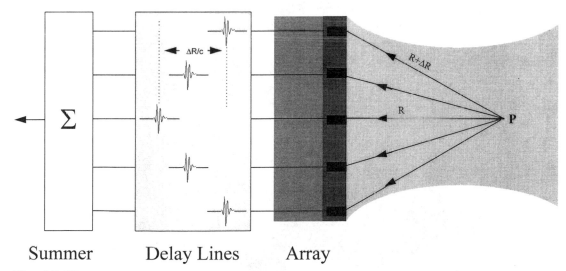

Summer   Delay Lines   Array

Fig. 15.10 Dynamic focusing during reception from a point target P can be achieved by delaying the pulses that arrive at the array elements depending on the differences in path lengths. In this example, the pulse arriving at the innermost element has to be delayed by $\Delta R/c$ relative to that arriving at the outermost element. Constructive interference results when these delayed pulses are summed.

A

B

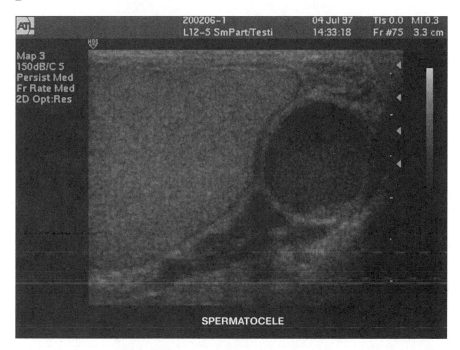

SPERMATOCELE

C

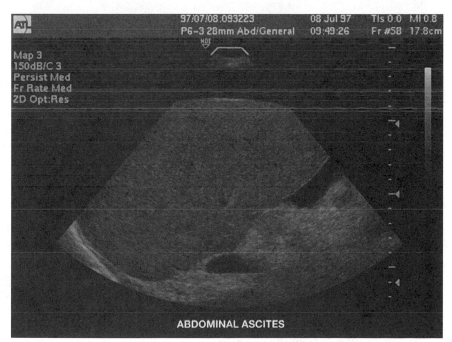

ABDOMINAL ASCITES

Fig. 15.11 (A) Photograph of a modern ultrasonic scanner. (B) The round anechoic mass in this ultrasonic image obtained by a linear array indicates a spermatocele, which is a cystic dilation of a duct in the testicle. The large homogeneous structure on the left of the image is the testicle. (C) An image of abdominal ascites, which are pockets of fluid that accumulate in the peritoneal cavity. The large homogeneous organ is the liver (courtesy of ATL, Bothell, WA).

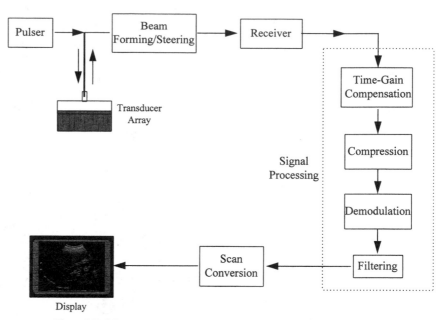

Fig. 15.12 Block diagram of a B-mode ultrasonic imaging system.

forms of TGC that are available on the console. Following signal processing, the signal may be digitized by an A/D converter. In high-end systems, A/D conversion is done in the beam former. This makes the system much more expensive since many more A/D converters of a higher sampling rate are required. The scan converter is a digital memory device that stores the data which have been converted from the format in which they were collected into a format that is displayable by a monitor. Before display, the video data may be processed again via band-pass filtering, high-pass filtering, low-pass filtering, gray-scale mapping, etc. Signal processing performed before and after the scan converter is called pre- and postprocessing, respectively.

The B-mode ultrasonic images exhibit a granular appearance, called speckle pattern, which is caused by the constructive and destructive interferences of the wavelets scattered by the tissue components as they arrive at the transducer surface. This speckle appearance very much resembles the speckle pattern that results from laser scattering by a rough surface. If the incident ultrasound beam is totally coherent like a laser beam, the speckle carries no information about the microstructure of the tissues. Fortunately, ultrasonic scanners use partially coherent incident waves, i.e., pulses. Thus, the speckle patterns exhibited by tissue do contain useful information about the structures of the tissues that can be used clinically for tissue differentiation.

For most of the B-mode scanners, only one ultrasound pulse is being transmitted at any one instant of time. The time needed to form one frame of image, t_f, can be readily calculated from the following equation:

$$t_f = \frac{2D \times N}{c}$$ (15.17)

where D is the depth of penetration determined by the pulse repetition frequency of the pulser, N is the number of scan lines in the image, and c is the sound speed in tissue. Rearranging this equation gives

$$F \times N \times D = \frac{c}{2} \tag{15.18}$$

where $F = 1/t_f$ is the frame rate.

Example Problem 15.6

A B-mode scanner is designed for obstetrical applications. The frequency of the scanner is specified to be 3.5 MHz. The scanner is also specified to have a frame rate of 30/s and the probe is a linear array of 128 elements with a subaperture of 32 elements that are fired simultaneously to form one scan line. Calculate the depth of penetration. What can be done to increase the depth of penetration?

Solution

There will be $128 - 32 + 1 = 97$ scan lines in one image since 32 elements are fired at the same time. From Eq. (15.18),

$$D = \frac{1.54 \times 10^5 \text{ cm/s}}{2 \times 30\text{/s} \times 97} = 26.5 \text{ cm}$$

The depth of penetration is therefore 26.5 cm. The number of scan lines or the frame rate must be reduced to increase the depth of penetration. ∎

15.3.3 Resolution of B-Mode Ultrasonic Imaging Systems

The resolution of a B-mode imaging system in the imaging or the azimuthal plane is determined by the duration of the pulse in the axial direction (i.e., in the direction of the beam) and the width of the ultrasonic beam in the lateral direction (i.e., in the direction perpendicular to the beam). The reason for this will become clear by examining Fig. 15.13, in which the echo waveform, or an A-line, from two targets separated by a distance x is shown. Scanner A with a shorter pulse duration will be able to resolve the two targets, whereas scanner B with a longer pulse duration will not. A similar argument can be made to understand why beamwidth determines the resolution in the lateral direction.

The axial and lateral resolutions have been conventionally defined as the spatial extents of the ultrasound pulse in the axial direction and beam profile in the lateral direction within -3 or -6 dB of the maxima. The axial resolution is determined by the pulse duration, which is primarily related to the bandwidth of the imaging system. The higher the bandwidth, the better the axial resolution. It is for this reason that system and transducer engineers strive for the widest bandwidth possible in system and transducer design. Currently, the bandwidth of an ultrasonic imaging system is for the most part limited by the performance of the transducer. In an optimally designed system, the bandwidth can be as high as 80–100%. If the medium

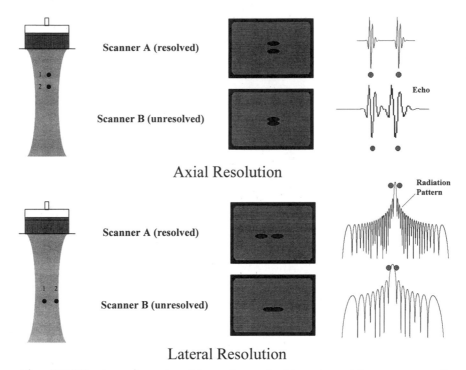

Fig. 15.13 Two targets, 1 and 2, can be resolved by scanner A but not scanner B. Therefore, scanner A has a better axial resolution than scanner B. Both B-mode images and echo waveforms or A-lines are shown to illustrate the concept of axial resolution. (Bottom) scanner A has a narrower radiation pattern and also has a better lateral resolution that is capable of resolving targets 1 and 2.

in which the beam is propagating causes significant frequency-dependent attenuation, as is the case in tissues, the bandwidth is reduced. Thus, the axial resolution of the scanner worsens as the beam penetrates deeper into the tissue because signal components at the higher frequencies are attenuated more.

Just like an optical lens, the lateral resolution of a transducer has been shown to be proportional to the product of the f number ($f_{\#}$), defined as the ratio of focal distance divided by the aperture size, and λ. Either a higher frequency or a larger aperture must be used to improve the lateral resolution at a fixed focal distance.

Ultrasound imaging is diffraction limited. This means that ultrasound can only be focused within a limited field of view or depth. Because of the wave diffraction effect, in the region near and very far from the transducer the beam cannot be properly focused. The field of view within which the beam can be properly focused and yield acceptable lateral resolution is limited to a $f_{\#}$ ranging from 1 to 10, given the aperture size and the frequency. In addition to diffraction, imaging in the far field is also affected to a great degree by the attenuation of the medium, which reduces the signal-to-noise ratio.

15.3.4 M-Mode and C-Mode Imaging

In M-mode display, one intensity-modulated A-line or one B-line is swept across the monitor as a function of time at a rate much slower than the pulse repetition frequency (PRF) of the A-line. In this format, the ultrasound beam is fixed at a certain position or angle and the displacement of a target relative to the probe along the beam direction is displayed as a function of time. This type of display is most useful for monitoring the motion of anatomical structures, for example, valves in the heart.

C-mode is a form of display similar to conventional radiography in which a second transducer is used to detect the pulse after traversing a medium. The image obtained is then a 2-D map of the ultrasonic attenuation experienced by the pulse. Reflection-type C-mode ultrasonography is also possible by using time gating to select only the echoes that originate from a certain plane or at a constant depth relative to the transducer.

Ultrasonic computed tomography, which may be useful for imaging organs such as breast and testicles, has also been studied extensively but has not found any practical applications to date because of the divergence and refraction of an ultrasonic beam as it penetrates into the body. These problems are difficult to compensate for or to overcome.

15.3.5 Doppler Methods for Flow Measurement

As was discussed earlier, the Doppler effect provides the unique capability for using ultrasound to measure blood flow. Conventionally, two different approaches have been used for ultrasonic Doppler flow measurements: continuous wave (CW) and pulsed wave (PW) Doppler. A CW system is shown in Fig. 15.14. A probe consist-

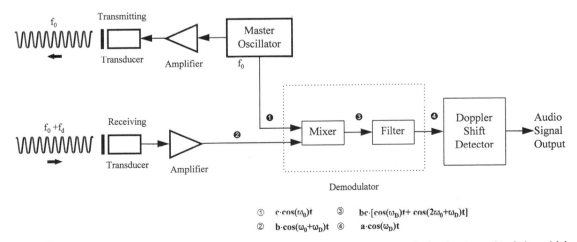

Fig. 15.14 Block diagram of a CW Doppler flowmeter. Points 1–4 are, respectively, the transmitted sinusoidal signal at ω_o, the Doppler shifted returned signal, the signal following mixing, and the demodulated signal at ω_d, the Doppler signal. Here, a–c are constants.

ing of two piezoelectric elements, one for transmitting the ultrasound signal and one for receiving the echoes returned from blood, is excited by an oscillator. The returned Doppler shifted echoes are amplified, demodulated, and low-pass filtered to remove the carrier frequency. The information contained in the output signal from the demodulator can be detected in different ways. It can be heard by a speaker since the Doppler shift is in the audible range. Alternatively, a zero-crossing counter can be used to estimate the mean Doppler frequency, or a spectrum analyzer can be used to display the spectrum. The spectrum is usually displayed in the format shown in Fig. 15.15, in which the vertical axis indicates Doppler frequency or velocity, the horizontal axis time, and the gray scale the intensity of the Doppler signal at that frequency or velocity. At each instant of time, the displayed line represents the Doppler spectrum calculated for that time within a 5–10-ms time window. From the Doppler spectrum, the mean frequency or other frequencies, e.g., median frequency and mode frequency, can easily be estimated.

A problem with a CW Doppler is its inability to differentiate the origins of the Doppler signals produced within the ultrasound beam. Signals coming from blood flowing in two blood vessels in the same vicinity, e.g., an artery and a vein, may overlap. To alleviate this problem a pulsed wave Doppler may be used. As illustrated in Fig. 15.16, ultrasound bursts of long duration consisting of many cycles are used to excite the probe. The returned echoes received by the same transducer are amplified and demodulated. The demodulated signal is then sampled and held by a sample and hold circuit that is triggered by the delayed pulses. The time delay allows the selection of the location at which the Doppler shift frequency is monitored. Following low-pass filtering, the Doppler signal can be displayed or heard as the CW Doppler.

A drawback of the pulsed Doppler is the limit of the highest Doppler frequency or maximal velocity that it can measure. This is determined by the PRF of the device, which has to be at least twice as large as the maximal Doppler frequency. This

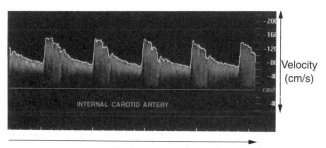

Time (s)

Fig. 15.15 A spectrasonogram delineating the Doppler spectrum from blood flowing in a human artery as a function of time in seconds. The Doppler power within a time window of 5–10 ms is represented by the gray scale. The ordinate represents Doppler frequency or velocity (cm/s). The bright white line indicates the traced maximal Doppler frequency (courtesy of ATL). Both the spectral content and the maximal Doppler frequency will be altered by diseases associated with the blood vessel such as arterial stenosis.

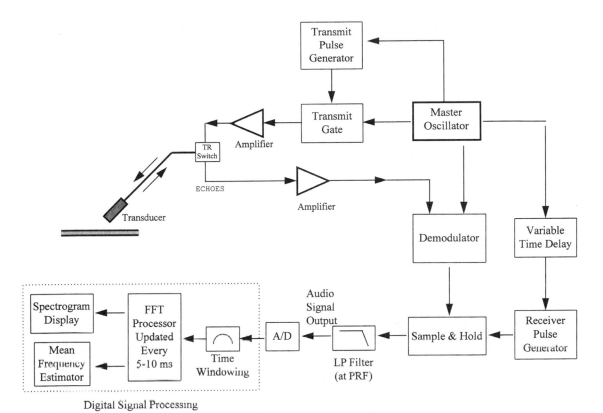

Fig. 15.16 The block diagram of a pulsed Doppler system. The audio signal output can be heard via a speaker or digitized following time windowing to minimize spectral distortion for Fourier analysis.

may pose a problem when measuring high velocities in the body, e.g., outflow tracts of cardiac valves and stenosis in a blood vessel. Modern high-end ultrasonic imaging machines are equipped with both CW and PW capabilities.

Duplex imaging combines B-mode real-time imaging with Doppler for measuring blood flow. Both sets of information are displayed simultaneously. A cursor line is typically superimposed on the B-mode image to indicate the direction of the Doppler beam. A fast Fourier transform algorithm is used to compute the Doppler spectrum, which is displayed in real time.

15.3.6 Color Doppler Flow Imaging

Color Doppler flow imaging systems are duplex scanners capable of displaying both B-mode and Doppler blood flow data simultaneously in real time. The Doppler information is encoded in color. Typically, a color such as red is assigned to represent flow toward the transducer, and a color such as blue is assigned to represent flow away from the transducer. The magnitude of the velocity is indicated by different shades of the color. Typically, the lighter the color, the higher the velocity. The color Doppler image is superimposed on the gray-scale B-mode image.

The basic concept of the color Doppler is similar to that of the pulsed Doppler instruments which extract the mean Doppler shift frequency from a sample volume that is defined by the beam width and the gate width. The only exception is that the color Doppler instruments are capable of estimating the mean Doppler shifts of many sample volumes along a scan line in a very short period of time, on the order of 30–50 ms. To be able to do so, fast algorithms had to be developed. One such algorithm was based on the well-known Wiener–Khinchine theorem, which indicates that the autocorrelation function $H(\tau)$ of a function $f(t)$ is the Fourier transform of the power spectrum $P(\omega)$ of $f(t)$. Mathematically this is given by

$$H(\tau) = \int_{-\infty}^{\infty} P(\omega)e^{j\omega\tau}d\omega \tag{15.19}$$

It can be easily shown that

$$j<\omega> = \frac{\dot{H}(0)}{H(0)} \tag{15.20}$$

where $<\omega>$ denotes the mean of ω and the dot operation represents the first derivative and

$$\sigma^2 = \left[\frac{\dot{H}(0)}{H(0)}\right]^2 - \frac{\ddot{H}(0)}{H(0)} \tag{15.21}$$

where σ^2 is the variance of ω and the double-dot operation denotes the second derivative. Since the magnitude and phase of $H(\tau)$,

$$H(\tau) = |H(\tau)|e^{j\phi(\tau)} \tag{15.22}$$

are respectively an even function and an odd function, Eqs. (14.20) and (14.21) can be further simplified to

$$<\omega> = \frac{\phi(T)}{T} \tag{15.23}$$

and

$$\sigma^2 = \frac{1}{T^2}\left[1 - \frac{|H(T)|}{H(0)}\right] \tag{15.24}$$

where T is the pulse repetition period. These are all simple arithmetic operations which require little time for computation if the autocorrelation function $H(\tau)$ is known.

In a color Doppler system, the signal received by a probe is divided into three paths — one for constructing the gray-scale B-mode image, one for calculating the flow information from Doppler data using a hard-wired autocorrelator, and one for conventional Doppler measurements. Once the autocorrelation function is known, the mean frequency and the variance can be readily calculated as described. Eight or more shades are used in these systems to depict the magnitude of the velocity. Since the basic idea of Doppler flow mapping is similar to that of pulsed Doppler, the maximal Doppler frequency that can be detected without aliasing is one-half of the PRF. A color Doppler image is shown in Fig. 15.17.

Many clinical applications have been found for this device, including diagnosing tiny shunts in the heart's wall and valvular regurgitation and stenosis. It considerably reduces the examination time in many diseases associated with flow disturbance. Problematic regions can be quickly identified first from the flow mapping. More quantitative conventional Doppler measurements are then made on these areas.

15.4 NEW DEVELOPMENTS

There have been many new and interesting developments in diagnostic ultrasound in recent years. Only a selected few of particular importance are discussed here.

15.4.1 Color Doppler "Power Mode" Imaging

Recently, a new way of displaying the color Doppler information, i.e., power mode or "energy mode" imaging, was introduced and has been well accepted by the clinical community. A majority of the top-of-the-line scanners now have this option. In-

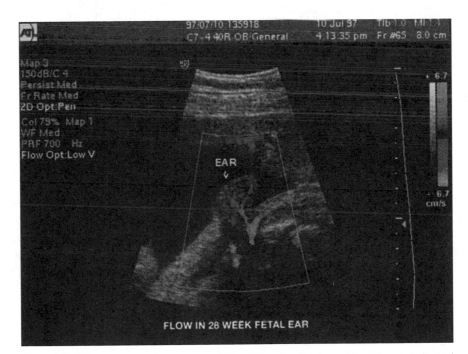

Fig. 15.17 A color Doppler image of blood flow in a fetal ear. Blood flow is indicated in color, whereas the fetal ear and surrounding anatomical structures are in gray scale (courtesy of ATL).

stead of the mean Doppler shift, the power contained in the Doppler is displayed in this approach. There are several advantages to doing this. First, a threshold can be set to minimize the effect of noise. Second, the data can be averaged to achieve a better signal-to-noise ratio. Third, the images are less dependent on the Doppler angle. Finally, aliasing is no longer a problem since only the power is detected. As a result, signals from blood flowing in much smaller vessels can be detected. The images so produced have an appearance similar to that of X-ray angiography. The disadvantages of this approach are that it is more susceptible to motion artifacts because of frame averaging and the image contains no information on flow velocity and direction. The motion problem, however, can be partially overcome by harmonic imaging in conjunction with the injection of an air-containing contrast agent that will be discussed later. A "power" Doppler image is shown in Fig. 15.18.

Power Doppler imaging is easier to implement than conventional color Doppler. In fact, the parameter, Doppler power, is readily available in conventional color Doppler systems. $H(0)$ in Eq. (15.24) that is needed to calculate the variance of Doppler frequencies is the power contained in the Doppler spectrum. This becomes apparent when one considers the definition of the autocorrelation function given by Eq. (15.19), i.e.,

$$H(\tau = 0) = \int_{-\infty}^{\infty} P(\omega)d\omega \tag{15.25}$$

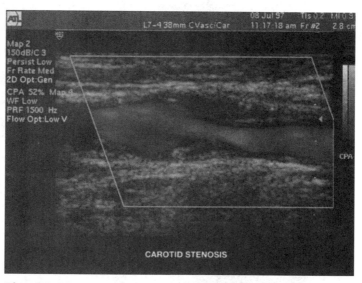

Fig. 15.18 A power color Doppler image of a stenotic carotid artery in which the color indicates the power contained in the Doppler signal rather than the Doppler shift frequency (courtesy of ATL).

15.4.2 Time-Domain Flow Estimation

As was discussed in preceding sections, Doppler devices have several shortcomings, including that it is difficult to estimate the angle between the direction of blood flow and the ultrasound beam and the maximal flow that can be measured is limited by the pulse repetition frequency. Alternatives have been sought, including several approaches that estimate blood flow in the time domain. Both frame-to-frame tracking of movement of speckles generated by blood from B-mode images and direct correlation of radio frequency (RF) echoes from blood show great promise. The frame-to-frame speckle tracking is desirable in that the flow information can be directly obtained from the B-mode images that are acquired by the scanner without the need for additional hardware. Its drawbacks are that the signal level from blood in the frequency range of 3–10 MHz is too weak to obtain a reliable estimation and the frame rate of 30 per second is too low to estimate arterial blood flow velocity. The former problem may be overcome by injecting a contrast agent, whereas a high frame scanner has to be used to solve the latter problem.

Direct correlation of RF echoes scattered by blood cells yields flow information along the ultrasound beam. Consider the two RF echo trains shown in Fig. 15.19 that are obtained at times t_0 and t_1. The distance z is given by $ct/2$, where c is sound velocity in the medium and t is the time of flight. The distance that has been traversed by the scatterers, which are red blood cells in the blood, is $z_1 - z_0$ and can be estimated from cross-correlating the echo trains or waveforms, $f(z) = f(ct/2)$, col-

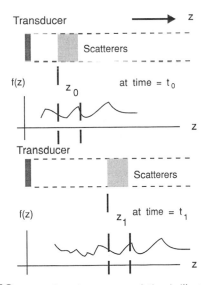

Fig. 15.19 Time domain cross-correlation is illustrated graphically. The shaded areas indicate a volume of scatterers which moves from z_0 at time = t_0 to z_1 at time = t_1 with a velocity of v. The echo waveform received by the transducer is denoted by $f(z)$.

lected at different times, t_0 and t_1. The cross-correlation function of the two time waveforms is given by

$$\varsigma(\tau) = \int_{-\infty}^{\infty} f_{t0}(t) f_{t1}(t - \tau) dt$$

where $\varsigma(\tau)$ is the cross-correlation function and $f_{t0}(t)$ and $f_{t1}(t)$ are echo waveforms obtained at t_0 and t_1. A cross-correlation time can be estimated from computing the cross-correlation function as τ is changed and is, in fact, given by the value when ς is maximal. Once $z_1 - z_0$ is known, the velocity, v, of the scatterers moving in the direction of the beam can be calculated from the following equation if $t_1 - t_0$ is set to T, the pulse repetition period of the transmitted pulses:

$$v = \frac{z_1 - z_0}{T} \tag{15.27}$$

This approach has been implemented in at least one commercial scanner. It has several advantages:

1. It can track slower flow since no clutter rejection filter, the high pass filter used in Doppler systems to reject large slowly moving echoes, is needed.
2. Its demand on signal-to-noise ratio is lower, thus requiring less averaging and yielding a higher frame rate.
3. There is no aliasing. It gives no estimation if the PRF is too low to track the motion.
4. The spatial resolution is superior because it uses shorter pulses.

Example Problem 15.7

An ultrasonic time domain flow estimation system was designed to have a PRF of 5 kHz. It was used to estimate blood flow in a carotid artery. The cross-correlation time, estimated from computing the cross-correlation functions between two successive echo waveforms returned from the blood in the vessel over a finite record of the waveform, was 0.05 µs. Assuming that the ultrasound beam emitted by the transducer is aimed in the same direction as the flow, what is the blood flow velocity?

Solution

The relationship between the time of flight and the distance from the tranducer is $z = ct/2$. This means that for a correlation time of 0.05 µs, the volume of blood from which the echo waveform is collected has moved a distance of

$$\Delta z = z_2 - z_1 = (1.54 \times 10^5 \text{ cm/s}) (0.05 \times 10^{-6} \text{ s})/2 = 0.0039 \text{ cm}$$

within the time interval given by the pulse repetition period equal to

$$\frac{1}{\text{Pulse repetition frequency}} = \frac{1}{5 \text{ kHz}} = 0.2 \text{ ms}$$

Blood flow velocity in the vessel can be found from Eq. (15.27):

$$v = 0.0039 \text{ cm}/0.2 \text{ ms} = 19.5 \text{ cm/s.} \qquad \blacksquare$$

15.4.3 Multidimensional Arrays

The linear arrays used by current scanners are only capable of dynamic focusing in one dimension in the azimuthal plane, the imaging plane. Beam focusing in the elevational plane (the plane perpendicular to the imaging plane), which determines the slice thickness, has to be achieved mechanically either by shaping or shading the elements or by a lens. In other words, the focused region of a linear array probe is fixed in the elevational plane. Thus, the slice thickness varies in the field of view and is smallest only near the focus. To further improve the image quality, it is necessary to control the beam in the elevational plane. The ultimate approach would be to use 2-D arrays that allow 2-D beam focusing and steering. Unfortunately, 2-D arrays with resolution comparable to that of 1-D arrays are difficult to fabricate and extremely expensive. A prototype 2-D array of more than 40 × 40 elements has been investigated for 3-D real-time volumetric imaging. Intensive investigations are under way to use very large-scale integration technology to miniaturize electronic components associated with the array elements and to develop novel interconnection technologies.

Alternatives have been sought because of the prohibitive cost and technical difficulty in developing 2-D arrays. One possibility is to use fewer rows in the elevational plane to achieve limited focusing at different depths and to reduce the slice thickness in the near field and far field of the transducer. These are called 1.5-D arrays. One such array is shown in Fig. 15.20. A corresponding image is shown in Fig. 15.21 in which better image uniformity is seen.

15.4.4 3-D Imaging and Parallel Processing

Since ultrasonic images are 2-D tomograms, 3-D reconstruction can be readily achieved off-line from multiple slices of ultrasonic images if the imaging planes are spatially encoded (Fig. 15.22). The most intriguing aspect of 3-D ultrasonic imaging is its potential for 3-D real-time imaging using a 2-D array that allows beam steering and dynamic focusing in 3-D. Parallel processing of the data is imperative

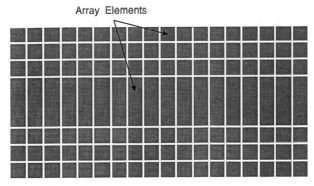

Array Elements

Fig. 15.20 A 1.5-D array with several rows of elements in the elevational plane to control slice thickness.

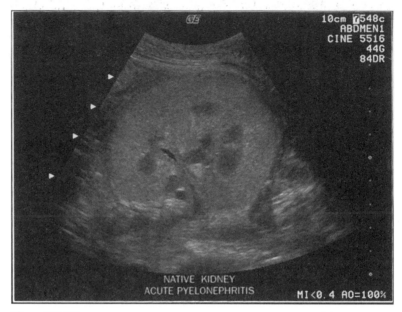

Fig. 15.21 An ultrasonic image of a kidney with acute infection of the parenchyma and pelvis, produced by a 1.5-D array (courtesy of GE Medical Systems, Milwaukee, WI).

in order to accomplish this. Only one acoustic pulse is propagated in the tissue at any one time in conventional ultrasonic image processing. The scanning time can be considerably shortened if the acoustic pulse can be sent into several directions simultaneously or if several scan lines of information can be collected and processed at the same time. In one parallel processing scheme, data were acquired in eight or more directions for each transmitted pulse with a wider than normal ultrasonic beam. Parallel processing inevitably increases the complexity and cost of an ultrasonic imaging system.

15.4.5   Intracavity and Very High Frequency Imaging

Since the attenuation of ultrasound in a tissue is approximately linearly proportional to frequency, the frequency range that can be used for imaging internal organs over the body surface is limited by a concern for patient exposure. Intracavity imaging, in which the probe is placed closer to the organ to allow application of higher frequencies, offers an alternative. Novel specialty probes in the form of mechanical sector, linear array, or phased array are available for transesophageal, transrectal, transvaginal, and intravascular scanning. Intravascular imaging is technically more challenging in that it uses frequencies higher than 20 MHz. At these frequencies, the resolution is sufficiently high to resolve the internal structures of the arterial wall. There has been a growing interest in developing devices operating in the 30–100 MHz range for applications in imaging superficial structures in the eye and skin.

a

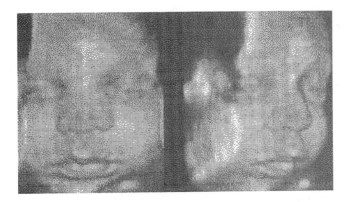

b

c

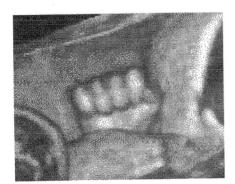

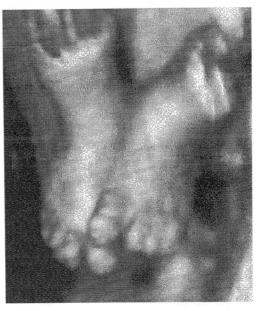

Fig. 15.22 3-D ultrasound images of a fetus. a, face; b, hands; c, feet (courtesy of Dr. Bernard Benoit).

Ultrasonic imaging and Doppler devices have also found applications in endoscopic surgery. These devices are useful because optical endoscopic instruments allow only the visualization of the surface of an anatomic structure. Since the surgeons lose tactile sense during endoscopic surgery, the capability of visualizing tissues beneath the surface becomes critical in diagnosing the state of the structure or organ.

15.4.6.   Ultrasound Contrast Media and Harmonic Imaging

Contrast media have been used extensively in radiology to enhance certain structures in an X-ray image. For instance, iodinated compounds can be injected into the coronary artery to better visualize the coronary vasculature in the heart. Similar ultrasonic agents have been successfully developed. A majority of these agents utilize microscopic air bubbles that are extremely strong ultrasound scatterers because of the acoustic impedance mismatch. They are generated as a result of chemical action or mechanical agitation. Moreover, air bubbles resonate when insonified by an ultrasonic wave. The echoes from the bubbles can be further enhanced if the incident wave is tuned to the resonant frequency of the bubbles. The resonant frequency, f_r, for a free air bubble, i.e., a bubble without a shell, is related to the radius of the bubble, a, by

$$f_r = \frac{1}{2\pi a} \sqrt{\frac{3\varepsilon P_0}{\rho_W}} \tag{15.28}$$

where ε is the ratio of the specific heats of gases and equals 1.4 for air, P_0 is the ambient pressure and equals 1.013×10^5 Pa or 1.013×10^6 dyns/cm^2 at 1 atm, and ρ_W is the density of the surrounding medium, e.g., water.

Example Problem 15.8

Calculate the scattering cross section at 2 MHz and resonant frequency of an air bubble of 3-μm radius in water.

Solution

The compressibility and density of air are respectively 6.9×10^{-7} cm^2/dyn and 1.3×10^{-3} g/cm^3. The compressibility of water is 4.6×10^{-11} cm^2/dyn. From Table 15.1, the speed of sound in water is 1480 m/s, so λ at 2 MHz is 740 μm. The scattering cross section of an air bubble can be calculated from Eq. (15.15). The compressibility term dominates so that the density term can be neglected. Also, G_s in the present case is $\gg G$:

$$\sigma_s = \frac{4}{9}\pi\left(\frac{2\pi}{740 \times 10^{-6}}\right)^4 (3 \times 10^{-6})^6 \left|\frac{6.9 \times 10^{-7}}{4.6 \times 10^{-11}}\right|^2$$

$$= 1.19 \times 10^{-7} \text{ cm}^2$$

This is much greater than the scattering cross section of a red cell at 2 MHz (1.5×10^{-12} cm^2). From Eq. (15.28), the resonant frequency for an air bubble of 3 μm radius is

$$f_r = \frac{1}{2\pi \times 3 \times 10^{-4}} \sqrt{\frac{3 \times (1.4) \times (1.013 \times 10^6)}{1.013}}$$

$$= 3.3 \times 10^6 \text{ Hz or 3.3 MHz}$$

If the bubble radius is increased so it resonates at 2 MHz, the scattering cross section can be further increased. ■

Current problems in contrast sonography include producing bubbles of suitable size and bubble lifetime. It is preferable to have the contrast medium injected intravenously since a significant amount of the contrast agent must be able to traverse the pulmonary circulation. Only bubbles <8 μm are capable of passing through the pulmonary capillaries. The requirement for the bubbles to maintain their size stems from the fact that their size has to be known or maintained when being imaged or monitored in order to quantitate blood perfusion. To lengthen the bubble lifetime in blood, the bubbles are typically encapsulated in some sort of nontoxic shell. Albunex is a commercial contrast agent that consists of air bubbles with a mean diameter of 3–5 μm that are encapsulated in a shell of albumin. Current commercial agents can persist in a blood stream from a few seconds to a few hours.

A new exciting application of the air-containing agents is found in harmonic imaging and Doppler measurements in which the effect of the surrounding stationary structures on the image and results is minimized. Harmonic imaging or Doppler measurements following the injection of a gas-containing contrast agent are possible because only microbubbles resonate when impinged upon by ultrasound and emit ultrasound at harmonic frequencies. Ultrasound at harmonic frequencies will only be produced by anatomic structures that contain these agents. Tissues that do not contain gaseous contrast agents presumably will not produce harmonic signals. A good example is blood flowing in a blood vessel. Blood containing the contrast agent will produce harmonic signals but the blood vessel will not. The contrast between the blood and the blood vessel will therefore be significantly improved in the resultant harmonic image if only the echoes at the harmonic frequency are detected by the transducer. Various types of oral contrast media that absorb or displace stomach or bowel gas for ultrasonic imaging have also been developed for better visualization of the abdominal organs such as the stomach and pancreas.

15.4.7 Elastography

Elastography is a new mode of ultrasonic imaging in which tissue elasticity is displayed. Two different approaches, static and dynamic, have been used. The static approach is illustrated in Fig. 15.23, in which a displacement, x_t, of the transducer would result in a displacement, x, of the tissue. The displacement of a region may be estimated by cross-correlating the speckle pattern. An elastic constant may then be calculated from the ratio of the strain and the stress, where the strain and the stress are defined as the displacement per unit distance and the force per unit area, respectively. This number can be estimated for each pixel or voxel and expressed by either a color or gray scale. This type of image displays information that is different from that of traditional pulse-echo imaging. Therefore, an object that has little contrast in a B-mode image may have significantly improved contrast in an elastogram. Dynamic elastography or sonoelasticity imaging uses an external vibrator. The motion of the tissue related to tissue elastic properties is monitored by Doppler methods. This technique has been shown to be useful to diagnose prostate cancer.

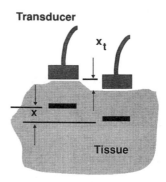

Fig. 15.23 An elastogram is obtained by pressing a transducer against a tissue. A displacement of x_t of the transducer results in a displacement of x of a volume of the tissue underneath the transducer.

15.5   BIOLOGICAL EFFECTS OF ULTRASOUND

Although ultrasound is known as a form of noninvasive radiation, biological effects inevitably occur if its intensity is increased beyond a certain limit. Ultrasound biological effects can be classified into two categories: thermal and mechanical. The **thermal effect** results from an increase in temperature when ultrasound energy is absorbed by tissue. The **mechanical effect** is caused by the mechanical disturbance produced by ultrasound and is manifested by the generation of microscopic air bubbles or **cavitation.** Currently, no biological effects have been found for ultrasound with a spatial peak temporal average intensity <100 mW/cm^2. A consortium of professional and manufacturing organizations has developed a standard for labeling acoustic output levels. These quantities are the thermal index and the mechanical index, which have been determined to have critical relevance to the thermal and mechanical biological effects, respectively. Manufacturers are encouraged to display these indices on the scanner monitor.

The thermal index (TI) is defined as the ratio of the acoustical power produced by the transducer to the power required to raise the tissue temperature by 1°C. TI depends on the tissue type. Obviously, bone absorbs heat differently than does soft tissues, so the TI for bone is different from that of soft tissues. The value should be displayed on the monitor when the TI produced by a transducer is higher than 0.4. The mechanical index (MI) is related to the probability of the generation of cavitation and is proportional to the peak rarefactional pressure and inversely proportional to the square root of the ultrasound frequency. Therefore, the likelihood for cavitation to occur increases as the ultrasound intensity is increased; however, it decreases as the frequency is increased. The MI should not be higher than 1.9. These indices are introduced to encourage the operators to follow the ALARA (as low as reasonably achievable) principle during patient scanning.

15.6   THERAPEUTIC ULTRASOUND

Ultrasound has been used for therapeutic purposes for many years since high-intensity ultrasound produces biological effects. Based on the duration and intensity level of the exposure during therapy, ultrasonic therapeutic applications can be categorized into hyperthermia and noninvasive surgery. In general, tissues are exposed to ultrasound of longer duration at lower intensity levels in ultrasound hyperthermia. The duration is typically from 10 to 30 min at an intensity level from a few to a few tens of Watts/cm^2, causing tissue temperature to be elevated to 43–45°C. For noninvasive surgery, the duration is shorter (on the order of a few seconds), and intensity level is higher (on the order of a few hundreds Watts/cm^2). The purpose is to raise the temperature in the focused region of the transducer to 70–90°C within a short period of time. Cell death results immediately at this temperature. High-intensity focused ultrasound surgery has been shown to be clinically efficacious in the treatment of many diseases, including benign prostate hyperplasia, ocular tumor, and retinal tear.

Although ultrasonic therapeutic devices are commercially available, there are still some technical difficulties that must be overcome to achieve wide acceptance. The foremost problem is in noninvasive temperature monitoring. Both ultrasonic imaging and MRI are currently being investigated to estimate temperature rise *in situ*. Another problem lies in proper focusing of the beam. The phased array technology that has been used in imaging is now being adapted to therapeutic ultrasound. Admittedly, designing transducers capable of handling high ultrasound power is as challenging as finding an appropriate methodology for monitoring temperature.

EXERCISES

1. The plane wave solutions for the wave equation are of the forms $e^{j(\omega t - kz)}$ and $e^{j(\omega t + kz)}$, one representing a wave going in the $+z$ direction and one in the $-z$ direction. Explain why.
2. A plane ultrasound wave of 3 MHz with peak pressure equal to 2 MPa is incident upon a flat boundary between two media of acoustic impedances, 1.5 and 2.1 MRayl. (a) Calculate the transmitted and reflected pressures and intensities. (b) Repeat (a) if the impedances are 1.5 and 1.3 MRayl. (c) Can p_t be greater than p_i? Why? (d) From the answers for (a) and (b), show that $I_i = I_t + I_r$.
3. Calculate the distance that 1 MHz ultrasound will travel before its intensity drops to one-half of its incident value for (a) air, (b) water, and (c) liver. Repeat the calculation with the frequency increased to 5 MHz.
4. A 96-element linear array is used to image a fetus *in utero* in real-time. The maternal wall plus a full-term fetus may need a penetration depth of 20 cm. Suppose that 16 elements are fired as a group to form one scan line. (a) How many scan lines are there in one frame of image? (b) What is the frame rate of the scanner?

5. A 5-element subaperture of a 128-element linear array is used to form a beam as shown in Fig. 15.10. The gap width between two elements is 0.2 mm. The element width is 0.5 mm. During reception, the beam is focused to point P, located 1 cm in front of the subaperture. Calculate the time delays that should be applied to the center elements relative to the two elements at the edges to achieve appropriate focusing.

6. A 5-MHz CW Doppler flowmeter is used to monitor blood flow in a carotid artery. The probe makes a 60° angle relative to the flow. The velocity is 40 cm/s. What is the Doppler shift?

7. Can tissue attenuation affect the axial resolution and lateral resolution of an ultrasonic imaging system? Justify your answer.

8. To avoid range ambiguity in pulsed Doppler and pulse echo imaging, the echoes from the deepest structures in the body must be received before the next pulse is transmitted. Consequently, the depth of penetration, z_{max}, determines the PRP of the pulsed Doppler or pulse echo imaging system. Given that the PRP of a pulsed Doppler system has to be at least twice as large as the maximum Doppler frequency, show that $z_{max} v_{max} \leq c^2/8f$, where v_{max} is the maximum velocity, c is the ultrasound velocity in blood, and f is the ultrasound frequency.

9. From Eqs. (15.19) and (15.25), derive Eqs. (15.20) and (15.21).

10. From Eqs. (15.20)–(15.22), derive Eqs. (15 23) and (15.24).

11. Calculate the scattering cross section of a collagen sphere in water, assuming that collagen has a compressibility of 2.3×10^{-11} cm^2/dyn. Show why collagen is less desirable than air bubbles as an ultrasonic contrast agent.

12. Calculate the radius of an air bubble that resonates at 2 MHz in blood and in water.

Suggested Reading

Cho, Z. H., Jones, J. P., and Singh, M. (1993). *Foundations of Medical Imaging.* Wiley, New York.

Christensen, D. A. (1988). *Ultrasonic Bioinstrumentation.* Wiley, New York.

Evans, D. H., McDicken, W. N., Skidmore, R., and Woodcock, J. P. (1989). *Doppler Ultrasound: Physics, Instrumentation, and Clinical Applications.* Wiley, New York.

Jensen, J. A. (1996). *Estimation of Blood Velocities Using Ultrasound.* Cambridge Univ. Press, Cambridge, UK.

Kino, G. S. (1987). *Acoustic Waves: Devices, Imaging, & Analog Signal Processing.* Prentice Hall, Englewood Cliffs, NJ.

Krestel, E. (1990). *Imaging Systems for Medical Diagnosis.* Siemens, Aktiengesellschaft, Germany.

McDicken, W. N. (1991). *Diagnostic Ultrasonics: Principles and Use of Instruments.* Churchill Livingstone, Edinburgh, UK.

Shung, K. K., Smith, M. B., and Tsui, B. M. W. (1992). *Principles of Medical Imaging.* Academic Press, San Diego.

Webb, S. (1990). *The Physics of Medical Imaging.* Hilger, Bristol.

Well, P. N. T. (1977). *Biomedical Ultrasonics.* Academic Press, London.

Zagzebski, J. A. (1996). *Essentials of Ultrasound Physics.* Mosby, St. Louis.

16 NUCLEAR MAGNETIC RESONANCE AND MAGNETIC RESONANCE IMAGING

Chapter Contents

At the conclusion of this chapter, the reader will be able to:

- Describe the principle of nuclear magnetism

- Describe the mechanism by which a nuclear magnetic resonance signal, a free-induction decay, is created

- Explain the procedure and mathematics for creating magnetic resonance images using two-dimensional Fourier imaging

- Draw and describe gradient echo and spin-echo pulse sequences

- Explain the contrast mechanism in magnetic resonance images and select pulse parameters to maximize contrast between two tissue types

- Draw and explain a block diagram of a magnetic resonance imaging system

- Calculate the signal-to-noise ratio obtained from a sample in a simple magnetic resonance experiment

16.1 INTRODUCTION

Nuclear magnetic resonance (NMR) has been used for many years as a spectroscopy technique for analytical chemistry. During the early 1970s, Paul Lauterbur, Ray Damadian, and Peter Mansfield independently pioneered the use of NMR as a technique for *in vivo* imaging. Magnetic resonance imaging (MRI) provides a "window" into the body that had not previously been available by actually interrogating the nuclear environment of selected nuclei. The most common nucleus studied is the hydrogen proton. MRI has undergone extremely rapid development, in terms of both the technology and the sophistication of the applications, because of the richness of the information available and the unparalleled soft-tissue contrast it provides. From initial clinical demonstrations in the late 1970s, MRI had become the gold standard for most neurological procedures by the late 1980s. A volume reconstruction made in the early 1990s is shown in Fig. 16.1, demonstrating the remarkable improvements in capability during a relatively short time. MR images such as that in Fig. 16.1, demonstrating the remarkable improvements in capability during a relatively short time. MR images such as that in Fig. 16.1 are becoming commonly used for surgical planning.

Recently, researchers have begun to expand the applications of MRI. Two of the many applications that have seen tremendous interest are MR angiography and functional imaging. A MR angiogram of the brain, depicting an aneurysm, a defect in a blood-vessel wall, is shown in Fig. 16.2a. MR angiography represents a signif-

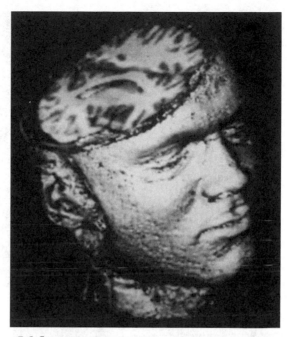

Fig. 16.1 A three-dimensional reconstruction of a head made using MRI. Images such as these are becoming widely used for surgical planning.

icant advance over previous X-ray methods in that it does not require the injection of contrast agents, which always carry some risk of complication. Figure 16.2b illustrates an example of the imaging of physiological function. In a technique known as blood oxygenation level-dependent imaging, increases in brain activity can be mapped due to changes in the magnetic susceptibility caused by changes in the blood oxygenation during task-induced activation of brain tissue. The image in Fig. 16.2b demonstrates the ability to identify the motor cortex by obtaining and processing a series of images made with and without the subject moving his hands. The activated pixels are overlaid in white over a reference image.

While there are over 10,000 clinical MR systems in use, the technology continues to experience intensive research and development. Because MR imaging and spectroscopy involve the interaction between two complex systems, the MR spectrometer and the human sample, bioengineers are often uniquely suited for research in this important diagnostic imaging modality. The goal of this chapter is to provide an introduction to MR imaging. Section 16.2 provides a brief overview of the basic mechanism upon which MRI is based, nuclear magnetism. NMR, the phenomenon which allows different nuclear species to be distinguished, is described in Section 16.3. The basic principles of Fourier imaging, as applied to NMR, are discussed in Section 16.4, and a brief overview of a typical MRI system is given in the final section.

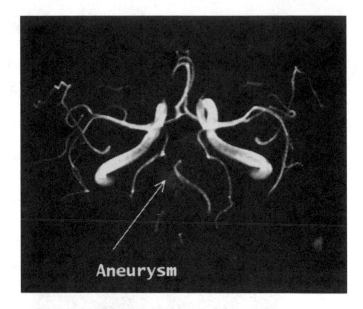

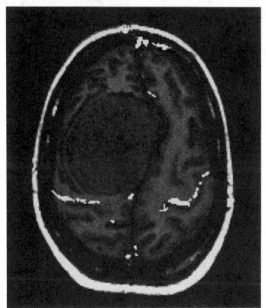

Fig. 16.2 (a) An MR angiogram of the brain depicting an aneurysm without the use of external contrast agents, (courtesy of Siemens Medical Systems, Iselin, NJ). (b) A functional brain image localizing the motor cortex.

16.2 NUCLEAR MAGNETISM

All nuclei with an odd number of protons, neutrons, or both exhibit nuclear magnetism. This includes approximately two-thirds of all stable nuclei. Table 16.1 provides some of the most important nuclei used for NMR. The nuclei can be characterized by two quantum numbers, I and m. I determines the number of nuclear spin states, and m defines the different states. m can take on $2I + 1$ integer values ranging from $-I$ to I.

From quantum mechanics, if I is integral or half-integral, the nuclei will exhibit nuclear magnetism. The vast majority of all clinical MRI uses the hydrogen proton, primarily due to the large numbers present in most tissues. Of particular interest for NMR are two properties that are exhibited when one of these nuclei is placed in a magnetic field: the creation of distinct spin states with a known energy difference between the states and the magnetic dipole moments, commonly referred to as "spins."

16.2.1 Energy Levels and the Larmor Equation

When placed in a static magnetic field, assumed to be z-directed and denoted B_0, the nuclei will exhibit $2I + 1$ distinct energy states specified by the integer m and given by

$$E_m = -\gamma \hbar B_0 m \qquad (16.1)$$

where $m = \pm I, \pm(I - 1),...,$ γ is the Larmor ratio with units of radians/Tesla or Hz/Tesla, B_0 is the magnetic flux density in Tesla, and $\hbar = h/2\pi$ is Planck's constant, 1.055×10^{-34} J·s. For a spin 1/2 nucleus such as the hydrogen proton, the energy difference between the two permissible states, ΔE, is given by

$$(16.2)$$

This energy difference, known as Zeeman splitting, is proportional to the magnetic field, as illustrated in Fig. 16.3.

Because of Zeeman splitting, placing a sample in a known magnetic field creates a well-defined energy level difference between spin states. An incident photon of the

TABLE 16.1 NMR Properties for Some Common Nuclei

Nucleus	Spin (I)	Natural abundance (%)	γ(MHz/T)
^{1}H	1/2	99.99	42.5759
^{2}H	1	1.5×10^{-2}	6.5357
^{13}C	1/2	1.1	10.7054
^{14}N	1	99.6	3.0756
^{19}F	1/2	100	40.0541
^{23}Na	3/2	100	11.262
^{31}P	1/2	100	17.235

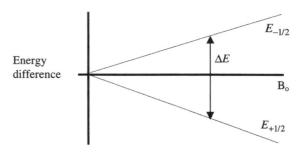

Fig. 16.3 Zeeman splitting of nuclear energy states due to a static magnetic field.

correct energy, or alternatively frequency, can induce a transition between the two states. This is the interaction that is fundamental to NMR.

Combining the expressions for the transition energy, $\Delta E = (E) = \gamma \hbar B_0$, and the corresponding photon frequency, given by $E = hf$, gives an expression for the frequency of the electromagnetic fields which will cause energy level transitions in a specified magnetic field. This equation is known as the **Larmor equation** and is given by

$$f = \frac{\gamma B_0}{2\pi} \quad \text{or, equivalently,} \quad \omega = \gamma B_0 \quad (16.3)$$

It should be noted at this point that the Larmor equation is often written as $f = \gamma B_0$. γ is different by a factor of 2π in the two forms of the equations.

Example Problem 16.1

Find the energy difference and corresponding frequency between the two spin states for ^{1}H if $B_0 = 0.1$ T.

$$\gamma = 2\pi \times 42.577 \times 10^6 \ \text{s}^{-1} \ \text{T}^{-1}$$

$$I = 1/2$$

Solution

For ^{1}H

Using $\quad \hbar = \dfrac{h}{2\pi} = \dfrac{6.63 \times 10^{-34}}{2\pi} \ \text{J} \cdot \text{s}$ (Planck's constant)

and $\quad \Delta E = (E_{-1/2} - E_{1/2}) = \gamma \hbar B_0$, gives

$$\Delta E = 2\pi \cdot 42.577 \times 10^6 \left(\frac{1}{\text{T s}}\right) \cdot \frac{6.63 \times 10^{-34}}{2\pi} \ (\text{J s})$$

$$\Delta E = 2.8228 \times 10^{-27} \ \text{J}$$

Note that this is a very small amount of energy separating these two energy states. A single photon of green light has an energy of 3.98×10^{-19} J, approximately eight orders of magnitude higher than the energy contained in the photons in this problem. Additionally, the frequency calculated is the energy of a photon which is matched to the transition energy between the two states.

The corresponding frequency is found from $E = hf$:

$$f = \frac{E}{h} = \frac{2.8228 \times 10^{-27}\text{J}}{6.63 \times 10^{-34}\text{J s}} = 4.257 \text{ MHz}$$

The majority of clinical MRI is done at magnetic field strengths between 0.5 and 1.5 T, corresponding to Larmor frequencies for hydrogen of 21.3–63.9 MHz. For comparison, the earth's magnetic field is on the order of 0.5 G or 50 μT. ∎

16.2.2 The Nuclear Magnetic Dipole Moment

The behavior of the spin states can be described using a classical approach, but descriptions based on quantum mechanics, which are not discussed here, are required for a more complete description. In a classical approach, each spin state is considered to have a corresponding nuclear magnetic dipole moment. The magnitude of the dipole moment is denoted by μ

$$\mu = \gamma\hbar\sqrt{I(I + 1)} \tag{16.4}$$

For reasons which will be discussed later, the z component, μ_z, is of interest here. μ_z takes on the discrete values given by

$$\mu_z = \gamma\hbar m \quad m = \pm I, \pm (I - 1), \cdots \tag{16.5}$$

For a spin 1/2 nucleus, such as hydrogen, Eq. (16.5) defines two states, usually denoted as the spin-up, or +1/2, and spin-down, or −1/2, states, with z components of their magnetic dipole moment given by

$$\mu_z^+ = \frac{+\gamma\hbar}{2} \ (m = 1/2, \text{ parallel or spin-up}) \tag{16.6}$$

$$\mu_z^- = \frac{-\gamma\hbar}{2} \ (m = -1/2, \text{ antiparallel or spin-down}) \tag{16.7}$$

Since the magnitude of the magnetic dipole moment for each state is the same, a magnetic vector for each state, $\vec{\mu}^+$ and $\vec{\mu}^-$ can be described. The relationship between the magnitude and z component of the dipole moments leads to an orientation of each state with respect to the static field B_0 (assumed to be along the z axis), as illustrated in Fig. 16.4. Note that $\vec{\mu}$ can only have discrete orientations, each of which can be viewed as having an orientation θ from the axis of the static magnetic field.

Motion of an Isolated Spin

Consider an isolated magnetic dipole moment, $\vec{\mu}$, in an applied field, $\vec{B}_0 = B_0\hat{a}_z$. A torque $\vec{T}$ is exerted on the magnetic moment by the static field, given by

$$\mu = \gamma\hbar\sqrt{I(I + 1)} \tag{16.8}$$

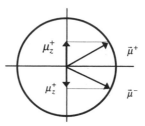

Fig. 16.4 Permissible states of the magnetic dipole moment vectors for a spin 1/2 nucleus.

The torque causes the dipole moment to move in the direction of the applied force. Since the magnetic field $\vec{B}_0$ is always in the z direction, the torque and the change in the momentum are always in the x–y plane, causing the isolated magnetic dipole moment, or spin, to precess about the applied field. Note that the precession is in a left-handed sense, i.e., the spin precesses in the direction of the fingers of the left hand when the thumb is aligned along the z axis, the direction of the static magnetic field.

What is the precessional frequency? Classical mechanics relates the applied torque to the change in the angular momentum vector, $\vec{P}$, $d\vec{P}/dt = \vec{T}$. The angular momentum vector is related to the nuclear magnetic dipole moment $\vec{\mu}$, through $\vec{\mu} = \gamma\vec{P}$. Combining these equations, the time rate of change of the magnetic dipole moment, $d\vec{\mu}/dt$, is obtained:

$$\frac{d\vec{\mu}}{dt} = \gamma\frac{d\vec{P}}{dt} = \gamma\vec{T} = \gamma\vec{\mu} \times \vec{B}_0 \tag{16.9}$$

Using the vector identity $\vec{A} \times \vec{B} = -\vec{B} \times \vec{A}$,

$$\frac{d\vec{\mu}}{dt} = -\gamma\vec{B}_0 \times \vec{\mu} \tag{16.10}$$

Equation (16.10) is the equation of motion for a rotating vector $\vec{\mu}$ rotating in the left-handed sense about the vector $\vec{B}_0$, with angular velocity $\omega_0 = \gamma B_0$ (the Larmor relation). The angular velocity is often written signed to indicate the left or right handedness of the rotation, as either $\vec{\omega}_0 = -\gamma\vec{B}_0$ or simply $\omega_0 = -\gamma B_0$. The solution to this equation is easily found by expanding into three scalar equations. The solution is found to be

$$\begin{aligned} \mu_z(t) &= \mu_z \\ \mu_x(t) &= \mu_{xy}\cos\gamma B_0 t \\ \mu_y(t) &= -\mu_{xy}\sin\gamma B_0 t \end{aligned} \tag{16.11}$$

where μ_{xy} is the magnitude of the transverse component of $\vec{\mu}$.

Equation (16.11) clearly shows that the dipole moment of the spin rotates around the magnetic field in a left-handed sense with frequency $\omega = \gamma B_0$.

Effect of an Ensemble of Spins: M_0, the Magnetization Vector

While it is conceptually useful to consider a single spin, MR imaging is inherently a low-sensitivity experiment and benefits from having a huge number of spins with which to work. For example, 1 mm^3 of water contains 6.7×10^{19} hydrogen protons. The behavior of the aggregate of spins must be considered. The net magnetization, i.e., the sum of all the dipole moments, is denoted $\vec{M}_0$, where

$$\vec{M}_0 = \sum_n \vec{\mu}_n \tag{16.12}$$

In the absence of an external field, the number of protons in the two spin states ($m = +\frac{1}{2}$ and $m = -\frac{1}{2}$) are essentially equal. However, once an external field is applied, the spin $+1/2$, ($m = +\frac{1}{2}$) state is preferred. Figure 16.5 illustrates the behavior of a set of dipole moments in an applied static magnetic field.

Since the individual spins are precessing about the z axis with no preferred phase, they will form two cones at the angles $\theta_{1/2}$ and $\theta_{-1/2}$. Without any phase coherence, the transverse, or x–y, component of the net magnetism will be zero. However, because the spin-up, or spin $+1/2$, state is preferred (as the lower energy state), there will be a net z component, or longitudinal magnetic moment, $\vec{M}_0$, aligned with the static field. This net magnetization vector $\vec{M}_0$, is the basis of NMR. Unfortunately, even though a huge number of hydrogen protons may be available, the excess in the spin-up state is ordinarily very small. In order to estimate the signal-to-noise ratio attainable in a NMR experiment, the number of spins contributing to $\vec{M}_0$ must be quantified. Consider a sample with a total of N spins/unit volume. The ratio of the populations in the two states is specified by the Boltzman distribution:

$$\frac{n^-}{n^+} = \exp^{(-\Delta E/kT)} \tag{16.13}$$

where n^+ and n^- are the spin populations in the $+1/2$ and $-1/2$ states per unit volume, k is Boltzman's constant, 1.38×10^{-23} J/K, and T represents the temperature

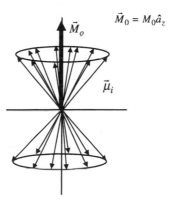

$$\vec{M}_0 = M_0 \hat{a}_z$$

$$\vec{\mu}_i$$

Fig. 16.5 An ensemble of magnetic dipole moments in a static magnetic field. By convention, individual dipoles, or spins, precess in a left-handed sense about the static field.

in degrees Kelvin. Using $\Delta E = \gamma \hbar B_0$, and small argument approximations for the exponential, an expression for the population excess in the spin-up, or $+1/2$, state can be obtained:

$$(n^+ - n^-) = \frac{N \gamma \hbar B_0}{2kT} \tag{16.14}$$

The net magnetic moment is simply the product of the excess population multiplied by the z component of the individual dipole moments, $\mu_z = \gamma \hbar m$:

$$|\vec{M_0}| = M_0 = \mu_z \, (n^+ - n^-) \tag{16.15}$$

For a spin 1/2 nucleus, such as the hydrogen proton,

$$M_0 = \frac{N \gamma^2 \hbar^2 B_0}{4kT} \tag{16.16}$$

Example Problems 16.2 and 16.3 illustrate that the magnetization and the excess population in the spin up state are extremely small.

Example Problem 16.2

Compute the magnetization in a cubic millimeter of water at 0.22 T.

Solution

As mentioned previously, there are 6.7×10^{19} protons in a cubic millimeter of water. Using

$$M_0 = \frac{N \gamma^2 \hbar^2 B_0}{4kT}$$

Substite the remaining values,

$$M_0 = \frac{6.7 \times 10^{19} \cdot \left(2\pi \cdot 42.57 \times 10^6 \, \frac{1}{T \, s}\right)^2 \cdot \left(\frac{6.63}{2\pi} \times 10^{-34} \, J \, s\right)^2 \cdot 0.22 \, T}{4 \cdot 1.38 \times 10^{-23} \, J/K \cdot 300 \, K}$$

$$= 7.09 \times 10^{-13} \, J/T \qquad \blacksquare$$

Example Problem 16.3

Compute $n^+ - n^-/n^+$ for hydrogen at room temperature in a 0.1 T magnetic field.

Solution

Exercise Problem 16.4 states that

$$\frac{n^+ - n^-}{n+} \cong \frac{\gamma \hbar B_0}{kT}$$

for spin 1/2. Thus,

$$\frac{n^+ - n^-}{n+} = \left[\frac{2\pi \cdot 42.57 \times 10^6 \, \frac{1}{T \, s} \cdot \left(\frac{6.63}{2\pi} \times 10^{-34} \right) \text{J s 0.1 T}}{1.38 \times 10^{-23} \, \text{J/K} \cdot 290 \, \text{K}} \right]$$

or $n^+ - n^-/n^+ = 0.7 \times 10^{-6}$, a very small number. ∎

As illustrated by the previous example, the spin-up excess is about 0.7 ppm/T, a very small number. This can be increased by either lowering the temperature or raising the magnetic field strength. Unfortunately, in human imaging, the former is impractical, whereas the latter increases system complexity and the concern for bioeffects. Nevertheless, systems are being considered with fields as high as 10 T for human imaging.

At equilibrium, there is no time variation, i.e., $\vec{M}_0$ is static. If $\vec{M}_0$ were displaced from the z axis, it would also be influenced by the torque caused by the magnetic field and precess around $\vec{B}_0$, in turn creating its own time-varying magnetic field. This time-varying field can induce a voltage in a detector coil, as will be discussed in the next section. The next step is to describe the perturbation of $\vec{M}_0$ (the response to a stimulus, which will be a radiofrequency signal at or near the Larmor frequency) and its return to equilibrium following removal of the stimulus.

16.3 NMR

16.3.1 The Bloch Equation

Since the magnetization vector $\vec{M}_0$ is simply the sum of all the individual nuclear magnetic dipole moments $\vec{M}_0 = \sum_n \vec{\mu}_n$, the magnetization vector will follow the same equations of motion as a single spin, or magnetic dipole moment, $\vec{\mu}$, which was found earlier to be

$$\frac{d\vec{\mu}}{dt} = -\gamma \vec{B}_0 \times \vec{\mu} \qquad (16.17)$$

Just as with the isolated magnetic dipole moment, a torque is created in the direction perpendicular to the magnetization vector and the magnetic field. At equilibrium, these two vectors are aligned, no force is created, and the system simply remains at equilibrium. In the process of performing an NMR experiment, $\vec{M}_0$ is disturbed from its equilibrium state along the z axis. At that point it will precess around the existing magnetic field, again in a left-handed sense. The time-varying magnetization can be denoted by $\vec{M}(t)$. However, the variation with time will simply be understood and will be omitted. The equation of motion for the magnetization vector can be obtained by analogy to the equation governing the single dipole moment:

$$\frac{d\vec{M}}{dt} = -\gamma \vec{B}_0 \times \vec{M} = \gamma \, \vec{M} \times \vec{B}_0 \qquad (16.18)$$

This equation is the Bloch equation for the magnetization vector in a static magnetic field. Just as with the equation for the isolated spin, the solution to the Bloch equation can be easily found by separating it into three scalar components, giving

$$M_z(t) = M_z^0$$

$$M_x(t) = M_x^0 \rightarrow M_{xy}^0 \, \cos\gamma B_0 t \qquad (16.19)$$

$$M_y(t) = -M_x^0 \rightarrow M_{xy}^0 \, sin\gamma B_0 t$$

where $M_z^0 = M_z(t = 0)$, and the same applies for the other components. These equations are simple to solve for the equilibrium magnetization. In this case, $\vec{M}(t = 0) = \vec{M}_0 = M_0\hat{a}_z$. The x and y components are zero and remain so. There is no time variation of the magnetization vector. However, the solution clearly shows that, if the magnetization vector is "tipped" into the x–y plane, the resulting vector would remain in the x–y plane, precessing around the static field at the Larmor frequency, $\omega = \gamma B_0$. This time-varying magnetization, which precesses in a left-handed sense, creates a time-varying magnetic flux density which can be detected by a sensor, a radiofrequency coil. This is the signal that is used to create NMR images, and it will be discussed in more detail later.

How is the magnetization vector tipped away from the equilibrium position along the z axis? Based on the discussion in Section 16.2, this can be done by applying photons of the appropriate energy. Transitions between states can be caused by controlling the strength and duration of the incident photons. The application of photons is accomplished by turning on a second, time-varying magnetic field called the B_1 field. Although it is time varying, its influence is still described by the Bloch equation. The magnetization will be affected by any magnetic field, not just the static field $\vec{B}_0$. To reflect this, the equation of motion for the magnetization can be written:

$$\frac{d\vec{M}}{dt} = \gamma\vec{M} \times (\vec{B}_0 + \vec{B}_1) = \gamma\vec{M} \times \vec{B} \qquad (16.20)$$

where $\vec{B}$ is the total magnetic field present. Again, the variation with time is understood.

16.3.2 The Rotating Frame

In addition to the effects of the static and time-varying magnetic fields discussed previously, the effect on the spin system due to such factors as blood flow, relaxation effects, magnetic field inhomogeneities, and many other perturbations needs to be included in a complete model of the MR experiment. Analyzing these other effects is much easier if the precession due to the ideal, static magnetic field is separated. This is achieved by introducing a new frame of reference known as the "rotating frame." The standard, fixed Cartesian frame, denoted by fixed x, y, and z axes, is referred to as the "laboratory frame." The rotating frame rotates about the z axis in the same direction as the precessing magnetic dipoles, or spins, with a rotational rate ω_{rf} typically equal to the Larmor frequency. In the case of an ideal, homogeneous static field B_0, with no other impressed fields, the precessing spins will appear motionless in the rotating frame. Any other motions due to additional perturbations on the magnetic field can be analyzed separately from the Larmor precession, which has effectively been removed. Additionally, a radiofrequency (RF) field at the Larmor frequency can become very simple to analyze in the rotating frame. In the remainder of the chapter, primes will be used to designate quantities expressed in the rotating frame. The rotating frame is defined by primed axes, x', y', z' and unit vectors $\hat{a}'_x$, $\hat{a}'_y$, $\hat{a}'_z$.

It will be convenient to use the rotating frame for most discussions of MRI. Conversion from one frame to the other is done using the transformation matrix $[J]$, given by

$$[J] = \begin{bmatrix} \cos\omega t & -\sin\omega t & 0 \\ \sin\omega t & \cos\omega t & 0 \\ 0 & 0 & 1 \end{bmatrix} \tag{16.21}$$

Using $[J]$ or its inverse $[J]^{-1}$ any vector in one frame can be converted to the other frame using $\vec{A}' = [J]\vec{A}$ or $A = [J]^{-1}A'$. Example Problem 16.4 demonstrates the conversion of a radiofrequency field from the laboratory to the rotating frame.

Example Problem 16.4

Express $\vec{B}_1(t) = -\hat{a}_y B_1 \cos\omega t$ in the rotating frame.

Solution

Using the transformation matrix above gives

$$\vec{B}'_1 = [J]\vec{B}_1 \quad \text{or} \quad \begin{bmatrix} B_{x'} \\ B_{y'} \\ B_{z'} \end{bmatrix} = \begin{bmatrix} \cos\omega t & -\sin\omega t & 0 \\ \sin\omega t & \cos\omega t & 0 \\ 0 & 0 & 1 \end{bmatrix} \begin{bmatrix} B_x \\ B_y \\ B_z \end{bmatrix}$$

For this problem,

$$\begin{bmatrix} B_{x'} \\ B_{y'} \\ B_{z'} \end{bmatrix} = \begin{bmatrix} \cos\omega t & -\sin\omega t & 0 \\ \sin\omega t & \cos\omega t & 0 \\ 0 & 0 & 1 \end{bmatrix} \begin{bmatrix} 0 \\ -B_1\cos\omega t \\ 0 \end{bmatrix} = \begin{bmatrix} B_1 \sin\omega t \cos\omega t \\ -B_1 \cos^2\omega t \\ 0 \end{bmatrix}$$

Using $\cos^2\omega t = \frac{1}{2}(1 = \cos 2\omega t)$ and $\sin\omega t \cos\omega t = \frac{1}{2}\sin(2\omega t)$ results in

$$\vec{B}'_1(t) = \hat{a}_{x'}\frac{B_1}{2}\sin 2\omega t - \hat{a}_{y'}\frac{B_1}{2} - \hat{a}_{y'}\frac{B_1}{2}\cos 2\omega t$$

The first and third terms are components that are rotating at twice the Larmor frequency, have little effect on the spins, and can be ignored for most purposes. This gives a very simple form for the applied B_1 field in the rotating frame:

$$\vec{B}'_1(t) \cong \hat{a}y, \frac{B_1}{2} \qquad \blacksquare$$

With a little effort, the Bloch equation of Eq. (16.20) can be written in the rotating frame by a simple change of coordinates:

$$\frac{d\vec{M}}{dt} = \gamma\vec{M} \times \vec{B}_{\text{eff}} \tag{16.22}$$

where $\vec{B}_{\text{eff}}$ is the effective magnetic field in the rotating frame, $\vec{B}_{\text{eff}} = \vec{B}_0 + \vec{\omega}_{\text{rf}}/\gamma$. If the rotational frequency of the frame is equal to the Larmor frequency, $\vec{\omega}_{\text{rf}} = \vec{\omega}_0 = -\gamma\vec{B}_0$, then there is zero effective field, and $d\vec{M}/dt = 0$ in the rotating frame. There is no net precession in the rotating frame in the absence of other field perturbations. Equation (16.22) is a very general expression. $\vec{B}_{\text{eff}}$ can include a variety of effects, including the magnetic field gradients used for imaging and magnetic field variations caused by the nuclear environment used for spectroscopy. It is perhaps the key equation for describing NMR.

16.3.3. RF Pulses

Now consider the effect of applying an external RF field denoted by $\vec{B}_1(\vec{r}, t)$. For convenience, the space and time variations will be suppressed. $\vec{B}_1$ is applied through a radiofrequency coil similar to an antenna although it is not designed to radiate energy. The term resonator is more applicable. The magnetic flux produced by this resonator adds to the static field $\vec{B}_0$. From the Bloch equation, $\vec{M}_0$ will rotate about this total field. Expressed in the rotating frame, the effective magnetic flux field becomes

$$\vec{B}_{\text{eff}} = \vec{B}_0 + \frac{\vec{\omega}_{\text{rf}}}{\gamma} + \vec{B}_1' \tag{16.23}$$

As before, if the frame rotational frequency is equal to the Larmor frequency, $\omega_{\text{rf}} = \omega_0$, then $\vec{B}_{\text{eff}} = \vec{B}_1'$ and the Bloch equation becomes

$$\frac{d\vec{M}}{dt} = \gamma\vec{M} \times \vec{B}_1' \tag{16.24}$$

Note that this expression is in the rotating frame. Clearly, the magnetization vector will precess about the RF field $\vec{B}_1'$. The rotation will again be in a left-handed sense, as illustrated in Fig. 16.6.

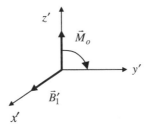

Fig. 16.6 The net magnetization vector precessing in the rotating frame of reference about an x-directed radiofrequency field $\vec{B}_1' = B_1\hat{a}_{x'}$. Note that the vector spins in the left-handed sense about the RF fields, just as with the static magnetic field.

Rectangular or "Hard" RF Pulses

The simplest pulse to consider is one of constant amplitude for a duration τ_p, expressed in the laboratory frame as $\vec{B}_1(t) = -\hat{a}_y B_1 \cos\omega t$. This represents a linearly polarized field of amplitude B_1 T/m along the y axis. As shown in Example Problem 16.4, in the rotating frame this RF pulse becomes

$$\vec{B}_1'(t) = \begin{array}{cc} -\hat{a}_{y'} \dfrac{B_1}{2} & 0 \le t \le \tau_p \\ 0 & \text{otherwise} \end{array} \tag{16.25}$$

If applied to the net magnetization vector at equilibrium, $M_{x'}(0) = 0$, $M_{y'}(0) = 0$, and $M_{z'}(0) = M_0$, the solution for $\vec{M}(t)$ is

$$\begin{aligned} M_{x'}(t) &= M_0 \sin \Omega t \\ M_{y'}(t) &= 0 \\ M_{z'}(t) &= M_0 \cos \Omega t \end{aligned} \tag{16.26}$$

The rotational frequency, Ω, of the magnetization vector is given by

$$\Omega = \gamma B_1' \tag{16.27}$$

For a linearly polarized RF pulse, $B_1' = B_1/2$, and

$$\Omega = \gamma \frac{B_1}{2} \tag{16.28}$$

where B_1 is the intensity of the linearly polarized RF pulse. A pulse of length τ_p, will rotate $\vec{M}$ an angle α, called the "tip angle":

$$\alpha = \int_0^{\tau_p} \Omega dt = \int_0^{\tau_p} \gamma B_1' dt \tag{16.29}$$

Since the applied field is constant,

$$\alpha = \gamma B_1' \tau_p \tag{16.30}$$

π and $\pi/2$ RF Pulses

Two of the most important RF pulses are the π and $\pi/2$ pulses, which tip $\vec{M}$ through a tip angle α of 90° and 180°, respectively. A $\pi/2$ pulse moves the net magnetization vector, initially at equilibrium (aligned along the $+z$ axis), to the transverse, or x–y, plane, as illustrated in Fig. 16.7a. Since the magnetization is precessing about the z axis, it is now time varying and will induce a voltage in a receiver coil. A π pulse completely inverts the magnetization. If it is initially along the $+z$ axis, it will end up on the $-z$ axis. Similarly, if it is initially on the $+x'$ axis, such as following a $\pi/2$ pulse, it would be on the $-x'$ axis following a π pulse, as shown in Fig. 16.7b. Both pulses have important applications in MRI. Additionally, pulses with tip angles <90° are often used in rapid imaging sequences. RF pulses provide the primary method for manipulating the RF spins. Magnetic field gradients provide the second major method.

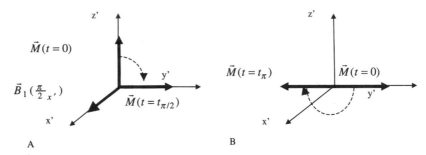

Fig. 16.7 (a) A $\pi/2$, or 90, pulse along the $+x'$ axis rotates the equilibrium magnetization 90° in a left-hand sense to the $+y'$ axis. (b) A π, or 180, pulse rotates the magnetization 180°.

Example Problem 16.5.

Consider an RF coil which produces a linearly polarized magnetic field of 10 μT/A. A continuous 10 A current is applied. Calculate the pulse length required to produce a $\pi/2$ pulse in distilled water.

Solution

The tip angle is given by Eq. (16.30), $\alpha = \gamma\, B'_1 \tau_p$. Solving for τ_p,

$$\tau_p = \frac{\alpha}{\gamma B'_1}$$

For ^{1}H, $\gamma = 42.57$ MHz/T, and, as shown earlier, $\vec{B}'_1(t) \cong \hat{a}_x$, $B_{1/2}$, or $B'_1(t) = \frac{(10\ \mu\text{T/A})(10\ \text{A})}{2} = 50\ \mu\text{T}$:

$$\tau_{\pi/2} = \frac{\pi/2}{2\pi \cdot 42.57 \times 10^6\, \frac{1}{Ts} \cdot 50 \times 10^{-6}\ T} = 117\ \mu s$$

Note that this is quite a long time compared to the precessional frequency of the spins about $\vec{B}_0$. In a 1 T field, ^{1}H spins precess about the static field once every $1/f_0$ s, or about once every 23.5 ns. In 117 μs, the spins rotate 90° around the x axis in the rotating frame and approximately 5000 times about the z axis in the laboratory frame. ∎

16.3.4 Relaxation Processes

Following a $\pi/2$ RF pulse, the magnetization vector is placed in the x–y plane, where, in the absence of a net magnetic field, it will remain. In the laboratory frame the magnetization precesses about the z axis, creating a time-varying field that can be detected with a radio frequency sensor coil. Clearly, this cannot continue indefinitely. Instead, the magnetization undergoes a number of relaxation processes and returns to equilibrium. For many materials, the relaxation process can be approximated by two first-order processes called the longitudinal or T_1 relaxation, and the transverse or T_2 relaxation.

T_1 Relaxation

T_1 relaxation refers to the return of the longitudinal component of the magnetization, M_z, to its equilibrium value, M_0. As illustrated in Fig. 16.8, T_1 relaxation is approximated by an exponential process with a time constant T_1:

$$M_z(t) = M_0\left[1 - \left(1 - \frac{M_z^0}{M_0}\right)e^{-t/T_1}\right] \tag{16.31}$$

where M_z^0 is the z component of the magnetization following the RF pulse or, alternatively, at some time denoted $t = 0$. The time constant T_1 is the "longitudinal" or "spin-lattice" time constant, which describes the return of the spin state populations to equilibrium.

T_2 Relaxation

A second relaxation process is known as transverse or T_2 relaxation and is similarly characterized by an exponential process with a time constant T_2. T_2 relaxation is caused by a loss of phase coherence in the spins forming the net magnetization vector $\vec{M}$. Recall that $\vec{M}$ is composed of the sum of many individual spins, $\vec{M} = \sum_n \vec{\mu}_n$, each of which has a transverse component at arbitrary phase, yielding only a net z component at equilibrium. Ideally, when $\vec{M}$ is tipped away from the z axis, the individual spins precess about $\vec{B}_0$ at the Larmor frequency, yielding the evolution of magnetization shown previously due to the T_1 magnetization. However, due to the complex nuclear environment, the spins experience a distribution of magnetic field strengths due to the effects of interactions with neighboring atoms. Each spin will precess at a frequency corresponding to the magnetic field it experiences. Some will precess faster than the Larmor frequency, some slower, and some at exactly the Larmor frequency. The net result is a dephasing of the spins in the transverse plane. The spins appear to "fanout," as illustrated in Fig. 16.9.

Once the phase of the spins has reached a random distribution, there will be no transverse magnetization. This can happen very rapidly, even before significant T_1

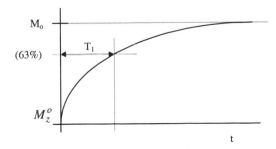

Fig. 16.8 The T_1 relaxation time is the time required for 63% of the magnetization to recover following a 90° RF pulse.

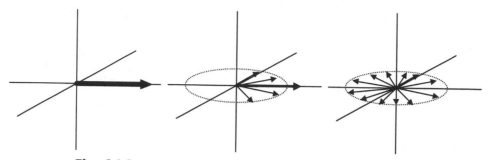

Fig. 16.9   Transverse magnetization dephasing following a 90° pulse.

relaxation has occurred. This process can again be described in many cases as a simple first-order process, illustrated in Fig. 16.10, and expressed as

$$M_{xy}(t) = M_{xy}^0 e^{-t/T_2} \qquad (16.32)$$

T_2^* and T_2^{**} Relaxation

T_2 is a fundamental parameter of the sample that is due to spin interactions. There are other causes of dephasing that can affect the apparent time constant for the transverse magnetization. For example, magnetic field inhomogeneities due to air bubbles, tissue interfaces, and imperfections in the static magnetic field will cause spatial variations in the precession frequency. Thus, inhomogeneities lead to a more rapid dephasing, decreasing the apparent relaxation constant. The effective relaxation constant which includes the dephasing due to magnetic field inhomogeneities is termed T_2^* and is commonly given as

$$\frac{1}{T_2^*} = \frac{1}{T_2} + \frac{\gamma \Delta B_0}{2} \qquad (16.33)$$

where ΔB_0 is the range of the magnetic field over the excited spins. Additionally, the application of magnetic field gradients to spatially encode the MR signal introduces

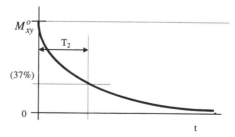

Fig. 16.10   The T_2 relaxation time is the time required for the transverse magnetization to decay to 37% of its value immediately following an excitation pulse.

a significant magnetic field inhomogeneity and further reduces the effective transverse relaxation time constant. The effective time constant which includes dephasing due to magnetic field gradients is termed T_2^{**} and is given by

$$\frac{1}{T_2^{**}} = \frac{1}{T_2} + \frac{\gamma \Delta B_0}{2} + \gamma Gr \qquad (16.34)$$

where G is the gradient strength (T/m) and r is the object radius (m). Note that $T_2^{**} \leq T_2^{*} \leq T_2 \leq T_1$

Modification of the Bloch Equation to Include Relaxation

The previous relaxation processes satisfy the modified Bloch equation:

$$\frac{d\vec{M}}{dt} - \gamma \vec{M} \times \vec{B} - \frac{M_x \hat{a}_x + M_y \hat{a}_y}{T_2} - \frac{(M_z - M_z^0)\hat{a}_z}{T_1} \qquad (16.35)$$

If it is assumed that the transverse magnetization initially lies along the x' axis, the solution to the modified Bloch equation is

$$M_z(t) = M_0 \left[1 - \left(1 - \frac{M_z^0}{M_0}\right)e^{-t/T_1} \right] \qquad (16.36)$$
$$M_x(t) = M_{xy}^0 \cos\omega_0 t \; e^{-t/T_2}$$
$$M_y(t) = -M_{xy}^0 \sin\omega_0 t \; e^{-t/T_2}$$

While the longitudinal (z) component of the magnetization decays with time constant T_1, the transverse components vary much more rapidly due to the precession about the static field. Thus, it is the transverse component of magnetization that induces a signal in the detection coil. It is useful to adopt the symbol $\vec{M}_{xy}$ to represent the transverse components of magnetization,

$$\vec{M}_{xy}(t) - M_{xy}^0(\hat{a}_x \cos\omega_0 t - \hat{a}_y \sin\omega_0 t)e^{-t/T_2} \qquad (16.37)$$

or, alternatively,

$$\vec{M}_{xy}(t) = M_{xy}^0 e^{-t/T_2} \, \text{Re}\{(\hat{a}_x + j\hat{a}_y)e^{+j\omega_0 t}\} \qquad (16.38)$$

It is useful to define a polarization unit vector for the magnetization, $\hat{p}_M = (\hat{a}_x + j\hat{a}_y)/\sqrt{2}$, and to write the transverse magnetization in complex notation that will be more convenient:

$$\vec{M}_{xy}(t) = \sqrt{2} \, M_{xy}^0 e^{-t/T_2} \, \hat{p}_M e^{+j\omega_0 t} \qquad (16.39)$$

The real part of Eq. (16.39) is taken to recover the standard form of the transverse magnetization.

16.3.5 Signal Detection: The Free Induction Decay

Once a transverse magnetization has been created by tipping $\vec{M}_0$ from the z axis, a time-varying magnetic flux and electric field, $\vec{B}_M(\vec{r}, t)$ and $\vec{E}_M(\vec{r}, t)$, are created by the time-varying transverse magnetization, $\vec{M}_{xy}(\vec{r}, t)$. It is known from basic electromagnetics (Faraday's law) that a time-varying flux will induce an electromotive force. If the flux passes through a detector coil, a voltage is produced at the wire terminals and is given by

$$v(t) = \oint_l \vec{E} \cdot \vec{dl} = -\frac{d}{dt}\int_s \vec{B} \cdot \vec{ds} \tag{16.40}$$

This voltage can be amplified and demodulated to provide the MR signal. Calculation of this voltage is relatively difficult but can be simplified, at least conceptually, by using reciprocity relations. Reciprocity relations allow Eq. (16.40) to be rewritten as

$$v(t) = \frac{1}{I_1}\int_v \vec{B}_1 \cdot \frac{d\vec{M}_{xy}}{dt}dv \tag{16.41}$$

where $\vec{B}_1$ is the flux density which would be produced by the detector coil if it were used as a transmitter, and I_1 is the current used to produce $\vec{B}_1$. Assuming that the sample is small enough that it and the flux of the detector coil are constant over the region of integration, Eq. (16.41) becomes

$$v(t) = \frac{\vec{B}_1}{I_1} \cdot \frac{d\vec{M}_{xy}}{dt}\Delta V \tag{16.42}$$

where ΔV is the voxel (volume element) size containing the magnetization. Finally, performing the derivative and ignoring the terms proportional to $1/T_2$, an expression for the voltage induced in a coil following an RF pulse is obtained:

$$v(t) = j\sqrt{2}\omega_0\Delta VM^0_{xy}B_{1t}e^{-t/T_2}e^{+j\omega_0 t} \tag{16.43}$$

where B_{1t} is the effective coil sensitivity, given by

$$B_{1t} = \frac{\vec{B}_1}{I_1} \cdot \hat{p}_M \tag{16.44}$$

$v(t)$ is the free-induction decay (FID) generated by an initial excitation pulse.

Example Problem 16.6.

Find the maximum voltage detected from 1 cm^3 of distilled water in a 2.0 T magnet using a radiofrequency coil that is linearly polarized in the x direction and produces a flux density of 20 µT per amp of current.

Solution

From Eq. (16.43), the magnitude of the induced voltage immediately following the RF pulse is given by

$$|v(t)| = \left| \sqrt{2}\omega_0\Delta VM^0_{xy}B_{1t} \right|$$

At 2.0 T, $\omega_0 = 2\pi\gamma B_0 = 2\pi(42.57 \times 10^6)2.0 = 5.35 \times 10^8$ 1/s. The magnetization of water in a 0.22 T field was found to be 7.09×10^{-13} J/(T mm^3) in Example Problem 16.3. At 2.0 T,

$$M_0 = 7.09 \times 10^{-13} \left(\frac{2.0}{0.22}\right) = 6.45 \times 10^{-12} \text{ J/(T } mm^3)$$

The maximum transverse magnetization would occur immediately following a 90° pulse, in which case $M_{xy}^0 = M_0$. B_{1t} is obtained from Eq. (16.44):

$$B_{1t} = \frac{\vec{B}_1}{I_1} \cdot \hat{p}_M = \frac{20\mu\text{T}}{1\text{A}}\,\hat{a}_x \cdot \left(\frac{\hat{a}_x + j\hat{a}_y}{\sqrt{2}}\right) = \frac{20}{\sqrt{2}}\frac{\mu\text{T}}{\text{A}}$$

Combining these terms, the voltage is obtained as

$$|v(t)| = \left| \sqrt{2} \cdot 5.35 \times 10^8 \frac{1}{s} \cdot 1 \text{ cm}^3 \cdot \left(\frac{10^3 \text{ mm}^3}{1 \text{ cm}^3}\right) \cdot 6.45 \times 10^{-17} \text{ J/(T } mm^3) \cdot \frac{2.0 \times 10^{-5}}{\sqrt{2}}\frac{\text{T}}{\text{A}} \right|$$

$$|v(t)| = 69 \ \mu V$$

While this is a relatively small voltage, it is certainly easily detectable. Resonating the coil will increase the voltage considerably, and low-noise preamplifiers can amplify the voltage further. However, it should be noted that the signal will decay rapidly, losing 67% of its amplitude in one time constant T_2^*, which is on the order of milliseconds. The signal-to-noise ratio obtained is easily estimated from the resistance of the receiver coil, which will be discussed in Section 16.5. ■

Signal Detection

The voltage $v(t)$ is a radiofrequency signal. In order to digitize this signal, it is practical to demodulate it to a lower frequency, typically an audio frequency. One method for achieving this is to use a quadrature demodulator, which will be discussed in Section 16.5. In a quadrature demodulator, the input signal is combined or "mixed" with a reference signal at a second frequency, ω_{mix}. The mixing results in the original frequency being shifted to the sum and difference frequencies of the two signals. In this manner, it is possible to convert the radiofrequency signal detected by the coil to a signal $S(t)$ at a frequency that can be easily digitized:

$$S(t) = \frac{1}{\sqrt{2}}\,j\omega_0\Delta V M_{xy}^0 e^{-t/T_2}\, B_{1t} e^{\,j(\omega_0 - \omega_{mix})t} \tag{16.45}$$

This is the starting point for developing the imaging equations.

16.3.6   Spin Echoes: General Concept

The signal that has been discussed so far is called a FID. It is an induced magnetization vector that decays in time. Due to the application of gradients, the FID may decay very quickly, limiting the ability to encode information (imaging). A very useful concept is that of the "spin echo," which was introduced by Erwin L. Hahn in 1950. The spin echo is a recalled net magnetization vector resulting from a reversal of the dephasing of the T_2^* decay that is caused by magnetic field inhomogeneities.

Hahn demonstrated that the application of a π RF pulse following the initial $\pi/2$ pulse would cause refocusing of the magnetization, forming a spin echo. A classic analogy is that of two runners, one fast and one slow, racing down a straight track. If the runners suddenly reverse their direction (and assuming neither gets tired) they will return to the starting line at the same time. Two variations of the spin echo will be considered, the Hahn and the Carr–Purcell, Meiboom–Gill echo.

Inside the sample, the magnetic field varies with position for a variety of reasons. Perturbations due to the sample itself, the parameter of interest, lead to T_2 decay. Magnetic field inhomogeneities due to imperfect magnets and susceptibility differences between sample types lead to T_2^* decay. Finally, the application of magnetic field gradients for information encoding, described in the next section, leads to T_2^{**} decay. For simplicity, consider two spins, one in a field less than B_0, and the other in a field greater than B_0. Furthermore, assume that the rotating frame precession rate is $\omega_{rf} = \omega_0$; thus, one spin is precessing faster than ω_0 and the other slower.

Hahn Spin Echoes

The Hahn spin-echo sequence consists of a 90° (or $\pi/2$) pulse, a delay, and then a 180° pulse along the same axis as the 90° pulse. If the RF pulses are applied along the x' axis, this simple "pulse sequence" can be denoted as $\pi/2_{x'} \rightarrow \tau \rightarrow \pi_{x'}$. The behavior of the two spins at various points in the pulse sequence is illustrated in Fig. 16.11. The spins are initially at equilibrium along the z axis. In Fig. 16.11a, the 90° RF pulse, oriented along the x' axis, rotates the spins in a left-handed sense to the y' axis. Immediately following the RF pulse, the two spins precess according to the local magnetic field. As described previous spin 1, in a field slightly lower than B_0, precesses slower than the rotating frame rate, assumed to be equal to the Larmor frequency. Thus, in the rotating frame, spin 1 appears to slowly precess in a counterclockwise sense (looking from the $+z$ axis). Similarly, spin 2, in a slightly higher magnetic field, precesses clockwise away from the $^+y'$ axis in the rotating frame (Fig. 16.11b). After a time delay τ, a 180° pulse is applied along the $+x'$ axis. Both spins rotate 180° again in a left-handed sense, about the applied field, as shown in Fig. 16.11c. Following the 180° pulse, the spins continue to precess in the same directions as before because they still experience the same magnetic fields. However, because of the 180° pulse, the angle that each spin had rotated away from the $+y'$ axis becomes the angle it needs to rotate to reach the $-y'$ axis. Hence, a spin echo is formed along the $-y'$ axis at an echo time (TE) equal to 2τ, as illustrated in Fig. 16.11d.

Although only two spins are considered here, it should be clear that the spins are dephasing due to the magnetic field inhomogeneities. If many spins are included in the analysis, the signal induced in a receiver coil will quickly diminish as the spins dephase, as discussed earlier. When an echo is formed, the amplitude will only reach a maximum value of $M_{xy}^0 e^{-t/T_2}$ because while the spin echo compensates for magnetic field inhomogeneities due to imperfect magnets and local magnetic susceptibility variations, it does not rephase signal decay due to T_2 relaxation in the sample. Thus, the echo is "T_2 weighted," meaning that it carries information on the T_2 relaxation time of the sample. The echo reforms and decays with time constant T_2^* or, if gradients are present, T_2^{**} The process can be repeated, forming multiple

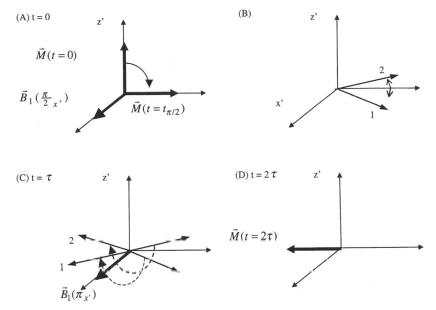

Fig. 16.11 Behavior of two isolated spins during a Hahn spin-echo pulse sequence.

echoes. The amplitude of each echo is governed by the T_2 decay curve. Note that the polarity of the echo is reversed from the FID since it forms on the $-y'$ axis. A second echo would rephase on the $+y'$ axis having the same sign as the FID. Successive echoes will alternate in sign.

Another form of spin echo commonly used is the Carr–Purcell, Meiboom–Gill (CPMG) echo. The CPMG sequence is identical to the Hahn echo sequence except that the second pulse is applied along the y' axis in the rotating frame. The pulse sequence can be expressed $\pi/2_{x'} \rightarrow \tau \rightarrow \pi_{y'}$. Unlike the Hahn echo, the CPMG echo refocuses along the original axis, y' in this case. Just as with the Hahn echo, repeated refocusing 180° pulses can be applied, forming an echo train decaying as e^{-t/T_2}. Typically, the phase of these pulses is alternated, called phase cycling, to reduce the effect of imperfections in the RF pulses.

16.4 MRI

The preceding section described how to create FIDs by tipping the magnetization vector $\vec{M}_0$ with an RF pulse. Using additional RF pulses, the FID can be recalled in the form of an echo. As will be discussed later, simple pulse sequences such as these can be used to measure T_1 and T_2 in bulk samples. What if the sample is not homogeneous, such as in a human body? Clearly, the next step is to spatially localize the signal.

The key ingredient in spatial localization came in 1972 when Paul Lauterbur developed the idea of "linking" the spatial position to the frequency using gradients in the static magnetic field. The use of magnetic field gradients allows one to spatially encode information from the object into the FID or echo. Undoubtedly, the most common MR imaging method in use today is two-dimensional (2-D) Fourier imaging, which is the method considered in this chapter. In 2-D Fourier imaging, the data collected from the sample in taking an MR image form the Fourier transform of the image. The image is obtained by a 2-D inverse Fourier transform. Of course, the technique can be extended to three or more dimensions of encoding. Before discussing the imaging procedure, it will be useful to discuss magnetic field gradients and to review the Fourier transform.

16.4.1 Fourier Transforms

Recall the definition of the Fourier transform: Given a function $f(x)$, its Fourier transform $\tilde{f}(k_x)$ is given by

$$\tilde{f}(k_x) = \int_{-\infty}^{\infty} f(x)e^{+jk_x x}dx \tag{16.46}$$

The inverse relationship is given by

$$f(x) = \frac{1}{2\pi}\int_{-\infty}^{\infty} \tilde{f}(k_x)e^{-jk_x x}dk_x \tag{16.47}$$

The Fourier transform will be denoted by

$$F\{f(x)\} = \tilde{f}(k_x) = \text{Fourier transform of } f(x)$$

$$F^{-1}\{\tilde{f}(k_x)\} = f(x) = \text{inverse Fourier transform of } f(x)$$

This definition can be extended to two dimensions using conjugate variable pairs (x, y) and (k_x, k_y). In two dimensions the forward and inverse transforms are given by

$$\tilde{f}(k_x,k_y) = \int_{-\infty}^{\infty}\int_{-\infty}^{\infty} f(x,y)e^{+jk_x x}e^{+jk_y y}\,dxdy \tag{16.48}$$

$$f(x,y) = \frac{1}{4\pi^2}\int_{-\infty}^{\infty}\int_{-\infty}^{\infty} \tilde{f}(k_x,k_y)e^{-jk_x x}e^{-jk_y y}\,dk_xdk_y \tag{16.49}$$

16.4.2 Magnetic Field Gradients

Consider a homogeneous field, $\vec{B}(\vec{r}) = B_0\hat{a}_z$, to which a second field, a gradient, is added:

$$\vec{G}_z(z) = G_z z\hat{a}_z \tag{16.50}$$

The gradient is a z-directed magnetic field that varies in the z direction only, has odd symmetry about $z = 0$, and has value 0 at $z = 0$. In practice, this gradient will be created by a small electromagnet, a gradient coil, inserted into the bore of the main magnet. Although the gradient coil will produce a linear gradient over a limited region, and must produce gradients in other directions as well to obey Maxwell's equations, these secondary effects will be ignored. Similarly, one can create gradients in the static field in the x and y directions so that the total static field is described by

$$\vec{B}(\vec{r}) = B_0\hat{a}_z + (G_x x + G_y y + G_z z)\hat{a}_z \qquad (16.51)$$

It is convenient to express Eq. (16.51) using vector notation for the gradient and the spatial position:

$$\vec{B}(\vec{r}) = (B_0 + \vec{G} \cdot \vec{r})\hat{a}_z \qquad (16.52)$$

where $\vec{G} = G_x\hat{a}_x + G_y\hat{a}_y + G_z\hat{a}_z$ and $\vec{r} = x\hat{a}_x + y\hat{a}_y + z\hat{a}_z$. Finally, as discussed in Section 16.3.2, in the rotating frame the static field term drops out if $\omega_{rf} = \omega_0$. Adding a radiofrequency field, the effective magnetic field becomes

$$\vec{B}_{eff}(\vec{r}) = \vec{B}_1'(\vec{r}) + \vec{G} \cdot \vec{r}\hat{a}_z \qquad (16.53)$$

Equation (16.53) can be used with the Bloch equation to describe the motion of a spin system in the presence of an RF and a gradient magnetic field.

The gradient links position to frequency through the Larmor relationship which, in the presence of a spatially varying magnetic field, must be generalized to $\omega = \gamma B$, where $B = |\vec{B}(\vec{r})|$. A linear z gradient such as in Eq. (16.50) creates a linear variation in the Larmor frequency with position. For example, consider two small objects, one at $z = +a$ and the other at $z = -a$. By tuning the RF signal to the appropriate Larmor frequency in a z-directed gradient, it would be possible to selectively tip the spins of one of the objects while leaving the other unaffected. Alternatively, both samples could be excited simultaneously by turning off the gradient and tuning the radiofrequency signal to $\omega = \omega_0 = \gamma B_0$. If the gradient is applied immediately following the RF pulse, the FID will become a sum of two signals, with amplitudes and frequencies corresponding to the position and volume of water in each sample. This is the principle of frequency encoding.

16.4.3 Slice Selection

While there are many techniques, or "pulse sequences," which can be used for generating a cross-sectional MR image, most begin by selecting a "slice" of the object to be imaged. Slice selection can be accomplished by applying a shaped RF pulse in conjunction with a magnetic field gradient, as illustrated in Fig. 16.12. In the following discussion, the slice-select gradient will be assumed to be along the z direction, but it will be denoted G_s for generality.

In the presence of a gradient, the total static field becomes a function of position, $\vec{B}(z) = \hat{a}_z(B_0 + G_z z)$. Additionally, since $\omega = \gamma B$, the Larmor frequency can be plotted as a function of position (Fig. 16.13).

This link between position and frequency can be used to select a slice through an object from which an image will be made. Consider selecting an imaging plane in a right cylindrical phantom along z, as illustrated in Fig. 16.14. If an RF field of fre-

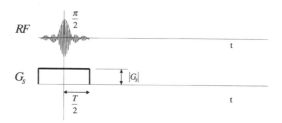

Fig. 16.12 A simple pulse sequence for selecting a rectangular slice of a sample.

quency $\omega = (\omega_0 + \omega')$ is applied, it will interact with a plane at the position z defined by

$$z = \frac{\omega'}{\gamma G_z} \qquad (16.54)$$

where ω' is the offset from the Larmor frequency determined by the gradient strength and the position

$$\omega' = \gamma G_z z \qquad (16.55)$$

where G_z is in T/m. Clearly, by varying the offset frequency, ω', the selected plane, or slice, can be moved.

To interact with one plane requires the excitation of a single frequency, which in turn requires a pulse length of infinite duration. A more realistic and useful approach is to apply an RF pulse containing a bandwidth of RF energy, $\omega_1 \leq \omega' \leq \omega_2$, where $\omega' = \omega - \omega_0$. This selects a slice of the cylinder between $z = z_1$ and $z = z_2$, as illustrated in Fig. 16.15. The edges of the slice are obtained from Eq. (16.54), $z_1 = \omega_1/(\gamma G_z)$, $z_2 = \omega_2/(\gamma G_z)$. Note that the slice thickness is dependent on the gradient strength and the frequency bandwidth: A narrower slice can be excited either by increasing the gradient strength or by decreasing the bandwidth of the RF pulse.

Slice Profile

From the previous discussion, excitation of a finite-thickness slice of the sample to a desired tip angle requires an RF pulse with a frequency spectrum that is con-

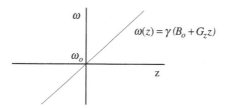

Fig. 16.13 The Larmor frequency is linked to position by the frequency-encoding gradient.

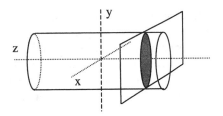

Fig. 16.14 A slice selected from a cylindrical object.

stant over a finite frequency bandwidth. The bandwidth of the RF pulse, $B_1(\omega)$ is related to the time waveform $B_1(t)$ by the Fourier transform relationship. In the MR literature, it is common to use time as the Fourier domain variable, with the spectrum of an RF pulse given by the inverse Fourier transform of the time waveform:

$$B_1(\omega) = F^{-1}\{B_1(t)\} \tag{16.56}$$

The gradients link each frequency component in the spectrum to a position z through the relationship $\omega' = \gamma G_z z$. For small tip angles, it is reasonable to assume that the tip angle is linearly related to the RF pulse power. Under this assumption, known appropriately as the "small tip-angle" approximation, the transverse magnetization at the end of the RF pulse, $t = T/2$ can be shown to be given by the Fourier transform of the RF pulse:

$$M_{xy'}(z) \propto M_z^0 F^{-1}\{B_1(t)\} \tag{16.57}$$

Note that the time reference has been set to the center of the RF pulse for convenience. The actual expression includes a phase term accumulated by the spins precessing in the gradients during the RF pulse:

$$M_{xy'}(z) = 2\pi M_z^0 e^{+i\gamma G_z z T/2} F^{-1}\{B_1(t)\} \tag{16.58}$$

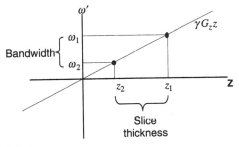

Fig. 16.15 An RF pulse with a finite bandwidth excites a slice of proportional thickness determined by the slice-select gradient amplitude.

The magnetization in Eq. (16.58) exhibits a z-dependent phase shift following the RF pulse, $\phi_1(z) = \gamma G_z z\,T/2$. This can be removed or "rewound" by applying a gradient pulse of appropriate duration with the opposite sign as the slice-selection gradient. During a gradient pulse of amplitude G_{rw} and of duration T_{rw}, the spins will precess faster or slower than the Larmor frequency, depending on their position in the gradient. This results in a position-dependent phase shift, $\phi_2(z) = \gamma G_{rw} z T_{rw}$. If the rewind pulse is chosen such that $G_{rw} T_{rw} = -G_s T/2$ then the phase across the slice profile will be canceled. This discussion presumes even, symmetrical RF pulses such as a $\sin(t)/t$ or a "sinc" pulse, commonly used for slice selection, and ideal gradient pulses. To first order, nonideal gradient pulses can be accounted for by working with the integrals of the gradient pulse time waveforms. A pulse sequence for a slice-selective pulse, including a phase-rewind pulse with $T_{rw} = T/2$ and $G_{rw} = -G_s$, is illustrated in Fig. 16.16.

A sinc pulse, or $\sin(x)/x$, in time can be used to obtain a rectangular slice profile in z. The slice thickness can be modified by either changing the gradient strength or scaling the RF pulse in time. A more exact treatment of the effects of truncation and the nonlinearities of the Bloch equation can be done in a number of ways, including a finite-difference solution of the Bloch equation.

Following the excitation of an imaging plane with an RF pulse, the next step is to encode spatial information into the ensuing FID. While there are many ways to do this, by far the most popular method in MR imaging is 2-D Fourier imaging. In this method, orthogonal gradients are used to spatially encode the FID or echo in such a way that the received signal is one line through the 2-D Fourier transform of the encoded image. After obtaining enough independently encoded FIDs or echos, the image is obtained by an inverse Fourier transform on the received and stored data set. The image is encoded by two mechanisms called "frequency" and "phase" encoding.

16.4.4 Frequency Encoding

Consider an excited slice at $z = z_0$, with a spin density distribution $\rho(x,y,z_0)$. It is useful to again consider a disk in the x–y plane, which is one slice from a right cylinder oriented along the z axis, as shown in Fig. 16.14. If no gradient is present, then following the excitation pulse, a voxel of size $\Delta V = \Delta x \Delta y \Delta z$ produces a signal given

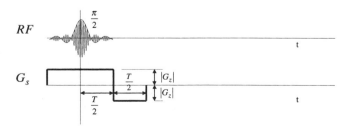

Fig. 16.16 Slice-selection pulse sequence with a phase-rewind pulse.

by Eq. (16.45). However, a frequency-encoding gradient along the x axis creates a frequency variation in the x direction, so Eq. (16.45) becomes

$$S(t) = j\frac{1}{\sqrt{2}}\omega_0\Delta V M_{xy}^0 e^{-t/T_2}B_{1t}e^{+j(\omega(x)-\omega_{mix})t} \qquad (16.59)$$

The variation in ω is small enough that it can be ignored in the magnitude terms but not in the phase term. The signal $S(t)$ received from an entire slice is obtained by integrating in x, y, and z. For a thin slice, it is reasonable to assume homogeneity in z and simply multiply by the slice thickness, Δz:

$$S(t) = \iint_{\text{slice}} \frac{1}{\sqrt{2}}\, j\omega_0 M_{xy}^0 B_{1t}e^{-t/T_2}\Delta z e^{+j(\omega(x)-\omega_{mix})t}dxdy \qquad (16.60)$$

In general, B_{1t}, T_2 and M_{xy}^0 are functions of position. In fact, these are factors that are included in the eventual cross-sectional image that will be formed. It is convenient to simply denote these factors $I(x,y)$, where $I(x,y)$, is the image, a map of the signal obtained from the object in the selected plane. For the case of a FID following a slice-selective excitation pulse as considered here,

$$I(x,y) = j\frac{\omega_0\Delta z}{\sqrt{2}}\, M_{xy}^0(x,y,z_0)B_{1t}(x,y,z_0)e^{-t/T_2} \qquad (16.61)$$

Using this notation, the received signal becomes

$$S(t) = \iint_{\text{slice}} I(x,y)e^{+j(\omega(x)-\omega_{mix})t}dxdy \qquad (16.62)$$

Note that $I(x,y)$ is dependent on the acquisition timing through the term e^{-t/T_2}. Analogously to the z-directed slice-select gradient, the Larmor frequency as a function of x is given by $\omega(x) = \gamma(B_0 + G_x x) = \omega_0 + G_x x$, and $S(t)$ becomes

$$S(t) = \iint_{\text{slice}} I(x,y)e^{+j(\omega_0+\gamma G_x x)-\omega_{mix})t}dxdy \qquad (16.63)$$

where G_x is the magnitude of the x gradient in T/m. If the mixing frequency $\omega_{mix} = \omega_0$, then $S(t)$ becomes

$$S(t) = \iint_{\text{slice}} I(x,y)e^{+j\gamma G_x xt}dxdy \qquad (16.64)$$

It is very useful to use the notation $k_x = \gamma G_x t$, which has units of rad/m, in which case Eq. (16.64) can be expressed as

$$S(k_x) = \iint_{\text{slice}} I(x,y)e^{+jk_x x}dxdy \qquad (16.65)$$

$S(k_x)$ can be interpreted as the Fourier transform of a "projection" of the image map $I(x,y)$ onto the x axis. The projection of an object onto an axis is the integral of the object in the variable orthogonal to the projection axis. For example, for the projection of $I(x,y)$ onto the x axis, $I_p(x)$ is given by

$$I_p(x) = \int I(x,y)dy \tag{16.66}$$

One can observe that $S(k_x)$ is the Fourier transform of $I_p(x)$ by writing

$$S(k_x) = \int_x \left\{ \int_y I(x,y)dy \right\} e^{+jk_x x}dx = \int_x I_p(x)e^{+jk_x x}dx \tag{16.67}$$

$$S(k_x) = F\{I_p(x)\}$$

Example Problem 16.7

What signal $S(t)$ is obtained from a cubic phantom in a 3.0 T field using a 10 G/cm x-directed gradient? Assume the sample is homogeneous.

Solution

The signal is given by the expression

$$S(k_x) = \int\int_{slice} I(x,y)e^{+jk_x x}dxdy$$

$I(x,y)$ was defined in Eq. (16.61). For a small phantom of distilled water, the flux density of the coil and the transverse magnetization can be assumed to be constant over the entire object:

$$I(x,y) = j\Delta z\frac{\omega_0}{\sqrt{2}} M_{xy}^0(x,y,z_0)B_{1r}(x,y,z_0)e^{-t/T_2} = C$$

The T_2 decay has simply been lumped in with the constant C because it is usually much slower than the gradient-induced signal decay. The limits of integration are the edges of the sample (no signal is contributed if there are no spins), and $S(k_x)$ becomes

$$S(k_x) = C \int_{-a}^{a}\int_{-a}^{a} e^{+jk_x x}dx$$

$$= C(2a)^2 \, \text{sinc}(k_x a)$$

Since $k_x = \gamma G_x t$, the form of the detected signal is given by

$$S(t) = C(2a)^2 \, \text{sinc}(\gamma G_x t a) \qquad \blacksquare$$

It is interesting to examine $S(t)$ from the previous example. $S(t)$ is plotted in Fig. 16.17 for the values $a = 0.5$ cm, $G_x = 10$ G/cm or 0.001 T/cm, $k_x = \gamma G_x t$, and $\gamma = 2\pi \cdot 42.57 \times 10^6$ Hz/T for the ^{1}H proton. These are typical values for a MRI experiment. Note that the signal has died out significantly after only 300 μs. To form an image, it will be necessary to sample $N/2$ points in this time, where N is the image resolution. This is illustrated in Fig. 16.18. If $N = 256$, one needs to sample at a rate on the order of two points (real and imaginary) every few microseconds.

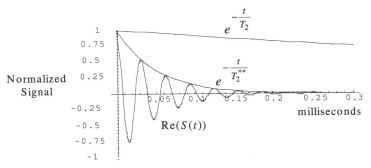

Fig. 16.17 Simulated free-induction decay in the presence of a 10 G/cm gradient. Decay envelopes due to T_2 and T_2^{**} are also indicated. The signal $S(t)$ decays due to the gradient-induced dephasing, the T_2^{**} curve.

Additionally, note that k_x increases with time. By sampling the signal $S(t)$ at intervals of $n\Delta t$ or $n\Delta k_x$, one is sampling a line in the $k_x - k_y$ plane. The $k_x - k_y$ plane is commonly referred to as "k space."

Sampling the FID fills only one-half of one line in k space, at $k_y = 0$, as illustrated in Fig. 16.18. To obtain the data to reconstruct an image, it is necessary to sufficiently sample a 2-D grid of k space. Once k space has been sufficiently sampled, the encoded image $I(x,y)$ can be obtained by inverse Fourier transform. The other half of a line in k space can be acquired by sampling an echo rather than a FID. In this case, "negative" values of k_x correspond to time values before the echo formation at $t = 2\tau = \text{TE}$, the echo time. However, k space must be sampled sufficiently in both the k_x and k_y directions. While there are a variety of methods, in 2-D Fourier imaging this is accomplished using a phase-encoding gradient.

16.4.5 Phase Encoding

Application of an x gradient allowed the sampling of a line in k space during the FID. Different lines in the $k_x - k_y$ plane (typically called k space) can be sampled

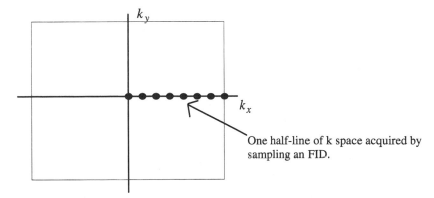

One half-line of k space acquired by sampling an FID.

Fig. 16.18 One half-line of k space, at $k_y = 0$, is sampled by a single FID. To form a 2-D image, k space must be sampled sufficiently in both the k_x and k_y directions.

by applying an additional phase-encoding gradient pulse prior to the application of the frequency-encoding gradient. Denote the phase-encoding gradient G_p. It will be assumed to be along the y axis:

$$\vec{G}_p = \hat{a}_z G_y y \tag{16.68}$$

If G_p is applied at time t_1 following the slice-select and phase-refocusing gradient pulses, the signal at time t becomes

$$S(t) = \int_{-\infty}^{\infty} \int_{-\infty}^{\infty} I(x,y,z_0) e^{+j\gamma(B_0 + G_y y)(t-t_1)} dx dy \tag{16.69}$$

where t_1 is the start of the phase-encode gradient pulse. The effect of the gradient pulse is to create a y-dependent phase shift, proportional to the product of the gradient amplitude and the duration of the pulse. It is convenient to use k space notation, where $k_y = \gamma G_y T_p$. T_p is the duration of the phase-encoding gradient pulse. The magnetization can be moved to any location on the k_y axis by the application of a gradient pulse of the appropriate duration. After the phase-encoding pulse the FID can be sampled as discussed previously. Following a phase-encoding gradient pulse, the sampled FID signal during the frequency-encoding gradient becomes

$$S(t,T_p) = \int\int_{\text{slice}} I(x,y) e^{+j\gamma G_x x t_f} e^{+j\gamma G_y y T_p} dx dy \tag{16.70}$$

By substituting $k_x = \gamma G_x t_f$ and $k_y = \gamma G_y T_p$, Eq. (16.70) becomes

$$S(k_x, k_y) = \int\int_{\text{slice}} I(x,y) e^{+jk_x x} x e^{+jk_y y} dx dy \tag{16.71}$$

from which it is apparent that each line acquired is one line in the 2-D Fourier transform of the object. A pulse sequence that would acquire an arbitrary line in k space is shown in Fig. 16.19. Clearly, all of k space could be interrogated by repeating the sequence with different values of G_y and/or T_p and by changing the sign of G_x or using an echo to obtain an entire line parallel to the k_x. The image $I(x,y)$ can then be obtained by inverse Fourier transform after suitably sampling k space.

16.4.6   MRI Pulse Sequences

A pulse sequence is a combination of RF and gradient pulses, delay times, and signal acquisition periods designed to encode particular information into the signal detected from an object. A wide variety of pulse sequences have been developed for different applications and needs. For example, one can design pulse sequences to encode different physiological parameters into $I(x,y)$, to perform dynamic imaging with frame rates of tens of images per second, or for MR microscopy. For some experiments, such as a one-pulse MR spectroscopy experiment, a single RF pulse can constitute an entire pulse sequence. For imaging applications, it is necessary to design a pulse sequence that acquires or "samples" k space sufficiently to form the desired image. In the following sections, three of the most common pulse sequences will be discussed.

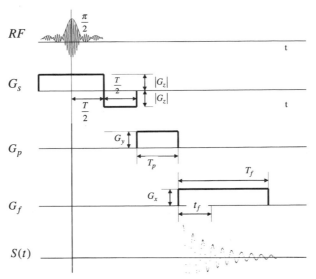

Fig. 16.19 A pulse sequence that could be used for sampling any line of k space by proper selection of G_x and G_y.

Partial Saturation

Perhaps the simplest MR imaging sequence is the partial saturation (PS) sequence, formed by simply repeating the sequence of Fig. 16.19 multiple times with a different phase-encode gradient amplitude during each repetition. A PS imaging sequence is illustrated in Fig. 16.20. One "line" of the sequence is illustrated, with this basic building block being repeated, each time incrementing the amplitude of the phase-encode gradient by ΔG_y. The repetition time, denoted TR, can vary from milliseconds to seconds depending on the application. Each repetition of this line acquires one line of k space. The k_y coordinate is given by $k_y = \gamma G_{y,n} T_p$, where $G_{y,n}$ is the value of the phase-encode gradient during the nth repetition. Taking N samples of the FID would sample N points of k space at this value of k_y and with N values of k_x, given by $k_{x,n} = n\gamma G_x \Delta t$, where Δt is the sampling rate.

One difficulty with the PS sequence as described here is that acquisition of a single FID only obtains values of k space on one side of the k_y axis. In order to obtain an entire set of k-space samples there are a number of options, one of which would be to repeat the pulse sequence with the sign of the frequency-encoding gradient, G_x, reversed. This would double the length of the sequence. Another option is to vary both G_x and G_y with each repetition in such a manner as to scan k space on a polar grid. This technique leads to projection reconstruction imaging, which will not be discussed here. Another option is to use echoes for data collection.

Gradient-Echo Sequences

Consider the modified PS sequence illustrated in Fig. 16.21. The only difference between this sequence and the one considered previously is the addition of one prephasing pulse in the frequency-encoding gradient. As discussed earlier, the gradi-

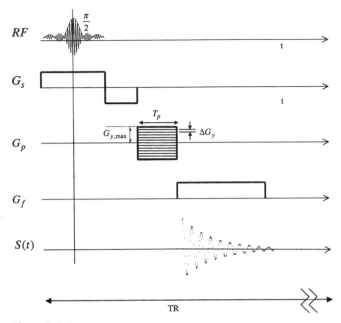

Fig. 16.20 A partial saturation imaging sequence. The sequence line is repeated with a repetition time TR, incrementing the phase-encode gradient by ΔG_y with each repetition. All other parameters are unchanged.

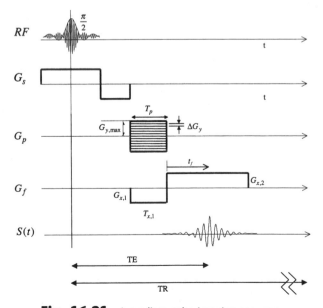

Fig. 16.21 A gradient-echo imaging sequence.

ent causes spins at different x coordinates to precess at different rates, creating a phase distribution along the x axis quantified by the spatial frequency k_x. A pulse of duration $T_{x,1}$ and amplitude $-G_{x,1}$ moves the spins to $k_x = -\gamma G_{x,1} T_{x,1}$. Just as before, the second gradient pulse causes the magnetization to move in the $+k_x$ direction. A gradient echo is formed with a peak when $k_x = 0$, which occurs when $G_{x,1} T_{x,1} = G_{x,2} t_f$. The length of time from the beginning of each repetition of the pulse sequence to the center of the echo is termed TE for echo time. Note that it is common to designate the time reference as the center of the initial RF pulse that creates transverse magnetization.

With this modification, one gradient echo samples one entire line of k space, as shown in Fig. 16.22. The line shown is at $k_y = 0$, which would be acquired during the repetition in which the amplitude of the phase-encoding gradient was zero. Other lines in k space would be acquired during different repetitions. For example, acquiring 256 samples of each of 256 echoes, each with a different phase-encode gradient value, would provide 256×256 data points. These points can then be transformed into an image using a discrete Fourier transform. Obviously, the choices of gradient values and pulse lengths affect both the image resolution and the contrast (discussed later).

Spin-Echo Sequences

A final variation to be discussed is the spin-echo sequence. This sequence has formed the basis of clinical MRI for most of its existence and is only recently seeing significant competition from other, faster sequences. A disadvantage of the gradient-echo sequence described previously is that the echo is diminished by magnetic field inhomogeneities, i.e., it reflects T_2^*. Images formed using a gradient sequence have contrast that reflects the local value of T_2^*. In many applications, such as some types of functional brain imaging, this can be very useful. However, in many cases it is desired to form images which reflect the T_2 values of the sample. As discussed previously, this can be achieved by using a spin echo rather than a gradient echo. The most significant difference between a spin-echo sequence and a gradient-echo sequence is the inclusion of a 180° RF pulse at TE/2. A basic spin-echo sequence is

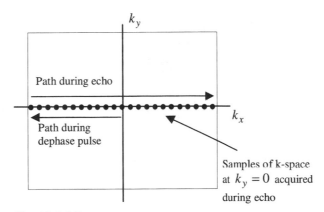

Fig. 16.22 One repetition of the basic gradient echo pulse sequence samples one line of k space.

shown in Fig. 16.23. The 180° pulse will form a spin echo at time TE. Ordinarily, the sequence is designed so that $G_{x,1}T_{x,1} = G_{x,2}t_f$ at $t =$ TE.

The 180° refocusing pulse is slice selective in most instances. The phase of the pulse can be chosen to obtain either a CPMG or a Hahn spin echo. Additionally, the sign of the initial frequency-encoding gradient dephasing pulse is reversed from that used with the gradient echo. In the gradient echo example shown in Fig. 16.21, the dephasing pulse moved the magnetization to the left in k space and then the frequency-encoding gradient moved it back through the middle, creating an echo. In the case of the spin echo, the π pulse causes the magnetization to flip across the y axis. Therefore, in order to create an echo moving through the center of k space in the same direction, it is necessary to begin with a dephasing pulse of the same sign as the frequency-encoding gradient pulse.

16.4.7 Resolution and Field of View

If all of k space were known, one could exactly reconstruct the image data $I(x,y)$. Unfortunately, it is impossible to characterize all of k space. In practice the information represented by k space, which is the echo or FID, is discretely sampled and reconstructed using the discrete Fourier transform. The sampling rate determines the resolution with which k space is observed. This resolution is denoted by Δk_x and Δk_y for 2-D Fourier imaging with a z-directed slice selection. Additionally, in practice one is limited in the extent, or range, to which k space can be sampled, $|k_{x,max}|$ and $|k_{y,max}|$, as indicated in Fig. 16.24. This limitation is due to the need to keep acquisition time low (sampling more points at a given acquisition rate increases the ac-

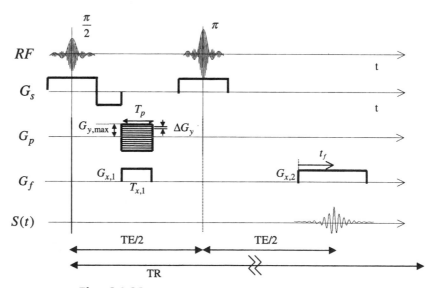

Fig. 16.23 A basic spin-echo imaging pulse sequence.

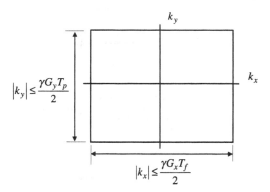

$$|k_y| \le \frac{\gamma G_y T_p}{2}$$

$$|k_x| \le \frac{\gamma G_x T_f}{2}$$

Fig. 16.24 A finite portion of k space sampled to form an image. The image resolution and field of view are determined by the sampling rate and extent of k space.

quisition time), memory limitations, (more points requires larger amounts of memory), and the inevitable result that eventually the signal will decay to below the noise, rendering further sampling meaningless. Typical sample rates are on the order of microseconds per sample, with between 64 and 1024 samples being collected for imaging experiments and more for spectroscopy experiments in which greater resolution is needed.

Assuming symmetric sampling of the echo, and a phase encode and frequency dephasing time of T_p and T_f, the portion of k space within $|k_x| \le \gamma G_x T_f/2$, $|k_y| \le \gamma G_y T_p/2$, is sampled as shown in Fig. 16.24.

How can the appropriate range and resolution to be used when sampling k space be found? The answer is determined by the required field of view and resolution in the reconstructed image and the available gradient strengths and pulse sequence parameters. The range of the reconstructed image, or the field of view (FOV), and the sampling rate of the k space data are inversely related as shown in Fig. 16.25. The FOV in the x and y directions of the image are given by the sampling resolution in k space:

$$\text{FOV}_x = \frac{2\pi}{\Delta k_x}, \qquad \text{FOV}_y = \frac{2\pi}{\Delta k_y} \qquad (16.72)$$

The remaining task is to relate Δk_x, Δk_y, $k_{x,\text{max}}$, and $k_{y,\text{max}}$ to the pulse sequence parameters in order to design pulse sequences. For 2-D Fourier-encoded MR imaging, it is standard to sample a line of k space by sampling the echo during a constant frequency-encoding gradient. The location in k space is given by the amount of dephasing experienced by the spins due to the gradient:

$$k_x(t) = \int_0^t \gamma G_x(t') dt' \qquad (16.73)$$

In a constant gradient, the increment in k_x during one time period Δt is simply

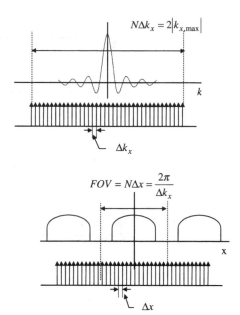

Fig. 16.25 The field of view and the resolution of the reconstructed image are determined by the sampling rate and number of samples of k space obtained.

$$\Delta k_x = \gamma G_x \Delta t \tag{16.74}$$

Thus, the resolution in k space in the frequency-encode direction can be controlled either by changing the sampling time or by changing the gradient strength. The range of sampling in k space is simply $N\Delta k_x$ and can be varied by the gradient strength, sampling time, or number or samples acquired. From Eq. (16.74), and assuming N equally spaced points in the FOV, the resolution in the frequency-encode direction is obtained as

$$\Delta x = \frac{2\pi}{N\gamma G_x \Delta t} \tag{16.75}$$

For 2-D Fourier imaging in which the phase encode is accomplished by varying the phase-encode gradient amplitude with each repetition, the increment in k_y is determined by the amount of increment of this gradient, ΔG_y. The duration of the phase-encode gradient pulse could be varied instead of the amplitude, but this is generally considered more complicated because another time must be varied to compensate in order to maintain fixed TE and TR times. Denoting the length of the phase-encode gradient pulse T_p, the increment or sampling rate in the phase-encode direction becomes

$$\Delta k_y = \gamma \Delta G_y T_p \tag{16.76}$$

and the resolution in the phase-encode direction is given by

$$\Delta y = \frac{2\pi}{N\gamma\Delta G_y T_p} \qquad (16.77)$$

Example Problem 16.8.

Determine the gradient parameters needed to obtain an image with 256×256 pixels and a 10-cm FOV in both frequency- and phase-encode directions. The sampling rate is 25 μs, and the maximum value of the phase- and frequency-encoding gradients should be the same.

Solution

Frequency-encoding gradient: The resolution in both directions is specified by the FOV and number of points:

$$\Delta x - \Delta y - \frac{FOV}{N} - \frac{10 \text{ cm}}{256} - 0.039 \text{ cm}$$

The frequency-encoding gradient strength is determined from the resolution and the sampling rate, specified as 25 μs:

$$\Delta x = \frac{2\pi}{N\gamma G_x \Delta t} \Rightarrow G_x = \frac{2\pi}{N\gamma\Delta x \Delta t}$$

$$G_x = \frac{2\pi}{256 \cdot 2\pi \cdot \left(42.57 \times 10^6 \frac{1}{s \cdot T}\right) \cdot (0.039 \text{ cm}) \cdot (25 \times 10^{-6} \text{ sec})}$$

$$= 9.40 \times 10^{-5} \frac{T}{cm} = 0.940 \frac{G}{cm}$$

where the conversion $1 \text{ T} = 10^4$ G has been used.

The total sampling time, or acquisition time T_{acq}, for one echo is simply

$$T_{acq} = N\Delta t = 256 \cdot 25 \text{ μs} = 6.4 \text{ ms}$$

Assuming ideal gradients, this is the length of the frequency-encoding gradient pulse.

Phase-encoding gradient: The problem statement specified identical maximum values of phase- and frequency-encoding gradients. This is commonly done but not necessary in general. In this case, the increment in the phase-encode gradient, Δk_y, becomes

$$\Delta Gy = \frac{2G_{y,max}}{N}$$

as the phase-encoding gradient is incremented from negative to positive of its maximum value to cover k space. Since $G_{y,max} = G_x = 0.940$ G/cm,

$$\Delta Gy = \frac{2 \cdot 0.940 \text{ G/cm}}{256} = 7.34 \times 10^{-3} \text{G/cm}$$

The phase-encode time, phase-encode increment, and spatial resolution are related by

$$\Delta y = \frac{2\pi}{N\gamma\Delta G_y T_p}$$

Solving for the phase-encode gradient pulse length, T_p

$$T_p = \frac{2\pi}{N\gamma\Delta G_y \Delta y}$$

$$= \frac{2\pi}{256 \cdot (2\pi \cdot 42.57 \times 10^6 \text{ 1/s T}) \cdot (73.4 \times 10^{-3} \text{ G/cm})(1\text{T}/10^4\text{G})(0.039 \text{ cm})} = 3.2 \text{ ms}$$

Note that the phase-encoding gradient pulse is exactly one-half the length of the frequency-encoding pulse. This is a direct result of choosing the maximum value of the frequency- and phase-encoding gradients to be the same. Note that the frequency-dephasing pulse, if turned on simultaneously with the phase-encoding gradient, would have the amplitude $-G_x$ computed previously.   ∎

16.4.8   Contrast in MRI

The previous sections introduced the general concepts of spatial encoding of an image using magnetic field gradients. It was shown that through the use of frequency-and-phase encode gradients, it is possible to collect a set of data which represents the Fourier transform of an image of the sample. A separate issue is the content of the image. For the partial saturation pulse sequence shown in Fig 16.20, the encoded information as a function of x and y was given by Eq. (16.61), repeated here:

$$I(x,y) = \frac{j\omega_0\Delta z}{\sqrt{2}} M_{xy}^0 (x,y,z_0)B_{1t}(x,y,z_0)e^{-t/T_2} \qquad (16.78)$$

Additionally, for the partial saturation sequence the T_2 relaxation term was ignored. Different pulse sequences can be designed to encode various physiological parameters into the image. Certainly, the most common parameters that are encoded are the relaxation times. Through proper pulse sequence design, one can produce images in which the intensity of the image is predominantly proportional to the T_1 or T_2 time of the tissue in each pixel of the image. Because the T_1 and T_2 times can be changed by different disease or injury processes, and because they act as natural contrast agents, forming separate T_1 and T_2 weighted images can be a useful diagnostic technique. Table 16.2 lists relaxation times for a few common tissues. For example, T_1 and T_2 weighted images of a spine are shown in Fig. 16.26a and 16.26b respectively. In this example, the T_2 weighted image more clearly demonstrates the encroachment of several vertebral bodies into the spinal cord canal. In some cases it is of interest to quantify the relaxation times. In order to understand the mechanism of image contrast in simple pulse sequences, the starting point is the development of equations for the transverse magnetization created by a simple pulse sequence.

Steady-State Transverse Magnetization

Consider a general object having T_1 and T_2 spin density and distributions $T_1(x, y)$, $T_2(x, y)$, and $M_z^0(x,y)$. The objective is to develop an equation for the steady-state

TABLE 16.2 Typical T_1 and T_2 Values for Some Common Tissue Types

Tissue	T_1 (ms)	T_2 (ms)
Gray matter	520	95
White matter	380	85
Typical edema or infarction	600	150
Malignant tumor	800	200
Fat	160	100
CSF	2000	1000

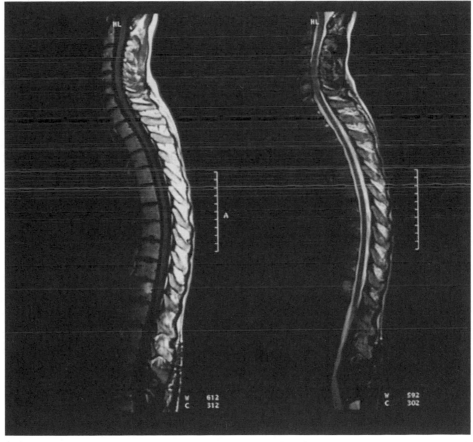

Fig. 16.26 Examples of T_1 and T_2 "weighted" images of the spine, in which the image intensity is primarily proportional to the local value of T_1 (a) and T_2 (b) (images courtesy of Siemens Medical Systems, Iselin, NJ).

value of the transverse magnetization in typical pulse sequences, from which the image can be computed or, alternatively, appropriate values of TE and TR which optimize contrast between two tissue types can be selected. The simplest image sequence to consider is a repeated partial saturation sequence consisting of a 90° pulse and a readout, or frequency-encoding gradient, illustrated in Fig. 16.27. This basic sequence is repeated at an interval of TR, the repetition time. Presumably a phase-encode pulse would be applied simultaneously with the gradient-refocusing pulse to enable imaging. At time $t = 0$, immediately prior to the RF pulse, the transverse magnetization is zero because the net magnetization vector is along the static field axis:

$$M_{xy}(0) = M_{xy}^0 = 0$$
$$M_z(0) = M_z^0 \qquad\qquad (16.79)$$

The RF pulse tips the spins toward the transverse plane. Because the time scale of the RF pulse is usually short in comparison to the relaxation effects, it is common to ignore T_1 and T_2 effects during the RF pulses. With this approximation, immediately following the RF pulse and the refocusing gradient the transverse and longitudinal magnetization become

$$M_{xy}(0^+) = M_z^0 \sin\alpha$$
$$M_z(0^+) = M_z^0(1 - \sin\alpha) \qquad\qquad (16.80)$$

where the time point following the RF pulse is denote 0^+ for simplicity. Furthermore, if $\alpha = 90°$, this simplifies to

$$M_{xy}(0^+) = M_z^0$$
$$M_z(0^+) = 0 \qquad\qquad (16.81)$$

Immediately prior to the second RF pulse, at approximately time TR, the transverse magnetization has been decreased by both T_1 and T_2 effects. Similarly, the longitudinal (z-directed) magnetization has been partially restored by T_1 relaxation:

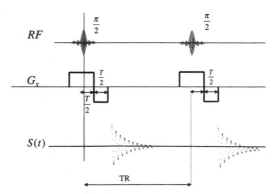

Fig. 16.27 A simple pulse sequence for the purpose of understanding contrast in MR pulse sequences.

$$M_{xy}(\text{TR}) = M_z^0 e^{-\text{TR}/T_1} e^{-\text{TR}/T_2}$$
$$M_z(\text{TR}) = M_z^0 (1 - e^{-\text{TR}/T_1}) \tag{16.82}$$

Again assuming a 90° RF pulse and ignoring relaxation effects during the RF pulse, the transverse and longitudinal magnetization following the second pulse become

$$M_{xy}(\text{TR}^+) = M_z^0 (1 - e^{-\text{TR}/T_1})$$
$$M_z(\text{TR}^+) = 0 \tag{16.83}$$

The 90° RF pulse simply transfers the longitudinal magnetization into the transverse plane. Continuing the same process, one can find the magnetization after the end of the second repetition, $t = 2\text{TR}$. The transverse magnetization has again undergone T_1 and T_2 relaxation:

$$M_{xy}(2 \cdot \text{TR}) = M_z^0 (1 - e^{-\text{TR}/T_1}) e^{-\text{TR}/T_1} e^{-\text{TR}/T_2} \tag{16.84}$$

The longitudinal component consists of two parts — the relaxing transverse magnetization, and the magnetization still returning to equilibrium from the effects of the first pulse. This latter part is the reservoir of $M_{xy}(\text{TR})$ relaxed by one more factor of TR:

$$M_z(2 \cdot \text{TR}) = M_z^0 (1 - e^{-\text{TR}/T_1})(1 - e^{-\text{TR}/T_1}) + M_z^0 e^{-\text{TR}/T_1}(1 - e^{-\text{TR}/T_1}) \tag{16.85}$$

After simplifying, this term becomes simply $M_z(2\text{TR}) - M_z^0(1 - e^{-\text{TR}/T_1})$, identical to the longitudinal magnetization at the end of the first repetition period. Indeed, one can show that the spins have reached steady state following a single RF pulse:

$$M_z(N \cdot \text{TR}) = M_z^0 (1 - e^{-\text{TR}/T_1}) \quad N \geq 1 \tag{16.86}$$

The $(N + 1)$th RF pulse simply transfers this magnetization into the transverse plane. Including the x–y dependence on the T_1, following every 90° pulse but the first, the transverse magnetization is simply

$$M_{xy}(x, y, N \cdot \text{TR}^+) = M_z^0 (1 - e^{-\text{TR}/T_1(x,y)}) \tag{16.87}$$

Since $I(x, y) = 1/\sqrt{2} j\omega_0 \Delta z M_{xy}^0(x, y, z_0) B_{1t}(x, y, z_0) e^{-t/T_2}$, it is apparent that for values of t that are small in comparison with T_2 and T_2^* the intensity of the image is weighted by the T_1 relaxation time distribution. Any spatial variations in the receiver coil sensitivity and the spin density are also included in the image intensity, but these factors are often much less significant for most tissues and well-designed RF coils. For this reason a partial saturation sequence can be used to produce T_1 weighted images. Equations for the steady-state magnetization in other pulse sequences can be obtained similarly, although the analysis becomes more complicated.

T_1 and T_2 Weighting

A common diagnostic procedure in MRI is to obtain T_1 and T_2 "maps" of the area of interest. Such a pair of images through the pathology of interest can demonstrate relative changes in both relaxation times, which is diagnostic for a wide range of diseases. Rather than providing direct quantitative measurements of the T_1 and T_2, images are obtained where the effect on image intensity is primarily determined by only one relaxation constant. Such an image is called a weighted image to reflect

TABLE 16.3 Steady-State Transverse Magnetizations for Standard Imaging Sequences

Sequence	Steady-state magnetization
Partial saturation	$M_z^0(1 - e^{-TR/T_1})$
Inversion recovery	$M_z^0(1 - 2e^{-TI/T_1} + e^{-TR/T_1})$
Spin echo	$M_z^0(1 - e^{-TR/T_1})e^{-TE/T_2}$
Gradient echo	$M_z^0 \dfrac{(1 - e^{-TR/T_1})e^{-TE/T_2^*}\sin\alpha}{1 - \cos\alpha \cdot e^{-TR/T_1}}$

the qualitative nature of the mapping. Consider the partial saturation sequence from Fig. 16.27. As discussed previously, as long as the acquisition delay is small relative to the T_2 times in the slice, the effect of T_2 relaxation can be ignored. The intensity equation from Table 16.3 can be written to specifically include the spatial distribution of equilibrium magnetization and T_1 times over the imaging slice:

$$M_{xy}(x,y) = M_z^0(x,y)(1 - e^{-TR/T_1(x,y)}) \tag{16.88}$$

$M_z^0(x, y)$, the equilibrium magnetization distribution, is a measure primarily of the density of nuclear dipole moments and can be referred to as the "spin-density" distribution. For many images, there will be relatively little variation in the spin density. If TR is on the order of the T_1 times contained in the slice, then the image contrast in the PS sequence will be predominately determined by the T_1 times. Hence, the PS sequence can be used to create a T_1 weighted image. Alternatively, if TR is made much larger than the T_1 times, the exponential approaches zero, and the image contrast becomes a function of the spin-density distribution.

A spin-echo sequence can be used to obtain either T_1 or T_2 images. The transverse magnetization in a spin-echo sequence, including spatial variations, was given by

$$M_{xy}(x,y) = M_z^0(x,y)(1 - e^{-TR/T_1(x,y)})e^{-TE/T_2(x,y)} \tag{16.89}$$

A T_1 weighted image can be obtained by selecting a very short TE, typically below 25 ms, and a TR on the order of the T_1 times of interest. Alternatively, a very short TE and a very long TR leaves only the spin-density distribution to determine contrast, allowing spin-density weighting. Finally, using very long TR times to eliminate the first exponential and TE comparable to the T_2 times results in images that are primarily T_2 weighted.

T_2^* Weighting

Gradient-echo imaging, in addition to allowing much faster image acquisition times, provides the ability to form T_2^* weighted images. The spatial distribution of transverse magnetization, which determines the contrast, is approximately

$$M_{xy}(x,y) = M_z^0 \frac{(1 - e^{-TR/T_1(x,y)})e^{-TE/T_2^*(x,y)}\sin\alpha}{1 - \cos\alpha \cdot e^{-TR/T_1(x,y)}} \tag{16.90}$$

Unlike spin-echo sequences in which one must make $\text{TR} \gg T_1(x, y)$ to remove the $e^{-\text{TR}/T_1}$ term, using small tip angles causes the effect of T_1 contrast to drop out, reducing Eq. (16.90) to

$$M_{xy}(x,y) \approx M_z^0 e^{-\text{TE}/T_2^*(x,y)}\sin\alpha \qquad (16.91)$$

This is an interesting result because it indicates that while the contrast will be largely determined by the T_2^* times, the transverse magnetization is necessarily low due to the need to use small tip angles. The amount of transverse magnetization can be improved by using a higher tip angle, increasing the signal-to-noise ratio. The need to balance the trade-off between a pure T_2^* map and higher signal-to-noise leads to the subject of contrast optimization, discussed in the next section. Finally, it is also possible to create T_1 and spin-density maps using gradient-echo sequence through proper choices of parameters. All the parameters discussed previously are summarized in Table 16.4. Note that the parameters listed are chosen to give strongly weighted images. In practice these choices are usually influenced by system limitations, particularly in terms of minimum echo times, and the need to reduce imaging time to make the MR system available for more studies and to increase patient comfort.

T_1 and T_2 Measurement

Direct measurement of the relaxation times is often desirable, although there is wide variation in normal values in biological samples. To the extent that a spin-echo image can be modeled by Eq. (16.89), it is possible to obtain quantitative T_1 and T_2 images using a basis set of three spin-echo images, all of the same slice. Two of the images are made with the same TR and different TE, and two of the images have the same TE and different TR values. Two images are used to fit a T_2 relaxation curve ($e^{-\text{TE}/T_2}$), and the other two are used to fit a T_1 relaxation curve ($1 - e^{-\text{TR}/T_1}$).

Given three images taken with TE and TR values as shown in Table 16.5, consider one pixel at the same location in each image, denoted by I_1, I_2, and I_3. All other settings must be identical between the three images, so that the system dependent constant K is identical. The pixel values from images 1 and 2 can be used to obtain an estimate of the T_2 value at that pixel, and images 1 and 3 can be used to obtain an estimate for T_1. To obtain an estimate for T_1, one uses the ratio of the pixel intensities in the two images:

$$\frac{I_1}{I_2} = \frac{\text{K}M_z^0(1 - e^{-\text{TR}_1/T_1})e^{-\text{TE}_1/T_2}}{\text{K}M_z^0(1 - e^{-\text{TR}_1/T_1})e^{-\text{TE}_2/T_2}} \qquad (16.92)$$

TABLE 16.4 Pulse Sequence Parameters for Weighted Images

Image	Partial saturation	Spin echo	Gradient echo
T_1 weighting	$\text{TR} \approx T_1$	$\text{TR} \approx T_1, \text{TE} \ll T_2$	$\text{TE} \ll T_2^*, \alpha$ large
T_2 weighting	—	$\text{TR} \gg T_1, \text{TE} \approx T_2$	—
T_2^* weighting	—	—	$\text{TE} \approx T_2^*, \text{TR}$ long
Spin-density weighting	$\text{TR} \gg T_1$	$\text{TR} \gg T_1, \text{TE} \ll T_2$	$\text{TE} \ll T_2^*, \text{TR}$ long

TABLE 16.5 Relaxation and Echo Times in Three Images to Obtain T_1 and T_2 Maps

Image No.	Relaxation time (TR)	Echo time (TE)
1	TR_1	TE_1
2	TR_1	TE_2
3	TR_2	TE_1

Canceling common terms and setting the equation to zero gives an equation which can be solved for the value of T_2 at the pixel, with the result

$$T_2 = \frac{TE_2 - TE_1}{\ln(I_1/I_2)} \tag{16.93}$$

The solution for T_1 is similar, but the resulting equation must be solved iteratively for T_1 and in most cases:

$$\frac{I_1}{I_3} - \frac{1 - e^{-TR_1/T_1}}{1 - e^{-TR_2/T_1}} = 0 \tag{16.94}$$

Keep in mind two significant limitations of this method. First, the pixel values in any given image are affected by noise. When only two images are used to fit a curve, the sensitivity of the solution to noise can be quite high. Additionally, it is always an approximation when one assumes that any pixel is governed by a single exponential decay. The signal from most pixels consists of contributions from multiple tissue types, both due to the size of the pixel and due to the intricacies of the imaging procedure, which "blurs" the actual pixel by the point spread function of the experiment. Additionally, even a homogeneous tissue is quite complicated. Intracellular and extracellular water, for example, have different relaxation times. Typically, however, treating a pixel as a monoexponential decay curve is a good approximation for the echo times and repetition times commonly used.

16.5 INSTRUMENTATION FOR MRI

It is often said that an MRI system is a perfect playground for engineers. A typical block diagram of an MR system is shown in Fig. 16.28. The system consists of a receiver and transmitter, current amplifiers for the gradients, a pulse sequence generator (PSG), and a host computer to provide image display and pulse sequence development and control. This section will provide a brief overview of the major systems as well as a brief discussion of the RF coils, gradient coils, and main magnet, not shown in the block diagram of Fig. 16.28.

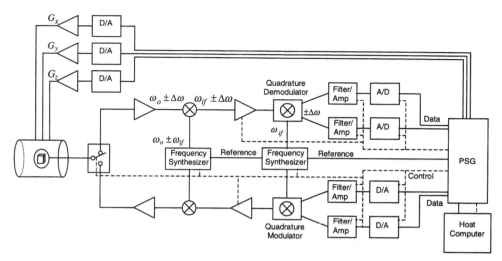

Fig. 16.28 Block diagram of a typical MRI system.

16.5.1 The MR Spectrometer

Because of the critical importance of accurate timing in MR experiments, a separate PSG is used to control the experiment. Data such as timing intervals, gradient amplitudes, RF pulse waveforms, and phase-encode gradient tables are downloaded from the computer to the PSG prior to beginning a pulse sequence. Once the user begins the sequence, the PSG is in control of the process. For most experiments the PSG simply executes the pulse sequence as scripted, but some level of interactivity is usually available. One example of interactive control of a pulse sequence is cardiac gating. The PSG will execute one or more lines in the sequence, stop, and wait for a trigger before continuing. The trigger is provided by an electrocardiogram device and ensures that data acquisition occurs during a specific point in the cardiac cycle, usually when the heart is in end systole. Of concern in a PSG is the number of control words available for controlling digital-to-analog converters and other control lines and the amount of memory, both for the pulse sequence and for data buffering on receive. Additionally, because the timing of the pulse sequence is controlled by the PSG, one needs programmability in very fine step sizes, typically 10 ns or better. The PSG is driven from a master clock, often derived from a reference signal in the frequency synthesizer. It is very important that this clock be very stable in order to allow signals to be averaged over long periods of time to improve the signal-to-noise ratio. The host computer provides the user interface for image display and pulse sequence generation. The data link to the PSG can be a standard serial or parallel link unless data buffering is done in the host computer.

Receiver

The upper data path in Fig. 16.28, after the RF switch, represents the receiver. A low-noise preamplifier, typically <1.0 dB noise figure and 20–40 dB gain, is used in the first stage. After amplification the signal is mixed to a intermediate frequency. A mixer, indicated by the multiplication symbol after the preamplifier, combines the detected signal, centered at the Larmor frequency ω_0, and a second signal at frequency $\omega_0 \pm \omega_{IF}$. The mixer effectively multiplies these two signals, producing the sum and difference of the two frequencies. A filter (not shown) attenuates the undesired frequency component. While the MR signal may be at a variety of frequencies depending on field strength and nucleus, mixing to an intermediate frequency allows the quadrature demodulation to always be done at a single frequency. The value of this frequency depends on manufacturer and system and can be higher or lower than the original MR frequency (up or down conversion). An IF amplifier is generally provided to ensure optimal signal amplitude into the quadrature demodulator. Although not shown, a variable gain stage is also often provided at the RF frequency stage.

The next stage in the receiver of Fig. 16.28 is to demodulate the signal, now at the IF, to baseband, or audio frequencies for digitizing. A quadrature demodulator mixes the IF bandwidth with the IF frequency, resulting in two signals, one centered at $2\omega_{IF}$ and the other at DC. Consider two magnetic dipole moments, or spins, one on either side of the magnet isocenter, in the direction of the frequency-encoding gradient. One of the spins creates a signal at $\omega_1 = \omega_0 + \gamma G_x \Delta x$ and the other at $\omega_2 = \omega_0 - \gamma G_x \Delta x$. After two stages of mixing and filtering, these signals are at frequencies of $\pm \gamma G_x \Delta x$. Assuming both singnals are at zero phase, they can be represented by

$$\begin{aligned} S_1(t) &= C \cos(+\gamma G_x \Delta x) \\ S_2(t) &= C \cos(-\gamma G_x \Delta x) \end{aligned} \tag{16.95}$$

Unfortunately, because $\cos(x) = \cos(-x)$, these two signals are indistinguishable. One solution is to mix to a center frequency sufficiently above DC so that all frequency components of the MR signal are above DC. One drawback of this solution is that it will require a digitization rate equal to the full bandwidth of the MR signal, $2\Delta\omega = \gamma G_x \text{FOV}$. Quadrature demodulation provides an alternate approach. In quadrature demodulation, illustrated in Fig. 16.29, the MR signal at IF is split and

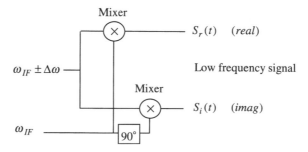

Fig. 16.29 A quadrature mixer resolves a complex RF signal into real and imaginary components at the sum and difference frequencies of the input signals.

supplied to two mixers. Each mixes with the IF to provide baseband signals centered at DC, but the two IF signals supplied for mixing are phase shifted by 90°.

Now, following the mixer, $S_r(t)$

$$S_r(t) = C\{\cos((\omega_{IF} + \gamma G_x \Delta x)t) \cdot \cos(\omega_{IF}t)\} = \frac{C}{2}\cos((2\omega_{IF} + \gamma G_x \Delta x)t) + \frac{C}{2}\cos(\gamma G_x \Delta xt)$$

$$S_i(t) = C\{\cos((\omega_{IF} + \gamma G_x \Delta x)t) \cdot \sin(\omega_{IF}t)\} = \frac{C}{2}\sin((2\omega_{IF} + \gamma G_x \Delta x)t) - \frac{C}{2}\sin(\gamma G_x \Delta xt)$$

(16.96)

Following low-pass filtering to remove the higher frequency component, the two signals become

$$S_r(t) = \frac{C}{2}\cos(\gamma G_x \Delta xt)$$

$$S_i(t) = -\frac{C}{2}\sin(\gamma G_x \Delta xt)$$

(16.97)

It can be shown that the spin shifted from center by $-\Delta x$ provides signals

$$S_r(t) = \frac{C}{2}\cos(\gamma G_x \Delta xt)$$

$$S_i(t) = \frac{C}{2}\sin(\gamma G_x \Delta xt)$$

(16.98)

The quadrature demodulator has removed the ambiguity of the signals. The real and imaginary channels are separately digitized using the analog-to-digital converters (ADC) shown in Fig. 16.28. For MR imaging, 16-bit ADCs are typically employed, commonly with digitization rates up to a few microseconds per point. Because the signal was mixed down to DC using quadrature demodulation, the low-pass, antialiasing filters are set for a bandwidth of $\Delta\omega - \frac{1}{2}\gamma G_x FOV$, one-half that required previously for a single mixer demodulation. Control signals for triggering the ADCs, setting the antialiasing filters, and controlling the system gain are provided by the PSG under control of the host computer.

Transmitter

The transmitter is, in most respects, the reverse of the receiver. For slice-selective imaging, the PSG provides a complex RF waveform for slice excitation. This might be a sinc pulse of appropriate bandwidth to excite a slice of the desired bandwidth. In order to shift the slice a phase is applied to the RF waveform. Two digital-to-analog converters create the complex waveform at baseband. The amplitude of the waveform can be adjusted at this point to control the tip angle. Additionally, low-pass filters are often included to remove undesired high-frequency components in the waveform. The quadrature modulator essentially performs the same function as the demodulator but in reverse. It combines the real and imaginary baseband waveform and mixes it to the intermediate frequency. An IF amplifier provides further flexibility in setting the system gain. After mixing to the Larmor frequency, the signal is amplified with a linear amplifier to provide the necessary RF power to the coil. Multiple-kilowatt RF amplifiers are commonly used in clinical systems.

Transmit/Receive Switch

It is often desirable to use the same RF coil for transmit and receive. A switch is needed to alternately connect the coil to the transmitter during RF excitation pulses and to the receiver during signal acquisition. A passive transmit/receive (T/R) switch is shown in Fig. 16.30. The crossed diode pairs act as switches, limiting the voltage across them to approximately 0.7 V. They behave as open circuits for the small signals experienced during receive but as short circuits to the high-voltage transmit pulses. During receive, they effectively open the path to the transmitter. During transmit, the first pair of diodes is shorted, providing a path to the coil. The second pair behaves as a short circuit to ground, protecting the receiver from the high-voltage transmit pulses. The quarter-wavelength transmission line is used to prevent the short-circuited second diode pair from also shorting out the RF coil. A quarter-wavelength transmission line acts as an impedance transformer, transforming the short circuit at the output to an open circuit. One disadvantage of the passive switch discussed here is that its performance is frequency dependent due to the requirement that the coaxial cable be a quarter wavelength.

16.5.2 Radiofrequency Coils

The RF coil is used to couple energy into the sample during transmit and to detect the time-varying magnetic flux density during signal acquisition. There are a wide variety of coils, loosely grouped into volume coils and local coils. Volume coils encompass the entire sample and are intended to provide a uniform B_1 field over the entire region of interest. Local coils are smaller and are designed to couple more tightly to a specific area, primarily to improve the signal-to-noise ratio (SNR). In the MR literature the SNR is usually defined as the ratio of the signal voltage to the noise voltage. At the beginning of signal acquisition the magnitude of the signal voltage can be obtained from Table 16.3 and Eq. (16.43):

$$|V_{sig}| = \sqrt{2}\omega_0 \Delta V M_{xy}|B_{1t}| \qquad (16.99)$$

where ω_0 is the Larmor frequency in rad/s, ΔV is the volume of interest, M_{xy} is the steady-state transverse magnetization per unit volume, and B_{1t} is the effective RF flux density given in Eq. (16.44).

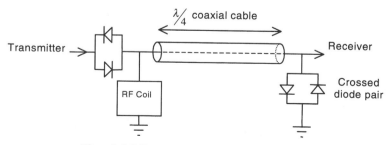

Fig. 16.30 A passive transmit/receive switch.

The noise voltage, V_{noise}, produced by a coil with impedance $Z_{coil} = R_{coil} + jX_{coil}$, is given by

$$V_{noise} = \sqrt{4kT\Delta f R_{coil}} \tag{16.100}$$

where k is Boltzman's constant, T is the temperature in degrees K, and Δf is the receiver bandwidth in Hz. The ratio of these two voltages gives the voltage signal to noise, SNR_v:

$$SNR_v = \frac{|V_{sig}|}{V_{noise}} = \frac{\sqrt{2}\omega_0 \Delta V M_{xy}|B_{1t}|}{\sqrt{4kT\Delta f R}} \tag{16.101}$$

For purposes of this analysis, the constants which remain unchanged from voxel to voxel can be ignored, giving a "coil signal-to-noise ratio," SNR_c:

$$SNR_c = \frac{\sqrt{2}|B_{1t}|}{\sqrt{R}} \tag{16.102}$$

Equation (16.102) describes the voltage SNR of a receive coil in terms of the flux density it would produce as a transmit coil. The constant factor of $\sqrt{2}$ is included for convenience and can be ignored. Thus, the relative SNR provided by two coils can be evaluated if the resistance of each coil and the transverse flux density produced by 1A into each coil is known. Of course, there are other measures of a coil, depending on the applications. For volume coils in particular, homogeneity of the field produced by the coil is very important.

Many methods are available to model coils and coils systems. In theory the calculation of impedances and flux densities of coils over lossy samples is mildly complex. A highly accurate calculation would need to include the inhomogeneities and physical extent of the sample, the current distribution on the coil in the presence of the load, and the shielding effects on the electric and magnetic fields due to eddy currents induced in the conducting sample. Several approaches are available that can include some or all of these complications. Examples include analytical solutions for canonical geometries, finite element approaches, integral equation solvers, and finite-difference time-domain approaches.

Fortunately, for a very high percentage of clinical applications, the magnetic flux density can be calculated relatively accurately using the Biot–Savart law. Assuming a known current distribution, magnetic flux density $\vec{B}(\vec{r})$ produced by a known current distribution $I(\vec{r})$ can be obtained in a straightforward manner as

$$\vec{B}(\vec{r}) = \frac{\mu}{4\pi} \int_l \frac{I(\vec{r}')\vec{dl} \times \hat{a}_R}{R^2} \tag{16.103}$$

where $\vec{r}'$ is the vector from the integration point on the conductor to the observation point $\vec{r}$, $\vec{dl}$ is the differential path length along the conductor at the point of integration, and $\hat{a}_R$ is a unit vector in the direction of the vector $\vec{R} = \vec{r} - \vec{r}'$, which is the vector from the source point to the observation point. R is the magnitude of the vector $\vec{R}$. The geometry is illustrated in Fig. 16.31. The current distribution $I(\vec{r})$ is generally known to a high degree of accuracy for practical, well-performing coils,

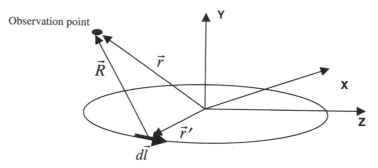

Fig. 16.31   Geometry for calculating the field of a circular loop using the Biot–Savart law.

such as simple loops, planar pairs, and birdcage designs. Once the current begins to substantially diverge from being constant over a given segment, performance of the coil, as well as this simple model, begins to suffer.

Only in a very few cases can Eq. (16.103) be evaluated analytically, such as in the case of the flux on axis of a circular or solenoid, each of which are discussed later. In general, the integral is performed numerically by discretizing the loop into a number of straight segments and summing the flux densities created by each straight segment. Fortunately, many experiments are done with variations of three simple coils — the circular loop, solenoids, and "birdcage" coils. Approximate formulas exist for the fields and impedances of these coils. Although not accurate in all cases, particularly as the coil dimensions approach a significant fraction of a wavelength, these formulas are accurate enough for most applications.

The Circular Loop

The flux density on axis at a distance y from a circular coil of radius a, located in the x–z plane, is given by

$$\vec{B}(0,y,0) = \frac{\mu_0 N I a^2}{2(a^2 + y^2)^{3/2}} \, \hat{a}_y \tag{16.104}$$

where N is the number of turns in the coil, I is the (complex) current, and μ_0, is the permeability of free space, $\mu_0 = 4\pi \times 10^{-7}$ H/m. The inductance of a circular loop of radius a, and with wire diameter d, is given approximately as

$$L = a\mu_0\left[\ln\left(\frac{16a}{d}\right) - 1.75\right] \tag{16.105}$$

This formula can be used to estimate the inductance of any small coil by computing an equivalent radius for a circular coil with the same area.

To estimate the SNR, the coil resistance is required. This can be measured or calculated using one of the methods mentioned previously or other methods such as quasistatic approximations. For purposes of estimating the SNR, the resistance can be estimated from the Q factor of the coil, typically on the order of 100 for a loaded coil. The resistance can be obtained from the definition of Q:

$$Q = \frac{\omega L}{R} \qquad (16.106)$$

yielding $R = \omega L/Q$.

Example Problem 16.9

Estimate the impedance of a circular loop of radius 5 cm, with a wire radius of 2.5 mm and a Q factor of 75.

Solution

The inductance is found from Eq. 16.105 as

$$L - 0.05 \cdot 4\pi \times 10^{-7} \left[\ln\left(\frac{16 \cdot 0.05}{2.5 \times 10^{-3}} \right) \quad 1.75 \right]$$

$$= 0.252 \ \mu H$$

From the value of Q, the resistance can be calculated:

$$R = \frac{2\pi \cdot 63.9 \times 10^6 \cdot 0.252 \times 10^{-6}}{75}$$

$$= 1.35\Omega$$

Thus, the impedance of this coil is

$$Z = R + j\omega L$$

$$= 1.35 + j101\Omega$$

The Solenoid Coil

Another common coil for MRI experiments is the solenoid, illustrated in cross section in Fig. 16.32. The solenoid is along the y axis, and the flux density produced by a solenoid of radius a, length l, with N turns, and carrying a current of I amps, is given by

$$\vec{B}(0,y,0) = \frac{\mu_0 NI}{2l} (\cos\alpha_1 - \cos\alpha_2)\hat{a}_z \qquad (16.107)$$

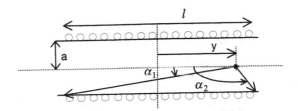

Fig. 16.32 Cross-sectional illustration of a solenoid coil of length l and radius a. The flux density at observation point y is determined by the angles α_1 and α_2.

where the angles α_1 and α_2 are as illustrated in Fig. 16.32. The inductance of this coil is approximated by

$$L = \frac{\mu_0 N^2 \pi a^2}{l^2} [(l^2 + a^2)^{1/2} - a] \qquad (16.108)$$

Just as with the circular loop, the resistance can be computed but is easily obtained from a Q measurement. The Q for the solenoid is also on the order of 100 in most applications.

Birdcage Coils

The solenoid coil is unsuitable for use as a body or head coil on most clinical magnets because the axis of the solenoid must be approximately orthogonal to the static field direction. A solenoid wrapped around the head, for example, creates a field that is aligned with the static field direction in a standard superconducting magnet. Various alternatives have been suggested which create fields in the x–y plane but which can encompass a large sample. The most popular of these volume coils is known as a birdcage coil, illustrated in Fig. 16.33. The coil consists of a set of "legs" or "rungs" connected to a set of end rings. Capacitors are inserted into the legs or, alternatively, between the legs in the end rings to create a resonant structure that supports a sinusoidal current distribution in the azimuthal direction. To first order, the current on the nth leg can be assumed constant, with an amplitude determined by its phi coordinate, ϕ_n:

$$I_n = I_0 \sin(\phi_n) \qquad (16.109)$$

The current on the end rings is obtained by enforcing current continuity. The birdcage coil creates a very homogeneous field and has some very useful properties. For example, it will easily support a second current mode, rotated 90° from the other. Each mode can be independently fed at points 90° apart, allowing the birdcage coil to create circular polarization, which can result in an increase in SNR of approxi-

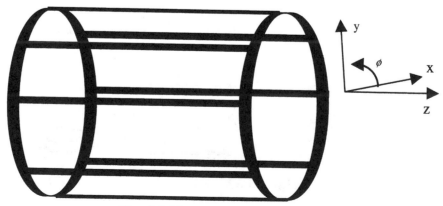

Fig. 16.33   An eight-leg birdcage coil.

mately 40%. Additionally, there are a variety of techniques to resonate the coil to multiple frequencies, allowing simultaneous studies of multiple nuclei.

16.5.3 Magnets, Shims, and Gradient Coils

The selection of a static field magnet for MRI is governed by many factors, including cost, field strength, safety, installation concerns, and image quality. The superconducting magnet has a clear advantage in field strength, capable of creating much larger fields than resistive and permanent magnets. While the higher the field the better the SNR, the relaxation times are also a function of the magnetic field strength. In particular, the increase of T_1 times with magnetic field strength decreases the transverse magnetization obtained for a given TR. For this reason the increase in SNR is not as rapid as would be indicated by the increase in the equilibrium magnetization, M_0. Additionally, an increase in chemical shift effects, RF penetration effects, and potential bioeffects such as RF heating make the choice of filed strength more complicated than would first appear.

Another consideration is the presence of fringe fields from the magnets which affect the performance of nearby electronic equipment in particular. Patients with pacemakers and other electronic implants are generally excluded from areas with fringe fields 5 G, which can be many meters from an unshielded clinical magnet. This problem is further complicated by the fact that the majority of clinical MR scanners are superconducting magnets operated in the persistent mode, in which the field is present at all times. The fringe fields from superconducting magnets can be greatly reduced by including iron shielding or active shielding, in which additional, opposed windings are included outside the main field winds to oppose and reduce the fringe fields created by the magnet. An actively shielded superconducting magnet is shown in Fig. 16.34.

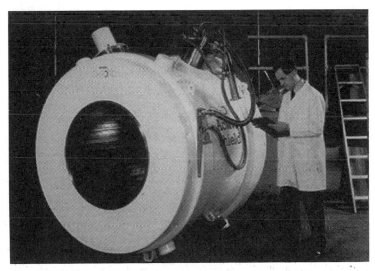

Fig. 16.34 An actively shielded superconducting magnet. Additional windings, opposed to the main field coils, are included within the cryostat to reduce the fringe fields (courtesy of Oxford Magnet Technology, Ltd.).

Field homogeneity and stability are very important concerns. The static field magnets used in MRI must be capable of creating highly homogeneous fields, constant to within at least a few tens of parts per million over the imaging volume. Some experiments, in particular spectroscopy and gradient-echo experiments, can require even more stringent control over the static field homogeneity. The field produced by the static magnet is refined by the use of static field shims, which can be either passive or electronic. Passive shimming involves small mechanical adjustments of the magnet or small pieces of ferrous material to compensate for imperfections in the main magnet. Electronic shimming uses a set of coils, placed outside the gradient coils, each designed to produce a particular field variation. Depending on the number of shim coils present, a very good fit to the magnetic field inhomogeneity can be created with the shims, effectively enabling one to cancel the variations to a very high order.

Operating and purchase costs of the magnets can be a significant factor. Superconducting magnets and permanent magnets are relatively expensive compared to resistive magnets. The operating costs of a superconducting magnet are principally dependent on the cost and availability of cryogens, whereas resistive magnets can entail high energy costs. Permanent magnets have essentially no operating costs.

Recently there has been increased interest in magnets of more open design, prompting renewed interest in resistive and permanent magnets. An open-design resistive magnet, suitable for interventional procedures under MR guidance, is shown in Fig. 16.35.

Fig. 16.35 A resistive magnet designed to provide a large, unrestricted imaging region (courtesy of Oxford Magnet Technology, Ltd.).

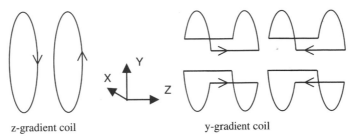

z-gradient coil

y-gradient coil

Fig. 16.36 General design of gradient coils for a cylindrical magnet with a static field oriented along the z axis. The x-gradient coil is obtained by rotating the y-gradient design by 90°.

Most superconducting and resistive magnets can be modeled (to first order) as a set of two or more circular loops on the same axis with a different number or turns, radii, and currents, designed to optimize the magnetic flux density homogeneity over a given distance along the z axis. In the absence of ferrous materials, the flux density created by the loops can be analyzed using the Biot–Savart law.

The gradient coils produce the variations in the z-directed static fields in the x, y, and z directions, $\partial B_z/\partial x$, $\partial B_z/\partial y$, and $\partial B_z/\partial z$. In addition to being able to produce strong gradients on the order of 1 G/cm, they need to be able to switch very rapidly, on the order of 100 μs, to support rapid imaging sequences. Keeping switching times short without dramatically increasing the power requirements, leads to low-inductance designs, minimizing the number of turns used in the coils. In addition to these design factors, the gradients must be linear to avoid image distortion and very stable to prevent phase errors which can lead to image artifacts. This latter issue is primarily a gradient power supply concern. Eddy currents produced in the cryostat and RF shields by the rapidly switching gradients can significantly distort the gradient waveforms, requiring compensation to the waveforms input to the gradient power supplies. Clearly, the design of high-quality gradient systems and power supplies is a very complex task. However, in principal at least, gradient coils for a cylindrical bore can be designed using the Biot–Savart law. In its simplest form, a z-gradient consists of two opposing sets of circular windings, spaced approximately $\sqrt{3}a$, where a is the radius of the coils. A y-gradient coil can be built from two saddle coils, as illustrated in Fig. 16.36. The optimization of the spacing between the coils in the z-gradient set is considered in an exercise at the end of the chapter.

EXERCISES

1. Compute the resonant frequency of ^{31}P, ^{1}H, and ^{2}H at 2.0 and 0.22 T.
2. Calculate the energy level separation in Joules between the two spin states of ^{1}H at 2.0 T.

3. Show that $n^+ - n^- = N\dfrac{1 - e^x}{1 + e^x}$, where $x = -\gamma\hbar B_0/kT$ for a spin 1/2 nucleus.

4. Show that $(n^+ - n^-)/n^+ \approx -\gamma\hbar B_0/kT$ for a spin 1/2 nucleus.

5. Calculate $(n^+ - n^-)/N$ for ^{31}P and ^1H at 2.0 and 0.22 T.

6. Transform the RF field $\vec{B}_1(t) = B_1\cos(\omega t)\hat{a}_x$ to the rotating frame, keeping all terms. Next, determine the effective RF field in the rotating frame using reasonable approximations.

7. Write an expression for M_{xy}^0, the transverse magnetization immediately following an alpha degree RF pulse.

8. Explain the difference between a CPMG and a Hahn spin echo.

9. Consider the following experimental apparatus for NMR: main magnet with $B_0 = 0.5$ T, sample = ^1H, and a 10-turn solenoid coil with length 1 cm and radius 5 mm. The coil has a Q of 50. Excite an object 1 cm long with a radius of 4 mm using a pulse length of 10 μs, a tip angle of 90°, and a receiver bandwidth of 10 kHz. Assume a uniform field is produced by the solenoid coil.

    a. How much power is required?

    b. Estimate the signal-to-noise ratio.

10. Find the Fourier transform of a pulse function defined as

$$\Pi\left(x; \frac{1}{2}\right) = \begin{array}{cc} 1 & -\dfrac{1}{2} \le x \le \dfrac{1}{2} \\ 0 & \text{otherwise} \end{array}$$

11. (a) Given a RF pulse with a bandwidth of 1000 Hz, centered at the Larmor frequency for a 2.0 T field, what slice-select gradient strength is needed in order to excite a 4-mm slice in a homogeneous water phantom? (b) How would you excite a 20-mm slice with the same RF pulse?

12. Draw a standard, spin-echo imaging sequence. Label all pertinent details. Include the RF, gradients, and signal lines.

13. Derive and plot the signal (FID) from a 0.5-cm square phantom in a 8.7 T magnet using a 5 G/cm frequency-encoding gradient (zero phase-encoding gradient). Assume perfect magnetic field homogeneity, and assume that the square phantom is aligned along the gradient. Scale the plot so the FID has a maximum value of 1.0.

14. What sample time, acquisition window (total time to digitize the echo), and phase-encoding time would be needed to make a 256 × 256 image with a FOV of 5 × 5 mm using maximum gradient amplitudes of 20 G/cm.

15. Given the following data obtained from two regions (pixels) of three images of the same slice made with differing values of TE and TR, estimate T_1 and T_2 for each pixel.

Image No.	TE	TR	S_1 (signal in pixel 1)	S_2 (signal in pixel 2)
1	35	600	430	281
2	70	600	251	195
3	35	1300	543	448

16. (a) Plot the z-directed magnetic field of a solenoid 5 cm in diameter by 20 cm long from -20 to 20 cm along the axis of the solenoid. (b) Find the distance from the center over which the field has a total variation of no more than ± 50 parts per million (ppm).

17. Consider a magnet made from two concentric loops of wire in the x–y plane, located at $\pm d$ on the z axis. Assume 1 A of current in each, flowing in the same direction. The coils are 1 m in diameter.

    a. Find the separation that gives the largest homogeneous region, defined as the largest region in which the magnetic field varies by no more than ± 50 ppm from a nominal value.

    b. Plot the field along the axis of the magnet, showing the "homogeneous" region. Label and specify the homogeneous region of the magnet.

SUGGESTED READING

Abragham, A. (1983). *Principles of Nuclear Magnetism.* Oxford Univ Press, New York.

Callaghan, P. T. (1993). *Principles of Nuclear Magnetic Resonance Spectroscopy.* Oxford Univ. Press, New York.

Canet, D. (1996). *Nuclear Magnetic Resonance Concepts and Methods.* Wiley, New York.

Cho, Z. H., Jones, J. P., and Singh, M. (1993). *Foundations of Medical Imaging.* Wiley, New York.

Farrar, T. C. (1989). *Introduction to Pulse NMR Spectroscopy.* Farrugut, Madison, WI.

Fukushima, E., and Roeder, S. B. W. (1981). *Experimental Pulse NMR: A Nuts and Bolts Approach.* Addison-Wesley, Reading, MA.

Harris, R. K. (1986). *Nuclear Magnetic Resonance Spectroscopy.* Longman, Avon, UK.

Hinshaw, W. S., and Lent, A. H. (1983, March). An introduction to NMR imaging: From the Bloch equation to the imaging equation. *Proc. IEEE* 71(3), 338–350.

Lauterbur, P. C. (1973). Image formation by induced local interactions: Examples employing nuclear magnetic resonance. *Nature,* **242,** 190–191.

Liang, Z. P., and Lauterbur, P. C. *Principles of Magnetic Resonance Imaging: A Signal Processing Approach,* in press.

Mansfield, P., and Morris, P. G. (1982). *NMR Imaging in Biomedicine.* Academic Press, New York.

Matwiyoff, N. A. (1990). *Magnetic Resonance Workbook.* Raven Press, New York.

Partain, C. L., Price, R. R., Patton, J. A., Kulkarni, M. V., and James, A. E., Jr. (1988). *Magnetic Resonance Imaging,* 2nd ed., Saunders, Philadelphia.

Pykett, I. L. (1982, May), NMR imaging in medicine. *Sci. Am.* 78–88.

Shellock, F. G., and Kanal, E. (1994). *Magnetic Resonance Bioffects, Safety, and Patient Management.* Raven Press, New York.

17 BIOMEDICAL OPTICS AND LASERS

Chapter Contents

At the conclusion of this chapter, the reader will be able to:

- Understand essential optical principles and fundamentals of light propagation in tissue as well as other biological- and biochemical-based media

- Utilize basic engineering principles for developing therapeutic, diagnostic, sensing, and imaging devices with a focus on the use of lasers and optical technology

- Describe the biochemical and biophysical interactions of optic and fiber-optic systems with biological tissue

- Understand the photothermal interactions of laser systems with biological tissue

17.1   INTRODUCTION

The use of light for therapeutic and diagnostic procedures in medicine has evolved from illumination of sunlight for heat therapy and a simple inspection of eyes, skin, and wounds for diagnosis to the current use of lasers and endoscopy in medical procedures. Indeed, the introduction of photonic technology into medicine during the past decade has revolutionized many procedures, and because of its low cost, relative to many other technologies, photonic technology has the potential to greatly impact health care. For example, coherent fiber-optic bundles have been applied in laproscopy cholecystectomy (minimally invasive removal of the gallbladder) to transform a once painful and expensive surgery into virtually an outpatient procedure. Biomedical engineers have been instrumental in defining and demonstrating the engineering fundamentals of the interaction of light and heat with biological media resulting in advances in dermatology, ophthalmology, cardiology, and urology.

Optical engineering has traditionally been taught as part of an electrical engineering curriculum and was primarily applied in defense technology. In biomedical optics applications, however, there are many issues related to the interaction of light with participating biological media that classical optics books and curricula do not cover. Therefore, the goal of this chapter is to provide students with a strong foundation in the growing field of biomedical optics based on the assimilation of fundamental principles for the understanding and design of optically based therapeutic, diagnostic, and monitoring devices.

17.2 ESSENTIAL OPTICAL PRINCIPLES

Throughout the ages great minds, including those of Galileo, Newton, Huygens, Fresnel, Maxwell, Planck, and Einstein, have studied the nature of light. Initially, it was believed that light was either corpuscular or particle in nature or it behaved as waves. After much debate and research, however, it has been concluded that there are circumstances in which light behaves like waves and those in which it behaves like particles. In this section, an overview of the essential principles of the behavior of light is described beginning with the wave equation, which is commonly derived in many introductory optics texts from Maxwell's equations. In addition, the optical spectrum and the fundamental governing equations for light absorption, scattering, and polarization are discussed.

17.2.1 Electromagnetic Waves and the Optical Spectrum

It was in the late 1800s that J. Clerk Maxwell showed conclusively that light waves were electromagnetic in nature. This was accomplished by expressing the basic laws of electromagnetism so as to allow the derivation of the wave equation. The key to validating this derivation was that the free space solutions to the wave equation corresponded to electromagnetic waves with a velocity equal to the known experimental value of the velocity of light. Many introductory optics texts, such as the work of Hecht (1987) and Pedrotti and Pedrotti (1987), begin with the vectorial form for Maxwell's equations to derive the wave equation, namely,

$$\nabla^2 \mathbf{E} = \varepsilon_0 \mu_0 \partial^2 \mathbf{E} / \partial t^2 \tag{17.1}$$

where $\nabla^2 \mathbf{E}$ is the partial second derivative of the electric field with respect to position ($\partial^2 \mathbf{E}/\partial x^2 + \partial^2 \mathbf{E}/\partial y^2 + \partial^2 \mathbf{E}/\partial z^2$), ε is the electric permittivity of the medium, and μ is the magnetic permeability of the medium. In free space $\mu = \mu_0$ and $\varepsilon = \varepsilon_0$. From symmetry, there is a similar solution for $\mathbf{H}$, the magnetic field, and, as mentioned previously, the wave velocity is equal to the known value of the velocity of light, namely,

$$c_0 = 1/(\varepsilon_0 \mu_0)^{1/2} \tag{17.2}$$

Therefore, the change in the electric or magnetic field in space is related to the velocity and the change with respect to time. Given Cartesian axes O_x, O_y, O_z shown in Fig. 17.1, a typical solution to this wave equation is the sinusoidal solution given by

$$E_x = E_0 \exp[j(\omega t - kz)] \tag{17.3}$$

$$H_y = H_0 \exp[j(\omega t - kz)] \tag{17.4}$$

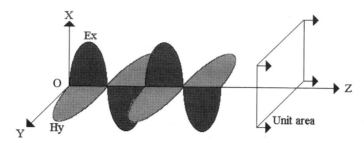

Fig. 17.1   Sinusoidal electromagnetic wave.

which states that the electric field (e-field) oscillates sinusoidally in the x–z plane, the magnetic field oscillates in the y–z plane (orthogonal to the e-field and in phase), and the wave propagates in the O_z direction.

The frequency, f, and wavelength, λ, of the wave are related to the velocity, c, and given by

$$f = \omega/2\pi \tag{17.5}$$

$$\lambda = 2\pi/k \tag{17.6}$$

$$c = f\lambda = \omega/k \tag{17.7}$$

The ratio of the velocity of the wave in free space, c_0, to that in the medium through which the light propagates, c, is known as the index of refraction of the medium, $n = c_0/c$.

The intensity or power per unit area of the wave can be expressed in terms of the Poynting vector, Π, defined in terms of the vector product of the electric and magnetic field as

$$\Pi = \mathbf{E} \times \mathbf{H} \tag{17.8}$$

The intensity of the wave will be the average value of the Poynting vector over one period of the wave; thus, if E and H are spatially orthogonal and in phase as shown in Fig. 17.1, then

$$I = <|\Pi|> = \text{power/unit area} = c\varepsilon E^2 \tag{17.9}$$

where $<\Pi>$ represents the average value of Π and it is clear that the intensity is proportional to the square of the electric field. This is true for the propagation of a wave through an isotropic dielectric medium and clearly from Maxwell's equations the intensity is also proportional to the square of the magnetic field. If the electric and magnetic field were in phase quadrature (90° out of phase) then the intensity would be zero since the average value of a sine wave times a cosine wave is zero:

$$I = <|\Pi|> = <E_0 \cos(\omega t)> = 0 \tag{17.10}$$

In addition to intensity, the light can be characterized by the wavelength or electromagnetic spectrum which ranges from low-frequency radio waves at 10^3 Hz to gamma radiation at 10^{20} Hz. As depicted in Fig. 17.2, the wavelength region of light

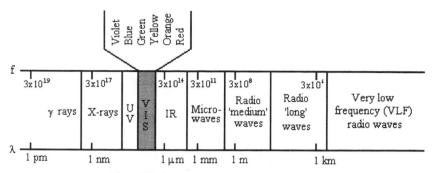

Fig. 17.2 The electromagnetic spectrum.

can be divided into ultraviolet, which is 0.003–0.4 μm, the visible region from 0.4 to 0.7 μm (7.5×10^{14} to 4.3×10^{14} Hz frequency range), the near-infrared region from 0.7 to 2.5 μm, the mid-infrared region from 2.5 to 12 μm, and the far infrared beyond 12 μm.

17.2.2 Optical Polarization

In the previous section the sinusoidal solution in terms of F_x and H_y is only one of an infinite number of such sinusoidal solutions to Maxwell's equations. The general solution for a sinusoid with angular frequency ω can be written as

$$\mathbf{E}(\mathbf{r}, t) = \mathbf{E}(\mathbf{r}) \exp(j\omega t) \qquad (17.11)$$

in which $\mathbf{E}(\mathbf{r}, t)$ and $\mathbf{E}(\mathbf{r})$ are complex vectors and $\mathbf{r}$ is a real radius vector.

For simplicity, consider the plane monochromatic (single wavelength) waves propagating in free space in the O_z direction; then the general solution to the wave equation for the electric field can be written:

$$E_x = e_x \cos(\omega t - kz + \delta_x) \qquad (17.12)$$

$$E_y = e_y \cos(\omega t - kz + \delta_y) \qquad (17.13)$$

in which δ_x and δ_y are arbitrary phase angles. Thus, this solution can be described completely by means of two waves (e-field in the x–z plane and e-field in the y–z plane). If these waves are observed at a particular value of z (e.g., z_0) they take the oscillatory form

$$E_x = e_x \cos(\omega t + \delta'_x) \qquad \delta'_x = \delta_x - kz_0 \qquad (17.14)$$

$$E_y = e_y \cos(\omega t + \delta'_y) \qquad \delta'_y = \delta_y - kz_0 \qquad (17.15)$$

and the top of each vector appears to oscillate sinusoidally with time along a line. E_x is said to be linearly polarized in the direction O_x and E_y is said to be linearly polarized in the direction O_y.

It can be seen from Eq. (17.15) that if the light is fully polarized, it can be completely characterized as a 2×1 matrix in terms of its amplitude and phase

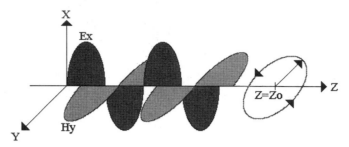

Fig. 17.3 Polarization propagation of an **E** wave.

[$e_x \exp(j\delta x')$ and $e_y \exp(j\delta y')$]. This vectoral representation is known as the Jones vector. When the optical system design includes the propagation of the light through nonscattering, and hence nondepolarizing, medium such as a lens or clear biological sample, the medium can be completely characterized by a 2 × 2 Jones matrix. Therefore, the output of the propagation of the polarized light can be modeled as a multiplication of the Jones matrix of the optical system and the input light vector. A system that yields both polarized and partially polarized light such as that obtained from tissue scattering can be characterized using 4 × 4 matrices known as Mueller matrices and a 4 × 1 matrix known as the Stokes vector.

Regarding Eq. (17.15), the tip of the polarized light vector is the vector sum of E_x and E_y which, in general, is an ellipse as depicted in Fig. 17.3 whose Cartesian equation in the x–y plane at $z = z_0$ is

$$E_x^2/e_x^2 + E_y^2/e_y^2 - 2(E_x E_y/e_x e_y) \cos \delta = \sin {}^2\delta \qquad (17.16)$$

in which $\delta = \delta_y' - \delta_x'$. It should be noted that the ellipse folds into a straight line producing linear light if (i) $E_x \neq 0$, and $E_y = 0$, (ii) $E_y \neq 0$ and $E_x = 0$, or (iii) $\delta = m\pi$, in which m is a positive or negative integer. The light becomes circular if $e_x = e_y$ and $\delta = (2m + 1) \pi/2$ since E_x and E_y would then have equal amplitudes and be in phase quadrature.

The polarization properties of light become particularly important for anisotropic media in which the physical properties are dependent on direction (i.e., E_x is different from E_y and thus values e_x, e_y, and δ will vary along the propagation path as depicted with a birefringent crystal or polarization preserving optical fiber).

Example Problem 17.1

Consider a general solution to the wave equation represented as a plane electromagnetic wave (in SI units) given by the expression

$$E_x = 0 \qquad E_z = 0 \qquad E_y = 2 \cos[2\pi \times 10^{14}(t - x/c) + \pi/2]$$

For the equations shown, determine the frequency (f), wavelength (λ), direction of motion (x, y, or z,) amplitude (A), initial phase angle (ϕ) and polarization of the wave (linear, circular, or elliptical).

Solution

The general form of a plane electromagnetic wave can be expressed as $E = A \cos(kx \pm \omega t + \phi)$, where $k = 2\pi/\lambda$, $\lambda = c/f$, $\omega = 2\pi f$, and $c = 3 \times 10^8 \text{m/s}$. The previous equation can therefore be written in general form as $E_y = 2 \cos[2\pi \times 10^{14} t - (2\pi \times 10^{14} x/c)] + \pi/2$. Thus, comparing the two equations yields the following values: $f = 1 \times 10^{14}$ Hz, $\lambda = 3 \times 10^{-6} = 3$ μm, motion is in the positive x direction, $A = 2$ V/m, $\phi = \pi/2$, and since $E_x = E_z = 0$ the wave must be linearly polarized in the y direction. ∎

17.2.3 Absorption, Scattering, and Luminescence

Other ways in which the light can change as it passes through biological media are now discussed. The optical properties of the light are defined in terms of the absorption, scattering, and anisotropy of the light. Knowing the absorption, scattering, and direction of propagation of the light has applications for both optical diagnostics and sensing as well as optical therapeutics. For instance, many investigators are currently trying to quantifiably measure certain body chemicals such as glucose noninvasively using the absorption of light. In addition, by knowing the absorption and scattering of the light, investigators are modeling the amount of coagulation or "cooking" of the tissue that is expected for a given wavelength of light, such as in the case of prostate surgery.

The optical penetration depth or volume of tissue affected by the light is a function of both the light absorption and scattering. If it is assumed that scattering is negligible or absorption is the dominant component, then the change in light intensity is determined by the number of absorbing species in the optical path and the illumination wavelength. For instance, for wavelengths above 1.5 μm the light propagation through all tissues, since they have a large water content, is absorption dominant. The intensity relationship as a function of optical path and concentration can be described by the Beer–Lambert law as

$$I(z) = I_0 \exp(-\mu_a z) \tag{17.17}$$

where I is the transmitted light intensity, I_0 is the incident light intensity, z is the path length of the light, and μ_a is the absorption coefficient. The absorption coefficient can be further broken down into two terms which include the concentration of the chemically absorbing species times the molar absorptivity, which is a function of wavelength.

Luminescence, known as fluorescence or phosphorescence depending on the fluorescent lifetime, is a process that occurs when photons of electromagnetic radiation are absorbed by molecules raising them to some excited state and then, on returning to a lower energy state, the molecules emit radiation or luminesce. Fluorescence does not involve a change in the electron spin and therefore occurs much faster. The energy absorbed in the luminescence process can be described in discrete photons as follows:

$$E = hc/\lambda \tag{17.18}$$

in which E is the energy, h is Planck's constant (6.626×10^{-34} J s), c is the speed of light, and λ is the wavelength. This absorption and reemission process is not

100% efficient since some of the originally absorbed energy is dissipated before photon emission. Therefore, the excitation energy is greater than that emitted:

$$E_{\text{excited}} > E_{\text{emitted}} \qquad (17.19)$$

and

$$\lambda_{\text{emitted}} > \lambda_{\text{excited}} \qquad (17.20)$$

The energy level emitted by any fluorochrome is achieved as the molecule emits a photon, and due to the discrete energy level individual fluorochromes excited by a given wavelength typically only fluoresce at a certain emission wavelength. Originally, fluorescence was used as an extremely sensitive, relatively low-cost technique for sensing dilute solutions. Only recently have investigators used fluorescence in turbid media such as tissue for diagnostics and sensing.

It is well known that tissues scatter as well as absorb light, particularly in the visible and near-infrared wavelength regions. Scattering, unlike absorption and luminescence, need not involve a transition in energy between quantized energy levels in atoms or molecules but rather is typically a result of random spatial variations in the dielectric constant. Therefore, the actual light distributions can be substantially different from distributions estimated using Beer's law. In fact, light scattered from a collimated beam undergoes multiple scattering events as it propagates through tissue. The transport equation that describes the transfer of energy through a turbid medium (a medium that absorbs and scatters light) is an approach which has proven to be effective. The transport theory has been used to describe scattering, absorption, and fluorescence, but until recently polarization was not included in part since relatively few scattering events are required to completely randomize the polarization of the light beam. The transport theory approach will be described later and it is a heuristic theory based on a statistical approximation of photon particle transport in a multiple-scattering medium.

17.3 FUNDAMENTALS OF LIGHT PROPAGATION IN BIOLOGICAL TISSUE

In this section the propagation of light through media such as tissue is discussed beginning with a simple ray optics approach for light traveling through a nonparticipating media or rather the effect of the absorption and scatter within the media is ignored. The addition of the effects of absorption and scatter on light propagation is then discussed along with the effect of boundary conditions and means of measuring the optical absorption and scattering properties.

17.3.1 Light Interactions with Nonparticipating Media

Previously, light was described as a set of electromagnetic waves traveling in the O_z direction. In this section the light will be treated as a set of "rays" traveling in straight lines that, when combined, make up the plane waves described using the complex exponentials discussed previously. Given this treatment of the light, the effect on the light at the boundaries between two different optical media will be investigated. In the treatment of geometrical or ray optics of the light it is first assumed that the incident, reflected, and refracted rays all lie in the same plane of incidence. The propagation at the boundary between these two interfaces as shown in Fig. 17.4 is based on two basic laws: The angle of reflection equals the angle of incidence and the sine of the angle of refraction bears a constant ratio to the sine of the angle of incidence (Snell's law). It should be noted that ray propagation can be very useful for a majority of the applications considered with the propagation of the light through bulk optics such as in the design of lenses and prisms. It has severe limitations, however, in that it cannot be used to predict the intensities of the refracted and reflected rays and it does not incorporate the effects of scatter and absorption. In addition, when apertures are used which are smaller than the bulk combined rays a wave process known as diffraction occurs which causes the geometrical theory to break down and the beams to diverge.

Wave theory is used to describe the phenomena of reflection and refraction in an effort to determine the intensity of the light as it propagates from one medium to another. If two nonconducting media are considered, as in Fig. 17.4, the boundary conditions follow from Maxwell's equations such that the tangential components of $\mathbf{E}$ and $\mathbf{H}$ are continuous across the boundary $\mathbf{E}_i + \mathbf{E}_p = \mathbf{E}_t$ and the normal components of $\mathbf{B}$ (the magnetic flux density) and $\mathbf{D}$ (the electric displacement) are continuous across the boundary. These conditions can only be true at all times and all places on the boundary if the frequencies of all waves are the same. The relative amplitudes of the waves will now be considered. Note that the value of $\mathbf{E}$, $\mathbf{H}$, $\mathbf{D}$, and $\mathbf{B}$ will depend on the direction of vibration of $\mathbf{E}$ and $\mathbf{H}$ fields of the incident wave relative to the planes of incidence. In other words, they depend on the polarization of the wave. The wave can be divided into two orthogonal polarization states and using Maxwell's equations, Snell's law, and the law of reflection as well as imposing the boundary conditions, the following Fresnel equa-

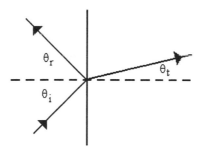

Fig. 17.4 Ray propagation of light at the boundary of two interfaces. Clearly the angle of incidence equals the angle of reflection ($\theta_i = \theta_r$) and from Snell's law the sine of the angle of refraction bears a constant ratio to the sine of the angle of incidence [$n_i \sin(\theta_i) = n_t \sin(\theta_t)$].

tions can be derived. The reflected and transmitted equations in the parallel configuration are

$$r_p = \left(\frac{E_r}{E_i}\right)_p = (-n_1 \cos \theta_t + n_2 \cos \theta_i)/(n_1 \cos \theta_t + n_2 \cos \theta_i) \qquad (17.21)$$

$$t_p = \left(\frac{E_t}{E_i}\right)_p = (2n_1 \cos \theta_i)/(n_1 \cos \theta_t + n_2 \cos \theta_i) \qquad (17.22)$$

The normal or perpendicular reflected and transmitted equations are

$$r_\perp = \left(\frac{E_r}{E_i}\right)_\perp = (n_1 \cos \theta_i - n_2 \cos \theta_t)/(n_1 \cos \theta_i + n_2 \cos \theta_t) \qquad (17.23)$$

$$t_\perp = \left(\frac{E_t}{E_i}\right)_\perp = (2n_1 \cos \theta_i)/(n_1 \cos \theta_i + n_2 \cos \theta_t) \qquad (17.24)$$

Using the previous equations, there are two noteworthy limiting cases. The
first case is as follows: If $n_2 > n_1$ then $\theta_i > \theta_t$, $r_\perp$ is always negative for all values
of θ_I, and r_p starts out positive at $\theta_i = 0$ and decreases until it equals 0 when $\theta_i +
\theta_t = 90°$. This implies that the refracted and reflected rays are normal to each other
and from Snell's law this occurs when $\tan \theta_i = n_2/n_1$. This particular value of θ_i
is known as the polarization angle or, more commonly, the Brewster angle. At
this angle only the polarization with E normal to the plane of incidence is reflected
and as a result this is a useful way to polarize a wave. As θ_i increases beyond θ_p,
r_p becomes progressively negative, reaching -1.0 at $90°$ which implies the surface performs as a perfect mirror at this angle. On the other hand, at normal
incidence $\theta_i = \theta_r = \theta_t = 0$, then $t_p = t_\perp = 2n_1/(n_1 + n_2)$ and $r_p = -r_\perp = (n_2 -
n_1)/(n_1 + n_2) = (n_1 - n_2)/(n_1 + n_2)$. The wave intensity, which can actually be mea-
sured by a detector, is defined proportional to the square of the electric field. Thus,
$(I_r/I_i) = (r_p)^2 = (E_r/E_i)^2 = (n_2 - n_1)^2/(n_1 + n_2)^2$ and $(I_t/I_i) = (t_p)^2 = (E_t/E_i)^2 =
4n_1^2/(n_1 + n_2)^2$. The previous expressions are useful since they represent the
amount of light lost by normal specular reflection when transmitting from one
medium to another.

The second limiting case is if $n_2 < n_1$, then $\theta_t > \theta_i$, $r_\perp$ is always positive for
values of θ_i, and $r_\perp$ increases from its initial value at $\theta_i = 0$ until it reaches 1.0 at
the critical angle θ_c. At this angle $\theta_i = \theta_c$, then $\theta_t = 90°$ and from Snell's law $\sin
\theta_t = n_1/n_2 \sin \theta_i$. Clearly, for $n_1 > n_2$ (less dense to more dense medium) $\sin\theta_t$ could
be greater than 1 according to the previous equation. This is not possible for any
real value of θ_t and thus at the angle for which $\theta_i = \theta_c$, $(n_1/n_2)\sin \theta_c = 1$ or rather
$\theta_t = 90°$. Thus, for all values of $\theta_i > \theta_c$ the light is totally reflected at the boundary.
This phenomenon is used for the creation of optical fibers in which the core of the
fiber (where the light is supposed to propagate) has an index of refraction slightly
higher than the cladding surrounding it so that the light launched into the fiber will
totally internally reflect, allowing for minimally attenuated propagation down the
fiber.

17.3.2 Light Interactions with Participating Media: The Role of Absorption and Scattering

When light is incident on tissue either from a laser or from other optical devices, tissue acts as a participating medium by reflecting, absorbing, scattering, and transmitting various portions of the incident wave of radiation. Ideally, an electromagnetic analysis of the light distribution in the tissue would be performed. Unfortunately, this can be quite cumbersome and a reliable database of electrical properties of biological tissues would be required. An alternate practical approach to the problem is to use transport theory, which starts with the construction of the differential equation for propagation of the light intensity. In the following section the equation for light intensity distribution in a purely absorbing medium will be derived. With this motivation, later we will introduce the general equation of transfer for a medium that scatters and absorbs the light.

The Case of Pure Absorption

In order to describe the distribution and transport of laser energy in a nontransparent participating medium, the medium can be viewed as having two coexisting "phases": a material phase for all the masses of the system and a photon phase for the electromagnetic radiation. Figure 17.5 shows the material phase as circles and the photon phase as curved arrows which strike the material phase. The energy balance equation for the material phase is introduced and discussed in the thermodynamic descriptions in Sections 17.4 and 17.6. The energy balance equation for the photon phase is discussed here.

Consider an infinitesimal volume of the material under irradiation (Fig. 17.6). The rate of change of radiative energy, $U^{(rad)}$, with time is the difference between the incoming and outgoing radiative fluxes in the element minus the rate of energy ab-

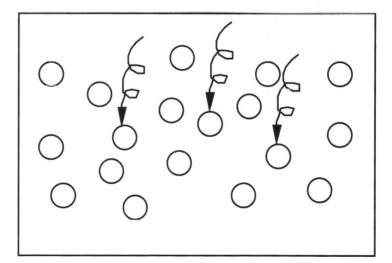

Fig. 17.5 Two coexisting phases — photon phase and material phase.

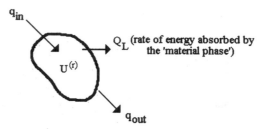

Fig. 17.6 Control volume of the photon phase with radiative fluxes in and out, radiative energy, and rate of energy lost to the material phase by absorption.

sorption by the material phase. The difference between the incoming and outgoing fluxes is, in the limit, the negative of the divergence of the radiative flux. Therefore, denoting the energy absorbed by the material phase, which is the laser heat source term, by Q_L and the radiative flux by $\vec{q}^{(rad)}$, the governing differential equation for energy rate balance in the photon phase can be written as

$$\partial U^{(rad)}/\partial t = -\boldsymbol{\nabla} \cdot \mathbf{q}^{(rad)} - Q_L \tag{17.25}$$

Any possible photon emission or scattering by the material phase has been ignored.

For a steady-state system in which radiation is collimated and monochromatic, laser light travels only in the positive z direction, assuming only one forward flux, Eq. (17.25) reduces to

$$dq_z^{(rad)}/dz = -Q_L \tag{17.26}$$

Considering the speed of light and the dimensions of a biological tissue, the assumption of steady-state condition is a reasonable asumption for most applications except when very fast light sources are used and/or when time-resolved analyses are considered. The important step at this point is to use the phenomenological relation

$$Q_L = \mu_a I \tag{17.27}$$

where μ_a is the absorption coefficient and I is the total light intensity. The total intensity at a point is the sum of radiative fluxes received at that point. In the case presented here, using purely absorbing tissue, a single radiative flux is used; thus,

$$I = q_z^{(rad)} \tag{17.28}$$

Combining Eqs. (17.27) and (17.28) into Eq. (17.26), the following differential equation is obtained:

$$\frac{dI}{dz} = -\mu_a I \tag{17.29}$$

which has the simple solution

$$I(z) = I_0 \exp(-\mu_a z) \tag{17.30}$$

where I_0 denotes the intensity at the surface, $z = 0$. This equation is the well-known Beer–Lambert law of absorption for a purely absorbing medium.

The laser heat source term can now be written (Eqs. 17.27 and 17.30) as

$$Q_L = \mu_a I_0 \exp(-\mu_a z) \tag{17.31}$$

This equation can be generalized to an axisymmetric three-dimensional case to include the effect of radial beam profile of a laser light incident orthogonally on a slab by writing it as

$$Q_L(r,z) = \mu_a I_0 \exp(-\mu_a z) f(r) \tag{17.32}$$

where $f(r)$ is the radial profile of an axisymmetric laser beam. For a Gaussian beam profile, which is a common mode of laser irradiation,

$$f(r) = \exp\left(-\frac{2r^2}{\omega_0^2}\right) \tag{17.33}$$

where ω_0 is known as the $1/e^2$ radius of the beam since at $r = \omega_0$, $f(r) = 1/e^2$.

Example Problem 17.2

For a Gaussian laser beam irradiating at a wavelength of 2.1 μm the absorption coefficient is estimated to be 25 cm^{-1} and scattering is negligible. The radial profile of light intensity and the rate of heat generation at the tissue surface, $z = 0$, and its axial profile along the center axis of the beam, $r = 0$, can be found using the previous equations. Graph $I(r, 0)$ for r ranging from $-2\omega_0$ to $2\omega_0$ and $I(0, z)$ for $z = 0$ to $z = 5/\mu_a$ as well as graph of $Q_L(r, 0)$ and $Q_L(0, z)$. For simplicity, assume $I_0 = 1$ W/cm^2.

Solution

Plot of the radial and axial profile of light intensity, I, and volumetric rate of absorption, Q_L, in a purely absorbing tissue.

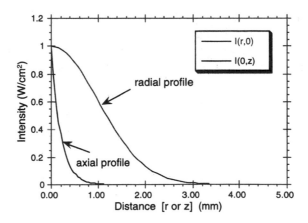

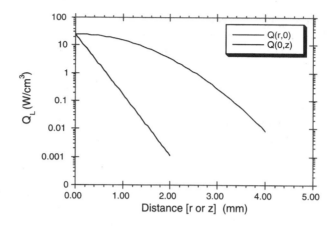

Methods for Scattering and Absorbing Media

The fundamental quantity in the transport theory approach is the total specific intensity, L, which is a function of position $\vec{r}$ for light in the direction given by a unit vector $\hat{s}$ and its units are W cm^{-2} s^{-1}. The equation of transfer satisfied by L can be written as

$$\frac{dL(\mathbf{r}, \hat{s})}{ds} = -(\mu_a + \mu_s)L(\mathbf{r}, \hat{s}) + \mu_s \int_{4\pi} p(\hat{s}, \hat{s}')L(\mathbf{r}, \hat{s}')d\omega' + S(\mathbf{r}, \hat{s}) \qquad (17.34)$$

This equation shows the decrease in L due to scattering (μ_s), and absorption (μ_a) and the increase in L due to scattering from L coming from another direction($\hat{s}'$) (Fig. 17.7). The sum of scattering and absorption, $\mu_t = \mu_a + \mu_s$, is defined as the attenuation coefficient. The function $p(\hat{s}, \hat{s}')$ is called the "phase function" and is related to the scattering amplitude of a particle, a scaled form of the probability distribution of scattering angles. Note that by ignoring scattering and assuming a collimated light source Eq. (17.34) reduces to Eq. (17.29). $S(\mathbf{r}, \hat{s})$ is the source term which could be irradiation on the surface, fluorescence generated inside the tissue, or an internal volumetric light source.

The equation of transfer is an integrodifferential equation for which a general solution does not exist. However, several approximate solutions have been found, such as the two-flux and multiflux models, the discrete ordinate finite element method, the spherical harmonic method, the diffusion approximation, and the Monte Carlo method. Each of these is subject to certain limitations and assump-

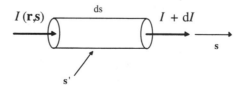

Fig. 17.7 Light intensity change in the transport theory approach including scattered light from other directions.

tions. In this section the focus will be on the diffusion approximation, which is one of the more widely used solutions.

Diffusion Approximation

The diffusion approximation is a second-order differential equation that can be derived from the radiative transfer equation (Eq. 17.34) under the assumption that the scattering is "large" compared with absorption. The solution to this equation provides a useful and powerful tool for the analysis of light distribution in turbid media. The governing differential equation for the diffusion approximation is

$$\nabla^2 \phi_d - 3\mu_a \mu_{t'} \phi_d = -\frac{\mu_{s'}}{D} \phi_c \tag{17.35}$$

where ϕ_d is the diffuse fluence rate and the parameters of the equation are $\mu_{s'} = \mu_s (1 - g)$, $\mu_{t'} = \mu_a + \mu_{s'}$, and $D = 1/3 \ \mu_{t'}$, in which g is defined as the anisotropy of the medium.

The total light fluence rate, ϕ (W/cm^2), is the sum of the collimated part, ϕ_c, and the diffuse part, ϕ_d. The total fluence rate as given by the following equation is a key parameter in laser–tissue interaction. It can be thought of as the total light received at a point in space, or within the medium, through a small sphere divided by the area of that sphere:

$$\phi(r, z) = \phi_c (r, z) + \phi_d (r, z) \tag{17.36}$$

The collimated fluence rate is given by

$$\phi_c = I_0(r)(1 - r_{sp}) \exp(-\mu_{t'} z) \tag{17.37}$$

where I_0 is the surface flux density (W/cm^2) of the incident beam and r_{sp} is the specular reflectance. The optical boundary condition at the beam axis ($r = 0$) is

$$\frac{\partial \phi_d}{\partial r}\bigg|_{r = 0} = 0 \tag{17.38}$$

The boundary condition elsewhere is

$$\phi_d - 2AD\nabla\phi_d \cdot \tilde{n} = 0 \tag{17.39}$$

where A is the internal reflectance factor and $\tilde{n}$ is the inward unit normal vector. The internal reflectance factor A can account for the effect of mismatch in the index of refraction between the boundary and the surrounding medium and is given by

$$A = \frac{1 + r_i}{1 - r_i} \tag{17.40}$$

where r_i is evaluated by an empirical formula

$$r_i = -1.440 \ n_{rel}^{-2} + 0.710 \ n_{rel}^{-1} + 0.688 + 0.0636 \ n_{rel} \tag{17.41}$$

and n_{rel} is the ratio of the refractive indices of the tissue and the medium. The internal reflectance factor A reduces to 1 in the case in which the boundary is matched, i.e., when $n_{rel} = 1$.

Equations (17.35)–(17.41) provide the governing differential equations and boundary condition for the diffusion approximation. A solution of these either by analytical or numerical methods allows the calculation and analysis of fluence rate within a scattering and absorbing media such as biological tissues.

For example, for an isotropic point source inside an infinite medium the solution for fluence rate as measured by a detector imbedded inside the medium at a large distance r from the fiber can be derived using the Green's functions solution of the previous equations as follows:

$$\phi(r) = \frac{\phi}{4\pi D} \frac{e^{-r/\delta}}{r} \tag{17.42}$$

where $\delta = \sqrt{D/\mu_a}$ is the penetration depth.

Example Problem 17.3

It is desired to measure the concentration of an absorber in a scattering medium with known scattering coefficient. If the reduced scattering coefficient $\mu_{s'}$ is known and the relative intensity at a distance r_0 from an isotropic source can be measured, an algebraic equation may be solved for the absorption coefficient based on the diffusion approximation given that r_0 is large enough for diffusion approximation to be valid.

Solution

Given scattering coefficient $\mu_{s'}$, r_0, and $\phi(r_0)/\phi_0$, an algebraic equation for μ_a based on the diffusion approximation is required:

$$\phi(r) = \phi_0 \exp(-r/\delta)/(4\pi D r)$$

$$\log[\phi(r)/\phi_0] = -r/\delta - [\log 4 + \log \pi + \log D + \log r]$$

$$D = 1/3\mu_{t'} \tag{ex. 1}$$

$$\delta = \sqrt{D/\mu_a}$$

The left-hand side of Eq. (ex. 1) is a known value, e.g., k_1,

$$\therefore \ k_1 = -r/\delta - [\log 4 + \log \pi + \log D + \log r]$$

$$\therefore \ k_1 + \log 4 + \log \pi + \log r = -r/\delta - \log D \tag{ex. 2}$$

The left-hand side of Eq. (ex. 2) is again a known value, e.g., k_2,

$$\therefore \ k_2 = -r/(\sqrt{D/\mu_a}) - \log D$$

$$\therefore \ k_2 = -r(\sqrt{\mu_a/D}) - \log D$$

$$\therefore \ k_2 = -r(\sqrt{\mu_a/1/(3\mu_{t'})}) - \log(1/3\mu_{t'}) \tag{ex. 3}$$

$$\therefore \ k_2 = -r(\sqrt{3\mu_a \mu_{t'}}) - \log(1/3\mu_{t'})$$

$$\therefore \ k_2 = -r(\sqrt{3\mu_a(\mu_a + \mu_{s'})}) - \log(1/3(\mu_a + \mu_{s'}))$$

Solving Eq. (ex. 3), the value of the coefficient of absorption can be found.   ∎

17.3.3 Measurement of Optical Properties

The measurement of optical properties, namely, absorption coefficient (μ_a), scattering coefficient (μ_s), and scattering anisotropy (g), of biological tissues remains a central problem in the field of biomedical optics. Knowledge of these parameters is important in both therapeutic and diagnostic applications of light in medicine. For example, optical properties are necessary to make accurate assessments of localized fluence during irradiation procedures such as photodynamic therapy, photocoagulation, and tissue ablation. Also, in addition to having a profound impact on *in vivo* diagnostics such as optical imaging and fluorescence spectroscopy, the optical properties can potentially be used to provide metabolic information and diagnose diseases.

To date, a number of methods have been developed to measure tissue optical properties. The collimated transmission technique can be used to measure the total interaction coefficient ($\mu_a + \mu_s$). In this technique, a collimated light beam illuminates a thin piece of tissue. Unscattered transmitted light is detected while the scattered light is rejected by use of a small aperture. The unscattered transmitted light can be calculated based on the Beer–Lambert law, which is an extension of Eq. (17.30). The Beer–Lambert law including scattering media is $I(z) = I_0 \exp[-(\mu_a + \mu_s) z]$, where $I(z)$ is the unscattered transmitted light intensity after penetrating a depth of z. In collimated transmission measurements, I_0, $I(z)$, and z are measured. Therefore, $\mu_a + \mu_s$ can be deduced.

Example Problem 17.4

A 5-mW collimated laser beam passes through a 4-cm nonabsorbing, scattering medium. The collimated transmission was measured through a small aperture to be 0.035 mW. Calculate the scattering coefficient.

Solution

Based on Beer's law, $\mu_s = \ln(5/0.035)/4 = 1.2$ cm^{-1}. ∎

Probably the most common technique is the integrating sphere measurement. In this technique, a thin slice of tissue is sandwiched between two integrating spheres (spheres with an entrance and exit port whose inner surface is coated with a diffuse reflecting material). A collimated beam is incident upon the tissue sample. Both the diffuse reflectance and the transmittance are measured by integrating the diffusely reflected and transmitted light, respectively. These two measurements are used to deduce the absorption coefficient (μ_a) and the reduced scattering coefficient ($\mu_{s'}$) with a model. The model could be based on the adding–doubling method, delta-Eddington method, Monte Carlo method, or other light transport theories.

Another technique is normal incidence video reflectometry. A collimated light beam is normally incident upon a tissue. The spatial distribution of diffuse reflectance is collected using either a CCD camera or an optical fiber bundle. Diffusion theory is used to fit the measured spatial distribution of diffuse reflectance to determine the optical properties. The measured spatial distribution of diffuse reflectance must be in absolute units unless total diffuse reflectance is measured with the diffuse reflectance profile. Calibration to absolute units is a sensitive procedure and hence this method is not ideal for a clinical setting.

It is possible to use time-resolved or frequency-domain techniques to measure optical properties. However, these techniques require instrumentation that may not be cost-effective for nonresearch applications.

A recent promising approach for *in vivo* optical property measurements is fiber-optic-based oblique incidence reflectometry. It is a fairly simple and accurate method for measuring the absorption and reduced scattering coefficients, μ_a and $\mu_{s'}$ providing the sample can be regarded as a semi-infinite turbid media, as is the case for most *in vivo* tissues. Therefore, this approach will be discussed in more detail in this section.

Obliquely incident light produces a spatial distribution of diffuse reflectance that is not centered about the point of light entry. The amount of shift in the center of diffuse reflectance is directly related to the medium's diffusion length, D. A fiber-optic probe may be used to deliver light obliquely and sample the relative profile of diffuse reflectance. Measurement in absolute units is not necessary. From the profile, it is possible to measure D, perform a curve fit for the effective attenuation coefficient, μ_{eff}, and then calculate μ_a and $\mu_{s'}$. Here, μ_{eff} is defined as

$$\mu_{eff} = \sqrt{\mu_a/D} \tag{17.43}$$

The spatial distribution of diffuse reflectance of normally incident light has previously been modeled using two isotropic point sources, one positive source located 1 mean free path' (mfp') below the tissue surface and one negative image source located above the tissue surface. The positive source represents a single scatter source in the tissue, and the height in z of the image source depends on the boundary condition. The transport mean free path is defined as

$$\text{mfp}' = 1/(\mu_a + \mu_{s'}) \tag{17.44}$$

With oblique incidence, the buried source should be located at the same path length into the tissue, with this distance now measured along the new optical path as determined by Snell's law. It is assumed that (i) the angle of incidence and (ii) the indexes of refraction for both the tissue and the medium through which the light is delivered are known. The net result is a change in the positions of the point sources, particularly a shift in the x direction. These two cases are diagrammed in Figs. 17.8a and 17.8 b.

The diffuse reflectance profile for oblique incidence is centered about the position of the point sources, so the shift, Δx, can be measured by finding the center of diffuse reflectance relative to the light entry point.

The two-source model gives the following expression:

$$R(x) = \frac{1}{4\pi}\left[\frac{\Delta z(1 + \mu_{eff}\rho_1)\exp(-\mu_{eff}\rho_1)}{\rho_1^3} + \frac{(\Delta z + 2z_b)(1 + \mu_{eff}\rho_2)\exp(-\mu_{eff}\rho_2)}{\rho_2^3}\right] \tag{17.45}$$

where $\Delta z = 3D\cos\theta_t$. Equation (17.45) can be scaled arbitrarily to fit a relative reflectance profile that is not in absolute units. ρ_1 and ρ_2 are the distances from the two sources to the point of interest (the point of light collection; see Fig.

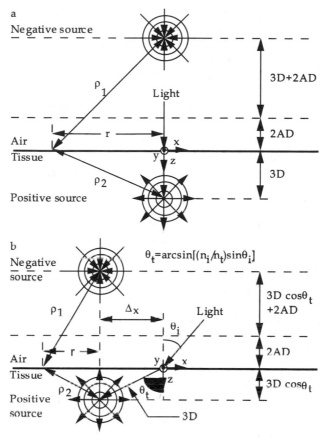

Fig. 17.8 (a) Positions of point sources in diffusion theory model for normal incidence. (b) Positions of point sources in diffusion theory model for oblique incidence. The y axis points out of the page. r_1 and r_2 are the distances from the positive and negative point sources to the point of interest on the tissue surface, at a radius of r from the axis of the sources. θ_t is determined from Snell's law.

17.8). As can be seen in Fig. 17.8b, the diffusion coefficient can be calculated from Δx:

$$D = \Delta x/(3 \sin \theta_t) \tag{17.46}$$

Optical properties of biological tissues depend on the tissue type and optical wavelength. For instance, the liver, with its reddish color, would have a much higher absorption coefficient from a green light source such as an argon laser than a tan piece of tissue such as chicken breast. Depending on the tissue type and exact wavelength, a typical set of optical properties for visible or near-infrared (NIR) light would be 0.1 cm^{-1} for the absorption coefficient, 100 cm^{-1} for the scattering coefficient, and 0.9 for the anisotropy. In the UV region, light absorption is dominated by proteins. In the IR region, light absorption is dominated by water. The NIR (~800 nm) is considered a diagnostic window because of the minimal absorption and relatively low scattering in this region.

An important application of optical properties is the measurement of hemoglobin oxygenation saturation, which is a critical physiological parameter. Because the oxygenated and deoxygenated hemoglobin have different absorption spectra, the relative concentration ratio between the two forms of hemoglobin can be calculated once the optical properties of the tissue are measured.

17.4   PHYSICAL INTERACTION OF LIGHT AND PHYSICAL SENSING

From the previous section, it is quite apparent that most of the tissues in our body are not transparent to light; that is, light is typically either absorbed or scattered in tissues. In this section the physical interaction of light with tissue is described and its use toward both sensing and therapeutics is discussed. The four fundamental physical interactions that will be described are the thermal changes induced or measured using light, pressure changes induced or measured using light, velocity changes manifested as Doppler frequency shifts in the light, and path-length changes in the sample which cause an interference pattern between two or more light beams.

17.4.1   Temperature Generation and Monitoring

All tissues in our body absorb light at various wavelengths and the absorption process converts light energy into heat. In addition, our body tissues, as with any object above absolute zero temperature, also generate light radiation known as blackbody radiation. Thus, light can be used to heat tissues for therapy or be measured from tissues to determine temperature.

Temperature Monitoring

The measurement of temperature, in general, has been traditionally performed reliably and fairly inexpensively using electrically based sensors such as thermistors and thermocouples. However, these sensors are not as useful when strong electromagnetic interference is present, which is common in a hospital-based setting. Also, when attempting to monitor temperature rise due to laser radiation, these sensors are inappropriate since they absorb the laser radiation. Therefore, to measure temperature in the body several indirect optic and fiber-optic approaches have been reported. For example, liquid crystal optrodes show a dramatic change with color due to temperature differences, interferometric sensors change phase or fringes with a change in the path length due to temperature variations, and the luminescence of many materials depends strongly on temperature.

The direct type of temperature sensor initially described is known as a thermographic or radiometer system. The primary science and engineering of this type of system has been done by the military for tasks such as detecting vehicles, personnel, and even ships in total darkness. The formula for the total emission from a black-body at temperature T is

$$I = \varepsilon \sigma T^4 \tag{17.47}$$

where ε is the emissivity and σ is the Stefan–Boltzmann constant. At room temperature objects emit mainly in the far-IR region of the wavelength spectrum, but as temperature rises the emission appears in the NIR and finally in the visible spectrum. The emissivity is 1 ($\varepsilon = 1$) for biological tissue and hence the total emission is dependent only on the temperature T. For the military applications a thermographic picture generated from the IR emission can be used to display the temperature, and using this technique the surface temperature of the human body can be monitored for medical applications. For example, a thermal camera may be used with a TV monitor to determine the temperature distribution on the human chest in an effort to reveal the pathologic condition of breast cancer since the effected breast tissue will be slightly warmer than the healthy tissue. Although IR imaging is less reliable than X-ray mammography for breast screening, it is totally passive and patients are not exposed to ionizing radiation. The thermographic systems used for this IR emission measurement require a line of sight between the warm surface and the detector. When no direct line of sight exists, an IR-transmitting optical fiber can sometimes be used to make the connection to the detector. This type of fiber-optic radiometer has been proposed for treatments using microwave heating and for laser tissue heating treatments.

Light-Induced Heating

As mentioned previously, in addition to monitoring light radiation from the body, light can be used to heat tissues that absorb light. Light-induced heating of tissue can be used for a variety of applications, including biostimulation, sealing or welding blood vessels, tissue necrosis, and tissue vaporization. Biostimulation has been determined to be a result of minute heating of the tissue whereby light-induced heat stimulates nerves and/or accelerates wound healing. Higher energy absorbed laser light can facilitate the joining of tissues, in particular blood vessels (i.e., anastomosis), and can be used to coagulate blood to stop bleeding during a surgical procedure. If light-induced heating causes the temperature to rise above 45°C, tissue necrosis and destruction occurs as would be desired for the treatment of cancer or enlarged prostate tissues. At even higher power densities ablation or vaporization of tissue occurs, as is the case for the corrective eye surgery known as radial keratectomy.

Light-induced heating is typically performed with laser light. The lasers used have different wavelengths, power densities (i.e., ratio between beam power and irradiated area), and duration times. The amount of energy imparted to the tissue and hence temperature rise can be changed by varying either the power density or the duration of the time pulse of the laser. For high-power densities, coagulation, necrosis, and vaporization of tissue can occur, whereas at low-power densities minimal heating is observed. The amount of absorption by the tissue is a function of the wavelength of the light used. For some wavelengths, for instance, those of the known strong absorption bands of water, the laser beam is highly absorbed since tissue is primarily made of water. At these wavelengths the energy is highly absorbed in a relatively thin layer near the surface where rapid heating occurs (i.e., radial keratectomy). Less absorption occurs for other wavelengths and this results in slower heating of a larger volume of the tissue (i.e., for prostate coagulation).

Temperature Generation and Rate of Photon Absorption

A thermodynamically irreversible mode of interaction of light with materials is the process of absorption in which the photon energy is absorbed by the material phase. In the absence of conduction, the temperature rise in tissue is governed by a thermodynamic equation of state. The equation of state requires that the change in internal energy of the system is proportional to temperature rise. The change in internal energy of the system, in the absence of conduction and other heat transfer processes, is equal to the rate of energy deposition by the laser. Expressing this relation in terms of time derivatives gives

$$\Delta U/\Delta t = \rho C \Delta T/\Delta t \approx Q_L \qquad (17.48)$$

where ΔT (K) is temperature rise, Q_L (W/m^3) is volumetric rate of photon absorption by the tissue, ρ (kg/m^3) is mass density, C (J/kg K) is specific heat, ρC (J/m^3 K) is volumetric heat capacity, and t is the duration. A very important factor in temperature rise by photons is the rate of photon absorption Q_L which is also known as the light/laser source term and as the rate of energy deposition. For ordinary interaction processes, the rate of absorption of photon by the material is proportional to irradiance, and the constant of proportionality is the absorption coefficient:

$$Q_L = \mu_a \phi \qquad (17.49)$$

Irradiance ϕ is related to the spatial distribution of photons within the tissue. For multiphoton light–tissue interaction processes, the exponent of ϕ is increased. (For example, for a two-photon process $Q_L = \mu_a \times \phi^2$.) For purely absorbing materials, for instance, the Beer–Lambert law applies, and if a Gaussian beam profile is additionally assumed then

$$Q_L = \mu_a \times \phi_0 \exp(-\mu_a z) \exp(-2r^2/\omega^2) \qquad (17.50)$$

where ϕ_0 is incident intensity, z is depth, and r is radial distance. In the presence of scattering, one of the scattering models discussed previously can be used to describe ϕ_0.

Equation (17.48) assumes no other thermal interaction processes such as conduction, convection, or perfusion. If these processes can be ignored, as for example for very short laser pulses, then temperature rise can be estimated as

$$\Delta T = Q_L \Delta t/\rho C \qquad (17.51)$$

where Δt is exposure duration. Other thermal effects of laser–tissue interaction will be considered in more detail in Section 17.6. Equation (17.51) is valid for very short irradiations during which heat diffusion is negligible. A criteria or an estimate of upper limit for validity of Eq. (17.51) is

$$\Delta t_{max} = \frac{1}{\sqrt{4\mu_a{}^2\alpha}} \qquad (17.52)$$

where α is the thermal diffusivity of the material. For water the value of α is about 1400 cm²/s.

17.4.2 Laser Doppler Velocimetry

In addition to temperature, another physical interaction of light is the Doppler phenomenon which is based on a frequency shift due to the velocity of a moving object. For medical applications these objects are typically the moving red blood cells with a diameter of approximately 7 μm. The Doppler approach is used for measuring blood flow velocity for a variety of medical applications, including heart monitoring, transluminar coronary angioplasty, coronary arterial stenosis, tissue blood flow on the surface of the body, and monitoring blood flow on the scalp of a fetus during labor.

When a light wave of frequency f and velocity c impinges on a stationary object, it is reflected at the same frequency. However, if the object moves with velocity, v, the reflected frequency, f', is different from f. This difference or shift in frequency, δf, from the original light wave is known as the Doppler effect or Doppler shift and can be described as

$$\delta f = f - f' \qquad (17.53)$$

The Doppler shift is described in Eq. (17.54) in terms of the velocity, v, of the moving red blood cells, the refractive index of the medium, n, the speed of light in the tissue, c_0, the input frequency, f, and the angle between the incident beam and the vessel, θ:

$$\delta f/f = 2v\, n\, \cos\theta/c_0 \qquad (17.54)$$

Furthermore, it is well known that the wavelength, λ, is equal to the speed of light divided by the frequency so that

$$\delta f = 2\, v\, n\, \cos\theta/\lambda_0 \qquad (17.55)$$

A comparable analogy for the Doppler effect with sound waves is a train moving toward an observer. As the train gets closer the whistle sounds like a higher pitch.

Laser Doppler is a good method for monitoring velocities, but one reason that it has not gained wide use clinically is that the flow (cm³/s) of blood or the average velocities of all the particles in the fluid is the physical quantity of diagnostic value and flow cannot be directly measured using the Doppler approach. Calculating the flow rate for a rigid tube filled with water by knowing the velocity is a straightforward problem to solve. However, determining the flow rate of blood in the body by simply using the Doppler measured velocity is a much more difficult problem, particularly for narrow vessels. Blood is thicker than water and the flow characteristics are more complex. Also, blood vessels are not rigid, straight tubes, and the flow of blood is pulsatile. Lastly, a fiber-optic probe is often inserted in the blood vessel and scanned to get a series of velocity measurements to determine flow, but the probe can alter the flow measurement.

Overall, laser Doppler velocimetry is a simple concept which can, with some effort, be used to measure relative changes in flow, but it is very difficult to calibrate this approach for absolute flow measurements. Thus, the standard for blood mass-flow measurements has been thermodilation and this is widely used in clinical practice. The thermodilation method is used to determine flow by inducing a predetermined change in the heat content of the blood at one point of the circulation and detecting the resultant change in temperature at some point downstream, after the flow has caused a controlled degree of mixing across the vessel diameter.

17.4.3 Interferometry

The phenomenon of interference also enables physical light interaction to take place and it depends on the superposition of two or more individual waves, typically originating from the same source. Since light consists of oscillatory electric and magnetic fields which are vectors, they add vectorally. Thus, when two or more waves emanating from the same source are split and travel along different paths, they can reunite and interfere constructively or destructively. When constructive and destructive interferences are seen to alternate in a spatial display, the interference is said to produce a series or pattern of fringes. If one of the paths in which the light travels is altered by any small perturbation, such as temperature, pressure, or index of refraction changes, then, once recombined with the unaltered reference beam, the perturbation causes a shift in the fringe pattern which can be readily observed with optoelectronic techniques to about 10^{-4} of the fringe spacing. The useful information regarding the changing variable of interest can be measured quite accurately as a path-length change on the order of 1/100 of a wavelength or 5×10^{-9} m for visible light.

There are several ways to produce light interference, including the Rayleigh refractometer, one of the first instruments developed from which came the Mach–Zehnder interferometer. For the sake of brevity, this section will focus on two more sophisticated variations of the Rayleigh idea; the dual-beam Michelson interferometer and the multiple-beam Fabry–Perot interferometer. As depicted in Fig. 17.9, the Michelson interferometer begins with a light source that is split into two beams by means of a beam-splitting mirror or fiber optics which also serves to recombine the light after reflection from fully silvered mirrors. A compensating plate is sometimes needed to provide equal optical paths before introduction of any perturbation or sample to be measured. The perturbation can take the form of a pressure or temperature change causing a strain and hence path-length change in the sample arm of the fiber optic or as a change due to the addition of a tissue sample or replacement of the mirror with a tissue sample. For example, the Michelson interferometer has been investigated for measuring tissue thickness, particularly for corneal tissue as feedback for the radial keratectomy procedure (laser removal or shaving of the

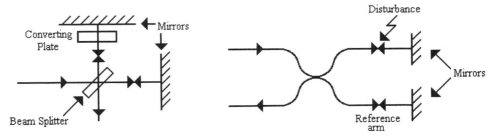

Fig. 17.9 Michelson interferometer (a) bulk optics and (b) fiber optics.

cornea to correct vision). This interferometer, when used with a low-coherent light source, has been researched for use in optical coherence tomographic imaging of superficial tissues. The governing equation for the irradiance of the fringe system of circles concentric with the optic axis in which the two interfering beams are of equal amplitude is given as

$$I = 4I_0 \cos^2(\delta/2) \tag{17.56}$$

where I_0 is the input intensity and the phase difference δ is defined as

$$\delta = 2\pi\Delta/\lambda \tag{17.57}$$

In this equation, λ is the wavelength and the net optical path difference, Δ, is defined as

$$\Delta = 2d\cos(\theta) + \lambda/2 = (m + 0.5)\lambda \tag{17.58}$$

Clearly, from Eq. (17.58) the following equation can be derived:

$$2d\cos(\theta) = m\lambda \tag{17.59}$$

where $2d$ is the optical path difference or the difference in the two paths from the beam splitter, m is the number of fringes, and θ is the angle of incidence ($\theta = 0$ is a normal or on-axis beam). When a plate, gas, or if a thin minimally absorbing or scattering tissue slice is assumed with constant index of refraction and is inserted in one of the paths then $d = (n_s - n_{air})L$, where L is the actual length of the substance, n_s is the index of refraction of the substance, and n_{air} is the index of refraction of the air.

Example Problem 17.5

A thin sheet of clear tissue, such as a section of the cornea of the eye ($n \approx 1.33$), is inserted normally into one beam of a Michelson interferometer. Using 589 nm of light the fringe pattern is found to shift by 50 fringes. Determine the thickness of the tissue section.

Solution

From Eq. (17.59), $2d\cos(\theta) = m\lambda$, and thus $d = (50 \times 0.589)/2 = 14.72$ μm, which is the calculated optical path length. However, the physical length of the tissue must take into account the index of refraction of the sample and the air. Thus, as described previously, the equation for physical path length is $L = d/(n_s - n_{air}) = 14.72/(1.33 - 1.0) = 44.6$ μm. ∎

The dual-beam interferometry techniques, such as the Michelson and Mach–Zehnder approaches, suffer from the limitation that the accuracy depends on the location of the maxima (or minima) of a sinusoidal variation as shown in Eq. (17.59). For very accurate measurements, such as precision spectroscopy, this limitation is severe. Rather than dual beam, if the interference of many beams is utilized the accuracy can be improved considerably. A Fabry–Perot interferometer uses a multiple-beam approach as depicted in Fig. 17.10. As can be seen in Fig. 17.10, the interferometer makes use of a plane parallel plate to produce an interference pattern by combining the multiple beams of the transmitted light. The parallel plate is typically composed of two thick-glass or quartz plates which enclose a plane parallel plate of air between them. The flatness and reflectivity of the inner surfaces are important and are polished generally better than $\lambda/50$ and coated with a highly reflective layer of silver or aluminum. The silver is good for wavelengths above 400 nm in the visible but aluminum has better reflectivity below 400 nm. These film coatings must also be thin enough to be partially transmitting (~50-nm thickness for silver coatings). In many instances the outer surfaces of the glass plate are purposely formed at a small angle relative to the inner faces (several minutes of arc) to eliminate spurious fringe patterns that can arise from the glass acting as the parallel plate interferometer. When L is fixed, the instrument is referred to as a Fabry–Perot etalon. The nature of the superposition at point P is determined by the path difference between successive parallel beams; thus,

$$\Delta = 2n_f L \cos\theta_t = m\lambda_{\text{(minimum)}} = 2L \cos\theta_t \qquad (17.60)$$

using $n_f = 1$ for air.

Other beams from different points of the source but that are in the same plane and make the same angle θ_t with the axis satisfy the same path difference and also arrive at P. With L fixed, the previous equation for Δ is satisfied for certain angles θ_t and the fringe system is composed of the familiar concentric rings due to the focusing of fringes of equal inclination as depicted in Figs 17.11a and 17.11b. If the

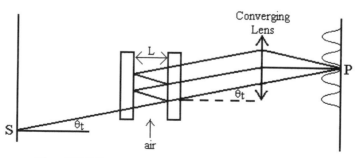

Fig. 17.10 Fabry–Perot multiple-beam interferometer.

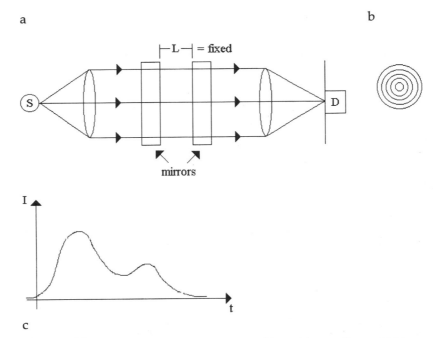

Fig. 17.11 (a) F-petalon with L fixed showing (b) spacial separation and (c) temporal separation.

thickness of L varies with time, a detector D will record an interferogram as a function of time as depicted in Fig. 17.11c.

Variation in the fringe pattern irradiance of the Fabry–Perot as a function of phase or path difference is known as the fringe profile. The sharpness of the fringes is important to the ultimate resolving power of the instrument. Using the trig identity $\cos \delta = 1 - 2 \sin^2 (\delta/2)$, the transmittance can be written as

$$T = I_T/I_i = 1/(1 + [4r^2/(1 - r^2)^2] \sin^2(\delta/2)) \quad \text{(Airy function)} \quad (17.61)$$

where r is the reflectivity and the term in the square brackets in the denominator is known as the coefficient of finesse, F; thus,

$$T = 1/(1 + F \sin^2(\delta/2)) \quad (17.62)$$

It should be noted that as $0 < r < 1$ then $0 < F < \infty$. The coefficient of finesse, F, also represents a certain measure of fringe contrast:

$$F = ((I_T)_{max} - (I_T)_{min})/I_{Tmin} = (T_{max} - T_{min})/T_{min} \quad (17.63)$$

It should be noted that $T_{max} = 1$ when $\sin(\delta/2) = 0$ and $T_{min} = 1/(1 + F)$ when $\sin(\delta/2) = \pm 1$. Given r, the fringe profile can be plotted as shown in Fig. 17.12. As seen in Fig. 17.12, $T = T_{max} = 1$ at $\delta = m2\pi$ and $T = T_{min} = 1/(1 + F)$ at $\delta = (m + \frac{1}{2}) 2\pi$. As can be seen $T_{max} = 1$ regardless of r, and T_{min} is never zero but approaches zero as r approaches 1. Also, the transmittance peaks sharply at higher val-

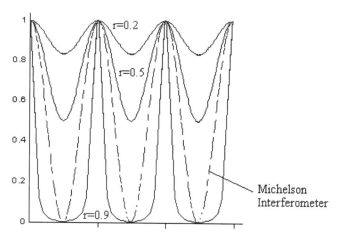

Fig. 17.12 The difference in resolution between the multiple-beam Fabry–Perot interferometer with different reflectivities and the dual-beam Michelson interferometer.

ues of r, remaining near zero for most of the region between fringes. It should be noted that the Michelson interferometer produces broad fringes when it is normalized to the same maximum value since it has a simple $\cos^2(\delta/2)$ dependence on the phase (dashed lines in Fig. 17.12). An example of a fiber-optic-based Fabry–Perot interferometer is the commercially available intracranial fluid pressure monitoring system for patients with severe head trauma or hydrocephalus (increased amount of cerebral spinal fluid in the ventricles and/or subarachnoid spaces of the brain).

17.5 BIOCHEMICAL MEASUREMENT TECHNIQUES USING LIGHT

In the past few years, there has been much enthusiasm as well as a strong effort by medical device companies and universities to perform diagnostic procedures such as cancer detection and to quantifiably monitor blood chemicals, such as glucose, lactic acid, albumin, and cholesterol, using various optical approaches. The most well-known optical monitoring approach used clinically is the blood oxygenation sensor, which detects qualitative changes in the strong optical absorption peaks of oxygenated and deoxygenated hemoglobin. The most common optical approaches for diagnostic and sensing applications include absorption, scattering, luminescence, and polarimetry. The primary variable for each of these approaches is a change in light intensity as it passes through the medium, which can change as a function of the wavelength or polarization of light. For absorption and luminescence or fluorescence, the intensity will change nearly linearly for moderate analyte concentrations and nonlinearly for high concentrations.

17.5.1 Spectroscopic Measurements Using Light Absorption

Many investigators have suggested IR absorption as a potential route to blood chemical monitoring and diagnostic sensing, in particular for glucose monitoring and cancer detection. The governing equation (Eq. 17.17) for purely absorbing media as well as fluorescence was described previously using the Beer–Lambert law. Expressed logarithmically, the equation becomes

$$A = \ln(T) = \ln(I_0/I) = \mu z = z\Sigma\varepsilon_i \, C_i \qquad (17.64)$$

where A is absorbance, T is transmittance, I_0 is the incident light, I is the transmitted light, z is the path length, and μ is the absorption coefficient or the sum of the multiplication of the molar absorptivity times the concentration of all the different components in the analyte. For tissue and blood this equation is valid in the mid-IR wavelength region (wavelengths of 2.5–12 µm) in which the absorption peaks due to various chemicals are distinguishably sharp and the scatter is weak. Unfortunately, the absorption of light due to water within tissue in this region is orders of magnitude stronger than that of any of the blood chemicals which results in the possibility of only short path-length sample investigations (on the order of micrometers). Thus, surface or superficial investigations on skin or, with the use of fiber optics, on internal body cavities are possible. Unfortunately, the optical fibers available in this region tend to be toxic, hydroscopic, and/or rigid. Consequently, the combination of water absorption and lack of good fibers makes *in vivo* diagnostics and sensing very difficult in the mid-IR region. The light sources used in this region include the broad-based tungsten bulb, nernst glower, nichrome wire, globar rod, and narrowly tunable, typically liquid nitrogen-cooled laser diodes. The detectors used include the cooled mercury cadmium telluride, thermopile, and thermistor. The optics include sodium chloride and potassium bromide, mirrors (typically gold coated), and gratings.

In the NIR wavelength region the spectrum is not affected by water to the same degree as is the mid-IR region allowing for path lengths of 1 mm to 1 cm to be used. In addition, low OH silica fibers are quite transparent across this range and are non-toxic and nonhydroscopic. The NIR region (700–2500 nm) exhibits absorptions due to low-energy electronic vibrations (700–1000 nm) as well as overtones of molecular bond stretching and combination bands (1000–2500 nm). These bands result from interactions between different bonds (−CH, −OH, and −NH) to the same atom. Typically, only the first, second, and third overtones of a molecular vibration are detectable, and are broad in nature. Thus, only at high concentrations of the chemical species are these overtones qualitatively detectable, with the intensity dropping off rapidly as overtone order increases. The NIR absorption bands are also influenced by temperature, pressure, and hydrogen bonding effects and can overlap significantly. In the 700- to 1200-nm region they are also influenced by scattering effects. Thus, unlike mid-IR spectroscopy, NIR spectroscopy is primarily empirical and not well suited for qualitative work. However, using techniques such as multivariate statistics, quantitative analysis is possible with the NIR spectrum. The light sources used in this region include the broad-based tungsten bulb, light-emitting diodes, laser-pumped solid-state lasers such as the tunable titanium sapphire laser, and laser diodes. The detectors used include silicon (good to 1 μm), germanium (good to 1.7 μm), and indium antiminide (good to 5.5 μm). The optics includes low OH glasses, quartz glass, gratings, and mirrors.

The primary methods for varying the wavelength of the light are dispersive and nondispersive methods. The dispersive approach uses either a ruled reflective grating or transmissive prism to separate the wavelengths of light. The nondispersive systems include a series of wavelength selection filters or a Fourier transform infrared (FTIR)-based instrument. The FTIR method uses an interferometer similar to the Michelson interferometer depicted in Fig. 17.9 to collect the entire spectrum and then deconvolutes it using Fourier transform techniques. Both approaches can be configured to cover the NIR and mid-IR ranges. The advantages of the dispersive systems include higher resolution and separation of closely spaced wavelength bands, whereas the nondispersive systems generally have better throughput since all the light passes through the sample.

17.5.2   Monitoring Approaches Using Scattered Light

There are fundamentally two types of optical scattering for diagnostics and monitoring, elastic and inelastic. Elastic scattering can be described using Mie theory (or Rayleigh scattering for particles small compared to the wavelength) in which the intensity of the scattered radiation can be related to the concentration, size, and shape of the scattering particles. Inelastic scattering in which the polarization of the particle is not constant can be described as Raman scattering. Fluorescence is also inelastic.

When using the light scattering phenomenon for sensing, the intensity of the reflected light is usually considered. The reflection of light, however, can be divided into two forms. The first, specular reflection or a "mirror" type of reflection, occurs at the interface of a medium. The returned light yields no information about the material since it never penetrates the medium and thus the specularly reflected light is typically minimized or eliminated with the design of the optical sensor. Diffuse reflection, however, occurs when light penetrates into a medium, becomes absorbed and multiply scattered, and makes its way back to the surface of the medium. The model describing the role of diffuse scattering in tissue is based on the radiative transfer theory as was described in Section 17.3. The same theory applies for sensing.

The use of elastically scattered light has been suggested for diagnostic procedures such as cancer detection and for monitoring analytes such as glucose noninvasively for diabetics. For use as a monitoring application of chemical changes such as glucose, researchers have used an intensity-modulated frequency domain NIR spectrometer capable of separating the reduced scattering and absorption coefficients to detect changes in the reduced scattering coefficient showing correlation with blood glucose in human muscle. This approach may be promising as a relative measure over time since clearly an increase in glucose concentration in the physiologic range decreases the total amount of tissue scattering. However, the drawbacks of the light elastic scattering approach for analyte monitoring are still quite daunting. The specificity of the elastic scattering approach is the biggest concern with this method since other physiologic effects unrelated to glucose concentration could produce similar variations of the reduced scattering coefficient with time and, unlike the absorption approach, elastic light scattering is nearly wavelength independent. The measurement precision of the reduced scattering coefficient and separation of scattering and absorption changes is another concern with this approach. It is difficult to measure such small changes and be insensitive to some of the larger absorption changes in the tissue, particularly hemoglobin. This approach also needs to take into account the different refractive indices of tissue. Tissue scattering is caused by a variety of substances and organelles (membranes, mitochondria, nuclei, etc.) and all of them have different refractive indices. The effect of blood glucose concentration and its distribution at the cellular level is a complex issue that needs to be investigated before this approach can be considered viable. An instrument of this type would require *in vivo* calibration against a gold standard since the reduced scattering coefficient is dependent on additional factors such as cell density. Lastly, there is a need to account for factors that might change the reduced scattering coefficient, such as variations in temperature, red blood cell concentration, electrolyte levels, and movements of extracellular and intracellular water. As a diagnostic screening tool for cancer detection, measurement of the scatter in thin tissues or cells may hold promise. Many of the changes in tissue due to cancer are morphologic rather than chemical and thus occur with changes in the size and shape of the cellular and subcellular components. Thus, the changes in elastic light scatter should be larger with the morphologic tissue differences. If the wavelength of the elastic light scattering is carefully selected so as to be outside the major absorption areas due to water and hemoglobin, and if the diffusely scattered light is measured as a function of angle of incidence, there is potential for this approach to aid in pathologic diagnosis of disease.

Inelastic Raman spectroscopic scattering has been utilized during the past 30 years mainly by physicists and chemists. Raman spectroscopy has become a powerful tool for studying a variety of biological molecules, including proteins, enzymes, and immunoglobulins, nucleic acids, nucleoproteins, lipids and biological membranes, and carbohydrates, but with the advent of more powerful laser sources and more sensitive detectors it has also become useful as a diagnostic and sensing tool. The phenomenon of Raman scattering is observed when monochromatic (single wavelength) radiation is incident upon the media. In addition to the transmitted light, a portion of the radiation is scattered. Some of the incident light of frequency w_0 exhibits frequency shifts $\pm w_m$, which is associated with transitions between rotational, vibrational, and electronic levels. In general, the intensity and polarization of the scattered radiation is dependent on the position of observation relative to the incident energy. Most studies used the Stokes type of scattering bands that correspond to the $w_0 - w_m$ scattering. Therefore, the Raman bands of interest are shifted to longer wavelengths relative to the excitation wavelength.

As with IR spectroscopic techniques, Raman spectra can be utilized to identify molecules since these spectra are characteristic of variations in the molecular polarizability and dipole moments. Raman spectroscopy can be considered as complementary to absorption spectroscopy because neither technique alone can resolve all the energy states of a molecule. In fact, for certain molecules, some energy levels may not be resolved by either technique. Due to the anharmonic oscillator model for dipoles, overtone frequencies exist in addition to fundamental vibrations. An advantage of Raman spectroscopy is that the overtones are much weaker than the fundamental tones, thus contributing to simpler spectra compared to absorption spectroscopy. One advantage to using Raman spectroscopy in biological investigations is that the Raman spectrum of water is weaker which, unlike IR spectroscopy, only minimally interferes with the spectrum of the solute and thus the spectrum can be obtained from aqueous solutions. However, the Raman signal is also weak and only recently, with the replacement of slow photomultiplier tubes with faster CCD arrays as well as the manufacture of higher power NIR laser diodes, has the technology become available to allow researchers to consider the possibility of distinguishing normal and abnormal tissue types as well as quantifying blood chemicals in near real time. In addition, investigators have applied statistical methods such as partial least squares to aid in the estimation of biochemical concentrations from Raman spectra.

As with elastic scatter, Raman spectroscopy has been used for both diagnostics and monitoring. The diagnostic approaches search for the presence of different spectral peaks and/or intensity differences in the peaks due to different chemicals present in, for instance, cancerous tissue. For quantifiable monitoring it is the intensity difference that is investigated. In tissue, a problem in addition to water absorption is the high fluorescence background signal as a result of autofluorescence incurred in heavily vascularized tissue due to the high concentration of proteins and other fluorescent components. Instrumentation to excite in the NIR wavelength range has been proposed to overcome the autofluorescence problem since the fluorescence component falls off with increasing wavelength. Excitation in the NIR region offers the added advantage of longer wavelengths that pass through larger tissue samples with lower absorption and scatter than other spectral regions, such as visible or ultraviolet. However, in addition to fluorescence falling off with wavelength, the Raman signal also falls off to the fourth power as wavelength increases. Thus, there is a trade-off between minimizing fluorescence and maintaining the Raman signal. The eye has been suggested as a site for analyte concentration measurements using Raman spectroscopy to minimize autofluorescence; however, the disadvantage of using the eye for Raman spectroscopy is that the laser excitation powers must be kept low to prevent injury, but this significantly reduces the signal-to-noise ratio. Lastly, like IR and NIR absorption, to quantifiably determine the inherently low concentrations of analytes *in vivo* different chemicals must be accounted for that yield overlapping Raman spectra.

17.5.3 Use of the Luminescence Property of Light for Measurement

As described previously, luminescence is the absorption of photons of electromagnetic radiation (light) at one wavelength and reemission of photons at another wavelength. The photons are absorbed by the molecules in the tissue or medium raising them to some excited energy state and then, upon returning to a lower energy state, the molecules emit radiation or luminesce. The luminscent effect can be referred to as fluorescence or phosphorescence. Fluorescence is luminescence that has energy transitions that do not involve a change in electron spin and therefore the reemission occurs much faster. Consequently, fluorescence occurs only during excitation, whereas phosphorescence can continue after excitation. For example, a standard television while turned on produces fluorescence but for a very short time after it is turned off the screen will phosphoresce.

The measurement of fluorescence has been used for both diagnostic and monitoring purposes. Obtaining diagnostic information, particularly with respect to cancer diagnosis or the total plaque in arteries, has been attempted using the extrinsic and intrinsic fluorescence of tissue. The intrinsic fluorescence is due to the naturally occurring proteins, nucleic acids, and nucleotide coenzymes, whereas extrinsic fluorescence is induced by the uptake of certain "impurities" or dyes in the tissue. Extrinsic fluorescence has also been investigated, for instance, to monitor such analytes as glucose, intracellular calcium, proteins, and nucleotide coenzymes. Unlike the use of fluorescence in chemistry on dilute solutions, the intrinsic or autofluorescence of tissue as well as the scattering and absorption of the tissue act as a noise source for the extrinsic approach.

The response of a fluorescence sensor can be described in terms of the output intensity as

$$I_f = \Phi_f (I_0 - I) \tag{17.65}$$

where I_f is the radiant intensity of fluorescence, Φ_f is the fluorescence efficiency, I_0 is the radiant intensity incident on the sample, and I is the radiant intensity emerging from the sample. The fluorescence efficiency can be described as the combination of three factors. First is the quantum yield, which is the probability that an excited molecule will decay by emitting a photon rather than by losing its energy nonradiatively. This parameter varies from 1.00 to 0.05 and varies with time on the order of nanoseconds. Thus, in addition to intensity measurements, time-resolved fluorescence measurements are possible using pulsed light sources and fast detectors. The second parameter is the geometrical factor which is the solid angle of fluorescence radiation subtended by the detector and depends on probe design. Lastly is the efficiency of the detector for the emitted fluorescence wavelength.

Since fluorescence is an absorption/reemission technique, it can be described in terms of Beer's law as

$$I_f = \Phi_f I_0 [1 - \exp(-\varepsilon C l)] \tag{17.66}$$

where C is the concentration of the analyte, l is the path length, and ε is the molar absorptivity. The previous equation can be described in terms of a power series, and for weakly absorbing species ($\varepsilon C l < 0.05$) only the first term in the series is significant. Therefore, under these conditions, the response of the fluorescence sensor becomes linear with analyte concentration and can be described as

$$I_f = \Phi_f I_0 \varepsilon C l \tag{17.67}$$

The primary fluorescent sensors are based on the measurement of intensity; however, lifetime measurements in the time or frequency domain are also possible. To gain the most information, particularly in a research or teaching setting, dual monochromators (grating-based wavelength separation devices) are used with either a photomuliplier tube as the detector or a CCD array detector. In a typical bench-top fluorimeter a broad, primarily ultraviolet/visible xenon bulb is used as the light source. The light is coupled first through a monochromator, which is a wavelength separator that can be set for any excitation wavelength within the range of the source. The light then passes through the sample and is collected by a second monochromator. The light reflected from the grating within the second monochromator can be scanned so that a photomultiplier tube (PMT) receives the different wavelengths of light as a function of time. Alternatively, all the wavelengths from the grating can be collected simultaneously on a CCD detector array. The advantage of the CCD array is that it provides for realtime collection of the fluorescence spectrum. The advantage of the PMT is that it is typically a more sensitive detector. In many systems a small portion of the beam is split at the input and sent to a reference detector to allow for correction of fluctuations in the light source. Once the optimal configuration for a particular biomedical application, such as cervical cancer detection or glucose sensing, has been investigated using the bench-top machine, an intensity measurement system can be designed with a simpler, more robust configuration. Such a system can be designed with wavelength-specific filters instead of monochromators and can be made to work at two or more discrete wavelengths. In addition, optical fibers can be used for delivery and collection of the light to the remote area. Since the excitation wavelength and fluorescent emission wavelengths are different, the same fiber or fibers can be used to both deliver and collect the light. In any configuration it is important to match the spectral characteristics of the source, dye if used, sample, and detector. For instance, depending on the tissue probed, strong absorbers, scatterers, and autofluorescence need to be factored into any fluorescent system design.

17.5.4 Measurements Made Using Light Polarization Properties

Since we have discussed the fundamental electromagnetic theory of polarized light generation, some applications of polarized light shall now be discussed. Two of the emerging applications of polarized light are for biochemical quantification such as glucose and tissue characterization, in particular to aid in cancer identification or for use in the measurement of the nerve fiber layer of normal and glaucomatous eyes.

The rotation of linearly polarized light by optically active substances has been used for many years to quantify the amount of the substance in solution. A variety of both polarimeters, adapted to the examination of all optically active substances, and saccharimeters, designed solely for polarizing sugars, have been developed. The concept behind these devices is that the amount of rotation of polarized light by an optically active substance depends on the thickness of the layer traversed by the light, the wavelength of the light used for the measurement, the temperature, the pH of the solvent, and the concentration of the optically active material. Historically, polarimetric measurements have been obtained under a set of standard conditions. The path length typically employed as a standard in polarimetry is 10 cm for liquids, the wavelength is usually that of the green mercury line (5461 Å), and the temperature is 20°C. If the layer thickness in decimeters is L, the concentration of solute in grams per 100 ml of solution is C, α is the observed rotation in degrees, and $[\alpha]$ is the specific rotation or rotation under standard conditions which is unique for all chiral molecules, then

$$C = \frac{100\,\alpha}{L[\alpha]} \qquad (17.68)$$

In the previous equation the specific rotation $[\alpha]$ of a molecule is dependent on temperature (T), wavelength (λ), and the pH of the solvent.

For polarimetry to be used as a noninvasive technique, for instance, in blood glucose monitoring, the signal must be able to pass from the source, through the body, to a detector without total depolarization of the beam. Since the skin possesses high scattering coefficients, which causes depolarization of the light, maintaining polarization information in a beam passing through a thick piece of tissue (i.e., 1 cm), including skin, would not be feasible. Tissue thickness of <4 mm may potentially be used but the polarimetric sensing device must be able to measure millidegree rotations in the presence of >95% depolarization of the light due to scattering from the tissue. As an alternative to transmitting light through the skin, the aqueous humor of the eye has been investigated as a site for detection of *in vivo* glucose concentrations since this sensing site is a clear biological optical media. It is also known that glucose concentration of the aqueous humor of the eye correlates well with blood glucose levels, with a minor time delay (on the order of minutes), in rabbit models. The approximate width of the average anterior chamber of a human eye is 1 cm. Therefore, an observed rotation of about 4.562 millidegrees per optical pass can be expected for a normal blood glucose level of 100 mg/100 ml, given a specific rotation of glucose at $\lambda = 633$ nm of $45.62°$ cm^2 g^{-1} and a thickness of 1 cm. The eye as a sensing site, however, is not without its share of potential problems. For instance, potential problems with using the eye include corneal birefringence and eye motion artifact. As shown in Eq. (17.68), the rotation is directly proportional to the path length and thus it is critical that this length be determined or at least kept constant for each individual subject regardless of the sensing site. If the eye is used as the sensing site, the angle of incidence on the surface of the cornea must be kept relatively constant for each patient so that not only the path length but also alignment remain fixed each time a reading is taken. In most tissues, including the eye, the change in rotation due to other chiral molecules such as proteins needs to be accommodated in any final instrument. In addition, most other tissues also have a birefringence associated with them that would need to be accounted for in a final polarimetric glucose sensor.

The birefringence and retardation of the polarized light as well as polarized scattering of the tissue is the signal rather than the noise when using polarized light for tissue characterization. For example, a scanning laser polarimeter has been used to measure changes in retardation of the polarized light impinging on the retinal nerve fiber layer. It has been shown that scanning laser polarimetry provides statistically significant higher retardation for normal eyes in certain regions over glaucoma eyes. Images generated from the scattering of various forms of polarized light have also been shown to be able to differentiate between cancer and normal fibroblast cells.

17.6 FUNDAMENTALS OF PHOTOTHERMAL THERAPEUTIC EFFECTS OF LASERS

Therapeutic application of a laser is mediated by conversion of photonic energy to absorbed energy within the material phase of the tissue. The primary mode of this energy conversion manifests itself as a nonuniform temperature rise which leads to a series of thermodynamic processes. These thermodynamics processes can then be exploited as a means to affect therapeutic actions, such as photothermal coagulation and ablation of tissue. Another mode of interaction is the utilization of the absorbed energy in activation of endogenous or exogenous photosensitizing agents in a photochemical process known as photodynamic therapy.

Laser interaction with biological tissue is composed of a combination of optical and thermodynamical processes. These processes can be summarized as shown in Fig. 17.13. Once laser light is irradiated on tissue, the photons penetrate into the tissue and, depending on the tissue optical properties such as scattering coefficient, absorption coefficient, and refractive index, the energy is distributed within the tissue. A portion of this energy is absorbed by the tissue and converted into thermal energy, making the laser act as a distributed heat source. This laser-induced heat source in turn initiates a nonequilibrium process of heat transfer manifested by a temperature rise in tissue. The combined mechanisms of conduction, convection, and emissive radiation distribute the thermal energy in the tissue, resulting in a time- and space-

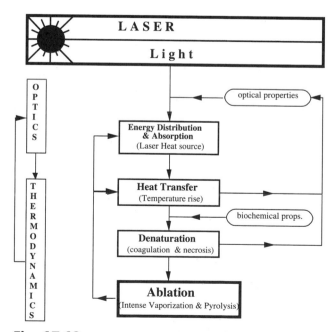

Fig. 17.13 Flowchart of photothermochemical processes in thermal interaction of laser light with biological tissues.

dependent temperature distribution in the tissue. The temperature distribution depends on the thermal properties, conductivity, heat capacity, convective coefficients, and emissivity of the tissue.

As heat deposition and transfer continues, a certain threshold can be reached above which a process of irreversible thermal injury initiates. This process leads to coagulation of tissue caused by denaturation of enzymes and proteins and finally leads to necrosis of constituent cells. As a result of this thermochemical process of injury, properties of the tissue, especially the optical properties, start to change. The change in properties in turn influences the process of energy absorption and distribution in the tissue.

The next stage in these processes is the onset of ablation. As the temperature continues to rise, a threshold temperature is reached at which point, if the rate of heat deposition continues to exceed the rate at which the tissue can transport the energy, an intensive process of vaporization of the water content of the tissue combined with pyrolysis of tissue macromolecules is initiated which results in ablation or removal of tissue.

17.6.1 Temperature Field during Laser Coagulation

In this section the governing equations will be described for a thermodynamic analysis of laser heating of biological tissues up to the onset of ablation. First, the heat conduction equation and typical boundary and initial conditions are described. Next, the Arrhenius–Henriques model for quantitative analysis of irreversible thermal injury to biological tissue will be introduced.

In Section 17.3, how the laser energy is absorbed by a participating medium such as biological tissue was described. As laser energy with a rate Q_L is absorbed by a material under irradiation, there is an immediate thermal energy flux traveling in different directions. This is caused by nonuniformity of Q_L in both the radial direction due to beam profile and the axial direction due to absorption. The energy rate balance equation in the material can be found as follows. Consider an infinitesimal element of the material under laser irradiation (Fig. 17.14). The rate of thermal energy storage in the element, U, depends on the difference between incoming and outgoing thermal fluxes, the rate of laser energy absorption, Q_L, and other energy rate

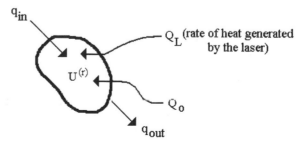

Fig. 17.14 Thermal fluxes and energy in a control volume in the material phase.

interactions which are lumped together and labeled Q_o. The difference between influx and efflux is, in the limit, equal to the negative of the divergence of the thermal flux vector. Therefore, the energy rate balance can be written as

$$\partial U/\partial t = -\mathbf{\nabla} \cdot \mathbf{q} + Q_L + Q_o \tag{17.69}$$

where $\nabla \cdot \mathbf{q}$ is the change in thermal flux. The equation of state for thermal energy is

$$U = \rho\, CT \tag{17.70}$$

where T is an absolute temperature (K), ρ (kg/m^3) is the density, and C (J/kg/K) is the specific heat of the material.

At this point, as in the derivation of the photon phase, a phenomenological relation for the thermal flux is needed. The commonly used relation is known as Fourier's law and is written as

$$\mathbf{q} = -\kappa\, \mathbf{\nabla}\, T \tag{17.71}$$

where κ (W/m K) is the thermal conductivity of the material. Proceeding now by substituting Eqs. (17.71) and (17.70) into Eq. (17.69), the heat conduction equation is obtained:

$$\partial(\rho CT)/\partial t = \mathbf{\nabla} \cdot \kappa\, \mathbf{\nabla}\, T + Q_L + Q_o \tag{17.72}$$

This equation, solved with application of proper initial and boundary conditions, can in theory predict the temperature distribution in a material under laser irradiation.

In the case of biological tissue with other energy rates involved, Q_o can be written as

$$Q_o = Q_b + Q_m \tag{17.73}$$

where Q_b is the heat rate removed by blood perfusion, and Q_m is the rate of metabolic heat generation. These terms are not considered here. They are in general much smaller than the laser heat source term Q_L, except when lower laser powers are used for longer times as in hyperthermia treatments.

The material properties κ and ρC have been measured for many biological tissues including human tissues. In general, these properties may depend on temperature and water content which could cause a nonlinear effect in Eq. (17.72). Nonlinearity may also be due to temperature dependence of the optical properties which, when present, has a more dominant role.

17.6.2   Thermal Coagulation and Necrosis: The Damage Model

The idea of quantifying the thermal denaturation process was first proposed in the 1940s. Using a single-constituent kinetic rate model this early work grossly incorporates the irreversible biochemical processes of coagulation, denaturation, and necrosis associated with thermal injury to biological tissue in terms of a single function. Defining an arbitrary nondimensional function, Ω, as an index for the severity of damage, the model assumes that the rate of change of this function follows an Arrhenius relation:

$$\frac{d\Omega}{dt} = A \exp\left(-\frac{E}{RT}\right) \tag{17.74}$$

where R is the universal gas constant (cal/mol), T is the absolute temperature (K), and A (1/s) and E (cal/mol) are constants to be determined experimentally. Total damage accumulated over a period t can be found by rearranging and integrating Eq. (17.74):

$$\Omega = A \int_0^t \exp\left(-\frac{E}{RT}\right) dt \tag{17.75}$$

The experimental constants A and E for pig skin were determined such that $\Omega = 1.0$ corresponded to complete transepidermal necrosis and $\Omega = 0.53$ indicated the minimum condition to obtain irreversible epidermal injury. The reported values are $A = 3.1 \times 10^{98}$ (1/s) and $E = 150,000$ (cal/mol). These values result in a threshold temperature of about 45°C. Different values for other tissues have been reported. For instance, for human arterial vessel walls, the values are $A = 4.1 \times 10^{44}$ (1/s) and $E = 74,000$ (cal/mol). For these values $\Omega = 1.0$ was defined as the threshold for histologically observed coagulation damage. It should be noted that the coagulation process, i.e., collagen denaturation, is a different damage process than that seen in skin burns.

Once the heat conduction (Eq. 17.72) is solved for the temperature field, the temperature history at every point in the tissue can be used in Eq. (17.75) to predict the accumulated damage at that point.

17.6.3 Thermodynamics: Ablation

The temperature rise in a biological tissue under laser irradiation cannot continue indefinitely. A threshold temperature can be reached beyond which, if the rate of heat deposition by the laser continues to be higher than the rate at which heat can be transported by heat transfer mechanisms, intense vaporization of the water content of the tissue occurs. This results in creation of vacuoles, vacuolization, followed by pyrolysis at higher temperatures. The combined processes of intense vaporization, vacuolization, and pyrolysis of tissue macromolecule constitutents is the laser ablation process. This is the primary underlying mechanism for removal of tissue by laser surgery.

The large water content of most biological tissues suggests that the dominant part of the ablation process is the intense vaporization process. Apart from a short discussion of pyrolysis, intense vaporization is the main process of concern.

In the following section, a derivation is given for the ablation interface energy balance equation which introduces the mathematical nonlinearity of the problem of ablation as a moving boundary problem. The section is concluded by a discussion of pyrolysis.

Ablation Interface Energy Balance Equation

The heat conduction equation alone cannot provide a mathematical model for the determination of the dynamic behavior of the ablation front as a function of time. In fact, the motion of the ablation front influences the heat transfer process. An energy balance at the ablation front is needed in order to determine the motion of the front of ablation and its influence on temperature field.

Consider a portion of a material under laser irradiation with unit cross-sectional area and thickness Δs as shown in Fig. 17.15 which is at the ablation threshold temperature and is to be vaporized in the next interval of time Δt. Using Fourier's law for heat flux (Eq. 17.71) and after mathematical manipulations the following equation for energy balance at the ablation interface can be derived:

$$\kappa \frac{\partial T(s,\,t)}{\partial z} = \rho f_L L \frac{ds}{dt} \qquad (17.76)$$

where f_L is the water fraction parameter. Inherent nonlinearity can be realized from this equation in that it couples the temperature gradient at the ablation interface to the rate of change of the front position which is not known a priori. The solution requires an iteration procedure, which requires the solution of the heat conduction equation simultaneously with the ablation front equation, with application of proper boundary and initial conditions.

Equation (17.76) is essential in providing information on the dynamics and thermodynamics of ablation. When solved simultaneously with the heat conduction equation and with the application of proper boundary and initial conditions, this equation provides the information on the position and velocity of the front of abla-

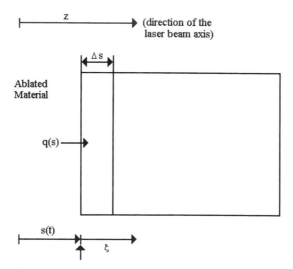

Fig. 17.15 Ablation interface energy balance which is depicted as the amount of energy required to ablate a portion Δs of the tissue.

tion. Furthermore, an important observation can be readily made about the temperature profile within the tissue as discussed later.

Note that on the right-hand side of Eq. (17.76) ds/dt must be positive for the ablation process to proceed. This means that on the left-hand side the temperature, $\partial T/\partial z$, must be positive. Consequently, the subsurface temperature must be higher than the surface temperature. That is, subsurface tissue must be superheated to temperatures higher than the surface ablation temperature so that thermal energy can be provided to the surface for the ablation process to proceed. This is in accordance with one of the corollaries of the second law of thermodynamics which states that heat can be transported only from hotter points to colder points. In fact, the higher than ablation threshold temperature puts the subsurface tissue in a "metastable" equilibrium condition which may be perturbed by internal tissue conditions and results in nucleation and vaporization below the surface and manifests itself by an "explosion" and mechanical tearing of the tissue surface. Initiation of the ablation process involves nonequilibrium nucleation processes which are not considered in the current approach.

17.6.4 Analytical Solution of Laser Heating in a Purely Absorbing Medium

The governing equations for laser irradiation with ablation were described in Section 17.6.3. Here, the governing equations will be repeated for a one-dimensional semi-infinite and purely absorbing medium. The governing equation for the preablation heating stage can be written as

$$\frac{\partial \rho c T}{\partial t} = \frac{\partial^2 T}{\partial z^2} + \mu_a I e^{-\mu_a z} \qquad (17.77)$$

where μ_a is the absorption coefficient. This equation is valid up to the onset of ablation which is assumed to occur when the surface temperature reaches the ablation threshold temperature, T_{ab}. The details of analysis of ablation are beyond the scope of this chapter and will not be considered.

In the late 1970s a nondimensionalization of the heat conduction equation for an axisymmetric threedimensional case of preablation laser heating of tissue by a Gaussian beam in an absorbing medium was solved. The following relations will transform the governing equations into a dimensionless moving frame:

$$\theta = \frac{T - T_0}{T_{ab} - T_0} \quad \xi = (I_0 c/kL)z \quad \tau = (I_0 c/kL)(I_0/\rho L)t \qquad (17.78)$$

The variables introduced are, respectively, dimensionless temperature θ, dimensionless coordinate ξ in the moving frame with origin at the ablation front, and dimensionless time τ. A dimensionless absorption parameter, B, and a dimensionless heating parameter λ are also defined as

$$B = (kL/I_0 c)\alpha$$
$$\lambda = c(T_{ab} - T_0)/L \qquad (17.79)$$

Analytical solutions of the nondimensional form of the governing equations can be found by Laplace transformation of the space variable ξ. The solution is as follows:

$$\theta(\xi,\tau) = \frac{1}{B\lambda}\left\{ 2B\sqrt{\tau}\,\text{ierfc}\left[\frac{\xi}{2\sqrt{\tau}}\right] - e^{-B\xi} \right.$$
$$\left. + \frac{1}{2}e^{B^2\tau}\left(e^{-B\xi}\text{erfc}\left[B\sqrt{\tau} - \frac{\xi}{2\sqrt{\tau}}\right] + e^{B\xi}\text{erfc}\left[B\sqrt{\tau} + \frac{\xi}{2\sqrt{\tau}}\right]\right)\right\} \tag{17.80}$$

This equation describes the temperature field as a function of the dimensionless space variable ξ and dimensionless time τ and is valid up to the onset of ablation when $\theta = 1$. The symbol ierfc indicates the integral of the function erfc, ierfc$(z) = \int_z^\infty \text{erfc}(t)dt$. This equation can be implemented with relative ease using modern user-friendly computer tools such as MATLAB or Mathematica.

An example of the graph of this solution, which is the progression in time of the temperature profile as a function of depth, is shown in Fig. 17.16.

By letting $\theta = 1$ (i.e., $T = T_{ab}$) at $\xi = 0$ in Eq (17.78), the following transcendental algebraic equation is obtained, which can be solved numerically for the time for the onset of ablation τ_{ab}:

$$B\lambda = \frac{2}{\sqrt{\pi}}B\sqrt{\tau_{ab}} + e^{B^2\tau_{ab}}\text{erfc}[B\sqrt{\tau_{ab}}] - 1 \tag{17.81}$$

Note that for large values of $B\lambda$ the behavior is almost linear with a slope of $2\backslash\sqrt{\pi}$.

The current analysis did not consider the case of scattering tissue. An analytical solution for Q_L could be obtained using the diffusion approximation approach described in Section 17.3. However, implementation of that solution into Eq. (17.72) to solve analytically for temperature would be highly cumbersome at best. There-

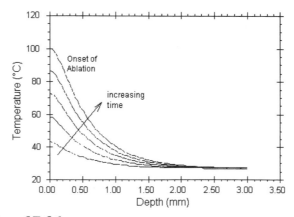

Fig. 17.16 Nondimensional temperature as a function of nondimensional depth at various times prior to the onset of ablation.

fore, a solution of temperature field would normally require a numerical approach, such as the finite difference or the finite element method.

17.6.5 Biochemical Damage Analysis

The Arrhenius model for prediction of damage was introduced previously, and the damage function was defined in Section 17.6.2. For numerical computation it is more convenient (due to large values of A and E) to rewrite the integrand of this function in an equivalent form so that Eq. (17.83) becomes

$$\Omega(z,t) = \int_0^t \exp\left(\ln A - \frac{E}{RT(z,t)} \right) dt \tag{17.82}$$

Two sets of values were used for A and E — one for pig skin and another for human vessel wall. The value of $\Omega(z, t)$ can be determined by replacing for $T(z,t)$ from the analytical solution of the previous section. The preablation solution should be used up to the onset of ablation, thereafter, the ablation-stage solution should be used.

The extent of damage or the position of the damage front at a given time t can be found by setting $\Omega = 1$ in Eq. (17.82) and solving for z. Alternatively, for every point z, Eq. (17.82) can be integrated by a controlled variable time-step procedure up to the time when $\Omega = 1$. The latter method was used to determine the position of damage front as a function of time and the difference in the value integrand at the upper and lower bounds of integration was monitored and used to change the time step. An example of such a calculation was done for a 5 W laser pulse of 500 ms duration. The calculation was done using values for both A and E from pig skin and human vessel wall. The results, together with the position of ablation front as a function of time (details are beyond the scope of this chapter), are shown in Fig. 17.17.

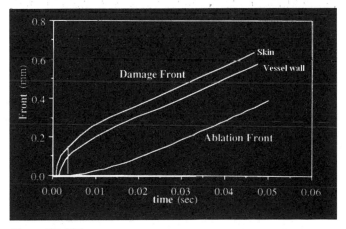

Fig. 17.17 Ablation and damage front propagation. For the same irradiation conditions, there would be more thermal damage beyond the ablation front in skin than in the vessel wall.

The vertical line at about 0.0035 s indicates the onset of ablation which indicates that there had been biochemical damage in the tissue before the onset of ablation. After the onset of ablation, the difference between the damage front and ablation front is the actual extent of damage in the tissue intact. The important observation from Fig. 17.17 is the difference in the prediction of the extent of damage using the two different sets of values for A and E. For the same irradiation conditions, thermal damage would propagate faster in skin than it would in vessel wall. This is expected since the damage mechanisms for vessel wall and skin are different.

Example Problem 17.6

Consider the skin subjected to a constant temperature heating T. For exposure durations $t = 1$ μs, $t = 1$ ms, $t = 1$ s, $t = 60$ s, and $t = 30$ min, find the critical temperature (in °C) T_c required to achieve thermal damage, $\Omega = 1$, during the exposure. Plot T_c vs exposure duration.

Solution

Given that the exposure durations are 1 μs, 1 ms, 1 s, 60 s, and 30 min and R is the universal gas constant $= 2$ cal/gm °C, T_c (critical temperature at which point $\Omega = 1$) must be calculated for each case.

Assumptions: Values of ln A and E are approximated to pig tissue values of 102.72 and 74,000, respectively.

Equation:

$$\Omega(z, t) = \int_0^t \exp\left(\ln A - E/(RT(z,t))\right)dt$$

$$\therefore \; 1 = \int_0^t \exp\left(\ln A - E/(RT_c)\right)dt$$

$$\therefore \; 1 = \int_0^t \exp\left(102.72 - 74000/2 \times T_c)\right)dt \qquad \text{(ex.1)}$$

$$\therefore \; 1 = \exp\left(102.72 - 74000/2 \times T_c)\right) \int_0^t dt \qquad \text{(ex.2)}$$

$$\therefore \; 1 = \exp\left(102.72 - 74000/2 \times T_c)\right)t \qquad \text{(ex.3)}$$

Solving Eq. (ex.3) for all values of t, we can get T_c for those values. Tabulated values of t and T_c are as follows:

Exposure time t (s)	Critical temperature T_c (°C)
$1\,e^{-6}$	143.1548
$1\,e^{-3}$	113.153
1	87.2
60	74.56
1800	72

17.6.6 Effect of Vaporization and Ablation Temperature

The coefficient f_L was introduced previously in relation to the heat of ablation as the water fraction parameter. Another nonvaporization phenomenon which can affect this coefficient is ejection of material due to subsurface nucleation. As discussed previously, chunks of the material may be ejected without vaporization. Another factor that can affect the ablation rate is the excess energy required for pyrolysis which, as discussed previously, manifests itself as a higher ablation threshold temperature.

Although the whole solution of the ablation problem in this chapter was not considered, in order to understand how these phenomena affect the ablation rate the steady-state ablation velocity may be used. This equation can be directly derived to be

$$v_{ss} = \frac{I}{\rho c \Delta T_{ab} + \rho f_L L} \tag{17.83}$$

Fixing the values of I, ρ, c, and L, the effect of the parameters f_L and ΔT_{ab} on v_{ss} can be determined. In the following example the physical properties of water are used and I is chosen to be 267.8 W/cm^2 so that $v_{ss} = 1$ mm/s for $\Delta T_{ab} = 100°C$ and $f_L = 1$. Figure 17.18 shows the effect of ablation temperature on the ablation ve-

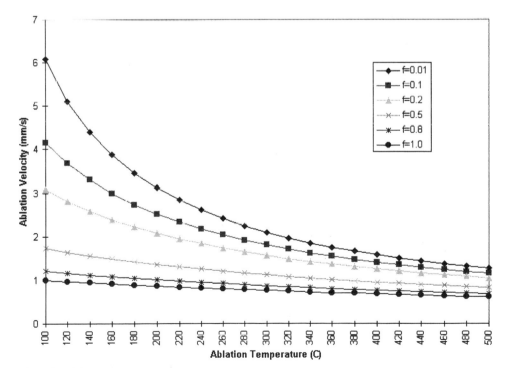

Fig. 17.18 The effect of ablation temperature on ablation velocity for various water content levels.

locity for various values of f_L from 0.01 to 1.0. The general effect of an increase either in ablation temperature or in f_L is a decrease in ablation velocity as is also obvious from Eq. (17.83). Observe, however, that the larger the value of f_L, the less effect the ablation temperature has on the ablation velocity. For instance, whereas for $f_L = 0.1$ ablation velocity drops from 4 to 1.15 mm/s over a temperature change from 100 to 500°C, for the same range and $f_L = 0.8$ the change is only from 1.20 to 0.69 mm/s. Therefore, if f_L indicates the fraction of water in tissue, for example, a decrease in ablation rate should be of concern as a result of higher pyrolysis/ablation temperatures only for tissue with low water content (e.g., 30% or lower). A cross section of families of curves in Fig. 17.18 at $\Delta T_{ab} = 200°C$ for more values of f_L is shown in Fig. 17.19. This figure shows that a change of f_L from 0 to 1 for $\Delta T_{ab} = 200°C$ results in a change in the ablation velocity from 3.2 to 0.86 mm/s; that is, the ablation velocity can be more than three times as large as its value of $f_L = 1.0$. The families of curves in Fig. 17.18 are more closely packed at higher values of ablation temperature. This means that the effect of a change in f_L on the ablation velocity is less significant at higher ablation temperatures.

17.7   FIBER OPTICS AND WAVEGUIDES IN MEDICINE

Rigid tubes for the examination of body cavities were used many centuries before the birth of Christ; however, it was not until the 1800s that illumination was used in the tube by means of a candle and 45° mirror. The introduction in the early 1900s of multiple lenses to transmit images led to semiflexible tubes for insertion into the body. The use of fiber-optic probes based on thin and transparent threads of glass

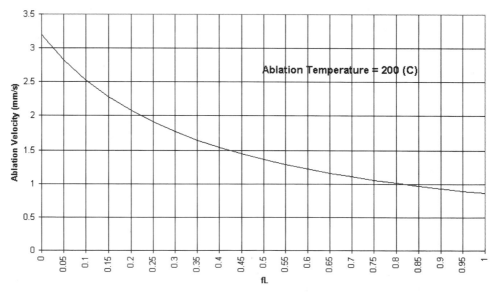

Fig. 17.19   The effect of water content, f_L, on ablation velocity.

dates back to the late 1920s but was dormant for two decades until the idea was revived in the 1950s. The first medical instrument, a flexible fiber-optic gastroscope, was developed and first used on patients in 1959. In the 1960s the first lasers were developed, and in the early 1970s there was a rapid development in the field of fiber optics for communications. All these events contributed to the modern fiber-optic probes and endoscopes used today.

17.7.1 Principles of Fiber Optics and Waveguides

In Section 17.3.1, the interaction of light with a nonparticipating medium was described. Light was described as rays and the Fresnel equations for the interaction of the light rays at the boundaries of two media were derived. It was determined that, depending on the index of refraction of two slab-type materials and the angle of incidence of the light, the amount of reflection and refraction could change. Although transmission of light through an optical fiber is a complex problem, the phenomenon can be understood with a simple geometrical model. As depicted in Fig. 17.20, the fiber can be thought of as a long rod of transparent material in which the rod or core of the fiber has a higher index of refraction (n_1) than the surrounding cylindrical shell material or cladding (n_2). The propagation of the light occurs down the core because of the total internal reflection of the light from the core–cladding interface which, from Snell's law, occurs when the angle of reflection at this interface is greater than the critical angle. In order for the light to be internally reflected at the core–cladding interface, the light injected into the end of the fiber must be smaller than a cone with some limiting angle θ_0 defined by

$$n_0\sin(\theta_0) = (n_1^2 - n_2^2)^{1/2} = \text{NA} \tag{17.84}$$

where NA, which stands for numerical aperture, is defined by this limiting value. In silica fibers the NA is generally between 0.2 and 0.4. As can be determined in the previous equation, increasing the difference in the index of refraction of the core and the cladding will increase the NA and also the acceptance angle. A large NA and acceptance angle gives rise to a large number of rays with different reflection angles or different transmission modes. A different zigzag of the beam path is thus

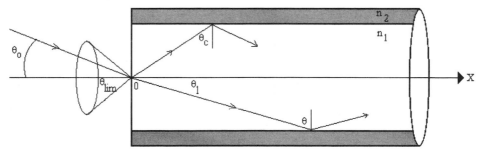

Fig. 17.20 Basic model of an optical fiber with a cylindrical core with index of refraction (n_1) and cladding index (n_2) where $n_2 < n_1$.

made for each mode. For a small core diameter, a single mode propagation can be generated in a fiber.

Many different kinds of optical fibers have been created with different structure, geometry, and materials depending on the ultimate application. For instance, the tips of the fibers can be changed to produce side-firing beams for therapeutic applications such as coagulation of prostate tissue. The fiber tips can also be tapered for pinpoint application of the light beam or made as a diffuse tip for broad uniform application of the light. In general, optical fibers have been classified in terms of the refractive index profile of the core and whether there are single modes or multi-modes propagating in the fiber. For instance, as depicted in Fig. 17.21, if the fiber core has a uniform or constant refractive index it is called a step-index fiber; if it has a nonuniform, typically parabolic, refractive index that decreases from the center to the cladding interface it is known as a graded-index fiber; and if the fiber core is small with a low NA only a single mode will propagate. The graded-index fibers have been shown to reduce the modal dispersion by a factor of 100 times and increase the bandwidth over a comparably sized step-index fiber. These general classifications apply to most fibers, but special fiber geometries are often required, as discussed later, for endoscopic coherent fiber bundle fiber imaging, single fibers or noncoherent bundles for sensing and diagnostics, and optical fibers made for high-powered therapeutic applications.

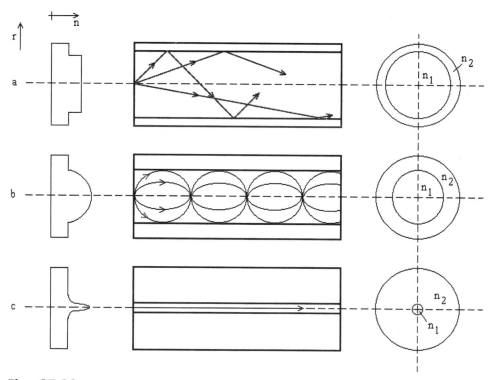

Fig. 17.21 Typical geometry of the main types of optical fibers including a (a) step index, (b) graded index, and (c) single-mode fiber.

17.7.2 Coherent Bundles for Imaging

Optical fibers can be bundled together in an aligned fashion such that the orders of the fibers at both ends are identical. Such a fiber bundle is called a coherent bundle or an ordered bundle. Of course, coherence means a correlation between the spatial positions of the fibers at both ends and has nothing to do with the light coherence. The important property of coherent bundles is their capability of transmitting images through a flexible channel. If an image is projected onto one end of a bundle, a replicate of the image is produced at the other end. Coherent bundles of optical fibers are the key components in endoscopes.

In endoscopic applications, an internal organ is imaged and viewed outside the body in a minimally invasive fashion. An incoherent (nonordered) bundle of optical fibers is used to illuminate the portion to be imaged inside human body. A coherent (ordered) bundle of optical fibers is used to transmit an image of the target portion. A white light source is usually used for the illumination such that a real color image of the tissue can be obtained.

The quality of image transmission is mainly determined by two factors: light collection and image resolution. The collection power of each individual optical fiber is limited by the diameter of its core and the NA. A large NA and core diameter allows a good transmission of light from the illuminated object to the eye of the physician. The image resolution indicates how fine details can be seen and is limited by the core diameter, d, of the cores of the individual fibers. The resolution, in number of discernible lines per millimeter, is approximately $d/2$. The smaller the core diameter, the better the image resolution. Due to the limited resolution, a straight line in an object may appear zigzagged in the image.

Cladding of each individual optical fiber is required to minimize or avoid cross talking among the fibers. Unclad fibers were used in early endoscopes and had poor imaging quality. Light in unclad fibers may leak from the core and cross into other fibers. The cross talk causes an overlay of various portions of an image during transmission and hence produces a blurred image.

The requirements of high light collection and good image resolution conflict. A large core diameter allows good light transmission but gives poor resolution. A thick cladding layer avoids cross talk but limits light collection and image resolution. A trade-off has to be made. In practice, the core diameter is usually 10–20 μm and the cladding thickness is approximately 1.5–2.5 μm.

Several other factors may deteriorate the image obtained with a coherent bundle. Some stray light may transmit through the cladding layers into the cores. The stray light would add an undesirable background that reduces the image contrast. Defective fibers in a coherent bundle would cause a serious problem. If a defective fiber does not transmit any light, a static dark spot appears in the image. If the ordering of the fibers is not identical at both ends, image distortion will degrade the images.

Lenses may be added at both ends of an imaging coherent bundle to adjust the magnification. Although the distance between the objective lens and the fiber bundle can be adjusted in principle to adjust the focusing of the coherent bundle, a fixed focus with a large depth of focus is often used for simplicity. A VCR may be used to record the images in real time while the images are displayed on a monitor.

Endoscopes may be specially built for imaging of various organs. The following are commonly used endoscopes: angioscopes for veins and arteries, arthroscopes (or orthoscopes) for the joints, bronchoscopes for the bronchial tubes, choledoschoscopes for the bile duct, colonoscopes for the colon, colposcopes for the vagina, cystoscopes for the bladder, esophagoscopes for the esophagus, gastroscopes for the stomach and intestines, laparoscopes for the peritoneum, laryngoscopes for the larynx, and ventriculoscopes for the ventricles in the brain.

17.7.3   Diagnostic and Sensing Fiber Probes

In terms of sensors for minimally invasive measurements into the body, fiber-optic probes are relatively new to the field (1980s). To understand the potential capability of these probes a sensor is first defined and some of the requirements for a good sensor are discussed. A sensor is a device that transforms an input parameter or measurand into another parameter known as the signal. For instance, a displacement membrane on the tip of a fiber probe could transform a pressure signal into a light intensity change which is then depicted as a change in voltage from a light detector. The requirements for any good sensor are specificity or the ability to pick out one parameter without interference of the other parameters, sensitivity or the capability to measure small changes in a given measurand, accuracy or closeness to the true measurement, and low cost. As with most sensors, fiber optics have to trade-off these parameters to within some limit, for example, one may be able to get 95% accuracy at a reasonable cost but to obtain 99% accuracy might require a huge increase in cost (i.e., military specifications of components). In addition to the previous requirements, fiber-optic probes also offer the potential for miniaturization, good biocompatability for the visible and NIR wavelengths, speed since light is used, and safety since no electrical connections to the body are required.

All fiber-optic probes transmit light into the body and the light, directly or indirectly, interacts with the biological parameter of interest, be it a biological fluid, tissue, pressure, or body temperature. The interaction causes a change in the light beam or beams that travel back to the detection system. The returning signal can be physically separated from the input signal in a separate fiber or fibers, or it can be separated if the output beam is at a separate wavelength from the input beam, as is the case for fluorescence probes. Fiber-optic sensors can be used for each of the previous chemically and physically based measurements described in Sections 17.4 and 17.5. However, rather than discuss each fiber-optic probe for each application, the rest of this section will focus on separating the fiber probes into two classes — indirect and direct fiber-optic sensors. Direct fiber-optic sensors are defined as those probes in which the light interacts directly with the sample. For instance, absorption measurements can theoretically be made to determine such things as blood oxygenation or to quantify blood analytes with either two side-firing fiber-optic probes separated at a fixed distance as depicted in Fig. 17.22a or an evanscent wave fiber optic which has the cladding stripped off so that some of the light traveling down the core travels out of the fiber, interacts with the sample, and then returns to the fiber core as depicted in Fig. 17.22b. In Fig. 17.22c a fiber-optic bundle is depicted which could be used to transmit and receive light to and from tissue to distinguish

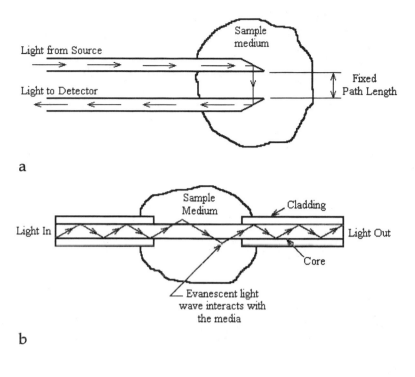

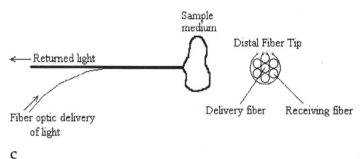

Fig. 17.22 Direct fiber-optic probe designs including (a) absorption probe using two side-firing fibers, (b) an evanescent wave probe in which the light transmits from the core to the sample and back into the core, and (c) a multi-fiber design for use in distinguishing normal from cancerous tissue by directly measuring autofluorescence or Raman spectra.

between normal and cancerous or precancerous lesions using direct measurement of tissue autofluorescence or Raman spectrum. Interferometers such as the Fabrey–Perot type described in Section 17.4.4 have been designed using partially reflecting mirrors built into a fiber optic, and any physical change imparted on the fiber, such as that due to fluctuations in body temperature or pressure, can be directly measured by the fiber in the form of a change in the interference pattern of the light.

Lastly, the fiber-optic probe for measurement of tissue optical properties described in Section 17.3.4 is another example of a direct probe which measures reflections from the sample. It should be noted that these descriptions are somewhat oversimplified in that a great deal of design must be done to gain the specificity, accuracy, and sensitivity required to measure these various low-level parameters in the noisy environment of the body. For indirect fiber-optic measurements a miniaturized transducer (sometimes referred to as an optode) is attached to the distal end of the fiber so that the light interacts with this transducer and it is the transducer that interacts with the sample. A displacement optode such as that depicted in Fig. 17.23a can be used to monitor pressure or temperature changes by simply causing less light to be specularly reflected into the fiber as the tip is moved. In addition, fluorescent chemistry can be placed within a membrane on the distal tip of the fiber, as depicted in Fig. 17.23b, in which the amount of fluorescent light produced is a measure of the concentration of a particular blood analyte. These indirect methods of measurement are also simplified and the actual final design to obtain an accurate and sensitive signal can be quite complicated. Overall, the indirect fiber-optic measurements typically have higher specificity over direct measurement approaches but at the expense of requiring a more complicated probe.

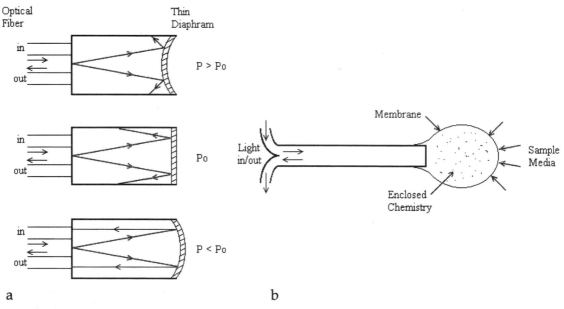

Fig. 17.23 Indirect fiber-optic probe designs including (a) a thin reflectance diaphragm for pressure or temperature measurements and (b) an optode with some chemistry enclosed within a membrane at the distal tip of the fiber, which can be used to determine a particular analyte concentration using fluorescent chemistry specific to the analyte of interest.

17.8 BIOMEDICAL OPTICAL IMAGING

Medical imaging has revolutionized the practice of medicine in the past century. Physicians are empowered to "see" through the human body for abnormalities noninvasively and to make diagnostic decisions rapidly, which has impacted the therapeutic outcomes of the detected diseases.

Medical imaging dates back to 1895 when X rays were discovered serendipitously by German physicist W. C. Roentgen, who received the first Nobel prize in physics in 1901 (*http://www.almaz.com/nobel/*) for this important work). Human anatomy can be easily imaged by simple X-ray projections, such as chest X rays, which are still being used in clinics. Contemporary medical imaging began with the invention of computerized tomography (CT) in the 1970s. G. N. Hounsfield in England produced the first computer-reconstructed images experimentally based on the theoretical foundation laid by A. M. Cormack in the United States. Both of them were awarded the Nobel prize in medicine in 1979. The essence of CT is that if an object is viewed from a number of different angles, then a cross-sectional image of it can be computed, i.e., "reconstructed."

The advent of CT has inspired other new tomographic and even 3-D imaging techniques. The application of reconstruction to conventional nuclear medicine imaging led to positron emission tomography and single photon emission computed tomography. A similar application to the technique of nuclear magnetic resonance led to magnetic resonance imaging (MRI).

Different imaging modalities are used to detect different aspects of biological tissues through a variety of contrast mechanisms. Using X-ray imaging electron density is primarily detected. In MRI, proton density and its associated relaxation properties are detected. Using ultrasonography acoustic impedance is detected. In nuclear imaging, nuclear radiation emitted from the body is detected after introducing a radiopharmaceutical inside the body to tag a specific biochemical function. The advantages and disadvantages of different imaging modalities may be illustrated using breast cancer detection. Breast cancer is the most common malignant neoplasm and the leading cause of cancer deaths in women in the United States. A means for prevention of breast cancer has not been found, and early detection and treatment is the best solution to improve cure rate. Currently, X-ray mammography and ultrasonography are clinically used for breast cancer detection. Mammography is currently the only reliable means of detecting nonpalpable breast cancers. As a supplementary tool, ultrasound is used to evaluate the internal matrix of circumscribed masses found using mammography or of palpable masses that are obscured by radiographically dense parenchyma using mammography. However, X-ray mammography is ionizing radiation, and imaging of radiographically dense breasts is difficult. Ultrasonography cannot detect many of the nonpalpable cancers that are not visible on mammograms of good quality. Most of the studies claiming that ultrasonography is useful in this regard compare it to poor-quality mammography or fail to compare it to mammography.

Several other techniques are under investigation for breast cancer imaging. MRI offers great promise for imaging of the radiographically dense breast. Breast MRI is superior to mammography in differentiating solid from cystic lesions and is equivalent to mammography in providing information regarding different parenchymal patterns. Injection of intravenous contrast material with MRI increases cancer detectability despite the fact that breast cancer and glandular tissues have similar magnetic resonance tissue characteristics. However, breast MRI is expensive, has inferior spatial resolution to mammography, and cannot image microcalcifications. Breast CT has been investigated for the differentiation of benign from malignant solid masses. Breast CT involves the use of intravenous injection of iodinated contrast material and has limited spatial resolution and high cost; hence, it is not suited for routine breast cancer screening.

17.8.1 Optical Tomographic Imaging

Nonionizing optical tomography is a new and active research field although projection light imaging was investigated as early as 1929. The optical properties of normal and diseased tissues are usually different despite the large variation of values in optical properties of the normal tissues alone. Therefore, it is possible to detect some breast cancers based on measurements of optical properties.

The optical difference is not surprising because cancerous tissues manifest significant architectural changes in the cellular and subcellular levels, and the cellular components that cause elastic scattering have dimensions typically on the order of visible to NIR wavelengths. Some tumors are associated with vascularization, where blood causes increased light absorption. The use of optical contrast agents may also be exploited to enhance the optical contrast between normal and abnormal tissues. Because the optical information is determined by the molecular conformations of biological tissues, optical imaging is expected to provide sensitive signatures for early cancer detection and monitoring.

Because tissues are optically turbid media that are highly scattering, light is quickly diffused inside tissues as a result of frequent scattering. The strong scattering has made optical detection of biological tissues challenging. A typical scattering coefficient for visible light in biological tissues is 100 cm^{-1} in comparison with 0.2 cm^{-1} for X rays used in medical diagnostics. Light transmitted through tissues is classified into three categories: ballistic light, quasi-ballistic light, and diffuse light. Ballistic light experiences no scattering by tissue and thus travels straight through the tissue. Ballistic light carries direct imaging information as does X-ray radiation. Quasi-ballistic light experiences minimal scattering and carries some imaging information. Multiply scattered diffuse light carries little direct imaging information and overshadows ballistic or quasi-ballistic light.

One of the techniques used for optical tomography is called "early photon imaging." If diffuse light is rejected, and ballistic or quasi-ballistic light is collected, buried objects can be detected much like X-ray projection. This technique uses a short-pulse laser (<1 ps pulse width) to illuminate the tissue. Only the initial portion of transmitted light is allowed to pass to a light detector, and the late-arriving light is gated off by a fast optical gate. Because the ballistic or quasi-ballistic photons travel along the shortest path length, they arrive at the detector sooner than diffuse photons. If only ballistic light is detected, the technique is called ballistic imaging. It has been shown that ballistic imaging is feasible only for tissue of thickness <1.4 mm or 42 mfps (mean free paths). Most ballistic imaging techniques reported in the literature have achieved approximately 30 mfps. Therefore, this approach is suitable for thin tissue samples but suffers loss of signal and resolution for thick tissues as a result of the strong scattering of light by the tissue.

Example Problem 17.7

Calculate the decay of ballistic light after penetrating a biological tissue 30 mfps thick. If the scattering coefficient of the tissue is 100 cm^{-1}, calculate the corresponding thickness in centimeters.

Solution

Based on Beer's law, the decay is $\exp(-30) = 9.4 \times 10^{-14}$. The thickness is 30/100 = 0.3 cm. ∎

For tissue of clinically useful thickness (5–10 cm), scattered light must be used to image breast cancers. It has been shown that for a 5-cm-thick breast tissue with an assumed absorption coefficient of 0.1 cm^{-1} and reduced scattering coefficient of 10 cm^{-1}, the detector must collect transmitted light that has experienced at least 1100 scattering events in the tissue to yield enough signal. Therefore, ballistic light or even quasi-ballistic light does not exist for practical purposes. However, if a 10-mW visible or NIR laser is incident on one side of the 5-cm-thick breast tissue, it has been estimated, using diffusion theory, that the diffuse transmittance is on the order of 10 nW/cm^2 or 10^{10} photons/s cm^2, which is detectable using a photomultiplier tube capable of single-photon counting. Similarly, the diffuse transmittance through a 10-cm-thick breast tissue would be on the order of 1 pW/cm^2 or 10^6 photons/s cm^2. The significant transmission of light is due to the low absorption coefficient despite the high scattering coefficient.

Imaging resolution of pure laser imaging degrades linearly with increased tissue thickness. The temporal profiles of the scattered light may be detected using a streak camera. The early portion of the profiles was integrated to construct the images of buried objects in a turbid medium. This time-domain technique requires expensive short-pulse lasers and fast light detectors.

Optical-coherence tomography (OCT) has emerged as a useful clinical tool. This technique is based on Michelson interferometer (see Fig. 17.9) with a short-coherence length light source. One arm of the interferometer leads to the sample of interest, and the other leads to a reference mirror. The reflected optical beams are detected at the photodetector. The two beams coherently interfere only when the sample and the reference path lengths are equal to within the source coherence length. Heterodyne detection is performed by taking advantage of the direct Doppler frequency shift that results from the uniform high-speed scan of the reference path length. Recording the interference signal magnitude as a function of the reference mirror position profiles the reflectance of the sample, which produces an image similar to an ultrasonic A scan. OCT has achieved 10-μm resolution but a penetration depth of <1 mm.

OCT was recently extended to image blood flow in superficial vessels based on

Doppler shift. The blood flow causes a Doppler shift on the frequency of the light. Frequency analysis with fast Fourier transform of the optical interference signal yields the Doppler shift. The Doppler shift is used to calculate the velocity of the blood flow.

A technique for optical imaging of thick tissue is the frequency-domain technique, which is based on photon-density waves. The governing equation of photon-density waves is the following:

$$\partial\Phi(r,t)/c\partial t + \mu_a\Phi(r,t) - D\nabla^2\Phi(r,t) = S(r,t) \tag{17.85}$$

where $\Phi(r,t)$ is light fluence rate at point r and time t (W/cm^2), c is the speed of light in the medium (cm/s), μ_a is the absorption coefficient (cm^{-1}), and $S(r,t)$ is the source intensity (W/cm^{-3}). The frequency-domain imaging technique requires the use of complex inverse algorithms for image reconstruction. Amplitude-modulated (at approximately 100 MHz) laser light is used to illuminate the tissue at multiple sites. At each illumination site, an optical detector measures the amplitude and phase of diffuse light at multiple locations around the tissue. The measured diffuse light can be estimated by the diffusion equation (Eq. 17.85) if the optical properties of the tissue sample are known. Conversely, the optical properties can be calculated by use of the measured diffuse light, which is the image reconstruction. A sample reconstructed image is shown in Fig. 17.24, which was based on a theoretically generated data set of diffuse light.

17.8.2 Hybrid Optical Imaging

Several emerging imaging techniques, coined hybrid optical imaging, are being developed by combining relatively transparent acoustic energy with strongly scatter-

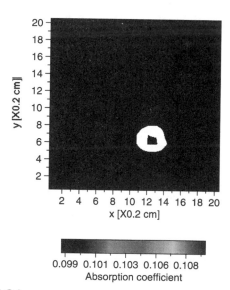

Fig. 17.24 Sample reconstructed image using photon-density waves. A simulated tumor with elevated absorption coefficient is visible in the lower right quadrant.

ing light. Ultrasound-modulated optical tomography, photoacoustic tomography, and sonoluminescence tomography will be briefly discussed.

In ultrasound-modulated optical tomography, an ultrasonic wave is focused into a scattering medium to modulate the laser light passing through the medium containing buried objects. The modulated laser light collected by a photomultiplier tube is related to the local mechanical and optical properties in the zone of ultrasonic modulation. If the buried objects have optical properties that are different from those of the background scattering medium, an image can be obtained by raster scanning the device.

The inverse of ultrasound-modulated optical tomography (acoustooptical tomography) is photoacoustic tomography. In photoacoustic tomography, a short-pulse light beam is injected into the scattering medium. The light is diffused in the medium and partially absorbed. If there is a strong optical absorber such as a tumor in the middle of the medium, more light will be absorbed by this optical absorber than by its neighbor. The absorbed optical energy is converted into heat. Due to thermal elastic expansion, an acoustic wave is generated. Stronger heat generation will produce a stronger acoustic wave. Therefore, a strong optical absorber emanates a strong acoustic wave. If several acoustic transducers are used to measure the acoustic signal around the medium, the absorber can be located based on the temporal distribution of the acoustic signals and hence produce an image of the medium.

The ultrasonic generation of light known as sonoluminescence (SL) was first reported in 1934 as a multiple-bubble sonoluminescence (MBSL). SL has attracted an extraordinary amount of attention in this decade since single-bubble sonoluminescence (SBSL) was reported in 1990. Although the full explanation of SL is still in development, it is well known that light is emitted when tiny bubbles driven by ultrasound collapse. The bubbles start out with a radius of several micrometers and expand to $\sim$50 μm due to a decrease in acoustic pressure in the negative half of a sinusoidal period. After the sound wave reaches the positive half of the period, the situation rapidly changes. The resulting pressure difference leads to a rapid collapse of the bubbles accompanied by emission of light. The flash time of SL has been measured to be in the tens of picoseconds. SBSL is so bright that it can be seen by the naked eye even in a lighted room, whereas MBSL is also visible in a darkened room. Researchers have envisioned possible applications of SL in sonofusion, sonochemistry, and building ultrafast lasers using the ultrafast flash of light in SL.

Sonoluminescent tomography (SLT) is a an application of SL that has been developed for cross-sectional imaging of strongly scattering media noninvasively. Sonoluminescence, which is generated internally in the medium by exposure of the medium to external ultrasound, is used to produce images of a scattering medium by raster scanning the medium. The spatial resolution is limited by the focal spot size of the ultrasound and can be improved by tightening the focus.

Optical imaging techniques may also measure the optical spectra of biological tissues noninvasively. In addition to the absorption spectra that are usually measured for other materials, the scattering spectra may also be measured. The optical spectra may be used to quantify important physiological parameters such as the saturation of hemoglobin oxygenation.

In summary, optical imaging and biomedical optics techniques in general have the following advantages: (i) the use of nonionizing radiation, (ii) the capability of measuring functional (physiological) parameters, (iii) the potential high sensitivity to pathologic state of biological tissues, and (iv) low cost. However, biomedical optics is a challenging research field in its infancy, and as it grows the participation of many diverse talented physicians, scientists, and engineers will be required.

EXERCISES

1. (a) Name at least three reasons why it might be desirable to use optic or fiber-optic sensors for biomedical applications. (b) Name three potential drawbacks to using optic or fiber-optic biomedical sensors.

2. In one paragraph, explain the similarities and differences between polarization, luminescence, absorption, scatter, and Raman scattering of light.

3. Given an interface between clear tissue (i.e., water) ($n_t = 1.33$) and glass ($n_g = 1.5$), compute the transmission angle for a beam incident in the water at 45°. If the transmitted beam is reversed so that it impinges on the interface from the glass side, calculate what the transmitted angle will be through the tissue. Does this make sense? Explain.

4. Using Matlab or some other software, plot ($R_\perp = r_\perp^2$) and ($R_P = r_P^2$) as a function of the input angle θ_i for a glass ($n_i = 1.5$) and air ($n_t = 1.0$) interface. Explain what it means.

5. An Nd:YAG laser operating at 1064 nm with a Gaussian beam is to be used for performing a Portwine stain treatment. Assume a heuristic model for light distribution in a homogeneous material such that $I(r,z) = I_0 \exp(-\mu z) \exp(-2(r/\omega_b)^2)$, where $\mu = \mu_a + \mu_s$, and μ_a and μ_s are the absorption and scattering coefficients. For the multilayer case below at $r = 0$ give the expressions and graph intensity I and the rate of heat generation Q in the tissue by the laser, $I(r = 0, z)$ and $Q(r = 0, z)$. Assume $I_0 = 1$. The absorption and scattering coefficients of each skin layer and blood vessel for this wavelength are given below. Discuss how selective photothermolysis of the blood layer is achieved.

	$\mu_a(cm^{-1})$	$\mu_s{'}(cm^{-1})$
Blood	20	20
Epidermis	0.1	10
Dermis	0.5	10

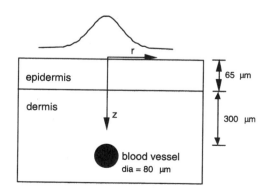

6. Consider a turbid biological tissue with an absorption coefficient $\mu_a = 0.1$ cm^{-1}. Using the diffusion approximation, graph on the same plane the logarithm of relative intensity vs radius for four different values of reduced scattering coefficient: $\mu_s' = 1, 10, 50,$ and 100 cm^{-1}; r varying from 0.1 to 2 cm. Describe the result in terms of the effect of scattering on the intensity profile.

7. For liver tissue irradiated by the pulsed Ho:YAG laser, determine the exposure duration during which the equation for heat conduction may be ignored. Assume the absorption coefficient at the Ho:YAG laser wavelength is 30 cm^{-1} and assume the thermal diffusivity of the tissue is the same as that of water. If the threshold temperature of ablation is 150°C, determine the threshold fluence (J/cm^2) for ablation.

8. Given a Helium–Neon laser-based fiber-optic Doppler probe at an angle of 60° with a blood vessel that registers a frequency shift of 63 KHz, what is the velocity of the blood? Is this a reasonable number from a physiologic point of view when compared to the average velocity in the major and minor blood vessels in the human body?

9. Assume you want to measure the thickness of a piece of tissue ($n_t = 1.33$) bounded by air ($n_t = 1.0$) using either the Michelson or Fabry–Perot interferometric approach. Your light source has a wavelength of 780 nm and your instrument can count 20 fringes. The tissue is ablated (or cut) with a high powered pulsed laser which removes approximately 2 μm of tissue per pulse. The maximum tissue thickness before it is all removed is 20 μm. (a) Determine if each of the interferometer systems will have a long enough dynamic range, without moving the system, to be used with this piece of tissue. Explain your result. (b) Assume you want to use either of your interferometric systems as a feedback unit for the laser removal of tissue. What is the range in the number of fringes (m_{min} and m_{max}) for each of your systems in order to measure both the minimum slice thickness and the maximum tissue thickness? Explain.

10. In general, when is a Fabry–Perot interferometer preferred over a standard Michelson interferometer? Can you make a Fabry–Perot interferometer which performs worse than a Michelson interferometer? Explain.

11. For Raman spectroscopy, what are the Stokes and anti-Stokes bands? Which bands are typically used for sensing? What are the main challenges that must be overcome for Raman spectroscopy to be used for biomedical sensing?

12. Assume you want to measure glucose through the anterior chamber of the eye as a means of noninvasively quantifying blood glucose. Given glucose has a specific rotation of 41.89°/(dm g/dL) at a wavelength of 656 nm and the anterior chamber of the eye has a path length of approximately 0.8 cm, calculate the concentration of glucose for a rotation of 15 millidegrees. Is this a reasonable value from a physiologic point of view? Would the patient be considered normal or diabetic?

13. In the NIR region of the optical spectrum between 600 and 1100 nm there are well-known absorbance peaks for oxygenated and deoxygenated hemoglobin. An assumption can be made for the moment that in this re-

gion the dominant optical signal is due to absorption (which is not generally the case). Given that you propagate through a 2-cm sample of tissue, you need to calculate three parameters: the concentration of oxyhemoglobin and deoxyhemoglobin and a background blood absorbance. At three wavelengths (758, 798, and 898 nm), you can measure the extinction coefficients for oxygenated (0.444, 0.702, and 1.123 mM^{-1} cm^{-1}) and deoxygenated hemoglobin (1.533, 0.702, and 0.721 mM cm^{-1}), respectively. The total absorption coefficients at these three wavelengths in the tissue can be measured using a time-resolved system as 0.293, 0.1704, and 0.1903 cm^{-1}, respectively. Calculate the oxygenated and deoxygenated hemoglobin levels given these parameters. Is this reasonable from a physiologic point of view? Explain.

14. Using the parameters of the equation for problem 7, create a graph of temperature vs time at $z = 0$, for t varying from 0 to 1 ms, using the analytical solution of the heat conduction equation given in the chapter. On the same graph plot temperature ignoring the effect of conduction, which assumes linear variation with time. Comment on your finding in relation to the diffusion time found in Exercise 10.

15. Consider skin tissue which is exposed to linear heating, for example, as in pulsed laser coagulation, such that $T = T_0 + mt$, where T_0 is initial temperature (e.g., 37°C) and m is the rate of heating (e.g., 10^3°C/s). Describe how you would find the critical temperature for thermal damage. Do you expect that to be lower or higher than the constant temperature of the same exposure duration?

16. For the study of ablation onset time, plot a graph of $B\sqrt{\tau_{ab}}$ vs $B\lambda$. Measure the slope for large values of $B\lambda$; it should be $2/\sqrt{\pi}$. For absorption coefficient of 100/cm and a laser intensity of 1000 W/cm², find the time for the onset of ablation using water thermal properties, assuming ablation initiates at 150°C.

17. In the visible region, tissue scattering dominates absorption. Assume the scattering coefficient of a 5-cm-thick tissue is 100 cm^{-1} and the wavelength of light is 0.5 μm. In order to detect on average a single ballistic photon transmitted through the tissue, what is the energy in Joules that is required for the incident light? Compare the energy of the incident light with the rest energy of the earth using Einstein's mass–eneregy equivalence equation $E = mc^2$, where the mass of the earth is 6×10^{24} kg.

18. Use diffusion theory to estimate light penetration in biological tissues. A 10-mW isotropic point source is buried in an infinite turbid medium. Assume that the absorption coefficient is 0.1 cm^{-1}, the scattering coefficient of a 5-cm-thick tissue is 100 cm^{-1}, the scattering anisotropy is the 0.9, and the wavelength of light is 0.5 μm. Calculate the light fluence rate 5 cm from the source.

19. Use diffusion theory to estimate sonoluminescence light transmission in biological tissues. A unit-power isotropic point source is buried in an infinite turbid medium. Assume that the absorption coefficient is 0.1 cm^{-1}, the scattering coefficient of a 5-cm-thick tissue is 100 cm^{-1}, and the scattering

anisotropy is 0.9. Calculate the light fluence rate integrated over a sphere of a 5-cm radius centered at the source.

SUGGESTED READING

Boisdé, G., and Harmer, A. (1996). *Chemical and Biochemical Sensing with Optical Fibers and Waveguides.* Artech House, Boston.

Culshaw, B., and Dakin, J. (1989). *Optical Fiber Sensors: Vol. I & II.* Artech House, Boston.

Hecht, E. (1987). *Optics.* Addison-Wesley, Reading, MA.

Katzir, A. (1993). *Lasers and Optical Fibers in Medicine.* Academic Press, San Diego.

Pedrotti, F. L., and Pedrotti, L. S. (1987). *Introduction to Optics, Second Edition.* Prentice Hall, Upper Saddle River, NJ.

Welch, A. J., and van Gemert, M. J. C. (1995). *Optical–Thermal Response of Laser Irradiated Tissue.* Plenum, New York.

Willard, H. H., Merritt, L. L. Jr., Dean, J. A., and Settle, F. A., Jr. (1988). *Instrumental Methods of Analysis,* 7th ed. Wadsworth, Belmont, CA.

18 REHABILITATION ENGINEERING AND ASSISTIVE TECHNOLOGY

Chapter Contents

At the conclusion of this chapter, the reader will be able to:

- Understand the role played by rehabilitation engineers and assistive technologists in the rehabilitation process

- Be aware of the major activities in rehabilitation engineering
- Be familiar with the physical and psychological consequences of disability
- Know the principles of assistive technology assessment and its objectives and pitfalls
- Be cognizant of key engineering and ergonomic principles of the field
- Be aware of the career opportunities and information sources available

18.1   INTRODUCTION

Since the late 1970s, there has been major growth in the application of technology to ameliorate the problems faced by people with disabilities. Various terms have been used to describe this sphere of activity, including prosthetics/orthotics, rehabilitation engineering, assistive technology, assistive device design, rehabilitation technology, and even biomedical engineering applied to disability. With the gradual maturation of this field, several terms have become more widely used, bolstered by their use in some federal legislation.

The two most frequently used terms today are **assistive technology** and **rehabilitation engineering.** While used somewhat interchangeably, they are not identical. In the words of James Reswick (1982), a pioneer in this field, "rehabilitation engineering is the application of science and technology to ameliorate the handicaps of individuals with disabilities." In contrast, assistive technology can be viewed as a product of rehabilitation engineering activities. Such a relationship is analogous to health care being the product of the practice of medicine.

One widely used definition for assistive technology is found in Public Law 100-407. It defines assistive technology as "any item, piece of equipment or product system whether acquired commercially off the shelf, modified, or customized that is used to increase or improve functional capabilities of individuals with disabilities." Notice that this definition views assistive technology as a broad range of devices, strategies, and/or services that help an individual to better carry out a functional activity. Such devices can range from "low"-technology devices that are inexpensive and simple to make to "high"-technology devices that are complex and expensive to fabricate. Examples of low-tech devices include dual-handled utensils and mouth sticks for reaching. High-tech examples include computer-based communication devices, reading machines with artificial intelligence, and externally powered artificial arms (Fig. 18.1).

Other terms often used in this field include **rehabilitation technology** and **orthotics** and **prosthetics.** Rehabilitation technology is that segment of assistive technology that is designed specifically to rehabilitate an individual from his or her present set of limitations due to some disabling condition, permanent or otherwise. In a classical sense, orthotics are devices that augment the function of an extremity, whereas prosthetics replace a body part both structurally and functionally. These two terms now broadly represent all devices that provide some sort of functional replacement. For example, an augmentative communication system is sometimes referred to as a speech prosthesis.

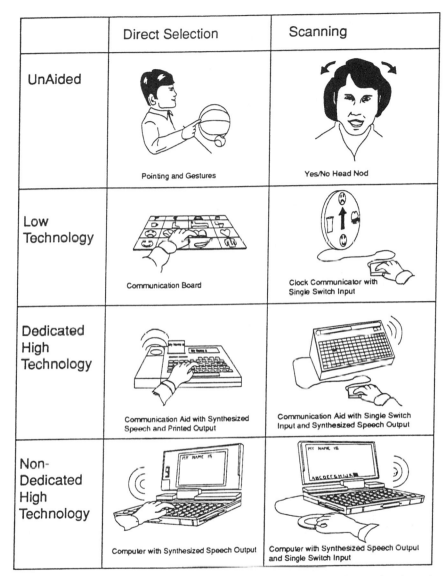

Fig. 18.1 Augmentative communication classification system. (from Church and Glennen, 1992).

18.1.1 History

A brief discussion of the history of this field will explain how and why so many different yet similar terms have been used to denote the field of assistive technology and rehabilitation. Throughout history, people have sought to ameliorate the impact of disabilities by using technology. This effort became more pronounced and concerted in the United States after World War II. The Veterans Administration (VA) realized that something should be done for the soldiers who returned from war with numerous and serious handicapping conditions. There were too few well-trained artificial limb and brace technicians to meet the needs of the returning soldiers. To train these much-needed providers, the federal government supported the establishment of a number of prosthetic and orthotic schools in the 1950s.

The VA also realized that the state of the art in limbs and braces was primitive and ineffectual. The orthoses and prostheses then available were uncomfortable, heavy, and offered limited function. As a result, the federal government established the Veterans Administration Prosthetics Research Board, whose mission was to improve the orthotics and prosthetic appliances that were available. Scientists and engineers formerly engaged in defeating the Axis powers now turned their energies toward helping people, especially veterans, with disabilities. As a result of their efforts, artificial limbs, electronic travel guides, and wheelchairs that were more rugged, lighter, cosmetically appealing, and effective were developed.

The field of assistive technology and rehabilitation engineering was nurtured by a two-prong approach in the federal government. One approach directly funded research and development efforts that would utilize the technological advances created by the war effort toward improving the functioning and independence of injured veterans. The other approach helped to establish centers for the training of prosthetists and orthotists, forerunners of today's assistive technologists.

In the early 1960s, another impetus to rehabilitation engineering came from birth defects found in infants born to expectant European women who took Thalidomide to combat "morning sickness." The societal need to enable children with severe deformities to lead productive lives broadened the target population of assistive technology and rehabilitation engineering to encompass children as well as adult men. Subsequent medical and technical collaboration in research and development produced externally powered limbs for people of all sizes and genders, automobiles that could be driven by persons with no arms, sensory aids for the blind and deaf, and various assistive devices for controlling a person's environment.

Rehabilitation engineering received formal governmental recognition as an engineering discipline with the passage of the federal Rehabilitation Act of 1973. The act specifically authorized the establishment of several centers of excellence in rehabilitation engineering. The formation and supervision of these centers were under the jurisdiction of the National Institute for Handicapped Research, which later became the National Institute on Disability and Rehabilitation Research. By 1976, about 15 rehabilitation engineering centers (RECs), each focusing on a different set of problems, were supported by grant funds totaling about $9 million per year.

The REC grants initially supported university-based rehabilitation engineering research and provided advanced training for graduate students. Beginning in the mid-1980s, the mandate of the RECs was broadened to include technology transfer and service delivery to persons with disabilities. During this period, the VA also established three of its own RECs in order to focus attention on some unique rehabilitation needs of veterans. Areas of investigation by VA and non-VA RECs include prosthetics and orthotics, spinal cord injury, lower and upper limb functional electrical stimulation, sensory aids for the blind and deaf, effects of pressure on tissue, rehabilitation robotics, technology transfer, personal licensed vehicles, accessible telecommunications, and vocational rehabilitation.

Another milestone, the formation of the Rehabilitation Engineering Society of North America (RESNA) in 1979, gave greater focus and visibility to rehabilitation engineering. Despite its name, RESNA is an inclusive professional society that welcomes everyone who is involved with the development, manufacturing, provision, and usage of technology for persons with disabilities. Members of RESNA include occupational and physical therapists, allied health professionals, special educators, and users of assistive technology. RESNA has grown to become an adviser to the government, a developer of standards and credentials, and, via its annual conferences and its journal, a forum for exchange of information and the display of state-of-the-art rehabilitation technology. In recognition of its expanding role and members who were not engineers, RESNA modified its name in 1995 to signify the Rehabilitation Engineering and Assistive Technology Society of North America.

Despite the need for and the benefits of providing rehabilitation engineering services, reimbursement for such services by third-party payers (e.g., insurance companies, social service agencies, and government programs) remained very difficult to obtain during much of the 1980s. Reimbursements for rehabilitation engineering services often had to be subsumed under more "accepted" categories of care, such as client assessment, prosthetic/orthotic services, or miscellaneous evaluation. For this reason, the number of practicing rehabilitation engineers remained relatively static despite a steadily growing demand for their services.

The shortage of rehabilitation engineers with suitable training and experience was specifically addressed in the Rehab Act of 1986 and the Technology-Related Assistance Act of 1988. These laws mandated that rehabilitation engineering services had to be available and funded for disabled persons. They also required an individualized work and rehabilitation plan (IWRP) for each vocational rehabilitation client. These two laws were preceded by the original Rehab Act of 1973 that mandated reasonable accommodations in employment and secondary education as defined by a least restrictive environment (LRE). Public Law 95-142 in 1975 extended the reasonable accommodation requirement to children 5–21 years of age and mandated an individual educational plan (IEP) for each eligible child. Table 18.1 summarizes the major United States federal legislation that has impacted the field of assistive technology and rehabilitation engineering.

In concert with federal legislation, several federal research programs have attempted to increase the availability of rehabilitation engineering services for persons with disabilities. The National Science Foundation (NSF), for example, initiated a program called the Bioengineering and Research to Aid the Disabled. The program's goals were (i) to provide student-engineered devices or software to disabled individuals that would improve their quality of life and degree of independence, (ii) to enhance the education of student engineers through real-world design experiences, and (iii) to allow the university an opportunity to serve the local community. The Office of Special Education and Rehabilitation Services in the U.S. Department of Education funded special projects and demonstration programs that addressed iden-

TABLE 18.1   Recent Major U.S. Federal Legislation That Has
Impacted Assistive Technologies

Legislation	Major assistive technology impact
Rehabilitation Act of 1973	Mandated reasonable accommodation, LRE in federally funded employment and higher education; requires both assistive technology devices and services
Education for All Handicapped Children Act of 1975 (PL 94-142)	Extended reasonable accommodation and LRE to ages 5–21 education; mandated IEP for each child: assistive technology plays major role in gaining access to educational programs
Handicapped Infants and Toddlers Act (PL 99-457)	Extended PL 94-142 to infants and to ages 3–5 years; expanded emphasis on educationally related assistive technologies
1986 Amendments to the Rehabilitation Act of 1973 (PL 99-506)	Required all states to include provision for assistive technology services in both state plan and IWRP for each client; Section 508 mandated equal access to electronic office equipment for all federal employees
Technology-Related Assistance for Individuals with Disabilities Act of 1988 (PL 100-407)	First legislation specifically related to assistive technologies; extends Section 508 to all funded states and mandates consumer-driven assistive technology services and statewide system change
Americans with Disabilities Act (ADA) of 1990 (PL 101-336)	Civil rights act for disabled; extends Sections 503, 504, and 508 and other provisions to all citizens in terms of public accommodation, private employment, transportation, and telecommunications
Individuals with Disabilities Education Act (IDEA)	Reauthorization of PL 94-102; extends assistive technology device and service definitions to education
Reauthorization of the Rehabilitation Act of 1973	Brings rehabilitation act language and mandates in line with ADA; defines rehabilitation technology as rehabilitation engineering and assistive technology devices and services; mandates rehabilitation technology as primary benefit to be included in IWRP

Note. From Cook and Hussey (1995).

tified needs such as model assessment programs in assistive technology, the application of technology for deaf–blind children, interdisciplinary training for students of communicative disorders (speech pathologists), special education, and engineering. In 1993, the National Institute on Disability Related Research committed $38.6 million to support Rehabilitation Engineering Centers that would focus on the following areas: adaptive computers and information systems, augmentative and alternative communication devices, employability for persons with low back pain, hearing enhancement and assistive devices, prosthetics and orthotics, quantification of physical performance, rehabilitation robotics, technology transfer and evaluation, improving wheelchair mobility, work site modifications and accommodations, geriatric assistive technology, personal licensed vehicles for disabled persons, rehabilitation technology services in vocational rehabilitation, technological aids for blindness and low vision, and technology for children with orthopedic disabilities.

18.1.2 Sources of Information

Like any other emerging discipline, the knowledge base for rehabilitation engineering was scattered in disparate publications in the early years. Owing to its interdisciplinary nature, rehabilitation engineering research papers appeared in such diverse publications as the *Archives of Physical Medicine & Rehabilitation, Human Factors, Annals of Biomedical Engineering, IEEE Transactions on Biomedical Engineering,* and *Biomechanics.* Some of the papers were very practical and application specific, whereas others were fundamental and philosophical. In the early 1970s, many important papers were published by the Veterans Administration in its *Bulletin of Prosthetic Research,* a highly respected and widely disseminated peer-reviewed periodical. This journal was renamed the *Journal of Rehabilitation R&D* in 1983. In 1989, RESNA began *Assistive Technology,* a quarterly journal that focused on the interests of practitioners engaged in technological service delivery rather than the concerns of engineers engaged in research and development. The IEEE Engineering in Medicine and Biology Society founded the *IEEE Transactions on Rehabilitation Engineering* in 1993 to give scientifically based rehabilitation engineering research papers a much needed home. This journal is published quarterly and covers the medical aspects of rehabilitation (rehabilitation medicine), its practical design concepts (rehabilitation technology), and its scientific aspects (rehabilitation science).

18.1.3 Major Activities in Rehabilitation Engineering

The major activities in this field can be categorized in many ways. Perhaps the simplest way to grasp its breadth and depth is to categorize the main types of assistive technology that rehabilitation engineering has produced (Table 18.2). The development of these technological products required the contributions of mechanical, material, and electrical engineers, orthopedic surgeons, prosthetists and orthotists, allied health professionals, and computer professionals. For example, the use of voice in many assistive devices, as both inputs and outputs, depends on digital signal processing chips, memory chips, and sophisticated software developed by electrical and computer engineers. Figures 18.2–18.4 illustrate some of the assistive technologies that are currently available. As explained in subsequent sections of this chapter, the proper design, development, and application of assistive technology devices require the combined efforts of engineers, knowledgeable and competent clinicians, informed end-users or consumers, and caregivers.

TABLE 18.2   Categories of Assistive Devices

Prosthetics and orthotics
    Artificial hand, wrist, and arms
    Artificial foot and legs
    Hand splints and upper limb braces
    Functional electrical stimulation orthoses
Assistive devices for persons with severe visual impairments
    Devices to aid reading and writing (e.g., closed-circuit TV magnifiers, electronic Braille, reading
        machines, talking calculators, and auditory and tactile vision substitution systems)
    Devices to aid independent mobility (e.g., laser cane, binaural ultrasonic eyeglasses, handheld
        ultrasonic torch, electronic enunciators, and robotic guide dogs)
Assistive devices for persons with severe auditory impairments
    Digital hearing aids
    Telephone aids (e.g., TDD and TTY)
    Lipreading aids
    Speech to text converters
Assistive devices for tactile impairments
    Cushions
    Customized seating
    Sensory substitution
    Pressure relief pumps and alarms
Alternative and augmentative communication devices
    Interface and keyboard emulation
    Specialized switches, sensors, and transducers
    Computer-based communication devices
    Linguistic tools and software
Manipulation and mobility aids
    Grabbers, feeders, mounting systems, and page turners
    Environmental controllers
    Robotic aids
    Manual and special-purpose wheelchairs
    Powered wheelchairs, scooters, and recliners
    Adaptive driving aids
    Modified personal licensed vehicles
Recreational assistive devices
    Arm-powered cycles
    Sports and racing wheelchairs
    Modified sit-down mono-ski

18.2   THE HUMAN COMPONENT

In order to knowledgeably apply engineering principles in design and fabrication technology that will help persons with disabling conditions, it is necessary to have a perspective on the human component and the consequences of various impairments. One way to view a human being is as a receptor, processor, and responder of information (Fig. 18.5). The human user of assistive technology perceives the envi-

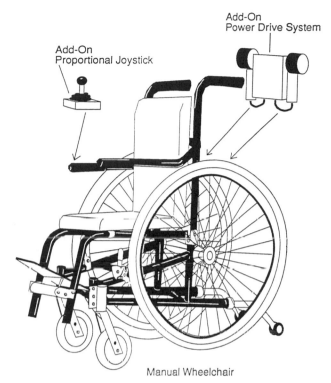

Add-On
Power Drive System

Add-On
Proportional Joystick

Manual Wheelchair

Fig. 18.2 Add-on wheelchair system (from Church and Glennen, 1992).

ronment via senses and responds or manipulates the environment via effectors. Interposed between the sensors and effectors are central processing functions that include perception, cognition, and movement control. **Perception** is the way in which the human being interprets the incoming sensory data. The mechanism of perception relies on the neural circuitry found in the peripheral nervous system and central psychological factors such as memory of previous sensory experiences. **Cognition** refers to activities that underlie problem solving, decision making, and language formation. **Movement control** utilizes the outcome of the processing functions described previously to form a motor pattern that is executed by the effectors (nerves, muscles, and joints). The impact of the effectors on the environment is then detected by the sensors, thereby providing feedback between the human and the environment. When something goes wrong in the information processing chain, disabilities often result. Table 18.3 lists the prevalence of various disabling conditions in terms of anatomic locations.

Interestingly, rehabilitation engineers have found a modicum of success when trauma or birth defects damage the input (sensory) end of this chain of information processing. When a sensory deficit is present in one of the three primary sensory

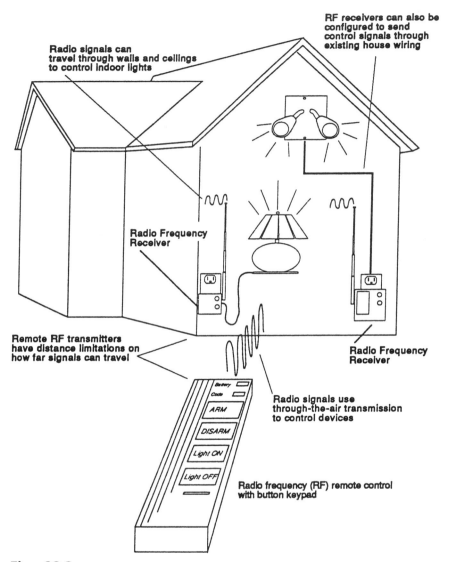

Fig. 18.3 Environmental control unit using radio frequency (RF) control (from Church and Glennen, 1992).

channels (vision, hearing, and touch), assistive devices can detect important environmental information and present it via one or more of the other remaining senses. For example, sensory aids for severe visual impairments utilize tactile and/or auditory outputs to display important environmental information to the user. Examples of such sensory aids include the laser cane, ultrasonic glasses, and the robotic guide dog. Rehabilitation engineers also have been modestly successful at replacing or

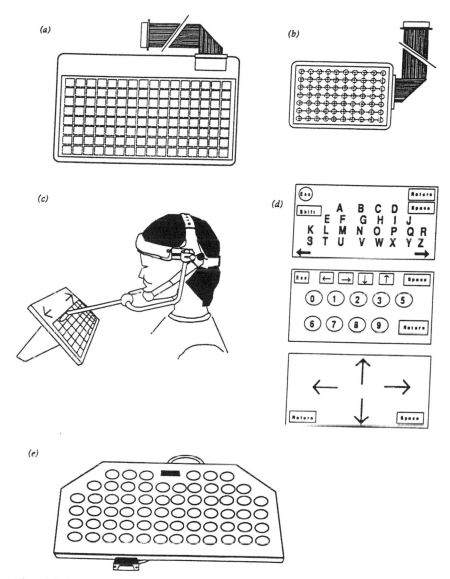

Fig. 18.4 Alternative keyboards can replace or operate in addition to the standard keyboard. (a) Expanded keyboards have a matrix of touch-sensitive squares that can be grouped together to form larger squares. (b) Minikeyboards are small keyboards with a matrix of closely spaced touch-sensitive squares. (c) The small size of a minikeyboard ensures that a small range of movement is needed to access the entire keyboard. (d) Expanded and minikeyboards use standard or customized keyboard overlays. (e) Some alternative keyboards plug directly into the keyboard jack of the computer, needing no special interface or software (from Church and Glennen, 1992).

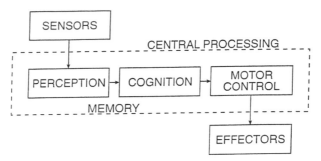

Fig. 18.5 An information processing model of the human operator of assistive technologies. Each block represents a group of functions related to the use of technology.

augmenting some motoric (effector) limitations (Fig. 18.6). As listed in Table 18.2, these include artificial arms and legs, wheelchairs of all types, environmental controllers, and, in the future, robotic assistants.

However, when dysfunction resides in the "higher information processing centers" of a human being, assistive technology has been much less successful ameliorating the resultant limitations. For example, rehabilitation engineers and speech pathologists have been unsuccessful in enabling someone to communicate effectively when that person has difficulty formulating a message (aphasia) following a stroke. Despite the variety of modern and sophisticated alternative and augmentative com-

TABLE 18.3 Prevalence of Disabling Conditions in the United States

45–50 million persons have disabilities that slightly limit their activities
    32% hearing
    21% sight
    18% back or spine
    16% leg and hip
    5% arm and shoulder
    4% speech
    3% paralysis
    1% limb amputation
7–11 million persons have disabilities that significantly limit their activities
    30% back or spine
    26% leg and hip
    13% paralysis
    9% hearing
    8% sight
    7% arm and shoulder
    4% limb amputation
    3% speech

Note. Data from Stolov and Clowers (1981).

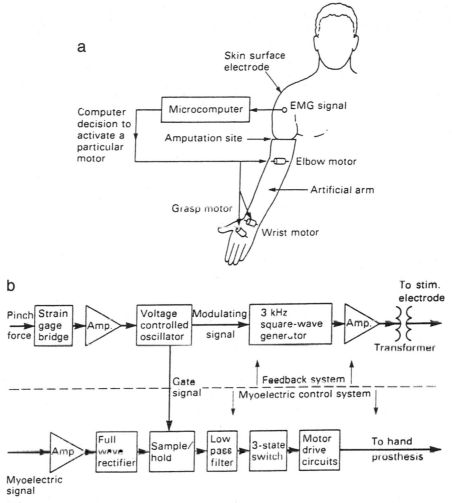

Fig. 18.6 (a) This system generates temporal signatures from one set of myoelectric electrodes to control multiple actuators. (b) Electrical stimulation of the forearm to provide force feedback may be carried out using a system like this one (from Webster *et al.,* 1985).

munication devices that are available, none has been able to replace the volitional aspects of the human being. If the user is unable to cognitively formulate a message, an augmentative communication device is often not helpful.

An awareness of the psychosocial adjustments to chronic disability is desirable because rehabilitation engineering and assistive technology seek to ameliorate the consequences of disabilities. Understanding the emotional and mental states of the person who is or becomes disabled is necessary so that offers of assistance and recommendations of solutions can be appropriate, timely, accepted, and, ultimately, used.

One of the biggest impacts of chronic disability is the minority status and socially devalued position that a disabled person experiences in society. Such loss of social status may result from the direct effects of disability (social isolation) and the indirect effects of disability (economic setbacks). Thus, in addition to the tremendous drop in personal income, a person who is disabled must battle three main psychological consequences of disability — the loss of self-esteem, the tendency to be too dependent on others, and passivity.

For individuals who become disabled through traumatic injuries, the adjustment to disability generally passes through five phases — shock, realization, defensive retreat or denial, acknowledgment, and adaptation or acceptance. During the first days after the onset of disability, the individual is usually in shock, feeling and reacting minimally with little awareness of what has happened. Counseling interventions or efforts of rehabilitation technologists are typically not very effective at this time.

After several weeks, the individual usually begins to acknowledge the reality and seriousness of the disability. Anxiety, fear, and even panic may be the predominant emotional reactions. Depression and anger may also occasionally appear during this phase. Because of the individual's emotional state, intense or sustained intervention efforts are not likely to be useful during this time as well.

In the next phase, the individual makes a defensive retreat in order to not be psychologically overwhelmed by anxiety and fear. Predominant among these defenses is denial — claiming that the disability is only temporary and that full recovery will occur. Such denial may persist or reappear occasionally long after the onset of disability.

Acknowledgment of the disability occurs when the individual achieves an accurate understanding of the nature of the disability in terms of its limitations and likely outcome. Persons in this phase may exhibit a thorough understanding of the disability but may not possess a full appreciation of its implications. The gradual recognition of reality is often accompanied by depression and a resultant loss of interest in many activities previously enjoyed.

Adaptation, or the acceptance phase, is the final and ultimate psychological goal of a person's adjustment to disability. An individual in this phase has worked through the major emotional reactions to disability described earlier. Such a person is realistic about the likely limitations and is psychologically ready to make the best use of his or her potential. Intervention by rehabilitation engineers or assistive technologists during the acknowledgment and acceptance phases of the psychosocial adjustment to disability is usually appropriate and effective. Involvement of the disabled individual in identifying needs, planning the approach, and choosing among possible alternatives can be very beneficial psychologically as well as physically.

18.3   PRINCIPLES OF ASSISTIVE TECHNOLOGY ASSESSMENT

Rehabilitation engineers not only need to know the physical principles that govern their designs but they also must adhere to some key principles that govern the applications of technology for people with disabilities. To be successful, the needs, preferences, abilities, limitations, and even environment of the individual seeking the assistive technology must be carefully considered. There are at least five major misconceptions that exist in the field of assistive technology:

1: Assistive technology can solve all the problems. Although assistive devices can make accomplishing tasks easier, technology alone cannot mitigate all the difficulties that accompany a disability.

2: Persons with the same disability need the same assistive devices. Assistive technology must be individualized because similarly disabled persons can have very different needs, wants, and preferences.

3: Assistive technology is necessarily complicated and expensive. Sometimes low-technology devices are the most appropriate and even preferred for their simplicity, ease of use and maintenance, and low cost.

4: Assistive technology prescriptions are always accurate and optimal. Past experience clearly demonstrates that the application of technology for persons with disabilities is inexact and will change with time. Changes in the assistive technology user's health, living environment, preferences, and circumstances will require periodic reassessment by the user and those rehabilitation professionals who are giving assistance.

5: Assistive technology will always be used. According to data from the 1990 U.S. Census Bureau's National Health Interview Survey, about one-third of the assistive devices not needed for survival are unused or abandoned just 3 months after they are initially acquired.

In addition to avoiding common misconceptions, a rehabilitation engineer and technologist should follow several principles that have proven to be helpful in matching appropriate assistive technology to the person or consumer. Adherence to these principles will increase the likelihood that the resultant assistive technology will be welcomed and fully utilized:

Principle 1: The user's goals, needs, and tasks must be clearly defined, listed, and incorporated as early as possible in the intervention process. To avoid overlooking needs and goals, checklists and premade forms should be used. Many helpful forms can be found in the suggested reading list at the end of this chapter.

Principle 2: Involvement of rehabilitation professionals with differing skills and know-how will maximize the probability for a successful outcome. Depending on the purpose and environment in which the assistive technology device will be used, a number of professionals should participate in the process of matching technology to a person's needs. Table 18.4 lists various technology areas and the responsible professionals.

Principle 3: The user's preferences, cognitive and physical abilities and limitations, living situation, tolerance for technology, and probable change in the future must be thoroughly assessed, analyzed, and quantified. Rehabilitation engineers will find that the highly descriptive vocabulary and qualitative language

TABLE 18.4 Professional Areas in Assistive Technology

Technology area	Responsible professionals[a]
Academic and vocational skills	Special education
	Vocational rehabilitation
	Psychology
Augmentative communication	Speech–language pathology
	Special education
Computer access	Computer technology
	Vocational rehabilitation
Daily living skills	Occupational therapy
	Rehabilitation technology
Specialized adaptations	Rehabilitation engineering
	Computer technology
	Prosthetics/orthotics
Mobility	Occupational therapy
	Physical therapy
Seating and positioning	Occupational therapy
	Physical therapy
Written communication	Speech–language pathology
	Special education

[a] Depending on the complexity of technical challenges encountered, an assistive technologist or a rehabilitation engineer can be added to the list of responsible professionals.

used by nontechnical professionals needs to be translated into attributes that can be measured and quantified. For example, whether a disabled person can use one or more upper limbs should be quantified in terms of each limb's ability to reach, lift, and grasp.

Principle 4: Careful and thorough consideration of available technology for meeting the user's needs must be carried out to avoid overlooking potentially useful solutions. Electronic databases (e.g., assistive technology websites and websites of major technology vendors) can often provide the rehabilitation engineer or assistive technologist with an initial overview of potentially useful devices to prescribe, modify, and deliver to the consumer.

Principle 5: The user's preferences and choice must be considered in the selection of the assistive technology device. Surveys indicate that the main reason why assistive technology is rejected or poorly utilized is inadequate consideration of the user's needs and preferences. Throughout the process of searching for appropriate technology, the ultimate consumer of that technology should be viewed as a partner and stakeholder rather than as a passive, disinterested observer.

Principle 6: The assistive technology device must be customized and installed in the location and setting where it primarily will be used. Often seemingly minor or innocuous situations at the usage site can spell success or failure in the application of assistive technology.

Principle 7: Not only must the user be trained to use the assistive device, but also the attendants or family members must be made aware of the device's intended purpose, benefits, and limitations. For example, an augmentative communication device usually will require that the communication partners adopt a different mode of communication and modify their behavior so that the user of the device can communicate a wider array of thoughts and even assume a more active role in the communication paradigm, such as initiating a conversation or changing the conversational topic. Unless the attendants or family members alter their ways of interacting, the newly empowered individual will be dissuaded from utilizing the communication device, regardless of how powerful it may be.

Principle 8: Follow-up, readjustments, and reassessments of the user's usage patterns and needs are necessary at periodic intervals. During the first 6 months following the delivery of the assistive technology device, the user and others in that environment learn to accommodate the new device. As people and the environment change, what worked initially may become inappropriate, and the assistive device may need to be reconfigured or reoptimized. Periodic follow-up and adjustments will lessen technology abandonment and the resultant waste of time and resources.

18.4 PRINCIPLES OF REHABILITATION ENGINEERING

Knowledge and techniques from different disciplines must be utilized in order to design technological solutions that can alleviate problems caused by various disabling conditions. Since rehabilitation engineering is intrinsically multidisciplinary, identifying universally applicable principles for this emerging field is difficult. Often the most relevant principles depend on the particular problem being examined. For example, principles from the fields of electronic and communication engineering are paramount when designing an environmental control system that is to be integrated with the user's battery-powered wheelchair. However, when the goal is to develop an implanted functional electrical stimulation orthosis for an upper limb impaired by spinal cord injury, principles from neuromuscular physiology, biomechanics, biomaterials, and control systems would be the most applicable.

Whatever the disability to be overcome, however, rehabilitation engineering is inherently design oriented. Rehabilitation engineering design is the creative process of identifying needs and then devising a product to fill those needs. A systematic approach is essential in order to successfully complete a rehabilitation project. Key elements of the design process involve the following major steps: analysis, synthesis, evaluation, decision, and implementation.

Analysis

Inexperienced but enthusiastic rehabilitation engineering students often respond to a plea for help from someone with a disability by immediately thinking about possible solutions. They overlook the important first step of doing a careful analysis of the problem or need. What they discover after much ineffectual effort is that a thorough investigation of the problem is necessary before any meaningful solution can be found. Rehabilitation engineers first must ascertain where, when, and how often the problem arises. What is the environment or the task situation? How have others performed the task? What are the environmental constraints (cost, size, speed, weight, location, physical interface, etc.)? What are the psychosocial constraints (user preferences, support of others, gadget tolerance, cognitive abilities, and limitations)? What are the financial considerations? Answers to these questions will require diligent investigation and quantitative data such as the weight and size to be lifted, the shape and texture of the object, and the operational features of the desired device. An excellent endpoint of problem analysis would be a list of operational features or performance specifications that the "ideal" solution should have. Such a list of performance specifications can serve as a valuable guide for choosing the best solution during later phases of the design process.

Example Problem 18.1

Develop a set of performance specifications for an electromechanical device to raise and lower the lower leg of a wheelchair user (to prevent edema).

Solution

A sample set of performance specifications of the ideal mechanism might be written as follows:

- Be able to raise or lower leg in 5 s
- Independently operable by the wheelchair occupant
- Have an emergency stop switch
- Compatible with existing wheelchair and its leg rests
- Quiet operation
- Entire adaptation weighs no more than 5 pounds ∎

Synthesis

A rehabilitation engineer who is able to describe in writing the nature of the problem is likely to have some ideas for solving the problem. While not strictly sequential, the synthesis of possible solutions usually follows the analysis of the problem. The synthesis of possible solutions is a creative activity that is guided by previously learned engineering principles and supported by handbooks, design magazines, catalogs, and consultation with other professionals. While making and evaluating the list of possible solutions, a deeper understanding of the problem usually is reached and other, previously not apparent, solutions arise. A recommended endpoint for the synthesis phase of the design process include sketches and technical descriptions of each possible solution.

Evaluation

Depending on the complexity of the problem and other constraints, such as time and money, the two or three most promising solutions should undergo further evaluation, possibly via field trials with mock-ups, computer simulations, and/or detailed mechanical drawings. Throughout the evaluation process, the end user and other stakeholders in the problem/solution should be consulted. Experimental results from field trials should be carefully recorded, possibly on videotape, for later review. One useful method for evaluating promising solutions is to use a quantitative comparison chart to rate how well each solution meets or exceeds the performance specifications and operational characteristics based on the analysis of the problem.

Decision

The choice of the final solution is often made easier when it is understood that it always involves a compromise. After comparing the various promising solutions, more than one may appear equally satisfactory. At this point, the final decision may be made based on the preference of the user or some other intangible factor that is difficult to anticipate. Sometimes choosing the final solution may involve consulting with someone else who may have encountered a similar problem. What is most important, however, is that careful consideration to the user's preference (Principle 5 of assistive technology) be given.

Implementation

To fabricate, fit, and install the final (or best) solution requires additional project planning that, depending on the size of the project, may range from a simple list of tasks to a complex set of scheduled activities involving many people with different skills.

Example Problem 18.2

List the major technical design steps needed to build the automatic battery-powered leg raiser described in Example Problem 18.1.

Solution

The following are some of the key design steps:

- Mechanical design of the linkages to raise the wheelchair's leg rests
- Static determination of the forces needed to raise the occupant's leg
- Determination of the gear ratios and torque needed from the electric motor
- Estimation of the power drain from the wheelchair batteries
- Purchase of the electromechanical components
- Fabrication of custom parts and electronic components
- Assembly, testing, and possible redesign
- Field trials and evaluation of prototype device   ∎

18.4.1 Key Engineering Principles

Each discipline and subdiscipline that contributes to rehabilitation engineering has its own set of key principles that should be considered when a design project is begun. For example, a logic family must be selected and a decision whether to use synchronous or asynchronous sequential circuits must be made at the outset in digital design. A few general hardware issues are applicable in a wide variety of design tasks, including worst-case design, computer simulation, temperature effects, reliability, and product safety. In worst-case design, the electronic or mechanical system must continue to operate adequately even when variations in component values degrade performance. Computer simulation and computer-aided design (CAD) software can often be used to predict how well an overall electronic system will perform under different combinations of component values or sizes.

The design should also take into account the effects of temperature and environmental conditions on performance and reliability. For example, temperature extremes can reduce a battery's capacity. Temperature may also affect reliability, so proper venting and use of heat sinks should be employed to prevent excessive temperature increases. For reliability and durability, proper strain relief of wires and connectors should be used in the final design.

Product safety is another very important design principle, especially for rehabilitative or assistive technology. An electromechanical system should always incorporate a panic switch that will halt a device's operation if an emergency arises. Fuses and heavy-duty gauge wiring should be employed throughout for extra margins of safety. Mechanical stops and interlocks should be incorporated to ensure proper interconnections and to prevent dangerous or inappropriate movement.

When the required assistive device must lift or support some part of the body, an analysis of the static and dynamic forces (biomechanics) that are involved should be performed. The simplest analysis is to determine the static forces needed to hold the object or body part in a steady and stable manner. The basic engineering principles needed for static and dynamic analysis usually involve the following steps: (i) Determine the force vectors acting on the object or body part, (ii) determine the moment arms, and (iii) ascertain the centers of gravity for various components and body segments. Under static conditions, all the forces and moment vectors sum to zero. For dynamic conditions, the governing equation is Newton's second law of motion in which the vector sum of the forces equals mass times an acceleration vector ($F = ma$).

Example Problem 18.3

Suppose a 125-lb person lies supine on a board resting on knife edges spaced 72 in. apart (Fig. 18.7). Assume that the center of gravity of the lower limb is located through the centerline of the limb and 1.5 in. above the knee cap. Estimate the weight of this person's right leg.

Solution

Record the scale reading with both legs resting comfortably on the board and when the right leg is raised almost straight up. Sum the moments about the left knife edge pivot to yield the following static equation:

$$WD = L(S_1 - S_2)$$

where W is the weight of the right limb, L is the length of the board between the supports, S_1 is the scale reading with both legs resting on the board, S_2 is the scale reading with the right leg raised, and D is the horizontal distance through which the limb's center of gravity was moved when the limb was raised. Suppose the two scale readings were 58 lbs for S_1 and 56 lbs for S_2 and $D = 7$ in. Substituting these

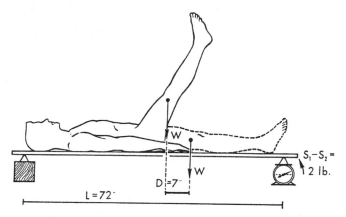

Fig. 18.7 Method of weighing body segments with board and scale (from Le Veau, 1976).

values into the equation above would yield an estimate of 20.6 lbs as the weight of the right leg. ∎

Example Problem 18.4

A patient is exercising his shoulder extensor muscles with wall pulleys (Fig. 18.8). Weights of 20, 10, and 5 lbs are loaded on the weight pan, which weighs 4 lbs. The patient is able to exert 45 lbs on the pulley. What is the resultant force of the entire system? What are the magnitude and direction of acceleration of the weights?

Solution

All the weights and the pan act straight down, whereas the 45 lbs of tension on the pulley's cable exerts an upward force. The net force (F) is 6 lbs upward. Apply Newton's second law of motion, $F = ma$, where m is the mass of the weights and the pan and a is the acceleration of the weights and pan. The mass, m, is found by dividing the weight of 39 lbs by the acceleration of gravity of 32.2 ft/s^2 to yield $m = 1.21$ slugs. Substituting these values into $a = F/m$ yields an acceleration of 4.96 ft/s^2 in the upward direction. ∎

18.4.2 Key Ergonomic Principles

Ergonomics or human factors is another indispensable part of rehabilitation engineering and assistive technology design. Applying information about human behavior, abilities, limitations, and other characteristics to the design of tools, adaptations, electronic devices, tasks, and interfaces is especially important when designing assistive technology because persons with disabilities generally will be less able to accommodate poorly designed or ill-fitted assistive devices. Several ergonomic principles that are especially germane to rehabilitation engineering are discussed in the following sections.

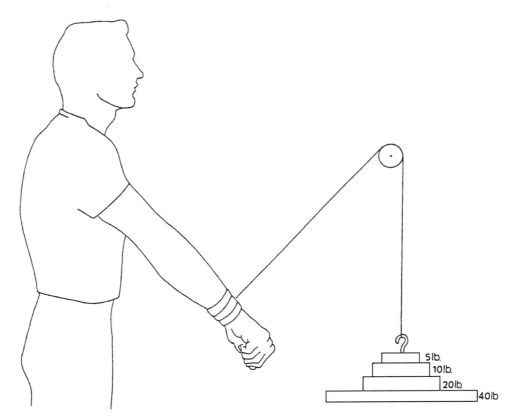

Fig. 18.8 Patient exercising his shoulder extensor muscles with wall pulleys (from Le Veau, 1976).

Principle of Proper Positioning

Without proper positioning or support, an individual who has lost the ability to maintain a stable posture against gravity may appear to have greater deformities and functional limitations than truly exist. For example, the lack of proper arm support may make the operation of even an enlarged keyboard unnecessarily slow or mistake prone. Also, the lack of proper upper trunk stability may unduly limit the use of an individual's arms because the person is relying on them for support.

During all phases of the design process, the rehabilitation engineer must ensure that whatever adaptation or assistive technology is being planned, the person's trunk, lower back, legs, and arms will have the necessary stability and support at all times (Fig. 18.9). Consultation with a physical therapist or occupational therapist familiar with the focus individual during the initial design phases should be considered if postural support appears to be a concern. Common conditions that require considerations of seating and positioning are listed in Table 18.5.

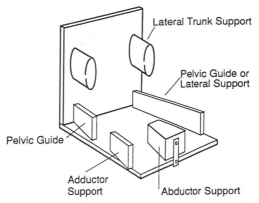

Fig. 18.9 Chair adaptations for proper positioning (from Church and Glennen, 1992).

Principle of the Anatomical Control Site

Since assistive devices receive command signals from the users, users must be able to reliably indicate their intent by using overt, volitional actions. Given the variety of switches and sensors that are available, any part of the body over which the user has reliable control in terms of speed and reliability can serve as the anatomical control site. Once the best site has been chosen, an appropriate interface for that site can be designed by using various transducers, switches, joysticks, and keyboards. In addition to obvious control sites such as the finger, elbow, shoulder, and knee, subtle movements such as raising an eyebrow or tensing a particular muscle can also be employed as the control signal for an assistive device. Often, the potential control sites can and should be analyzed and quantitatively compared for their relative speed, reliability, distinctiveness, and repeatability of control actions. Field trials using mock-ups, stopwatches, measuring tapes, and a video camera can be very helpful for collecting such performance data.

Principle of Simplicity and Intuitive Operation

The universal goal of equipment design is to achieve intuitively simple operation, and this is especially true for electronic and computer-based assistive devices. The key to intuitively simple operation lies in the proper choice of compatible and optimal controls and displays. Compatibility refers to the degree to which relationships between the control actions and indicator movements are consistent, respectively, with expectations of the equipment's response and behavior. When compatibility relationships are incorporated into an assistive device, learning is faster, reaction time is shorter, fewer errors result, and the user's satisfaction is higher. Although people can and do learn to use adaptations that do not conform to their expectations, they do so at a price. Hence, the rehabilitation engineer needs to be aware of and follow some common compatibility relationships and basic ergonomic guidelines:

TABLE 18.5   Conditions That Require Consideration of Seating and Positioning

Condition	Description and Characteristics	Seating Considerations
Cerebral palsy	Nonprogressive neuromuscular	
Increased tone (high tone)	Fixed deformity, decreased movements, abnormal patterns	Correct deformities, improve alignment, decrease tone
Decreased tone (low tone)	Subluxations, decreased active movement, hypermobility	Provide support for upright positioning, promote development of muscular control
Athetoid (mixed tone)	Excessive active movement, decreased stability	Provide stability, but allow controlled mobility for function
Muscular dystrophies	Degenerative neuromuscular	
Duchenne	Loss of muscular control proximal to distal	Provide stable seating base, allow person to find balance point
Multiple sclerosis	Series of exacerbations and remissions	Prepare for flexibility of system to follow needs
Spina bifida	Congenital anomaly consisting of a deficit in one or more of the vertebral arches, decreased or absent sensation	Reduce high risk for pressure concerns, allow for typically good upper extremity and head control
Spinal cord injury	Insult to spinal cord, partial or complete loss of function below level of injury, nonprogressive once stabilized, decreased or absent sensation, possible scoliosis/kyphosis	Reduce high risk for pressure concerns, allow for trunk movements used for function
Osteogenesis imperfecta	Connective tissue disorder, brittle bone disease, limited functional range, multiple fractures	Provide protection
Orthopedic impairments	Fixed or flexible	If fixed, support if flexible, correct
Traumatic brain injury	Severity dependent on extent of central nervous system damage, may have cognitive component, nonprogressive once stabilized	Allow for functional improvement as rehabilitation progresses, establish a system that is flexible to changing needs
Elderly		
Typical aged	Often, fixed kyphosis, decreased bone mass, and decreased strength, incontinence	Provide comfort and visual orientation, moisture-proof, accommodate kyphosis
Aged secondary to primary disability	Example—older patients with cerebral palsy may have fixed deformities	Provide comfort, support deformities

Adapted with permission from *Evaluating, Selecting, and Using Appropriate Assistive Technology*, J. C. Galvin, M. J. Scherer, p. 66, © 1996 Aspen Publishers, Inc.

■ The display and corresponding control should bear a physical resemblance to each other.
■ The display and corresponding control should have similar physical arrangements and/or be aided by guides or markers.

- The display and corresponding control should move in the same direction and within the same spatial plane (e.g., rotary dials matched with rotary displays and linear vertical sliders matched with vertical displays).
- The relative movement between a switch or dial should follow population stereotypic expectations (e.g., an upward activation to turn something on, a clockwise rotation to increase something, and scale numbers that increase from left to right).

Additional guidelines for choosing among various types of visual displays are given in Table 18.6.

Principle of Display Suitability

In selecting or designing displays for transmission of information, the selection of the sensory modality is sometimes a foregone conclusion, such as when designing a warning signal for a visually impaired person. When there is an option, however, the rehabilitation engineer must take advantage of the intrinsic advantages of one sensory modality over another for the type of message or information to be conveyed. For example, audition tends to have an advantage over vision in vigilance types of warnings because of its attention-getting qualities. A more extensive comparison of auditory and visual forms of message presentation is presented in Table 18.7.

Principle of Allowance for Recovery from Errors

Both rehabilitation engineering and human factors or ergonomics seek to design assistive technology that will expand an individual's capabilities while minimizing errors. However, human error is unavoidable no matter how well something is designed. Hence, the assistive device must provide some sort of allowance for the possibility of errors without seriously compromising system performance or safety. Errors can be classified as errors of omission, errors of commission, sequencing errors, and timing errors.

A well-designed computer-based electronic assistive device will incorporate one or more of the following attributes:

- The design makes it inherently impossible to commit the error (e.g., using jacks and plugs that can fit only one way or the device automatically rejects inappropriate responses while giving a warning).
- The design makes it less likely, but not impossible, to commit the error (e.g., using color-coded wires accompanied by easily understood diagrams).
- The design reduces the damaging consequences of errors without necessarily reducing the likelihood of errors (e.g., using fuses and mechanical stops that limit excessive movement or speed).

Principle of Adaptability and Flexibility

One fundamental assumption in ergonomics is that devices should be designed to accommodate the user and not vice versa. As circumstances change and/or as the user gains greater skill and facility in the operation of an assistive device, its operational characteristics must adapt accordingly. In the case of an augmentative electronic communication device, its vocabulary set should be changed easily as the user's needs, skills, or communication environment change. The method of selection

TABLE 18.6 General Guide to Visual Display Selection

To display	Select	Because	Example
Go, no go, start, stop, on, off	Light	Normally easy to tell if it is on or off.	
Identification	Light	Easy to see (may be coded by spacing, color, location, or flashing rate; may also have label for panel applications).	
Warning or caution	Light	Attracts attention and can be seen at great distance if bright enough (may flash intermittently to increase conspicuity)	
Verbal instruction (operating sequence)	Enunciator light	Simple "action instruction" reduces time required for decision making	
Exact quantity	Digital counter	Only one number can be seen, thus reducing chance of reading error	
Approximate quantity	Moving pointer against fixed scale	General position of pointer gives rapid clue to the quantity plus relative rate of change.	
Set-in quantity	Moving pointer against fixed scale	Natural relationship between control and display motions.	
Tracking	Single pointer or cross pointers against fixed index	Provides error information for easy correction.	
Vehicle attitude	Either mechanical or electronic display of position of vehicle against established reference (may be graphic or pictorial)	Provides direct comparison of own position against known reference or base line.	

Abstracted from Human Factors in Engineering and Design, 7th ed., by Sanders and McCormick, 1993.

TABLE 18.7 Choosing between Auditory and Visual Forms of Presentation

Use auditory presentation if	Use visual presentation if
The message is simple	The message is complex
The message is short	The message is long
The message will not be referred to later	The message will be referred to later
The message deals with events in time	The message deals with location in space
The message calls for immediate action	The message does not call for immediate action
The visual system of the person is overburdened	The auditory system of the person is overburdened
The message is to be perceived by persons not in the area	The message is to be perceived by someone very close by
Use artificially generated speech if the listener cannot read	Use visual display if the message contains graphical elements

Note. Adapted and modified from Saunders and McCormick (1993, p. 53, Table 3.1).

and feedback also should be flexible, perhaps offering direct selection of the vocabulary choices in one situation while reverting to a simpler row–column scanning in another setting. The user should also be given the choice of having auditory, visual, or a combination of both as feedback indicators.

Principle of Mental and Chronological Age Appropriateness

When working with someone who has had lifelong and significant disabilities, the rehabilitation engineer cannot presume that the mental and behavioral age of the individual with disabilities will correspond closely with that person's chronological age. In general, people with congenital disabilities tend to have more limited variety, diversity, and quantity of life experiences. Consequently, their reactions and behavioral tendencies often mimic those of someone much younger. Thus, during assessment and problem definition, the rehabilitation engineer should ascertain the functional age of the individual to be helped. Behavioral and biographical information can be gathered by direct observation and by interviewing family members, teachers, and social workers.

Special human factor considerations also need to be employed when designing assistive technology for very young children and elderly individuals. When designing adaptations for such individuals, the rehabilitation engineer must consider that they may have a reduced ability to process and retain information. For example, generally more time is required for very young children and older people to retrieve information from long-term memory, to choose among response alternatives, and to execute responses. Studies have shown that elderly persons are much slower in searching for material in long-term memory, in shifting attention from one task to another, and in coping with conceptual, spatial, and movement incongruities.

The previous findings suggest that the following design guidelines be incorporated into any assistive device intended for an elderly person:

- Strengthen the displayed signals by making them louder, brighter, larger, etc.
- Simplify the controls and displays to reduce irrelevant details that could act as sources of confusion.
- Maintain a high level of conceptual, spatial, and movement congruity.
- Reduce the requirements for monitoring and responding to multiple tasks.
- Provide more time between the execution of a response and the need for the next response. Where possible, let the person set the pace for the task.
- Allow more time and practice for learning the material or task to be performed.

18.5   PRACTICE OF REHABILITATION ENGINEERING AND ASSISTIVE TECHNOLOGY

18.5.1   Career Opportunities

As efforts to constrain health care costs intensify, it is reasonable to wonder whether career opportunities will exist for rehabilitation engineers and assistive technologists. Given an aging population, the rising number of children born with cognitive and physical developmental disorders, the impact of recent legislative mandates (Table 18.1), and the proven cost benefits of successful rehabilitation, the demand for assistive technology (new and existent) will likely increase rather than decrease. Correspondingly, employment opportunities for technically oriented persons interested in the development and delivery of assistive technology should steadily increase as well.

In the early 1980s, the value of rehabilitation engineers and assistive technologists was unappreciated and thus required significant educational efforts. While the battle for proper recognition may not be entirely over, much progress has been made in the intervening years. For example, Medi-Cal, the California version of the federally funded medical assistance program, now funds the purchase and customization of augmentative communication devices. Many states routinely fund technology devices that enable people with impairments to function more independently.

Career opportunities for rehabilitation engineers and assistive technologists currently can be found with hospital-based rehabilitation centers, public schools, vocational rehabilitation agencies, manufacturers, and community-based rehabilitation technology suppliers; opportunities also exist as independent contractors. For example, one recent job announcement for a rehabilitation engineer contained the following job description (Department of Rehabilitative Services, Commonwealth of Virginia, 1997):

Provide rehabilitation engineering services and technical assistance to persons with disabilities, staff, community agencies, and employers in the area of employment and reasonable accommodations. Manage and design modifications and manufacturer of adaptive equipment. . . . Requires working knowledge of the design, manufacturing techniques, and appropriate engineering problem-solving techniques for persons with disabilities. Skill in the operation of equipment and tools and the ability to direct others involved in the manufacturing of assistive devices. Ability to develop and effectively present educational programs related to rehabilitation engineering. Formal training in engineering with a concentration in rehabilitation engineering, mechanical engineering, or biomedical engineering or demonstrated equivalent experience a requirement.

The salary and benefits of the job in the previous announcement were competitive with those of other types of engineering employment opportunities. Similar announcements regularly appear in trade magazines such as *Rehab Management* and *Team Rehab* and in newsletters of RESNA.

An example of employment opportunities in a hospital-based rehabilitation center is provided by the Bryn Mawr Rehabilitation Center in Malvern, Pennsylvania. The center is part of the Jefferson Health System, a nonprofit network of hospitals and long-term, home care, and nursing agencies. Bryn Mawr's assistive technology center provides rehabilitation engineering and assistive technology services. Its newly opened geriatric rehabilitation clinic brings together several of the facility's departments to work at keeping senior citizens in their own homes longer. This clinic charges Medicare for assessments and the technology needed for independent living. Support for this new program stems from the potential cost savings related to keeping older people well and in their own homes.

Rehabilitation engineers and assistive technologists can also work for school districts that need to comply with the Individuals with Disabilities Education Act. A rehabilitation engineer working in such an environment would perform assessments, make equipment modifications, customize assistive devices, assist special education professionals in classroom adaptations, and advocate to funding agencies for needed educationally related technologies. An ability to work well with nontechnical people, such as teachers, parents, students, and school administrators, is a must.

One promising employment opportunity for rehabilitation engineers and assistive technologists is in community-based service providers such as the local United Cerebral Palsy Association or the local chapter of the National Easter Seals Society. Through the combination of fees for service, donations, and insurance payments, shared rehabilitation technology engineering services in a community service center can be financially viable. The center would employ assistive technology professionals to provide information, assessments, customized adaptations, and training.

Rehabilitation engineers can also work as independent contractors or as employees of companies that manufacture assistive technology. Because rehabilitation engineers understand technology and the nature of many disabling conditions, they can serve as the liaison between the manufacturer and its potential consumers. In such a capacity, they could help identify and evaluate new product opportunities. Rehabilitation engineers, as independent consultants, could also offer knowledgeable and trusted advice to consumers, funding agencies, and worker compensation insurance companies. Such consultation work often involves providing information, offering client evaluations, and assessing the appropriateness of an assistive device. It is important that a rehabilitation engineer who wishes to work as an independent consultant be properly licensed as a Professional Engineer (PE) and be certified through RESNA as described later. The usual first step in attaining the PE license is to pass the Fundamentals of Engineering Examination given by each state's licensing board.

18.5.2 Rehabilitation Engineering Outlook

Rehabilitation engineering has reached adolescence as a separate discipline. It has a clearly defined application. It has been recognized as an activity worthy of support by many governments, and many universities offer formal graduate programs in this field. Fees for such services have been reimbursed by public and private insurance policies. Job advertisements for rehabilitation engineers appear regularly in newsletters and employment notices. In 1990, the Americans with Disabilities Act granted civil rights to persons with disabilities and made reasonable accommodations mandatory for all companies having more than 25 employees. Archival journals publish research papers that deal with all facets of rehabilitation engineering. Student interest in this field is increasing. What is next?

Based on recent developments, the following trends will likely dominate the practice of rehabilitation engineering and its research and development activities during the next decade:

- Certification of rehabilitation engineers will be fully established in the United States. Certification is the process by which a nongovernmental agency or professional association validates an individual's qualifications and knowledge in a defined functional or clinical area. RESNA is leading such a credentialing effort for providers of assistive technology. RESNA will certify someone as a Professional Rehabilitation Engineer if they are a registered Professional Engineer (a legally recognized title), possess the requisite relevant work experience in rehabilitation technology, and pass an examination that contains some 200 multiple-choice questions. For nonengineers, certification as an Assistive Technology Practitioner (ATP) or Assistant Technology Supplier (ATS) is also available. Sample questions from RESNA's credentialing examination are provided at the end of the chapter.
- Education and training of rehabilitation technologists and engineers will expand worldwide. International integration and exchange of information has been occurring informally. Recent initiatives by government entities and professional associations such as RESNA have given impetus to this trend. For example, the U.S. Department of Education recently announced its intention to support a consortium of several American and European universities in the training of rehabilitation engineers. One indirect goal of this initiative is to foster formal exchanges of information, students, and investigators.
- Universal access and universal design of consumer items will become commonplace. Technological advances in the consumer field have greatly benefited people with disabilities. Voice-recognition systems have enabled people with limited movement to use their computer as an interface to their home and the world. Telecommuting permits gainful employment without requiring a disabled person to be physically at a specified location. Ironically, benefits are beginning to flow in the opposite direction. Consumer items that once were earmarked for the disabled population (e.g., larger knobs, easy to use door and cabinet handles, curb cuts, closed-caption television programming, and larger visual displays) have become popular with

everyone. In the next decade, the trend toward universal access and products that can be used easily by everyone will expand as the citizenry ages and the number of people with limitations increases.

- Ergonomic issues will play a more visible role in rehabilitation engineering. When designing for people with limitations, ergonomics and human factors play crucial roles, often determining the success of a product. In recognition of this, the *IEEE Transactions on Rehabilitation Engineering* published a special issue on "Rehabilitation Ergonomics and Human Factors" in September 1994. The Human Factors and Ergonomics Society has a special interest group on "Medical Systems and Rehabilitation." In the next decade, more rehabilitation engineering training programs will offer required courses in ergonomics and human factors. The understanding and appreciation of human factors by rehabilitation engineers will be commonplace. The integration of good human factors design into specialized products for people with disabilities will be expected.

- Cost–benefit analysis regarding the impact of rehabilitation engineering services will become imperative. This trend parallels the medical field in that cost containment and improved efficiency have become everyone's concern. Econometric models and socio-economic analysis of intervention efforts by rehabilitation engineers and assistive technologists will soon be mandated by the federal government. It is inevitable that health maintenance organizations and managed care groups will not continue to accept anecdotal reports as sufficient justification for supporting rehabilitation engineering and assistive technology services. Quantitative studies in rehabilitation, performed by unbiased investigators, will likely be the next major initiative from funding agencies.

- Quality assurance and performance standards for categories of assistive devices will be established. As expenditures for rehabilitation engineering services and assistive devices increase, there will undoubtedly be demand for some objective assurance of quality and skill level. One example of this trend is the ongoing work of the Wheelchair Standards Committee jointly formed by RESNA and American National Standards Institute. Another example of this trend is the drive for certifying assistive technology providers and assistive technology suppliers as mentioned earlier.

- Technology will become a powerful equalizer as it reduces the limitations of manipulation, distance, location, mobility, and communication that are the common consequences of disabilities. Sometime in the next 20 years, rehabilitation engineers will utilize technologies that will enable disabled individuals to manipulate data and information and to alter system behavior remotely through their voice-controlled, internet-based, wireless computer workstation embedded in their nuclear-powered wheelchairs. Rather than commuting daily to work, persons with disabilities will or can work at home in an environment uniquely suited to their needs. They will possess assistive technology that will expand their abilities. Their dysarthric speech will be automatically recognized and converted into intelligible speech in real time by a powerful voice-recognition system. Given the breathtaking speed at which technological advances occur, these futuristic devices are not mere dreams but realistic extrapolations of the current rate of progress.

Students interested in rehabilitation engineering and assistive technology R&D will be able to contribute to making such dreams a reality shortly after they complete their formal training. The overall role of future practicing rehabilitation engineers, however, will not change. They will still need to assess someone's needs and limitations, apply many of the principles outlined in this chapter, and design, prescribe, modify, or build assistive devices.

EXERCISES

Like the engineering design process described earlier, answers to the following study questions may require searching beyond this textbook for the necessary information. A good place to begin is the Suggested Reading section. You also may try looking for the desired information using the World Wide Web (internet).

1. As a school-based rehabilitation engineer, you received a request from a teacher to design and build a gadget that would enable an 8-year-old second-grade student to signal her desire to respond to questions or make a request in class. This young student uses a powered wheelchair, has multiple handicaps, cannot move her upper arms very much, and is unable to produce understandable speech. Prepare a list of quantitative and qualitative questions that will guide your detailed analysis of this problem. Produce a hypothetical set of performance specifications for such a signaling device.

2. Write a sample set of performance specifications for a voice-output oscilloscope to be used by a visually impaired electrical engineering student for a laboratory exercise having to do with operational amplifiers. What features would be needed in the proposed oscilloscope?

3. Sketch how the leg raiser described in Example Problem 18.1 might fit onto a battery-powered wheelchair.

4. Do a careful search of commercially available electronic communication devices that meet the following performance specifications: speech output, icon-based membrane keyboard, portable, weighs less than 7.5 lbs, no more than 2.5 in. thick, no larger than a standard three-ring binder, and able to be customized by the user to quickly produce frequently used phrases. Hint: Consult "The Closing the Gap Product Directory" and the "Cooperative Electronic Library on Disability." The latter is available from the Trace Research and Development Center at the University of Wisconsin, Madison.

5. (a) A person is known to have spinal cord injury at the C5-C6 level. What does this mean in terms of this person's probable motoric and sensory abilities and limitations? (b) Repeat (a) for a person with multiple sclerosis. Include the prognosis for this individual.

6. To be portable, an electronic assistive device must be battery powered. Based on your study of technical manuals and battery handbooks, list the pros and cons of using disposable alkaline batteries versus lithium–hydride

rechargeable batteries. Include in your comparison an analysis of the technical issues (e.g., battery capacity, weight, and charging circuitry), cost issues, and practicality issues (e.g., user preferences, potential for misplacement or improper usage, and user convenience).

7. A young person with paraplegia wishes to resume his love of skiing, canoeing, and golfing. For each of these sports, list four or five adaptations or equipment modifications that are likely to be needed. Sketch and briefly describe these adaptations.

8. A 21-year-old female who has muscular dystrophy requested assistance with computer access, particularly for writing, using spreadsheets, and playing computer games. She lacks movement in all four extremities except for some wrist and finger movements. With her left hand, she is able to reach about 6 in. past her midline. With her right hand, she is able to reach only 2 in. past her midline. Both her hands can reach out about 8 in. from the body. If given wrist support, she has good control of both index fingers. Based on this description, sketch the work area that she appears able to reach with her two hands. Describe the adaptations to a standard or contracted keyboard that she would need to access her home computer. For additional information, consult "The Closing the Gap Product Directory," the "Cooperative Electronic Library on Disability," and the suggested reading materials listed at the end of this chapter.

9. The two main computer user interfaces are the command line interface (CLI), as exemplified by MS-DOS commands, and the graphical user interface (GUI), as exemplified by the Windows operating system. For someone with limited motoric abilities, each type of interface has advantages and disadvantages. List and compare the advantages and disadvantages of CLI and GUI. In what circumstances and for what kinds of disabilities would the CLI be superior or preferred over the GUI?

10. What would the second scale reading (S_2) be if the person in Example Problem 18.3 raised both of his legs straight up and D was known to be 14 in.?

11. How much tension would be exerted on the pulley in Example Problem 18.4 if the weights were observed to be falling at 1.5 ft/s^2?

12. How much contraction force must the flexor muscles generate in order for a person to hold a 25-lb weight in his hand, 14 in. from the elbow joint? Assume that the flexor muscle inserts at 90° to the forearm and 2 in. from the elbow joint and that his forearm weighs 4.4 lbs. Use the equilibrium equation, $\Sigma F_x = \Sigma F_y = \Sigma M = 0$, and Fig. 18.10 to aid your analysis.

13. For persons with good head control and little else, the Head Master (by Prentke Romich Co., Wooster, OH) has been used to emulate the mouse input signals for a computer. The Head Master consists of a headset connected to the computer by a cable. The headset includes a sensor that detects head movements and translates such movements into a signal interpreted as movements of the mouse. A puff-and-sip pneumatic switch is also attached to the headset and substitutes for clicking of the mouse. Based on this brief description of the Head Master, draw a block diagram of how this device might work and the basic components that might be needed in

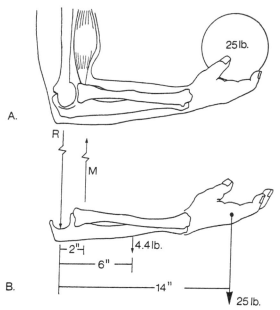

Fig. 18.10   Static forces about the elbow joint during an elbow flexor exercise (from Le Veau, 1976).

the Head Master. Include in your block diagram the ultrasonic signal source, detectors, timers, and signal processors that would be needed.

14. An electronic guide dog has been proposed as an electronic travel aid for a blind person. List some of the things that such a device must be able to do and the information processing steps needed to realize such an assistive device. List as many items and give as much detail as possible. Hint: Consider the problems of obstacle detection, information display, propulsion system, inertial guidance, route recall, power supply, etc.

15. How much force will the head of the femur experience when a 200-lb person stands on one foot? Hint: Apply the equilibrium equation, $\Sigma F_x = \Sigma F_y = \Sigma M = 0$, to the skeletal force diagram (Fig. 18.11) in your analysis.

16. Analyze the difference between visual-pursuit tracking and visual scanning in terms of the oculomotor mechanisms that underlie these two activities.

Sample Multiple-Choice Questions from RESNA's Credentialing Examination in Assistive Technology

1. Which of the following abilities is necessary for development of skilled upper-extremity movements?
   a. Equilibrium reactions in the standing position
   b. Ability to cross midline

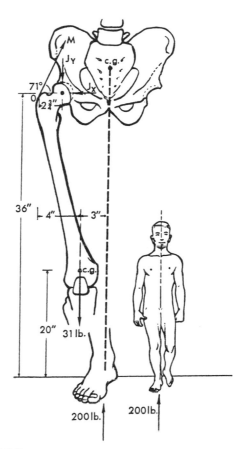

Fig. 18.11 Determination of the compression force on the supporting femoral head in unilateral weight bearing (from Le Veau, 1976).

   c. Good postural control of the trunk and head
   d. Pincer grasp
2. A 12-year-old male with Duchenne's muscular dystrophy is being evaluated for a mobility system. The therapist notes that he has lateral bending of the trunk and leans to the left. The most appropriate next step is assessment for
   a. Kyphosis
   b. Lordosis
   c. Left-sided weakness
   d. Scoliosis
3. The most appropriate location for training and instruction in functional use of an assistive technology device is
   a. A quiet area with few distractions
   b. The individual's home environment

    c. The environment in which the device will be used

    d. A training center where several therapists are available

4. An architect with C4-C5 quadriplegia would like to use a computer-assisted design (autoCAD) system when he returns to work. The most appropriate first step is assessment of the client's ability to use

    a. Mouth stick

    b. Eye-blink switch

    c. Alternate mouse input

    d. Sip-and-puff switch

5. Under the Individuals with Disabilities Education Act, assistive technology is defined as a device that

    a. Increases functional capability

    b. Improves mobility or communication

    c. Compensates for physical or sensory impairment

    d. Is considered durable medical equipment

6. In addition to the diagnosis, which information must be included in a physician's letter of medical necessity?

    a. Cost of assistive technology requested

    b. Client's prognosis

    c. Client's range of motion

    d. Client's muscle tone

7. Plastic is an ideal seat base for the person with incontinence because it is

    a. Light weight

    b. Less costly than wood

    c. Nonabsorbent

    d. Detachable from wheelchair

8. When considering structural modification of a newly purchased commercial device, which of the following is the *most* important concern?

    a. Future use by other individuals

    b. Voidance of warranty

    c. Resale value

    d. Product appearance

9. A client is interested in using a voice-recognition system to access the computer. Which of the following factors is *least* critical to success with this method?

    a. Hand function

    b. Voice clarity

    c. Voice-recognition system training

    d. Type of computer system used

10. A 9-year-old is no longer able to drive her power-base wheelchair. Training was provided following delivery of the wheelchair 2 years ago. Which of the following is the first step in evaluation?

    a. Interview the parents and child

    b. Perform a cognitive evaluation

    c. Reevaluate access in the wheelchair

    d. Contact the wheelchair manufacturer

SUGGESTED READING

Adams, R. C., Daniel, A. N., McCubbin, J. A. and Rullman L. (1982). *Games, Sports, and Exercises for the Physically Handicapped,* 3rd ed. Lea & Febiger, Philadelphia.

American Academy of Orthopaedic Surgeons (1975). *Atlas of Orthotics: Biomechanical Principles and Application.* Mosby, St. Louis.

Bailey, R. W. (1989). *Human Performance Engineering,* 2nd ed. Prentice Hall, Englewood Cliffs, NJ.

Blackstone, S. (Ed.) (1986). *Augmentative Communication: An Introduction.* American Speech–Language–Hearing Association, Rockville, MD.

Childress, D. S. (1993). Medical technical collaboration in prosthetics research and development. *J. Rehab. R&D* **30**(2), vii–viii.

Church, G., and Glennen, S. (1992). *The Handbook of Assistive Technology.* Singular, San Diego.

Cook, A. M., and Hussey, S. M. (1995). *Assistive Technology.* Mosby, St. Louis.

Cooper, R. A., Robertson, R. N., VanSickel, D. P., Stewart, K. J., and Albright, S. J. (1994) Wheelchair impact response to ISO test pendulum and ISO standard curb. *IEEE Trans. Rehab. Eng.* **2**(4), 240–246.

Daus, C. (1996, April/May). Credentialing rehabilitation engineers. *REHAB Manag.,* 115–116.

Deatherage, B. (1972). Auditory and other sensory forms of information presentation. In *Human Engineering Guide to Equipment Design* (H. Van Cott and R. Kincade, Eds.). Government Printing Office, Washington, DC.

Department of Rehabilitative Services, Commonwealth of Virginia (1997, Spring). Job announcement for a rehabilitation engineer.

Galvin, J. C., and Scherer, M. J. (1996). *Evaluating, Selecting, and Using Appropriate Assistive Technology.* Aspen, Gaithersburg, MD.

Goldenson, R. M., Dunham, J. R., and Dunham, C. S. (1978). *Disability and Rehabilitation Handbook.* McGraw Hill, New York.

Guthrie, M. (1997, July). Where will jobs be? *TeamRehab Rep.,* 14–23.

Hammer, G. (1993, September). The next wave of rehab technology. *TeamRehab Rep.,* 41–44.

Jones, M., Sanford, J., and Bell, R. B. (1997, October). Disability demographics: How are they changing? *TeamRehab Rep.* 36–44.

Lange, M. (1997, September). What's new and different in environmental control systems. *TeamRehab Rep.* 19–23.

Le Veau (1976). *Biomechanics of Human Motion,* 2nd ed.

McLaurin, C. A. (1991, August/September). A history of rehabilitation engineering. *REHAB Manag.* 70–77.

Mercier, J., Mollica, B. M., and Peischl, D. (1997, August). Plain talk: A guide to sorting out AAC devices. *TeamRehab Rep.* 19–23.

Reswick, J. (1981). What is a rehabilitation engineer? *Annu. Rev. Rehab* pp. 254–258.

Sanders, M. S., and McCormick, E. J. (1993). *Human Factors in Engineering and Design,* 7th ed. McGraw-Hill, New York.

Scherer, M. (1993). *Living in the State of Stuck: How Technology Impacts the Lives of People with Disabilities.* Brookline, Cambridge, MA.

Smith, R. V., and Leslie, J. H., Jr. (Eds.) (1990). *Rehabilitation Engineering.* CRC Press, Boca Raton, FL.

Stolov, W. C., and Clowers, M. R. (Eds.) (1981). *Handbook of Severe Disability.* U.S. Department of Education, Washington, DC.

U.S. Census Bureau (1990). *National Health Interview Survey on Assistive Devices (NHIS-AD)*. U.S. Census Bureau, Washington, DC.

Webster, J. G., Cook, A. M., Tompkins, W. J., and Vanderheiden, G. C. (Eds.) (1985). *Electronic Devices for Rehabilitation*. Wiley, New York.

Wilcox, A. D. (1990). *Engineering Design for Electrical Engineers*. Prentice Hall, Englewood Cliffs, NJ.

19 CLINICAL ENGINEERING AND ELECTRICAL SAFETY

Chapter Contents

At the conclusion of this chapter, the reader will be able to:

- Understand the role played by clinical engineers in the modern health care system
- Know the physiological effects of electrical currents
- Understand the methodologies/procedures used to protect patients and health care professionals from electrical hazards
- Be familiar with procedures to follow to establish an electrical safety program for a clinical facility
- Understand how to prepare for a career in the field of clinical engineering

19.1   INTRODUCTION

As mentioned in Chapter 1, biomedical engineers assist in the struggle against illness and disease by providing materials and tools that can be utilized for research, diagnosis, and treatment by health care professionals. One particular subset of the biomedical engineering community, namely clinical engineers, has become an integral part of the health care delivery team. They have achieved this position by assuming responsibility for managing the use of medical equipment within the hospital environment. Exactly what is clinical engineering? Over the years, a number of organizations have attempted to provide an appropriate definition. Consider the following:

- The American Hospital Association defines a clinical engineer as "a person who adapts, maintains, and improves the safe use of equipment and instruments in the hospital."
- The American College of Clinical Engineering (ACCE) defines a clinical engineer as "a professional who supports and advances patient care by applying engineering and managerial skills to health-care technology."
- The Association for the Advancement of Medical Instrumentation describes a clinical engineer as "a professional who brings to health-care facilities a level of education, experience, and accomplishment which will enable him[/her] to responsibly, effectively, and safely manage and interface with medical devices, instruments, and systems and the user thereof during patient care."
- The Board of Examiners for Clinical Engineering Certification considers a clinical engineer to be "an engineer whose professional focus is on patient–device interfacing; one who applies enginering principles in managing medical systems and devices in the patient setting."

Taking these definitions into consideration, we define a clinical engineer as an engineer who has graduated from an accredited academic program in engineering or who is licensed as a professional engineer or engineer-in-training and is engaged in the application of scientific and technological knowledge developed through engineering education and subsequent professional experience within the health-care environment in support of clinical activities.

With this definition in mind, the purpose of this chapter is to provide a brief historical perspective regarding the development of clinical engineering, a review of the basic functions performed by clinical engineers within a hospital environment, some insights into a typical day of a clinical engineering professional, a detailed description of their activities regarding electrical safety in the clinical environment, a brief summary of the future of clinical engineering, and some suggestions regarding how individuals might prepare themselves to enter this field.

19.2 A HISTORICAL PERSPECTIVE

Engineers were first encouraged to enter the clinical scene during the late 1960s in response to concerns about electrical safety of hospital patients. This safety scare reached its peak when consumer activists, most notably Ralph Nader (in the *Ladies Home Journal,* April 24, 1970) claimed that "at the very least, 1200 Americans are electrocuted annually during routine diagnostic and therapeutic procedures in hospitals." This concern was based primarily on the supposition that catheterized patients with a low-resistance conducting pathway from outside the body into blood vessels near the heart could be electrocuted by voltage differences well below the normal level of sensation. Despite the lack of statistical evidence to substantiate these claims, this outcry served to raise the level of consciousness of health care professionals with respect to the safe use of medical devices.

In response to this concern, a new industry — hospital electrical safety — arose almost overnight. Organizations such as the National Fire Protection Association (NFPA) wrote standards addressing electrical safety in hospitals. Electrical safety analyzer manufacturers and equipment safety consultants became eager to serve the needs of various hospitals that wanted to provide a "safety fix," and some companies, particularly those specializing in power distribution systems (most notably isolation transformers), deliberately preyed on the fears of hospitals. To alleviate these fears, the Joint Commission on the Accreditation of Healthcare Organizations (then known as the Joint Commission on Accreditation of Hospitals) turned to NFPA codes as the standard for electrical safety and further specified that hospitals must inspect all equipment used on or near a patient for electrical safety at least every 6 months. To meet this new requirement, hospital administrators considered a number of options, including (i) paying medical device manufacturers to perform these electrical safety inspections, (ii) contracting for the services of shared-services organizations, and (iii) providing these services with in-house staff. As discussed in Chapter 1, when faced with this decision, most large hospitals opted for in-house service and created whole departments to provide the technological support necessary to address these electrical safety concerns.

As a result, a new engineering discipline — clinical engineering — was born. Many hospitals established centralized clinical engineering departments. Once these departments were in place, however, it soon became obvious that electrical safety failures represented only a small part of the overall problem posed by the presence of medical equipment in the clinical environment. At the time, this equipment was neither totally understood nor properly maintained. Simple visual inspections often revealed broken knobs, frayed wires, and even evidence of liquid spills. Many devices did not perform in accordance with manufacturers' specifications and were not maintained in accordance with manufacturers' recommendations. In short, electrical safety problems were only the tip of the iceberg. By the mid-1970s, complete performance inspections before and after equip-ment use became the norm and sensible inspection procedures were developed. In the process, these clinical engineering pioneers began to play a more substantial role within the hospital. As new members of the hospital team, they

- Became actively involved in developing cost-effective approaches for using medical technology
- Provided hospital administrators advice regarding the purchase of medical equipment based on their ability to meet specific technical specifications
- Started utilizing modern scientific methods and working with standards-writing organizations
- Became involved in the training of health care personnel regarding the safe and efficient use of medical equipment.

During the 1970s and 1980s, a major expansion of clinical engineering occurred, primarily due to the following events:

- The Veterans Administration (VA), convinced that clinical engineers were vital to the overall operation of the VA hospital system, divided the country into biomedical engineering districts, with a chief biomedical engineer overseeing all engineering activities in the hospitals in that district.
- Throughout the United States, clinical engineering departments were established in most large medical centers and hospitals and in some smaller clinical facilities with at least 300 beds.
- Health care professionals, i.e., physicians and nurses, needed assistance in utilizing existing technology and incorporating new innovations.
- Certification of clinical engineers became a reality to ensure the continued competence of practicing clinical engineers.

During the 1990s, the evaluation of clinical engineering as a profession has continued with the establishment of the ACCE and the Clinical Engineering Division within the International Federation of Medical and Biological Engineering.

Clinical engineers today provide extensive engineering services for the clinical staff and serve as a significant resource for the entire hospital. Possessing in-depth knowledge regarding available in-house technological capabilities as well as the technical resources available from outside firms, the modern clinical engineer enables the hospital to make effective and efficient use of all of its technological resources.

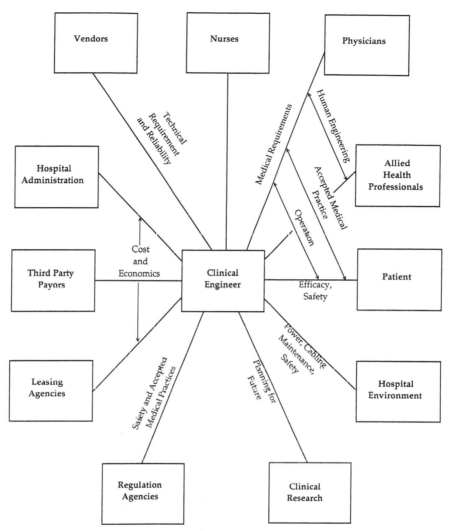

Fig. 19.1 Diagram illustrating the range of interactions in which a clinical engineer may be required to engage in a hospital setting.

19.3 THE ROLE OF THE CLINICAL ENGINEER

Figure 19.1 illustrates the multifaceted role played by modern clinical engineers. It is important to point out that for clinical engineers to be successful today, they must interface with clinical staff, hospital administrators, regulatory agencies, etc.

Examples Problem 19.1

Consider the following simple organizational chart of a major medical institution:

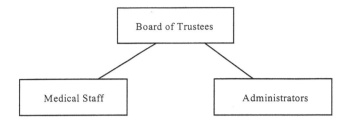

Specify the major responsibilities of each group.

Solution

- The Board of trustees determines the overall mission of the institution and is responsible for overseeing the fiscal operation of the facility.
- The medical staff advises the board regarding medical affairs and accepts responsibility for the quality of care rendered to patients in the hospital.
- Administrators submit plans of operation to the board of trustees for their approval and are responsible for the proper allocation of resources to achieve the mission of the institution.   ■

Example Problem 19.2

Draw a diagram that illustrates how a clinical engineering department interacts with the medical staff, nursing, and other allied health professionals in the institution.

Solution

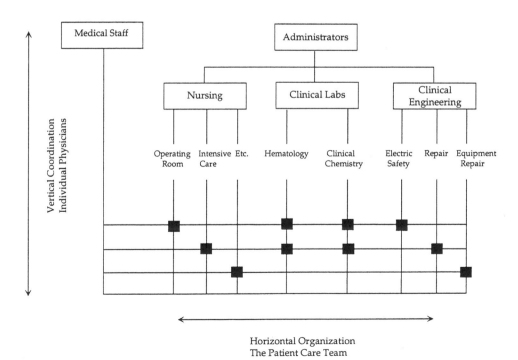

Patient care requires both a vertical and a horizontal (lateral) reporting relationship. Each nurse and clinical engineer report vertically to their respective supervisor as well as laterally by following directions given to them by the medical staff, i.e., physicians and physician assistants, directing the patient's clinical management. ∎

What do clinical engineers do? To answer this question, consider the following list of activities of typical clinical engineers. They

- Engage in clinical applications engineering, i.e., design and/or modify medical devices for clinical research, evaluate new noninvasive monitoring systems, etc.
- Repair sophisticated medical instruments or systems
- Supervise the performance of safety testing of medical equipment by biomedical equipment technologists (BMETs)
- Inspect all incoming equipment
- Establish performance benchmarks for all equipment
- Organize medical equipment inventory control
- Manage medical equipment calibration and repair service in a cost-effective manner
- Coordinate use of outside technical services (e.g., maintenance on radiology or imaging equipment) and vendors
- Train medical personnel in the safe and effective use of medical devices and systems
- Provide input to the design of clinical facilities where medical technology is used, e.g., operating rooms (ORs) and intensive care units
- Develop and implement documentation protocols required by external accreditation and licensing agencies.

The previous list is not complete, but it is sufficient to provide an understanding of the scope of activities in which clinical engineers are engaged.

With these tasks in mind, consider a typical day in the life of a clinical engineer:

4:00–5:00 AM: Presurgical equipment testing by nursing staff uncovers faulty respirator. Clinical engineering department is alerted and sends BMETs on call to service equipment.

8:00–9:00 AM: Clinical engineering director meets with other clinical engineers and BMETs to review work assignments for the day.

9:00 AM–12:00 PM: Clinical engineering director has meetings with hospital administrators to discuss the status of their equipment needs for the future. Other members of the department work on repairs, preventative maintenance, inspection of incoming equipment, and special clinical application projects.

12:00–1:00 PM: Clinical engineering director has lunch with other administrators or staff to discuss general topics related to department function.

1:00–3:00 PM: Review of work orders, inventory, etc. using medical information system. Engage in planning for short- and long-term goals. Prepare for inservice training session.

3:00–4:00 PM: Provide inservice training for nursing staff on various pieces of medical equipment.

4:00–5:00 PM: Clinical engineering director meets with capital equipment acquisition committee to review requests for purchasing new equipment. Other members of staff evaluate equipment needed for the next day's surgical procedures.

5:00–6:00 PM: Clinical engineering director usually engages in some employee evaluations at the end of each day.

6:00 PM–8:00 AM: On call for any emergency.

As one can see, the role of the modern clinical engineer is multifaceted. It ranges from the mundane tasks of managing the inspection, testing, and repair of medical equipment to higher level skills of equipment evaluation and planning. For clinical engineers to perform their tasks in the most effective and efficient manner, they must be properly organized.

In many hospitals, administrators have established clinical engineering departments to make effective use of the talents of these professionals and provide a broad-based engineering program that addresses all aspects of medical instrumentation and systems support. Figure 19.2 illustrates the organizational chart of the medical support services division of a typical major medical facility. It is worth noting that within this organizational structure, the director of clinical engineering reports directly to the vice president of medical support services. This administrative relationship is extremely important since it recognizes the important role that clinical engineering departments play in delivering quality care. In another common organizational structure, clinical engineering is placed under "facilities," "materials management," or even "support services." In either case, the success of clinical engineering departments depends on the leadership and administrative skills of the director or head of the department. Therefore, it is critical that clinical engineering directors develop sound managerial skills.

As a result of the wide-ranging scope of interrelationships within the medical setting outlined in Fig. 19.1, the duties and responsibilities of clinical engineering directors are extremely diversified. However, there is a common thread that is provided by the very nature of the technology they manage. Directors of clinical engineering departments are usually involved in the following areas:

- Developing, implementing, and directing equipment management programs: Specific tasks include evaluating and selecting new technology, accepting and installing new equipment, and managing the inventory of medical instrumentation, all in keeping with the responsibilities and duties defined by the administrator. The director advises the administrator of the budgetary, personnel, space, and test equipment requirements necessary to support this equipment management program. Directors also advise administration and medical and nursing staffs in areas such as safety, the purchase of new medical instrumentation and equipment, and the design of new clinical facilities.
- Evaluating and taking appropriate action on incidents attributed to equipment malfunction or misuse: The director summarizes the technological significance of each incident and documents the findings of the investigation.

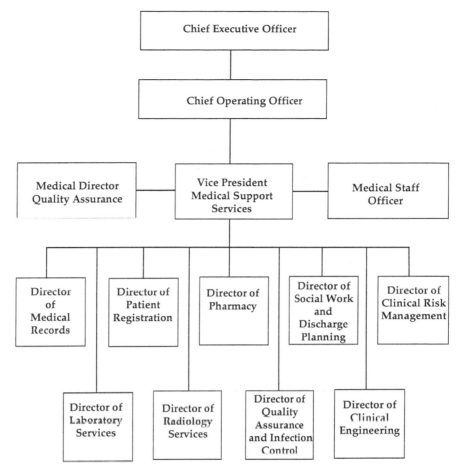

Fig. 19.2 Organizational chart of medical support services division for a typical major medical facility. The organizational structure indicates the critical interrelationship between the clinical engineering department and the other primary services provided by the medical facility.

He or she submits a report to the appropriate hospital authority and, according to the Safe Medical Devices Act of 1990, to the device manufacturer, the Food and Drug Administration, or both.

- Selecting departmental staff and training them to perform their functions in a professional manner.
- Establishing departmental priorities, developing and enforcing departmental policies and procedures, and supervising and directing departmental activities: The director takes an active role in leading the department to achieve its overall technical goals.

Example Problem 19.3

Provide a general organizational structure for a clinical engineering department of a large hospital.

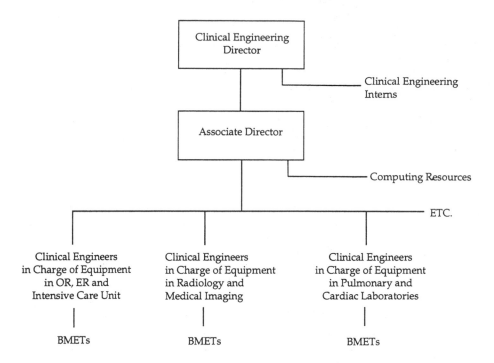

Solution

Although hospitals vary widely in administrative structure, size, and complexity, the basic structure of the clinical engineering program remains the same. ■

In reviewing the activities of clinical engineering departments, their core functions usually include the following (for a detailed review, see Bronzino, 1992):

- Technology management
- Safety and risk management
- Technology assessment
- Facilities design and project management
- Quality assurance
- Training

Since it is not possible to discuss all these functions in detail, I focus on one major activity — safety in the clinical environment.

19.4 SAFETY IN THE CLINICAL ENVIRONMENT

Patients enter hospitals today with much the same feelings as they did years ago — that is, with a mixture of hope and fear. They hope that modern medicine is capable of alleviating their physical distress and entrust themselves to the care and ministrations of physicians and nurses. At the same time, however, they try to suppress the apprehension and fear they feel as they enter a complex and technologically sophisticated health care delivery system. These concerns are real and appropriate since patients today are likely to encounter a host of medical devices and systems that may be both potentially beneficial and hazardous to their well-being. Of particular concern here are the hazards associated with medical devices, including the following:

- Electrical hazards: Patients are surrounded by and often connected to electrical devices. Some patients, because of their physical condition and the invasive nature of certain medical procedures, are particularly susceptible to low-level electrical hazards. Furthermore, in oxygen-enriched environments and in those clinical areas in which flammable anesthetic gases are used, electrical energy can be an ignition source.

- Mechanical hazards: Mechanical devices used for clinical purposes in the modern hospital include mobility aids, transfer devices, prosthetic devices, mechanical-assist devices, and patient-support equipment. Since each of these devices embodies numerous threats to patients as well as to hospital staff, they must be subjected to careful design review and failure indication, and specifications for safe use must be established.

- Environmental hazards: The hospital has its own ecosystem. The environment within the hospital includes such elements as solid wastes, noise, utilities (natural gas, water, etc.), and building structures which must be carefully controlled and managed to reduce hazards ranging from the spread of infection to injury from physical objects.

- Biological hazards: Infection control is a major factor in hospital safety today, especially in light of the AIDS epidemic. Here, the term "safety" is used in its broadest sense because infection threatens the safety of staff members as well as that of patients. Infection control is accomplished by instituting and managing programs that identify exposures to and eliminate sources of infection. Isolation, decontamination, sterilization, and appropriate methods of biological waste disposal are the primary tools of infection control.

- Radiation hazards: Radiation and radioactive materials have assumed a broad role in medical treatment. However, their use also may be hazardous to both patients and the medical staff. The health hazards from these devices and substances require the establishment of programs to control the use of diagnostic machines and therapeutic devices that release ionizing radiation and to monitor exposure of patients and medical staff. Control of radioactive waste is especially critical because of the extended periods during which these substances remain hazardous.

The presence of these and other hazards in the medical environment has made it essential for modern health care facilities to have a clinical engineering department that is intimately involved in and responsible for establishing and maintaining a safe environment. While physicians, nurses, and other medical personnel are busily serving patients under their care, there is a distinct need for the clinical engineer to focus on those hazards that may disrupt or destroy the life-saving efforts of the medical team.

Since a detailed presentation of all aspects of safety within the hospital is not possible here, this chapter emphasizes those safety issues that are of particular relevance to clinical engineering departments, with special emphasis on electrical and medical device safety.

19.5 ELECTRICAL SAFETY

Due to the prevalence of electrical devices, electrical safety is a vital component of a hospital's comprehensive safety program. As previously discussed, concern for electrical safety in the patient care environment has a long history and was the stimulant for the development of the clinical engineering profession.

Since the 1980s, however, dramatic changes have affected the hospital's approach to electrical safety. Consider the following:

- Significant improvements have been made regarding the safe operation, better performance, and greater reliability of medical equipment.
- Codes that are more realistic and standards have been developed that make more efficient use of hospital resources in minimizing electrical hazards.
- Clinical engineering departments have been effectively utilized to assist in the purchase, inspection, and preventive maintenance of medical equipment.

All these factors have made it possible for clinical engineering departments to relegate the electrical safety issue to the status of an important but routine matter that is reasonably controlled by simple periodic measures. Nonetheless, it is an important topic with significant engineering considerations.

19.5.1 Physiological Effects of Electrical Currents

For electricity to have any effect on the body, certain conditions must be satisfied:

- An electrical potential difference (or voltage) must be present.
- The individual must be part of the electrical circuit, permitting electrical current to flow through the body.

When these requirements are met, the actual effect of the electrical current passing through the body depends on the amount of current passed and the pathway the current takes through the body.

Figure 19.3 summarizes the physiological effects of 1-s exposures to various levels of alternating current [root mean squared (RMS) values] at 60 Hz applied arm-to-arm. The let-go current is defined as the maximum current at which the subject can voluntarily withdraw from the stimulus. As can be seen, very little current (1 mA) is needed to produce an awareness of the presence of current within the body, whereas 10–100 mA may result in a hazardous situation for the individual.

Since the magnitude of the current flowing in any circuit depends not only on the voltage between the connections but also on the electrical resistance of the various pathways through the body, it is important to consider the values of resistance of bi-

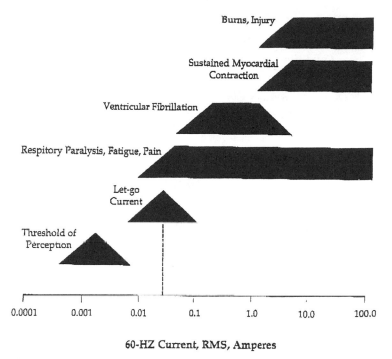

Fig. 19.3 Physiological responses to electricity at 60Hz. Redrawn from Webster, J.G., *Medical Instrumentation: Application and Design*, 1992.

ological tissues and how they may be changed. Most body tissues contain a high percentage of water (containing ions). Therefore, they are usually considered to be fairly good conductors of electricity. However, the surface resistance — that is, the condition of the actual contact point — also affects the amount of current that flows through the body. For example, if contact is made directly on moist skin, there will be little surface resistance, and a considerable amount of current will flow directly into the body.

On the other hand, if there is no direct contact with body fluids (e.g., when the skin is dry or when contact is made against protective clothing), then the surface resistance will be quite high, and much less current will be able to find its way into body tissue. Table 19.1 indicates, for example, that the skin impedance at 60 Hz may vary from 200 Ω to 93 kΩ, depending on the condition of the skin. Thus, the human body's first line of defense against electrical shock is its own protective covering — the skin.

Example Problem 19.4

What is the RMS voltage (at 60 Hz) required to deliver a current of 10 µA between two gelled surfaces each having a cross-sectional area of 1.5 cm^2? Assume that the skin impedance is 200 Ω.

TABLE 19.1 Impedance per cm^2 of Skin Tissue at 60 Hz[a]

Condition	Skin impedance
Dry skin	93 kΩ cm^2
Electrode gel on skin	10.8 kΩ cm^2
Penetrated skin	200 Ω cm^2

[a] Note that values of impedance depend on the electrode gel used, skin properties, and the cross-sectional area of the contact.

Solution

The impedance at each electrode is:

$$Z_{electrode} = \frac{10.8 \text{ k}\Omega - cm^2}{1.5 \text{ cm}^2} = 7.2 \text{ k}\Omega$$

The total current impedance is:

$$Z_{total} = 2Z_{electrode} + Z_{skin} = 14.6 \text{ k}\Omega$$

The voltage required to deliver 10 µA is

$$V = Z_{total} \times I = 146 \text{ mV}$$

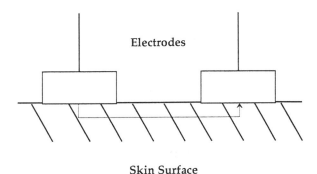

Electrodes

Skin Surface

Keep in mind, however, that the physiological effect of this current will depend on the path the current takes through the body. Since the distribution of current flow is determined by the specific resistance of local tissue (i.e., a nonhomogeneous volume conductor), the electrical current tends to spread out. Consequently, there are a variety of pathways that the current may follow in its course through the body, depending on the resistance of the tissues involved. It should be apparent that it is possible for situations to arise in which an individual comes into contact with a relatively large voltage source but experiences only a minor electrical shock. Conversely, application of a relatively low voltage directly to the heart (e.g., during surgey or via the invasive components of certain diagnostic and therapeutic equipment) can produce ventricular fibrillation, a life-threatening arrhythmia. The danger in one's exposure to various electrical devices is not the voltage but the magnitude and pathway of electrical current that flows through body tissue and major organs. In addition, for electrically sensitive organs such as the heart, physiological effects are also dependent on the frequency and waveform of the current.

If electrical current enters the body, it may also burn the tissue through which it passes. Whenever electrical current passes through any resistive element, some of the electrical energy is dissipated in the form of heat. As a result, damage to the biological tissue in the form of a burn will result when the temperature produced by this phenomenon is high enough to affect the tissue. This is especially true at the point of contact where the current intensity is usually quite high and the pathway is limited to a relatively small area.

Consider the case of skin burns that may result when electrosurgical units are used in surgery to cut tissue or stop bleeding of small blood vessels. When these devices are used, a dispersive electrode is usually placed under the patient's shoulder or buttocks to ensure that high current densities occur only at the point of contact for the "active" electrode that is controlled by the surgeon. If the dispersive electrode is too small, there is a significant risk of skin burn at the contact point. The absence or presence of burns is directly related to the current density ($J = I$/area). For any given amount of current, the current density will vary depending on the cross-sectional area of the dispersive electrode.

Example Problem 19.5

What is the difference in current density in two ground return pads that have the following dimensions:

Ground return pad 1: l (8 cm) $\times$ w (10 cm)

Ground return pad 2: 1 cm in diameter

Solution

$$J_1 = \frac{I_1}{A_1} \quad J_2 = \frac{I_2}{A_2}$$

Assuming $I_1 = I_2$

$$\frac{J_1}{J_2} = \frac{A_1}{A_2} = \frac{\pi r^2}{1 \times w} = \frac{0.25\pi}{80} = \frac{\pi}{320}$$

$$\frac{J_1}{J_2} = \frac{3.14}{320} \cong 0.01$$

or J_2 is 100 times greater than J_1.

As noted in Chapters 2 and 3 and in Fig. 19.3, electrical current can also affect the functioning of nerve and muscle tissue (e.g., stimulating the nerve or muscle into activity). When individual neurons are activated, an action potential is generated that travels along the axon to its synaptic connection with other cells. This action potential can be stimulated, with some startling results, by an electrical current passing through the nerve. For example, if sensory nerves are stimulated, the result is a prickling sensation that at sufficient intensity becomes unpleasant and even painful. If motor nerves or muscles are affected, severe muscular contraction can result. If an electrical current interferes with the functioning of heart muscle, ventricular fibrillation could result. If this occurs, the pumping action of the heart will stop, circulation of the blood will be interrupted, and the patient will be in critical condition. Thus, the introduction into the body of conductive materials, such as fluid-filled catheters and pacemaker leads, can drastically alter the electrical environment of the patient by creating direct low-impedance pathways to electrically sensitive organs.

Electrical shock hazards are usually classified as either macroshock or microshock. Macroshock is the term applied to high currents (approximately milliamperes and above) that pass through the skin, typically from limb to limb (Fig. 19.4). This type of electrical shock can result in burns and a wide variety of physiological consequences. Macroshock is a hazard to healthy people in everyday life as

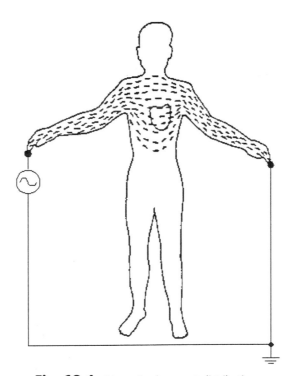

Fig. 19.4 Macroshock current distribution.

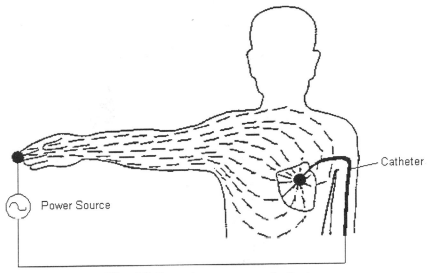

Fig. 19.5 Microshock current distribution.

well as to patients in the clinical environment. Microshock is the term applied to low currents (approximately microamperes) that present a fibrillation hazard to patients who are electrically susceptible because of invasive diagnostic and therapeutic procedures (Fig. 19.5). These categories of electrical shock differ not only in magnitude of electrical current but also in their etiology and control.

19.5.2 Protection against Electrical Hazards

All electrical and electronic devices are sources of potentially harmful current. Most of this equipment is powered by electricity obtained from the power distribution system of the building. In practice, electricity is usually generated at a power generator station, transmitted at very high voltages, and subsequently transformed to a much lower voltage for local distribution. This lower voltage is further reduced for local use to 110–120 or 220–240 V by power distribution transformers located at the electrical entrances to hospitals, industrial plants, and residential districts for local use.

Heavy-duty electrical devices requiring larger amounts of current, such as large air conditioners, ovens, and X-ray equipment, operate on 230 V, whereas most wall receptacles receive 115 V. Today, within the hospital, electrical power is provided by a three-wire system in which one of the wires (G) is at ground potential, another is called the "hot" wire (H), and the last is the return or neutral wire (N), which is also connected to ground (Fig. 19.6). The line voltage between the hot and neutral wires is either 115 or 230 V. The hot and neutral wires, respectively, carry current to and from equipment, whereas the ground wire does not normally carry substantial currents. However, it serves as a first line of defense against electrical hazards.

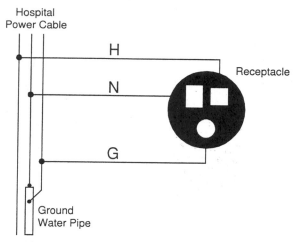

Fig. 19.6 Conventional (grounded) electrical power distribution system.

Grounding and Leakage Current Limits

If electrical devices were perfect, the conventional (ground-referenced) electrical power distribution system described previously would require only two conductors. All power current would be confined to those conductors and the electrical equipment that they supply. However, in practical applications, there are two major departures from this ideal case:

- If a fault occurs (i.e., through miswiring, insulation failure, or electrical component failure), it is possible for an electrical potential to exist between an exposed conductive surface (e.g., the metal case of a device) and a grounded surface (a wet floor, the grounded metal case of another device, etc.). Any person who bridges these surfaces is subject to macroshock.
- Even if a fault does not occur, imperfect insulation and electromagnetic coupling (capacitive or inductive) may produce an electrical potential relative to ground. An electrically susceptible patient who provides a path for this "leakage current" to flow to ground is subject to microshock.

Example Problem 19.6

Consider the following situation: A nurse touches a defective monitor having a leakage current of 100 μA while at the same time touching a sink fixture tied to the ground. Is the nurse safe?

Solution

The leakage current is sufficient to cause paralysis and death, but most of it flows through the ground wire. The following electrical circuit serves as a good model of the situation:

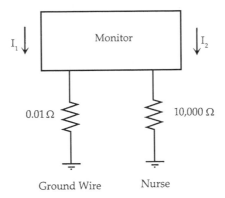

In this leaky monitor circuit, a battery of voltage V drives the current through two resistors in parallel. The important question is How does the 100 μA divide between the two resistors $R_1 = 10,000$ Ω for the nurse and $R_2 = 0.01$ Ω for the ground wire? Since there are two resistors in parallel:

$$R_{total} = \frac{R_1 \times R_2}{R_1 + R_2} = \frac{0.01 \times 10,000}{0.01 + 10,000} \approx 0.01 \ \Omega$$

Thus, $R_{total} \approx R_1$

$$V_{total} = V_1 = I_{total} \times R = (100 \ \mu A) \times (0.01 \ \Omega) = 1 \ \mu V$$

$$I_2 (\text{through the nurse}) = \frac{V_2}{R_2} = \frac{1 \times 10^{-6}}{10,000} = 1 \times 10^{-10} \ A$$

$$I_{nurse} = 10^{-10} A$$

Therefore, almost all the leakage current goes through the ground wire and only 10^{-10} A goes through the nurse. The nurse is safe. ∎

To address these problems, all electrical power distribution systems in the clinical environment are required to have a third conductor (ground wire) that is directly connected to ground. All line-operated electrical devices (except those meeting strict "double-insulation" standards) use this ground wire to ground all external conductive surfaces. This produces the following effects:

- If an electrical fault occurs (e.g., a short circuit between the hot power conductor and the metal case of the device), the resulting fault current will cause the circuit breaker to open (after a short delay) and remove power from the device. This method of reducing the macroshock hazard depends on the ability of the grounding system (ground connections within the device, the device power cord and the plug, and the power distribution system receptacle and wiring) to conduct high currents to ground.
- If there is no fault, the ground wire serves to conduct leakage current safely back to the electrical power source. As long as the grounding system

provides a low-resistance pathway to ground, the microshock hazard caused by leakage current is greatly reduced.

To meet both objectives, the grounding system must have large-gauge, low-resistance conductors as well as high-integrity connections along the entire electrical pathway. The size and type of conductors are specified by the design standards for buildings and equipment. On the other hand, high-integrity connections require good design (e.g., the use of hospital-grade plugs and receptacles) but also good maintenance (as discussed later).

Because they are unique to the clinical environment, microshock hazards created by equipment leakage currents require additional consideration. Leakage current may occur in the following ways:

- Capacitive coupling: Since it is possible to establish a potential difference between any two metal conductors through the process of "separation of charge," leakage current will flow whenever anyone makes contact with both metal surfaces. This type of fault is frequency sensitive because as the frequency of the voltage increases or decreases, so does the capacitive coupling.
- Inductive coupling: In this case, the hazardous voltage difference is established by "transformer action" via the close proximity of motors, transformers, and other inductive devices. Inductive coupling is also frequency dependent.

Leakage current limits recommended by the Emergency Care Research Institute (ECRI) were established so that if the grounding of a device should fail, injuries would be avoided if a patient touched the chassis of that device while in contact with a grounded conductor (Table 19.2). Thus, these leakage current limits repre-

TABLE 19.2 ECRI-Recommended Risk Current Limits for Line-Powered, Cord-Connected, Equipment[a]

Type of equipment	Patient applied parts or leads	Chassis (μA)
Patient equipment w/o patient connection	NA	100[b]
Patient contact equipment with nonisolated patient connections	50 A each lead or 100 A all leads together	100[b]
Patient contact equipment with isolated patient connections	10 A each lead[c]	100[b]
Equipment located outside the patient vicinity and nonmedical appliances	NA	500

[a] These values are based on limited experimentation on humans and on the results of similar tests on animals using currents required to cause ventricular fibrillation. (Reprinted with permission of Emergency Care Research Institute, 1990, pp. 27–28.

[b] The chassis leakage current criterion in NFPA 99-1990 is under appeal and may be modified to 500 μA.

[c] An isolated patient connection must also pass an isolation test. The NFPA 99-1990 criterion for lead-to-ground measurements of isolated patient connections is 10 A with the ground intact and 50 μA with the ground open.

sent a redundant protection system. Therefore, in any electrical safety program, the values of leakage current for medical equipment in these areas must be verified to be below the limits specified in applicable codes and standards.

Example Problem 19.7

Consider the case of a patient in a grounded bed in which the bed is connected to ground at the circuit breaker. Is the patient safe?

Solution

As long as the patient touches only the grounded bed, the patient is "properly grounded" via the circuit breaker (A), and the patient is safe. A problem arises only if the patient comes in contact with the ground wire of another circuit (B) while circuit B is draining off leakage current from some distant appliance, such as a vacuum cleaner. ∎

Isolated Patient Connections

Most devices, such as electrocardiogram and blood pressure monitors, that are connected directly to the patient have patient input isolation. An isolated patient connection not only limits leakage current levels from the patient contact points to very low levels but also reduces the amount of current that can flow into the patient connection from another (defective) device. It is recommended that equipment with isolated patient leads be used whenever the device is to be in contact with the heart or with a fluid-filled catheter located within the heart or great vessels.

A device designated as having an isolated patient connection must pass an isolation test from the patient connection (electrode). The test consists of applying a 120-V, 60-Hz, ground-referenced potential to each patient connection and measuring the resulting current (Fig. 19.7). The maximum allowable current is 20 A when the voltage is applied to the distal end of the patient cable (connected to the device under test) and 10 A when the voltage is applied directly to the patient cable con-

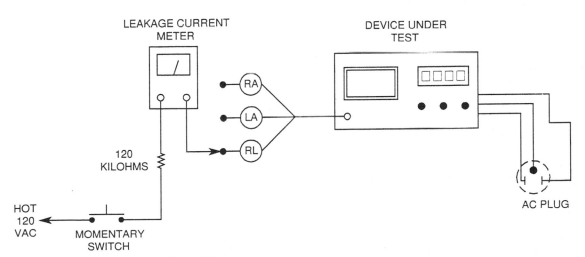

Fig. 19.7 Patient lead input isolation test circuit.

nector (with the patient cable removed) on the chassis of the device. The leads are also tested for leakage currents. The standards for isolated and nonisolated inputs differ.

Isolated Power Systems

The conventional electrical power distribution system described previously is "ground referenced" in that the neutral wire (N) is connected to ground, and therefore the hot wire (H) can serve as a source of current to any grounded conductor. An alternative scheme is to isolate both power-carrying lines from ground. In this scheme, called an isolated power system, the two power-carrying wires are labeled L1 and L2 (rather than H and N) and are electrically identical. It is important to note, however, that isolated power systems still include a ground wire (G) that functions just like the ground wire in a conventional power distribution system (Fig. 19.8).

Although isolated power systems are expensive to install and maintain, they have the following beneficial features:

- The perfect isolated power system would allow no current flow from either L1 or L2 to ground. However, because of imperfections in the transformers used to convert a ground-referenced system into an isolated system, some current does flow. In the early days of microshock hysteria, isolated power systems were recommended for use in intensive care units to reduce microshock hazards. However, the degree of isolation is inadequate to be of significant benefit in this regard, and the use of isolated power systems in intensive care units is no longer recommended.
- Isolated power systems, in conjunction with explosion-proof electrical outlets, can reduce the likelihood of arcing that could ignite flammable anesthetic gases. This hazard has become rare since the introduction of nonflammable inhalation anesthetics. In fact, explosive anesthetic agents are no longer used in the United States and Canada. Although many operating rooms have isolated power systems, they are not required in locations where flammable inhalation anesthetics are banned by institutional policy.
- In the ground-referenced power distribution system, certain equipment faults result in opened circuit breakers and loss of power. In an isolated power system, such equipment may continue to operate without interruption. For

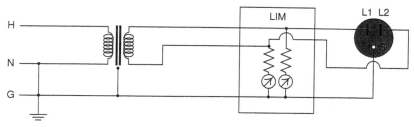

Fig. 19.8 An isolation transformer connected to a line-isolation monitor (LIM).

example, in a ground-referenced system, a short circuit between the hot wire (H) to the grounded case of a device would cause the circuit breaker to open and the device to cease functioning. However, in an isolated system, a short circuit from L1 (or L2) to ground would not cause a current overload. In fact, the effect would be to functionally transform the isolated power system into a ground-referenced system. Thus, the isolated power system can allow equipment to function under certain "single-fault" conditions. Faults of this type are rare but sometimes occur in wet environments such as operating rooms. Coupled with the critical nature of equipment in the operating theater, this may justify the installation of an isolated power system.

In addition to special transformers, isolated power systems include line-isolation monitors (LIMs) to check the degree of electrical isolation in the system. In effect, LIMs "predict" the maximum amount of current that can flow from L1 or L2 to ground. This value increases as a result of either a degradation within the isolated power system or faulty equipment connected to the isolated power system. When the LIM indication exceeds a designated set point, visible and audible alarms are activated. It is important to note that these alarms indicate only a reduction in isolation, which is not a hazard in itself. The alarm warns that an additional equipment fault could create a hazard. It is important for clinical engineers to understand the various types of LIMs and the appropriate responses to LIM alarms.

Ground Fault Circuit Interrupters

Another protective device that can contribute to electrical safety is the ground fault circuit interrupter (GFCI). In certain circumstances, this inexpensive device can provide protection against macroshock hazards. It is usually installed as part of the electrical wiring that feeds a receptacle or a cluster of receptacles, or it may even be integrated into the power outlet. There are also portable, stand-alone GFCIs and devices with built-in GFCIs.

Figure 19.9 illustrates the circuit of a generic GFCI. It consists of a magnetic toroid around which the hot lead and the neutral wire are wound, each with the same number of turns but in opposite directions. The GFCI continually senses the

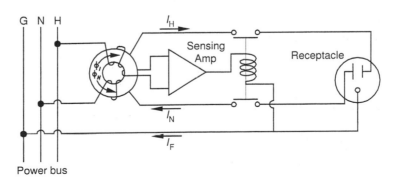

Fig. 19.9 Schematic diagram of a generic ground fault circuit interrupter (GFCI) (redrawn with permission from Aston, 1990).

difference in current flow between the hot and neutral wires. Normally, the difference between these two currents is small, and the magnetic fields created by them are opposite and approximately equal and will therefore cancel. However, under fault conditions, in which some or all of the current returns to the generating source by some route other than the neutral wire (i.e., through a person in contact with the electrical device), the corresponding fluxes in the coil will be unequal, thereby producing a voltage that will be detected by the sensing amplifier. When this potential difference (proportional to the current imbalance) exceeds some critical value, power to the GFCI receptacle or device is shut off. It is important to note that although GFCIs protect against excessive exposure to ground fault currents, they do not limit these currents (i.e., a substantial fault current may flow for a brief period of time). Therefore, GFCIs are not intended to protect against leakage currents in the microshock range but will help protect against macroshock. In some instances, they may be used to meet requirements for special protection against electrical shock in locations where fluids are frequently used or in designated wet locations, such as hydrotherapy tanks.

Example Problem 19.8

Discuss what happens when one of the transformer secondary leads touches the OR table by accident.

Solution

A ground fault detector (GFD) is often built into the OR power system to provide a warning in case something goes wrong with the isolation of the power system. When one of the transformer secondary leads touches the OR table, there is no chance, because of the isolated power system, that either sparking or patient injury will occur.

However, if the other secondary output lead from the transformer should also accidentally touch ground or the patient, then a dangerous situation will be created. The GFD alerts the operating room staff that the isolation has been lost, thereby permitting the situation to be corrected before something else goes wrong. If the second fault occurs before the first is corrected, then the patient and staff may be in jeopardy.

If a GFCI was used instead of a GFD, the GFCI would operate by constantly measuring the current going out through the hot lead and comparing it with the current returning via the neutral lead. If they are equal, all is well. If not, the GFCI opens the circuit, thereby limiting leakage current to a level that is considered to be harmless.   ■

The choice of locations for the use of GFCIs in the hospital must be carefully considered because, in the interest of preventing shock, GFCIs may cause a far greater hazard by shutting off power to critical or life-support patient-care equipment. For this reason, it is not recommended that such equipment be powered from a GFCI circuit. Instead, an isolated power system may be necessary. Isolated power systems and GFCIs are required by code only in areas designated as wet locations.

19.6 ELECTRICAL SAFETY PROGRAMS

Electrical safety in hospitals today clearly requires that appropriate attention be paid to the electrical environment of the patient. To achieve this goal, it is necessary to develop a comprehensive safety plan. This involves implementing a periodic maintenance check of all line-operated equipment and establishing a good record-keeping scheme to keep track of the tests that have been conducted and the relative status of the equipment that has been tested. In providing such a safety program, the following steps are required:

- Checking the environment
- Checking the equipment before bringing it into the environment
- Providing education and training in safety procedures

19.6.1 Environmental Inspection

The hospital environment consists of a wide range of facilities, each of which requires special consideration from a safety perspective. Consider the following locations:

Anesthetizing locations: These are areas normally used for the administration of an inhalation anesthetic agent. Traditionally, this term refers to operating rooms and delivery rooms; however, with the advent of changing patient populations and new technologies, delivery of anesthesia occurs in other areas, such as emergency and medical imaging sites (e.g., X ray and magnetic resonance imaging).

Anesthetizing location — nonflammable refers to anesthetizing locations that, by issuance of a hospital policy, are posted to be used only for nonflammable inhalation anesthetizing agents.

Anesthetizing location — flammable must be labeled as such and be specifically designed with safeguards to allow for safe administration of flammable inhalation anesthetic agents.

Anesthetizing locations — mixed refers to all anesthetizing sites that allow either flammable or nonflammable inhalation anesthetic agents. Hospital policies must ensure that only properly designed equipment is used in these locations and must require appropriate posting of signs on each room and labeling of equipment.

Critical care locations: These areas include all patient care areas classified as critical care sites by hospital policy or where patients are subjected to invasive procedures and directly connected to line-operated medical devices. Typically, these include operating rooms, delivery rooms, and intensive care, cardiac care, and cardiac catheterization areas.

General care locations: These are the patient care locations where patients are expected to come in contact with ordinary electrical appliances (lamps, beds, televisions, etc.) or be connected nonelectrically to medical devices. This does not include nursing stations.

Nonpatient areas: These are areas in which patients are not normally cared for or treated, such as administrative offices, laboratories, nursing stations, storage areas, kitchens, or plant equipment areas.

Wet locations: These include patient care areas normally subject to wet conditions, including standing water on the floor or routine dousing or drenching of the work area. Routine housekeeping procedures and incidental spillage do not constitute cause for designation as a wet location. Generally, this classification includes only hydrotherapy tanks in the physical therapy department, garbage storage areas, and certain food service areas.

In all these areas of the hospital, the use of proper grounding procedures to remove hazardous electrical currents from the patient environment is critical. A proposed or expected ground point must be verified as actually being at ground potential, thereby presenting a low-resistance path to the common reference point at the service panel. The reason for this precaution is clear when one considers the numerous discontinuities that exist in the ground connections located throughout the building. It is at these discontinuities that electrical hazards may arise. A loose or disconnected wire is all that is required to remove a particular outlet or room from ground. Therefore, one must never simply assume continuity to ground — it must be tested. Isolated power systems and GFCIs must also be checked periodically in accordance with some appropriate schedule.

19.6.2   Equipment Inspection

Once the environment has been checked, it is necessary to test all electrical equipment brought into the patient environment. This means that the quality and continuity of ground between the chassis of each instrument and the ground lug of its line cord must be verified. This is extremely important because defective ground lugs and wires in the plug are among the most common causes of shock hazard. Equipment grounding and leakage currents can be quickly measured by safety analyzers designed for this purpose.

19.6.3   Education of Clinical Staff

One of the key factors in preventing electrical accidents is the education of the hospital staff. Individuals who are well informed and knowledgeable about the causes and consequences of electrical hazards represent the best preventive measure any institution can have. This does not imply that any of the steps just discussed may be discarded, but rather that only with proper training can these safety measures be effectively implemented. The extent to which a safe electrical environment is established within a hospital today is directly related to the support given by the clinical staff. These individuals can easily carry out the simple steps of electrical safety by using and interacting with user-friendly monitors and test equipment. Consequently, programs directed at the in-house training of medical personnel in the basics of electrical safety and safety instrumentation are absolutely necessary.

In the modern hospital, patient safety is considered by many to be a medical problem that requires the coordinated effort of the entire health care delivery team. It is essential that medical personnel be properly trained to look for possible electrical defects before using line-operated equipment on patients. When the previously mentioned steps (checking the environment, checking the equipment, and educating the necessary personnel) are followed, patients will not only receive the best medical care possible but also be protected from the potential electrical hazards that may surround them.

19.7 THE FUTURE OF CLINICAL ENGINEERING

From its early days when electrical safety testing and basic preventive maintenance were the primary concerns to the present, the practice of clinical engineering has changed enormously. However, it is appropriate to use the cliché "The more things change, the more they stay the same." Today, hospital-based clinical engineers still have the following as their primary concerns: patient safety and good hospital equipment management. However, these basic concerns are being supplemented by new areas of responsibility, making the clinical engineer not only the chief technology officer but also an integral part of the hospital management team.

In large part, these demands are due to the economic pressures that hospitals face. To further complicate this financial situation, state-of-the-art, high-level instruments, such as magnetic resonance imaging systems, surgical lasers, and other sophisticated devices, are now used as a matter of course in patient care. Because of the high cost and complexity of such instrumentation, the institution needs to plan carefully — at both a technical and a managerial level — for the assessment, acquisition, and use of this new technology.

With these needs in mind, hospital administrators have begun to turn to their clinical engineering staffs for assistance in operational areas. Clinical engineers now provide assistance in the application and management of many other technologies that support patient care, e.g., computer support, telecommunications, and facilities operations.

19.7.1 Computer Support

The use of personal computers (PCs) has grown enormously in the past decade. PCs are now commonplace in every facet of hospital operations, including data analysis for research, use as a teaching tool, and many administrative tasks. PCs are also increasingly used as integral parts of local area networks and hospital information systems.

Because of their technical training and experience with computerized patient record systems and inventory and equipment management programs, clinical engineers have extended their scope of activities to include PC support. In the process, the hospital has accrued several benefits from this involvement in computer servicing. The first is time: Whether computers are used in direct clinical applications or in administrative work, downtime is expensive. In-house servicing can provide faster and, often, more dependable repairs than an outside group. Second, with in-house service, there is no need to send a computer out for service, reducing the possibility that computer equipment will be damaged or lost. Finally, in-house service reduces costs by permitting the hospital to avoid expensive service contracts for computers and peripheral equipment.

With all these benefits, it might seem that every clinical engineering department should carry out computer servicing. However, the situation is not so simple. At the most basic level, the clinical engineering program must have the staff, money, and space to do the job well. To assist in making this determination, several questions should be asked: Will computer repair take too much time away from the department's primary goal — patient care instrumentation? Is there enough money and space to stock needed parts, replacement boards, diagnostic software, and peripheral devices? Should personnel with computer expertise be in the department? For those hospitals that do commit the resources needed to support computer repair, clinical engineers have found that their departments can provide these services very efficiently, and they subsequently receive added recognition and visibility within the hospital.

19.7.2   Telecommunications

Another area of increased clinical engineering involvement is hospital-based telecommunications. In the modern health care institution, telecommunications covers many important activities, the most visible of which is telephone service. Broadly speaking, however, telecommunications includes many other capabilities.

Up to the 1970s, telephone systems were basically electromechanical, using switches, relays, and other analog circuitry. During the 1970s, digital equipment began to be used. This development allowed the introduction of innovations such as touch-tone dialing, call forwarding, conference calling, improved call transferring, and other advanced services. The breakup of the Bell system also changed the telecommunications field enormously by opening it to competition and diversification of services. Today, many hospitals use a private branch exchange (PBX) in which all switching equipment is located on-site. In comparison to the old PBX systems, which could fill an entire room, newer systems offer several advantages. They take up less space, they can be serviced in-house, and they are relatively easy to customize — all cost-saving features. For example, it is possible to limit long-distance capabilities for specific telephones. PBX systems also handle more traffic, serve local area network functions, and tap into data networks.

Data transmission capability allows the hospital to send scans and reports to physicians at their offices or at other remote locations. Data such as patient ECGs can be transmitted from the hospital to a data analysis system at another location, and the results can be transmitted back. Hospitals are also making increased use of facsimile transmission. This equipment allows documents such as patient charts to be sent via telephone line from a remote location and reconstructed at the receiving site in a matter of minutes. Of particular importance is the utilization of telemedicine to provide medical service to remote sites where health care specialists are in short supply.

Modern telecommunications equipment also allows the hospital to conduct educational conferences. This is accomplished through microwave links that allow video transmission of a conference taking place at a separate site. Pictorial information such as patient slides can also be digitally transmitted via a phone line and then electronically reassembled to produce a video image.

Clinical engineers can play an important role in helping hospital administrators develop plans for a new telecommunications system. They can provide technical support during the planning stage, assist in the development of requests for proposals for a new system, and help resolve any physical plant issues associated with installing the new system. Thereafter, the clinical engineer can assist in reviewing responses to the hospital's requests for quotations and can help with the installation of the system.

19.7.3 Facilities Operations

Recently, some hospital administrators have begun to tap the high-level technical skills available within their clinical engineering departments to assist in other operational areas. One of these is facilities operations, which includes heating, ventilation, and air conditioning; electrical supply and distribution (including isolated power); central gas supply and vacuum systems; and other physical plant equipment. This has occurred for a number of reasons. First, the modern physical plant contains microprocessor-driven control circuits, sophisticated circuitry, and other high-level technology. In many instances, facilities personnel lack the training needed to understand the engineering theory underlying these systems. Thus, clinical engineers can effectively work in a consultation role to help correct malfunctions in the hospital's physical plant systems. Another reason is cost. The hospital may be able to avoid expensive service contracts by performing this work in-house.

Clinical engineers can assist in facilities operations in other ways. For example, they can lend assistance when compressed gas or vacuum delivery systems need to be upgraded or replaced. While the work is taking place, the clinical engineer can serve as the administration's technical arm, ensuring that the job is done properly, is in compliance with code, and is completed with minimal disruption.

19.7.4 Strategic Planning

Today's emphasis on health care cost control requires that clinical engineers assist in containing costs associated with medical instrumentation. To accomplish this objective, clinical engineers are increasingly becoming involved in strategic planning, technology assessment, and purchase review. During assessment and purchase review, the clinical engineer studies a request to buy a new system or device and ensures that the purchase request includes (i) needed accessories, (ii) warranty information, and (iii) user and service training. This review process ensures that the device is, in fact, needed (or whether there exists a less costly unit that meets the clinician's requirements) and will be properly integrated into the hospital's existing equipment and physical environment.

Clinical engineers can provide valuable assistance during the planning for and financial analysis of potential new services. If one considers the steps involved in planning for a new patient care area, many design questions immediately come to mind: What is the best layout for the new area? What equipment will be used there? How many suction, air, oxygen, and electrical outlets will be needed? Is there a need to provide for special facilities, such as those needed for hemodialysis treatment at the bedside? With their knowledge of instrumentation and user needs, clinical engineers can help select reliable, cost-effective equipment to help ensure that the hospital obtains the best possible plan.

In the future, clinical engineering departments will need to concentrate even more heavily on management, with the goals of increased productivity and reduced costs. Clinical engineers will also need to keep up with the latest developments in medical instrumentation, computers and telecommunications systems, and other technologies that affect the financial operation of the hospital. By expanding their horizons, clinical engineers can further contribute to the overall operation of the hospital and ensure high-quality patient care at a reasonable cost.

19.8 PREPARATION FOR CLINICAL ENGINEERS

Having identified the roles clinical engineers play and the types of knowledge and experience they must possess, it is appropriate to ask, "How can an individual prepare to enter the field of clinical engineering?" As we approach the twenty-first century, students interested in clinical engineering should begin their education by obtaining a bachelor's degree in an engineering discipline, such as biomedical, chemical, electrical, or mechanical engineering. Upon completion of the undergraduate degree program, the individual should proceed to work in a hospital and gain direct professional experience. An appropriate alternative would be to enter a master's degree program in biomedical engineering that includes a clinical engineering internship.

It is important to emphasize that the academic component of a clinical engineer's knowledge base is essentially found within the biomedical engineering curriculum. Therefore, a typical plan of graduate study in clinical engineering should follow that of a normal master's degree program in biomedical engineering. In addition to the engineering courses required to provide the necessary fundamentals in specific biomedical engineering topics, it is necessary that students interested in clinical engineering be exposed to and work in the hospital environment. Such an experience permits these students to observe both the operation of specific medical instruments and the environment in which they are used and the people who use them. The nature of this hospital experience may vary in terms of its duration and specificity, but an internship of some type is essential.

One way to obtain this experience is through an internship established as part of the individual's overall educational program. In the past, several biomedical engineering programs have attempted to meet this objective by initiating a required or optional internship program with a clinical institution. Most of these programs have been based primarily within an academic institution and have provided internships of varying duration. For example, some institutions require students to engage in a hospital project for 1 or 2 months, whereas others insist on a 1- or 2-year experience.

Any internship experience, however, should be designed to permit budding clinical engineers to assume a role analogous to that of their medical counterparts and learn the intricacies of their profession in a personal manner. Since specific clinical, organizational, and managerial skills needed by clinical engineers cannot be taught in a formal academic program, such an internship experience can provide insights into the profession that might otherwise be unattainable. Programs that provide such an educational opportunity offer the participating student excellent exposure to the real world in which the clinical engineer must function.

EXERCISES

1. Select a medical device that was invented between 1945 and 1970. Describe the fundamental principles of how it operates, and discuss its impact on health care delivery.
2. Repeat Exercise 1 for a device that was invented after 1970.
3. Select two key words from the definition of "clinical engineering" and discuss their significance.
4. List the areas of knowledge necessary to practice clinical engineering. Identify where in the normal educational process one can acquire this knowledge. How best can the administrative skills be acquired?
5. Summarize some of the major roles of clinical engineers. Provide examples of the activities that are associated with each role.
6. Provide an organizational chart for the clinical engineering department of a major medical facility in your community. Discuss the advantages and disadvantages of such an organizational layout.
7. What is meant by the term, "clinical engineering shared services"? How are they organized? What do they provide to the hospital?
8. Briefly describe what you consider to be the most important duty of the clinical engineering director and discuss why you chose that duty as the most important.
9. When and why are patients susceptible to electrical shock hazards?
10. If the RMS voltage delivered in Example Problem 19.4 is 73 mV, how much current will be flowing through these electrodes?
11. If the cross-sectional area of each electrode in Example Problem 19.4 was decreased to 1.0 cm^2 and the RMS voltage was 146 mV, what would be the current?
12. What is the difference in current density in Example Problem 19.5 if ground return pad No. 2 is 0.1 cm in radius?
13. Assume that 120 V (RMS) is delivered between two gelled surface electrodes with cross-sectional areas of 3 cm^2 and that the impedance across

the skin between the two electrodes is 200 Ω. (a) What is the resultant current? (b) What will be the physiological effect of this current? (c) What will be the effect on the current when the cross-sectional area of the electrodes is reduced to 0.3 cm²?

14. A normal ground return pad is the size of an 8 1/2 × 11-in. sheet of paper. Discuss what happens if only one-fifth of the ground pad comes into contact with the patient. What would be the difference in current density in these circumstances?

15. Distinguish between macroshock and microshock hazards. Give examples of each.

16. Define leakage current. Provide a circuit diagram to illustrate how to test for leakage current. What types of devices are available to test for and protect against leakage current? For two specific pieces of medical equipment, list the leakage current specified by the manufacturer.

17. If in Example Problem 19.6, the nurse is stepping in a pool of water so that the resistance is reduced to 0.001 Ω, what would be the new current flow? Is the nurse still safe?

18. In Example Problem 19.6, at what current would the nurse be at risk?

19. If the leakage current was 100 µA and the resistance of the nurse was Ω, what portion of the current would go through the nurse?

20. About an hour after plugging an electrical device into a wall receptacle, smoke was observed rising from behind the receptacle coverplate. Name several faults that may have contributed to this condition. For each fault, design an inspection program to preclude future faults.

21. Describe a situation in which an electronically susceptible patient, i.e., in the operating room, would be exposed to an electrical hazard that would lead to ventricular fibrillation.

22. In example problem 19.7, suppose the patient's foot is also grounded by an old-fashioned ECG monitor to ground via circuit breaker B, and this circuit breaker also happens to be on the ground circuit for the receptacle in the hall. If the janitor uses that receptacle for a vacuum cleaner that has its motor short-circuited when the cleaner picks up a paper clip, will the patient be in danger?

23. As a clinical engineer observing an open-heart procedure in the OR, you note that when a mechanical respirator is plugged in for possible use, the GFCI alarm goes off. The surgeon says, "What is that?" What is your response?

24. Isolated power systems have many beneficial features. Discuss two of them.

25. What is the purpose of line-isolation monitors? Explain how they work.

26. Discuss the steps required in establishing an electrical safety program.

27. Describe whether a danger in a patient's exposure to electrical devices lies in (a) the voltage, (b) resistance of local tissue, (c) magnitude of current, and (d) pathway that electrical current takes.

28. Describe whether the absence or presence of skin burns is directly related to (a) magnitude of voltage, (b) skin resistance, (c) current density, and (d) size of electrodes.

29. Describe whether the electrical environment of the patient can be adversely affected if (a) conductive materials are placed in the body, (b) high currents are supplied to the body, (c) the skin is punctured, and (d) high voltages are present.

30. Describe whether a ground within a structure can be made to (a) a cold-water pipe, (b) grounding conductor, and (c) an electrical conduit, if used.

31. Describe whether the ultimate ground for a structure is (a) the cold-water intake pipe to the structure, (b) the neutral wire of the electrical service to the structure, (c) the structural metal frame of the building, and (d) a copper pipe, at least 20 ft in length, driven into damp earth.

32. Describe whether a ground fault (as used in the National Electrical Code) refers to (a) a short between the neutral and ground conductors, (b) a current return to ground by any path other than the neutral line, and (c) a break in the grounding system.

33. Is there a relationship between the physiological consequences of current flow through the body and frequency? Hint: See Reilly (1992).

SUGGESTED READING

American Hospital Association Resource Center (1986). *Hospital Administration Terminology,* 2nd ed. American Hospital Publishing, Washington, DC.

Aston, R. (1990). *Principles of Biomedical Instrumentation and Measurement.* Merrill, Columbus, OH.

Bauld, T. J. (1991). The definition of a clinical engineer. *J. Clin. Eng.* **16,** 403–405.

Bronzino, J. D. (1992). *Management of Medical Technology.* Butterworth, Boston.

Emergency Care Research Institute (1990). Electrical Safety Document 1.4.

Goodman, G. (1989). The profession of clinical engineering. *Clin. Eng.* **14,** 27–37.

International Certification Commission (ICC) (1991).

International Certification Commission's Definition of a Clinical Engineer, International Certification Commission Fact Sheet. ICC, Arlington, VA.

Reilly, J. P. (1992). *Electrical Stimulation and Electropathology.* Cambridge Univ. Press, New York.

Webster, J. G. (1992). *Medical Instrumentation, Applications and Design.* Wiley, New York.

20 MORAL AND ETHICAL ISSUES

Chapter Contents

After completing this chapter, the reader will be able to:

- Define and distinguish between the terms morals and ethics
- Present the rationale underlying two major philosophical schools of thought: utilitarianism and nonconsequentialism
- Present the codes of ethics for the medical profession, nursing profession, and biomedical engineering
- Identify the modern moral dilemmas, including redefining death, deciding how to care for the terminally ill, and experimentation on humans, which arise from the two moral norms: beneficence (the provision of benefits) and nonmaleficence (the avoidance of harm)
- Discuss the moral judgments associated with current policies regarding the regulation of the development and use of new medical devices

20.1   INTRODUCTION

The tremendous infusion of technology into the practice of medicine during the past 50 years has created a new medical era. Advances in material science have led to the production of artificial limbs, heart valves, and blood vessels, thereby permitting "spare-parts" surgery. Numerous patient disorders are now routinely diagnosed using a wide range of highly sophisticated imaging devices, and the lives of many patients are being extended through significant improvements in resuscitative and supportive devices, such as respirators, pacemakers, and artificial kidneys.

These technological advances, however, have not been benign. They have had significant moral consequences. Provided with the ability to develop cardiovascular assist devices, perform organ transplants, and maintain the breathing and heartbeat of terminally ill patients, society has been forced to reexamine the meaning of such terms as death, quality of life, heroic efforts, and acts of mercy and consider such moral issues as the right of patients to refuse treatment (living wills) and to participate in experiments (informed consent). As a result, these technological advances have made the moral dimensions of health care more complex and have posed new and troubling moral dilemmas for medical professionals, the biomedical engineer, and society at large.

The purpose of this chapter is to examine some of the moral questions related to the use of new medical technologies. The objective, however, is not to provide solutions or recommendations for these questions. Rather, the intent is to demonstrate that each technological advance has consequences that affect the very core of human values.

Technology and ethics are not foreigners; they are neighbors in the world of human accomplishment. Technology is a human achievement of extraordinary ingenuity and utility and is quite distant from the human accomplishment of ethical values. They face each other rather than interface. The personal face of ethics looks at the impersonal face of technology in order to comprehend technology's potential and its limits. The face of technology looks to ethics to be directed to human purposes and benefits.

In the process of making technology and ethics face each other, it is our hope that individuals engaged in the development of new medical devices, as well as those responsible for the care of patients, will be stimulated to examine and evaluate critically "accepted" views and to reach their own conclusions. This chapter, therefore, begins with some definitions related to morality and ethics, and a more detailed discussion of some of the moral issues of special importance to biomedical engineers follows.

20.2 MORALITY AND ETHICS: A DEFINITION OF TERMS

From the beginning, individuals have raised concerns about the nature of life and its significance. Many of these concerns have been incorporated into the four fundamental questions posed by the German philosopher Immanuel Kant (1724–1804): What can I know? What ought I do? What can I hope? and What is man? Evidence that early societies raised these questions can be found in the generation of rather complex codes of conduct embedded in the customs of the earliest human social organization, the tribe. By 600 BC, the Greeks were successful in reducing many primitive speculations, attitudes, and views on these questions to some type of order or system and integrating them into the general body of wisdom called *philosophy*. Being seafarers and colonizers, the Greeks had close contact with many different peoples and cultures. In the process, struck by the variety of customs, laws, and institutions that prevailed in the societies that surrounded them, they began to examine and compare all human conduct in these societies. They called this part of philosophy *ethics*.

The term ethics is derived from the Greek *ethos*, meaning "custom." On the other hand, the Latin word for custom is *mos*, and its plural, *mores*, is the equivalent of the Greek ethos and the root of the words moral and morality. Although both terms (ethics and morality) are often used interchangeably, there is a distinction between them that should be made.

Philosophers define ethics as a particular kind of study and use morality to refer to its subject matter. For example, customs that result from some abiding principal human interaction are called *morals*. Some examples of morals in our current society are telling the truth, paying one's debts, honoring one's parents, and respecting the rights and property of others. Most members of society usually consider such conduct not only customary but also correct or right. Thus, morality encompasses what people believe to be right and good and the reasons they give for it.

Most of us follow these rules of conduct and adjust our lifestyles in accordance with the principles they represent. Many even sacrifice life itself rather than diverge from them, applying them not only to their own conduct but also to the behavior of others. Individuals who disregard these accepted codes of conduct are considered deviants and, in many cases, are punished for engaging in an activity that society as a whole considers unacceptable. For example, individuals committing "criminal acts" (defined by society) are often "outlawed" and, in many cases, severely punished. These judgments regarding codes of conduct, however, are not inflexible; they must continually be modified to fit changing conditions and thereby avoid the trauma of revolution as the vehicle for change.

While morality represents the codes of conduct of a society, ethics is the study of right and wrong and of good and evil in human conduct. Ethics is not concerned with providing any judgments or specific rules for human behavior but rather with providing an objective analysis about what individuals "ought to do." Defined in this way, it represents the philosophical view of morals and therefore is often referred to as moral philosophy.

Consider the following questions: (i) Should badly deformed infants be kept alive? (ii) Should treatment be stopped to allow a terminally ill patient to die? and (iii) Should humans be used in experiments? Are these questions of morality or ethics? In terms of the definitions just provided, all three of these inquiries are questions of moral judgment.

Philosophers argue that all moral judgments are considered to be "normative judgments," i.e., they can be recognized simply by their characteristic evaluative terms such as "good," "bad," "right," and "wrong." Typical normative judgments include the following:

- Stealing is wrong.
- Everyone should have access to an education.
- Voluntary euthanasia should not be legalized.

Each of these judgments expresses an evaluation, i.e., conveys a negative or positive attitude toward some state of affairs. Each, therefore, is intended to play an action-guiding function.

Arriving at moral judgments, however, requires knowledge of valid moral standards in our society. Nevertheless, how is such knowledge obtained? The efforts to answer this question lie in two competing schools of thought that currently dominate normative ethical theory: **utilitarianism,** a form of consequentialism, and **Kantianism,** a form of nonconsequentialism. Consequentialism holds that the morally right action is always the one among the available options that has the best consequences. An important implication of consequentialism is that no specific actions or courses of conduct are automatically ruled out as immoral or ruled in as morally obligatory. The rightness or wrongness of an action is wholly contingent upon its effects.

According to utilitarianism, there are two steps to determining what should be done in any situation. First, determine which courses of action can be taken. Second, determine the consequences of each alternative. When this has been accomplished, the morally right course of action is the one that maximizes pleasure, minimizes pain, or both — the one that does the "greatest good for the greatest number." Because the central motivation driving the design, development, and use of medical devices is improvement of medicine's capacity to protect and restore health, an obvious virtue of utilitarianism is that it assesses medical technology in terms of what many believe makes health valuable — the attainment of well-being and the avoidance of pain.

Utilitarianism, therefore, advocates that the end justifies the means. As long as any form of treatment maximizes good consequences, it should be used. Many people, though, believe that the end does not always justify the means and that individuals have rights that are not to be violated no matter how good the consequences might be.

In opposition to utilitarianism is the school of normative ethical thought known as nonconsequentialism. Proponents of this school deny that moral evaluation is simply and wholly a matter of determining the consequences of human conduct. They agree that other considerations are relevant to moral assessment and reject the view that morally right conduct is whatever has the best consequences. Based largely on the views of Immanuel Kant, this ethical school of thought insists that there is something uniquely precious about human beings from the moral point of view. According to Kant's theory, humans have certain "rights" that do not apply to any other animal. For example, the moral judgments that we should not kill and eat each other for food, hunt each other for sport, or experiment on each other for medical science are all based on this view of human rights. In short, humans are owed a special kind of respect simply because they are people.

To better understand the Kantian perspective, it may be helpful to recognize that Kant's views are an attempt to capture in secular form a basic tenet of Christian morality. What makes human beings morally special entities deserving a unique type of respect? Christianity answers this question in terms of the doctrine of ensoulment. This doctrine holds that only human beings are divinely endowed with an eternal soul. According to Christian ethics, the soul makes humans the only beings with intrinsic value. Kant's secular version of the doctrine of ensoulment asserts that human beings are morally unique and deserve special respect because of their autonomy. Autonomy is taken by Kant to be the capacity to make choices based on rational deliberation. The central task of ethics is to specify what human conduct is required to respect the unique dignity of human beings. For most Kantians, this means determining what limits human beings must observe in the way they treat each other and this, in turn, is taken to be a matter of specifying each individual's fundamental moral rights.

These two ethical schools of thought, therefore, provide some rationale for moral judgments. However, when there is no clear moral judgment, one is faced with a dilemma. In medicine, moral dilemmas arise in situations that raise fundamental questions about right and wrong in the treatment of sickness and the promotion of health in patients. In many of these situations, the health professional usually faces two alternative choices, neither of which seems to be a satisfactory solution to the problem. For example, is it more important to preserve life or prevent pain? Is it right to withhold treatment when doing so may lead to a shortening of life? Does an individual have the right to refuse treatment when refusing it may lead to death? All these situations seem to have no clear-cut imperative based on our current set of convictions about right and wrong. This is the dilemma first raised by Kant: What ought I to do?

In the practice of medicine, moral dilemmas are certainly not new. They have been present throughout medical history. As a result, over the years there have been efforts to provide a set of guidelines for those responsible for patient care. These efforts have resulted in the development of specific "codes of professional conduct." We next examine some of these codes or guidelines.

For the medical profession, the World Medical Association adopted a version of the *Hippocratic Oath* titled the Geneva Convention Code of Medical Ethics in 1949. This declaration contains the following statements:

I solemnly pledge myself to consecrate my life to the services of humanity;

I will give to my teachers the respect and gratitude which is their due;

I will practice my profession with conscience and dignity;

The health of my patient will be my first consideration;

I will respect the secrets which are confided in me;

I will maintain by all the means in my power, the honour and the noble traditions of the medical profession;

My colleagues will be my brothers;

I will not permit considerations of religion, nationality, race, party politics, or social standing to intervene between my duty and my patient;

I will maintain the utmost respect for human life from the time of conception; even under threat;

I will not use my medical knowledge contrary to the laws of humanity;

I make these promises solemnly, freely, and upon my honour.

In 1980, the House of Delegates of the American Medical Association adopted the following Principles of Medical Ethics:

PREAMBLE: The medical profession has long subscribed to a body of ethical statements developed primarily for the benefit of the patient. As a member of this profession, a physician must recognize responsibility not only to patients, but also to society, to other health professionals, and to self.

I. A physician shall be dedicated to providing competent medical service with compassion and respect for human dignity.

II. A physician shall deal honestly with patients and colleagues, and strive to expose those physicians deficient in character or competence, or who engage in fraud or deception.

III. A physician shall respect the law and also recognize a responsibility to seek changes in those requirements which are contrary to the best interests of the patient.

IV. A physician shall respect the rights of patients, of colleagues, and of other health professionals, and shall safeguard patient confidences within the constraints of the law.

V. A physician shall continue to study, apply, and advance scientific knowledge, make relevant information available to patients, colleagues, and the public, obtain consultation, and use the talents of other health professionals when indicated.

VI. A physician shall, in the provision of appropriate patient care, except in emergencies, be free to choose whom to serve, with whom to associate, and the environment in which to provide medical services.

VII. A physician shall recognize a responsibility to participate in activities contributing to an improved community.

For the nursing profession, the American Nurses Association formally adopted in 1976 the Code for Nurses, the statements and interpretations of which provide guidance for conduct and relationships in carrying our nursing responsibilities.

PREAMBLE: The Code for Nurses is based on belief about the nature of individuals, nursing, health, and society. Recipients and providers of nursing services are viewed as individuals and groups who possess basic rights and responsibilities, and whose values and circumstances command respect at all times. Nursing encompasses the promotion and restoration of health, the prevention of illness, and the alleviation of suffering. The statements of the Code and their interpretation provide guidance for conduct and relationships in carrying out nursing responsibilities consistent with the ethical obligations of the profession and quality in nursing care.

1. The nurse provides services with respect for human dignity and the uniqueness of the client unrestricted by considerations of social or economic status, personal attributes, or the nature of health problems.
2. The nurse safeguards the client's right to privacy by judiciously protecting information of a confidential nature.
3. The nurse acts to safeguard the client and the public when health care and safety are affected by the incompetent, unethical, or illegal practice of any person.
4. The nurse assumes responsibility and accountability for individual nursing judgments and actions.
5. The nurse maintains competence in nursing.
6. The nurse exercises informed judgment and uses individual competence and qualifications as criteria in seeking consultation, accepting responsibilities, and delegating nursing activities to others.
7. The nurse participates in activities that contribute to the ongoing development of the profession's body of knowledge.
8. The nurse participates in the profession's efforts to implement and improve standards of nursing.
9. The nurse participates in the profession's efforts to establish and maintain conditions of employment conducive to high-quality nursing care.
10. The nurse participates in the profession's effort to protect the public from misinformation and misrepresentation and to maintain the integrity of nursing.
11. The nurse collaborates with members of the health professions and other citizens in promoting community and national efforts to meet the health needs of the public.

These codes take as their guiding principle the concepts of service to humankind and respect for human life. It is difficult to imagine that anyone could improve on these codes as summary statements of the primary goals of individuals responsible for the care of patients. However, some believe that such codes fail to provide answers to many of the difficult moral dilemmas confronting health professionals today. For example, in many situations, all the fundamental responsibilities of the nurse cannot be met at the same time. When a patient suffering from a massive insult to the brain is kept alive by artificial means and this equipment is needed elsewhere, it is not clear from these guidelines how "nursing competence is to be maintained to conserve life and promote health." Although it may be argued that the decision to treat or not to treat is a medical and not a nursing decision, both professions are so intimately involved in the care of patients that they are both concerned with the ultimate implications of any such decision.

For biomedical engineers, an increased awareness of the ethical significance of their professional activities has also resulted in the development of codes of professional ethics. Typically consisting of a short list of general rules, these codes express both the minimal standards to which all members of a profession are expected to conform and the ideals for which all members are expected to strive. Such codes provide a practical guide for the ethical conduct of the profession's practitioners. Consider, for example, the code of ethics endorsed by the American College of Clinical Engineers:

> As a member of the American College of Clinical Engineering, I subscribe to the established Code of Ethics in that I will:
>
> - Accurately represent my level of responsibility, authority, experience, knowledge, and education.
> - Strive to prevent a person from being placed at risk due to the use of technology.
> - Reveal conflicts of interest that may affect information provided or received.
> - Respect the confidentiality of information.
> - Work toward improving the delivery of health care.
> - Work toward the containment of costs by the better management and utilization of technology.
> - Promote the profession of clinical engineering.

Although these codes can be useful in promoting ethical conduct, such rules obviously cannot provide ethical guidance in every situation. A profession that aims to maximize the ethical conduct of its members must not limit the ethical consciousness of its members to knowledge of their professional code alone. It must also provide them with resources that will enable them to determine what the code requires in a particular concrete situation, and thereby enable them to arrive at ethically sound judgments in situations in which the directives of the code are ambiguous or simply do not apply.

20.3   TWO MORAL NORMS: BENEFICENCE AND NONMALEFICENCE

Two moral norms have remained relatively constant across the various moral codes and oaths that have been formulated for health care providers since the beginnings of Western medicine in classical Greek civilization: **beneficence,** the provision of benefits, and **nonmaleficence,** the avoidance of doing harm. These norms are traced back to a body of writings from classical antiquity known as the *Hippocratic Corpus.* Although these writings are associated with the name of Hippocrates, the acknowledged founder of Western medicine, medical historians remain uncertain whether any, including the *Hippocratic Oath,* were actually his work. Although portions of the *Corpus* are believed to have been authored during the sixth century BC, other portions are believed to have been written as late as the beginning of the Christian era. Medical historians agree that many of the specific moral directives of the *Corpus* represent neither the actual practices nor the moral ideals of the majority of physicians of ancient Greece and Rome.

Nonetheless, the general injunction, "As to disease, make a habit of two things — to help or, at least, to do no harm," was accepted as a fundamental medical ethical norm by at least some ancient physicians. With the decline of Hellenistic civilization and the rise of Christianity, beneficence and nonmaleficence became increasingly accepted as the fundamental principles of morally sound medical practice. Although beneficence and nonmaleficence were regarded merely as concomitant to the craft of medicine in classical Greece and Rome, the emphasis on compassion and the brotherhood of humankind, central to Christianity, increasingly made these norms the only acceptable motives for medical practice. Even today, the provision of benefits and the avoidance of doing harm are stressed just as much in virtually all contemporary Western codes of conduct for health professionals as they were in the oaths and codes that guided the health care providers of past centuries.

Traditionally, the ethics of medical care has given greater prominence to nonmaleficence than to beneficence. This priority was grounded in the fact that, historically, medicine's capacity to do harm far exceeded its capacity to protect and restore health. Providers of health care possessed many treatments that posed clear and genuine risks to patients and that offered little prospect of benefit. Truly effective therapies were all too rare. In this context, it is surely rational to give substantially higher priority to avoiding harm than to providing benefits.

The advent of modern science changed matters dramatically. Knowledge acquired in laboratories, tested in clinics, and verified by statistical methods has increasingly dictated the practice of medicine. This ongoing alliance between medicine and science became a critical source of the plethora of technologies that now pervade medical care. The impressive increases in therapeutic, preventive, and rehabilitative capabilities that these technologies have provided have pushed beneficence to the forefront of medical morality. Some have even gone so far as to hold that the old medical ethic of "Above all, do no harm" should be superseded by the new ethic that "The patient deserves the best." However, the rapid advances in medical technology capabilities have also produced great uncertainty as to what is most beneficial or least harmful for the patient. In other words, along with increases in the ability to be beneficent, medicine's technology has generated much debate about what actually counts as beneficent or nonmaleficent treatment. Having reviewed some of the fundamental concepts of ethics and morality, we next discuss several specific moral issues posed by the use of medical technology.

20.4 REDEFINING DEATH

Although medicine has long been involved in the observation and certification of death, many of its practitioners have not always expressed philosophical concerns regarding the beginning of life and the onset of death. Since medicine is a clinical and empirical science, it would seem that health professionals had no medical need to consider the concept of death: The fact of death was sufficient. The distinction between life and death was viewed as the comparison of two extreme conditions separated by an infinite chasm. With the advent of technological advances in medicine to assist health professionals to prolong life, this view has changed.

There is no doubt that the use of medical technology has in many instances warded off the coming of the "grim reaper." One need only look at the trends in average life expectancy for confirmation. For example, in the United States today, the average life expectancy for males is 74.3 years and for females 76 years, whereas at the turn of the century the average life expectancy for both sexes was only 47 years. Infant mortality has been significantly reduced in developed nations in which technology is an integral part of the culture. Premature births no longer constitute a threat to life because of the artificial environment that medical technology can provide. Today, technology has not only helped individuals avoid early death but also been effective in delaying the inevitable. Pacemakers, artificial kidneys, and a variety of other medical devices have enabled individuals to add many more productive years to their lives. Technology has been so successful that health professionals responsible for the care of critically ill patients have been able to maintain their "vital signs of life" for extensive periods of time. In the process, however, serious philosophical questions concerning the quality of the life provided these patients have arisen.

Consider the case of the patient who sustains a serious head injury in an automobile accident. When the attendants in the ambulance reached the scene of the accident, the patient was unconscious but still alive with a beating heart. After the victim was rushed to the hospital and into the emergency room, the resident in charge verified the stability of the vital signs of heartbeat and respiration during examination and ordered a computerized tomography scan to indicate the extent of the head injury. The results of this procedure clearly showed extensive brain damage. When the electroencephalogram (EEG) was obtained from the scalp electrodes placed about the head, it was noted to be significantly abnormal. In this situation, obvious questions arise: What is the status of the patient? and Is the patient alive?

Alternatively, consider the events encountered during one open-heart surgery. During this procedure, the patient was placed on the heart bypass machine while the surgeon attempted to correct a malfunctioning valve. As the complex and long operation continued, the EEG monitors that had indicated a normal pattern of electrical activity at the onset of the operation suddenly displayed a relatively straight line indicative of feeble electrical activity. However, since the patient's so-called vital signs were being maintained by the heart–lung bypass, what should the surgeon do? Should the medical staff continue on the basis that the patient is alive or is the patient dead?

The increasing occurrence of these situations has stimulated health professionals to reexamine the definition of death. In essence, advances in medical technology that delay death actually hastened its redefinition. This should not be surprising because the definition of death has always been closely related to the extent of medical knowledge and available technology. For many centuries, death was defined solely as the absence of breathing. Since it was believed that the spirit of the human being resided in the spiritus (breath), its absence was indicative of death. With the continuing proliferation of scientific information regarding human physiology and the development of techniques to revive a nonbreathing person, attention turned to the pulsating heart as the focal point in determination of death. However, this view was to change as a result of additional medical and technological advances in supportive therapy, resuscitation, cardiovascular assist devices, and organ transplantation.

As understanding of the human organism increased, it became obvious that one of the primary constituents of the blood is oxygen and that any organ deprived of oxygen for a specified period of time will cease to function and die. The higher functions of the brain are particularly vulnerable to this type of insult since the removal of oxygen from the blood supply even for a short period of time (3 min) produces irreversible damage to the brain tissues. Consequently, the evidence of "death" began to shift from the pulsating heart to the vital, functioning brain. Once medicine was provided with the means to monitor the brain's activity (i.e., the EEG), another factor was introduced in the definition of death. Advocates of the concept of brain death argued that the human brain is truly essential to life. When the brain is irreversibly damaged, so are the functions that are identified with self and "humanness": memory, feeling, thinking, knowledge, etc.

As a result, it became widely accepted that the meaning of clinical death implies that the spontaneous activity of the lungs, heart, and brain is no longer present. The irreversible cessation of functioning of all three major organs was required before anyone was pronounced dead. Although damage to any other organ system such as the liver or kidney may ultimately cause the death of the individual through a fatal effect on the essential functions of the heart, lungs, or brain, this aspect was not included in the definition of clinical death.

With the development of modern respirators, however, the medical profession encountered an increasing number of situations in which a patient with irreversible brain damage could be maintained almost indefinitely. Once again, a new technological advance created the need to reexamine the definition of death.

The movement toward redefining death received considerable impetus with the publication of a report sponsored by the ad hoc committee of the Harvard Medical School in 1968 in which the committee offered an alternative definition of death based on the functioning of the brain. The report of this committee was considered a landmark attempt to deal with death in light of technology.

In summary, the criteria for death established by this committee included the following: (i) The patient must be unreceptive and unresponsive, that is, in a state of irreversible coma; (ii) the patient must have no movements of breathing when the mechanical respirator is turned off; (iii) the patient must not demonstrate any reflexes; and (iv) the patient must have a flat EEG for at least 24 h indicating no electrical brain activity. When these criteria are satisfied, then death may be declared.

At the time, the committee also strongly recommended that the decision to declare the person dead and then to turn off the respirator should not be made by physicians involved in any later efforts to transplant organs or tissues from the deceased individual. In this way, a prospective donor's death would not be hastened merely for the purpose of transplantation. Thus, complete separation of authority and responsibility for the care of the recipient from the physician or group of physicians who are responsible for the care of the prospective donor is essential.

The shift to a brain-oriented concept involved deciding that much more than just biological life is necessary to be a human person. The brain-death concept was essentially a statement that mere vegetative human life is not personal human life. In other words, an otherwise intact and alive but brain-dead person is not a human person. Many of us have taken for granted the assertion that being truly alive in this world requires an "intact functioning brain." However, precisely this issue was at stake in the gradual movement from using heartbeat and respiration as indices of life to using brain-oriented indices.

Indeed, total and irreparable loss of brain function, referred to as "brain stem death," "whole brain death," or simply "brain death," has been widely accepted as the legal standard for death. By this standard, an individual in a state of brain death is legally indistinguishable from a corpse and may be legally treated as one even though respiratory and circulatory functions may be sustained through the intervention of technology. Many take this legal standard to be the morally appropriate one, noting that once destruction of the brain stem has occurred, the brain cannot function at all, and the body's regulatory mechanisms will fail unless artificially sustained. Thus, mechanical sustenance of an individual in a state of brain death is merely postponement of the inevitable and sustains nothing of the personality, character, or consciousness of the individual. It is simply the mechanical intervention that differentiates such an individual from a corpse, and a mechanically ventilated corpse is a corpse nonetheless.

Even with a consensus that brain stem death is death, and thus that an individual in such a state is indeed a corpse, difficult cases remain. Consider the case of an individual in a persistent vegetative state, the condition known as "neocortical death." Although severe brain injury has been suffered, enough brain function remains to make mechanical sustenance of respiration and circulation unnecessary. In a persistent vegetative state, an individual exhibits no purposeful response to external stimuli and no evidence of self-awareness. The eyes may open periodically and the individual may exhibit sleep–wake cycles. Some patients even yawn, make chewing motions, or swallow spontaneously. Unlike the complete unresponsiveness of individuals in a state of brain stem death, a variety of simple and complex responses can be elicited from an individual in a persistent vegetative state. Nonetheless, the chances that such an individual will regain consciousness are remote. Artificial feeding, kidney dialysis, and the like make it possible to sustain an individual in a state of neocortical death for decades. This sort of condition and the issues it raises are exemplified by the famous case of Karen Ann Quinlan.

In April 1975, Karen Quinlan suffered severe brain damage and was reduced to a chronic vegetative state in which she no longer had any cognitive function. Accepting the doctors' judgment that there was no hope of recovery, her parents sought permission from the courts to disconnect the respirator that was keeping her alive in the intensive care unit of a New Jersey hospital.

The trial court, and then the Supreme Court of New Jersey, agreed that Karen's respirator could be removed, so it was disconnected. However, the nurse in charge of her care in the Catholic hospital opposed this decision and, anticipating it, had begun to wean her from the respirator so that by the time it was disconnected she could remain alive without it. Therefore, Karen did not die. She remained alive for 10 years. In June 1985, she finally died of acute pneumonia. Antibiotics, which would have fought the pneumonia, were not given.

If brain stem death is death, is neocortical death also death? Again, the issue is not a straightforward factual matter because it, too, is a matter of specifying which features of living individuals distinguish them from corpses and so make treatment of them as corpses morally impermissible. Irreparable cessation of respiration and circulation, the classical criterion for death, would entail that an individual in a persistent vegetative state is not a corpse and so, morally speaking, must not be treated as one. The brain stem death criterion for death would also entail that a person in a state of neocortical death is not yet a corpse. Using this criterion, it is crucial that brain damage be severe enough to cause failure of the regulatory mechanisms of the body.

Is an individual in a state of neocortical death any less in possession of the characterstics that distinguish the living from cadavers than one whose respiration and circulation are mechanically maintained? It is a matter that society must decide. Until society decides, it is not clear what is considered beneficent or nonmaleficent treatment of an individual in a state of neocortical death.

20.5 THE TERMINALLY ILL PATIENT AND EUTHANASIA

Terminally ill patients today often find themselves in a strange world of institutions and technology devoted to assisting them in their fight against death. However, at the same time, this modern technologically oriented medical system may cause patients and their families considerable economic, psychological, and physical pain. In enabling medical science to prolong life, modern technology has in many cases made dying slower and more undignified. As a result of this situation, there is a moral dilemma in medicine: Is it right or wrong for medical professionals to stop treatment or administer a lethal dose to terminally ill patients?

This problem has become a major issue for our society. Although death is all around us in the form of accidents, drug overdose, alcoholism, murder, and suicide, for most of us, the end lies in growing older and succumbing to some from of chronic illness. As the aged approach the end of life's journey, they may eventually wish for the day when all troubles can be brought to an end. Such a desire, frequently shared by a compassionate family, is often shattered by therapies provided with only one concern: to prolong life regardless of the situation. As a result, many claim a dignified death is often not compatible with today's standard medical view.

Consider the following hypothetical version of the kind of case that often confronts contemporary patients, their families, health care workers, and society as a whole. Suppose a middle-aged man suffers a brain hemorrhage and loses consciousness as a result of a ruptured aneurysm. Suppose that he never regains consciousness and is hospitalized in a state of neocortical death, a chronic vegetative state. He is maintained by a surgically implanted gastronomy tube that drips liquid nourishment from a plastic bag directly into his stomach. The care of this individual takes 7.5 h of nursing time daily and includes shaving, oral hygiene, grooming, attending to his bowels and bladder, and so forth. Suppose further that his wife undertakes legal action to force his caregivers to end all medical treatment, including nutrition and hydration, so that complete bodily death of her husband will occur. She presents a preponderance of evidence to the court to show that her husband would have wanted just this result in these circumstances.

The central moral issue raised by this sort of case is whether the quality of the individual's life is sufficiently compromised to make intentional termination of that life morally permissible. While alive, he made it clear to both family and friends that he would prefer to be allowed to die rather than be mechanically maintained in a condition of irretrievable loss of consciousness. Deciding whether his judgment in such a case should be allowed requires deciding which capacities and qualities make life worth living, which qualities are sufficient to endow it with value worth sustaining, and whether their absence justifies deliberate termination of a life, at least when this would be the wish of the individual in question. Without this decision, the traditional norms of medical ethics, beneficence and nonmaleficence, provide no guidance. Without this decision, it cannot be determined whether termination of life support is a benefit or a harm to the patient.

For many individuals, the fight for life is a correct professional view. They believe that the forces of medicine should always be committed to using innovative ways of prolonging life for the individual. However, this cannot be the only approach to caring for the terminally ill. Certain moral questions regarding the extent to which physicians engaged in heroic efforts to prolong life must be addressed if the individual's rights are to be preserved. The goal of those responsible for patient care should not solely be to prolong life as long as possible by the extensive use of drugs, operations, respirators, hemodialyzers, pacemakers, and soon but rather to provide a reasonable quality of life for each patient. It is out of this new concern that euthanasia has once again become a controversial issue in the practice of medicine.

The term euthanasia is derived from two Greek words meaning "good" and "death." Euthanasia was practiced in many primitive societies in varying degrees. For example, on the island of Cos, the ancient Greeks assembled elderly and sick people at an annual banquet to consume a poisonous potion. Even Aristotle advocated euthanasia for gravely deformed children. Other cultures acted in a similar manner toward their aged by abandoning them when they felt these individuals no longer served any useful purpose. However, with the spread of Christianity in the Western world, a new attitude developed toward euthanasia. Because of the Judeo-Christian belief in the biblical statements "Thou shalt not kill" (Exodus 20:13) and "He who kills a man should be put to death" (Leviticus 24:17), the practice of euthanasia decreased. As a result of these moral judgments, killing was considered a sin, and the decision about whether someone should live or die was viewed solely as God's responsibility — not humans.

In today's society, euthanasia implies to many "death with dignity," a practice to be followed when life is merely being prolonged by machines and no longer seems to have value. In many instances, it has been considered a contract for the termination of life in order to avoid unnecessary suffering at the end of a fatal illness and, therefore, has the connotation of relief from pain.

Discussions of the morality of euthanasia often distinguish active from passive euthanasia, a distinction that rests upon the difference between an act of commission and an act of omission. When failure to take steps that could effectively forestall death results in an individual's demise, the resultant death is an act of omission and a case of letting a person die. When a death is the result of doing something to hasten the end of a person's life (e.g., giving a lethal injection), that death is caused by an act of commission and is a case of killing a person. The important difference between active and passive euthanasia is that in passive euthanasia the physician does not do anything to bring about the patient's death. The physician does nothing and death results due to whatever illness already afflicts the patient. In active euthanasia, however, the physician does something to bring about the patient's death.

The physician who gives the patient with cancer a lethal injection has caused the patient's death, whereas if the physician merely ceases treatment, the cancer is the cause of death.

In active euthanasia, someone must do something to bring about the patient's death, and in passive euthanasia the patient's death is caused by illness rather than by anyone's conduct. Is this notion correct? Suppose a physician deliberately decides not to treat a patient who is terminally ill, and the patient dies. Suppose further that the physician were to attempt to exonerate himself by saying, "I did nothing. The patient's death was the result of illness. I was not the cause of death." Under current legal and moral norms, such a response would have no credibility. The physician would be blameworthy for the patient's death as surely as if he or she had actively killed the patient. Thus, the actions taken by a physician to continue treatment to the very end are understood.

Euthanasia may also be classified as involuntary or voluntary. An act of euthanasia is involuntary if it hastens the individual's death for his or her own good but against his or her wishes. Involuntary euthanasia, therefore, is no different in any morally relevant way from unjustifiable homicide. However, what happens when the individual is incapable of agreeing or disagreeing? Suppose that a terminally ill person is unconscious and cannot make his or her wishes known. Would hastening his or her death be permissible? It would if there were substantial evidence that the individual had given prior consent. The individual may have told friends and relatives that, in certain circumstances, efforts to prolong his or her life should not be undertaken or continued and may even have recorded his or her wishes in the form of a living will or an audio- or videotape. When this level of substantial evidence of prior consent exists, the decision to hasten death would be morally justified. A case of this sort would be case of voluntary euthanasia.

For a living will to be valid, the person signing it must be of sound mind at the time the will is made and shown not to have altered their opinion in the interim between its signing and the onset of the illness. In addition, the witnesses must not be able to benefit from the individual's death. As the living will states, it is not a legally binding document. It is essentially a passive request and depends on moral persuasion. Proponents of the will, however, believe that it is valuable in relieving the burden of guilt often carried by health professionals and the family in making the decision to allow a patient to die.

Those who favor euthanasia point out the importance of individual rights and freedom of choice and view euthanasia as a kindness ending the misery endured by the patient. The thought of a dignified death is much more attractive than the process of continuous suffering and gradual decay into nothingness. Viewing each person as a rational being possessing a unique mind and personality, proponents argue that terminally ill patients should have the right to control the ending of their own life.

On the other hand, those opposed to euthanasia demand to know who has the right to end the life of another. Some use religious arguments, emphasizing that euthanasia is in direct conflict with the belief that God, and God alone, has the power to decide when a human life ends. Their view is that anyone who practices euthanasia is essentially acting in the place of God, and that no human should ever be considered omnipotent.

A Living Will

TO MY FAMILY, MY PHYSICIAN, MY CLERGYMAN, MY LAWYER:

If the time comes when I can no longer take part in decisions about my own future, let this statement stand as testament of my wishes: If there is no reasonable expectation of my recovery from physical or mental disability, I request that I be allowed to die and not be kept alive by artificial means or heroic measures. Death is as much a reality as birth, growth, maturity, and old age — it is the one certainty. I do not fear death as much as I fear the indignity of deterioration, dependence, and hopeless pain. I ask that drugs be mercifully administered to me for the terminal suffering even if they hasten the moment of death. This request is made after careful consideration. Although this document is not legally binding, you who care for me will, I hope, feel morally bound to follow its mandate. I recognize that it places a heavy burden of responsibility upon you, and it is with the intention of sharing that responsibility and of mitigating any feelings of guilt that this statement is made.

Signed_____

Date_____

Witnessed by_____

Others turn to the established codes, reminding those responsible for the care of patients that they must do whatever is in their power to save a life. Their argument is that health professionals cannot honor their pledge and still believe that euthanasia is justified. If terminally ill patients are kept alive, there is at least a chance of finding a cure that might be useful to them. Opponents of euthanasia feel that legalizing it would destroy the bonds of trust between doctor and patient. How would sick individuals feel if they could not be sure that their physician and nurse would try everything possible to cure them, knowing that if their condition worsened they would just lose faith and decide that death would be better? Opponents of euthanasia also question whether it will be truly beneficial to the suffering person or will only be a means to relieve the agony of the family. They believe that destroying life (no matter how minimal) merely to ease the emotional suffering of others is indeed unjust.

Many fear that if euthanasia is legalized, it will be difficult to define and develop clear-cut guidelines that will serve as the basis for euthanasia to be carried out. Furthermore, once any form of euthanasia is accepted by society, its detractors fear that many other problems will arise. Even the acceptance of passive euthanasia could, if carried to its logical conclusion, be applied in state hospitals and institutions for the mentally retarded and the elderly. Such places currently house thousands of people who have neither hope nor any prospect of a life that even approaches normality.

Legalization of passive euthanasia could prompt an increased number of suits by parents seeking to end the agony of incurably afflicted children or by children seeking to shorten the suffering of aged and terminally ill parents. In Nazi Germany, for example, mercy killing was initially practiced to end the suffering of the terminally ill. Eventually, however, the practice spread so that even persons with the slightest deviation from the norm (e.g., the mentally ill and minority groups such as Jews and others) were terminated. Clearly, the situation is delicate and thought provoking.

20.6 TAKING CONTROL

Medical care decisions can be tremendously difficult. They often involve unpleasant topics and arise when we are emotionally and physically most vulnerable. Almost always these choices involve new medical information that seems alien and overwhelming. In an attempt to assist individuals to make these decisions, it is often helpful to follow the pathway outlined here:

1. Obtain all the facts, i.e., clarify the medical facts of the situation.
2. Understand all options and their consequences.
3. Place a value on each of the options based on your own set of personal values.

The three-step facts/options/values path concerns the "how" of decisions, but equally important is the "who." Every medical decision must be made by somebody. Ideally, decisions will be made by the person most intimately involved — the patient. Very often, however, it is made by someone else — spouse, family, or physician — or a group of people acting on behalf of the patient. It is therefore important to recognize the four concentric circles of consent:

- The first, and primary, circle is the patient.
- The second circle is the use of advance directives, that is, choosing in advance through the use of such documents as the living will.
- The third circle is others deciding for the patient, i.e., the move from personal control to surrogate control.
- The fourth and final circle is the courts and bureaucrats. It is the arena of last resort in which our society has decreed that we go when the patient may be incapacitated, where there is no clear advance directive, and where it is not clear who should make the decision.

These three steps and four circles are simply attempts to impose some order on the chaos that is medical decision making. They can help individuals take control.

20.7 HUMAN EXPERIMENTATION

Medical research has long held an exalted position in our modern society. It has been acclaimed for its significant achievements that range from the development of the Salk and Sabin vaccines for polio to the development of artificial organs. In order to determine their effectiveness and value, however, these new drugs and medical devices eventually are used on humans. The issue is therefore not only whether humans should be involved in clinical studies designed to benefit themselves or their fellow humans but also clarifying or defining more precisely the conditions under which such studies are to be permitted.

For example, consider the case of a 50-year-old female patient suffering from severe coronary artery disease. What guidelines should be followed in the process of experimenting with new drugs or devices that may or may not help her? Should only those procedures viewed as potentially beneficial to her be tried? Should experimental diagnostic procedures or equipment be tested on this patient to evaluate their effectiveness when compared to more accepted techniques, even though they will not be directly beneficial to the patient?

On the other hand, consider the situation of conducting research on the human fetus. This type of research is possible as a result of the legalization of abortion in the United States as well as the technological advances that have made fetal studies more practical than in the past. Under what conditions should medical research be conducted on these subjects? Should potentially hazardous drugs be given to women planning to have abortions to determine the effect of these drugs on the fetus? Should the fetus, once aborted, be used in any experimental studies?

Although these questions are difficult to answer, clinical researchers responsible for the well-being of their patients must face the moral issues involved in testing new equipment and procedures and at the same time safeguard the individual rights of their patients.

20.8   DEFINITION AND PURPOSE OF EXPERIMENTATION

One may ask, what exactly constitutes a human experiment? Although experimental protocols may vary, it is generally accepted that human experimentation occurs whenever the clinical situation of the individual is consciously manipulated to gather information regarding the capability of drugs and devices. In the past, experiments involving human subjects have been classified as either therapeutic or nontherapeutic. A therapeutic experiment is one that may have direct benefit for the patient, whereas the goal of nontherapeutic research is to provide additional knowledge without direct benefit to the person. The central difference is a matter of intent or aim rather than results.

Throughout medical history, there have been numerous examples of therapeutic research projects. The use of nonconventional radiotherapy to inhibit the progress of a malignant cancer, of pacemakers to provide the necessary electrical stimulation for proper heart function, or of artificial kidneys to mimic nature's function and remove poisons from the blood were all, at one time, considered novel approaches that might have some value for the patient. In the process, they were tried and found to be beneficial not only for the individual patient but also for humankind.

Nontherapeutic research has been another important vehicle for medical progress. Experiments designed to study the impact of infection from the hepatitis virus or the malarial parasite or the procedures involved in cardiac catheterization have had significant impacts on the advancement of medical science and the ultimate development of appropriate medical procedures for the benefit of all humans.

In the mid-1970s, the National Commission for the Protection of Human Subjects of Biomedical and Behavioral Research offered the terms "practice" and "research" to replace the conventional therapeutic and nontherapeutic distinction mentioned previously. Quoting the commission, Alexander Capron in his chapter on "Human Experimentation" (1986) wrote

[T]he term "practice" refers to interventions that are designed solely to enhance the well-being of an individual patient or client and that have a reasonable expectation of success." In the medical sphere, practices usually involve diagnosis, preventive treatment, or therapy; in the social sphere, practices include governmental programs such as transfer payments, education, and the like.

By contrast, the term "research" designates an activity designed to test a hypothesis, to permit conclusions to be drawn, and there by to develop or contribute to generalizable knowledge (expressed, for example, in theories, principles, statements of relationships). In the polar cases, then, practice uses a proven technique in an attempt to benefit one or more individuals, while research studies a technique in an attempt to increase knowledge.

Although the practice/research dichotomy has the advantage of not implying that therapeutic activities are the only clinical procedures intended to benefit patients, it is also based on intent rather than outcome. Interventions are "practices" when they are proven techniques intended to benefit the patient, whereas interventions aimed at increasing generalizable knowledge constitute research. What about interventions that do not fit into either category?

One such intervention is "nonvalidated practice" which may encompass prevention as well as diagnosed therapy. The primary purpose of the use of a nonvalidated practice is to benefit the patient while emphasizing that it has not been shown to be safe and efficacious. For humans to be subjected to nonvalidated practice, they must be properly informed and give their consent.

20.9 INFORMED CONSENT

Informed consent has long been considered by many to be the most important moral issue in human experimentation. It is the principal condition that must be satisfied in order for human experimentation to be considered both lawful and ethical. All adults have the legal capacity to give medical consent (unless specifically denied through some legal process). As a result, issues concerning legal capability are usually limited to minors. Many states, if not all, have some exceptions that allow minors to give consent.

Informed consent is an attempt to preserve the rights of individuals by giving them the opportunity for self-determination, that is, to determine for themselves whether they wish to participate in any experimental effort. In 1964, the World Medical Association in Finland endorsed a code of ethics for human experimentation as an attempt to provide guidelines in this area. In 1975, the 29th World Medical Assembly in Tokyo revised these guidelines.

Because it is often essential to use the results obtained in human experiments to further scientific knowledge, the World Medical Association (1975) prepared the following recommendations to serve as a guide to physicians all over the world. However, it is important to note that these guidelines do not relieve physicians, scientists, and engineers from criminal, civil, and ethical responsibilities dictated by the laws of their own countries.

Basic principles

- Biomedical research involving human subjects must conform to generally accepted scientific principles and should be based on adequately performed laboratory and animal experimentation and on a thorough knowledge of the scientific literature.
- The design and performance of each experimental procedure involving human subjects should be clearly formulated in an experimental protocol which should be transmitted to a specially appointed independent committee for consideration, comment, and guidance.
- Biomedical research involving human subjects should be conducted only by scientifically qualified persons and under the supervision of a clinically competent medical person. The responsibility for the human subject must always rest with a medically qualified person and never rest on the subject of the research, even though the subject has given his or her consent.
- Biomedical research involving human subjects cannot legitimately be carried out unless the importance of the objective is in proportion to the inherent risk to the subject.
- Every biomedical research project involving human subjects should be preceded by careful assessment of predictable risks in comparison with foreseeable benefits to the subject or to others. Concern for the interests of the subject must always prevail over the interests of science and society.
- The right of the research subject to safeguard his or her integrity must always be respected. Every precaution should be taken to respect the privacy of the subject and to minimize the impact of the study on the subject's physical and mental integrity and on the personality of the subject.
- Doctors should abstain from engaging in research projects involving human subjects unless they are satisfied that the hazards involved are believed to be predictable. Doctors should cease any investigation if the hazards are found to outweigh the potential benefits.
- In publication of the results of his or her research, the doctor is obliged to preserve the accuracy of the results. Reports of experimentation not in accordance with the principles laid down in this Declaration should not be accepted for publication.
- In any research on human beings, each potential subject must be adequately informed of the aims, methods, anticipated benefits, and potential hazards of the study and the discomfort it may entail. He or she should be informed

that he or she is at liberty to abstain from participation in the study and that he or she is free to withdraw his or her consent to participation at any time. The doctor should then obtain the subject's free-given informed consent, preferably in writing.

- When obtaining informed consent for the research project, the doctor should be particularly cautious if the subject is in a dependent relationship to him or her or may consent under duress. In that case, the informed consent should be obtained by a doctor who is not engaged in the investigation and who is completely independent of this official relationship.

- In the case of legal incompetence, informed consent should be obtained from the legal guardian in accordance with national legislation. Where physical or mental incapacity makes it impossible to obtain informed consent, or when the subject is a minor, permission from the responsible relative replaces that of the subject in accordance with national legislation.

- The research protocol should always contain a statement of the ethical considerations involved and should indicate that the principles enunciated in the present Declaration are complied with.

Medical research combined with professional care (clinical research)

- In the treatment of the sick person, the doctor must be free to use a new diagnostic and therapeutic measure, if in his or her judgment it offers hope of saving life, reestablishing health, or alleviating suffering.

- The potential benefits, hazards, and discomfort of a new method should be weighed against the advantages of the best current diagnostic and therapeutic methods.

- In any medical study, every patient — including those of a control group, if any — should be assured of the best proven diagnostic and therapeutic method.

- The refusal of the patient to participate in a study must never interfere with the doctor–patient relationship.

- If the doctor considers it essential not to obtain informed consent, the specific reasons for this proposal should be stated in the experimental protocol for transmission to the independent committee.

- The doctor can combine medical research with professional care, the objective being the acquisition of new medical knowledge, only to the extent that medical research is justified by its potential diagnostic or therapeutic value for the patient.

Nontherapeutic biomedical research involving human subjects (nonclinical biomedical research)

- In the purely scientific application of medical research carried out on a human being, it is the duty of the doctor to remain the protector of the life and health of that person on whom biomedical research is being carried out.

- The subjects should be volunteers — either healthy persons or patients from whom the experimental design is not related to the patient's illness.

■ The investigator or the investigating team should discontinue the research if in his/her or their judgment it may, if continued, be harmful to the individual.

■ In research on man, the interest of science and society should never take precedence over considerations related to the well-being of the subject.

These guidelines generally converge on six basic requirements for ethically sound human experimentation. First, research on humans must be based on prior laboratory research and research on animals as well as on established scientific fact, "so that the point under inquiry is well-focused and has been advanced as far as possible by nonhuman means." Second, research on humans should use tests and means of observation that are reasonably believed to be able to provide the information being sought by the research. Methods that are not suited for providing the knowledge sought are pointless and rob the research of its scientific value. Third, research should be conducted only by persons with the relevant scientific expertise. Fourth,

All foreseeable risks and reasonably probable benefits, to the subject of the investigation and to science, or more broadly to society, must be carefully assessed, and . . . the comparison of those projected risks and benefits must indicate that the latter clearly outweighs the former. Moreover, the probable benefits must not be obtainable through other less risky means.

Fifth, participation in research should be based on informed and voluntary consent. Sixth, "participation by a subject in an experiment should be halted immediately if the subject finds continued participation undesirable or a prudent investigator has cause to believe that the experiment is likely to result in injury, disability, or death to the subject." Conforming to conditions of this sort probably does limit the pace and extent of medical progress, but society's insistence on these conditions is its way of saying that the only medical progress truly worth having must be consistent with a high level of respect for human dignity. Of these conditions, the requirement to obtain informed and voluntary consent from research subjects is widely regarded as one of the most important protections.

A strict interpretation of the criteria mentioned previously for subjects automatically rules out whole classes of individuals from participating in medical research projects. Children, the mentally retarded, and any patient whose capacity to think is affected by illness are excluded on the grounds of their inability to comprehend exactly what is involved in the experiment. In addition, individuals having a dependent relationship to the clinical investigator, such as the investigator's patients and students, are eliminated based on constraint. Since mental capacity also includes the ability of subjects to appreciate the seriousness of the consequences of the proposed procedure, this means that even though some minors have the right to give consent for certain types of treatments, they must be able to understand all the risks involved.

Any research study must clearly define the risks involved. The patient must receive a total disclosure of all known information. In the past, the evaluation of risk and benefit in many situations belonged to the medical professional alone. Once made, it was assumed that this decision would be accepted at face value by the patient. Today, this assumption is not valid. Although the medical staff must still weigh the risks and benefits involved in any procedure they suggest, it is the patient that has the right to make the final determination. The patient cannot, of course, decide whether the procedure is medically correct since that requires more medical expertise than the average individual possesses. However, once the procedure is recommended, the patient must have enough information to decide whether the hoped-for benefits are sufficient to risk the hazards. Only when this is accomplished can a valid consent be given.

Once informed and voluntary consent has been obtained and recorded, the following protections are in place:

- It represents legal authorization to proceed. The subject cannot later claim assault and battery.
- It usually gives legal authorization to use the data obtained for professional or research purposes. Invasion of privacy cannot later be claimed.
- It eliminates any claims in the event that the subject fails to benefit from the procedure.
- It is defense against any claim of an injury when the risk of the procedure is understood and consented to.
- It protects the investigator against any claim of an injury resulting from the subject's failure to follow safety instructions if the orders were well explained and reasonable.

Nevertheless, can the aims of research ever be reconciled with the traditional moral obligations of physicians? Is the researcher/physician in an untenable position? Informed and voluntary consent once again is the key only if subjects of an experiment agree to participate in the research. What happens to them during and because of the experiment is a product of their own decision. It is not something that is imposed on them but rather, in a very real sense, something that they elected to have done to themselves. Because their autonomy is thus respected, they are not made a mere resource for the benefit of others. Although they may suffer harm for the benefit of others, they do so of their own volition, as a result of the exercise of their own autonomy rather than as a result of having their autonomy limited or diminished.

For consent to be genuine, it must be truly voluntary and not the product of coercion. Not all sources of coercion are as obvious and easy to recognize as physical violence. A subject may be coerced by fear that there is no other recourse for treatment, by the fear that nonconsent will alienate the physician on whom the subject depends for treatment, or even by the fear of disapproval of others. This sort of coercion, if it truly ranks as such, is often difficult to detect and, in turn, to remedy.

Finally, individuals must understand what they are consenting to do. Therefore, they must be given sufficient information to arrive at an intelligent decision concerning whether or not to participate in the research. Although a subject need not be given all the information of a researcher, it is important to determine how much should be provided and what can be omitted without compromising the validity of the subject's consent. Another difficulty lies in knowing whether the subject is competent to understand the information given and to render an intelligent opinion based on it. In any case, efforts must be made to ensure that sufficient relevant information is given and that the subject is sufficiently competent to process it. These are matters of judgment that probably cannot be made with absolute precision and certainty, but rigorous efforts must be made in good faith to prevent research on humans from involving gross violations of human dignity.

20.10 REGULATION OF MEDICAL DEVICE INNOVATION

In the United States, responsibility for regulating medical devices falls on the Food and Drug Administration (FDA) under the Medical Device Amendment of 1976. This statute requires approval from the FDA before new devices are marketed and imposes requirements for the clinical investigation of new medical devices on human subjects. Although the statute makes interstate commerce of an unapproved new medical device generally unlawful, it provides an exception to allow interstate distribution of unapproved devices in order to conduct clinical research on human subjects. This investigational device exemption (IDE) can be obtained by submitting to the FDA.

> A protocol for the proposed clinical testing of the device, reports of prior investigations of the device, certification that the study has been approved by a local institutional review board, and an assurance that informed consent will be obtained from each human subject.

With respect to clinical research on humans, the FDA categorizes devices into two categories: devices that pose significant risk and those that involve insignificant risk. Examples of the former include orthopedic implants, artificial hearts, and infusion pumps. Examples of the latter include various dental devices and daily-wear contact lenses. Clinical research involving a significant risk device cannot begin until an institutional review board (IRB) has approved both the protocol and the informed consent form and the FDA has given permission. This requirement to submit an IDE application to the FDA is waived in the case of clinical research in which the risk posed is insignificant. In this case, the FDA requires only that approval from an IRB be obtained certifying that the device in question poses only insignificant risk. In deciding whether to approve a proposed clinical investigation of a new device, the IRB and the FDA must determine the following:

1. Risks to subjects are minimized.
2. Risks to subjects are reasonable in relation to anticipated benefit and knowledge to be gained.
3. Subject selection is equitable.
4. Informed consent materials and procedures are adequate.
5. Provisions for monitoring the study and protecting patient information are acceptable.

The FDA allows unapproved medical devices to be used without an IDE in three situations: feasibility studies, emergency use, and treatment use.

20.11 ETHICAL ISSUES IN FEASIBILITY STUDIES

In a feasibility study, or "limited investigation," human research involving the use of a new device would take place at a single institution and involve no more than 10 human subjects. The sponsor of a limited investigation is required to submit to the FDA a "Notice of Limited Investigation" which includes a description of the device, a summary of the purpose of the investigation, the protocol, a sample of the informed consent form, and a certification of approval by the responsible medical board. In certain circumstances, the FDA could require additional information or require the submission of a full IDE application or suspend the investigation.

Investigations of this kind are limited to (i) investigations of new uses of existing devices, (ii) investigations involving temporary or permanent implants during the early developmental stages, and (iii) investigations involving modification of an existing device.

To comprehend adequately the ethical issues posed by clinical use of unapproved medical devices outside the context of an IDE, it is necessary to use the distinctions between practice, nonvalidated practice, and research discussed previously. How do these definitions apply to feasibility studies?

Clearly, the goal of the feasibility study — generalizable knowledge — makes it an instance of research rather than practice. Manufacturers seek to determine the performance of a device with respect to a particular patient population in an effort to gain information about its efficacy and safety. Such information is important in order to determine whether further studies (animal or human) need to be conducted, whether the device needs modification before further use, and so on. The main difference between use of an unapproved device in a feasibility study and its use under the terms of an IDE is that the former would be subject to significantly less intensive FDA review than the latter. This, in turn, means that the responsibility for ensuring that the use of the device is ethically sound falls primarily on the IRB of the institution conducting the study.

The ethical concerns posed here can best be comprehended only with a clear understanding of what justifies research in the first place. Ultimately, no matter how much basic research and animal experimentation has been conducted on a given device, the risks and benefits it poses for humans cannot be adequately determined until it is actually used on humans. The benefit of research on humans lies primarily in the generalizable information that is provided. This information is crucial to medical science's ability to generate new modes of medical treatment that are both efficacious and safe. Therefore, one condition for experimentation to be ethically sound is that it must be scientifically sound.

Although scientific soundness is a necessary condition of ethically sound research on humans, it is not of and by itself sufficient. The human subjects of such research are at risk of being mere research resources, that is, having value only for the ends of the research. Human beings are not valuable wholly or solely for the uses to which they can be put. They are valuable simply because they are human beings. To treat them as such is to respect them as people. Treating individuals as people means respecting their autonomy. This requirement is met by ensuring that no competent person is subjected to any clinical intervention without first giving voluntary and informed consent. Furthermore, respect for people means that the physician will not subject a human to unnecessary risks and will minimize the risks to patients in required procedures.

Much of the scrutiny that the FDA imposes upon use of unapproved medical devices in the context of an IDE addresses two conditions of ethically sound research: Is the experiment scientifically sound? and Does it respect the rights of the human subjects involved? Medical ethicists argue that decreased FDA scrutiny will increase the likelihood that either or both of these conditions will not be met. This possibility exists because many manufacturers of medical devices are, after all, commercial enterprises, companies that are motivated to generate profit and thus to get their devices to market as soon as possible with as little delay and cost as possible. These self-interest motives are likely, at times, to conflict with the requirements of ethically sound research and thus to induce manufacturers to fail to meet these requirements. Profit is not the only motive that might induce manufacturers to contravene the requirements of ethically sound research on humans. A manufacturer may sincerely believe that its product offers great benefit to many people and be prompted to take shortcuts that compromise the quality of the research. Whether the consequences being sought by the research are desired for reasons of self-interest, altruism, or both, the ethical issue is the same: Research subjects may be placed at risk of being treated as mere objects rather than as people.

What about the circumstances in which feasibility studies would take place? Are these not sufficiently different from the "normal" circumstances of research to warrant reduced FDA scrutiny? As noted previously, manufacturers seek to engage in feasibility studies in order to investigate new uses of existing devices, to investigate temporary or permanent implants during the early developmental stages, and to investigate modifications to an existing device. As also noted previously, a feasibility study would take place at only one institution and would involve no more than 10 human subjects. Given these circumstances, is the sort of research that is likely to occur in a feasibility study less likely to be scientifically sound or to fail to respect people than normal research on humans in normal circumstances?

Research in feasibility studies would be done on a very small subject pool, and the harm of any ethical lapses would likely affect fewer people than if such lapses occurred in more usual research circumstances. However, even if the harm done is limited to 10 or fewer subjects in a single feasibility study, the harm is still ethically wrong. To wrong 10 or fewer people is not as bad as to wrong more than 10 people, but it is engaging in wrongdoing nonetheless.

Are ethical lapses more likely to occur in feasibility studies than in studies that take place within the requirements of an IDE? Although nothing in the preceding discussion provides a definitive answer to this question, it is a question to which the FDA should give high priority. The answer to this question might be quite different when the device at issue is a temporary or permanent implant than when it is a previously approved device being put to new uses or modified in some way. Whatever the contemplated use under the feasibility studies mechanism, the FDA would be ethically advised not to allow this kind of exception to IDE use of an unapproved device without a reasonably high level of certainty that research subjects would not be placed in greater jeopardy than in normal research circumstances.

20.12   ETHICAL ISSUES IN EMERGENCY USE

What about the mechanism for avoiding the rigors of an IDE for emergency use? The FDA has authorized emergency use where an unapproved device offers the only alternative for saving the life of a dying patient. However, either an IDE has not yet been approved for the device or its use or an IDE has been approved but the physician who wishes to use the device is not an investigator under the IDE.

The purpose of emergency use of an unapproved device is to attempt to save a dying patient's life in circumstances in which no other alternative is available: This sort of use constitutes practice rather than research. Its aim is primary benefit to the patient rather than provision of new and generalizable information. Because this sort of use occurs before the completion of clinical investigation of the device, it constitutes a nonvalidated practice. What does this mean?

First, it means that while the aim of the use is to save the life of the patient, the nature and likelihood of the potential benefits and risks engendered by use of the device are far more speculative than those in the sort of clinical intervention that constitutes validated practice. In validated practice, thorough investigation of a device, including preclinical studies, animal studies, and studies on human subjects, has established its efficacy and safety. The clinician thus has a well-founded basis on which to judge the benefits and risks such an intervention poses for the patient.

It is precisely this basis that is lacking in the case of a nonvalidated practice. Does this mean that emergency use of an unapproved device should be regarded as immoral? This conclusion would follow only if there were no basis on which to make an assessment of the risks and benefits of the use of the device. The FDA requires that a physician who engages in emergency use of an unapproved device must have substantial reason to believe that benefits will exist. This means that there should be a body of preclinical and animal tests allowing a prediction of the benefit to a human patient.

Thus, although the benefits and risks posed by use of the device are highly speculative, they are not entirely speculative. Although the only way to validate a new technology is to engage in research on humans at some point, not all nonvalidated technologies are equal. Some will be largely uninvestigated, and assessment of their risks and benefits will be wholly or almost wholly speculative. Others will at least have the support of preclinical and animal tests. Although this is not sufficient support for incorporating use of a device into regular clinical practice, it may represent sufficient support to justify use in the desperate circumstances in emergency situations. Desperate circumstances can justify desperate actions, but desperate actions are not the same as reckless actions; hence, the ethical soundness of the FDA's requirement that emergency use be supported by solid results from preclinical and animal tests of the unapproved device.

A second requirement that the FDA imposes on emergency use of unapproved devices is the expectation that physicians "exercise reasonable foresight with respect to potential emergencies and . . . make appropriate arrangements under the IDE procedures." Thus, a physician should not "create" an emergency in order to circumvent IRB review and avoid requesting the sponsor's authorization of the unapproved use of a device. From a Kantian point of view, which is concerned with protecting the dignity of people, this is a particularly important requirement. To create an emergency in order to avoid FDA regulations is to treat the patient as a mere resource whose value is reducible to service to the clinician's goals. Hence, the FDA is quite correct to insist that emergencies are circumstances that reasonable foresight would not anticipate.

Also especially important is the nature of the patient's consent. Individuals facing death are especially vulnerable to exploitation and deserve greater measures for their protection than might otherwise be necessary. One such measure would be to ensure that the patient, or his or her legitimate proxy, knows the highly speculative nature of the intervention being offered; that is, to ensure that it is clearly understood that the clinician's estimation of the intervention's risks and benefits is far less solidly grounded than in the case of validated practices. The patient's consent must be based on an awareness that the device whose use is contemplated has not undergone complete and rigorous testing on humans and that estimations of its potential are based wholly on preclinical and animal studies. Above all, the patient must not be led to believe that the risks and benefits of the intervention are better understood than they in fact are by the clinical. Another important point is to ensure that the patient understands all the options, not simply life or death but also a life with severely impaired quality. Although desperate circumstances may legitimate desperate actions, the decision to take such actions must rest on the informed and voluntary consent of the patient, certainly for an especially vulnerable patient.

It is important for a clinician involved in emergency use of an unapproved device to recognize that these activities constitute a form of practice, albeit nonvalidated, and not research. Hence, the primary obligation is to the well-being of the patient. The patient enters into the relationship with the clinician with the same trust that accompanies any normal clinical situation. To treat this sort of intervention as if it were an instance of research, and hence justified by its benefits to science and society, would be an abuse of this trust.

20.13 ETHICAL ISSUES IN TREATMENT USE

The FDA has adopted regulations authorizing the use of investigational new drugs in circumstances in which a patient has not responded to approved therapies. This "treatment use" of unapproved new drugs is not limited to life-threatening emergency situations; it is also available to treat "serious" diseases or conditions. The FDA has not approved treatment use of unapproved medical devices, but it is possible that a manufacturer could obtain such approval by establishing a specific protocol for this kind of use within the context of an IDE.

The criteria for treatment use of unapproved medical devices would be similar to criteria for treatment use of investigational drugs: (i) The device is intended to treat a serious or life-threatening disease or condition; (ii) there is no comparable or satisfactory alternative product available to treat the condition; (iii) the device is under an IDE or has received an IDE exemption, or all clinical trials have been completed and the device is awaiting approval; and (iv) the sponsor is actively pursuing marketing approval of the investigational device. The treatment-use protocol would be submitted as part of the IDE and would describe the intended use of the device, the rationale for use of the device, the available alternatives and why the investigational product is preferable, the criteria for patient selection, the measures to monitor the use of the device and to minimize risk, and technical information that is relevant to the safety and effectiveness of the device for the intended treatment purpose.

Were the FDA to approve treatment use of unapproved medical devices, what ethical issues would be posed? First, because such use is premised on the failure of validated interventions to improve the patient's condition adequately, it is a form of practice rather than research. Second, since the device involved in an instance of treatment use is unapproved, such use would constitute nonvalidated practice. As such, like emergency use, it should be subject to the FDA's requirement that prior preclinical tests and animal studies have been conducted that provide substantial reason to believe that patient benefit will result. As with emergency use, although this does not prevent assessment of the intervention's benefits and risks from being highly speculative, it does prevent assessment from being totally speculative. Here, too although desperate circumstances can justify desperate action, they do not justify reckless action. Unlike emergency use, the circumstances of treatment use involve serious impairment of health rather than the threat of premature death. Hence, an issue that must be considered is how serious such impairment must be to justify resorting to an intervention whose risks and benefits have not been solidly established.

In cases of emergency use, the FDA requires that physicians not create an exception to an IDE to avoid requirements that would otherwise be in place. As with emergency use of unapproved devices, the patients involved in treatment uses would be particularly vulnerable patients. Although they are not dying, they are facing serious medical conditions and are thereby likely to be less able to avoid exploitation than patients in less desperate circumstances. Consequently, it is especially important that patients be informed of the speculative nature of the intervention and of the possibility that treatment may result in little or no benefit to them.

20.14 THE SAFE MEDICAL DEVICES ACT OF 1990

On November 28, 1991, the Safe Medical Devices Act of 1990 (Public Law 101-629) went into effect. This regulation requires a wide range of health care institutions, including hospitals, ambulatory-surgical facilities, nursing homes, and outpatient treatment facilities, to report information that "reasonably suggests" the likelihood that the death, serious injury, or serious illness of a patient at that facility has been caused or contributed to by a medical device. When a death is device related, a report must be sent directly to the FDA and to the manufacturer of the device. When a serious illness or injury is device related, a report must be sent to the manufacturer or to the FDA in cases in which the manufacturer is not known. In addition, summaries of previously submitted reports must be submitted to the FDA on a semiannual basis. Prior to this regulation, such reporting was wholly voluntary. This new regulation was designed to enhance the FDA's ability to learn quickly about problems related to medical devices and supplements the medical device reporting (MDR) regulations promulgated in 1984. MDR regulations require that reports of device-related deaths and serious injuries be submitted to the FDA by manufacturers and importers. The new law extends this requirement to users of medical devices along with manufacturers and importers. This act gives the FDA authority over device-user facilities.

The new regulation is ethically significant because, by attempting to increase the FDA's awareness of medical device-related problems, it attempts to increase the agency's ability to protect the welfare of patients. The main controversy regarding this new regulation is essentially utilitarian in nature. Skeptics of the law are dubious about its ability to provide the FDA with much useful information. They worry that much of the information generated by the new law will simply duplicate information already provided under MDR regulations. If this is the case, little or no benefit to patients would accrue from compliance with the regulation. Furthermore, the new regulations, according to the skeptics, are likely to increase lawsuits filed against hospitals and manufacturers and will require device-user facilities to implement formal systems for reporting device-related problems and to provide personnel to operate those systems. This would, of course, add to the costs of health care and thereby exacerbate the problem of access to care, a situation that many believe to be of crisis proportions already. In short, the controversy over the new policy centers on the worry that its benefits to patients will be marginal and significantly outweighed by its costs.

EXERCISES

1. Explain the distinction between the terms ethics and morality. Provide examples that illustrate this distinction in the medical arena.
2. Provide three examples of medical moral judgments.
3. What do advocates of the utilitarian school of thought believe?
4. What does Kantianism expect in terms of the patient's rights and wishes?
5. Discuss how the code of ethics for clinical engineers provides guidance to practitioners in the field.
6. Discuss what is meant by brain stem death. How is this distinguished from neocortical death?
7. Distinguish between active and passive euthanasia as well as voluntary and involuntary euthanasia. In your view, which, if any, are permissible? Provide your reasoning and any conditions that must be satisfied to meet your approval.
8. If the family of a patient in the intensive care unit submits the individual's living will, should it be honored immediately or should there be a discussion between physicians and the family? Who should make the decision? Why?
9. What constitutes a human experiment? Under what conditions are human experiments permitted? What safeguards should hospitals have in place?
10. A biomedical engineer has designed a new sleep apnea monitor. Discuss the steps that should be taken before it is used in a clinical setting.

11. Discuss the distinctions between practice, research, and nonvalidated practice. Provide examples of each in the medical arena.
12. What are the two major conditions for ethically sound research?
13. Informed consent is one of the essential factors for permitting humans to participate in medical experiments. What ethical principles are satisfied by informed consent? What should be done to ensure it is truly voluntary? What information should be given to human subjects?
14. What are the distinctions between feasibility studies and emergency use?

SUGGESTED READING

Abrams, N., and Buckner, M. D. (Eds.) (1983). *Medical Ethics*. MIT Press, Cambridge.

Baker, S. P. (1980). On lobbies, liberties and the public good. *Am. J. Public Health* 70(6).

Beecher, H. K. (1970). Research and the individual. In *Human Studies*. Little, Brown, Boston.

Bronzino, J. D. (1992). *Management of Medical Technology*. Butterworth, Boston.

Bronzino, J. D., Smith, V.H., and Wade, M.L. (1990). *Medical Technology and Society*. MIT Press, Cambridge.

Campbell, A. V. (1972). *Moral Dilemmas in Medicine*. Churchill Livingston, Edinburgh, UK.

Capron, A. (1978). *The Encyclopedia of Bioethics Vol. II*. Free Press, Glencoe, IL.

Capron, A. (1986). *Human Experimentation* "in (J. F. Childers *et al.*, eds) University Publications of America.

Cutler, D. R. (Ed.) (1968). *Updating Life and Death*. Beacon, Boston.

Daniels, N. (1987). *Just Health Care*. Cambridge Univ. Press, Cambridge, UK.

Dubler, N., and Nimmons, D. (1992). *Ethics on Call*. Harmony, New York.

Fagothey, A. (1972). *Right and Reason: Ethics in Theory and Practice*. Mosby, St. Louis.

Fletcher, J. (1974). *The Ethics of Genetic Control*. Anchor, Garden City, NY.

Horan, D. J., and Mall, D. (1977). *Death, Dying and Euthanasia*. Washington Univ. Publications of America, Washington, DC.

Jonsen, A. R. (1990). *The New Medicine and the Old Ethics*. Harvard Univ. Press, Cambridge, MA.

Maguire, D. C. (1974). *Death by Choice*. Doubleday, Garden City, NY.

Mendelsohn, E. Swazey, J. P., and Traviss, I. (Eds.) (1976). *Human Aspects of Biomedical Innovation*. Harvard Univ. Press, Cambridge, MA.

Moskop, J. C., and Kopelman, L. (Eds.) (1985). *Ethics and Critical Care Medicine*. Reidel, Boston.

Murphy, J., and Coleman, J. (1984). *The Philosophy of Law*. Rowman & Allenheld. TOTAWA, NJ.

Pence, G. E. (1990). *Classic Cases in Medical Ethics*. McGraw-Hill, New York.

Rachels, J. (Ed.) (1978). *Moral Problems*, 3rd ed. Harper & Row, New York.

Rachels, J. (1986). *Ethics at the End of Life: Euthanasia and Morality*. Oxford Univ. Press, Oxford.

Valasquez, M. (1988). *Business Ethics*, 2nd ed. Prentice-Hall, Englewood Cliffs, NJ.

Vandeveer, D., and Regan, T. (Eds.) (1978). *Health Care Ethics: An Introduction*. Temple Univ. Press, Philadelphia.

Veatch, R. (1976). *Death, Dying and Biological Revolution*. Yale Univ. Press, New Haven, CT.

Weber, H. (Ed.) (1976). *Experiments with Man*. World Council of Churches, Geneva.

Weir, R. F. (1977). *Ethical Issues in Death and Dying*. Columbia Univ. Press, New York.

INDEX